This book explains the basic aspects of symmetry groups as applied to problems in physics and chemistry using an approach pioneered and developed by the author. The symmetry groups and their representations are worked out explicitly, eliminating the unduly abstract nature of group theoretical methods.

The author has systematized the wealth of knowledge on symmetry groups that has accumulated during the century since Fedrov discovered the 230 space groups. All space groups, unitary as well as anti-unitary, are reconstructed from the algebraic defining relations of the point groups. The matrix representations are determined through the projective representations of the point groups. The representations of the point groups are subduced by the representations of the rotation group. The correspondence theorem on basis functions belonging to a representation is introduced to form the general expression for the symmetry-adapted linear combinations of equivalent basis functions with respect to a point group. This is then applied to form molecular orbitals and symmetry coordinates of vibration of a molecule or a crystal and the energy band eigenfunctions of the electrons in a crystal. The book assumes only an elementary knowledge of quantum mechanics. Numerous applications of the theorems are described to aid understanding.

This work will be of great interest to graduate students and professionals in solid state physics, chemistry, mathematics and geology and to those who are interested in magnetic crystal structures.

To 'Eomma'

Group Theoretical Methods

and Applications to Molecules and Crystals

SHOON K. KIM
Temple University

CAMBRIDGE UNIVERSITY PRESS
Cambridge, New York, Melbourne, Madrid, Cape Town, Singapore, São Paulo

Cambridge University Press
The Edinburgh Building, Cambridge CB2 2RU, UK

Published in the United States of America by Cambridge University Press, New York

www.cambridge.org
Information on this title: www.cambridge.org/9780521640626

First published 1999
This digitally printed first paperback version 2005

A catalogue record for this publication is available from the British Library

Library of Congress Cataloguing in Publication data

Kim, Shoon Kyung, 1920–
Group theoretical methods and applications to molecules and crystals/by Shoon K. Kim.
p. cm.
Includes bibliographical references and index.
ISBN 0 521 64062 8
1. Group theory. 2. Chemistry, Physical and theoretical – Mathematics. 3. Crystallography, Mathematical.
I. Title. QD455.3.G75K56 1999
512'.2–dc21 98–26493 CIP

ISBN-13 978-0-521-64062-6 hardback
ISBN-10 0-521-64062-8 hardback

ISBN-13 978-0-521-02038-1 paperback
ISBN-10 0-521-02038-7 paperback

Contents

Preface

This book is written for graduate students and professionals in physics, chemistry and in particular for those who are interested in crystal and magnetic crystal symmetries. It is mostly based on the papers written by the author over the last 20 years and the lectures given at Temple University. The aim of the book is to systematize the wealth of knowledge on point groups and their extensions which has accumulated over a century since Schönflies and Fedrov discovered the 230 space groups in 1895. Simple, unambiguous methods of construction for the relevant groups and their representations introduced in the book may overcome the abstract nature of the group theoretical methods applied to physical chemical problems.

For example, a unified approach to the point groups and the space groups is proposed. Firstly, a point group of finite order is defined by a set of the algebraic defining relations (or presentation) through the generators in Chapter 5. Then, by incorporating the translational degree of freedom into the presentations of the 32 crystallographic point groups, I have determined the 32 minimum general generator sets (MGGSs) which generate the 230 space groups in Chapter 13. Their representations follow from a set of five general expressions of the projective representations of the point groups given in Chapter 12. It is simply amazing to see that the simple algebraic defining relations of point groups are so very far-reaching.

In almost all other textbooks or monographs on solid-state physics, the space groups may be tabulated, but without their derivations, as if they were 'god-given'. The main reason could have been the lack of a simple method for the derivations. As a result, the group theoretical methods have been unnecessarily abstract in an age when students are very familiar with non-commuting physical quantities in quantum mechanics.

The book is self-sufficient even though some elementary knowledge of quantum mechanics is assumed. No previous knowledge of group theory is required. In providing the basic essentials, introductory examples are given prior to the theorems. Effort has been made to provide the simplest and easiest but rigorous proofs for any theorem described in the book. Applications are fully developed. Each chapter contains something new or different in approach that cannot be found in any other monograph. For example, even in the basic theory on matrix transformation given in Chapter 2, I have introduced an involutional transformation into the Dirac theory of the electron and arrived at the Dirac plane wave solution in one step. This transformation is used frequently in later chapters. The transformation is further extended to a new general theory of matrix diagonalization that provides the transformation matrix as a polynomial of the matrix to be diagonalized. This theory is included for its usefulness even though it is somewhat mathematical.

Some further typical features of the book are worth mentioning here. In Chapter 5, I have introduced a faithful representation for a point group using the unit basis vectors of the coordinate system. This allows one to construct the multiplication table of any point group, e.g. the octahedral group, with ease. A new unified system of classifications for the improper point groups and anti-unitary (or magnetic) point groups is introduced, using the fact that both the inversion and the time-reversal operator commute with all the point operations. This system is quite effective for describing their isomorphisms, and thereby greatly simplifies the construction of their matrix representations and co-representations in its entirety. In Chapter 7, I have introduced a simple correspondence theorem on the basis functions of a point group G and thereby developed a general method of constructing the symmetry-adapted linear combinations (SALCs) of equivalent basis functions with respect to G. It is then applied to construct SALCs of equivalent atomic orbitals and the symmetry coordinates of vibration for molecules and later for crystals in Chapter 15. This theory requires only knowledge of the elementary basis functions of the irreducible representation and does not require the matrix representation. This is in quite a contrast to the conventional projection operator method. The correspondence theorem is further extended to form the energy band eigenfunctions of the electron in a solid in Chapter 15. By incorporating the time-reversal symmetry into point groups, anti-unitary (magnetic) point groups are formed in Chapter 16. Analogously, 38 assemblies of MGGSs for 1421 anti-unitary space groups are formed from the 32 MGGSs of space groups in Chapter 17. Their co-representations are introduced and applied to the selection rules under the anti-unitary groups.

Once a reader is familiar with the basic aspects of the group theoretical methods given in Chapters 3, 4 and 5, the reader can pick and choose to read any applications in later chapters using the rest of the book as the built-in references. This is possible because each chapter is as self-contained as possible and also an effective numbering system is introduced for referring to the theorems, equations and figures given in the book. Numerous examples of the applications of theorems are given as illustrations. In some chapters, I introduced a simplified special proof for a theorem to help understanding, even though its general proof had been given in an earlier chapter. In particular, those who are interested in the applications to inorganic chemistry may directly start from Chapter 7 with minimum knowledge of the group theoretical methods. One of my colleagues, Professor S. Jansen-Varnum, used the theory of symmetry-adapted linear combinations based on the correspondence theorem described in Chapter 7 of my manuscript for teaching both undergraduate and graduate courses in inorganic chemistry.

Acknowledgment

I am very much indebted to many friends and colleagues for their help while I was writing this book. Firstly, I am very grateful to the chairpersons of the chemistry and physics departments, Dr G. Krow, Dr S. Wunder, Dr R. Salomon and E. Gawlinski. Special thanks are due to Dr D. J. Lee, the late Dr C. W. Pyun, Dr S. I. Choi, Dr S. Jansen-Varnum and Dr L. Mascavage for reviewing parts of the manuscript and to Dr K. S. Yun for valuable advice. I am also very grateful to Dr D. Titus for his help in scanning the figures into the manuscript. Whenever I had difficulty with a delicate sentence, Dr D. Dalton assisted me, so I express here my sincere gratitude for his help. Ms G. Basmajian typed the entire manuscript single-handedly. I am truly indebted to her patience and her typing skills in dealing with complicated mathematical equations.

Philadelphia, Pennsylvania S. K. KIM

List of symbols

$\in$	belongs to, e.g. $g \in S$ means an element g belongs to a set S.
$\forall$	for all, e.g. $\forall\, g \in S$ means for all $g \in S$.
$*$	complex conjugate.
$\sim$	transpose, e.g. $A^{\sim}$ is the transpose of the matrix A.
$\dagger$	adjoint or Hermitian adjoint, i.e. $A^{\dagger} = A^{*\sim}$.
$\rightarrow$	is mapped onto.
$\leftrightarrow$	one-to-one correspondence.
$\otimes$	direct product.
$\oplus$	direct sum.
$\cap$	set-theoretic intersection, e.g., $S_1 \cap S_2$ is the set common to the two sets S_1 and S_2.
$\{\}$	set of all elements.
$H < G$	H is a subgroup of a group G.
$H \lhd G$	H is an invariant subgroup of a group G.
$G_1 \times G_2$	the direct product group of two groups G_1 and G_2.
$F \wedge H$	the semidirect product of two groups F and H, where F is invariant under H.
$F \simeq H$	Two groups F and H are isomorphic.

1
Linear transformations

1.1 Vectors

A sequence of n numbers that are complex in general, denoted $\boldsymbol{v} = (v_1, v_2, \ldots, v_n)$, is called a vector in n-dimensional linear vector space $V^{(n)}$ with the components v_i; $i = 1, 2, \ldots, n$. The coordinates of a point with respect to a coordinate system in n-dimensional space can also be considered as a vector in $V^{(n)}$. Vectors obey the following rules.

(1) Two vectors $\boldsymbol{v}$ and $\boldsymbol{u}$ are said to be equal if their corresponding components are equal, i.e. $\boldsymbol{v} = \boldsymbol{u}$, if $v_i = u_i$ for all i.
(2) Addition of two vectors $\boldsymbol{v}$ and $\boldsymbol{u}$ is also a vector with components $(\boldsymbol{v} + \boldsymbol{u})_i = v_i + u_i$.
(3) The product $c\boldsymbol{v}$ of a number c with a vector $\boldsymbol{v}$ is a vector whose components are c times the components of $\boldsymbol{v}$, i.e. $(c\boldsymbol{v})_i = cv_i$.

A vector is a null vector if all its components vanish.

Vectors of a set $\boldsymbol{v}^{(1)}, \boldsymbol{v}^{(2)}, \ldots, \boldsymbol{v}^{(k)}$ are said to be *linearly independent* if there exists no relationship of the form

$$a_1\boldsymbol{v}^{(1)} + a_2\boldsymbol{v}^{(2)} + \cdots + a_k\boldsymbol{v}^{(k)} = 0$$

excluding the trivial case in which all coefficients $a_1, a_2, \ldots, a_k$ are zero. If these vectors are linearly dependent, there exists a non-zero coefficient, say $a_1 \neq 0$, then $\boldsymbol{v}^{(1)}$ is expressed in terms of the remaining vectors as follows:

$$\boldsymbol{v}^{(1)} = -\frac{a_2}{a_1}\boldsymbol{v}^{(2)} - \frac{a_3}{a_1}\boldsymbol{v}^{(3)} - \cdots - \frac{a_k}{a_1}\boldsymbol{v}^{(k)}$$

There exist no more than n vectors in $V^{(n)}$ which are linearly independent; in fact, if we introduce n linearly independent vectors of the form

$$\begin{aligned} \boldsymbol{e}^{(1)} &= (1, 0, 0, \ldots, 0) \\ \boldsymbol{e}^{(2)} &= (0, 1, 0, \ldots, 0) \\ &\ldots \quad \ldots \quad \ldots \\ \boldsymbol{e}^{(n)} &= (0, 0, 0, \ldots, 1) \end{aligned} \tag{1.1.1}$$

then an arbitrary vector $\boldsymbol{v} = (v_1, v_2, \ldots, v_n)$ is expressed in terms of the set of n vectors as follows:

$$\boldsymbol{v} = v_1\boldsymbol{e}^{(1)} + v_2\boldsymbol{e}^{(2)} + \cdots + v_n\boldsymbol{e}^{(n)}$$

The set $e = \{\boldsymbol{e}^{(1)}, \boldsymbol{e}^{(2)}, \ldots, \boldsymbol{e}^{(n)}\}$ is called *the natural basis* of the n-dimensional vector space $V^{(n)}$.

Another important elementary concept is the scalar product of two vectors. *The scalar product* of two vectors $\boldsymbol{v}$ and $\boldsymbol{u}$ is a number defined by

$$(\boldsymbol{v}, \boldsymbol{u}) = v_1 u_1 + v_2 u_2 + \cdots + v_n u_n \tag{1.1.2}$$

There exists another scalar product called *the Hermitian scalar product* defined by

$$\langle \boldsymbol{v}, \boldsymbol{u} \rangle = v_1^* u_1 + v_2^* u_2 + \cdots + v_n^* u_n = (\boldsymbol{v}^*, \boldsymbol{u}) \tag{1.1.3}$$

where v_i^* denotes the complex conjugate of v_i.

Let c be a number. Then the Hermitian scalar product is linear in the second factor but is 'antilinear' in the first factor:

$$\langle \boldsymbol{v}, c\boldsymbol{u} \rangle = c\langle \boldsymbol{v}, \boldsymbol{u} \rangle, \qquad \langle c\boldsymbol{v}, \boldsymbol{u} \rangle = c^* \langle \boldsymbol{v}, \boldsymbol{u} \rangle \tag{1.1.4}$$

whereas the simple scalar product is linear in both factors.

When $(\boldsymbol{v}, \boldsymbol{u}) = 0$ or $\langle \boldsymbol{v}, \boldsymbol{u} \rangle = 0$ we say that two vectors $\boldsymbol{v}$ and $\boldsymbol{u}$ are orthogonal under the ordinary scalar product or the Hermitian scalar product, respectively. A vector that satisfies $(\boldsymbol{v}, \boldsymbol{v}) = 0$ but $\langle \boldsymbol{v}, \boldsymbol{v} \rangle \neq 0$ is said to be *self-orthogonal* or *isotropic.*

Isotropic vectors were introduced by Cartan (1913) to formulate the 'spinor algebra.' Obviously, if $\langle \boldsymbol{v}, \boldsymbol{v} \rangle = 0$, then $\boldsymbol{v}$ is a null vector.

1.2 Linear transformations and matrices

Consider a set of n linear equations

$$\begin{aligned} y_1 &= A_{11}x_1 + A_{12}x_2 + \cdots + A_{1n}x_n \\ y_2 &= A_{21}x_1 + A_{22}x_2 + \cdots + A_{2n}x_n \\ &\cdots \quad \cdots \quad \cdots \\ y_n &= A_{n1}x_1 + A_{n2}x_2 + \cdots + A_{nn}x_n \end{aligned}$$

which may be written in a more compact form

$$y_j = \sum_{i=1}^{n} A_{ji}x_i; \qquad j = 1, 2, \ldots, n$$

The set of equations is said to form a linear transformation in n variables. It may be regarded as a point transformation that brings a point x to another point y in $V^{(n)}$. The point transformation is a mapping of the n-dimensional space $V^{(n)}$ into itself and is completely defined by the set of n^2 quantities $\{A_{ji}\}$, which can be any complex numbers. We therefore associate with the transformation the array of numbers called an $n \times n$ *matrix,*[1] denoted $A = \|A_{ji}\|$. Then the set of equations is written as follows:

$$y = Ax \tag{1.2.1}$$

Two matrices $A = \|A_{ji}\|$ and $B = \|B_{ji}\|$ are said to be equal if and only if $A_{ji} = B_{ji}$ for all j and i.

We define addition of two $n \times n$ matrices by the rule

$$(A + B)_{ji} = A_{ji} + B_{ji}$$

[1] An $n \times n$ matrix is called a square matrix. Hereafter we are primarily concerned with square matrices unless otherwise specified.

in accordance with the vector addition $z = Ax + Bx = (A + B)x$. Then addition is commutative and associative, for $A + B = B + A$ and $A + (B + C) = (A + B) + C$.

Now let $z_k = \sum_j B_{kj} y_j$ be a second transformation by an $n \times n$ matrix $B = \|B_{kj}\|$. The effect of two transformations consecutively by A and then by B produces a third transformation

$$z_k = \sum_j B_{kj} A_j = \sum_{ji} B_{kj} A_{ji} x_i; \qquad \text{i.e. } z = BAx$$

Therefore we define the product P of two matrices by the rule

$$P_{ki} = \sum_j B_{kj} A_{ji} \qquad \text{or} \qquad P = BA$$

The rule is analogous to the rule for the product of two determinants. If we denote the determinant of a matrix A by $\det A$, we have

$$\text{if } P = BA, \qquad \text{then} \qquad \det P = \det B \det A \tag{1.2.2}$$

Note that $\det B \det A = \det A \det B$ since determinants are numbers. However, the matrix product BA need not equal AB in general; e.g. see (1.2.5b). Thus, matrix multiplication need not be commutative. Multiplication is, however, associative: $A(BC) = (AB)C$, so that the product is simply written ABC.

Another important characteristic property of a matrix A is the *trace* of A defined by the sum of the diagonal elements

$$\operatorname{tr} A = \sum_i A_{ii}$$

Then, the trace of a product ABC equals the trace of the product BCA, because

$$\operatorname{tr} ABC = \sum_{ijk} A_{ij} B_{jk} C_{ki} = \sum_{ijk} B_{jk} C_{ki} A_{ij} = \operatorname{tr} BCA$$

That is, the trace of a product of matrices is invariant under a cyclic permutation of the matrices.

The matrix with unity in each position in the leading diagonal and zero elsewhere is called the unit matrix, denoted $\mathbf{1} = \|\delta_{ij}\|$, where $\delta_{ij} = 0$ $(i \neq j)$ and $\delta_{ii} = 1$. The unit matrix corresponds to the identity transformation. A matrix of the form $D = \|d_i \delta_{ij}\|$ that has diagonal elements $d_1, d_2, \ldots, d_n$ and zero elsewhere is called a diagonal matrix. It is denoted by

$$D = \operatorname{diag}[d_1, d_2, \ldots, d_n] \tag{1.2.3a}$$

then

$$\det D = d_1 d_2 \ldots d_n, \qquad \operatorname{tr} D = d_1 + d_2 + \cdots + d_n$$

Two diagonal matrices always commute and their product gives a diagonal matrix; in fact, let $D' = \operatorname{diag}[d'_1, d'_2, \ldots, d'_n]$ be another diagonal matrix, then

$$DD' = \operatorname{diag}[d'_1 d_1, d'_2 d_2, \ldots, d'_n d_n] = D'D \tag{1.2.3b}$$

If the determinant of a matrix is not zero, then the matrix is said to be non-singular. If the coefficient matrix A of (1.2.1) is non-singular, then the equation (1.2.1) may be solved for x; i.e. there exists the inverse transformation from y to x, which may be written as

$$x = A^{-1}y$$

where A^{-1} is called the inverse of A and satisfies

$$A^{-1}A = AA^{-1} = 1 \tag{1.2.4}$$

If, however, $\det A = 0$, there exists no inverse of A and the matrix A is said to be singular. When $y = 0$, on the other hand, the equation (1.2.1) has non-null solutions for x, if and only if A is singular; this has an important application in the theory of matrix diagonalization. Note that, for a product of non-singular matrices, we have $(ABC)^{-1} = C^{-1}B^{-1}A^{-1}$.

The product of a number c and a matrix $A = \|A_{ij}\|$ is a matrix whose elements are c times the elements of A:

$$cA = \|cA_{ij}\|$$

Accordingly, a number commutes with any matrix. Note, however, that

$$\det cA = c^n \det A$$

where n is the dimensionality of A.

The well-known examples of matrices in two dimensions are the *Pauli spin* matrices defined by

$$\sigma_x = \begin{bmatrix} 0 & 1 \\ 1 & 0 \end{bmatrix}, \qquad \sigma_y = \begin{bmatrix} 0 & -\mathrm{i} \\ \mathrm{i} & 0 \end{bmatrix}, \qquad \sigma_z = \begin{bmatrix} 1 & 0 \\ 0 & -1 \end{bmatrix} \tag{1.2.5a}$$

Their determinants are all equal to -1, and their traces are all zero. They also anticommute with each other:

$$\sigma_x\sigma_y = -\sigma_y\sigma_x, \qquad \sigma_y\sigma_z = -\sigma_z\sigma_y, \qquad \sigma_z\sigma_x = -\sigma_x\sigma_z \tag{1.2.5b}$$

Their squares are all equal to the unit matrix **1**; i.e. the σ_i are all involutional satisfying $x^2 = 1$. Therefore, they form a set of anticommuting matrices satisfying, with $\sigma_x = \sigma_1$, $\sigma_y = \sigma_2$ and $\sigma_z = \sigma_3$,

$$[\sigma_\nu, \sigma_\mu]_+ = \sigma_\nu\sigma_\mu + \sigma_\mu\sigma_\nu = 2\delta_{\nu\mu}; \qquad \nu, \mu = 1, 2, 3 \tag{1.2.6}$$

Furthermore, $\sigma_1\sigma_2\sigma_3 = \mathrm{i}$ so that $\sigma_1\sigma_2 = \mathrm{i}\sigma_3$. A set of matrices that satisfy anti-commutation relations as given by (1.2.6) is said to form *a Clifford algebra*.

Now, consider a set G of all non-singular matrices M in n dimensions. Then a product of two members of G is also a non-singular matrix belonging to G. This property is called *the group property* of G or the closure of G. Summarizing the properties of G discussed in this section, we may conclude that the set G satisfies the following properties:

1. The group property:[2] if $M_1, M_2 \in G$ then $M_1M_2 \in G$.
2. The associative property for the product: $(M_1M_2)M_3 = M_1(M_2M_3)$.
3. Existence of the unit matrix **1**: $\mathbf{1}M = M\mathbf{1}$ for any $M \in G$.
4. Existence of the inverse M^{-1} for any $M \in G$: $M^{-1}M = MM^{-1} = \mathbf{1}$.

[2] $M \in G$ means M belongs to the set G.

A set of elements that satisfies these four properties is said to form *a group*. Thus the set of non-singular $n \times n$ matrices forms a group called *the group of general linear transformations* in n dimensions, denoted $GL(n)$. Full discussion of the axiomatic system of a group will be given later in Chapter 3. Here, we need the group axiom simply to characterize a set of matrices.

1.2.1 Functions of a matrix

Let us define powers of a non-singular matrix A by

$$A^0 = 1, \qquad A^1 = A, \qquad A^{n+1} = A^n A, \qquad A^{-n} = (A^{-1})^n$$

where n is an integer. Then, corresponding to a function $f(z)$ of a scalar variable z that can be expanded in powers of z,

$$f(z) = \sum_{n=-\infty}^{\infty} c_n z^n = c_0 + c_1 z + c_2 z^2 + \cdots + c_{-1} z^{-1} + c_{-2} z^{-2} + \cdots \tag{1.2.7a}$$

one can define a function of a matrix A as follows:

$$f(A) = \sum_{n=-\infty}^{\infty} c_n A^n \tag{1.2.7b}$$

Such a function $f(A)$ commutes with any other function $g(A)$ of A. An elementary example of a function of A is the exponential function

$$f(sA) = \exp(sA) = 1 + sA + s^2 A^2/2! + \cdots + s^n A^n/n! + \cdots$$

where s is a scalar parameter. One can differentiate the function $\exp(sA)$ with respect to s regarding A as a constant and obtain

$$f'(sA) = \mathrm{d}[\exp(sA)]/\mathrm{d}s = A \exp(sA)$$

where $f'(z)$ is the derivative of $f(z)$ with respect to z. This can be checked by the expansion.

In the special case of a diagonal matrix D

$$D = \mathrm{diag}\,[d_1, d_2, \ldots, d_n]$$

the function $f(D)$ of D is simply given by a diagonal matrix with elements $f(d_1)$, $f(d_2), \ldots, f(d_n)$, i.e.

$$f(D) = \mathrm{diag}\,[f(d_1), f(d_2), \ldots, f(d_n)] \tag{1.2.7c}$$

This follows simply from the definition (1.2.7b). Thus, for example,

$$\exp\left(\begin{bmatrix} d_1 & 0 \\ 0 & d_2 \end{bmatrix}\right) = \begin{bmatrix} \mathrm{e}^{d_1} & 0 \\ 0 & \mathrm{e}^{d_2} \end{bmatrix} \tag{1.2.7d}$$

For the Pauli spin σ_z defined by (1.2.5a), we have

$$\exp(\lambda \sigma_z) = \begin{bmatrix} \mathrm{e}^{\lambda} & 0 \\ 0 & \mathrm{e}^{-\lambda} \end{bmatrix} \tag{1.2.7e}$$

where λ is a constant.

1.2.2 *Special matrices*

From a given matrix $A = \|A_{ij}\|$ one can define a new matrix by complex conjugation or by transposition of the rows and columns. The complex conjuguate A^* and *the transpose* $A^\sim$ of A are defined by the elements

$$(A^*)_{ij} = A^*_{ij}, \qquad (A^\sim)_{ij} = A_{ji} \tag{1.2.8a}$$

where A^*_{ij} is the complex conjugate of the element A_{ij}. For a product of matrices we have

$$(ABC)^* = A^* B^* C^*, \qquad (ABC)^\sim = C^\sim B^\sim A^\sim$$

Note that a transformation of a vector x by a matrix A can be written in the following two ways:

$$\sum_j A_{ij} x_j = \sum_j x_j A^\sim_{ji}$$

If one combines the above two operations defined by (1.2.8a), one obtains *the adjoint* or *Hermitian conjugate matrix* defined by

$$A^\dagger = A^{\sim *} = A^{*\sim} \tag{1.2.8b}$$

These special matrices come in when we describe the transformations of the scalar products, $(\boldsymbol{v}, \boldsymbol{u})$ and $\langle \boldsymbol{v}, \boldsymbol{u} \rangle$,

$$(A\boldsymbol{v}, \boldsymbol{u}) = (\boldsymbol{v}, A^\sim \boldsymbol{u}), \qquad \langle A\boldsymbol{v}, \boldsymbol{u} \rangle = \langle \boldsymbol{v}, A^\dagger \boldsymbol{u} \rangle \tag{1.2.9}$$

which can be verified by writing out the respective scalar product. By assuming various relationships between a matrix A and its complex conjugate, transpose, adjoint and inverse, one can obtain special kinds of matrices. For example, if $A^\dagger = A$, A is said to be *self-adjoint* or *Hermitian*. Further examples will be introduced, however, later in Section 1.5 along with their transformation properties.

1.2.3 *Direct products of matrices*

Frequently, we encounter the concept of a direct product $P = A \otimes B$ of two square matrices. It is defined by their elements as follows: let $A = \|A_{ij}\|$ and $B = \|B_{ks}\|$ then

$$P = A \otimes B = \|A_{ij} B_{ks}\|$$

so that the elements of P may be expressed by

$$P_{ik,js} = A_{ij} B_{ks};$$
$$i, j = 1, 2, \ldots, d_A; \qquad k, s = 1, 2, \ldots, d_B \tag{1.2.10a}$$

where the respective dimensionalities d_A and d_B of A and B need not be the same and the dimensionality of the direct product P equals $d_A d_B$. By definition, the product of two similar direct products $P = A \otimes B$ and $P' = A' \otimes B'$ is given by

$$PP' = AA' \otimes BB'$$

where the dimensionalities of A and A' should be the same and so should those of B and B'. Note the trace of a direct product $A \otimes B$ equals the product of the traces of A and B:

$$\mathrm{tr}\, A \otimes B = (\mathrm{tr}\, A)(\mathrm{tr}\, B)$$

Moreover, the transpose of a direct product is given by

$$(A \otimes B)^{\sim} = A^{\sim} \otimes B^{\sim}$$

The direct product $P = A \otimes B$ can also be expressed by *a super matrix* (whose elements are matrices) as follows:

$$A \otimes B = \begin{bmatrix} A_{11}B & A_{12}B & \dots & A_{1n}B \\ A_{21}B & A_{22}B & \dots & A_{2n}B \\ \cdot & \cdot & \dots & \cdot \\ A_{n1}B & A_{n2}B & \dots & A_{nn}B \end{bmatrix} \quad (1.2.10b)$$

where $n = d_A$. For example, for the Pauli spin matrices,

$$\sigma_x \otimes \sigma_y = \begin{bmatrix} 0 & \sigma_y \\ \sigma_y & 0 \end{bmatrix}$$

1.2.4 Direct sums of matrices

Let A and B be two square matrices, then their direct sum is defined by

$$S = A \oplus B = \begin{bmatrix} A & 0 \\ 0 & B \end{bmatrix} \quad (1.2.11)$$

It is a generalization of a diagonal matrix: it is a super matrix whose diagonal elements are square matrices while the remaining elements are null matrices. The respective dimensionalities d_A and d_B of A and B need not be the same and the dimensionality of the direct sum $A \oplus B$ equals $d_A + d_B$. For example, when $d_A = 2$ and $d_B = 3$ we have

$$A \oplus B = \begin{bmatrix} A_{11} & A_{12} & 0 & 0 & 0 \\ A_{21} & A_{22} & 0 & 0 & 0 \\ 0 & 0 & B_{11} & B_{12} & B_{13} \\ 0 & 0 & B_{21} & B_{22} & B_{23} \\ 0 & 0 & B_{31} & B_{32} & B_{33} \end{bmatrix}$$

Let $S' = A' \oplus B'$ be another direct sum that has a similar shape to that of S of (1.2.11), then their matrix product is given by

$$SS' = (A \oplus B)(A' \oplus B') = AA' \oplus BB'$$

Thus, e.g. for a polynomial of S

$$\sum c_n S^n = \sum_n c_n (A \oplus B)^n = \left(\sum_n c_n A^n\right) \oplus \left(\sum_n c_n B^n\right)$$

More generally, let $f(z)$ be a function of z defined by (1.2.7a), then

$$f(A \oplus B) = f(A) \oplus f(B) \quad (1.2.12)$$

i.e. a function of a direct sum $A \oplus B$ equals the direct sum of the functions $f(A)$ and $f(B)$. This means that a functional notation f may be regarded as a linear operator for a direct sum. For example,

$$\exp(A \oplus B) = \exp A \oplus \exp B \quad (1.2.13a)$$

so that (1.2.7d) can be rewritten as follows:

$$\exp(d_1 \oplus d_2) = \exp d_1 \oplus \exp d_2 \tag{1.2.13b}$$

The determinant and trace of a direct sum are given by

$$\det(A \oplus B) = \det A \det B$$

$$\operatorname{tr}(A \oplus B) = \operatorname{tr} A + \operatorname{tr} B \tag{1.2.14}$$

1.3 Similarity transformations

Let us consider the effect of a change of the coordinate system in $V^{(n)}$ on the linear transformation $y = Ax$ given by (1.2.1). Let $x' = (x'_1, x'_2, \ldots, x'_n)$ and $y' = (y'_1, y'_2, \ldots, y'_n)$ be the new coordinates of the two points originally defined by x and y respectively. Then, there exists a non-singular matrix S such that

$$x = Sx', \qquad y = Sy' \tag{1.3.1}$$

Substitution of these into $y = Ax$ yields the new transformation

$$y' = S^{-1}ASx'$$

The matrix defined by

$$A' = S^{-1}AS \tag{1.3.2}$$

is called the transform of the matrix A by the matrix S. Then, A is also the transform of A' by S^{-1}. The matrices which are transforms of one another are called equivalent matrices, while the transformation itself is called *the similarity transformation* or *equivalent transformation* by S.

The equivalent matrices have many properties in common. For example, their determinants are equal and so are their traces:

$$\det S^{-1}AS = \det S^{-1} \det A \det S = \det A \tag{1.3.3}$$

$$\operatorname{tr} S^{-1}AS = \operatorname{tr} ASS^{-1} = \operatorname{tr} A \tag{1.3.4}$$

We say that the determinant and trace of a matrix are *invariant* under a similarity transformation. Another important property of the similarity transformation is that

$$(S^{-1}AS)^n = S^{-1}A^nS \tag{1.3.5}$$

Suppose that $f(A)$ is a function of A that can be expanded in powers of A. Then we have

$$S^{-1}f(A)S = f(S^{-1}AS) \tag{1.3.6}$$

1.3.1 Functions of a matrix (revisited)

Suppose that the matrix A is diagonalized by the similarity transformation

$$S^{-1}AS = \Lambda = \operatorname{diag}[\lambda_1, \lambda_2, \ldots, \lambda_n]$$

Then we have, from (1.3.6) and (1.2.7c),

$$S^{-1}f(A)S = f(\Lambda) = \operatorname{diag}[f(\lambda_1), f(\lambda_2), \ldots, f(\lambda_n)]$$

so that

$$f(A) = Sf(\Lambda)S^{-1} \tag{1.3.7}$$

This gives $f(A)$ explicitly as an $n \times n$ matrix. Even if $f(z)$ cannot be expanded in powers of z, we may define the function $f(A)$ of the matrix A by (1.3.7), provided that A can be diagonalized by a similarity transformation. According to this definition the functions like $\sqrt{A}$ and $\ln A$ become meaningful, even though they cannot be expanded in powers of A. In the next section we shall discuss the condition for a matrix to be diagonalized.

1.4 The characteristic equation of a matrix

Let A be an $n \times n$ matrix that can be diagonalized by a similarity transformation. Then there exists a non-singular matrix T such that

$$T^{-1}AT = \Lambda = \text{diag}\,[\lambda_1, \lambda_2, \ldots, \lambda_n] \tag{1.4.1a}$$

This can be written in the form

$$AT = T\Lambda \qquad \text{or} \qquad \sum_j A_{kj}T_{ji} = T_{ki}\lambda i \tag{1.4.1b}$$

Let us regard the matrix T as a set of n column vectors and write the matrix elements in the form $T^i_j \equiv T_{ji}$ and define the ith column vector by

$$\boldsymbol{t}^{(i)} = (T^i_1, T^i_2, \ldots, T^i_n); \qquad i = 1, 2, \ldots, n \tag{1.4.2}$$

Then we arrive at the eigenvalue problem of the matrix A

$$A\boldsymbol{t}^{(i)} = \lambda_i \boldsymbol{t}^{(i)}; \qquad i = 1, 2, \ldots, n \tag{1.4.3}$$

where λ_i is called *an eigenvalue* of A and $\boldsymbol{t}^{(i)}$ is called *the eigenvector* of A belonging to the eigenvalue λ_i.

Conversely, if the eigenvalue problem (1.4.3) of A is solvable, i.e. if it provides a set of n linearly independent eigenvectors of A, then one constructs a non-singular matrix T by

$$T = [\boldsymbol{t}^{(1)}, \boldsymbol{t}^{(2)}, \ldots, \boldsymbol{t}^{(n)}] \tag{1.4.4}$$

which diagonalizes the matrix A by (1.4.1a).

The condition for the existence of the non-null eigenvector $\boldsymbol{t}^{(i)}$ of A is that the coefficient matrix $[\lambda_i \mathbf{1} - A]$ of (1.4.3) is singular for each λ_i. This implies that λ_i is a root of the following nth-order polynomial equation of x:

$$\begin{aligned} D^{(n)}(x) &= \det(x\mathbf{1} - A) \\ &= x^n + a_1x^{n-1} + \cdots + a_n \\ &= (x - \lambda_1)(x - \lambda_2) \ldots (x - \lambda_n) = 0 \end{aligned} \tag{1.4.5}$$

where $\det(x\mathbf{1} - A)$ is called *the secular determinant* of A. The coefficients a_1, a_2, $\ldots$, a_n are determined by the elements of the matrix A. This equation is called *the characteristic (or secular) equation* of the matrix A (even if A cannot be diagonalized), and its (*characteristic*) roots provide the eigenvalues of A. For an $n \times n$ matrix A, there exist exactly n characteristic roots, some of which may be repeated. One sees immediately that equivalent matrices satisfy the same characteristic equation, since

$$\det(x\mathbf{1} - S^{-1}AS) = \det[S^{-1}(x\mathbf{1} - A)S] = \det(x\mathbf{1} - A)$$

This means that all the characteristic roots and hence every coefficient of the characteristic equation are invariant under any similarity transformation. Note, in particular, that two invariants of A that we know already are given by the coefficients as follows: using (1.4.1a) and (1.4.5),

$$\operatorname{tr} A = \sum_i \lambda_1 = -a_1, \qquad \det A = \prod_i \lambda_i = (-1)^n a_n$$

The eigenvectors $\boldsymbol{t}^{(i)}$ and $\boldsymbol{t}^{(j)}$ belonging to different eigenvalues λ_i and λ_j are linearly independent. Suppose that $\boldsymbol{t}^{(i)}$ and $\boldsymbol{t}^{(j)}$ are linearly dependent, then there exist non-null coefficients such that

$$c_i\boldsymbol{t}^{(i)} + c_j\boldsymbol{t}^{(j)} = 0$$

By applying $(A - \lambda_j\mathbf{1})$ to both sides of this equation from the left we obtain $c_i(\lambda_i - \lambda_j) = 0$, which yields $c_i = 0$ for $\lambda_i \neq \lambda_j$, and hence $c_j = 0$ also. Thus, if all eigenvalues of A are different then all eigenvectors of A are linearly independent. This means that the matrix T defined by (1.4.4) with these eigenvectors is non-singular so that A can be diagonalized by T. On the other hand, if some of the eigenvalues of A are degenerate, the eigenvalue problem need not provide n linearly independent eigenvectors to form a non-singular transformation matrix T.

Before establishing the necessary and sufficient condition for a matrix to be diagonalized by a similarity transformation we shall show the following fundamental theorem for a square matrix.

Theorem 1.4.1. Every matrix satisfies its own characteristic equation.

This theorem follows from the very definition of the characteristic equation given by (1.4.5): since x commutes with A, we can substitute $x = A$ into (1.4.5) and obtain $D^{(n)}(A) = \det(A - A) = 0$, i.e.

$$D^{(n)}(A) = A^n + a_1A^{n-1} + \cdots + a_n\mathbf{1} = 0 \tag{1.4.6}$$

1.4.1 Diagonalizability and projection operators

We shall now discuss the condition for a matrix to be diagonalized by a similarity transformation. Suppose that the characteristic equation of a matrix A has the form

$$D^{(n)}(x) = \prod_{i=1}^{r}(x - \lambda_i)^{n_i} = 0 \tag{1.4.7}$$

where $\lambda_1, \lambda_2, \ldots, \lambda_r$ are all distinct roots with degeneracies $n_1, n_2, \ldots, n_r$ respectively. If the matrix A is diagonalized by a similarity transformation, i.e. $T^{-1}AT = \Lambda$, then the diagonal matrix Λ can be written in the form

$$\Lambda = \operatorname{diag}[\overbrace{\lambda_1, \ldots \lambda_1}^{n_1 \text{ times}}, \overbrace{\lambda_2, \ldots \lambda_2}^{n_2 \text{ times}}, \ldots, \overbrace{\lambda_r, \ldots \lambda_r}^{n_r \text{ times}}] \tag{1.4.8}$$

Accordingly the diagonal matrix Λ satisfies the following rth-order polynomial equation which has no multiple root:

$$p^{[r]}(x) = \prod_{i=1}^{r}(x - \lambda_i) = 0 \tag{1.4.9}$$

and hence the matrix A also satisfies $p^{[r]}(A) = 0$.

The polynomial equation of the least order satisfied by the matrix A is called *the reduced characteristic equation* of A. The following theorem holds (Littlewood 1950).

Theorem 1.4.2. The condition for a matrix to be diagonalized by a similarity transformation is that the reduced characteristic equation has no multiple root.

We have already shown through (1.4.9) that the condition is necessary. Instead of giving the sufficiency proof directly, it is more profitable to introduce the '*projection operator*' method which provides a general method of constructing a transformation matrix T (which diagonalizes A) based on the reduced characteristic equation (1.4.9) which has no multiple root. For this purpose we define a set of matrix operators by

$$P_\nu(A) = (1/\chi_\nu)\prod_{\sigma\neq\nu}^{r}(A - \lambda_\sigma \mathbf{1}); \qquad \chi_\nu = \prod_{\sigma\neq\nu}^{r}(\lambda_\nu - \lambda_\sigma) \neq 0 \tag{1.4.10}$$

where $\nu = 1, 2, \ldots, r$. These are normalized such that $P_\nu(\lambda_\mu \mathbf{1}) = \delta_{\nu\mu}$. The operators satisfy the eigenvalue problem

$$AP_\nu = \lambda_\nu P_\nu; \qquad \nu = 1, 2, \ldots, r \tag{1.4.11}$$

because $(A - \lambda_\nu \mathbf{1})P_\nu$ is proportional to $p^{[r]}(A)$, which equals zero from (1.4.9). Each P_ν has n_ν linearly independent column vectors: this is seen most easily from Jordan's canonical form[3] of P_ν, which is a triangular matrix whose diagonal elements are all zero except for n_ν unit diagonal elements. Thus we obtain n_ν linearly independent eigenvectors belonging to the eigenvalue λ_ν from the column vectors of P_ν. Therefore, we can construct a non-singular transformation matrix T from a set of n $(= n_1 + n_2 + \cdots + n_r)$ linearly independent column vectors of $P_1, P_2, \ldots, P_r$. This provides the sufficiency proof.

Further algebraic properties of P_ν are as follows. The set $\{P_\nu\}$ satisfies the orthogonality relations

$$P_\nu P_\mu = \delta_{\nu\mu} P_\mu; \qquad \nu, \mu = 1, 2, \ldots, r \tag{1.4.12a}$$

and also the completeness relation

$$\sum_\nu P_\nu = \mathbf{1} \tag{1.4.12b}$$

These are seen most easily from the diagonalized form of P_ν

[3] Any square matrix M can be reduced to a triangular form $||M_{ij}||$, where $M_{ij} = 0$ for $i > j$ by a similarity transformation. It is called Jordan's canonical form, where the diagonal elements are the characteristic roots of M (Murnaghan 1938).

$$T^{-1}P_\nu T = (1/\chi_\nu)\prod_{\sigma\neq\nu}^{r}(\Lambda - \lambda_\sigma \mathbf{1})$$

$$= \mathrm{diag}\,[0, 0, \ldots 0, \overbrace{1, 1, \ldots 1}^{n_\nu \text{ times}}, 0, 0, \ldots, 0] \tag{1.4.13}$$

The operator P_ν is called *a projection operator*: for since it satisfies $P_\nu^2 = P_\nu$, the projection of a vector $\boldsymbol{v}$ with P_ν defined by $P_\nu \boldsymbol{v}$ remains the same for further projections, i.e. $P_\nu(P_\nu \boldsymbol{v}) = P_\nu \boldsymbol{v}$.

The projection operator method of determining the eigenvectors described above is effective if the dimension of the matrix is not very high (see Section 1.6 exercises). There exists, however, a more effective general method for the matrix diagonalization introduced by Kim (1979a, b) which will be discussed later in Section 2.2.

1.5 Unitary transformations and normal matrices

A wide variety of matrices which appear in physics can be diagonalized by *unitary transformations*, i.e. similarity transformations with unitary matrices. Suppose that a transformation U leaves invariant the Hermitian scalar product of two vectors $\boldsymbol{x}$ and $\boldsymbol{y}$ in $V^{(n)}$, then using (1.2.9) we have

$$\langle U\boldsymbol{x}, U\boldsymbol{y}\rangle = \langle \boldsymbol{x}, U^\dagger U\boldsymbol{y}\rangle = \langle \boldsymbol{x}, \boldsymbol{y}\rangle \tag{1.5.1}$$

so that the matrix U satisfies

$$U^\dagger U = \mathbf{1}$$

Such a matrix U is called *a unitary matrix*. It is non-singular, for $|\det U|^2 = 1$, and the inverse is given by $U^{-1} = U^\dagger$ so that

$$U^\dagger U = UU^\dagger = \mathbf{1} \tag{1.5.2}$$

which is written explicitly, in terms of the matrix elements,

$$\sum_s U^*_{is} U_{js} = \sum_s U^*_{si} U_{sj} = \delta_{ij} \tag{1.5.3}$$

The basic theorem on matrix diagonalization through a unitary transformation may be stated as follows.

Theorem 1.5.1. The condition for a matrix M to be diagonalized by a unitary transformation is that M is normal, i.e. it satisfies

$$M^\dagger M = MM^\dagger \tag{1.5.4}$$

where $M^\dagger$ is the Hermitian conjugate of M.

For simplicity, we shall give here the proof of the theorem for the necessary condition only and refer the proof for the sufficient condition to the existing literature; e.g. Murnaghan (1938). Suppose that M is diagonalized by a unitary transformation, then we have

$$U^\dagger M U = \Lambda \tag{1.5.5}$$

where Λ is a diagonal matrix. Then, by substituting this into the commutation relation

$\Lambda\Lambda^\dagger = \Lambda^\dagger\Lambda$ (which holds since any two diagonal matrices commute), we arrive at (1.5.4).

The importance of this theorem is due to the fact that most of the matrices which appear in physics are normal. For example, a Hermitian matrix H is normal: for since $H^\dagger = H$

$$H^\dagger H = HH = HH^\dagger \tag{1.5.6a}$$

and a unitary matrix is normal; for by definition

$$U^\dagger U = UU^\dagger = 1 \tag{1.5.6b}$$

The unitary transformation greatly simplifies the matrix diagonalization discussed in Section 1.4, because it eliminates the necessity of calculating the inverse transformation matrix, for $U^{-1} = U^\dagger$. Let us express U in terms of the column vectors in the form, analogous to (1.4.4),

$$U = [\boldsymbol{u}^{(1)}, \boldsymbol{u}^{(2)}, \ldots, \boldsymbol{u}^{(n)}] \tag{1.5.7}$$

Then the unitary transformation (1.5.5) is reduced to the eigenvalue problem, as in Section 1.4, through $MU = U\Lambda$ or through its adjoint $M^\dagger U = U\Lambda^*$

$$M\boldsymbol{u}^{(i)} = \lambda_i \boldsymbol{u}^{(i)} \quad \text{or} \quad M^\dagger \boldsymbol{u}^{(i)} = \lambda_i^* \boldsymbol{u}^{(i)}; \qquad i = 1, 2, \ldots, n \tag{1.5.8}$$

where the eigenvectors satisfy the orthogonality relations from the unitary condition

$$\langle \boldsymbol{u}^{(i)}, \boldsymbol{u}^{(j)} \rangle = \delta_{ij} \tag{1.5.9}$$

Conversely, starting from the eigenvalue problem (1.5.8) of the matrix M one can construct the unitary matrix U (which diagonalizes M) by (1.5.7). Note that two eigenvectors belonging to different eigenvalues are orthogonal, for since

$$\lambda_i \langle \boldsymbol{u}^{(j)}, \boldsymbol{u}^{(i)} \rangle = \langle \boldsymbol{u}^{(j)}, M\boldsymbol{u}^{(i)} \rangle = \langle M^\dagger \boldsymbol{u}^{(j)}, \boldsymbol{u}^{(i)} \rangle = \lambda_j \langle \boldsymbol{u}^{(j)}, \boldsymbol{u}^{(i)} \rangle$$

we have $\langle \boldsymbol{u}^{(j)}, \boldsymbol{u}^{(i)} \rangle = 0$ if $\lambda_i \neq \lambda_j$. This implies that when we solve the eigenvalue problem (1.5.8) for a non-degenerate set of the eigenvalues, it is only necessary to normalize the eigenvectors to obtain the correct column vectors of U. In the case of a degenerate eigenvalue, however, one has to orthonormalize the corresponding eigenvectors according to (1.5.9).

1.5.1 Examples of normal matrices

In addition to Hermitian matrices and unitary matrices we have the following examples of the normal matrices:

(1) the anti-Hermitian matrix K: $K^\dagger = -K$
(2) the real orthogonal matrix R: $R^\sim R = RR^\sim = \mathbf{1}$ and $R^* = R$ (it is a special case of the unitary matrix since $R^\dagger = R^\sim = R^{-1}$)
(3) the real symmetric matrix S: $S^\sim = S$ and $S^* = S$ (it is a special case of the Hermitian matrix since $S^\dagger = S$) and
(4) the real antisymmetric matrix A: $A^\sim = -A$ and $A^* = A$ (it is a special case of the anti-Hermitian matrix, for $A^\dagger = A^\sim = -A$).

Remark 1. The eigenvalues of a Hermitian matrix are all real whereas those of a unitary matrix are all unimodular: for from (1.5.5) and its adjoint

$$U^\dagger M U = \Lambda, \qquad U^\dagger M^\dagger U = \Lambda^*$$

it follows that if $M^\dagger = M$ then $\Lambda = \Lambda^*$ and if $M^\dagger = M^{-1}$ then $\Lambda^{-1} = \Lambda^*$ i.e. $\Lambda\Lambda^* = \mathbf{1}$.

Remark 2. A real symmetric matrix S can be diagonalized by a real orthogonal matrix and so is a unitary symmetric matrix N. This is because their eigenvectors can be chosen to be real. To see this let $M\boldsymbol{v} = \lambda\boldsymbol{v}$, where $\boldsymbol{v}$ is an eigenvector belonging to an eigenvalue λ of a matrix M, then by the complex conjugation $M^*\boldsymbol{v}^* = \lambda^*\boldsymbol{v}^*$ we have, for $M = S$ or N,

$$M\boldsymbol{v}^* = \lambda\boldsymbol{v}^*$$

because, on account of Remark 1, M and λ are real for $M = S$ while $M^* = M^{-1}$ and $\lambda^* = \lambda^{-1}$ for $M = N$. Thus the eigenvector $\boldsymbol{v}$ can be replaced by its real and/or imaginary parts. These special cases are frequently encountered in the theory of matrix diagonalization in physics.

1.6 Exercises

1. The two-dimensional planar rotation through an angle θ is described by a 2×2 matrix

$$R^{(2)}(\theta) = \begin{bmatrix} \cos\theta & -\sin\theta \\ \sin\theta & \cos\theta \end{bmatrix} = \cos(\theta)\mathbf{1} - \mathrm{i}\sigma_y \sin\theta \tag{1.6.1a}$$

with the Pauli spin matrix σ_y defined by (1.2.5a). Show the following.

(1a) The characteristic equation is given by

$$x^2 - 2\cos(\theta)x + 1 = 0 \tag{1.6.1b}$$

(1b) The eigenvalues are $\mathrm{e}^{\mathrm{i}\theta}$ and $\mathrm{e}^{-\mathrm{i}\theta}$.

(1c) By direct matrix multiplication verify that $R^{(2)}(\theta)$ satisfies (1.6.1b), cf. Theorem 1.4.1.

(1d) From (1.4.10), the projection operators P_1 and P_2 of $R^{(2)}$ corresponding to the eigenvalues $\lambda_1 = \mathrm{e}^{-\mathrm{i}\theta}$ and $\lambda_2 = \mathrm{e}^{\mathrm{i}\theta}$, respectively, are given by

$$P_1 = \frac{1}{2}\begin{bmatrix} 1 & -\mathrm{i} \\ \mathrm{i} & 1 \end{bmatrix}, \qquad P_2 = \frac{1}{2}\begin{bmatrix} 1 & \mathrm{i} \\ -\mathrm{i} & 1 \end{bmatrix} \tag{1.6.1c}$$

(1e) The transformation matrix T which diagonalizes $R^{(2)}$ is given by

$$T = \frac{1}{\sqrt{2}}\begin{bmatrix} 1 & \mathrm{i} \\ \mathrm{i} & 1 \end{bmatrix} = \frac{1}{\sqrt{2}}(1 + \mathrm{i}\sigma_x) \tag{1.6.1d}$$

of which the first column vector is from P_1 while the second column vector is from P_2. Note that T is unitary.

(1f) The diagonalized form of $R^{(2)}$ is given, using (1.2.7e) or (1.2.13b), by

$$T^\dagger R^{(2)} T = \begin{bmatrix} \mathrm{e}^{-\mathrm{i}\theta} & 0 \\ 0 & \mathrm{e}^{\mathrm{i}\theta} \end{bmatrix} = \mathrm{e}^{-\mathrm{i}\theta\sigma_z} \tag{1.6.1e}$$

2. Let H be a Hermitian matrix, then $K = \mathrm{i}H$ is an anti-Hermitian matrix, while $U = \exp(\mathrm{i}H)$ is a unitary matrix. (It will be shown in Section 4.1 that any unitary matrix of a finite order can be always expressed in this exponential form.)
3. Let A be an antisymmetric matrix, then $R = \exp A$ is an orthogonal matrix. (It will be shown that any real orthogonal matrix of a finite order can be expressed in this exponential form. See Section 4.2.)
4. Show that the planar rotation $R^{(2)}$ defined by (1.6.1a) can be written in an exponential form as follows:

$$R^{(2)} = \mathrm{e}^{\vartheta}; \qquad \vartheta = \begin{bmatrix} 0 & -\theta \\ \theta & 0 \end{bmatrix} \tag{1.6.2}$$

Hint. Transform (1.6.1e) back to $R^{(2)}$, using $T\sigma_z T^\dagger = \sigma_y$.

5. The vector product $[\boldsymbol{s} \times \boldsymbol{x}]$ of two vectors $\boldsymbol{s}$ and $\boldsymbol{x}$ in three dimensions is defined by the components

$$s_2x_3 - s_3x_2, \qquad s_3x_1 - s_1x_3, \qquad s_1x_2 - s_2x_1$$

Show that it can be written in the form

$$[\boldsymbol{s} \times \boldsymbol{x}] = \omega \boldsymbol{x} \tag{1.6.3a}$$

where ω is a 3×3 antisymmetric matrix defined by

$$\omega = \omega(\boldsymbol{s}) = \begin{bmatrix} 0 & -s_3 & s_2 \\ s_3 & 0 & -s_1 \\ -s_2 & s_1 & 0 \end{bmatrix} \tag{1.6.3b}$$

6. A three-dimensional rotation about a unit vector $\boldsymbol{s}$ through an angle θ is given by

$$R^{(3)}(\boldsymbol{\theta}) = \mathrm{e}^{\theta\omega}, \qquad \boldsymbol{\theta} = \theta \boldsymbol{s} \tag{1.6.4a}$$

where $\omega = \omega(\boldsymbol{s})$ has been defined by (1.6.3b). The beauty of this expression is that it is explicitly given by the rotation angle θ and the unit vector $\boldsymbol{s}$ (called the *axis-vector*, see Section 4.4).

Solution. An infinitesimal displacement $\mathrm{d}\boldsymbol{x}$ of a vector $\boldsymbol{x} \in \boldsymbol{V}^{(3)}$ caused by a counterclockwise rotation about $\boldsymbol{s}$ (viewed from the $+\boldsymbol{s}$ direction) through an infinitesimal angle $\mathrm{d}\theta$ is given by

$$\mathrm{d}\boldsymbol{x} = [\boldsymbol{s} \times \boldsymbol{x}]\,\mathrm{d}\theta = \omega\boldsymbol{x}\,\mathrm{d}\theta \tag{1.6.4b}$$

using the identity (1.6.3a) and the geometrical meaning of the vector product. Integration of the above yields

$$\boldsymbol{x} = \mathrm{e}^{\theta\omega}\boldsymbol{x}_0 = R^{(3)}(\boldsymbol{\theta})\boldsymbol{x}_0 \tag{1.6.4c}$$

where $\boldsymbol{x}_0$ is the initial vector. The antisymmetric matrix ω is called the *infinitesimal rotation* about the axis vector $\boldsymbol{s}$ (see Chapter 4).

7. Diagonalize the rotation matrix $R^{(3)}(\boldsymbol{\theta})$ defined by (1.6.4a) by means of the projection operator method.

Solution. Since $R^{(3)}(\boldsymbol{\theta})$ is a function of the matrix ω given by (1.6.4a), it suffices to diagonalize the infinitesimal rotation ω. It will be diagonalized by a unitary transformation since ω is a normal matrix. From the characteristic equation of ω,

$$\omega^3 + \omega = \omega(\omega - \mathrm{i})(\omega + \mathrm{i}) = 0 \tag{1.6.5}$$

the eigenvalues of ω are 0, i and $-$i. Let the corresponding eigenvectors be $\boldsymbol{u}_1$, $\boldsymbol{u}_2$ and $\boldsymbol{u}_3$, respectively. Then the eigenvector $\boldsymbol{u}_1$ belonging to zero eigenvalue is obvious: it is given by $\boldsymbol{s}$ since $\omega\boldsymbol{s} = \boldsymbol{s} \times \boldsymbol{s} = 0$ from (1.6.3a). The eigenvector $\boldsymbol{u}_2$ belonging to the eigenvalue i is given by a non-null column vector of the projection operator P_2 defined by

$$P_2 = -\omega(\omega + \mathrm{i})/2 \tag{1.6.6}$$

following (1.4.10). Substitution of (1.6.3b) into this yields

$$P_2 = -\frac{1}{2}\begin{bmatrix} s_1^2 - 1 & s_1 s_2 - \mathrm{i}s_3 & s_1 s_3 + \mathrm{i}s_2 \\ s_2 s_1 + \mathrm{i}s_3 & s_2^2 - 1 & s_2 s_3 - \mathrm{i}s_1 \\ s_3 s_1 - \mathrm{i}s_2 & s_3 s_2 + \mathrm{i}s_1 & s_3^2 - 1 \end{bmatrix}$$

which shows that either one of the column vectors may provide the required eigenvector $\boldsymbol{u}_2$, because any one of them is not null. We take the second column for $\boldsymbol{u}_2$

$$\boldsymbol{u}_2 = (s_1 s_2 - \mathrm{i}s_3,\ s_2^2 - 1,\ s_2 s_3 + \mathrm{i}s_1)/[2(1 - s_2^2)]^{1/2}$$

with proper normalization. The eigenvector $\boldsymbol{u}_3$ belonging to $-$i is simply given by $\boldsymbol{u}_2^*$ (the complex conjuguate of $\boldsymbol{u}_2$). These three eigenvectors, $\boldsymbol{s}$, $\boldsymbol{u}_2$ and $\boldsymbol{u}_2^*$, are orthonormal under the Hermitian scalar product so that from (1.5.7) a unitary matrix U defined by

$$U = [\boldsymbol{s},\ \boldsymbol{u}_2,\ \boldsymbol{u}_2^*] \tag{1.6.7}$$

diagonalizes ω as well as $R(\boldsymbol{\theta}) = \mathrm{e}^{\theta\omega}$ as follows:

$$\begin{aligned} U^\dagger \omega U &= \mathrm{diag}\,[0,\ \mathrm{i},\ -\mathrm{i}] \\ U^\dagger R(\boldsymbol{\theta}) U &= \mathrm{diag}\,[1,\ \mathrm{e}^{\mathrm{i}\theta},\ \mathrm{e}^{-\mathrm{i}\theta}] \end{aligned} \tag{1.6.8}$$

2
The theory of matrix transformations

We shall introduce a general theory of matrix transformation that connects two matrices A and B of order $n \times n$ satisfying the same reduced characteristic equation of degree r: $p^{(r)}(x) = 0$, $r \leqslant n$ (Kim 1979a,b). According to this theorem, an intertwining matrix T_{AB} that connects two matrices A and B via $AT_{AB} = BT_{AB}$ is explicitly given by a polynomial of degree $r - 1$ in A and B. In the most important special case in which the reduced characteristic equation of A has no multiple root, this formalism explicitly provides a transformation matrix that diagonalizes the matrix A. It also contains the method of the idempotent matrix (or the projection operator method) for constructing the eigenfunctions of A as a special case. We shall first discuss the transformation of involutional matrices that satisfy $x^2 = 1$, for its simplicity and for its wide application. In a first reading only the first section, Section 2.1, may be recommended.

2.1 Involutional transformations

An involutional matrix is defined as a matrix that satisfies a simple quadratic equation

$$x^2 = \text{constant} \times \mathbf{1} \tag{2.1.1}$$

where $\mathbf{1}$ is the unit matrix in an appropriate dimension.

The 2×2 general solution of this equation is given by a traceless matrix, excluding the trivial constant matrices,

$$M = \begin{bmatrix} f & g \\ h & -f \end{bmatrix}; \qquad M^2 = (f^2 + gh)\mathbf{1} \tag{2.1.2a}$$

where f, g and h are constants, which are complex in general. The determinant of M is given by

$$\det M = -(f^2 + gh) \tag{2.1.2b}$$

Accordingly, if M is unit involutional, i.e. $M^2 = \mathbf{1}$, then it is improper, i.e. $\det M = -1$. The well-known examples are the Pauli spin matrices

$$\sigma_1 = \begin{bmatrix} 0 & 1 \\ 1 & 0 \end{bmatrix}, \qquad \sigma_2 = \begin{bmatrix} 0 & -\mathrm{i} \\ \mathrm{i} & 0 \end{bmatrix}, \qquad \sigma_3 = \begin{bmatrix} 1 & 0 \\ 0 & -1 \end{bmatrix} \tag{2.1.3}$$

Note that they are unit involutional and hence improper.

Further examples are the Dirac γ-matrices and also the Dirac Hamiltonian (Dirac 1947) for the free electron in the momentum representation. The involutional matrix in any arbitrary dimension has been obtained by Kim (1969) via the matrix representations of the general homogeneous linear transformations in two dimensions.

It has been recognized that involutional matrices have deep roots in various problems of mathematical physics. The symmetry operations like inversion, time reversal and charge conjugation are all involutional satisfying (2.1.1). A convenient feature of an involutional matrix is that, if an involutional matrix is normalized by $x^2 = \mathbf{1}$ then its inverse is the same as the original matrix. In an important special case in which an involutional matrix is Hermitian, it is unitary as well. Such a matrix may be called an *IUH* (involutional, unitary and Hermitian) matrix, whereby any two of the three properties guarantee the third. The well-known examples are the Pauli spin matrices and Dirac's γ-matrices. The purpose of the present section is to show the effectiveness of an involutional transformation that converts an involutional matrix A into another involutional matrix satisfying the same quadratic equation (2.1.1). The following lemma holds:

Lemma 2.1.1. Let A and B be two involutional matrices of a given order satisfying $A^2 = B^2 = \mathbf{1}$. If their anticommutator is a c-number ($\neq -2$), i.e.

$$AB + BA = 2c\mathbf{1}, \qquad c \neq -1$$

then there exists an involutional transformation that interchanges A and B via

$$YAY = B \qquad \text{or} \qquad A = YBY; \qquad Y^2 = \mathbf{1} \tag{2.1.4}$$

where

$$Y = (A + B)/(2 + 2c)^{1/2} \tag{2.1.5}$$

The proof is elementary. From $A^2 = B^2$ it follows that the sum $A + B$ is *an intertwining matrix* of A and B:

$$A(A + B) = (A + B)B$$

Moreover, the matrix $A + B$ is involutional satisfying

$$(A + B)^2 = A^2 + AB + BA + B^2 = (2 + 2c)\mathbf{1}$$

Thus, Y defined by (2.1.5) interchanges A and B via (2.1.4).

Corollary 2.1.1. The most general transformation V which connects the involutional matrices A and B of the lemma via a similarity transformation $V^{-1}AV = B$ is given by

$$V = F_A Y = Y F_B \tag{2.1.6}$$

where F_A and F_B are the same function of A and B, respectively.

In the special case in which $F_A = A$ and $F_B = B$ we have, from (2.1.6) and (2.1.5),

$$V = (\mathbf{1} + AB)/(2 + 2c)^{1/2}, \qquad V^{-1} = (\mathbf{1} + BA)/(2 + 2c)^{1/2}, \qquad c \neq -1 \tag{2.1.7}$$

Remark. If both A and B are IUH matrices, then so is Y; however, V is unitary but not involutional in general, for $V^2 = AB$. In two dimensions, the involutional matrices A, B and Y are all improper whereas V is proper, being a product of two improper matrices.

Example 1. The Pauli spin matrices (σ_1, σ_2 and σ_3) anticommute with each other so that for the lemma the involutional transformation which converts σ_1 into σ_2 is given by

$$Y\sigma_1 Y = \sigma_2; \qquad Y = (\sigma_1 + \sigma_2)/\sqrt{2} \tag{2.1.8a}$$

while the transformation corresponding to (2.1.7) is given by

$$V^\dagger \sigma_1 V = \sigma_2; \qquad V = (\mathbf{1} + \mathrm{i}\sigma_3)/\sqrt{2} \tag{2.1.8b}$$

using $\sigma_1\sigma_2 = \mathrm{i}\sigma_3$. Since the Pauli spin matrices are IUH matrices, Y is an IUH matrix and V is unitary. Note also that Y is an improper matrix that transforms σ_3 into $-\sigma_3$, whereas V is a proper matrix that leaves σ_3 invariant. In fact, $V = \exp[\mathrm{i}(\pi/3)\sigma_3]$, which describes $-90°$ rotation of the spin vector about the z-axis (see Chapter 10 for the spinor transformation).

Example 2. Lemma 2.1.1 holds in general for two dimensions, because the anticommutator of two involutional matrices in two dimensions is always a c-number. In fact, let M and M' be two involutional matrices defined by

$$M = \begin{bmatrix} f & g \\ h & -f \end{bmatrix}, \qquad M' = \begin{bmatrix} f' & g' \\ h' & -f' \end{bmatrix} \tag{2.1.9}$$

then their anticommutator satisfies

$$MM' + M'M = [M, M']_+ = (2ff' + gh' + hg')\mathbf{1} \equiv 2c\mathbf{1}$$

If M and M' are normalized by $f^2 + gh = f'^2 + g'h' = 1$, then they are connected by

$$YMY = M'; \qquad Y = (M + M')/(2 + 2c)^{1/2}$$

$$V^{-1}MV = M'; \qquad V = (1 + MM')/(2 + 2c)^{1/2}$$

where $c \neq -1$.

Example 3. Bogoliubov transformation. It describes the diagonalization of an involutional matrix defined by

$$M = \begin{bmatrix} s & t \\ -t & -s \end{bmatrix}; \qquad M^2 = (s^2 - t^2)\mathbf{1} \tag{2.1.10}$$

which is a special case of (2.1.9).

Let us assume that s and t are real and $s^2 - t^2 > 0$, then the eigenvalues of M are given by $\pm\varepsilon$ where $\varepsilon = (s^2 - t^2)^{1/2}$. From the lemma, the involutional transformation which diagonalizes M is given by

$$YMY = \begin{bmatrix} \varepsilon & 0 \\ 0 & -\varepsilon \end{bmatrix}; \qquad Y = N\begin{bmatrix} s+\varepsilon & t \\ -t & -s-\varepsilon \end{bmatrix}$$

where $N = (2\varepsilon s + 2\varepsilon^2)^{1/2}$ with $s > 0$. This transformation is called *the Bogoliubov transformation*, because it occurs for the diagonalization of the Bogoliubov Hamiltonian which describes the phonon field in a super-fluid system (Kim 1979a).

2.2 Application to the Dirac theory of the electron

2.2.1 The Dirac γ-matrices

Let $\{\gamma_1, \gamma_2, \ldots, \gamma_d\}$ be a set of matrices that satisfies the anticommutation relations

$$[\gamma_\nu, \gamma_\mu]_+ \equiv \gamma_\nu\gamma_\mu + \gamma_\mu\gamma_\nu = 2\delta_{\nu\mu}; \quad \nu, \mu = 1, 2, \ldots, d \tag{2.2.1}$$

where $\delta_{\nu\mu}$ are Kronecker's delta. Such a set of matrices is called a set of *Dirac's γ-matrices*. It is said to form *the Clifford algebra* of the order d. The effectiveness of Lemma 2.1.1 in describing the transformations of the γ-matrices is obvious from the fact that any linear combination of the γ-matrices is involutional; in fact, let γ_x be a linear form of a vector $\boldsymbol{x} = \{x_1, x_2, \ldots, x_d\}$ in $V^{(d)}$ defined by

$$\gamma_x = x_1\gamma_1 + x_2\gamma_2 + \cdots + x_d\gamma_d \tag{2.2.2a}$$

then it satisfies

$$(\gamma_x)^2 = x_1^2 + x_2^2 + \cdots + x_d^2 = x^2 \tag{2.2.2b}$$

That is, a set of γ-matrices turns the scalar square x^2 of a vector $\boldsymbol{x} \in V^{(d)}$ into the square of the linear form γ_x of $\boldsymbol{x}$. In order to apply the γ-matrices to describe the Dirac theory of the spinning electron we shall first give some examples of the γ-matrices.

2.2.1.1 The Clifford algebra of the order four

We introduce two commuting sets of 4×4 matrices by the direct products of the Pauli spin matrices with the unit matrix $\mathbf{1}$ in two dimensions as follows:

(i)
$$\Sigma_i = \mathbf{1} \times \sigma_i = \begin{bmatrix} \sigma_i & 0 \\ 0 & \sigma_i \end{bmatrix}; \qquad i = 1, 2, 3 \tag{2.2.3a}$$

(ii)
$$\rho_i = \sigma_i \times \mathbf{1}; \qquad i = 1, 2, 3 \tag{2.2.3b}$$

where

$$\rho_1 = \begin{bmatrix} 0 & \mathbf{1} \\ \mathbf{1} & 0 \end{bmatrix}, \qquad \rho_2 = \begin{bmatrix} 0 & -\mathrm{i}\mathbf{1} \\ \mathrm{i}\mathbf{1} & 0 \end{bmatrix}, \qquad \rho_3 = \begin{bmatrix} \mathbf{1} & 0 \\ 0 & -\mathbf{1} \end{bmatrix}$$

Since by definition these two sets commute with each other and since each set provides an anticommuting set of 4×4 matrices, a set of *the Clifford algebra* of the order four $\{\gamma_1, \gamma_2, \gamma_3, \gamma_4\}$ may be formed by their products as follows:

$$\gamma_i = \rho_2\Sigma_i = \sigma_2 \times \sigma_i = \begin{bmatrix} 0 & -\mathrm{i}\sigma_i \\ \mathrm{i}\sigma_i & 0 \end{bmatrix}; \qquad i = 1, 2, 3$$
$$\gamma_4 = \rho_3 \tag{2.2.4}$$

This set is called *the standard representation* of the γ-matrices for $d = 4$. Another set of four matrices $(\alpha_1, \alpha_2, \alpha_3, \beta)$ that describes the Dirac Hamiltonian is defined by

$$\alpha_i = \rho_1 \Sigma_i = \sigma_1 \times \sigma_i = \begin{bmatrix} 0 & \sigma_i \\ \sigma_i & 0 \end{bmatrix}; \qquad i = 1, 2, 3$$

$$\beta = \rho_3 \tag{2.2.5}$$

This representation is, however, equivalent to the standard representation, because σ_1 and σ_2 are equivalent under a similarity transformation which leaves σ_3 invariant as was shown by (2.1.8b); in fact, we have $T^\dagger \alpha_i T = \gamma_i$ and $T^\dagger \beta T = \gamma_4$ under T defined by

$$T = [(\mathbf{1} + \mathrm{i}\sigma_3)/\sqrt{2}] \times \mathbf{1} \tag{2.2.6}$$

2.2.1.2 The Clifford algebra for $d = 5$

From $(\gamma_1, \gamma_2, \gamma_3, \gamma_4)$ introduced in (2.2.4) we introduce an additional element γ_5 by

$$\gamma_5 = \gamma_1 \gamma_2 \gamma_3 \gamma_4 = -\rho_1 \tag{2.2.7}$$

Then the set $\{\gamma_1, \gamma_2, \gamma_3, \gamma_4, \gamma_5\}$ provides the Clifford algebra for $d = 5$.

Remark. In general, from a $2n \times 2n$ matrix representation for the Clifford algebra of $d = 2n$, one can form the Clifford algebra of the order $d + 1$ by incorporating an additional element γ_{d+1} defined by

$$\gamma_1 \gamma_2 \ldots \gamma_d = \iota \gamma_{d+1} \tag{2.2.8}$$

where ι equals 1 or i depending on whether n is even or odd. By definition, γ_{d+1} anticommutes with every member of the set $\{\gamma_1, \gamma_2, \ldots, \gamma_d\}$ and $(\gamma_{d+1})^2 = 1$ so that the set $\{\gamma_1, \gamma_2, \ldots, \gamma_d, \gamma_{d+1}\}$ forms the Clifford algebra (Brauer and Weyl 1935, Kim 1980a). In addition to the example $\{\gamma_1, \gamma_2, \gamma_3, \gamma_4, \gamma_5\}$ considered above, there is another trivial example: the Pauli spin matrices σ_1 and σ_2 form the Clifford algebra of $d = 2$, whereas σ_1, σ_2 and σ_3 satisfy $\sigma_2\sigma_2 = \mathrm{i}\sigma_3$ and form the Clifford algebra of the order $2 + 1 = 3$.

2.2.2 The Dirac plane waves

The Dirac Hamiltonian (Dirac 1947) for a free particle in the momentum representation is given, in terms of the four matrices $(\alpha_1, \alpha_2, \alpha_3, \beta)$, by

$$H = \boldsymbol{\alpha} \cdot \boldsymbol{p} + m\beta; \qquad c = \hbar = 1 \tag{2.2.9a}$$

where $\boldsymbol{p} = (p_1, p_2, p_3)$ is the momentum and m is the rest mass of the particle, and we have assumed that the velocity of light $c = 1$ and Planck's constant $\hbar = 1$. In view of (2.2.2), the Hamiltonian H is involutional satisfying

$$H^2 = E^2; \qquad E = \pm(m^2 + p^2)^{1/2} \tag{2.2.9b}$$

where $p = |\boldsymbol{p}|$. The energy E can be positive or negative but its absolute value $|E|$ is larger than the rest mass m. From (2.1.4), the Hamiltonian H is diagonalized via an involutional transformation

$$Y_E H Y_E = \beta E; \qquad Y_E^2 = \mathbf{1} \tag{2.2.10}$$

where

$$Y_E = Y_E(\boldsymbol{p}) = N_E(H + \beta E) = N_E[\boldsymbol{\alpha} \cdot \boldsymbol{p} + (E + m)\beta]$$

with the normalization constant N_E

$$N_E^{-1} = E(2 + 2m/E)^{1/2}$$

If we rewrite (2.2.10) in the form

$$HY_E = Y_E \beta E$$

we see that the four column vectors of Y_E expressed by

$$Y_E = [u_1,\ u_2,\ u_3,\ u_4] \tag{2.2.11a}$$

provide the eigenvectors of H, where the νth eigenvector u_ν belongs to the eigenvalue $\varepsilon_\nu E$ such that

$$HU_\nu = \varepsilon_\nu E u_\nu; \qquad \nu = 1,\ 2,\ 3,\ 4 \tag{2.2.11b}$$

with

$$\varepsilon_\nu = \begin{cases} 1 & \text{for } \nu = 1,\ 2 \\ -1 & \text{for } \nu = 3,\ 4 \end{cases}$$

Since the Hamiltonian is Hermitian, Y_E is an IUH matrix. From the unitary property of Y_E, it follows that the set of four column vectors of Y_E provides a complete set of orthonormalized eigenvectors of H for the given momentum $\boldsymbol{p}$ of the particle.

Exercise 1. Write down the explicit form of the involutional matrix Y_E and obtain the four eigenvectors $[u_1,\ u_2,\ u_3,\ u_4]$ of H from the column vectors of Y_E:

$$Y_E = N_E \begin{bmatrix} E+m & 0 & p_z & p_x - \mathrm{i}p_y \\ 0 & E+m & p_x + \mathrm{i}p_y & -p_z \\ p_z & p_x - \mathrm{i}p_y & -(E+m) & 0 \\ p_x + \mathrm{i}p_y & -p_z & 0 & -(E+m) \end{bmatrix} \tag{2.2.12}$$

So far, we have considered the involutional transformation which diagonalizes the Hamiltonian H. According to Lemma 2.1.1, we can transform H into various other forms. An interesting special case is to transform H into a form $E\alpha_p$ ($\alpha_p = \boldsymbol{\alpha} \cdot \boldsymbol{p}/p$) that represents a massless particle with a momentum equal to $E\boldsymbol{p}/p$. In fact, from Lemma 2.1.1, we have

$$Y_\alpha H Y_\alpha = E\alpha_p; \qquad \alpha_p^2 = \mathbf{1} \tag{2.2.13}$$

where

$$Y_\alpha = (H + E\alpha_p)/[2E(E+p)]^{1/2}$$

which is again an IUH matrix.

Exercise 2. Transform the Dirac Hamiltonian into the Weyl Hamiltonian of a massless particle defined by, with $\boldsymbol{\Sigma} = (\Sigma_1,\ \Sigma_2,\ \Sigma_3)$ of (2.2.3a),

$$H_{\mathrm{W}} = E\rho_3\Sigma_p; \qquad \Sigma_p = \boldsymbol{\Sigma} \cdot \hat{\boldsymbol{p}}; \qquad \hat{\boldsymbol{p}} = \boldsymbol{p}/p \tag{2.2.14}$$

where Σ_p is called the *helicity operator.*

Solution. Since $\alpha_p = \rho_1 \Sigma_p$ and since $\boldsymbol{\rho}$ commutes with $\boldsymbol{\Sigma}$, we arrive at, by a further transformation of (2.2.13) which brings ρ_1 to ρ_3,

$$T_W^\dagger H T_W = H_W; \qquad T_W = Y_\alpha(\rho_1 + \rho_3)/\sqrt{2} \tag{2.2.15}$$

where T_W is a unitary matrix.

2.2.3 The symmetric Dirac plane waves

The Dirac plane waves described by $\{u_\nu\}$ of (2.2.11) are degenerate since u_1 and u_2 belong to the energy E while u_3 and u_4 belong to the energy $-E$. These degeneracies may be removed by introducing the simultaneous eigenvectors of the Hamiltonian H and *the helicity operator* Σ_p defined in (2.2.14), which obviously commutes with the Dirac Hamiltonian because Σ_p commutes with $\{\rho_1, \rho_2, \rho_3\}$ and hence with $\alpha_p = \rho_1 \Sigma_p$.

From Lemma 2.1.1, the involutional transformation which diagonalizes the helicity operator Σ_p is given by

$$Y_s \Sigma_p Y_s = s\Sigma_3 = \text{diag}\,[s, -s, s, -s] \tag{2.2.16a}$$

where $s\ (= \pm 1)$ is inserted for convenience and

$$Y_s = Y_s(\hat{\boldsymbol{p}}) = N_s(\Sigma_p + s\Sigma_3); \qquad N_s = (2 + 2s\hat{p}_3)^{-1/2} \tag{2.2.16b}$$

By definition, $Y_s(\hat{\boldsymbol{p}})$ satisfies the following symmetry property:

$$Y_s(\hat{\boldsymbol{p}}) = -Y_{-s}(-\hat{p})$$

Then a unitary transformation U that diagonalizes H and Σ_p simultaneously via

$$U^\dagger H U = \beta E, \qquad U^\dagger \Sigma_p U = s\Sigma_3 \tag{2.2.17a}$$

is given by the product

$$U = Y_E Y_s \tag{2.2.17b}$$

because Y_s commutes with $\beta = \rho_3$ and Y_E commutes with Σ_p. The unitary matrix U defined by (2.2.17b) is highly symmetric. To see this we rewrite Y_E in a form similar to that of the Hamiltonian itself,

$$Y_E = Y_E(\boldsymbol{p}) = p'\alpha_p + m'\beta, \qquad \alpha_p = \boldsymbol{\alpha} \cdot \hat{\boldsymbol{p}} \tag{2.2.18}$$

where

$$p'(E) = \text{sgn}\,(E)\left[\frac{1}{2}\left(1 - \frac{m}{E}\right)\right]^{1/2}, \qquad m'(E) = \left[\frac{1}{2}\left(1 + \frac{m}{E}\right)\right]^{1/2}$$

These two quantities are connected by

$$p'(-E) = -\text{sgn}\,(E)\, m'(E)$$

Now, we write U in the form

$$U = Y_s Y_s Y_E Y_s = Y_s(sp'\alpha_3 + m'\beta)$$

through calculation of $Y_s Y_E Y_s$. Then U takes the desired symmetric form:

$$U = N_s \begin{bmatrix} m'\begin{pmatrix} \hat{p}_3 + s & \hat{p}_- \\ \hat{p}_+ & -\hat{p}_3 - s \end{pmatrix} & sp'\begin{pmatrix} \hat{p}_3 + s & -\hat{p}_- \\ \hat{p}_+ & \hat{p}_3 + s \end{pmatrix} \\ sp'\begin{pmatrix} \hat{p}_3 + s & -\hat{p}_- \\ \hat{p}_+ & \hat{p}_3 + s \end{pmatrix} & -m'\begin{pmatrix} \hat{p}_3 + s & \hat{p}_- \\ \hat{p}_+ & -\hat{p}_3 - s \end{pmatrix} \end{bmatrix} \tag{2.2.19}$$

where $\hat{p}_+ = \hat{p}_1 + \mathrm{i}\hat{p}_2$, $\hat{p}_- = \hat{p}_1 - \mathrm{i}\hat{p}_2$ and N_s has been defined by (2.2.16b). For further properties of $U = Y_E Y_s$ under the parity $P = -\gamma_4$, time reversal $T = -\mathrm{i}\Sigma_2 K$ and the charge conjugation $C = \gamma_2 K$, where K is the complex conjugation, see Kim (1980b). An analogous treatment can be given for the Coulomb Dirac wave (Kim 1980c) and for the representations of the Lorentz group (Kim 1980a).

2.3 Intertwining matrices

Let A be a matrix of order $n \times n$ that satisfies a polynomial equation of degree r ($\leqslant n$):

$$P^{(r)}(x) = x^r + c_1 x^{r-1} + \cdots + c_r = 0 \tag{2.3.1}$$

where $c_1, c_2, \ldots, c_r$ are constant coefficients. In terms of these coefficients we define a kth degree polynomial of x by

$$x^{(k)} = x^k + c_1 x^{k-1} + \cdots + c_k; \qquad k = 0, 1, 2, \ldots, r \tag{2.3.2a}$$

with $c_0 = 1$, then the set satisfies the following recursion formulae:

$$x^{(k+1)} = x x^{(k)} + c_{k+1}; \qquad k = 0, 1, \ldots, r-1 \tag{2.3.2b}$$

with $x^{(0)} = 1$. According to this notation, we have $x^{(r)} = p^{(r)}(x) = 0$ so that $A^{(r)} = p^{(r)}(A) = 0$. Let $M(n \times n, p^{(r)}(x))$ be a set of $n \times n$ matrices, every member of which satisfies $P^{(r)}(x) = 0$, then the matrix A belongs to the set, i.e.

$$A \in M(n \times n, p^{(r)}(x)) \tag{2.3.3}$$

When $P^{(r)}(x) = 0$ is the equation of least degree satisfied by the matrix A, it is called *the reduced characteristic equation* of A. To avoid confusion, the characteristic polynomial of A defined by *the secular determinant* is denoted by

$$D^{(n)}(x) = \det[x\mathbf{1} - A] \tag{2.3.4}$$

where $\mathbf{1}$ is the $n \times n$ unit matrix. The multiplicity of a root λ_ν of this equation is called *the degeneracy of the eigenvalue* λ_ν of A. With these preparations we state the basic theorem of matrix transformations introduced by Kim (1979b) as follows.

Theorem 2.3.1. Let $A, B \in M(n \times n, p^{(r)}(x))$ with $p^{(r)}(x)$ of (2.3.1). Then A and B are connected via

$$A T_{AB} = T_{AB} B \tag{2.3.5}$$

with T_{AB} given by

$$T_{AB} = \sum_{k=0}^{r-1} A^{r-1-k} B^{(k)} = \sum_{k=0}^{r-1} A^{(k)} B^{r-1-k} \tag{2.3.6}$$

where $A^{(k)}$ and $B^{(k)}$ are the kth-degree polynomials defined by (2.3.2a). If T_{AB} is non-singular, then the matrices A and B are equivalent.

Proof. The first equality in (2.3.6) defines T_{AB}. To show the second equality we write out both sides explicitly, using (2.3.2a):

$$T_{AB} = \sum_{k=0}^{r-1}\sum_{h=0}^{k} c_h A^{r-1-k} B^{k-h} = \sum_{s=0}^{r-1}\sum_{h=0}^{s} c_h A^{s-h} B^{r-1-s}$$

where the dummy summation index k of the right-hand side of (2.3.6) is replaced by s for convenience. Then we change the orders of summations for both sides and obtain a new equality to be proven:

$$\sum_{h=0}^{r-1}\sum_{k=h}^{r-1} c_h A^{r-1-k} B^{k-h} = \sum_{h=0}^{r-1}\sum_{s=h}^{r-1} c_h A^{s-h} B^{r-1-s}$$

which is true, however, simply because the two sides are related by the transformation of the summation indices via $r-1-k = s-h$. Next, to show (2.3.5) we use the recursion formulae (2.3.2b) and the second expression of T_{AB} given in (2.3.6) to obtain

$$\begin{aligned} AT_{AB} &= \sum_{k=0}^{r-1} AA^{(k)} B^{r-1-k} \\ &= \sum_{k=0}^{r-1} (A^{(k+1)} - c_{k+1}) B^{r-1-k} \\ &= T_{AB} B + A^{(r)} - B^{(r)} \\ &= T_{AB} B \end{aligned}$$

where we have used $A^{(r)} = B^{(r)}$ and $A^{(0)} = B^{(0)} = c_0 = 1$. Q.E.D.

Note that the above theorem can be generalized by replacing the assumption $A^{(r)} = B^{(r)} = 0$ with $A^{(r)} = B^{(r)}$, because only the latter condition is used for the proof. This generalization may be useful but will not be considered here any further.

It is obvious that, if T_{AB} is non-singular, then the matrices A and B are equivalent. However, the equivalence of A and B does not guarantee that T_{AB} is non-singular. We may call T_{AB} *the characteristic intertwining matrix of A to B* even if T_{AB} is singular. In general, T_{AB} is different from its reverse T_{BA} unless A and B commute or $r = 2$: for $r = 2$,

$$T_{AB} = T_{BA} = A + B - (\lambda_1 + \lambda_2)\mathbf{1} \tag{2.3.7a}$$

where λ_1 and λ_2 are the characteristic roots of A and also of B. This simple special case has many important applications; e.g. the idempotent matrices and involutional matrices satisfying $x^2 = x$ and $x^2 = 1$, respectively, belong to this special case. Another important special case occurs when A and B are general involutional matrices satisfying $x^r = 1$. In this case, we have $x^{(k)} = x^k$ for $0 \leqslant k \leqslant r-1$, so that, from (2.3.6),

$$T_{AB} = \sum_{k=0}^{r-1} A^{r-1-k} B^k = A^{r-1} + A^{r-2}B + \cdots + B^{r-1} \tag{2.3.7b}$$

A more general intertwining matrix V_{AB} that connects A and B via $AV_{AB} = V_{AB}B$ is given by

$$V_{AB} = F_A T_{AB} = T_{AB} F_B \tag{2.3.8a}$$

where F_A and F_B are the same function of A and B respectively. This is analogous to (2.1.6). Thus, for example, for $r = 2$, we have

$$V_{AB} = AT_{AB} = AB - \lambda_1 \lambda_2 \mathbf{1} \tag{2.3.8b}$$

which is a slight generalization of (2.1.7).

For later consideration we shall describe the property of the product $T_{AB}T_{BA}$. From (2.3.5), it follows that the product commutes with A and B, because

$$AT_{AB}T_{BA} = T_{AB}BT_{BA} = T_{AB}T_{BA}A$$

and analogously for B. Moreover, if both A and B are Hermitian, then we have $T^{\dagger}_{AB} = T_{BA}$ for the Hermitian conjugate so that $T_{AB}T_{BA} = T_{AB}T^{\dagger}_{AB}$. Accordingly the product is also Hermitian and positive semidefinite, i.e. the eigenvalues of $T_{AB}T_{BA}$ are positive or zero.

As has been stated before, if the characteristic transformation matrix T_{AB} is non-singular, two matrices A and B are equivalent. It is most desirable to give some simple criteria for the existence of a non-singular characteristic transformation matrix T_{AB}. We shall establish such a criterion for the most important special case in which B is a diagonal matrix equivalent to A. However, even if T_{AB} is singular, Theorem 2.3.1 can provide very significant consequences in some cases. A typical example is the case in which T_{AB} becomes an idempotent matrix: this occurs when B is a constant matrix equal to $\lambda_\nu \mathbf{1}$, where λ_ν is an eigenvalue of B. We shall first discuss this special case.

2.3.1 *Idempotent matrices*

Let $A,\ B \in M(n \times n, p^{(r)}(x))$ and $\lambda_1, \lambda_2, \ldots, \lambda_r$ be the roots of the polynomial equation $p^{(r)}(x) = 0$, some of which could be equal. Let λ_ν be one of the roots and set $B = \lambda_\nu \mathbf{1}$, then from (2.3.6) we obtain

$$T_{AB} = \sum_{k=0}^{r-1} A^{r-1-k} \lambda_\nu^{(k)} = \sum_{k=0}^{r-1} A^{(k)} \lambda_\nu^{r-1-k}; \qquad B = \lambda_\nu \mathbf{1} \tag{2.3.9a}$$

where $T_{AB} = T_{BA}$ with $B = \lambda_\nu \mathbf{1}$, because a constant matrix commutes with any matrix. If we compare this with the following identity:

$$\prod_{\sigma \neq \nu}^{r} (x - \lambda_\sigma) = p^{(r)}(x)/(x - \lambda_\nu) = \sum_{k=0}^{r-1} x^{r-1-k} \lambda_\nu^{(k)} \tag{2.3.9b}$$

which can be shown by dividing the polynomial $p^{(r)}(x)$ by $(x - \lambda_\nu)$, then we obtain, using the recursion formulae (2.3.2b),

$$T_{A,\lambda_\nu \mathbf{1}} = \prod_{\sigma \neq \nu}^{r} (A - \lambda_\sigma \mathbf{1}) = T_{\lambda_\nu \mathbf{1}, A} \tag{2.3.9c}$$

If we assume further that the reduced characteristic equation of A has no multiple root, then T_{AB} with $B = \lambda_\nu \mathbf{1}$ is proportional to the idempotent matrix P_ν belonging to the eigenvalue λ_ν of A defined by (1.4.10), i.e.

$$T_{A,\lambda_\nu 1} = \chi_\nu P_\nu; \qquad \chi_\nu = \prod_{\sigma \neq \nu}^{r} (\lambda_\nu - \lambda_\sigma); \qquad \nu = 1, 2, \ldots, r \tag{2.3.10a}$$

where the matrices P_ν satisfy the orthogonality and completeness relations

$$P_\nu P_\mu = \delta_{\nu\mu} P_\mu, \qquad \sum_{\nu=1}^{r} P_\nu = 1 \tag{2.3.10b}$$

given by (1.4.12a) and (1.4.12b). These relations will be used to prove the existence of a non-singular characteristic transformation matrix $T_{A\Lambda}$ that transforms A into a diagonal matrix Λ equivalent to A.

2.4 Matrix diagonalizations

We shall now develop a general theory of matrix diagonalizations based on Theorem 2.3.1. The condition for a matrix A to be diagonalized by a similarity transformation is that the reduced characteristic equation of A has no multiple root. This condition may be expressed by

$$A \in M(n \times n,\ p^{[r]}(x))$$

where $p^{[r]}(x)$ denotes a polynomial of degree r with all distinct roots. With this preparation, the basic theorem for matrix diagonalizations is stated as follows.

Theorem 2.4.1. Let $A \in M(n \times n,\ p^{[r]}(x))$ and Λ be a diagonal matrix equivalent to A. Then, there exists at least one diagonal matrix Λ that makes the characteristic transformation matrices $T \equiv T_{A\Lambda}$ and/or $\hat{T} \equiv T_{\Lambda A}$ non-singular, so that

$$T^{-1} A T = \Lambda \qquad \text{and/or} \qquad \hat{T} A \hat{T}^{-1} = \Lambda \tag{2.4.1}$$[1]

The proof of this theorem is somewhat involved. Before proceeding with the proof

[1] Suppose that

$$A = \begin{bmatrix} 1 & 0 \\ \alpha & -1 \end{bmatrix}$$

then A can be transformed to

$$\Lambda = \begin{bmatrix} 1 & 0 \\ 0 & -1 \end{bmatrix}$$

by $T_{A\Lambda} = A + \Lambda$ but not to

$$\Lambda' = \begin{bmatrix} -1 & 0 \\ 0 & 1 \end{bmatrix}$$

by $T_{A\Lambda'}$, because

$$T_{A\Lambda'} = A + \Lambda' = \begin{bmatrix} 0 & 0 \\ \alpha & 0 \end{bmatrix}$$

is singular.

we shall give some preparations. From (2.3.6), the explicit forms of T and $\hat{T}$ are given by

$$T = T_{A\Lambda} = \sum_{k=0}^{r-1} A^{r-1-k}\Lambda^{(k)} = \sum_{k=0}^{r-1} A^{(k)}\Lambda^{r-1-k}$$

$$\hat{T} = T_{\Lambda A} = \sum_{k=0}^{r-1} \Lambda^{r-1-k}A^{(k)} = \sum_{k=0}^{r-1} \Lambda^{(k)}A^{r-1-k} \qquad (2.4.2)$$

In the special case in which $r = 2$, we have

$$T = \hat{T} = A + \Lambda - (\lambda_1 + \lambda_2)\mathbf{1} \qquad (2.4.3)$$

Hereafter, we shall call T and $\hat{T}$ *the characteristic transformation matrices* of A, even if they are singular. It is obvious from (2.3.5) that the column (row) vectors of $T(\hat{T})$ are eigenvectors of A even if $T(\hat{T})$ is singular, unless they are null vectors.

Now a diagonal matrix Λ equivalent to A is characterized by a sequence of the whole set of n characteristic roots of A. Let the distinct characteristic roots of A be $\lambda_1, \lambda_2, \ldots, \lambda_r$, and their respective degeneracies be $n_1, n_2, \ldots n_r$. Then $n_1 + n_2 + \cdots + n_r = n$. The matrix elements of Λ may be expressed by

$$\Lambda_{ij} = \lambda_{\nu=g(j)}\delta_{ij}; \qquad \nu = 1, 2, \ldots, r; \qquad i, j = 1, 2, \ldots, n \qquad (2.4.4)$$

where δ_{ij} is the Kronecker delta and $\nu = g(j)$ defines the sequence of the characteristic roots on the principal diagonal of Λ. The inverse $g^{-1}(\nu)$ is a multivalued function of ν that gives the n_ν indices of the columns or the rows belonging to λ_ν. It is obvious that the intersection of two sets of indices belonging to two different roots λ_ν and $\lambda_{\nu'}$ is null,

$$g^{-1}(\nu) \cap g^{-1}(\nu') = 0, \qquad \text{for } \nu \neq \nu' \qquad (2.4.5)$$

This simple and obvious property plays an important role in the proof. If a sequence $\nu = g(j)$ is such that the equal roots are placed on the consecutive positions, then the corresponding Λ is called a *standard form* of Λ.

Now, let us denote the jth column (row) vectors of any matrix M by $M_{\cdot j}$ $(M_{j\cdot})$. Then we can state from (2.4.1) that $T_{\cdot j}$ $(\hat{T}_{j\cdot})$ belongs to the eigenvalue λ_ν of A if $j \in g^{-1}(\nu)$. Accordingly, from (2.4.2) and (2.3.9a), we have

$$T_{\cdot j} = \sum_{k=0}^{r-1} [A^{(k)}]_{\cdot j}\lambda_\nu^{r-1-k} = [T_{A,\lambda_\nu \mathbf{1}}]_{\cdot j}; \qquad j = g^{-1}(\nu)$$

and an analogous expression for $\hat{T}_{j\cdot}$. Hence, from (2.3.10a),

$$T_{\cdot j} = \chi_\nu [P_\nu]_{\cdot j}, \qquad \hat{T}_{j\cdot} = \chi_\nu [P_\nu]_{j\cdot}; \qquad j \in g^{-1}(\nu) \qquad (2.4.6)$$

This means that there exists an $n_\nu \times n_\nu$ submatrix $T^{(\nu)}$ common to T and $\hat{T}$ for each λ_ν. These relations are the crucial relations for the proof of the existence of a non-singular $T(\hat{T})$. One has to show that all the n columns (rows) of $T_{\cdot j}$ $(\hat{T}_{j\cdot})$ can be made linearly independent with a proper choice of the sequence $\nu = g(j)$. Since, as has been shown in (1.4.13), there exist n_ν linearly independent column vectors for P_ν belonging to the eigenvalue λ_ν of A, provided that the reduced characteristic equation has no multiple root, it is only necessary to show that there exists a sequence $\nu = g(j)$ for Λ

such that a set of n_ν linearly independent vectors for each λ_ν is provided by the set of the column vectors $\{[P_\nu]_{\cdot j}\}$ specified by $j \in g^{-1}(\nu)$ (not by any column vectors of P_ν).

By assumption, there exists a non-singular matrix G that diagonalizes A. Then G also diagonalizes P_ν, i.e.

$$G^{-1} P_\nu G = D_\nu; \qquad \nu = 1, 2, \ldots, r \tag{2.4.7}$$

where D_ν is a diagonal matrix that has n_ν unit elements and $n - n_\nu$ zeros on the diagonal such that

$$[D_\nu]_{ij} = \delta_{\nu, h(i)} \delta_{ij} \tag{2.4.8}$$

with $\nu = h(i)$, which defines the sequence of λ_ν in the diagonal matrix $D = G^{-1}AG$. Substitution of $P_\nu = GD_\nu G^{-1}$ into the first of (2.4.6) yields

$$T_{\cdot j} = \chi_\nu \sum_{k \in h^{-1}(\nu)} G_{\cdot k}[G^{-1}]_{kj}; \qquad j \in g^{-1}(\nu) \tag{2.4.9}$$

where $G_{\cdot k}$ is the kth column vector of G.

Since G is non-singular, all the column vectors $G_{\cdot k}$ are linearly independent. Hence, if the $n_\nu \times n_\nu$ coefficient matrix $\Delta^{(\nu)}$ of (2.4.9) with elements

$$\Delta^{(\nu)}_{kj} = [G^{-1}]_{kj}; \qquad k \in h^{-1}(\nu), \qquad j \in g^{-1}(\nu) \tag{2.4.10}$$

is non-singular, then the n_ν column vectors $T_{\cdot j}$ ($j \in g^{-1}(\nu)$) are linearly independent. Now, $\det[\Delta^{(\nu)}]$ is nothing but a minor of $\det[G^{-1}]$. Thus, if we apply the Laplace theorem on expansion of a determinant to $\det[G^{-1}]$ with respect to n_ν columns given by $h^{-1}(\nu)$, changing ν successively from 1 to r, we must have at least one set of non-vanishing minors such that

$$\prod_{\nu=1}^{r} \det[\Delta^{(\nu)}] \neq 0 \tag{2.4.11}$$

with conditions

$$h^{-1}(\nu) \cap h^{-1}(\nu') = 0, \qquad g^{-1}(\nu) \cap g^{-1}(\nu') = 0, \qquad \text{for } \nu \neq \nu' \tag{2.4.12}$$

This means that there exists at least one sequence $\nu = g(j)$ of λ_ν for Λ that makes T non-singular for a given sequence $\nu = h(k)$.

An analogous proof holds for $\hat{T}$. In fact, since the row vector $\hat{T}_{i\cdot}$ is given by

$$\hat{T}_{i\cdot} = \chi_\nu \sum_{k \in h^{-1}(\nu)} G_{ik}[G^{-1}]_{k\cdot}, \qquad i \in g^{-1}(\nu) \tag{2.4.13}$$

by proceeding as before we can find a set of non-vanishing minors of $\det[G]$ that determines a sequence $\nu = g'(i)$ of λ_ν for Λ that makes $\hat{T}$ non-singular. Q.E.D.

It is to be noted that the proper sequences $\nu = g(i)$ and $\nu = g'(i)$ for T and $\hat{T}$, respectively, need not be the same except for $r = 2$ unless there exists a certain symmetry in A. In most cases, however, one can find such a sequence for which both T and $\hat{T}$ are non-singular. The number of ways of defining $\nu = g(i)$ that makes T non-singular is anything from 1 to $n!/(n_1!n_2! \ldots n_r!)$. The minimum number occurs in the exceptional case in which A is triangular. In this case, the transformation matrices T and $\hat{T}$ are also triangular, similar in shape to that of A, and the diagonal elements of T and $\hat{T}$ are given by, from (2.3.9c),

$$T_{ss} = \hat{T}_{ss} = \prod_{\sigma \neq g(s)}^{r} (A_{ss} - \lambda_\sigma); \qquad s \in g^{-1}(\nu) \tag{2.4.14}$$

where $g(s)$ describes the assumed sequence of the characteristic roots in Λ. Thus, to obtain non-singular $T\,(\hat{T})$ one must take Λ equal to the diagonal part of A, i.e.

$$A_{ss} = \lambda_{g(s)}$$

When we have found a proper sequence of λ_ν in Λ that makes $T\,(\hat{T})$ non-singular it is always possible to transform Λ into a standard form whereby the equal roots are placed in the consecutive positions by a unitary transformation of A and Λ that is merely a simultaneous renumbering of the columns and the rows of A and Λ. Hereafter, we shall assume that such a renumbering has been performed so that the proper Λ takes a standard form unless otherwise specified. It will be shown by the illustrative examples given in Section 2.7 that a standard sequence is frequently a proper sequence for a given matrix A without any simultaneous renumbering.

It follows from (2.4.6) that the present method of matrix diagonalization by T is simpler than the method of the projection operators P_ν, because the latter requires construction of r idempotent projection operators $\{P_\nu;\ \nu = 1, 2, \ldots, r\}$ that are linearly independent.

2.5 Basic properties of the characteristic transformation matrices

We shall discuss the basic properties of the characteristic transformation matrices T and $\hat{T}$ which stem from their relations with the idempotent matrices given by (2.4.6). Then, these properties will be used in the next section to construct a new transformation matrix U that is more effective than $T\,(\hat{T})$ in the sense that its inverse is written down immediately and it becomes unitary when A is Hermitian. Let $\nu = g(j)$ be any sequence of λ_ν in Λ equivalent to the matrix $A \in M(n \times n,\ p^{[r]}(x))$. Then from (2.4.6) and (2.3.10b) we have

$$[\hat{T}T]_{ij} = \begin{cases} 0, & \text{for } g(i) \neq g(j) \\ \chi_\nu T_{ij}^{(\nu)}, & \text{for } \nu = g(i) = g(j) \end{cases} \tag{2.5.1}$$

where $T^{(\nu)}$ is an $n_\nu \times n_\nu$ submatrix common to T and $\hat{T}$

$$T_{ij}^{(\nu)} \equiv T_{ij} = \hat{T}_{ij}; \qquad i, j \in g^{-1}(\nu) \tag{2.5.2}$$

These are remarkable properties of T and $\hat{T}$. If we assume that there is a standard form for Λ, we can express it in the form of a direct sum:

$$\Lambda = \lambda_1 \mathbf{1}_1 \oplus \cdots \oplus \lambda_r \mathbf{1}_r \tag{2.5.3}$$

where $\mathbf{1}_\nu$ is the $n_\nu \times n_\nu$ unit matrix. Then $\hat{T}T$ and $\hat{T}AT$ also take the forms of direct sums:

$$\hat{T}T = \chi_1 T^{(1)} \oplus \chi_2 T^{(2)} \oplus \cdots \oplus \chi_r T^{(r)} \tag{2.5.4}$$

$$\hat{T}AT = \lambda_1 \chi_1 T^{(1)} \oplus \lambda_2 \chi_2 T^{(2)} \oplus \cdots \oplus \lambda_r \chi_r T^{(r)} \tag{2.5.5}$$

with use of $\hat{T}AT = \hat{T}T\Lambda$.

These equations suggest a method by which one can avoid a direct calculation of the inverse T^{-1} $(\hat{T}^{-1})$ for the diagonalization of the matrix A when both T and $\hat{T}$ are non-singular. From (2.5.1), we may regard $\{\chi_\nu T_{ij}^{(\nu)}\}$ for a given ν as the scalar products of two sets of vectors $\{\hat{T}_{i\cdot}\}$ and $\{T_{\cdot j}\}$ belonging to λ_ν of A. Then the mutual orthogonalization of these two vector sets (like the Schmidt orthogonalization method) may lead to the desired diagonalization of A, as will be discussed in the next section.

As a preparation we shall first discuss the condition for the common submatrices $T^{(\nu)}$ to be non-singular. From (2.5.4), the determinant of $\hat{T}T$ is given by

$$\det \hat{T}T = \prod_{\nu=1}^{r} (\chi_\nu)^{n_\nu} \det T^{(\nu)} \tag{2.5.6}$$

Thus, if both T and $\hat{T}$ are non-singular, then all $T^{(\nu)}$, $\nu = 1, 2, \ldots, r$, are non-singular and vice versa. If one of $T^{(\nu)}$ is singular, at least one of T and $\hat{T}$ must be singular. Hereafter, we shall assume that all $T^{(\nu)}$ are non-singular unless otherwise specified.

If the matrix A is Hermitian, then $T^\dagger = \hat{T}$ and thus the common matrices $T^{(\nu)}$ are all Hermitian. Moreover, $T^\dagger T$ is positive definite, so that all the submatrices $\chi_\nu T^{(\nu)}$ are also positive definite from (2.5.4). In fact, from (2.5.1) the following Hermitian form of $\chi_\nu T^{(\nu)}$ with respect to a set of complex variables $\{x_i;\ i \in g^{-1}(\nu)\}$ for each ν is positive definite, because

$$\chi_\nu \sum_{i,j \in g^{-1}(\nu)} x_i^* T_{ij}^{(\nu)} x_j = \sum_{k=1}^{n} \left| \sum_i T_{ki} x_i \right|^2 > 0 \tag{2.5.7a}$$

where x_i^* is the complex conjugate of x_i. Since the variables x_i in (2.5.7a) are arbitrary as long as the indices i belong to $g^{-1}(\nu)$, any principal submatrices of $\chi_\nu T^{(\nu)}$ are also positive definite; hence, their eigenvalues are all positive. Thus, their determinants (called *the principal minors* of $\chi_\nu T^{(\nu)}$) are also positive: the first few of them are

$$\begin{aligned} &\chi_\nu T_{ii}^{(\nu)} = \sum_k |T_{ik}|^2 > 0; \qquad && i \in g^{-1}(\nu) \\ &(\chi_\nu)^2 \det \begin{bmatrix} T_{ii}^{(\nu)} & T_{ij}^{(\nu)} \\ T_{ji}^{(\nu)} & T_{jj}^{(\nu)} \end{bmatrix} > 0; \qquad && i, j \in g^{-1}(\nu) \end{aligned} \tag{2.5.7b}$$

and so on. The first inequality in (2.5.7b) means that the signs of all diagonal elements of $T^{(\nu)}$ are given by the sign of χ_ν. If any one of the diagonal elements, T_{ii}, is zero, then T becomes singular since the ith column and row become null. These properties, which come from (2.5.7a), will provide the condition for $T^{(\nu)}$ to be diagonalized by the Gaussian elimination procedure which will be discussed in the next section.

2.6 Construction of a transformation matrix

It follows from (2.5.5) that the complete diagonalizations of the matrix A can be achieved by diagonalizing each submatrix $T^{(\nu)}$ of order $n_\nu \times n_\nu$. Obviously, the present formalism can be used for this. However, it is more effective to use the method of successive elimination since $\chi_\nu T_{ij}^{(\nu)}$ can be regarded as the scalar products of two vector sets $\{\hat{T}_{i\cdot}\}$ and $\{T_{\cdot j}\}$ according to (2.5.4). The method is based on the well-known Gaussian procedure, which transforms a square matrix $T^{(\nu)}$ into a triangular form by using another triangular matrix of the opposite shape. It serves as an effective algorithm for the Schmidt orthogonalization process.

Let us introduce two non-singular matrices $C^{(\nu)}$ and $\overline{C}^{(\nu)}$ of order $n_\nu \times n_\nu$ and diagonalize $T^{(\nu)}$ via $C^{(\nu)} T^{(\nu)} \overline{C}^{(\nu)}$. To this end, we put

$$S^{(\nu)} = T^{(\nu)}\overline{C}^{(\nu)}, \qquad \overline{S}^{(\nu)} = C^{(\nu)}T^{(\nu)} \tag{2.6.1a}$$

and require $S^{(\nu)}$ to be, say, a lower triangular form assuming an upper triangular form with unit diagonal elements for $\overline{C}^{(\nu)}$. Likewise, we require $\overline{S}^{(\nu)}$ to be upper triangular assuming a lower triangular form with unit diagonal elements for $C^{(\nu)}$. This set of equations is self-sufficient to determine unique solutions for each pair of $\{S^{(\nu)}, \overline{C}^{(\nu)}\}$ and $\{\overline{S}^{(\nu)}, C^{(\nu)}\}$, if $T^{(\nu)}$ satisfies the condition that the principal minors of $T^{(\nu)}$ are non-vanishing as in (2.5.7b). To see this, let us write out the first equation of (2.6.1a):

$$S_{ij}^{(\nu)} = \sum_{k=1}^{j} T_{ik}^{(\nu)}\overline{C}_{kj}^{(\nu)}; \qquad i, j \in g^{-1}(\nu) \tag{2.6.1b}$$

Then from $S_{ij}^{(\nu)} = 0$ for $i < j$ and $\overline{C}_{jj}^{(\nu)} = 1$ we have

$$0 = \sum_{k=1}^{j-1} T_{ik}^{(\nu)}\overline{C}_{kj}^{(\nu)} + T_{ij}^{(\nu)}; \qquad i < j$$

which provides a set of $j - 1$ linear equations for $\overline{C}_{kj}^{(\nu)}$ $(k < j)$ for a given j. This set is solvable for each j if the determinant of the coefficient matrix is non-singular: the first few of them are given by

$$j = 1: \qquad S_{11}^{(\nu)} = T_{11}^{(\nu)} \neq 0; \qquad 1 \in g^{-1}(\nu)$$

$$j = 2: \qquad S_{11}^{(\nu)}S_{22}^{(\nu)} = \begin{vmatrix} T_{11}^{(\nu)} & T_{12}^{(\nu)} \\ T_{21}^{(\nu)} & T_{22}^{(\nu)} \end{vmatrix} \neq 0; \qquad 1, 2 \in g^{-1}(\nu) \tag{2.6.1c}$$

Here the equalities follow from (2.6.1a), because the determinants of the leading submatrices of $S^{(\nu)}$ and $T^{(\nu)}$ are equal and because the determinant of a triangular matrix equals the product of the diagonal elements. Note that the above condition holds for a Hermitian matrix A, as shown by (2.5.7b).

Now, from (2.6.1a) we obtain

$$C^{(\nu)}T^{(\nu)}\overline{C}^{(\nu)} = C^{(\nu)}S^{(\nu)} = \overline{S}^{(\nu)}\overline{C}^{(\nu)} \tag{2.6.2}$$

which is diagonal, since the second equality means that the product of lower triangular matrices is equal to the product of upper triangular matrices and since a product of upper (lower) triangular matrices is upper (lower) triangular: thus

$$[C^{(\nu)}T^{(\nu)}\overline{C}^{(\nu)}]_{ij} = \delta_{ij}S_{ii}^{(\nu)} = \delta_{ij}\overline{S}_{ii}^{(\nu)}; \qquad i, j \in g^{-1}(\nu) \tag{2.6.3}$$

It is obvious that the above argument holds if we interchange all the upper and lower triangular forms.

When A is Hermitian, $T^{(\nu)}$ is also Hermitian so that it is only necessary to construct one of $S^{(\nu)}$ and $\overline{S}^{(\nu)}$, since from (2.6.1a)

$$\overline{S}^{(\nu)} = [S^{(\nu)}]^{\dagger}, \qquad \overline{C}^{(\nu)} = [C^{(\nu)}]^{\dagger}$$

Now we define a pair of transformation matrices of order $n \times n$ by

$$S = T\overline{C}, \qquad \overline{S} = C\hat{T} \tag{2.6.4}$$

where $\overline{C}$ and C are the direct sums of $\overline{C}^{(\nu)}$ and $C^{(\nu)}$, respectively:

$$\overline{C} = \sum_{\nu=1}^{r} \oplus \overline{C}^{(\nu)}, \qquad C = \sum_{\nu=1}^{r} \oplus C^{(\nu)} \tag{2.6.5}$$

Then S and $\overline{S}$ coincide with $S^{(\nu)}$ and $\overline{S}^{(\nu)}$ respectively in each degenerate subspace of λ_ν and also satisfy the eigenvalue problems, from (2.6.4) and (2.4.1),

$$AS = S\Lambda, \qquad \overline{S}A = \Lambda\overline{S} \tag{2.6.6a}$$

because Λ commutes with $\overline{C}$ and C from the fact that the diagonal matrix Λ is a constant matrix in each subspace of λ_ν. Moreover, their product $\overline{S}S$ is diagonal, from (2.6.4), (2.5.1) and (2.6.3),

$$\overline{S}S = D = \text{diag}\,[d_1, d_2, \ldots, d_n] \tag{2.6.6b}$$

with the diagonal elements

$$d_j = \chi_{g(j)} S_{jj}; \qquad j = 1, 2, \ldots, n \tag{2.6.6c}$$

Consequently S^{-1} is simply given by $S^{-1} = D^{-1}\overline{S}$. Here, for the existence of S^{-1}, it is essential that $S_{jj} \neq 0$ for all j. This condition, however, is guaranteed to hold by the required condition (2.6.1c) for the Gaussian elimination procedure.

Now, we shall normalize S and arrive at the desired transformation matrix U defined by

$$U = SD^{-1/2}, \qquad U^{-1} = D^{-1/2}\overline{S} \tag{2.6.7}$$

where the argument of each normalization coefficient $d_j^{-1/2}$ has to be assigned appropriately once for each d_j so that $(d_j^{-1/2})^2 = d_j^{-1}$. Then equations (2.6.6a) are combined to one equation

$$U^{-1}AU = \Lambda \tag{2.6.8a}$$

Since $S_{jj} = \overline{S}_{jj}$ from (2.6.3), the diagonal elements of U and U^{-1} are the same and given by

$$U_{jj} = [U^{-1}]_{jj} = d_j^{-1/2} S_{jj} = d_j^{1/2}/\chi_{g(j)} \qquad \text{for all } j \tag{2.6.8b}$$

from (2.6.6c). In the important special case in which the matrix A is Hermitian, the transformation matrix U becomes unitary because $S^\dagger = \overline{S}$ and D becomes a positive definite matrix from (2.6.6b). In cases in which there exists no degeneracy for the eigenvalues of A, we simply have $S = T$ and $\overline{S} = \hat{T}$ (because C and $\overline{C}$ in (2.6.4) become the unit matrix) so that $U = TD^{-1/2}$ with $D = \text{diag}\,[\chi_1 T_{11}, \chi_2 T_{22}, \ldots, \chi_n T_{nn}]$.

Finally, if the Gaussian elimination procedure is not possible, we simply calculate T^{-1} by

$$T^{-1} = \left(\sum_{\nu=1}^{r} \oplus [\chi_\nu T^{(\nu)}]^{-1} \right) \hat{T} \tag{2.6.9}$$

from (2.5.4).

2.7 Illustrative examples

Example 1. Let us diagonalize a 4×4 matrix A defined by

$$A = \begin{bmatrix} 1 & 1 & 1 & 1 \\ 1 & 1 & 1 & 1 \\ 1 & 1 & 1 & 1 \\ 1 & 1 & 1 & 1 \end{bmatrix} \tag{2.7.1}$$

This problem arises when we construct the eigenvectors of the total spin angular momentum squared S^2 for a system of four electrons; see Kim (1979b).

Since A is real and symmetric, it can be diagonalized by a real orthogonal matrix. From $A^2 = 4A$, the reduced characteristic equation of A is given by

$$A^2 - 4A = 0$$

which has no multiple roots. The eigenvalues λ_ν, their degeneracies n_ν and χ_ν of (2.3.10a) are given by

$$\begin{array}{cccc} & \lambda_\nu & n_\nu & \chi_\nu \\ \nu = 1 & 0 & 3 & -4 \\ \nu = 2 & 4 & 1 & 4 \end{array}$$

Choosing the following standard form for the diagonal matrix:

$$\Lambda = \mathrm{diag}\,[0,\ 0,\ 0,\ 4] \tag{2.7.2}$$

we can immediately write down the transformation matrices T and $\hat{T}$, using (2.4.3),

$$T = \hat{T} = A + \Lambda - 4\mathbf{1} = \left[\begin{array}{ccc:c} -3 & 1 & 1 & 1 \\ 1 & -3 & 1 & 1 \\ 1 & 1 & -3 & 1 \\ \hdashline 1 & 1 & 1 & 1 \end{array}\right] \tag{2.7.3}$$

One can easily see from this result that T is non-singular for any sequence of the characteristic roots in Λ. The submatrices $T^{(1)}$ and $T^{(2)}$ corresponding to λ_1 and λ_2 are shown by the dotted lines in (2.7.3). They are symmetric since A is symmetric. Note that the signs of the diagonal elements of T coincide with those of χ_1 and χ_2, respectively, satisfying (2.5.7b). By successive elimination of the elements above the principal diagonal of T by the linear combinations of the first three columns leaving the first column intact, we obtain the matrix S defined by (2.6.4):

$$S = \begin{bmatrix} -3 & 0 & 0 & 1 \\ 1 & -8/3 & 0 & 1 \\ 1 & 4/3 & -2 & 1 \\ 1 & 4/3 & 2 & 1 \end{bmatrix}$$

Then, by normalizing each column of S directly or using (2.6.7), we obtain the unitary transformation matrix U

$$U = \begin{bmatrix} -\dfrac{3}{\sqrt{12}} & 0 & 0 & \dfrac{1}{2} \\ \dfrac{1}{\sqrt{12}} & -\dfrac{2}{\sqrt{6}} & 0 & \dfrac{1}{2} \\ \dfrac{1}{\sqrt{12}} & \dfrac{1}{\sqrt{6}} & -\dfrac{1}{\sqrt{2}} & \dfrac{1}{2} \\ \dfrac{1}{\sqrt{12}} & \dfrac{1}{\sqrt{6}} & \dfrac{1}{\sqrt{2}} & \dfrac{1}{2} \end{bmatrix} \tag{2.7.4}$$

which is also a real orthogonal matrix as expected since A is real symmetric. For further application of the present formalism to calculate the vector coupling coefficients for angular momenta, see Kim (1979b).

Example 2. A rotation matrix for rotation about a unit vector $\boldsymbol{s} = (s_1, s_2, s_3)$ through an angle θ is expressed by

$$R(\boldsymbol{\theta}) = \exp[\theta\omega]; \qquad \omega = \begin{bmatrix} 0 & -s_3 & s_2 \\ s_3 & 0 & -s_1 \\ -s_2 & s_1 & 0 \end{bmatrix} \tag{2.7.5}$$

which has been introduced in (1.6.4a). Diagonalization of this matrix has been discussed in Exercise 7 in the previous chapter on the basis of the projection operator method. Here we shall diagonalize it using the characteristic transformation matrix as a comparison. Since it is a function of the matrix ω it is only necessary to diagonalize the infinitesimal rotation ω for the diagonalization of $R(\boldsymbol{\theta})$. The characteristic equation of ω is given by

$$\omega^3 + \omega = 0$$

which has no multiple root. The characteristic roots λ_ν and χ_ν of (2.3.10a) are

ν	1	2	3
λ_ν	0	i	−i
χ_ν	1	−2	−2

Take $\Lambda = \text{diag}[0, \text{i}, -1]$, then the characteristic transformation matrix $T_{\omega\Lambda}$ is given, from (2.4.2), by

$$T_{\omega\Lambda} = \omega^2 + \omega\Lambda^{(1)} + \Lambda^{(2)}$$

Since $\Lambda^{(1)} = \Lambda, \Lambda^{(2)} = \Lambda^2 + \mathbf{1}$, and $\omega^2 = \boldsymbol{s}^2 - \mathbf{1}$, we have

$$T_{\omega\lambda} = \begin{bmatrix} s_1^2 & s_1s_2 - \text{i}s_3 & s_1s_3 - \text{i}s_2 \\ s_1s_2 & s_2^2 - 1 & s_2s_3 + \text{i}s_1 \\ s_1s_3 & s_3s_2 + \text{i}s_1 & s_3^2 - 1 \end{bmatrix} \tag{2.7.6}$$

Note that the signs of the diagonal elements coincide with those of χ_1, χ_2 and χ_3 because $\boldsymbol{s}^2 = 1$.

Since all eigenvalues of ω are distinct, all we need to do is to normalize each column vector of $T_{\omega\Lambda}$ using $\boldsymbol{s}^2 = 1$ to obtain the required U matrix (which must be unitary for ω being normal)

$$U = [\boldsymbol{u}_1, \boldsymbol{u}_2, \boldsymbol{u}_3] \tag{2.7.7}$$

where the column vectors are given by

$$\boldsymbol{u}_1 = \begin{bmatrix} s_1 \\ s_2 \\ s_3 \end{bmatrix}, \qquad \boldsymbol{u}_2 = \frac{1}{[2(1-s_2^2)]^{1/2}} \begin{bmatrix} s_1 s_2 - \mathrm{i}s_3 \\ s_2^2 - 1 \\ s_3 s_2 + \mathrm{i}s_1 \end{bmatrix},$$

$$\boldsymbol{u}_3 = \frac{1}{[2(1-s_3^2)]^{1/2}} \begin{bmatrix} s_1 s_3 - \mathrm{i}s_2 \\ s_2 s_3 + \mathrm{i}s_1 \\ s_3^2 - 1 \end{bmatrix}$$

From $\Lambda = \mathrm{diag}\,[0, \mathrm{i}, -\mathrm{i}]$ we obtain

$$U^\dagger R(\boldsymbol{\theta}) U = \mathrm{diag}\,[1, \mathrm{e}^{\mathrm{i}\theta}, \mathrm{e}^{-\mathrm{i}\theta}] \tag{2.7.8}$$

Exercise. Show that $\boldsymbol{u}_3$ is proportional to $\boldsymbol{u}_2^*$.

3

Elements of abstract group theory

3.1 Group axioms

A group is a set G on which a multiplication operation with the following properties is defined.

1) If a and b belong to G, their product ab also belongs to G. This property is called *closure* or *the group property*. This property may be expressed as follows:

$$\text{if } a, b \in G, \text{ then } ab \in G$$

where $\in$ means 'belongs to'.

2) *The associative law* holds for multiplication:

$$(ab)c = a(bc)$$

for every a, b and c in G.

3) There is *an identity element* e in G such that

$$ea = ae = a \qquad \forall\, a \in G$$

4) For every $a \in G$, there is an inverse element $a^{-1} \in G$ such that

$$aa^{-1} = a^{-1}a = e$$

We define the powers of $g \in G$ as follows: $g^0 = e$, $g^1 = g$, $g^n = g^{n-1}g$ and $g^{-n} = (g^{-1})^n$ for a positive integer n. When a group element g satisfies $g^2 = e$, it is called a binary element of G.

In general, a product ab of two elements a, $b \in G$ may but need not equal ba. If $ab = ba$, we say that a and b commute. If all elements in G commute with each other, the group is said to be *Abelian*.

Let $G = \{g_i\}$ and define $aG = \{ag_i\}$. If $a \in G$, then $aG = G$ because $ag_i \in G$ by closure and $ag_i \neq ag_j$ means $g_i \neq g_j$ since a^{-1} exists. Thus, multiplication of G by an element of G may rearrange the elements but will not change the group G as a whole (the *rearrangement theorem*). The number of distinct elements of G is called *the order* of G, denoted $|G|$. If $|G|$ is finite, the group is called *a finite group*, otherwise the group is *an infinite group*.

Let G be a finite group and let $g \in G$. Then a sequence e, g, g^2, ... must return to the identity after a finite number of elements. The smallest number n for which $g^n = e$ is called the order of g. The sequence

$$\langle g \rangle = e,\ g,\ g^2, \ldots,\ g^{n-1} \tag{3.1.1}$$

is called *the period* of g. The period $\langle g \rangle$ forms a *cyclic group* C_n of order n generated by the element g.

3.1.1 The criterion for a finite group

Let G be a finite set of elements on which a multiplication operation is defined. If it satisfies the first two requirements of a group, i.e. closure and the associative law, the set is a group: let $a \in G$ and its order by n, then $a^n = e$ the identity and $a^{n-1} = a^{-1}$ the inverse.

A finite group G is defined by a multiplication table of G that is described by the $n \times n$ matrix whose ijth entry is $a_i a_j \in G$.

G	a_1	a_2	$\dots$	a_n
a_1	a_1a_1	a_1a_2	$\dots$	a_1a_n
a_2	a_2a_1	a_2a_2	$\dots$	a_2a_n
.	.	.	$\dots$	.
.	.	.	$\dots$	.
a_n	a_na_1	a_na_2	$\dots$	a_na_n

Let G be the multiplicative group with elements $\{1, -1\}$, then the multiplication table is given by

G	1	−1
1	1	−1
−1	−1	1

In general, construction of the multiplication table is rather cumbersome unless the order of the group is small. As will be described later, however, one can define a finite group by means of the group generators and the defining relations of the group.

3.1.2 Examples of groups

1. The set of all non-singular $n \times n$ matrices forms a group called *the group of general linear transformations* denoted $GL(n)$.
2. The set of all $n \times n$ unitary matrices forms a group called *the unitary group* in n dimensions, denoted $U(n)$.
3. The set of all real numbers forms a group, with addition as the multiplication. Here zero serves as the identity. It is Abelian.
4. The set of all non-zero real numbers forms a group under ordinary numerical multiplication.
5. Let an nth primitive root of unity be $\omega = \exp(2\pi i/n)$, then a set of numbers

$$\langle\omega\rangle = 1, \omega^2, \omega^3, \dots, \omega^{n-1}$$

forms a cyclic group of order n under numerical multiplication
6. A set Z of n integers

$$0, 1, 2, \dots, n-1$$

forms a group under addition, modulo n. For example, if $n = 5$, then $3 + 4 = 5 + 2 \equiv 2 \pmod 5$.[1] Note that the elements of Z can be put into one-to-

[1] $x \equiv y \pmod 5$ reads 'x is congruent to y modulo 5' and means '$x - y$ is an integral multiple of 5'. Thus if $x \equiv 0 \pmod 1$, then x is any integer. For simplicity we frequently write $x = y \pmod 5$.

one correspondence to the elements of the set $\langle\omega\rangle$ by making k correspond to ω^k. This correspondence has the property that 'products correspond to products', i.e. $k_1 + k_2 = k_3$ implies that $\omega^{k_1}\omega^{k_2} = \omega^{k_3}$. Such a pair of groups is said to be *isomorphic*.

7. Group C_n, *the uniaxial group of order n*. The group C_n is a cyclic group of order n generated by an n-fold rotation c_n, which is a rotation about a given axis through an angle $2\pi/n$. It is a special case of (3.1.1) and its elements are given by

$$\langle c_n\rangle = e,\ c_n,\ c_n^2,\ \ldots,\ c_n^{n-1}$$

8. Group C_∞, *the uniaxial rotation group of order infinity*. Let $c_z(\theta)$ be a rotation about the z-axis through an angle θ, then $c_z(\theta_2)c_z(\theta_1) = c_z(\theta_2 + \theta_1)$ and $c_z(\theta + 2\pi) = c_z(\theta)$. The set $c_\infty = \{c_z(\theta);\ -\pi \leqslant \theta < \pi\}$ forms a group. Note that $c_z(\theta)^{-1} = c_z(-\theta)$.
9. Group D_∞, *the dihedral group of order infinity*. This group is formed by augmenting (adjoining) the uniaxial group $C_\infty = \{c_z(\theta)\}$ with a two-fold rotation (or binary rotation) $2_{\boldsymbol{v}}$ about a unit vector $\boldsymbol{v}$ on the x, y plane perpendicular to the z-axis. To describe the group structure we introduce the following relation (see Figure 3.1):

$$c_z(\theta)2_{\boldsymbol{v}} = 2_{\boldsymbol{h}}, \qquad \boldsymbol{h}\cdot\boldsymbol{v} = \cos(\theta/2) \tag{3.1.2a}$$

where $2_{\boldsymbol{h}}$ is a binary rotation about a unit vector $\boldsymbol{h}$ on the x, y plane that makes an angle $\theta/2$ with the unit vector $\boldsymbol{v}$. The proof is easily achieved by the stereographic projection diagram given in Figure 3.1. Note that $c_z(\theta)$ rotates a point in the space through the angle θ whereas it rotates an axis of binary rotation $2_{\boldsymbol{v}}$ perpendicular to $c_z(\theta)$ through half the angle $\theta/2$ upon simple multiplication according to (3.1.2a).

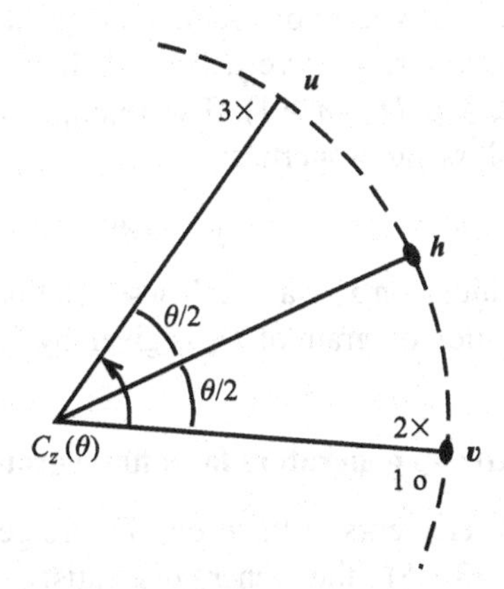

Figure 3.1. A stereographic projection diagram showing $2_{\boldsymbol{h}} = C_z(\theta)2_{\boldsymbol{v}}$. Consider a unit sphere centered at a point on the plane of the paper and denote the cross-section of the sphere in the paper by a dashed circle. A point on the sphere above (below) the paper is projected onto the paper and denoted by ○ (×). Place a binary axis of rotation $2_{\boldsymbol{v}}$ along a unit vector $\boldsymbol{v}$ on the paper shown in the figure. Then, applying $2_{\boldsymbol{v}}$ to an initial point 1 above the paper, we arrive at point 2 below the paper. Then applying $C_z(\theta)$ on point 2 we arrive at point 3. The resultant operation which brings point 1 to point 3 is given by a binary rotation $2_{\boldsymbol{h}}$, where $\boldsymbol{h}$ is a unit vector that makes the angle $\theta/2$ with the unit vector $\boldsymbol{v}$.

The elements of D_∞ are expressed by

$$D_\infty = \{c_z(\theta), 2_v\}$$

where $-\pi \leqslant \theta < \pi$ and 2_v represents any binary axis of rotation in the horizontal plane perpendicular to the z-axis. The group properties of D_∞ are understood from (3.1.2a); e.g.

$$2_h 2_v = c_z(\theta), \qquad 2_v c_z(\theta)^{-1} = 2_h \tag{3.1.2b}$$

Exercise. Show from (3.1.2b) that a similarity transformation of 2_v by $c_z(\theta)$ rotates 2_v through the angle θ, i.e.

$$c_z(\theta) 2_v c_z^{-1}(\theta) = 2_u \tag{3.1.2c}$$

where 2_u is a binary rotation about the unit vector $\boldsymbol{u}$ that makes an angle θ with $\boldsymbol{v}$ (see Figure 3.1).

10. Group D_n, *the dihedral group of order 2n*. This group is formed by augmenting (adjoining) the uniaxial group $C_n = \langle c_n \rangle$ with a two-fold axis of rotation u_o perpendicular to $c_n \in C_n$. Let $u_k = c_n^k u_o$; $k = 0, 1, 2, \ldots, n-1$, then the group elements of D_n are given by a total of $2n$ elements:

$$e,\ c_n,\ c_n^2,\ \ldots,\ c_n^{n-1},\ u_0,\ u_1,\ \ldots,\ u_{n-1} \tag{3.1.3}$$

where, from (3.1.2a), the u_k are binary axes of rotation perpendicular to c_n and their adjacent angles are π/n. See Figure 3.2 for the stereographic projections of the groups D_2 and D_3.

11. Group D_2 (or *the four group*). The group D_2 has four elements, which may be expressed by

$$e,\ u_x,\ u_y,\ u_z \tag{3.1.4}$$

where u_x, u_y and u_z are binary axes of rotation along the x-, y- and z-axes of the Cartesian coordinate system, respectively. It is Abelian. This is the only Abelian group in the dihedral group D_n ($n > 1$). The multiplication table of D_2 can be constructed using the following properties:

$$u_x^2 = e, \qquad u_x u_y = u_y u_x = u_z \tag{3.1.5}$$

and their cyclic permutations on x, y and z. These relations are verified by (3.1.2). The stereographic projection diagram of D_2 is given by Figure 3.2.

3.2 Group generators for a finite group

It is to be noted that the four elements of the group D_2 are generated by two elements $a = u_x$ and $b = u_y$. From (3.1.5), the generators satisfy the following algebraic relations:

$$a^2 = b^2 = (ab)^2 = e \tag{3.2.1}$$

These are called *the defining relations* of D_2, because from them we can construct the multiplication table of D_2. For simplicity, however, we simply derive the multiplicative properties (3.1.5) from (3.2.1). Firstly, the commutation relation $ba = ab$ is obtained by multiplying $(ab)^2 = abab = e$ by a from the left and by b from the right and using $a^2 = b^2 = e$. Set $u_x = a$, $u_y = b$ and $u_z = ab$, then $u_x u_y = u_y u_x = u_z$. Next, on multi-

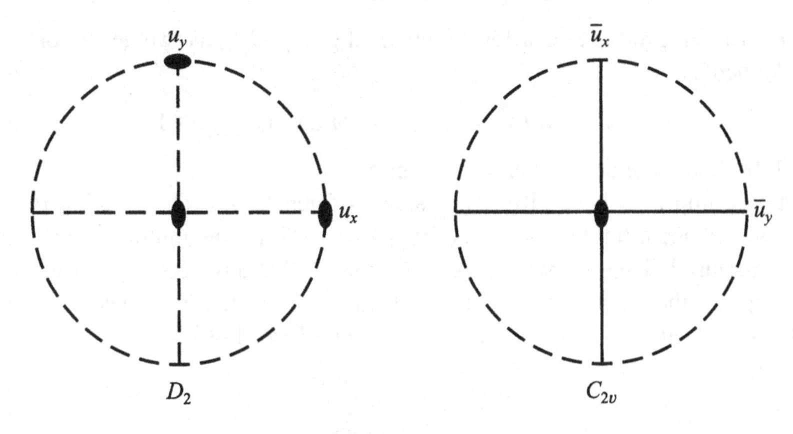

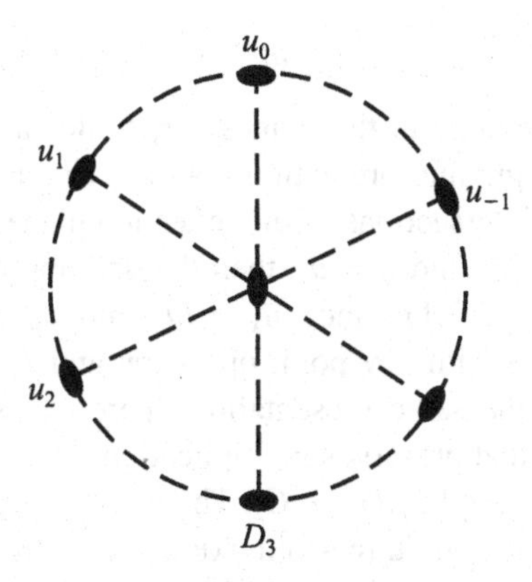

Figure 3.2. The stereographic projections for D_2, C_{2v} and D_3. Here ● denotes a two-fold axis while ▲ denotes a three-fold axis.

plying $u_x u_y = u_z$ from the left by u_x we obtain $u_y = u_x u_z$, and so on. In this way, we obtain all the necessary information for constructing the multiplication table; cf. Table 5.3 later.

In general, if all elements of a finite group G are generated by a subset of the elements of G through finite products of their powers (including negative powers), the subset is called the set of *group generators* of G. If the multiplication table of G is constructed by means of a set of algebraic relations satisfied by the generators g_1, $g_2, \ldots$,

$$g_1^{n_1} g_2^{n_2} \ldots = e; \qquad n_1, n_2, \ldots \text{ are integers} \tag{3.2.2}$$

then the set of relations is called the set of defining relations or the presentation of the group G. Every algebraic relation of the generators is a consequence of these relations.

For example, a cyclic group C_n of order n is defined by one generator a and one relation $a^n = e$. According to (3.1.3), the group D_n has two generators a $(= c_n)$ and b $(= u_o \perp c_n)$, which satisfy the defining relations

$$a^n = b^2 = (ab)^2 = e \tag{3.2.3}$$

Every proper point group (except for the uniaxial group C_n) has two generators a and b and is defined by

$$a^n = b^m = (ab)^2 = e; \qquad n,\ m \text{ are integers} > 1 \tag{3.2.4}$$

These will be studied in detail later in Chapter 5.

The same group may have different systems of generators. The defining relations will also differ accordingly. *The rank of a group G* is the minimum number of generators required. Then we will show in Chapter 5 that any proper point group has rank 2 except for the uniaxial group of rank 1. A set of defining relations of a group is also called *a presentation of the group* (Coxeter and Moser 1984).

3.2.1 Examples

1. Group C_{2v}, The group C_{2v} has four elements that are expressed, in the Cartesian coordinate system, by

$$e,\ \overline{u}_x,\ \overline{u}_y,\ u_z$$

where $\overline{u}_x$ and $\overline{u}_y$ are reflections in the planes perpendicular to the x- and y-axes respectively. In the stereographic projection diagram (Figure 3.2), the solid lines are the projections of the intersections of the reflection planes with the unit sphere. Let the generators be $a = \overline{u}_x$ and $b = \overline{u}_y$, then the defining relations are the same as those of D_2 given by (3.2.1), since $\overline{u}_x = Iu_x$ and $\overline{u}_y = Iu_y$, where I is the inversion which commutes with any point operation and $I^2 = e$. Now, when two groups G_1 and G_2 have the same presentation, there exists a one-to-one correspondence between them that preserves multiplication. In such a case they are said to be *isomorphic* and expressed by $G_1 \simeq G_2$. Then, $C_{2v} \simeq D_2$.

2. Group D_3. The group D_3 has one three-fold axis of rotation 3_z along the z-axis in the Cartesian coordinate system and three binary axes of rotations u_0, u_1 and u_2 in the x, y plane with the nearest neighbor angle $\pi/3$ (see Figure 3.2). It has six elements, e, 3_z, 3_z^2, u_0, $u_1 = 3_z u_0$ and $u_2 = 3_z u_1$. If we take $a = 3_z$ and $b = u_0$ as the generators, the defining relations are

$$a^3 = b^2 = (ab)^2 = e \tag{3.2.5}$$

It is not Abelian: from $abab = e$ we have $ab = ba^{-1}$ so that $ab = ba$ requires $a^{-1} = a$ or $a^2 = e$, which is incorrect.

3. Symmetric groups S (or permutation groups). A permutation of n different letters may by denoted by

$$P = \begin{pmatrix} 1 & 2 & \dots & n \\ m_1 & m_2 & \dots & m_n \end{pmatrix} \tag{3.2.6}$$

It displaces the letter originally in position 1 into position m_1, the letter in position 2 into position m_2, etc. The product P_1P_2 of two permutations P_1 and P_2 is defined as the permutation obtained by carrying out permutation P_2 first and then permutation P_1. The set of all permutations forms a group of order $n!$. It is seen easily that any permutation can be obtained by a product of transpositions, each of which interchanges two letters. Thus the set of generators may be given by $n-1$ transpositions $(1\ \ 2)$, $(2\ \ 3)$, $\dots$, $(n-1\ \ n)$. However, this is not the minimal set of the generators. In fact, all elements of S can be obtained from two group elements, (1 2) and the cyclic permutation $C_n = (1\ 2\ 3\ \ldots\ n)$ which shifts $1 \to 2$, $2 \to 3$ ... $n-1 \to n$ and $n \to 1$. Since $C_n(1\ 2)C_n^{-1} = (2\ 3)$, repeated conjugations by C_n bring (1 2) into any desired adjacent transposition. It is, however, more convenient to take $n-1$ adjacent transpositions as the generators.

4. The quaternion group Q. It has eight elements:

$$\pm\mathbf{1},\ \pm\mathbf{i},\ \pm\mathbf{j},\ \pm\mathbf{k} \tag{3.2.7}$$

where $\mathbf{1}$ is the unit element and the three elements $\mathbf{i}$, $\mathbf{j}$ and $\mathbf{k}$ are extensions of $\sqrt{-1}$ and satisfy

$$\mathbf{i}^2 = \mathbf{j}^2 = \mathbf{k}^2 = -1, \qquad \mathbf{ij} = -\mathbf{ji} = \mathbf{k} \qquad (\mathbf{i}, \mathbf{j}, \mathbf{k} \text{ cyclic}) \tag{3.2.8}$$

If we take $a = \mathbf{i}$ and $b = \mathbf{j}$ as the generators, the defining relations are given by

$$a^2 = b^2 = (ab)^2 = e', \qquad e'^2 = e \tag{3.2.9}$$

Conversely, from these we can reproduce (3.2.8) with $e' = -\mathbf{1}$. Multiplying $(ab)^2 = abab = -\mathbf{1}$ by a from the left and by b from the right we obtain $ba = -ab$. Next, we set $\mathbf{k} = ab$ and obtain the eight elements of Q which satisfy (3.2.8). Note that the above set (3.2.9) with $e' = e$ defines the dihedral group D_2.

It is interesting to construct a matrix group that is isomorphic to the quaternion group Q. Let σ_0 be a 2×2 unit matrix and σ_x, σ_y and σ_z be the Pauli spin matrices defined by

$$\sigma_0 = \begin{bmatrix} 1 & 0 \\ 0 & 1 \end{bmatrix}, \qquad \sigma_x = \begin{bmatrix} 0 & 1 \\ 1 & 0 \end{bmatrix}, \qquad \sigma_y = \begin{bmatrix} 0 & -\mathrm{i} \\ \mathrm{i} & 0 \end{bmatrix}, \qquad \sigma_z = \begin{bmatrix} 1 & 0 \\ 0 & -1 \end{bmatrix}$$

As was shown in (1.2.5b) or by direct matrix multiplication, these satisfy the following algebraic relations:

$$\sigma_x^2 = \sigma_y^2 = \sigma_z^2 = \sigma_0, \qquad \sigma_x\sigma_y = -\sigma_y\sigma_x = \mathrm{i}\sigma_z \qquad (x, y, z \text{ cyclic})$$

Thus, the set of eight elements $\{\pm\sigma_0, \pm\mathrm{i}\sigma_x, \pm\mathrm{i}\sigma_y, \pm\mathrm{i}\sigma_z\}$ forms a group. This group is isomorphic to Q by one-to-one correspondence:

$$\mathbf{1} \leftrightarrow \sigma_0, \qquad \mathbf{i} \leftrightarrow -\mathrm{i}\sigma_x, \qquad \mathbf{j} \leftrightarrow -\mathrm{i}\sigma_y, \qquad \mathbf{k} \leftrightarrow -\mathrm{i}\sigma_z \tag{3.2.10}$$

Such a matrix group is called *a matrix representation* of Q.

3.3 Subgroups and coset decompositions

If a subset H of a group G forms a group under original multiplication, then H is called *a subgroup* of G. If H is not one of the trivial subgroups, the identity group e and the group G itself, then H is called *a proper subgroup* of G. If H is a subgroup of G that is not G, we write $H < G$. If there is no subgroup of G that contains H, then H is called a maximal subgroup of G. Finding all the possible subgroups of a given group is very important in order to understand the group's structure. Examples of subgroups are

1. $U(n) < GL(n)$,
2. $C_\infty < D_\infty$, $\qquad C_n < D_n$,
3. The set Z of integers in the group of real numbers under addition,

4. The period $\langle x \rangle$ of an element x of a group G, and
5. The maximal subgroups of the cyclic group C_6 are C_2 and C_3.

3.3.1 *The criterion for subgroups*

If H is a non-empty subset of a group G, then H is a subgroup of G if and only if $y^{-1}x \in H$ when $x, y \in H$.

Proof. Since H is non-empty, it contains an element x. Taking $y = x$, we have $x^{-1}x = e$ is in H. If $y \in H$, then taking $x = e$ we have $y^{-1}e = y^{-1}$ in H. Finally, if x, $y \in H$, then $(y^{-1})^{-1}x = yx$ is in H. Thus, H is a subgroup of G.

This criterion is different from the criterion for a finite group given in Section 3.1 in that the present criterion applies to infinite groups as well; cf. Rotman (1973).

3.3.1.1 *Cosets*

Let G be a group and let $H = \{h_i\}$ be a subgroup of G. Let q_1 be an element of G that is not contained in H. Then the set of elements

$$q_1 H = \{q_1 h_i;\ \text{for all } h_i \in H\}$$

is called *the left coset* of H by q_1, and q_1 itself is called *the left coset representative* of H. Note that $q_1 h_i$ with any $h_i \in H$ can be chosen to be the coset representative. Then H may be considered also a coset of H by the identity. If there exists another element q_2 of G that is not contained both in H and in $q_1 H$, one can construct another left coset $q_2 H$. Then these cosets have no element in common because $q_1 h_i = q_2 h_j$ means that

$$q_1^{-1} q_2 = h_i h_j^{-1} \in H, \qquad \text{i.e. } q_2 \in q_1 H \tag{3.3.1}$$

The relation $q_1^{-1} q_2 \in H$ will be referred to as the condition for q_1 and q_2 to belong to the same coset. Continuing this process we obtain the left coset expansion of G:

$$G = H + q_1 H + q_2 H + \cdots = \{q_k H\} \tag{3.3.2a}$$

where + means simply the union of the disjoint sets. Obviously, one can define the right cosets of H and construct the right coset expansion. A convenient way of doing this is to take the inverses of all elements of G in (3.3.2a) and arrive at the right coset expansion given by

$$G + H + Hq_1^{-1} + \cdots = \{Hq_k^{-1}\} \tag{3.3.2b}$$

Thus the left and right cosets of H have equal numbers of elements. Since the order of each coset equals the order of H and all cosets are disjoint, we conclude that the order of a group G equals the number of the cosets multiplied by the order $|H|$ of the subgroup.

3.3.2 *Langrange's theorem*

Let H be a subgroup of a finite group G, then order $|H|$ of the subgroup is a divisor of the order $|G|$ of the group. The integer $|G|/|H| = J$ equals the number of the cosets of H in G and is called the *index* of H in G.

The above basic theorem is merely one of the many outcomes through coset expansions of a group by its subgroups. The coset expansion plays an essential role in constructing the matrix representations of a group G from those of its subgroups.

3.3.2.1 Examples

1. Let $H = \{e, u_x\}$ be a subgroup of D_2 given in (3.1.4), then
$$D_2 = H + u_y H = H + H u_y \tag{3.3.3}$$
2. Let $H = \{\pm\mathbf{1}, \pm\mathbf{i}\}$ in the quaternion group Q of (3.2.7), then
$$Q = H + \mathbf{j}H = H + H\mathbf{j} \tag{3.3.4}$$
In the above two examples, the left and right cosets coincide. Such a coincidence occurs whenever H is a so-called *invariant subgroup* of G, which will be discussed later in detail.
3. Let $H = \{e, u_0\}$ be a subgroup of the group D_3 defined in (3.2.5). Then the left coset decomposition of D_3 by H is given by
$$D_3 = H + 3_z H + 3_z^2 H$$
where
$$3_z H = \{3_z, u_1\}, \qquad 3_z^2 H = \{3_z^2, u_2\} \tag{3.3.5}$$
The corresponding right cosets are, using $3_z^{-1} = 3_z^2$,
$$H 3_z^2 = \{3_z^2, u_1\}, \qquad H 3_z = \{3_z, u_2\}$$
Thus, the left cosets do not coincide with the right cosets for this example.

3.4 Conjugation and classes

Let x be an element that may but need not be an element of a group G, but multiplication between x and the elements of G is allowed. Then, a conjugation of an element $g \in G$ by x is the similarity transformation of g by x
$$g^x = xgx^{-1} \tag{3.4.1}$$
The transform g^x is called *the conjugate* of g by x. Note that two successive conjugations by x and y are given by a single conjugation by yx
$$(g^x)^y = yxgx^{-1}y^{-1} = g^{yx} \tag{3.4.2}$$
In particular, $(g^x)^{-1} = (g^{-1})^x$. One of the most important properties of conjugation by a fixed element is that it preserves the multiplication
$$h^x g^x = xhx^{-1}xgx^{-1} = (hg)^x; \qquad h, g \in G \tag{3.4.3}$$
Consequently the group structure is not affected by a conjugation. A group defined by $G^x = \{g^x\}$ with a fixed x is called *a conjugate group* of $G = \{g\}$. It is isomorphic to G via the correspondence $g^x \leftrightarrow g$.

3.4.1 Normalizers

Let $H = \{h\}$ be a subset of a group G, then the conjugate of H by an element $x \in G$ is defined by $H^x = \{h^x\}$. If $H^x = H$ for an element $x \in G$, then the set H is said to be invariant under x of G. This does not mean that x commutes with every element of H. It simply means that x commutes with H as a whole. The subset N of G that leaves H invariant forms a subgroup of G, because the set N satisfies the criterion of a subgroup, i.e. if $x, y \in N$ then $y^{-1}x = \bar{y}x \in N$, for

$$H^{\bar{y}x} = (H^x)^{\bar{y}} = H^{\bar{y}} = H$$

where the last equality follows from $H^y = H$. This subgroup N is called *the normalizer* of the subset H in G, denoted

$$N_G(H) = \{x \in G | H^x = H\} \tag{3.4.4}$$

where the vertical line stands for 'with the property'. Obviously, $N_G(H) \leqslant G$.

3.4.1.1 Normal subgroups

If H is a subgroup of G and is invariant under all $g \in G$, then H is called an *invariant* (or *normal*) *subgroup* of G, denoted $H \triangleleft G$. In this case, the normalizer of H equals the group G itself. If there is no invariant subgroup of G that contains $H(\triangleleft G)$, then H is called a *maximal invariant subgroup* of G. For a normal subgroup H of G, the left coset decomposition of G by H coincides with the right coset decomposition. Obviously, every subgroup of an Abelian group is normal. A halving subgroup (subgroup of index 2) of a group G is always a normal subgroup of G. To see this, let H be a halving subgroup of G and let q be an element of G that is not contained in H. Then the left and right coset decompositions of G by H yield

$$G = H + qH = H + Hq$$

Accordingly, $qH = Hq$ or $qHq^{-1} = H$, i.e. H is a normal subgroup of G; cf. (3.3.3) and (3.3.4). Note that the normalizer of a normal subgroup of G is the group G itself.

3.4.1.2 Examples

1. The normalizer of a subset $H = \{\mathbf{i}, \mathbf{j}\}$ in the quaternion group Q of (3.2.7) is $\{\pm\mathbf{1}\}$. The normalizer of a subgroup $H = \{\pm\mathbf{1}, \pm\mathbf{i}\}$ in Q is Q itself so that H is a normal subgroup of Q. Note that $\mathbf{jij}^{-1} = -\mathbf{i}$. Here H is a halving subgroup of Q.
2. The cyclic groups C_2 and C_3 are normal subgroups of C_6. By definition, both C_2 and C_3 are maximal normal subgroups of C_6. Note that C_2 is not a subgroup of C_3.
3. The cyclic group C_n is a halving subgroup of the dihedral group D_n so that it is normal subgroup of D_n.
4. Let $C_n = \{c_n^m;\ m = 0, 1, \ldots, n\}$ be a uniaxial group contained in a proper point group P. Then the normalizer of a rotation c_n^m $(m \neq 0) \in C_n$ is the uniaxial group C_n except for the special case in which $m = n/2$ and P contains a binary rotation perpendicular to c_n. In this case the normalizer of the binary rotation $u = c_n^{n/2}$ is the dihedral group D_n, for $u^{-1} = u$.
5. Let G be the permutation group on n letters. Then the set of even permutations on n letters forms a subgroup of G, which is a halving subgroup of G.

3.4.2 The centralizer

The set of all elements in G that commute with every element of a subset H of G forms a subgroup of G called *the centralizer of the subset H in G*. It is denoted by

$Z_G(H) = \{g|h^g = h: \forall\, h \in H\}$. In the special case in which H is a single element h of G, the centralizer $Z_G(h)$ equals the normalizer $N_G(h)$.

The normalizer and centralizer or a subset H of a group G are often used to find subgroups of G. In general, $G \geqslant N_G(H) \geqslant Z_G(H)$.

3.4.2.1 Examples (continued)

6. The centralizer of a subset $\{\pm\mathbf{i}\}$ in the quaternion group Q is $\{\pm\mathbf{1}, \pm\mathbf{i}\}$, whereas the normalizer is Q itself.
7. The centralizer of the n-fold axis of rotation c_n in D_n $(n > 3)$ is the uniaxial group C_n $(< D_n)$. The centralizer of c_2 in D_2 is D_2.

3.4.3 The center

The center of a group G is the subgroup consisting of those elements which commute with every element of G. Thus, it is an Abelian invariant subgroup of G and denoted by $Z(G)$. By definition, the group G is the centralizer of $Z(G)$. For an Abelian group G, the center is G itself. The center of the dihedral group D_2 is D_2 itself whereas the center of D_3 is the identity. For the quaternion group Q, the center is $C_2 = \{\pm\mathbf{1}\}$.

3.4.4 Classes

Let $a, b \in G$. If there exists an element $x \in G$ such that

$$a = xbx^{-1} \qquad (= b^x)$$

we say that a is conjugate to b. Obviously, b is also conjugate to a, since $b = (x^{-1})a(x^{-1})^{-1}$. Thus we simply say that a and b are equivalent in G. Let this equivalence relation be denoted as $a \sim b$. If $a \sim b$ and $b \sim c$, then $a \sim c$ (transitivity). Thus one can decompose G into sets of conjugate elements, known as *conjugacy classes* or simply *classes*. Then each element of G appears in one and only one class. The conjugacy class $C(h)$ of an element h of G is the set of all elements of G conjugate to h. It is formed by

$$C(h) = \{ghg^{-1};\ \forall\, g \in G\} \qquad \text{(no repetition)}$$

where $\forall$ means for all; i.e. the class $C(h)$ is defined by all the different conjugates of h.

The identity e of G forms a class by itself. In an Abelian group, each class consists of a single element since $gag^{-1} = a$ for all $g \in G$. Now, the class $C(h)$ of an element h in G is formed by the conjugates of h by the coset representatives of the centralizer $Z_G(h)$ of h in G: since $Z_G(h)$ leaves h invariant under conjugation, each different coset representative of the centralizer defines a different element in the class $C(h)$. Thus we arrive at the following theorem.

3.4.4.1 The class order theorem

The order $|C(h)|$ of the class $C(h)$ of an element h in G is given by the index of the centralizer $Z_G(h)$ of h in G:

$$|C(h)| = |G|/|Z_G(h)| \tag{3.4.5}$$

Since $|Z_G(h)|$ is an integer, the order of a class is a divisor of $|G|$, i.e. $|Z_G(h)| = |G|/|C(h)|$.

This theorem is useful for calculating $|C(h)|$ since it is easier to calculate $|Z_G(h)|$ than it is to calculate $|C(h)|$ directly; cf. Example 4 in Section 3.4. It has an important application in analyzing the class structures of point groups. On account of this theorem we have the following useful relation: let $f(g)$ be any single-valued function on the group G, then for a given h in G

$$\sum_{g\in G} f(ghg^{-1}) = |Z_G(h)| \sum_{g\in C(h)} f(ghg^{-1}) \tag{3.4.6}$$

3.4.4.2 Ambivalent classes

The class $C(h)$ of h in G is said to be ambivalent if it contains h^{-1}. In this case every element of $C(h)$ has its inverse in the class, for $(h^x)^{-1} = (h^{-1})^x$. In a special case, when h is binary, i.e. $h^2 = e$, all the elements of the class are binary and hence every element of the class is its own inverse. If we exclude this special case and the class of the identity, the order of an ambivalent class of a finite point group is even, having an element and its inverse that are distinct.

3.4.4.3 Examples (continued)

8. The quaternion group Q has five classes:

$$\{\mathbf{1}\},\ \{-\mathbf{1}\},\ \{\pm\mathbf{i}\},\ \{\pm\mathbf{j}\},\ \{\pm\mathbf{k}\} \tag{3.4.7}$$

(use $\mathbf{jij}^{-1} = -\mathbf{i}$). These classes are all ambivalent and their orders satisfy the class order theorem (3.4.5), e.g. $Z_Q(\mathbf{1}) = Q$ and $Z_Q(\mathbf{i}) = \{\pm\mathbf{1}, \pm\mathbf{i}\}$, so that $|C(\mathbf{1})| = 8/8 = 1$ and $|C(\mathbf{i})| = 8/4 = 2$.

9. The group D_3 has three classes:

$$\{e\},\ \{c_3, c_3^{-1}\},\ \{u_0, u_1, u_2\} \tag{3.4.8}$$

because, in view of Figure 3.2, $u_k c_3 u_k = c_3^{-1}$ and $u_{k+2} = c_3 u_k c_3^{-1}$ from (3.1.2c). These classes are all ambivalent.

10. The group D_4 has five classes:

$$\{e\},\ \{c_4, c_4^{-1}\},\ \{c_4^2\},\ \{u_0, u_2\},\ \{u_1, u_3\}$$

which are all ambivalent.

3.5 Isomorphism and homomorphism

Two groups $G = \{g_i\}$ and $\overline{G} = \{\overline{g}_i\}$ are isomorphic if there exists a one-to-one correspondence $g_i \leftrightarrow \overline{g}_i$ between their elements such that products correspond to products, $g_i g_j \leftrightarrow \overline{g}_i \overline{g}_j$. Such a one-to-one correspondence (or mapping) between two groups is called isomorphism and is denoted by $G \simeq \overline{G}$. We have already seen that the groups D_2 and C_{2v} are isomorphic. In the abstract point of view, isomorphic groups are identical because they are represented by the same multiplication table. We have shown in Section 3.4 that a group $G = \{g\}$ and its conjugate group $G^x = \{g^x\}$ are isomorphic, i.e. $G \simeq G^x$ via the correspondence $g \leftrightarrow g^x$. A group is always isomorphic to itself. Such an isomorphism is called *automorphism*. The automorphism of a group $H = \{h\}$ through a conjugation $h \leftrightarrow h^x = xhx^{-1}$ by a fixed element x in H is called *inner automorphism*, whereas if H is invariant under x but x is not contained in H, then the mapping $h \leftrightarrow h^x$ is called *outer automorphism*.

A more general type of correspondence between two groups is *homomorphism*. Here, as opposed to isomorphism, the mapping is not required to be one-to-one. A group G is *homomorphic* to a group $\overline{G}$ if there exists a mapping $f\colon G = \{g_i\} \to \overline{G} = \{f(g_i)\}$ that brings G *onto* $\overline{G}$ and preserves multiplication, where $f(g)$ is a single-valued function on the group G that satisfies $f(g_i g_j) = f(g_i) \times f(g_j)$ for all g in G. Here, mapping G *onto* $\overline{G}$ means that every member of $\overline{G}$ corresponds to at least one element of G. Since more than one element of G may correspond to an element of $\overline{G}$, the homomorphism $f\colon G \to \overline{G}$ may be a many-to-one correspondence and hence is a directed relation. The larger group is *homomorphic* to the smaller. The smaller is called the *homomorph* of the larger. Homomorphism becomes isomorphism if the mapping is one-to-one.

In general a homomorphism $f\colon G \to \overline{G}$ means that $\overline{g} \in \overline{G}$ is a single-valued function of $g \in G$, i.e. $\overline{g} = f(g)$, where $\overline{g}$ and g are called the *image* and *preimage* respectively. Since many different g in G can correspond to one element $\overline{g}$ of $\overline{G}$, the inverse function $f^{-1}(\overline{g})$ may be a multivalued function of $\overline{g} \in \overline{G}$. The set of all preimages $\{g\}$ corresponding to a given image $\overline{g}$ is called the *fiber* of the preimages of $\overline{g}$, i.e. $\{g\} = f^{-1}(\overline{g})$ (see Figure 3.3). According to this terminology, for an isomorphism f, its inverse f^{-1} is unique.

3.5.1 Examples

1. Let $G = GL(n, r)$ be a group of non-singular $n \times n$ matrices with real entries. Then a mapping that brings every element M of G into determinant

$$M \to \det M \tag{3.5.1}$$

yields a set of real numbers that forms a multiplicative group Z_{R} of non-zero real numbers. The mapping preserves multiplication since

$$\det(AB) = \det A \times \det B$$

Thus $GL(n, r)$ is homomorphic to Z_{R}.

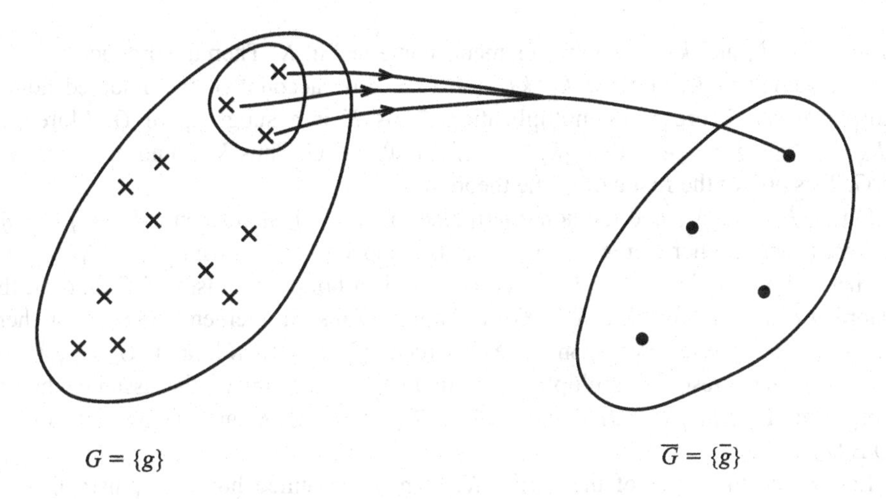

Figure 3.3. The fiber of $\overline{g}$: $\{g\} \in G^{-1}(\overline{g})$.

2. The quaternion group Q is homomorphic to group D_2 with the correspondence

$$\{\pm\mathbf{1}\} \to e, \qquad \{\pm\mathbf{i}\} \to u_x, \qquad \{\pm\mathbf{j}\} \to u_y, \qquad \{\pm\mathbf{k}\} \to u_z \tag{3.5.2}$$

which is two-to-one correspondence.

3.5.2 *Factor groups*

Let N be a normal subgroup of a group G. Then by definition $N = gNg^{-1}$ for all $g \in G$, so that $Ng = gN$. This means that left and right cosets of N coincide. Thus one can speak of cosets of N without specifying the left or right. Let the cosets of N be q_1N, $q_2N \ldots$, then the set of all cosets of N forms a group (with the cosets as the elements): closure is satisfied:

$$q_1Nq_2N = q_1q_2NN = q_1q_2N$$

and the associative law is satisfied:

$$(q_1Nq_2N)q_3N = q_1q_2q_3N = q_1N(q_2Nq_3N)$$

Moreover, the unit element is N itself, $N(q_iN) = (q_iN)N = (q_iN)$, while the inverse of q_iN is $q_i^{-1}N$. Accordingly, the set of all cosets of N forms a group, called a factor group of G by N, denoted G/N:

$$G/N = \{q_iN\}$$

Since the elements of a factor group G/N are cosets, its order equals the number of the cosets of N, i.e. the index of N in G. Thus $|G/N| = |G|/|N|$.

In terms of the factor group we shall discuss the structure of the homomorphism of a group $G = \{g\}$ onto a group $\overline{G} = \{\overline{g}\}$. For this purpose we introduce the concept of the *kernel*. It is defined by the subset K of G that is mapped onto the identity $\overline{e}$ of $\overline{G}$; it is the fiber of the identity $\overline{e}$ of $\overline{G}$. We have the following theorem.

3.5.2.1 *The isomorphism theorem*

The kernel K of the homomorphism of G onto $\overline{G}$ is a normal subgroup of G and the factor group G/K is isomorphic to $\overline{G}$, i.e. $G/K \simeq \overline{G}$.

Proof. Let k_1 and k_2 be any two elements contained in K. Then the product $k_1^{-1}k_2$ is also contained in K, because $k_1^{-1}k_2 \to \overline{e}^{-1}\overline{e} = \overline{e}$ on account of the assumed homomorphism which preserves multiplication. Thus K is a subgroup of G. Moreover, $gKg^{-1} \to \overline{g}\,\overline{e}\,\overline{g}^{-1} = \overline{e}$ so that $gKg^{-1} = K$ for all $g \in G$. Thus K is a normal subgroup of G. This proves the first half of the theorem.

Let $G/K = \{q_iK\}$ and the homomorphism be $q_i \to \overline{q}_i \in \overline{G}$, then $q_iK \to \overline{q}_i\overline{e} = \overline{q}_i$. If there exists another element p of G that is mapped onto $\overline{q}_i$, then $q_i^{-1}p \to \overline{q}_i^{-1}\overline{q}_i = \overline{e}$ so that $q_i^{-1}p \in K$, i.e. $p \in q_iK$. Since $G \to \overline{G}$ is a homomorphism of G onto $\overline{G}$, the mapping $q_iK \to \overline{q}_i$ for all q_iK in G/K should exhaust all elements of $\overline{G}$. Thus there exists a one-to-one correspondence between $G/K = \{q_iK\}$ and $\overline{G} = \{\overline{q}_i\}$ via $q_iK \leftrightarrow \overline{q}_i$ that preserves multiplication. In fact, on account of the assumed homomorphism, if $q_iKq_jK = q_kK$ then $\overline{q}_i\overline{q}_j = \overline{q}_k$. Thus G/K and $\overline{G}$ are isomorphic (Q.E.D.).

Let $|K|$ be the order of the kernel K, then the assumed homomorphism $G \to \overline{G}$ makes $|K|$-to-one correspondence.

3.5.2.2 Examples

1. For the homomorphism $GL(n, r) \to Z_R$ via the mapping $M \to \det M$ discussed in (3.5.1), the kernel K of the homomorphism is the special general linear transformation group $SGL(n, r)$ (with $\det M = 1$).
2. For the homomorphism $Q \to D_2$ discussed in (3.5.2), the kernel is $K = \{\pm 1\}$ and the correspondence is two-to-one. We have the isomorphism $Q/K \simeq D_2$.
3. A space group $\hat{G}$ (crystallographic group) is homomorphic to a point group G. The kernel of the homomorphism is the translational group T which leaves the crystal lattice invariant so the $\hat{G}/T \simeq G$ (see Chapter 13). The homomorphism is ∞-to-one, because $|T| = \infty$.

3.6 Direct products and semidirect products

Let $F = \{f_i\}$ and $H = \{h_j\}$ be two groups. The direct product of F and H is defined by a set of ordered pairs (f_i, h_j)

$$G = F \times H = \sum_{i,j} (f_i, h_j) \tag{3.6.1}$$

Then G is a new group under the multiplication $(f_1, h_1)(f_2, h_2) = (f_1 f_2, h_1 h_2)$. The number of elements in G equals the product of the orders of F and H. Furthermore, since

$$(f, h)(f_1, h_1)(f, h)^{-1} = (ff_1 f^{-1}, hh_1 h^{-1}) \tag{3.6.2}$$

the number of classes in G equals the product of the numbers of classes in F and in H.

There exists another form of group product, called the *semidirect product*. Consider two groups $F = \{f_i\}$ and $H = \{h_j\}$ that satisfy the following conditions:

1. their only common element is the identity e, i.e. $F \cap H = e$,
2. the product between two group elements is defined, i.e. $f_i h_j$ and $h_j f_i$ are meaningful and definite, and
3. F is invariant under H, i.e. $h_j f_i h_j^{-1} \in F$, for all $h_j \in H$. Then the set of the products of their elements denoted

$$F \wedge H = \sum_{i,j} f_i h_j \tag{3.6.3}$$

forms a group called the semidirect product of F and H; cf. Altmann (1997).

Proof

Identity: e (the only common element)
Closure: $f_i h_j f_s h_t = f_i (h_j f_s h_j^{-1}) h_j h_t \in F \wedge H$
The inverse: $(f_i h_j)^{-1} = (h_j^{-1} f_1^{-1} h_j) h_j^{-1} \in F \wedge H$
(Q.E.D.)

Note that the semidirect product symbol $\wedge$ is not commutative; by convention, the invariant group must always be placed first in the product $F \wedge H$. If H is also invariant under F, the semidirect product $F \wedge H$ becomes a direct product $F \times H$ with elements $f_i h_j$. This special situation arises most frequently when every element of F commutes with every element of H.

Examples

1. $C_{ni} = C_i \times C_n$, where $C_i = \{e, I\}$ is the group of inversion.
2. $D_n = C_n \wedge U$, where $U = \{e, u_0\}$ (see Section 3.1.2, Example 10).

4
Unitary and orthogonal groups

Two of the most important types of groups that occur in physics are unitary groups and orthogonal groups. We shall describe these groups as prototype examples of continuous groups in order to facilitate later discussion on the general theory of the continuous groups as well as of the finite groups, and, in particular, of the point groups.

4.1 The unitary group $U(n)$

4.1.1 Basic properties

A unitary transformation in n-dimensional vector space $V^{(n)}$ is defined by an $n \times n$ matrix U that satisfies

$$U^\dagger U = \mathbf{1} \tag{4.1.1}$$

which can be written also in the form $UU^\dagger = \mathbf{1}$ for $U^\dagger = U^{-1}$. According to (1.5.1), it leaves invariant the Hermitian scalar product of two vectors in $V^{(n)}$ over the complex field. The set of $n \times n$ unitary matrices forms a group under matrix multiplication. Firstly, the product of two unitary matrices is also a unitary matrix, i.e. if $U_2^\dagger U_2 = \mathbf{1}$ and $U_2^\dagger U_2 = \mathbf{1}$, then $(U_1U_2)^\dagger U_1U_2 = U_2^\dagger U_1^\dagger U_1U_2 = U_2^\dagger U_2 = \mathbf{1}$. Moreover, there exist the unit matrix $\mathbf{1}$ and the inverse given by $U^\dagger$, and obviously the matrix multiplication obeys the associative law. It is called *the unitary group in n dimensions* and denoted by $U(n)$. The matrix entries are complex so that an $n \times n$ unitary matrix is described by $2n^2$ real continuous parameters under the unitarity condition (4.1.1) which imposes n^2 linearly independent constraints (because the product $U^\dagger U$ is Hermitian). Accordingly, $U(n)$ is an n^2-parameter group, i.e. it is described by n^2 linear independent real continuous parameters. It is a *compact group* since its parameters vary over a continuously finite range due to the normalization conditions

$$\sum_{j=1}^{n} |U_{ij}|^2 = 1; \qquad i = 1, 2, \ldots, n$$

contained in (4.1.1).

The eigenvalues of a unitary matrix have modulus unity. Let Λ be the diagonalized form of a unitary matrix, then we have $\Lambda^\dagger \Lambda = \mathbf{1}$ from (4.1.1) so that every eigenvalue λ_j of a unitary matrix is unimodular satisfying

$$\lambda_j^* \lambda_j = |\lambda_j|^2 = 1; \qquad j = 1, 2, \ldots, n$$

Therefore, any eigenvalue can be expressed by $\lambda_j = \exp(\mathrm{i}\alpha_j)$ with a real parameter α_j. This means that a unitary matrix is periodic with respect to each angular parameter α_j.

In particular, the one-dimensional unitary group $U(1)$ is formed by the set $\{e^{i\alpha}; 0 \leqslant \alpha \leqslant 2\pi\}$.

The determinant of a unitary matrix U has modulus unity, i.e. $|\det U| = 1$, which follows by taking the determinants of both sides of $U^\dagger U = \mathbf{1}$. In the special case in which $\det U = 1$, the matrix U is called *a special unitary matrix*. The set of all special unitary matrices in n dimensions also forms a group called *the special unitary group* and is denoted as $SU(n)$. It is a subgroup of $U(n)$: $SU(n) < U(n)$, and is a $(n^2 - 1)$-parameter group due to the additional condition $\det U = 1$.

Example. The general element of $SU(2)$ is described by a 2×2 matrix

$$S(a, b) = \begin{bmatrix} a & b \\ -b^* & a^* \end{bmatrix}, \qquad |a|^2 + |b|^2 = 1 \tag{4.1.2}$$

which is a three-parameter group: let $a = a_1 + ia_2$ and $b = b_1 + ib_2$, then $a_1^2 + a_2^2 + b_1^2 + b_2^2 = 1$ so that only three of the real continuous parameters a_1, a_2, b_1 and b_2 are independent. From (4.1.2), it follows that a special unitary matrix S is expressed by the Pauli spin matrices as follows:

$$S = a_1\sigma_0 + ib_2\sigma_1 + ib_1\sigma_2 + ia_2\sigma_3$$

where $\sigma_0 = \mathbf{1}$, the 2×2 unit matrix. This expression is often very convenient for describing its transformation properties by those of the Pauli spin matrices. The parametrization of $SU(2)$ by (a, b) is called *the Cayley–Klein parametrization.*

4.1.2 The exponential form

Any unitary matrix U can be exponentiated as follows:

$$U = \exp K, \qquad K^\dagger = -K \tag{4.1.3}$$

where K is an anti-Hermitian matrix. If U is a special unitary matrix, we have $\operatorname{tr} K = 0 \pmod{2\pi i}$.

To show this, let the eigenvalues of a unitary matrix U be $\{e^{i\alpha_1}, e^{i\alpha_2}, \dots, e^{i\alpha_n}\}$, then U can be diagonalized by a unitary transformation T to the form

$$T^\dagger UT = e^{i\alpha_1} \oplus e^{i\alpha_2} \dots \oplus e^{i\alpha_n} = \exp K';$$
$$K' \equiv [i\alpha_1 \oplus i\alpha_2 \oplus \cdots \oplus i\alpha_n] \tag{4.1.4}$$

where we have used (1.2.12) for the second equality. On transforming this diagonalized form back to the original U, we obtain

$$U = T(\exp K')T^\dagger = \exp(TK'T^\dagger) \equiv \exp K$$

where $K = TK'T^\dagger$ is anti-Hermitian because K' is anti-Hermitian. This proves (4.1.3).

Next, taking the determinants of both sides of (4.1.4), we obtain

$$\det U = \exp(i\alpha_1 + i\alpha_2 + \cdots + i\alpha_n) = \exp(\operatorname{tr} K) \tag{4.1.5}$$

Accordingly, if U is a special unitary matrix, we have $\det U = 1$ by definition so that $\operatorname{tr} K = 0 \pmod{2\pi i}$. Q.E.D.

In general, we can set $K = K_0 + i\alpha\mathbf{1}$, where K_0 is the traceless part of K and α is a real number. Then, from (4.1.3),

$$U = e^{i\alpha} U_0, \qquad U_0 = \exp K_0 \tag{4.1.6}$$

where U_0 is a special unitary matrix defined by the traceless part K_0 of K satisfying $\det U_0 = 1$. The set of special unitary matrices in n dimensions forms the special unitary group $SU(n)$ whereas the set $\{e^{i\alpha}; 0 \leqslant \alpha \leqslant 2\pi\}$ forms the one-dimensional unitary group $U(1)$. Thus the n-dimensional unitary group $U(n)$ is expressed by a direct product group

$$U(n) = U(1) \times SU(n) \tag{4.1.7}$$

Previously, it has been shown that $U(n)$ is an n^2 parameter group[1] so that $SU(n)$ is an $(n^2 - 1)$-parameter group. For example, $SU(2)$ is a three-parameter group whereas $SU(3)$ is an eight-parameter group. The exponential form (4.1.3) is also very effective when we discuss the group generators.

Example. According to (4.1.6), the general element S of $SU(2)$ defined by (4.1.2) can be written in an exponential form as follows:

$$S(\boldsymbol{\theta}) = \exp[-iH(\boldsymbol{\theta})] \tag{4.1.8a}$$

where H is a traceless 2×2 Hermitian matrix that may be expressed by

$$H(\boldsymbol{\theta}) = \frac{1}{2}\begin{bmatrix} \theta_3 & \theta_1 - i\theta_2 \\ \theta_1 + i\theta_2 & -\theta_3 \end{bmatrix} \tag{4.1.8b}$$

Here, θ_1, θ_2 and θ_3 are three real parameters. These may be regarded as three components of a vector $\boldsymbol{\theta} = (\theta_1, \theta_2, \theta_3)$ that characterizes a rotation through an angle $\theta = |\boldsymbol{\theta}|$ around the unit vector $\boldsymbol{n} = \boldsymbol{\theta}/\theta$. It is called *the rotation vector* of the unitary transformation S. The numerical factor $\frac{1}{2}$ in (4.1.8b) is introduced to correlate $\boldsymbol{\theta}$ to the rotation vector of an ordinary rotation in three dimensions (cf. (4.3.2a)). The parametrization of $SU(2)$ by $\boldsymbol{\theta}$ is called *the Euler–Rodrigues parametrization.* Its relation with *the Cayley–Klein parametrization* introduced previously by (4.1.2) will be discussed in Section 10.1. In the theory of the spinning electron, an element $S(\boldsymbol{\theta})$ of $SU(2)$ is called *a spinor transformation* because it describes the transformation of the spin under a rotation characterized by the rotation vector $\boldsymbol{\theta} = \theta\boldsymbol{n}$. See Chapter 10 for further detail.

4.2 The orthogonal group $O(n, c)$

4.2.1 Basic properties

An orthogonal transformation in the n-dimensional vector space $V^{(n)}$ is defined by an $n \times n$ matrix R that satisfies the orthogonality relation

$$R^{\sim} R = \mathbf{1} \tag{4.2.1a}$$

which can be written also in the form $RR^{\sim} = \mathbf{1}$ for $R^{\sim} = R^{-1}$. In terms of the matrix elements, we have

$$\sum_s R_{is} R_{js} = \sum_s R_{si} R_{sj} = \delta_{ij}; \qquad i, j = 1, 2, \ldots, n$$

[1] This is most easily seen, through the exponential form (4.1.3), from the number of independent parameters of the anti-unitary group K, which equals n^2.

It is a transformation that leaves invariant the scalar product of two vectors x and y in $V^{(n)}$ over the complex field:

$$(Rx, Ry) = (x, R^{\sim}Ry) = (x, y)$$

The set of orthogonal transformations in $V^{(n)}$ over the complex field forms *the orthogonal group*, denoted $O(n, c)$. Since the matrix elements are complex, it is an $(n^2 - n)$-parameter group: $2n^2$ real continuous parameters with $n^2 + n$ orthonormality conditions imposed by (4.2.1a) (because $R^{\sim}R$ is a symmetric matrix).

Taking the determinants of the orthogonality relation (4.2.1a), we have $(\det R)^2 = 1$ so that

$$\det R = \pm 1 \tag{4.2.1b}$$

When $\det R = 1$ (-1), R is called *proper* (*improper*). The set of proper orthogonal transformations forms a group called the *special orthogonal group*, denoted $SO(n, c)$. It is a subgroup of $O(n, c)$, i.e. $SO(n, c) < O(n, c)$. An orthogonal transformation is called simply a *rotation*. Likewise, $O(n, c)$ may be called *the rotation group* (over the complex field) whereas $SO(n, c)$ is called the proper rotation group. The rotation group $O(n, r)$ over the real field is a subgroup of $O(n, c)$. The Lorentz group is a subgroup of $O(4, c)$; see Kim (1980a).

4.2.2 *Improper rotation*

The complement of the proper rotation group $SO(n, c)$ in $O(n, c)$ is the set of improper rotations in $V^{(n)}$; however, it does not form a group because a product of two improper rotations is a proper rotation. One of the simplest improper rotations is a reflection: a reflection m_h in a plane perpendicular to a unit vector $\boldsymbol{h}$ in $V^{(n)}$ is defined, in terms of *the diadic notation*,[2] by

$$m_h = \mathbf{1} - 2\boldsymbol{hh}; \qquad m_h^2 = \mathbf{1} \tag{4.2.2a}$$

or by the components

$$(m_h)_{ij} = \delta_{ij} - 2h_i h_j; \qquad i, j = 1, 2, \ldots, n$$

The reflection m_h transforms the vector $\boldsymbol{h}$ to $-\boldsymbol{h}$ and leaves invariant a vector $\boldsymbol{p}$ perpendicular to $\boldsymbol{h}$;

$$m_h \cdot \boldsymbol{h} = -\boldsymbol{h}, \qquad m_h \cdot \boldsymbol{p} = \boldsymbol{p}$$

for $\boldsymbol{h} \cdot \boldsymbol{h} = 1$ and $\boldsymbol{h} \cdot \boldsymbol{p} = 0$. It is involutional, satisfying $[m_h]^2 = 1$, so that its eigenvalues are 1 and -1. Directly from (4.2.2a), the trace is given by

$$\operatorname{tr} m_h = n - 2 \tag{4.2.2b}$$

from which the degeneracies of the eigenvalues 1 and -1 are $n - 1$ and 1, respectively; hence, $\det m_h = -1$ as expected. In the special case in which $\boldsymbol{h}$ is a natural basis vector $e^{(1)} = (1, 0, \ldots, 0)$ in $V^{(n)}$, we have

$$m_1 = \operatorname{diag}[-1, 1, \ldots, 1]$$

[2] A diad $D = \boldsymbol{uv}$ means that $D_{ij} = u_i v_j$, where $\boldsymbol{u} = \{u_i\}$ and $\boldsymbol{v} = \{v_j\}$ are vectors in $V^{(n)}$. By definition the transpose of D is given by $D^{\sim} = \boldsymbol{vu}$.

Let R be a proper rotation and m be a reflection in the full rotation group $O(n, c)$. Then any improper element can be expressed by *a rotation–reflection*: $R' = mR$. Thus the rotation group is decomposed into the cosets of the proper rotation group $SO(n, c)$ such that

$$O(n, c) = SO(n, c) + m\,SO(n, c) \tag{4.2.3a}$$

i.e. the proper rotation group $SO(n, c)$ is a halving subgroup of the full rotation group $O(n, c)$. When the dimensionality n is odd, however, it is simpler to define an improper rotation by *a rotation–inversion* $\overline{R} = IR$, where I is the inversion defined by

$$I \equiv \text{diag}\,[-1, -1, \ldots, -1] = -\mathbf{1} \tag{4.2.3b}$$

which commutes with every element of $O(n, c)$. Since $\det I = (-1)^n$, the inversion is proper (improper) when n is even (odd). Thus, if n is odd, the above decomposition of $O(n, c)$ is rewritten as

$$O(n, c) = SO(n, c) + I\,SO(n, c) \tag{4.2.3c}$$

This decomposition will be used for the three-dimensional rotation group.

It was Cartan (1913) who recognized the basic importance of the reflection as a fundamental foundation stone of the group of orthogonal transformations $O(n, c)$. His basic theorem may be stated as follows: 'Any proper (improper) rotation $R \in O(n, c)$ is given by a product of an even (odd) number ($\leqslant n$) of reflections.' Here a reflection means Cartan's reflection m_h defined by (4.2.2a); cf. Kim (1980a).

4.2.3 The real orthogonal group $O(n, r)$

If we assume that the matrix entries in $O(n, c)$ are real, we arrive at *the real orthogonal group*, denoted by $O(n, r)$. It is compact and a subgroup of $O(n, c)$. It is also a subgroup of the unitary group $U(n)$ because a real orthogonal matrix is a special case of a unitary matrix. Likewise, *the special real orthogonal group* $SO(n, r)$ is a subgroup of $SO(n, c)$ and $SU(n)$. Accordingly, all the theorems on $SU(n)$ and $SO(n, c)$ hold for $SO(n, r)$, with due modifications. The group–subgroup chains are

$$\begin{gathered} U(n),\ O(n, c) > O(n, r) > SO(n, r) \\ U(n) > SU(n) > SO(n, r) \\ O(n, c) > SO(n, c) > SO(n, r) \end{gathered} \tag{4.2.4}$$

Here, both $O(n, r)$ and $SO(n, r)$ are $[(n^2 - n)/2]$-parameter groups, because the numbers of the independent continuous parameters are the same for both. The real orthogonal group $O(n, r)$ is also called *the full (real) rotation group* in n dimensions, because its members are all real and leave invariant the distance between two points in the real linear vector space $V^{(n)}(r)$. A member of $O(n, r)$ is periodic with respect to angular parameters, because $O(n, r) < U(n)$. The special real orthogonal group $SO(n, r)$ is called *the proper (real) rotation group*.

The eigenvalues of a real orthogonal matrix $R \in O(n, r)$ are either 1 or -1 or in pairs $\{e^{i\theta_j}, e^{-i\theta_j}\}$ with real θ_j. To see this, let $u^{(j)}$ be an eigenvector belonging to an eigenvalue λ_j of R, then

$$Ru^{(j)} = \lambda_j u^{(j)}, \qquad Ru^{(j)*} = \lambda_j^* u^{(j)*} \tag{4.2.5}$$

where the second equation is the complex conjugate of the first. Since $O(n, r) < U(n)$, we can set $\lambda_j = e^{i\theta_j}$ with a real θ_j, then $\lambda_j^* = e^{-i\theta_j}$. In the case in which $\lambda_j^* = \lambda_j$ we have $\lambda_j^2 = 1$ so that $\lambda_j = 1$ or -1. Q.E.D.

4.2.4 *Real exponential form*

A real proper rotation R (i.e. $\det R = 1$) $\in SO(n, r)$ can be exponentiated in the form

$$R = \exp A; \qquad A^{\sim} = -A \tag{4.2.6}$$

where A is a real antisymmetric matrix which is traceless by definition. This is a special case of the special unitary matrix U_0 given by (4.1.6); cf. Littlewood (1950).

Previously we have shown that $SO(n, r)$ is an $[n(n-1)/2]$-parameter group; however, this is most easily seen from the number of the independent parameters of the real antisymmetric matrix A. Thus $SO(2, r)$ is a one-parameter group whereas $SO(3, r)$ is a three-parameter group.

Example. From (4.2.6), a plane rotation $R^{(2)}(\theta) \in SO(2, r)$ can be expressed by the following exponential form (see (1.6.2)):

$$R^{(2)}(\theta) = \begin{bmatrix} \cos\theta & -\sin\theta \\ \sin\theta & \cos\theta \end{bmatrix} = \exp\begin{bmatrix} 0 & -\theta \\ \theta & 0 \end{bmatrix} \tag{4.2.7}$$

4.3 The rotation group in three dimensions $O(3, r)$

Since there are many applications for the full rotation group in three dimensions $O(3, r)$ $(= O^{(3)})$ we shall discuss its properties in greater detail. According to (4.2.3c), the rotation group can be decomposed into

$$O(3, r) = SO(3, r) + I\, SO(3, r)$$

where SO(3, r) $(= O^{(+3)})$ is the proper rotation group (or the special real orthogonal group) in three dimensions and I is the inversion which commutes with any 3×3 matrix. We shall begin with the discussion of the proper rotation group $O^{(+3)}$.

4.3.1 *Basic properties of rotation*

According to (4.2.6), a real proper rotation in $V^{(3)}$ can be expressed by an antisymmetric matrix A that may be written in the following form:

$$R(\boldsymbol{\theta}) = e^A; \qquad A = \begin{bmatrix} 0 & -\theta_3 & \theta_2 \\ \theta_3 & 0 & -\theta_1 \\ -\theta_2 & \theta_1 & 0 \end{bmatrix} \tag{4.3.1}$$

where $(\theta_1, \theta_2, \theta_3)$ is a set of real parameters that completely determines the rotation and that may be regarded as a vector $\boldsymbol{\theta} = (\theta_1, \theta_2, \theta_3)$ whose transformation property must be determined by that of the rotation $R(\boldsymbol{\theta})$: it is an eigenvector of $R(\boldsymbol{\theta})$ belonging to the eigenvalue unity, i.e. $R(\boldsymbol{\theta})\boldsymbol{\theta} = \boldsymbol{\theta}$ for $A\boldsymbol{\theta} = 0$. Thus, $R(\boldsymbol{\theta})$ leaves invariant a line passing through the coordinate origin O and parallel to $\boldsymbol{\theta}$; the line is called the *axis of rotation*. The vector $\boldsymbol{\theta}$ is called the *rotation vector*: it must be an axial vector[3] because any rotation (through which $\boldsymbol{\theta}$ is defined) is invariant under the inversion I, i.e. $IR(\boldsymbol{\theta})I = R(\boldsymbol{\theta})$.

In the following, we shall show that $R(\boldsymbol{\theta})$ describes a rotation about the rotation vector $\boldsymbol{\theta}$ counterclockwise (viewed from the $+\boldsymbol{\theta}$ direction) through an angle θ defined by the magnitude $\theta = |\boldsymbol{\theta}|$. Let the unit vector $\boldsymbol{s} = \boldsymbol{\theta}/\theta$ along the rotation vector $\boldsymbol{\theta}$ be called the *axis-vector* of the rotation $R(\boldsymbol{\theta})$, then it is also axial since $\boldsymbol{\theta}$ is axial, and the rotation $R(\boldsymbol{\theta})$ is rewritten in the form (cf. (1.6.4c))

$$R(\theta \boldsymbol{s}) = \mathrm{e}^{\theta\omega}; \qquad \omega = \omega(\boldsymbol{s}) = \begin{bmatrix} 0 & -s_3 & s_2 \\ s_3 & 0 & -s_1 \\ -s_2 & s_1 & 0 \end{bmatrix} \tag{4.3.2a}$$

where the matrix $\omega = \omega(\boldsymbol{s})$ is called *the infinitesimal rotation* about $\boldsymbol{s}$ because $\partial R/\partial\theta|_{\theta=0} = \omega$. Let the plane perpendicular to the axis-vector $\boldsymbol{s}$ be called the *ω-plane*, then $R(\boldsymbol{\theta})$ is a plane rotation that leaves the ω-plane invariant. Let $\boldsymbol{x}$ be a vector that transforms according to $\boldsymbol{x} = \mathrm{e}^{\theta\omega}\boldsymbol{x}_0$, then the rate of rotation of $\boldsymbol{x}$ per unit angle is given by

$$\boldsymbol{x}' = \omega\boldsymbol{x} = [\boldsymbol{s} \times \boldsymbol{x}] \tag{4.3.2b}$$

From the geometric meaning of the vector product, the infinitesimal displacement of $\boldsymbol{x}$ under $R(\theta\boldsymbol{s})$ occurs counterclockwise about the axis-vector $\boldsymbol{s}$ (viewed from the $+\boldsymbol{s}$ direction), as shown in Figure 4.1.

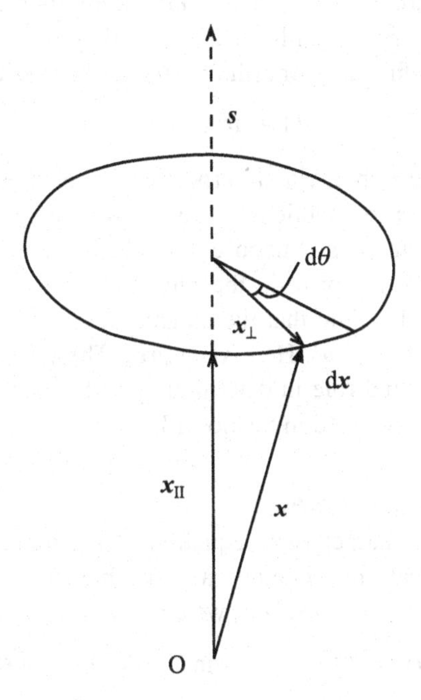

Figure 4.1. An infinitesimal rotation about $\boldsymbol{s}$ in $V^{(3)}$.

[3] A vector in $V^{(3)}$ that does not change its sign under the inversion I is called an axial vector whereas an ordinary vector changes its sign under I and is called a polar vector. A typical example of an axial vector is given by a vector product $[\boldsymbol{v} \times \boldsymbol{u}]$ of two vectors of the same parity.

Let us look into the vectorial nature of the axis-vector $\boldsymbol{s}$. Let $\boldsymbol{v}$ be a unit vector on the ω-plane and set $\boldsymbol{u} = \omega\boldsymbol{v} = [\boldsymbol{s} \times \boldsymbol{v}]$, then the set of three unit vectors $\{\boldsymbol{s}, \boldsymbol{v}, \boldsymbol{u}\}$ forms a right-handed orthogonal trio so that we can express $\boldsymbol{s}$ and ω in terms of $\boldsymbol{v}$ and $\boldsymbol{u}$ as follows:

$$\boldsymbol{s} = [\boldsymbol{v} \times \boldsymbol{u}], \qquad \omega = \boldsymbol{uv} - \boldsymbol{vu} \tag{4.3.2c}$$[4]

These expressions exhibit manifestly the axial nature of the axis-vector $\boldsymbol{s}$ and the tensorial nature of ω, because two vectors $\boldsymbol{v}$ and $\boldsymbol{u}$ ($= \omega\boldsymbol{v}$) are of the same parity. It has been shown by Kim (1980a) that a plane rotation belonging to the group $O(n, c)$ is also expressed by $R(\theta) = \exp(\theta\omega)$ with $\omega = \boldsymbol{uv} - \boldsymbol{vu}$, where $\boldsymbol{u}$ and $\boldsymbol{v}$ are n-dimensional vectors. This extension has an important application in the representation of the Lorentz group, which is a subgroup of $O(4, c)$.

4.3.1.1 *The characteristic equation of ω*

From (4.3.2a), the square of ω is given by

$$\omega^2 = \boldsymbol{ss} - 1 \tag{4.3.3a}$$

in the diadic notation. Since $\omega\boldsymbol{s} = [\boldsymbol{s} \times \boldsymbol{s}] = 0$, we have the following characteristic equation of ω,

$$\omega^3 + \omega = 0 \tag{4.3.3b}$$

From this, it follows that the operators ω and $(1 + \omega^2)$ are orthogonal in the sense that $\omega(1 + \omega^2) = 0$ and so are $(1 + \omega^2)$ and $-\omega^2$. Here, the last two operators are projection operators that are mutually dual and satisfy $X^2 = X$. Thus, for a given vector $\boldsymbol{x} \in V^{(3)}$ one can define a set of orthogonal trios by (see Figure 4.1)

$$\boldsymbol{x}' = \omega\boldsymbol{x} = [\boldsymbol{s} \times \boldsymbol{x}], \qquad \boldsymbol{x}_\parallel = (1 + \omega^2)\boldsymbol{x} = \boldsymbol{s}(\boldsymbol{s} \cdot \boldsymbol{x}), \qquad \boldsymbol{x}_\perp = -\omega^2\boldsymbol{x} \tag{4.3.4}$$

Here $\boldsymbol{x}'$ is a vector on the ω-plane and perpendicular to $\boldsymbol{x}$; $\boldsymbol{x}_\parallel$ is the projection of $\boldsymbol{x}$ onto $\boldsymbol{s}$ and hence is parallel to $\boldsymbol{s}$ while $\boldsymbol{x}_\perp$ ($= \boldsymbol{x} - \boldsymbol{x}_\parallel$) is the projection of $\boldsymbol{x}$ on the ω-plane, hence perpendicular to $\boldsymbol{s}$. Suppose that the vector $\boldsymbol{x}$ rotates according to $\boldsymbol{x} = \mathrm{e}^{\theta\omega}\boldsymbol{x}_0$, then $\boldsymbol{x}'$, $\boldsymbol{x}_\parallel$ and $\boldsymbol{x}_\perp$ provide the rate, axis and arm of the rotation of $\boldsymbol{x}$, respectively (see Figure 4.1). Note that the magnitudes of the rate and arm are equal, i.e. $\boldsymbol{x}_\perp^2 = \boldsymbol{x}'^2 = \boldsymbol{x}_0'^2$, for $\omega^4 = -\omega^2$. Here $\boldsymbol{x}_0' = \omega\boldsymbol{x}_0$. These projection operators introduced above play the essential role in determining all the possible lattice types of the crystal lattices, as will be discussed in Section 13.3.2.

4.3.1.2 *The matrix expression of $R(\boldsymbol{\theta})$*

Since $\omega^3 = -\omega$ from the characteristic equation of the infinitesimal rotation ω, the expansion of $\exp(\theta\omega)$ can not contain a term higher than ω^2. In fact, using $\omega^{2N+1} = (-1)^N\omega$ and $\omega^{2N+2} = (-1)^N\omega^2$, we arrive at the expansion

$$R(\boldsymbol{\theta}) = \mathrm{e}^{\theta\omega} = \mathbf{1} + \omega\sin\theta + \omega^2(1 - \cos\theta) \tag{4.3.5a}$$

Accordingly, the periodicity of the rotation $R(\boldsymbol{\theta})$ with respect to θ is described by

$$R((\theta + 2\pi)\boldsymbol{s}) = R(\theta\boldsymbol{s}) \tag{4.3.5b}$$

[4] The second expression is obtained via $\omega\boldsymbol{r} = [\boldsymbol{s} \times \boldsymbol{r}] = [\boldsymbol{v} \times \boldsymbol{u}] \times \boldsymbol{r} = \boldsymbol{u}(\boldsymbol{v} \cdot \boldsymbol{r}) - \boldsymbol{v}(\boldsymbol{u} \cdot \boldsymbol{r}) = (\boldsymbol{uv} - \boldsymbol{vu}) \cdot \boldsymbol{r}$, where $\boldsymbol{r}$ is an arbitrary vector.

Thus, the parameter domain Ω of the rotation group $O^{(+3)}$ is defined by a sphere of radius π with the cyclic boundary condition

$$\Omega: \quad 0 \leqslant |\boldsymbol{\theta}| \leqslant \pi; \qquad R(\pi \boldsymbol{s}) = R(-\pi \boldsymbol{s}) \tag{4.3.6}$$

where two opposite poles $\pi\boldsymbol{s}$ and $-\pi\boldsymbol{s}$ are regarded as one point in Ω. Then there exists a one-to-one correspondence between the rotation $R(\boldsymbol{\theta}) \in O^{(+3)}$ and the rotation vector $\boldsymbol{\theta}$ in the parameter domain Ω.

Now, we shall write down the 3×3 matrix representation of $R(\boldsymbol{\theta})$. Using (4.3.3a), we rewrite (4.3.5a) in the form

$$R(\theta\boldsymbol{s}) = \cos\theta\,\mathbf{1} + \omega\sin\theta + \boldsymbol{s}\boldsymbol{s}(1-\cos\theta) \tag{4.3.7}$$

In a simple special case in which the axis-vector $\boldsymbol{s}$ is parallel to the z-axis, i.e. $\boldsymbol{s} = (0, 0, 1)$, Equation (4.3.7) takes the following familiar form:

$$R(0, 0, \theta) = \begin{bmatrix} \cos\theta & -\sin\theta & 0 \\ \sin\theta & \cos\theta & 0 \\ 0 & 0 & 1 \end{bmatrix} \tag{4.3.8a}$$

which is the rotation in the x, y plane through an angle θ. In the general case for three dimensions, we have

$$R(\theta\boldsymbol{s}) = \begin{bmatrix} s_1^2 + \cos\theta\,(1 - s_1^2) & (1-\cos\theta)s_1 s_2 - \sin\theta\, s_3 & (1-\cos\theta)s_1 s_3 + \sin\theta\, s_2 \\ (1-\cos\theta)s_2 s_1 + \sin\theta\, s_3 & s_2^2 + \cos\theta\,(1 - s_2^2) & (1-\cos\theta)s_2 s_3 - \sin\theta\, s_1 \\ (1-\cos\theta)s_3 s_1 - \sin\theta\, s_2 & (1-\cos\theta)s_3 s_2 + \sin\theta\, s_1 & s_3^2 + \cos\theta\,(1 - s_3^2) \end{bmatrix} \tag{4.3.8b}$$

which is explicit with respect to the axis-vector $\boldsymbol{s}$ and the rotation angle θ; cf. Lomont (1959).

Next, we consider the inverse problem. Suppose that there is given a 3×3 real orthogonal matrix R, then the problem is to find its rotation vector $\boldsymbol{\theta}$. Firstly, the angle of rotation θ is determined by the trace of R

$$\operatorname{tr} R = 1 + 2\cos\theta, \qquad 0 \leqslant \theta \leqslant \pi \tag{4.3.9a}$$

Secondly, from (4.3.5a), the infinitesimal rotation ω is determined by

$$\omega = (R - \tilde{R})/2\sin\theta, \qquad 0 < \theta < \pi \tag{4.3.9b}$$

from which the axis-vector $\boldsymbol{s}$ follows via (4.3.2a). When $\theta = \pi$, ω becomes indefinite. In this case, R is a binary rotation given, directly from (4.3.7), by

$$R(\pi\boldsymbol{s}) = R(-\pi\boldsymbol{s}) = 2\boldsymbol{s}\boldsymbol{s} - \mathbf{1} \tag{4.3.10}$$

which determines $\boldsymbol{s}$ up to the sign. This corresponds simply to the fact that a binary rotation can be regarded as a rotation through angle π about $\boldsymbol{s}$ or $-\boldsymbol{s}$.

Exercise. Find the rotation vector $\boldsymbol{\theta}$ of a real orthogonal matrix defined by

$$R = \begin{bmatrix} 0 & 0 & 1 \\ 1 & 0 & 0 \\ 0 & 1 & 0 \end{bmatrix} \tag{4.3.11}$$

From (4.3.9a), we find $\theta = 120°$. Then from (4.3.9b) we have

$$\omega = \frac{1}{\sqrt{3}}\begin{bmatrix} 0 & -1 & 1 \\ 1 & 0 & -1 \\ -1 & 1 & 0 \end{bmatrix} \tag{4.3.12}$$

By comparison of this with (4.3.2a) we obtain $\boldsymbol{s} = (1/\sqrt{3}, 1/\sqrt{3}, 1/\sqrt{3})$. Thus R describes 3_{xyz}, the three-fold rotation about the diagonal direction with respect to the x-, y- and z-axes of the Cartesian coordinate system. For this simple case, the basis-vector representation 3_{xyz} is obtained directly from the matrix expression (4.3.11) through Lemma 5.1.1.

4.3.2 The conjugate rotations

The similarity transformation $TR(\boldsymbol{\theta})T^{\sim}$ of a rotation $R(\boldsymbol{\theta}) \in O^{(+3)}$ by a proper rotation $T \in O^{(+3)}$ is called the conjugate rotation of $R(\boldsymbol{\theta})$ by T. It is given by

$$R' = TR(\boldsymbol{\theta})T^{\sim} = R(T\boldsymbol{\theta}) \tag{4.3.13a}$$

That is, it simply rotates the rotation vector $\boldsymbol{\theta}$ by T.

This important relation follows from the fact that the rotation vector $\boldsymbol{\theta}$ is the eigenvector of $R(\boldsymbol{\theta})$ belonging to the eigenvalue unity, i.e. $R(\boldsymbol{\theta})\boldsymbol{\theta} = \boldsymbol{\theta}$. In fact, through $TR(\boldsymbol{\theta})T^{\sim}T\boldsymbol{\theta} = T\boldsymbol{\theta}$, the rotation vector of the conjugate rotation R' is given by $\boldsymbol{\theta}' = T\boldsymbol{\theta}$, because $\boldsymbol{\theta}' \to \boldsymbol{\theta}$ as $T \to 1$. Since the axis-vector $\boldsymbol{s}$ is parallel to $\boldsymbol{\theta}$, we have also

$$T\omega(\boldsymbol{s})T^{\sim} = \omega(T\boldsymbol{s}) \tag{4.3.13b}$$ [5]

The above relation (4.3.13a) is truly a life-saving result in the theory of rotation: instead of working out the similarity transform of $R(\boldsymbol{\theta})$, is is necessary only to rotate the rotation vector $\boldsymbol{\theta}$. Two rotation vectors connected by a rotation such that $\boldsymbol{\theta}' = T\boldsymbol{\theta}$ are said to be *equivalent*, and so are their axis-vectors $\boldsymbol{s}$ and $\boldsymbol{s}'$.

Obviously, Equation (4.3.13a) can be extended to an improper rotation $\overline{T} = IT$ belonging to the full rotation group $O(3, r)$ ($= O^{(3)}$), where I is the inversion. Since the inversion I commutes with any 3×3 matrix we have

$$\overline{T}R(\boldsymbol{\theta})\overline{T}^{\sim} = TR(\boldsymbol{\theta})T^{\sim} = R(T\theta) = R(\overline{T}\boldsymbol{\theta}) \tag{4.3.14}$$

where the last equality holds since the rotation vector is axial, i.e. $I\boldsymbol{\theta} = \boldsymbol{\theta}$.

The set of all conjugate rotations of a given rotation $R(\boldsymbol{\theta}) \in O^{(+3)}$ forms the conjugate class of $R(\boldsymbol{\theta})$ in the rotation group. Their rotation angles are all the same $|\boldsymbol{\theta}| = |T\boldsymbol{\theta}|$ because an orthogonal matrix T leaves the length of a vector invariant. Conversely, all rotations with the same rotation angle θ are in the same class of $SO(3, r)$, because their rotation vectors $\boldsymbol{\theta}$ lie on a concentric sphere with the radius θ in the parameter space Ω and any two vectors $\boldsymbol{\theta}_1$ and $\boldsymbol{\theta}_2$ on the sphere can be brought together by a rotation about an axis perpendicular to $\boldsymbol{\theta}_1$ and $\boldsymbol{\theta}_2$. Obviously the above discussion can easily be extended to the full rotation group $O^{(3)}$. We express an improper rotation by a rotation–inversion $\overline{R}(\boldsymbol{\theta}) = IR(\boldsymbol{\theta})$, and define the rotation vector of $\overline{R}(\boldsymbol{\theta})$ by that of the proper part $R(\boldsymbol{\theta})$. Then two improper rotations $\overline{R}(\boldsymbol{\theta}_1)$ and $\overline{R}(\boldsymbol{\theta}_2)$ belonging to $O^{(3)}$ are in the same class if their rotation angles $|\boldsymbol{\theta}_1|$ and $|\boldsymbol{\theta}_2|$ are the same simply because the inversion I commutes with any rotation in $O^{(3)}$.

[5] This relation may also be proven on the basis of (4.3.2c): $\omega(\boldsymbol{s}') = T\omega T^{\sim} = T(\boldsymbol{uv} - \boldsymbol{vu})T^{\sim} = (T\boldsymbol{u})(T\boldsymbol{v}) - (T\boldsymbol{v})(T\boldsymbol{u})$, so that $\boldsymbol{s}' = [T\boldsymbol{v} \times T\boldsymbol{u}] = T\boldsymbol{s}$.

Exercise 1. Let $R(\boldsymbol{\theta})$ be a plane rotation, then show that

$$\overline{2}_s R(\boldsymbol{\theta})\overline{2}_s = R(\boldsymbol{\theta}), \qquad \overline{2}_h R(\boldsymbol{\theta})\overline{2}_h = R(-\boldsymbol{\theta}) \tag{4.3.15a}$$

where $\boldsymbol{s} \parallel \boldsymbol{\theta} \perp \boldsymbol{h}$.

Exercise 2. Previously in (3.1.2a) we have shown the following relation via the stereographic projection diagram given in Figure 3.1:

$$2_h 2_v = c_z(\theta); \qquad \boldsymbol{h}\cdot\boldsymbol{v} = \cos(\theta/2) \tag{4.3.15b}$$

Show this relation using the conjugate transformation (4.3.13a).

Hint. $2_h = R(\boldsymbol{\theta}/2)2_v R(-\boldsymbol{\theta}/2)$; $\quad \boldsymbol{v} \perp \boldsymbol{\theta}$ so that

$$\begin{aligned} 2_h 2_v &= R(\boldsymbol{\theta}/2)2_v R(-\boldsymbol{\theta}/2)2_v \\ &= R(\boldsymbol{\theta}/2)R(\boldsymbol{\theta}/2) = R(\boldsymbol{\theta}) \end{aligned}$$

Exercise 3. Using the same notations as those in Exercise 2, show that

$$c_z(\theta)2_v = 2_v c_z(\theta)^{\sim} = 2_h \tag{4.3.15c}$$

Exercise 4. Show that two proper rotations $R_1 = R(\boldsymbol{\theta}_1)$ and $R_2 = R(\boldsymbol{\theta}_2)$ commute if and only if either their rotation axes are parallel or they are mutually perpendicular binary rotations, i.e.

$$\boldsymbol{\theta}_1 \parallel \boldsymbol{\theta}_2 \qquad \text{or} \qquad \boldsymbol{\theta}_1 \perp \boldsymbol{\theta}_2 \text{ and } |\boldsymbol{\theta}_1| = |\boldsymbol{\theta}_2| = \pi \tag{4.3.16}$$

Solution. Let $R_2 R(\boldsymbol{\theta}_1)R_2^{\sim} = R(\boldsymbol{\theta}_1')$ with $\boldsymbol{\theta}_1' = R_2\boldsymbol{\theta}_1$, then $R_2 R(\boldsymbol{\theta}_1) = R(\boldsymbol{\theta}_1')R_2$. Thus R_2 commutes with $R(\boldsymbol{\theta}_1)$ if and only if $R(\boldsymbol{\theta}_1') = R(\boldsymbol{\theta}_1)$. From the one-to-one correspondence between $R(\boldsymbol{\theta})$ and $\boldsymbol{\theta}$ in the parameter space Ω defined by (4.3.6), we have either $R_2\boldsymbol{\theta}_1 = \boldsymbol{\theta}_1' = \boldsymbol{\theta}_1$ so that $\boldsymbol{\theta}_1 \parallel \boldsymbol{\theta}_2$ or $R_2\boldsymbol{\theta}_1 = \boldsymbol{\theta}_1' = -\boldsymbol{\theta}_1$ with $|\boldsymbol{\theta}_1| = \pi$ so that both R_1 and R_2 are binary and $\boldsymbol{\theta}_1 \perp \boldsymbol{\theta}_2$. Q.E.D.

4.3.3 The Euler angles

Frequently, a rotation is described by a product of three elementary rotations about the coordinate axes (z-, y- and z-axes) through angles α, β and γ, respectively. These angles are called *the Euler angles*. It may be worthwhile to discuss these angles even though we seldom use them in this book. To understand the basic structure of the Euler construction, we shall first rewrite the conjugate transformation introduced by (4.3.13a) in the form

$$TR(\boldsymbol{\theta}) = R(\boldsymbol{\theta}')T; \qquad \boldsymbol{\theta}' = T\boldsymbol{\theta}$$

Repeated applications of this relation for a product of three successive rotations yield

$$\begin{aligned} R(\boldsymbol{\theta}_3)R(\boldsymbol{\theta}_2)R(\boldsymbol{\theta}_1) &= R(\boldsymbol{\theta}_2')R(\boldsymbol{\theta}_3)R(\boldsymbol{\theta}_1) \\ &= R(\boldsymbol{\theta}_2')R(\boldsymbol{\theta}_1')R(\boldsymbol{\theta}_3) \\ &= R(\boldsymbol{\theta}_1'')R(\boldsymbol{\theta}_2')R(\boldsymbol{\theta}_3) \end{aligned} \tag{4.3.17}$$

where

$$\theta_1'' = R(\theta_2')\theta_1', \qquad \theta_1' = R(\theta_3)\theta_1, \qquad \theta_2' = R(\theta_3)\theta_2$$

The above equation (4.3.17) relates a product of three successive rotations about three rotation vectors, θ_1, θ_2 and θ_3, fixed in the original coordinate system to a product of three successive rotations about θ_3 and the transformed rotation vectors θ_2' and θ_1''.

To describe the Euler construction, let e_x, e_y and e_z be three unit vectors along the x-, y- and z-axes of the Cartesian coordinate system fixed in space (see Figure 4.2) and choose the three rotation vectors as follows:

$$\theta_1 = \alpha e_z, \qquad \theta_2 = \beta e_y, \qquad \theta_3 = \gamma e_z$$

with the angles of rotation limited by

$$-\pi \leqslant \alpha < \pi, \qquad 0 \leqslant \beta \leqslant \pi, \qquad -\pi \leqslant \gamma < \pi$$

Then, (4.3.17) takes the form

$$R(\gamma e_z)R(\beta e_y)R(\alpha e_z) = R(\alpha e_z'')R(\beta e_y')R(\gamma e_z) \equiv R(\alpha, \beta, \gamma) \tag{4.3.18}$$

where $e_z' = e_z$ and

$$e_z'' = R(\beta e_y')e_z, \qquad e_y' = R(\gamma e_z)e_y$$

The angles α, β and γ of the three elementary rotations define the Euler angles. The elementary rotations in the first expression of (4.3.18) are with respect to the z-, y- and z-axes of the fixed coordinate system, whereas the elementary rotations in the second expression are with respect to the e_z, e_y' and e_z'' axes (see Figure 4.2). Here the first operation $R(\gamma e_z)$ transforms e_y to e_y', the second operation $R(\beta e_y')$ transforms e_z to e_z'' and the last operation $R(\alpha e_z'')$ describes a rotation about e_z'' through an angle α.

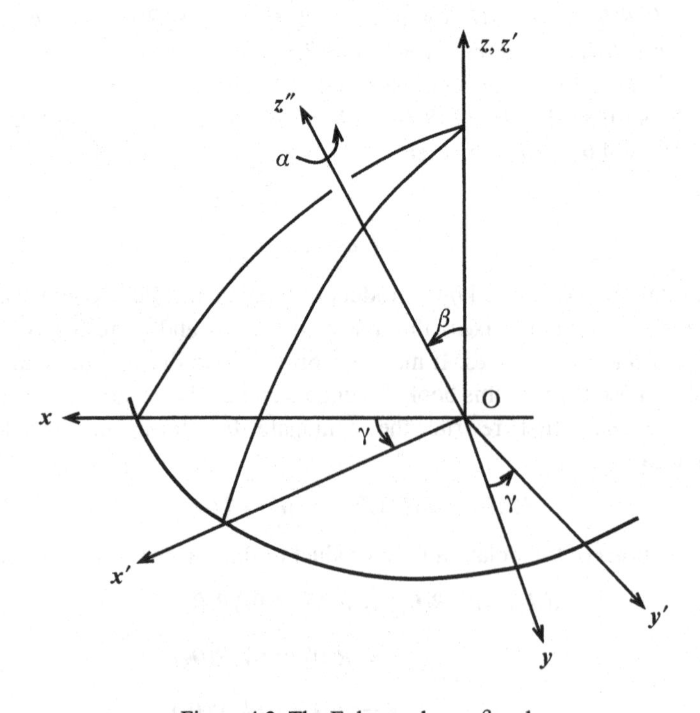

Figure 4.2. The Euler angles α, β and γ.

Depending on the situation, one can use either expression of equation (4.3.18) to calculate the nett rotation $R(\alpha, \beta, \gamma)$. When we describe the motion of a point (x, y, z) with respect to a fixed coordinate system as in the present approach, it is obvious that the first expression of (4.3.18) is more convenient, since each elementary rotation is referred to the fixed coordinate system. When we describe the coordinate transformation of a fixed point in space, however, the second expression of (4.3.18) is more convenient, since each elementary transformation is referred to the transformed coordinate system. On the basis of the first expression of (4.3.18), the explicit form of the rotation $R(\alpha, \beta, \gamma)$ is given by

$$R(\alpha, \beta, \gamma) = \begin{bmatrix} \cos\gamma & -\sin\gamma & 0 \\ \sin\gamma & \cos\gamma & 0 \\ 0 & 0 & 1 \end{bmatrix} \begin{bmatrix} \cos\beta & 0 & \sin\beta \\ 0 & 1 & 0 \\ -\sin\beta & 0 & \cos\beta \end{bmatrix} \begin{bmatrix} \cos\alpha & -\sin\alpha & 0 \\ \sin\alpha & \cos\alpha & 0 \\ 0 & 0 & 1 \end{bmatrix}$$

$$= \begin{bmatrix} \cos\gamma\cos\beta\cos\alpha - \sin\gamma\sin\alpha & -\cos\gamma\cos\beta\sin\alpha - \sin\gamma\cos\alpha & \cos\gamma\sin\beta \\ \sin\gamma\cos\beta\cos\alpha + \cos\gamma\sin\alpha & -\sin\gamma\cos\beta\sin\alpha + \cos\gamma\cos\alpha & \sin\gamma\sin\beta \\ -\sin\beta\cos\alpha & \sin\beta\sin\alpha & \cos\beta \end{bmatrix} \tag{4.3.19}$$

where each elementary rotation is obtained from (4.3.8b). This representation $R(\alpha, \beta, \gamma)$ for a rotation may be compared with the representation $R(\theta\boldsymbol{s})$ given by (4.3.8b).

5

The point groups of finite order

5.1 Introduction

Any subgroup of the full rotation group in three dimensions $O(3, r)$ $(= O^{(3)})$ is called a *point group*, because all the group elements leave at least one point (the coordinate origin) in space invariant. Since there are a great deal of applications for the point groups of finite order, we shall discuss their group structures in greater detail. There exist five types of proper and nine types of improper point groups of finite order. A new system of notations for the latter is introduced by expressing an improper operation with the inversion or a rotation–inversion. This system is very effective in describing the isomorphisms between proper and improper point groups, because the inversion commutes with any point operation.

Following the historical development, we shall introduce a point group by the symmetry point group of a geometric body such as a regular polyhedron. The symmetry of a geometric body is defined by the set of all symmetry transformations which brings the body into coincidence with itself. By definition, such a set of transformations forms a group, the symmetry group G of the body. More specifically, let p be a point on the body, then a set of transformations which leaves the point p invariant forms a subgroup H of G. Thus, the number n_p of the points which are equivalent to the point p is given by the index of H in G

$$n_p = |G|/|H| \qquad \text{or} \qquad |G| = n_p|H| \tag{5.1.0}$$

where $|G|$ and $|H|$ are the orders of the respective groups G and H. Since $|H| \geqslant 1$, the order $|G|$ of the symmetry group G is characterized by *the maximum number of mutually equivalent points on the body.* Equation (5.1.0) can be used to determine the order of the symmetry group of a geometric body or the number of the equivalent points n_p of the body. For example, a regular tetrahedron has four equivalent triangular faces and the mid-point of each face is invariant under the uniaxial group C_3. Hence its proper symmetry point group, the tetrahedral group T, has the order $|T| = 4 \times 3 = 12$ from (5.1.0). Analogously, $|O| = 8 \times 3 = 24$ for the octahedral group O and $|Y| = 20 \times 3 = 60$ for the icosahedral group Y. There exists no other proper polyhedral group besides these three groups T, O and Y (Section 5.3). The structure of the buckyball molecule C_{60} is interesting: it belongs to the icosahedral group Y and all 60 carbon atoms are on mutually equivalent vertices, but not along any proper axis of rotation (Section 5.3.6). This is consistent with $|Y| = 60$ in view of (5.1.0) with $|H| = 1$.

Guided by the geometric symmetry of a body, we shall arrive at *the generators* of the symmetric group G and *the set of the algebraic defining relations* or *the presentation* which contains the complete information on G. The truly exciting aspect is that the number of generators is very much limited ($\leqslant 3$) for any point group G of finite order. The algebraic approach based on the presentation is very effective in determining the subgroup structure and the representations of G and also for its extensions to the space groups and magnetic space groups.

5.1.1 The uniaxial group C_n

The set of all rotations about a fixed axis forms a uniaxial group of infinite order, denoted C_∞. It is Abelian. Let us consider a subgroup $C_n < C_\infty$ of a finite order n, then there exists the *minimum rotation* about the axis through an angle $2\pi/n$, which is called the *n-fold axis of rotation* and denoted by c_n. We say that the axis and axis-vector of c_n are *n-fold* or of *order n*. When we discuss the point symmetry of a body, the minimum angle of rotation about an axis which brings the body into coincidence with itself determines the order of the axis. Let c_4 be an axis of order 4, for example, then two successive four-fold rotations c_4^2 amount to a two-fold rotation: it is a two-fold rotation c_2 about an axis of order 4.

An n-fold axis of rotation c_n about an axis-vector $\boldsymbol{s}$ is represented by the orthogonal transformation $R(2\pi n^{-1}\boldsymbol{s})$ defined by (4.3.2a). Thus, a uniaxial group C_n of order n generated by c_n is expressed by

$$C_n = \{c_n^m = R(2\pi m n^{-1}\boldsymbol{s}); \qquad m = 0, 1, 2, \ldots, n-1\} \tag{5.1.1a}$$

It is a cyclic group and is defined by one generator $a = c_n$ and one *defining relation* (see (3.2.2))

$$a^n = e \tag{5.1.1b}$$

Since it is Abelian, the number of the classes equals the number of the elements n.

There is no further rotation about this axis besides the elements of C_n. If there is a rotation $R(\theta\boldsymbol{s})$ such that $\theta \neq 2\pi m/n$ for any m, then one can construct a rotation with an angle $(\theta - 2\pi m/n)$ that can be made smaller than the minimum angle $2\pi/n$.

5.1.2 Multiaxial groups. The equivalence set of axes and axis-vectors

Let P be a proper point group of a finite order in general. Then the group has a finite set of distinct rotation axes passing through the coordinate origin O, each of which is of a finite order and defines a uniaxial group C_n. These uniaxial groups are all distinct except for the identity element which they share.

Let $R(\theta\boldsymbol{s})$ be a rotation in P through an angle θ about an axis-vector $\boldsymbol{s}$. Then the conjugate transformation of $R(\theta\boldsymbol{s})$ by an element α in P rotates only the axis-vector $\boldsymbol{s}$, leaving θ fixed,

$$\alpha R(\theta\boldsymbol{s})\alpha^{-1} = R(\theta\alpha\boldsymbol{s}) \tag{5.1.2}$$

as has been shown in (4.3.13a). Here the transformed vector $\boldsymbol{s}' = \alpha\boldsymbol{s}$ should also represent one of the rotation axes in P. When axis-vectors are mapped onto each other by a group element of P, they are said to be equivalent in P and so are their axes of rotation. To a given axis of rotation there correspond two axis-vectors $\boldsymbol{s}$ and $-\boldsymbol{s}$ that are mutual inverses because $R^{-1}(\theta\boldsymbol{s}) = R(-\theta\boldsymbol{s})$. In the case in which an axis-vector $\boldsymbol{s}$ is equivalent to its inverse $-\boldsymbol{s}$, i.e. there exists a rotation α in P such that $-\boldsymbol{s} = \alpha\boldsymbol{s}$, we say that the corresponding axis is *two-sided* (otherwise it is *one-sided*). Obviously such an element α in P is a binary rotation c_2 whose axis is perpendicular to the axis-vector $\boldsymbol{s}$. Thus, if the two-sided axis is n-fold, the group P contains the dihedral group D_n as its subgroup. If one of the axes in an equivalence set is two-sided, then all its partners in the set are also two-sided (for $\alpha\boldsymbol{s} = -\boldsymbol{s}$ means $\beta\alpha\beta^{-1}(\beta\boldsymbol{s}) = -(\beta\boldsymbol{s})$, $\forall\beta \in P$. The class associated with an equivalence set of two-sided axes is *ambivalent*, containing a rotation $R(\theta\boldsymbol{s})$ and its inverse $R(-\theta\boldsymbol{s}) = R(\theta\boldsymbol{s})^{-1}$ in the same class. Conversely, if a class of rotations is ambivalent then the corresponding rotation axes are two-sided except for a class of binary rotations. A class of binary rotations is always ambivalent because a binary rotation equals its own inverse, i.e. $R(\pi\boldsymbol{s}) = R(-\pi\boldsymbol{s})$ irrespective of whether the axis is two-sided.

The definition of a two-sided axis can be extended to an improper point group as well. Here the axis-vector of an improper rotation $\overline{R}(\theta\boldsymbol{s}) = IR(\theta\boldsymbol{s})$ is defined by that of the proper part $R(\theta\boldsymbol{s})$, because the inversion I commutes with any point operation. In this case, an operation that brings $\boldsymbol{s}$ to $-\boldsymbol{s}$ can be a binary rotation c_2 or a reflection $m = Ic_2$; in either case, c_2 is perpendicular to $\boldsymbol{s}$, for $\boldsymbol{s}$ being an axial vector.

Wyle (1952) has shown on the basis of the allowed set of axis-vectors for a proper point group that there exist only five types of proper point groups of finite order: the uniaxial group C_n, dihedral group D_n, tetrahedral group T, octahedral group O and icosahedral group Y, as will be shown in Section 5.4.

5.1.3 Notations and the multiplication law for point operations

Let us denote by $n_{\boldsymbol{s}}$ an n-fold rotation $R(2\pi n^{-1}\boldsymbol{s})$ about an axis-vector $\boldsymbol{s}$ counter-clockwise (viewed from the $+\boldsymbol{s}$ direction). For example, in the Cartesian coordinate system, 3_{xyz}, $2_{x\bar{y}}$ and 4_x denote the three-fold, two-fold and four-fold axes of rotation about the respective axis-vectors $\boldsymbol{s}$ pointing toward points with the coordinates (1, 1, 1), (1, −1, 0) and (1, 0, 0), from the coordinate origin.[1] Moreover, the inversion I is denoted by $\bar{1}$ and a rotation–inversion $\overline{R}(2\pi n^{-1}\boldsymbol{s}) = IR(2\pi n^{-1}\boldsymbol{s})$ by $\bar{n}_{\boldsymbol{s}}$. This notation may be called *the axis-vector representation* of a rotation. The rotation–reflection axes, which are basic elements in the Schönflies notation, are completely replaced by rotation–inversion axes, see Section 5.5.

The axis-vector representation is very convenient for expressing the elements of crystallographic point groups. For a typical example, we consider *the octahedral group* O which is also the proper symmetry point group of a cube that has eight equivalent vertices with C_3 symmetry. The order $|O|$ of the group O equals 24, as has been shown via (5.1.0). Corresponding to six faces, eight vertices and 12 edges of a cube, there exist three four-fold axes c_4 joining the opposite faces of the cube, four three-fold axes c_3 joining the opposite vertices and six two-fold axes c_2 joining the mid-points of the opposite edges, see Figure 5.1. These axes are all two-sided. If we take the Cartesian coordinate axes along the three four-fold axes, then the 24 elements of the octahedral group O are expressed by

[1] By definition, $3_{xyz} = 3_{yxz}$, $2_{x\bar{y}} = 2_{\bar{y}x}$, etc. for the Cartesian coordinate system, i.e. the permutations of the subscripts are immaterial for the geometric meaning of the rotation. See Figure 5.1 for the graphical presentation of the axis-vectors. It is also to be noted that similar systems of notation have been introduced by Zak *et al.* (1969) and by Heine (1977).

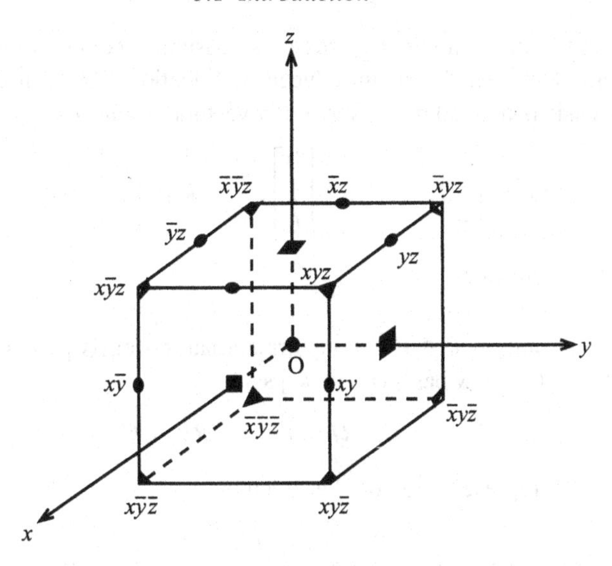

Figure 5.1. The graphical presentation of the axis-vectors for the elements of the octahedral group O and its subgroups. Note that each axis-vector is pointing toward a vertex, a mid-point of a face or a mid-point of an edge of a cube from the center of the cube.

$$
\begin{aligned}
&C(e): && 1\\
&C(c_4): && 4_z,\ 4_x,\ 4_y,\ 4_{\bar{z}},\ 4_{\bar{x}},\ 4_{\bar{y}}\\
&C(c_4^2): && 2_z,\ 2_x,\ 2_y\\
&C(c_3): && 3_{xyz},\ 3_{\bar{x}yz},\ 3_{x\bar{y}z},\ 3_{xy\bar{z}},\ 3_{\bar{x}\bar{y}\bar{z}},\ 3_{x\bar{y}\bar{z}},\ 3_{\bar{x}y\bar{z}},\ 3_{\bar{x}\bar{y}z}\\
&C(c_2): && 2_{xy},\ 2_{yz},\ 2_{zx},\ 2_{x\bar{y}},\ 2_{z\bar{y}},\ 2_{z\bar{x}}
\end{aligned}
\tag{5.1.3a}
$$

where $n_{\bar{s}} = n_s^{-1}$. Each set in (5.1.3a) forms a class, since all faces of a cube are equivalent and so are all the vertices and all the edges, respectively. This follows also from the conjugate relation (5.1.2), i.e. $\alpha n_s \alpha^{-1} = n_{\alpha s}$, which requires only the equivalence of the axis-vectors. Thus, in view of Figure 5.1, some typical conjugations are

$$
3_{xyz}4_z3_{xyz}^{-1} = 4_x, \qquad 4_z3_{xyz}4_z^{-1} = 3_{\bar{x}yz}, \qquad 3_{xyz}2_{xy}3_{xyz}^{-1} = 2_{yz} \tag{5.1.3b}
$$

Moreover, every class is ambivalent since all rotation axes are two-sided; e.g. $2_{x\bar{y}}3_{xyz}2_{x\bar{y}} = 3_{\bar{x}\bar{y}\bar{z}}$.

It remains to show that the set of these 24 elements given in (5.1.3a) closes under multiplication and thus forms a group. This will be proven by constructing the multiplication table of the group O through a faithful representation of the point group based on the three unit basis vectors of the Cartesian coordinate system.

5.1.3.1 Basis-vector representations

From the geometric meaning of the axis-vector representation n_s of an n-fold rotation, we shall introduce a faithful representation of rotation by which one can determine the 3×3 matrix representation as well as the product of two successive rotations, without using the general matrix expression given by (4.3.8b). It is based on the natural basis defined by the unit vectors along the axes of an assumed coordinate system in the linear vector space $V^{(n)}$, which was introduced by Equation (1.1.1) in Chapter 1. For $V^{(3)}$, the natural basis is defined by the three unit vectors $\boldsymbol{i}$, $\boldsymbol{j}$ and $\boldsymbol{k}$:

$$\boldsymbol{i} = \begin{bmatrix} 1 \\ 0 \\ 0 \end{bmatrix}, \qquad \boldsymbol{j} = \begin{bmatrix} 0 \\ 1 \\ 0 \end{bmatrix}, \qquad \boldsymbol{k} = \begin{bmatrix} 0 \\ 0 \\ 1 \end{bmatrix} \tag{5.1.4a}$$

The basic lemma is as follows.

Lemma 5.1.1. If a rotation $R \in V^{(3)}$ transforms a natural basis $[\boldsymbol{i}, \boldsymbol{j}, \boldsymbol{k}]$ of a coordinate system in $V^{(3)}$ into a new basis $[\boldsymbol{i}', \boldsymbol{j}', \boldsymbol{k}']$ such that

$$R\boldsymbol{i} = \boldsymbol{i}', \qquad R\boldsymbol{j} = \boldsymbol{j}', \qquad R\boldsymbol{k} = \boldsymbol{k}' \tag{5.1.4b}$$

then the 3×3 matrix representation of R is given by

$$R = [\boldsymbol{i}', \boldsymbol{j}', \boldsymbol{k}'] \tag{5.1.5}$$

that is, the three column vectors of R are given by the transformed basis vectors $\boldsymbol{i}'$, $\boldsymbol{j}'$ and $\boldsymbol{k}'$.

The proof is trivial. By uniting the three equations in (5.1.4b) into one single matrix formula, we obtain

$$R[\boldsymbol{i}, \boldsymbol{j}, \boldsymbol{k}] = [R\boldsymbol{i}, R\boldsymbol{j}, R\boldsymbol{k}] = [\boldsymbol{i}', \boldsymbol{j}', \boldsymbol{k}'] \tag{5.1.6a}$$

The lemma is proven, since the matrix $[\boldsymbol{i}, \boldsymbol{j}, \boldsymbol{k}]$ is the unit matrix in $V^{(3)}$

$$[\mathbf{i}, \mathbf{j}, \mathbf{k}] = \begin{bmatrix} 1 & 0 & 0 \\ 0 & 1 & 0 \\ 0 & 0 & 1 \end{bmatrix} \tag{5.1.6b}$$

in view of (5.1.4a). The expression (5.1.5) will be called *the basis-vector representation* of the rotation R. Obviously, the rotation–inversion is expressed by $\overline{R} = [-\boldsymbol{i}', -\boldsymbol{j}', -\boldsymbol{k}']$.

The lemma is easily extended to any dimensions, and applies for a rectilinear coordinate system as well as for an oblique coordinate system such as a rhombic or hexagonal coordinate system. Moreover, the basis-vector representation is very effective in calculating the product of two successive rotations, in particular, if the coordinate system is symmetric with respect to the rotations.

5.1.3.2 Examples

A rotation 3_{xyz} in the Cartesian coordinate system (right-handed) rotates the right-handed trio $\boldsymbol{i}, \boldsymbol{j}, \boldsymbol{k}$ counterclockwise such that $\boldsymbol{i} \to \boldsymbol{j} \to \boldsymbol{k} \to \boldsymbol{i}$, i.e.

$$3_{xyz}\boldsymbol{i} = \boldsymbol{j}, \qquad 3_{xyz}\boldsymbol{j} = \boldsymbol{k}, \qquad 3_{xyz}\boldsymbol{k} = \boldsymbol{i} \tag{5.1.7a}$$

Accordingly, from (5.1.5) and (5.1.4a),

$$3_{xyz} = [\boldsymbol{j}, \boldsymbol{k}, \boldsymbol{i}] = \begin{bmatrix} 0 & 0 & 1 \\ 1 & 0 & 0 \\ 0 & 1 & 0 \end{bmatrix} \tag{5.1.7b}$$

Analogously, some of the typical rotations in the group O are expressed by

Table 5.1. *The basis-vector representation of the octahedral group O based on the Cartesian coordinate system (here, $\bar{\boldsymbol{i}} = -\boldsymbol{i}$, for example)*

$$
\begin{aligned}
&1 = [\boldsymbol{i}, \boldsymbol{j}, \boldsymbol{k}] \\
&2_x = [\boldsymbol{i}, \bar{\boldsymbol{j}}, \bar{\boldsymbol{k}}],\ 2_y = [\bar{\boldsymbol{i}}, \boldsymbol{j}, \bar{\boldsymbol{k}}],\ 2_z = [\bar{\boldsymbol{i}}, \bar{\boldsymbol{j}}, \boldsymbol{k}] \\
&3_{xyz} = [\boldsymbol{j}, \boldsymbol{k}, \boldsymbol{i}],\ 3_{x\bar{y}\bar{z}} = [\bar{\boldsymbol{j}}, \boldsymbol{k}, \bar{\boldsymbol{i}}],\ 3_{\bar{x}y\bar{z}} = [\bar{\boldsymbol{j}}, \bar{\boldsymbol{k}}, \boldsymbol{i}],\ 3_{\bar{x}\bar{y}z} = [\boldsymbol{j}, \bar{\boldsymbol{k}}, \bar{\boldsymbol{i}}],\ 3_{\bar{x}\bar{y}\bar{z}} = [\boldsymbol{k}, \boldsymbol{i}, \boldsymbol{j}], \\
&\quad 3_{\bar{x}yz} = [\bar{\boldsymbol{k}}, \bar{\boldsymbol{i}}, \boldsymbol{j}],\ 3_{x\bar{y}z} = [\boldsymbol{k}, \bar{\boldsymbol{i}}, \bar{\boldsymbol{j}}],\ 3_{xy\bar{z}} = [\bar{\boldsymbol{k}}, \boldsymbol{i}, \bar{\boldsymbol{j}}] \\
&4_{\bar{x}} = [\boldsymbol{i}, \bar{\boldsymbol{k}}, \boldsymbol{j}],\ 4_x = [\boldsymbol{i}, \boldsymbol{k}, \bar{\boldsymbol{j}}],\ 4_{\bar{y}} = [\boldsymbol{k}, \boldsymbol{j}, \bar{\boldsymbol{i}}],\ 4_y = [\bar{\boldsymbol{k}}, \boldsymbol{j}, \boldsymbol{i}],\ 4_{\bar{z}} = [\bar{\boldsymbol{j}}, \boldsymbol{i}, \boldsymbol{k}],\ 4_z = [\boldsymbol{j}, \bar{\boldsymbol{i}}, \boldsymbol{k}] \\
&2_{z\bar{y}} = [\bar{\boldsymbol{i}}, \bar{\boldsymbol{k}}, \bar{\boldsymbol{j}}],\ 2_{zy} = [\bar{\boldsymbol{i}}, \boldsymbol{k}, \boldsymbol{j}],\ 2_{z\bar{x}} = [\bar{\boldsymbol{k}}, \bar{\boldsymbol{j}}, \bar{\boldsymbol{i}}],\ 2_{zx} = [\boldsymbol{k}, \bar{\boldsymbol{j}}, \boldsymbol{i}],\ 2_{x\bar{y}} = [\bar{\boldsymbol{j}}, \bar{\boldsymbol{i}}, \bar{\boldsymbol{k}}], \\
&\quad 2_{xy} = [\boldsymbol{j}, \boldsymbol{i}, \bar{\boldsymbol{k}}]
\end{aligned}
$$

$$
\begin{aligned}
&2_z = [-\boldsymbol{i}, -\boldsymbol{j}, \boldsymbol{k}], \qquad 4_z = [\boldsymbol{j}, -\boldsymbol{i}, \boldsymbol{k}], \qquad 3_{\bar{x}yz} = [-\boldsymbol{k}, -\boldsymbol{i}, \boldsymbol{j}] \\
&2_{z\bar{x}} = [-\boldsymbol{k}, -\boldsymbol{j}, -\boldsymbol{i}]
\end{aligned} \tag{5.1.7c}
$$

These are easily formed: 2_z is the 180° rotation about the z-axis, which brings $\boldsymbol{i} \to -\boldsymbol{i}$ and $\boldsymbol{j} \to -\boldsymbol{j}$, leaving $\boldsymbol{k}$ invariant; 4_z is a 90° rotation about the z-axis such that $\boldsymbol{i} \to \boldsymbol{j} \to -\boldsymbol{i}$ and leaving $\boldsymbol{k}$ invariant; $3_{\bar{x}yz}$ rotates the left-handed trio $\{-\boldsymbol{i}, \boldsymbol{j}, \boldsymbol{k}\}$ counterclockwise such that $-\boldsymbol{i} \to \boldsymbol{k} \to \boldsymbol{j} \to -\boldsymbol{i}$, from which there follow $\boldsymbol{i} \to -\boldsymbol{k}$, $\boldsymbol{j} \to -\boldsymbol{i}$ and $\boldsymbol{k} \to \boldsymbol{j}$; and, finally, $2_{z\bar{x}}$ is the two-fold rotation about the diagonal for the z- and $-x$-axes such that it inverts $\boldsymbol{j}$ to $-\boldsymbol{j}$ and exchanges $\boldsymbol{k}$ and $-\boldsymbol{i}$, thus there follow $\boldsymbol{i} \to -\boldsymbol{k}$, $\boldsymbol{j} \to -\boldsymbol{j}$ and $\boldsymbol{k} \to -\boldsymbol{i}$. Note also that $2_{z\bar{x}} = 2_{\bar{z}x}$, being a binary rotation. The basis-vector representation of the octahedral group O thus formed is presented in Table 5.1, which will be used to construct the multiplication table of the group O via the law of multiplication introduced below.

The product of two successive rotations based on the basis-vector representation is calculated by *the multiplication law*. Let $R = [\boldsymbol{a}, \boldsymbol{b}, \boldsymbol{c}]$ be a 3×3 matrix defined by three column vectors, then

$$
[\boldsymbol{a}, \boldsymbol{b}, \boldsymbol{c}]\boldsymbol{i} = \boldsymbol{a}, \qquad [\boldsymbol{a}, \boldsymbol{b}, \boldsymbol{c}]\boldsymbol{j} = \boldsymbol{b}, \qquad [\boldsymbol{a}, \boldsymbol{b}, \boldsymbol{c}]\boldsymbol{k} = \boldsymbol{c} \tag{5.1.8}
$$

which is obvious, since $\boldsymbol{i}$, $\boldsymbol{j}$ and $\boldsymbol{k}$ are the column vectors defined by (5.1.4a). For example, $3_{xyz}\boldsymbol{i} = [\boldsymbol{j}, \boldsymbol{k}, \boldsymbol{i}]\boldsymbol{i} = \boldsymbol{j}$. Thus, products of the elements of the group O are given by, e.g.,

$$
\begin{aligned}
4_z 3_{xyz} &= [\boldsymbol{j}, -\boldsymbol{i}, \boldsymbol{k}][\boldsymbol{j}, \boldsymbol{k}, \boldsymbol{i}] = [-\boldsymbol{i}, \boldsymbol{k}, \boldsymbol{j}] = 2_{zy} \\
2_{x\bar{y}} 2_{z\bar{x}} &= [-\boldsymbol{j}, -\boldsymbol{i}, -\boldsymbol{k}][-\boldsymbol{k}, -\boldsymbol{j}, -\boldsymbol{i}] = [\boldsymbol{k}, \boldsymbol{i}, \boldsymbol{j}] = 3_{\bar{x}\bar{y}\bar{z}} \\
2_z 3_{xyz} &= [-\boldsymbol{i}, -\boldsymbol{j}, \boldsymbol{k}][\boldsymbol{j}, \boldsymbol{k}, \boldsymbol{i}] = [-\boldsymbol{j}, \boldsymbol{k}, -\boldsymbol{i}] = 3_{x\bar{y}\bar{z}}
\end{aligned} \tag{5.1.9}
$$

Via the law of multiplication given by (5.1.8) and using the basis-vector representation of the group O given in Table 5.1, we have constructed the multiplication table of the group O as given by Table 5.6 later. This then proves that the set of point operations given in (5.1.3a) indeed closes under multiplication and hence forms a group.

Exercise. Show that the basis-vector representations for 3_{xyz} and $2_{z\bar{x}}$ given in Table 5.1 hold also for a rhombic coordinate system, and thereby show that $3_{xyz}2_{z\bar{x}} = 2_{z\bar{y}}$.

Table 5.2. *The Jones faithful representation of the octahedral group O based on the Cartesian coordinate system (here, $\bar{x} = -x$, for example)*

$1 = (x, y, z)$

$2_x = (x, \bar{y}, \bar{z}),\ 2_y = (\bar{x}, y, \bar{z}),\ 2_z = (\bar{x}, \bar{y}, z)$

$3_{xyz} = (z, x, y),\ 3_{x\bar{y}\bar{z}} = (\bar{z}, \bar{x}, y),\ 3_{\bar{x}y\bar{z}} = (z, \bar{x}, \bar{y}),\ 3_{\bar{x}\bar{y}z} = (\bar{z}, x, \bar{y}),\ 3_{\bar{x}\bar{y}\bar{z}} = (y, z, x),$
$\quad 3_{\bar{x}yz} = (\bar{y}, z, \bar{x}),\ 3_{x\bar{y}z} = (\bar{y}, \bar{z}, x),\ 3_{xy\bar{z}} = (y, \bar{z}, \bar{x})$

$4_{\bar{x}} = (x, z, \bar{y}),\ 4_x = (x, \bar{z}, y),\ 4_{\bar{y}} = (\bar{z}, y, x),\ 4_y = (z, y, \bar{x}),\ 4_{\bar{z}} = (y, \bar{x}, z),\ 4_z = (\bar{y}, x, z)$

$2_{z\bar{y}} = (\bar{x}, \bar{z}, \bar{y}),\ 2_{zy} = (\bar{x}, z, y),\ 2_{z\bar{x}} = (\bar{z}, \bar{y}, \bar{x}),\ 2_{zx} = (z, \bar{y}, x),\ 2_{x\bar{y}} = (\bar{y}, \bar{x}, \bar{z}),$
$\quad 2_{xy} = (y, x, \bar{z})$

5.1.3.3 Jones representations

Let $\boldsymbol{r}$ be an arbitrary point in $V^{(3)}$ with the coordinates (x, y, z) with respect to the natural basis of a coordinate system. Then it is expressed by a column vector

$$\boldsymbol{r} = x\boldsymbol{i} + y\boldsymbol{j} + z\boldsymbol{k} \tag{5.1.10}$$

Under a rotation R, the point $\boldsymbol{r}$ is transformed to a new point $\boldsymbol{r}' = (x', y', z')$ according to

$$R\boldsymbol{r} = x\boldsymbol{i}' + y\boldsymbol{j}' + z\boldsymbol{k}' = x'\boldsymbol{i} + y'\boldsymbol{j} + z'\boldsymbol{k} = \boldsymbol{r}' \tag{5.1.11}$$

Thus, one may represent the rotation by the coordinates of the transformed point $\boldsymbol{r}'$

$$R = (x', y', z') \tag{5.1.12}$$

which is known as *the Jones faithful representation* of the rotation R. For example, in the Cartesian coordinate system we have, from (5.1.10) and (5.1.7a),

$$3_{xyz}\boldsymbol{r} = x\boldsymbol{j} + y\boldsymbol{k} + z\boldsymbol{i}$$

so that $x' = z$, $y' = x$ and $z' = y$ from (5.1.11); hence, the Jones representation of 3_{xyz} is given by

$$3_{xyz} = (z, x, y) \equiv \begin{bmatrix} 0 & 0 & 1 \\ 1 & 0 & 0 \\ 0 & 1 & 0 \end{bmatrix} \tag{5.1.13}$$

Here, the matrix expression follows if we regard z, x and y in the column vector (z, x, y) as one-row matrices [0, 0, 1], [1, 0, 0] and [0, 1, 0], respectively. The Jones representation can also be used to calculate the product of two rotations.[2] In Table 5.2, we have provided the Jones representation for the octahedral group O, which will be used for the construction of the space groups in Chapter 13.

The Jones representation of a group G generated from an arbitrary initial point (x, y, z) is a faithful representation and thus describes the maximum set of the equivalent points with respect to G, cf. (5.1.0).

[2] Here, the multiplication law is given by $x(x', y', z') = x'$, $y(x', y', z') = y'$, $z(x', y', z') = z'$, for x, y and z being row vectors; e.g., $4_z 3_{xyz} = (\bar{y}, x, z)(z, x, y) = (\bar{x}, z, y) = 2_{zy}$.

5.2 The dihedral group D_n

The group D_n is the group of proper rotations which leaves a regular dihedral n-gon invariant. Since the dihedron has two equivalent faces (up and down) and the mid-point of each face is invariant under the uniaxial group C_n, the order of D_n is given by $|D_n| = 2 \times n = 2n$ according to (5.1.0). There exist an n-fold principal axis c_n normal to the face at the center of the dihedron and n binary axes evenly distributed on the face. Thus the angle between two adjacent binary axes is π/n. By definition the principal axis c_n is two-sided, whereas binary axes are two-sided only if n is even because $(c_n)^{n/2} = c_2$ for an even n. When n is odd, there exists only one equivalence set of binary axes, each of which joins a vertex and the mid-point of the opposite edge (see D_3 in Figure 5.2): there are all one-sided. When n is even, there exist two equivalence sets of binary axes: each axis in the first set joins two opposite vertices, whereas each axis in the second equivalence set joins the mid-points of the two opposite edges: these are all two-sided (see Figure 5.2 for D_4).

For any n, there exist three equivalence sets of *axis-vectors* for the symmetry group D_n of the regular dihedral n-gon: a set of two axis-vectors normal to the faces, a set of those pointing to n vertices and a set of those pointing to the mid-points of n edges.

An algebraic construction of D_n is to adjoin a two-fold axis u_0 to the uniaxial group C_n at a right angle to the principal axis of rotation c_n of C_n. Then, n two-fold rotations are generated from u_0 through

$$u_k = c_n^k u_0; \qquad k = 0, 1, 2, \ldots, n-1 \tag{5.2.1a}$$

where u_k makes an angle $\pi k/n$ with u_0; see (4.3.15c). Since $u_k c_n u_k = c_n^{-1}$ for any k, the uniaxial group C_n is a normal subgroup of D_n and the coset decomposition of D_n by C_n yields

$$D_n = C_n + C_n u_0 = C_n \wedge C_2' \tag{5.2.1b}$$

which may also be regarded as the semidirect product of C_n and $C_2' = \{e, u_0\}$.

The group D_n is generated by two generators $a = c_n$ and $b = u_0$, and *the defining relations* are given by

$$a^n = b^2 = (ab)^2 = e \tag{5.2.2}$$

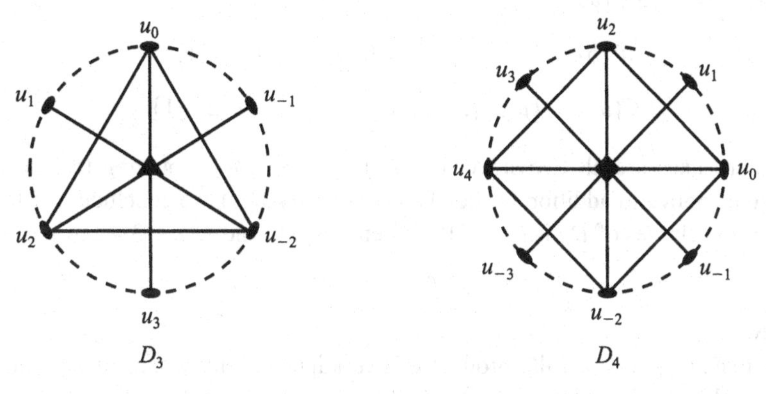

Figure 5.2. Rotation axes of dihedrons. All the binary axes $u_k \in D_3$ are one-sided, but all $u_k \in D_4$ are two-sided. Note that $u_{-k} = u_{n-k}$ for $c_n^n = e$.

Table 5.3. *The multiplication table of D_n (m and l are integers)*

	a^l	$a^l b$
a^m	a^{m+l}	$a^{m+l} b$
$a^m b$	$a^{m-l} b$	a^{m-l}

In fact, on the basis of these, one constructs the multiplication table, Table 5.3 of D_n expressed by the generators: The first line of the table is obvious. To understand the second line, we use $a^m b a^m = b$, which follows from the repeated use of $aba = b$ (obtained from $abab = e$ and $b^2 = e$). Then, for example, $a^m b a^l = a^{m-l}(a^l b a^l) = a^{m-l} b$.

The classes of D_n are also easily obtained, if we use Table 5.3. The class of a^m, denoted $C(a^m)$, is obtained through the conjugation $b a^m b = a^{-m}$:

$$C(e) = \{e\}; \qquad C(a^m) = \{a^m, a^{-m}\}, \qquad m = 1, 2, \ldots, [n/2] \tag{5.2.3}$$

where $a^{n/2} = a^{-n/2}$ for an even n. Here, $[n/2]$ is the integral part of $n/2$, i.e. $[n/2]$ equals $(n-1)/2$ or $n/2$ when n is odd or even, respectively. These are all ambivalent classes, containing an element and its inverse.

The classes of the binary axes of rotation $u_k \in D_n$ depend on the evenness or oddness of n. This is based on the equivalence relations

$$a u_k a^{-1} = u_{k+2} \qquad \text{for any } k \tag{5.2.4}$$

Note that, if k is odd (even), then $k+2$ is also odd (even). However, since $u_n = u_0$ from (5.2.1a) and $a^\nu u_1 a^{-\nu} = u_{2\nu+1}$, we have $u_1 \sim u_0$, if and only if $n = 2\nu + 1$ (odd). Thus, when n is odd, there should be only one class of binary rotations of order n denoted by $C(b)$:

$$C(b) = \{u_k; \qquad k = 0, 1, \ldots, n-1\} \tag{5.2.5}$$

These binary rotations are all one-sided because there is no binary axis perpendicular to them. On the other hand, when n is even there should be two classes of order $n/2$ denoted by $C(b)$ and $C(ab)$:

$$\begin{aligned} C(b) &= \{u_{2k}; \qquad k = 0, 1, \ldots, n/2 - 1\} \\ C(ab) &= \{u_{2k+1}; \qquad k = 0, 1, \ldots, n/2 - 1\} \end{aligned} \tag{5.2.6}$$

These binary axes are all two-sided, for $(c_n)^{n/2} = c_2 \perp u_k$. Counting the number of classes given above in addition to the class of identity $C(e)$, we conclude that the total number of the classes of D_n is $(n+3)/2$ when n is odd and $n/2 + 3$ when n is even.

Exercises

1. Show that $u_{k+1} u_k = n_z$: the product of two adjacent binary axes of D_n equals the principal axis of rotation.
2. Show that the group D_n is also defined by $x^2 = y^2 = (xy)^n = e$.

Table 5.4. *Regular polyhedra and the symmetry groups*

n	m	Regular polyhedra	Symmetry groups P_o	$\lvert P_o \rvert$
3	3	Tetrahedron	T	12
4	3	Octahedron	O	24
3	4	Cube	O	24
5	3	Icosahedron	Y	60
3	5	Dodecahedron	Y	60

5.3 Proper polyhedral groups P_o

A proper polyhedral group P_o is a group of proper rotations that leave a regular polyhedron invariant. In any regular polyhedron, there exist only three kinds of *characteristic symmetry points* under rotation: vertices, the mid-points of the faces and edges in addition to the center O of the body, which is the fixed point of all symmetry operations. All these symmetry points of the same kind are equivalent for a regular polyhedron. Correspondingly, *there exist three and only three equivalence sets of the axis-vectors of rotation in* P_o: each axis-vector is pointing toward each symmetry point of the body from the center O such that there exists a one-to-one correspondence between an equivalence set of axis-vectors in P_o and the corresponding equivalent set of symmetry points of the regular polyhedron.

There exist only five types of regular polyhedra. This follows simply from the condition that the sum of apex angles at a vertex in a regular polyhedron is less than 2π. Let n (>2) be the number of regular m-gons ($m>2$) that meet at any one vertex of a polyhedron. Then an internal angle of a regular m-gon given by $\pi - 2\pi/m$ must satisfy the inequality $n(\pi - 2\pi m) < 2\pi$. This is rewritten in the form

$$1/n + 1/m > 1/2 \tag{5.3.1}$$

Since the inequality (originally due to Diophantus) is symmetric with respect to n and m, we may first determine all the possible pairs (n, m) with the condition $n \geqslant m$ and then exchange n and m to obtain the final results. On combining $n \geqslant m$ with (5.3.1), we obtain $2/m > 1/2$. Thus, $4 > m > 2$, which has only one integral solution, $m = 3$. By substituting this back into (5.3.1), we obtain $n < 6$, which yields $n = 3, 4, 5 > 2$. Accordingly, we conclude that one of n and m equals 3 while the other equals 3, 4 or 5; hence, the possible regular polyhedra are the five types given in Table 5.4 and in Figure 5.3. Their proper symmetry point groups[3] are called the tetrahedral group T, octahedral group O and icosahedral group Y: there exist only three point groups because two pairs (n, m) and (m, n) belong to the same polyhedral group P_o, as will be shown later. Table 5.4 also contains the orders $|P_o|$ of P_o, which are obtained through (5.1.0) or preferably from (5.3.9) given later. Hereafter, a vertex, a regular face or an edge of a polyhedron characterized by an n-, m- or two-fold symmetry axis of

[3] In general, the symmetry point groups of regular polyhedra are improper and given by T_p, O_i and Y_i (see Section 5.5).

Table 5.5. *The characteristics of regular polyhedra*

	Number of vertices	Number of faces	Number of edges
Tetrahedron	$4 = 12/3$	$4 = 12/3$	$6 = 12/2$
Octahedron	$6 = 24/4$	$8 = 24/3$	$12 = 24/2$
Cube	$8 = 24/3$	$6 = 24/4$	$12 = 24/2$
Icosahedron	$12 = 60/5$	$20 = 60/3$	$30 = 60/2$
Dodecahedron	$20 = 60/3$	$12 = 60/5$	$30 = 60/2$

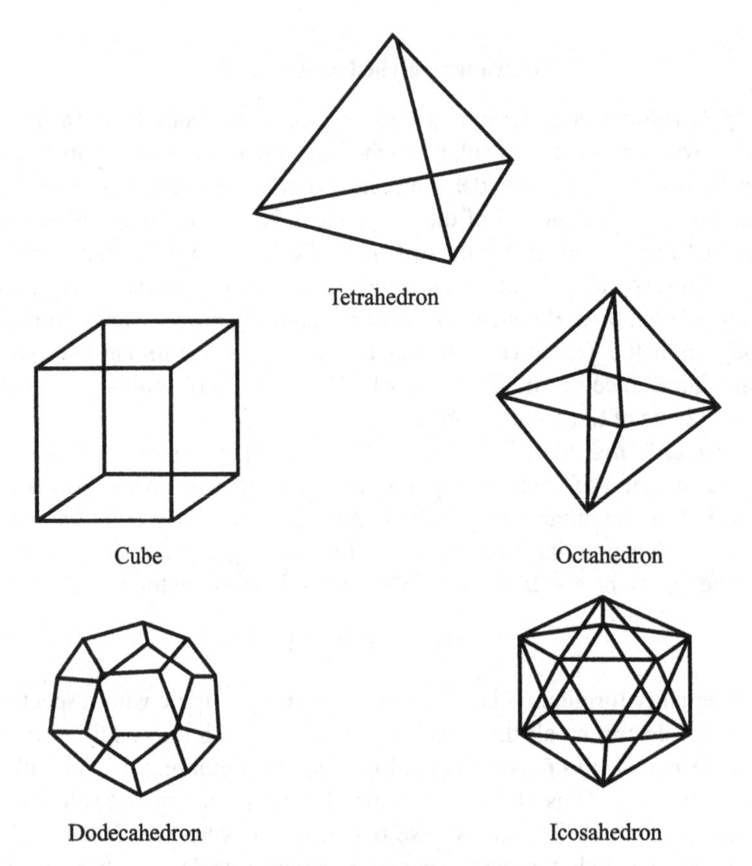

Figure 5.3. The five regular polyhedra.

rotation may be conveniently called an *n-fold vertex*, an *m-fold face* or a *two-fold edge* respectively. The numbers of these characteristic objects are given in Table 5.5.

From the correspondence between the characteristic symmetry points of a polyhedron and the axis-vectors of rotation, we see that most of the rotation axes of P_0 are two-sided since each of them connects a face to an opposite face, a vertex to an opposite vertex or an edge to an opposite edge. The only exceptions are the three-fold axes of the tetrahedral group T, which are one-sided, each of them connecting a three-fold face to a three-fold vertex. See Figure 5.3.

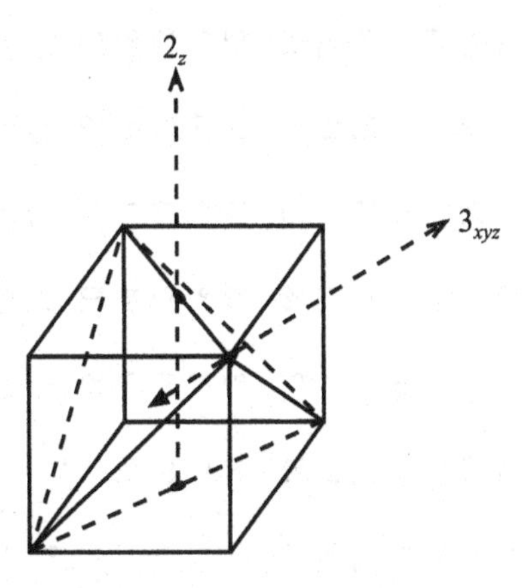

Figure 5.4. The generators of the group T. A regular tetrahedron is cut out from a cube.

5.3.1 Proper cubic groups, T and O

5.3.1.1 The tetrahedral group T

It is the proper symmetry point group of the regular tetrahedron. The group has three mutually perpendicular two-fold axes c_2 (two-sided) joining the mid-points of opposite edges of the tetrahedron and four three-fold axes c_3 (one-sided) joining each vertex of the tetrahedron to the mid-point of the opposite face, see Figure 5.3. Thus, the group T has 12 elements classified by four classes: $\{e\}$, $\{3c_2\}$, $\{4c_3\}$ and $\{4c_3^{-1}\}$. We shall show that all these elements of the group T are generated by two generators. To see this, let us introduce a Cartesian coordinate system along the three binary axes of T and introduce the following realizations of the generators (see Figure 5.4):

$$a = 2_z, \qquad b = 3_{xyz} \tag{5.3.2a}$$

then $ab = 2_z 3_{xyz} = 3_{x\bar{y}\bar{z}}$ from Table 5.6, so that $(ab)^3 = e$, which defines the mutual arrangement of the generators. Thus, the generators satisfy the defining relations of T

$$a^2 = b^3 = (ab)^3 = e \tag{5.3.2b}$$

From these follow all the symmetry properties of T. Firstly, the group T has a normal subgroup D_2 generated by $x = a$ and $y = bab^{-1}$, for $x^2 = y^2 = (xy)^2 = e$. Then, through the coset decomposition of T by D_2

$$T = D_2 + D_2 c_3 + D_2 c_3^{-1} = D_2 \wedge C_3 \tag{5.3.3a}$$

all the elements of T are generated by

$$\begin{aligned}
D_2\!: &\quad e,\ a = 2_z,\ bab^{-1} = 2_x,\ b^{-1}ab = 2_y \\
D_2 b\!: &\quad b = 3_{xyz},\ 2_z b = 3_{x\bar{y}\bar{z}},\ 2_x b = 3_{\bar{x}y\bar{z}},\ 2_y b = 3_{\bar{x}\bar{y}z} \\
D_2 b^{-1}\!: &\quad b^{-1} = 3_{\bar{x}\bar{y}\bar{z}},\ 2_z b^{-1} = 3_{x\bar{y}z},\ 2_x b^{-1} = 3_{xy\bar{z}},\ 2_y b^{-1} = 3_{\bar{x}yz}
\end{aligned} \tag{5.3.3b}$$

with use of the multiplication table of the octahedral group O, Table 5.6. This also shows that the group T is a subgroup of the group O. The above scheme of generating T by (5.3.3b) becomes very convenient when we construct the space groups in terms of the generators, see Section 13.9.1.

Table 5.6. *The multiplication table for the octahedral group O and its subgroups*

$i\,/\,j$	e	2_x	2_y	2_z	3_{xyz}	$3_{x\bar{y}\bar{z}}$	$3_{\bar{x}y\bar{z}}$	$3_{\bar{x}\bar{y}z}$	$3_{\bar{x}\bar{y}\bar{z}}$	$3_{\bar{x}yz}$	$3_{x\bar{y}z}$	$3_{xy\bar{z}}$	$4_{\bar{x}}$	4_x	$4_{\bar{y}}$	4_y	$4_{\bar{z}}$	4_z	$2_{z\bar{y}}$	2_{zy}	$2_{z\bar{x}}$	2_{zx}	$2_{x\bar{y}}$	2_{xy}
e	1	2	3	4	5	6	7	8	9	10	11	12	13	14	15	16	17	18	19	20	21	22	23	24
2_x	2	1	4	3	7	8	5	6	12	11	10	9	14	13	21	22	24	23	20	19	15	16	18	17
2_y	3	4	1	2	8	7	6	5	10	9	12	11	20	19	16	15	23	24	14	13	22	21	17	18
2_z	4	3	2	1	6	5	8	7	11	12	9	10	19	20	22	21	18	17	13	14	16	15	24	23
3_{xyz}	5	8	6	7	9	12	10	11	1	4	2	3	18	24	14	20	16	22	23	17	19	13	21	15
$3_{x\bar{y}\bar{z}}$	6	7	5	8	11	10	12	9	4	1	3	2	17	23	20	14	21	15	24	18	13	19	16	22
$3_{\bar{x}y\bar{z}}$	7	6	8	5	12	9	11	10	2	3	1	4	23	17	13	19	22	16	18	24	20	14	15	21
$3_{\bar{x}\bar{y}z}$	8	5	7	6	10	11	9	12	3	2	4	1	24	18	19	13	15	21	17	23	14	20	22	16
$3_{\bar{x}\bar{y}\bar{z}}$	9	11	12	10	1	3	4	2	5	7	8	6	22	15	24	17	20	13	21	16	23	18	19	14
$3_{\bar{x}yz}$	10	12	11	9	3	1	2	4	8	6	5	7	21	16	18	23	13	20	22	15	17	24	14	19
$3_{x\bar{y}z}$	11	9	10	12	4	2	1	3	6	8	7	5	15	22	23	18	14	19	16	21	24	17	13	20
$3_{xy\bar{z}}$	12	10	9	11	2	4	3	1	7	5	6	8	16	21	17	24	19	14	15	22	18	23	20	13
$4_{\bar{x}}$	13	14	19	20	16	15	22	21	24	23	18	17	2	1	8	7	9	10	4	3	6	5	11	12
4_x	14	13	20	19	22	21	16	15	17	18	23	24	1	2	6	5	12	11	3	4	8	7	10	9
$4_{\bar{y}}$	15	22	16	21	18	23	17	24	20	13	19	14	9	11	3	1	6	8	12	10	2	4	7	5
4_y	16	21	15	22	24	17	23	18	13	20	14	19	10	12	1	3	7	5	11	9	4	2	6	8
$4_{\bar{z}}$	17	23	24	18	14	20	19	13	22	16	15	21	7	6	9	12	4	1	8	5	10	11	3	2
4_z	18	24	23	17	20	14	13	19	15	21	22	16	8	5	11	10	1	4	7	6	12	9	2	3
$2_{z\bar{y}}$	19	20	13	14	21	22	15	16	23	24	17	18	3	4	7	8	11	12	1	2	5	6	9	10
2_{zy}	20	19	14	13	15	16	21	22	18	17	24	23	4	3	5	6	10	9	2	1	7	8	12	11
$2_{z\bar{x}}$	21	16	22	15	23	18	24	17	19	14	20	13	12	10	4	2	8	6	9	11	1	3	5	7
2_{zx}	22	15	21	16	17	24	18	23	14	19	13	20	11	9	2	4	5	7	10	12	3	1	8	6
$2_{x\bar{y}}$	23	17	18	24	19	13	14	20	21	15	16	22	6	7	10	11	2	3	5	8	9	12	1	4
2_{xy}	24	18	17	23	13	19	20	14	16	22	21	15	5	8	12	9	3	2	6	7	11	10	4	1

(1) The entries give the subscript k such that $a_i a_j = a_k$.
(2) The axis-vector $\boldsymbol{s}$ of each element n_s is given graphically in Figure 5.1.
(3) The multiplication table for the double octahedral group is given by Table 11.2.

The class structure of T may be characterized by the generators as follows:

$$C(e) = \{e\}$$

$$C(a) = \{2_z, 2_x, 2_y\}$$

$$C(b) = \{3_{xyz}, 3_{\bar{x}\bar{y}z}, 3_{x\bar{y}\bar{z}}, 3_{\bar{x}y\bar{z}}\} = D_2 3_{xyz}$$

$$C(b^{-1}) = \{3_{\bar{x}\bar{y}\bar{z}}, 3_{xy\bar{z}}, 3_{\bar{x}yz}, 3_{x\bar{y}z}\} = D_2 3_{\bar{x}\bar{y}\bar{z}} \tag{5.3.4}$$

where the class $C(a)$ is obtained by the repeated conjugations of a with b, whereas the class $C(b)$ is obtained by the conjugations of b with 2_z, 2_x and 2_y. Then, the class $C(b^{-1})$ is obtained by the inversion of the class $C(b)$, the c_3-axis being one-sided.

5.3.1.2 The octahedral group O (revisited)

This group has been discussed in Section 5.1.3 and its multiplication table has been constructed in Table 5.6. Here, we shall show that this group is also generated by two generators: a four-fold axis c_4 and a three-fold axis c_3. To see this, let us introduce the Cartesian coordinate along the three four-fold axes as in Figure 5.1 and set

$$a = 4_z, \qquad b = 3_{xyz} \tag{5.3.5a}$$

where their axis-vectors are chosen to be one of the closest pairs for the group O. Since $ab = 4_z 3_{xyz} = 2_{yz}$, as was shown in (5.1.9), the defining relations of the group O are given by

$$a^4 = b^3 = (ab)^2 = e \tag{5.3.5b}$$

where the powers 4, 3 and 2 are in accordance with the three kinds of order of rotation axes in the group O. Now, the subgroup T is generated by $a^2 = 2_z$ and $b = 3_{xyz}$ as before, and the remaining elements of O are provided by aT, because the left coset decomposition of O by T is given by

$$O = T + aT \tag{5.3.5c}$$

Note that T is a normal subgroup of O since it is a halving subgroup of O. For the class structure of O, see (5.1.3a).

Remark. If we had taken $a = 4_z$ and $b = 3_{x\bar{y}\bar{z}}$, for example, then $ab = 4_x$ so that $(ab)^4 = e$.

5.3.2 Presentations of polyhedral groups

There exist two generators for a polyhedral group P_o. Consider a regular polyhedron with n-fold vertices and m-fold faces. Then two generators may be defined by

$$a = c_n, \qquad b = c_m \tag{5.3.6a}$$

where c_n and c_m are the rotation axes passing through an n-fold vertex and the midpoint of an m-fold face from the center O of the polyhedron, respectively. If their axis-vectors are one of the closest pairs and thus connected to one regular face, then their product defines a two-fold axis c_2 closest to both axes and passing through the midpoint of an edge of the face:

$$c_2 = c_n c_m \tag{5.3.6b}$$

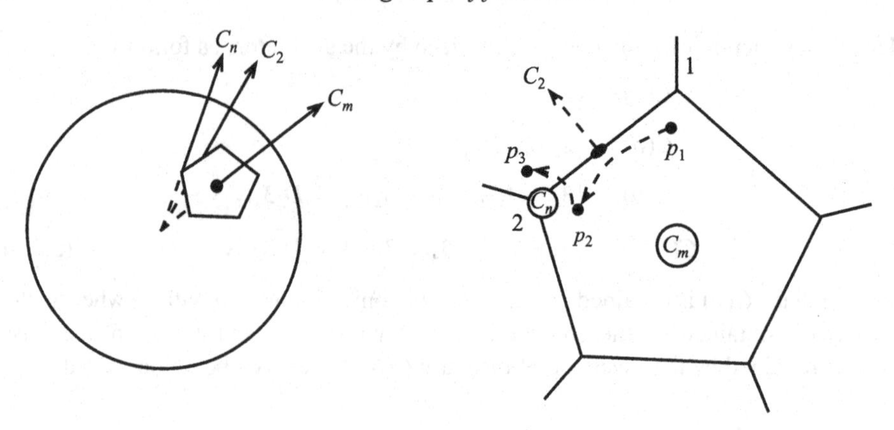

Figure 5.5. The geometric proof of the relation $c_n c_m = c_2$ through a dodecahedron.

We shall show this relation through simple geometric considerations. In Figure 5.5, let c_n and c_m be such a pair of rotation axes with closest axis-vectors, and let 1 and 2 be two adjacent vertices joined by one edge $\overline{12}$ of the face. Then the rotation c_m brings a point p_1 near vertex 1 to a point p_2 near vertex 2. Next, the rotation c_n passing through the vertex 2 brings the point p_2 to a point p_3, both of which are near vertex 2. The resultant displacement of the initial point p_1 to the final point p_3 by the product $c_n c_m$ can be brought about by the two-fold rotation c_2 perpendicular to the edge $\overline{12}$ and passing through its mid-point from the center O of the polyhedron. This proves (5.3.6b). Note that the axis-vectors of the three rotations in (5.3.6b) are nearest to each other and ordered clockwise from c_m to c_n and then to c_2. The remaining axes of rotation for the polyhedron are formed through mutual conjugations like

$$c_m c_n c_m^{-1} = c_n', \qquad c_m c_2 c_m^{-1} = c_2', \qquad c_n c_m c_n^{-1} = c_m' \tag{5.3.7}$$

which can be visualized through the equivalent transformation $a n_s a^{-1} = n_{as}$, where $a \in P_o$.

Now the defining relations of P_o characterized by a set (n, m) in Table 5.4 and those of D_n given by (5.2.2) are expressed by the following *standard presentation*:

$$a^n = b^m = (ab)^2 = e \tag{5.3.8a}$$

where $a = c_n$, $b = c_m$ and $ab = c_2$ are the rotations defined by the three closest axis-vectors of rotation in each group. Thus each group may be expressed by a set of three integers $\{n, m, 2\}$ that also refers to the axis orders of three equivalence sets of the axis-vectors in each group. Then, the abstract group represented by the presentation $\{n, m, 2\}$ is invariant under the permutations of the three axis orders n, m and 2. Firstly, the interchange of the axis orders (n, m) in (5.3.8a) defines a group isomorphic to the original group because $(ab)^2 = e$ means $(ba)^2 = e$. This then explains Table 5.4 in that both an octahedron and a cube belong to the same abstract group O whereas an icosahedral and a dodecahedron belong to the same abstract group Y. Secondly, let $x = ab$ and $y = b^{-1}$, then we have

$$x^2 = y^m = (xy)^n = e \tag{5.3.8b}$$

Thus, the permutation of $\{n, m, 2\}$ leaves the group invariant. For example, the presentation of the tetrahedral group T is also given by

$$x^2 = y^3 = (xy)^3 = e \tag{5.3.8c}$$

as in (5.3.2b).

There exist only five kinds of proper point groups P: C_n, D_n, T, O and Y, as was shown first by Wyle and will be shown in Section 5.4. Combining (5.1.1b) with (5.3.8a), the presentation of a proper point group P is expressed by

$$a^n = b^m = (ab)^l = e \tag{5.3.8d}$$

where the set of the axis orders $\{n, m, l\}$ for the set $\{a, b, ab\}$ is characteristic to P:

$$T = \{3, 3, 2\}, \quad O = \{4, 3, 2\}, \quad Y = \{5, 3, 2\}, \quad D_n = \{n, 2, 2\},$$
$$C_n = \{n, 0, 0\} \tag{5.3.8e}$$

which may be called the *axis-order notation* for P. The corresponding short notations are given by the generators in each standard presentation (5.3.8a):

$$T = (3, 3), \quad O = (4, 3), \quad Y = (5, 3), \quad D_n = (n, 2), \quad C_n = (n) \tag{5.3.8f}$$

These notations may be compared with the international notation of the point groups; cf. Table 5.7 later.

It is stressed here that the presentation of a proper point group P given by (5.3.8d) defines an abstract group P completely, independently from the geometric interpretation like (5.3.5a). In fact, by the presentation, we can determine all its subgroups (see Section 5.3.3) and identify all its group elements via the coset decomposition with respect to an appropriate subgroup of P. There exists also an algorithm called *the method of coset enumeration* by which we can construct the multiplication table of P solely by the presentation (5.3.8d), see Section 5.7. Moreover, on the basis of the presentation (5.3.8d), in Chapter 13 we shall construct all the space groups by adjoining the translational degrees of freedom to the respective point groups. It is simply surprising that the simple set of algebraic equations (5.3.8d) describes the whole symmetry properties of the point groups and their extensions. For the presentations of improper point groups see Section 5.5.

5.3.2.1 The Wyle relation

It has been shown by Wyle (1952) that the inequality (5.3.1) can be replaced by an equality

$$1/n + 1/m + 1/2 = 1 + 2/|P_o| \tag{5.3.9}$$

where $|P_o|$ is the order of the polyhedral group P_o of a regular polyhedron. Postponing its proof to Section 5.4, the orders of the polyhedral groups (including D_n) determined by (5.3.9) are

$$|T| = 12, \quad |O| = 24, \quad |Y| = 60, \quad |D_n| = 2n$$

which are also presented in Table 5.4. Note that all these orders are even.

5.3.2.2 One- or two-sidedness of a rotation axis

Let $\{j\}$ be an equivalence set of axes in a proper point group P. If the set is one-sided, then this corresponds to there being two equivalence sets of axis-vectors, $\{s_j\}$ and $\{-s_j\}$. If the set of axes $\{j\}$ is two-sided, then there is only one equivalence set of axis-vectors $\{s_j, -s_j\}$, which is *symmetric*. Since the equivalence sets of axis-vectors for one-sided axes must come in as a pair (if they exist), we arrive at the following rules for the three equivalence sets of axis-vectors in a group $\{n, m, 2\}$.

1. Either all three equivalence sets are symmetric (so that the corresponding axes are all two-sided) or one is symmetric and the other two are mutual inverses.
2. If one of the three orders $\{n, m, 2\}$ is different from the remaining two, the corresponding axis is two-sided. Thus, if three orders n, m and 2 are different, then the corresponding axes are all two-sided.
3. If two members of $\{n, m, 2\}$ are equal, the corresponding sets are mutual inverses or both symmetric.

Examples. For $O = \{4, 3, 2\}$ and $Y = \{5, 3, 2\}$, their axes are all two-sided. For $D_n = \{n, 2, 2\}$, the set of n-fold axes is two-sided whereas the remaining two sets of binary axis-vectors are either mutual inverses or both symmetric, depending on whether n is odd or even (for $c_n^{n/2} = c_2$ if n is even). For the tetrahedral group $T = \{3, 3, 2\}$, the binary axes are two-sided whereas the three-fold axes are all one-sided; if the latter were two-sided the group T would contain $D_3 = \{3, 2, 2\}$ as a subgroup, of which the binary axes are one-sided (which would be a contradiction).

5.3.3 Subgroups of proper point groups

The presentation (5.3.8d) of a proper point group P is very effective in finding its subgroups. We give here only the maximal subgroups of each proper point group P:

$$T > \underline{D}_2, C_3; \qquad O > D_4, D_3, \underline{T}; \qquad Y > D_5, D_3, T;$$
$$D_n > \underline{C}_n, C_2, D_{n/\nu}\ (\underline{D}_{n/2} \text{ if } n \text{ is even}); \qquad C_n > \underline{C}_{n/\nu} \tag{5.3.10}$$

where ν $(1 < \nu < n)$ is an integral devisor of n and those underlined are normal subgroups.

The dihedral subgroups of T, O and Y are due to the generators with two-sided axes. The subgroup T of O and Y may be shown as follows. In the presentation of O, $a^4 = b^3 = (ab)^2 = e$, we set $x = a^2$ and $y = b$ and obtain the presentation of T given by $x^2 = y^3 = (xy)^3 = e$, because $xy = a^2b = ab^{-1}a^{-1}$ using $ab = b^{-1}a^{-1}$. Analogously, in the presentation of Y, $a^5 = b^3 = (ab)^2 = e$, we set $x = ba$ and $y = aba^{-1}$ guided by the geometric interpretation (5.3.7b). Then we again arrive at $x^2 = y^3 = (xy)^3 = e$, because $xy = ba^2ba^{-1} = a^{-1}baa^2 = a^{-2}b^{-1}a^2$. Finally, from $a^n = b^2 = (ab)^2 = e$ of D_n it follows that $(a^\nu)^{n/\nu} = b^2 = (a^\nu b)^2 = e$ of $D_{n/\nu}$ because $a^\nu b$ is a binary rotation of D_n via $a^\nu ba^\nu b = abab = e$.

The normal subgroups given in (5.3.10) are mostly halving subgroups of the respective super-groups except for $T > \underline{D}_2$ and $C_n > \underline{C}_{n/\nu}$; the latter is obvious and the former follows from the fact that every c_2 of T belongs to D_2. It is also to be noted that the icosahedral group Y is the only point group of a finite order that does not have a proper normal subgroup. Such a group is called a *simple group*. It is a simple matter to extend (5.3.10) for the maximal subgroups of an improper point group, as will be shown in Section 5.5.3.

5.3.4 Theorems on the axis-vectors of proper point groups

The following lemma is basic to the Wyle theorem which determines all the possible proper point groups. It is also directly related to the characteristic properties of the regular polyhedra:

Lemma 5.3.1. Let $\boldsymbol{s}$ be an axis-vector of order n (> 1) in a proper point group P, then the number of distinct axis-vectors in the equivalence set of $\boldsymbol{s}$ with respect to P equals $|P|/n$.

Proof. The subgroup of P which leaves the axis-vector $\boldsymbol{s}$ invariant is a uniaxial group C_n, so that all distinct axis-vectors equivalent to $\boldsymbol{s}$ are given by

$$\{q_k \boldsymbol{s}\}; \qquad k = 1, 2, \ldots, |P|/n \tag{5.3.11}$$

where q_k is a coset representative of C_n in P. From this follows the Lemma 5.3.1.

Since there exists a one-to-one correspondence between an equivalence set of symmetry points of a regular polyhedron and the corresponding equivalence set of axis-vectors in P_o, the numbers of n-fold vertices, m-fold faces and two-fold edges and also the maximum number of the equivalent points on the regular polyhedron are given by

$$|P_o|/n, \qquad |P_o|/m, \qquad |P_o|/2, \qquad |P_o|/1 \tag{5.3.12}$$

Conversely, from these characteristics, we can determine the order $|P_o|$ of the group P_o.

Thus, from Table 5.4 and (5.3.12), we have the equivalence characteristics of the regular polyhedra given by Table 5.5.

Corollary 5.3.2. The number of n-fold axes of rotation c_n in an equivalence set for a proper point group P is given by

$$N(c_n) = \begin{cases} |P|/n, & \text{if the axes are one-sided} \\ |P|/(2n), & \text{if the axes are two-sided} \end{cases} \tag{5.3.13}$$

One minor result that follows from this corollary is that the order $|P|$ of a proper point group P that has a two-sided axis must be even. It turns out that all P except for a uniaxial group have a two-sided axis so that their orders are all even (cf. Table 5.4).

Theorem 5.3.3. (The class order theorem.) Let R be a rotation belonging to a uniaxial group C_n ($n \neq 1$) in P, then the order of the class $C(R)$ of R in P is given by

$$|C(R)| = \begin{cases} |P|/n, & \text{if } R \text{ is not a } \pi_2 \\ |P|/(2n), & \text{if } R \text{ is a } \pi_2 \end{cases} \tag{5.3.14}$$

where π_2 is a two-sided binary rotation.

Suppose that $R = c_n^m$; $m = 1, 2, \ldots, n-1$, then the theorem states that the class order of c_n^m equals $|P|/n$ independent of m except when $m = n/2$ and c_n is two-sided, i.e. R is a π_2.

Proof. According to the general class order theorem (3.4.5) the class order $|C(R)|$ of R is given by

$$|C(R)| = |P|/|Z_p(R)|$$

where $Z_p(R)$ is the centralizer of R in P. By assumption and from the condition for commutation (4.3.16), the centralizer of R equals C_n provided that R is not a binary rotation π_2 about a two-sided axis. If $R = \pi_2$, there exists in P a binary rotation perpendicular to π_2 that also commutes with π_2 so that $Z_p(\pi_2)$ equals the dihedral group D_n. Thus follows (5.3.14).

5.3.4.1 Examples

The group structure of each polyhedral group may be understood by considering the general overview of the proper point groups given below, based on (5.3.13) and (5.3.14).

$D_4 = \{4, 2, 2\}$. $|D_4| = 8$. All axes are two-sided so that the binary rotations are all π_2-type. Thus,

$$N(c_4) = 8/(4 \times 2) = 1, \qquad N(u_\nu) = 8/(2 \times 2) = 2$$
$$|C(c_4)| = 8/4 = 2, \qquad |C(c_4^2)| = 8/(2 \times 4) = 1, \qquad |C(u_\nu)| = 8/(2 \times 2) = 2 \tag{5.3.15a}$$

$D_3 = \{3, 2, 2\}$. $|D_3| = 6$. The c_3-axis is two-sided but the binary axes $\{u_\nu\}$ are one-sided. Thus,

$$N(c_3) = 6/(2 \times 3) = 1, \qquad N(u_\nu) = 6/2 = 3$$
$$|C(c_3)| = 6/3 = 2, \qquad |C(u_\nu)| = 6/2 = 3 \tag{5.3.15b}$$

$T = \{3, 3, 2\}$. $|T| = 12$. There exist two equivalence sets of the axis-vectors of order 3 that are one-sided and mutual inverses. The two-fold axes are two-sided and hence π_2-type. Thus,

$$N(c_3) = 12/3 = 4, \qquad N(c_2) = 12/(2 \times 2) = 3$$
$$|C(c_3)| = |C(c_3^{-1})| = 12/3 = 4, \qquad |C(c_2)| = 12/(2 \times 2) = 3 \tag{5.3.15c}$$

There exist in total four classes including the identity class.

$O = \{4, 3, 2\}$. $|O| = 24$. All axes are two-sided and hence all binary rotations are π_2-type. Thus,

$$N(c_4) = 24/(2 \times 4) = 3, \qquad N(c_3) = 24/(2 \times 3) = 4, \qquad N(c_2) = 24/(2 \times 2) = 6$$
$$|C(c_4)| = 24/4 = 6, \qquad |C(c_4^2)| = 24/(2 \times 4) = 3$$
$$|C(c_3)| = 24/3 = 8, \qquad |C(c_2)| = 24/(2 \times 2) = 6 \tag{5.3.15d}$$

There exists a total of five classes for O, all of which are ambivalent because all axes are two-sided.

$Y = \{5, 3, 2\}$. $|Y| = 60$. All axes are two-sided so that all binary rotations are π_2-type. Thus,

$$N(c_5) = 60/(2 \times 5) = 6, \qquad N(c_3) = 60/(2 \times 3) = 10, \qquad N(c_2) = 60/(2 \times 2) = 15$$
$$|C(c_5)| = |C(c_5^2)| = 60/5 = 12, \qquad |C(c_3)| = 60/3 = 20,$$
$$|C(c_2)| = 60/(2 \times 2) = 15 \tag{5.3.15e}$$

There exists a total of five classes for Y, all of which are ambivalent because every axis is two-sided.

5.4 The Wyle theorem on proper point groups

We have seen in the previous section that the structure of a proper point group P is effectively described by equivalence sets of axis-vectors of rotation. Wyle (1952) has shown on the basis of the allowed set of axis-vectors in a proper point group that the possible proper point groups of finite order are of the following five kinds: C_n, D_n, T, O and Y. The Wyle theorem summarizes all the findings on the point groups which we have described so far.

We shall count the total number of the axis-vectors of every non-null rotation in a proper point group P in the following two ways. Firstly, it is given by $2(|P| - 1)$ since there exists a total of $|P| - 1$ non-null rotations in P and also there exist two axis-vectors $\boldsymbol{s}$ and $-\boldsymbol{s}$ for each non-null rotation. Secondly, from Lemma 5.3.1, the number of the axis-vectors in the rth equivalence set in P is given by $|P|/n^{(r)}$, where $n^{(r)}$ is the order of the axis. Since there exist $n^{(r)} - 1$ non-null rotations about an axis of order $n^{(r)}$, we arrive at the equality

$$\sum_{r=1}^{H} (n^{(r)} - 1)|P|/n^{(r)} = 2(|P| - 1) \tag{5.4.1}$$

where H is the number of distinct equivalence sets of axis-vectors in P excluding the identity rotation. Note that on the right-hand side (rhs) of (5.4.1) we have twice the number of non-null rotations in P, whereas, on the left-hand side (lhs) we have every rotation axis twice through symmetric sets $\{\boldsymbol{s}_j, -\boldsymbol{s}_j\}$ for two-sided axes and through $\{\boldsymbol{s}_j\}$ and $\{-\boldsymbol{s}_j\}$ for one-sided axes.

Equation (5.4.1) was first derived by Wyle (1952) and is sufficient to determine all the possible point groups of finite order. For later developments, we rewrite the equation in a more convenient form

$$\sum_{r=1}^{H} (1 - 1/n^{(r)}) = 2(1 - 1/|P|) \tag{5.4.2}$$

Excluding the trivial case of the group of the identity, we may assume that $|P| \geqslant n^{(r)} \geqslant 2$. Then the rhs of (5.4.2) is less than 2 whereas each term on the lhs is larger than or equal to $1/2$. Therefore, we cannot have more than three equivalence sets of axis-vectors in P, i.e. $H \leqslant 3$.

Now, suppose that there exists only one equivalence vector set in P, then (5.4.2) gives $1/n^{(1)} = -1 + 2/|P|$, which is impossible to satisfy because the lhs > 0 while the rhs $\leqslant 0$. Next, for $H = 2$ we obtain, from (5.4.2),

$$|P|/n^{(1)} + |P|/n^{(2)} = 2$$

Since both $|P|/n^{(1)}$ and $|P|/n^{(2)}$ are integers for the indices of the uniaxial subgroups of order $n^{(1)}$ and $n^{(2)}$ being in P, the only solution for $H = 2$ is given by

$$|P|/n^{(1)} = |P|/n^{(2)} = 1. \tag{5.4.3}$$

which provides a uniaxial group of order $|P|$ with two axis-vectors s and $-s$ that are mutual inverses.

For the case of three equivalence vector sets one can rewrite (5.4.2) in the form

$$1/n^{(1)} + 1/n^{(2)} + 1/n^{(3)} = 1 + 2/|P|$$

where $n^{(r)} \geqslant 2$. For this case at least one of the orders $n^{(r)}$ must equal 2, because otherwise $n^{(r)} \geqslant 3$ for all r so that the lhs $\leqslant 1$, whereas the rhs > 1. For convenience we set $n^{(1)} = n$, $n^{(2)} = m$ and $n^{(3)} = 2$ and obtain

$$1/n + 1/m + 1/2 = 1 + 2/|P|; \qquad n,\ m \geqslant 2 \tag{5.4.4}$$

which is nothing other than the Wyle relation introduced in (5.3.9) without proof. It is emphasized here that the set $\{n, m, 2\}$ refers to three different equivalence sets of axis-vectors of the orders n, m and 2: this point had not been made very clear when we introduced the equation in (5.3.9). Any solution of this equation for the set $\{n, m, 2\}$ gives the possible order of the axis-vectors of rotation in P and thus determines the presentation (5.3.8a) of P. Once the presentation of P has been given, then, by purely algebraic manipulation such as the coset enumeration (see Section 5.7), we can determine the abstract structure of the point group P without any additional information.

Now, if one of n and m equals 2, say $m = 2$, we have a solution $\{n, 2, 2\}$ of (5.4.4) that yields $|P| = 2n$, where n is an arbitrary integer > 1. The set defines the group D_n through the defining relations (5.3.8a). When $n, m > 2$, we proceed as before in the case of a polyhedral group based on (5.3.1) (which follows obviously from (5.4.4)) and obtain three solutions of (5.4.4), $\{3, 3, 2\}$, $\{4, 3, 2\}$ and $\{5, 3, 2\}$, which define the groups T, O and Y, respectively, through the defining relations (5.3.8a).

As a conclusion, we state that there exist only five types of proper point groups of finite order: C_n, D_n, T, O and Y. It is stressed again that there exist no more than three equivalence sets of axis-vectors; in fact, with the exception of C_n, every proper point group has three equivalence sets of axis-vectors, either all three are symmetric (so that corresponding axes are all two-sided), or one is symmetric and the other two are mutual inverses.

5.5 Improper point groups

5.5.1 General discussion

An improper rotation in three dimensions is defined by a 3×3 real orthogonal matrix R with $\det R = -1$. Thus, an improper rotation can be expressed as the product of the inversion $\bar{1}$ and a proper rotation z, i.e. $\bar{z} = \bar{1}z$. Now, an improper point group is a subgroup of the real orthogonal group which contains improper rotations. It has a halving subgroup P that is proper. Accordingly, an improper point group is formed by augmenting a proper point group P with an improper element $\bar{z} = \bar{1}z$. Thus, an improper point group is defined by

$$P_{\bar{z}} = P + \bar{z}P \tag{5.5.1}$$

where $\bar{z}$ should be compatible with P such that

$$\bar{z}^2,\ \bar{z}h\bar{z}^{-1} \in P; \qquad \forall\, h \in P \tag{5.5.2}$$

because a halving subgroup of a group is an invariant subgroup. Thanks to the compatibility condition, there exist at most three kinds of augmentors $\overline{z}$ for a given proper point group P, obviously within a multiplicative element belonging to P. Let c_n be the principal axis[4] of rotation in P, then the possible augmentors, except for an arbitrary multiplicative factor $h \in P$, are expressed by

$$i = \overline{1}, \qquad p = \overline{c}_{2n}, \qquad v = \overline{c}_2{}' \tag{5.5.3}$$

where $c_{2n} \| c_n \perp c_2'$, i.e. $\overline{c}_{2n}$ is the $2n$-fold rotation–inversion with c_{2n} 'parallel' to c_n whereas $\overline{c}'$ is a reflection in a 'vertical' plane that contains c_n. Correspondingly, we have at most three kinds of improper point groups for a given proper point group P expressed by

$$P_i, \qquad P_p, \qquad P_v \tag{5.5.4}$$

For $P = C_n$, all three augmentors given in (5.5.3) are allowed. For $P = D_n$, we have only two alternative augmenting operators, $\overline{1}$ and $\overline{c}_{2n}$. Here $\overline{c}_2'$ is excluded[5] since it is equivalent either to $\overline{1}$ or to $\overline{c}_{2n}$ as an augmentor to D_n, because D_n already contains binary axes $c_2 \perp c_n$. Analogously, for $P = T$ we have two alternative augmentors $\overline{1}$ and $\overline{c}_4$. For $P = O$ or Y we have only one augmentor $\overline{1}$. Note that each augmentor introduced above is with a rotation axis of even order: the one with a rotation axis of odd order is reduced to the pure inversion.

Thus, from the five types of the proper point groups C_n, D_n, T, O and Y we obtain the nine types of improper point groups of finite order expressed by

$$C_{ni}, \quad C_{np}, \quad C_{nv}; \quad D_{ni}, \quad D_{np}; \quad T_i, \quad T_p; \quad O_i; \quad Y_i \tag{5.5.5}$$

These provide the complete set of the improper point groups of finite order. Extending $n \longrightarrow \infty$, we obtain three improper point groups of infinite order,

$$C_{\infty i}, \qquad C_{\infty v}, \qquad D_{\infty i} \tag{5.5.6}$$

Obviously, $C_{np} \longrightarrow C_{\infty i}$ and $D_{np} \longrightarrow D_{\infty i}$, in the limit $n \longrightarrow \infty$.

The system of notation (5.5.1) for improper point groups was introduced by the author; see Kim (1983b). As will be shown by (5.5.8a), it is very effective for describing their isomorphisms with proper point groups, which are essential for the systematic construction of their matrix representations and their extensions to the space groups and the magnetic groups. In Table 5.7, the present notation is compared with the Schönflies and the international notations. The author is fully aware of the confusion which an alternative set of notations might bring in. It is to be noted, however, that the present notation does not conflict with any existing notations and also that the present notation and the Schönflies notation are complementary to each other since the former is based on *inversion* and *rotation–inversion* whereas the latter is based on *reflection* and *rotation–reflection*. Note that, in the present notation, the existence of reflection planes in $P_{\overline{z}}$ can be easily seen from the fact that a

[4] The principal axes of D_n, T, O and Y are defined to be c_n, c_2, c_4 and c_5, respectively.

[5] The only allowed reflection plane $\overline{c}_2'$ for D_n as an augmentor that satisfies (5.5.2) is that which is either $\overline{u}_k = \overline{1}u_k$, where u_k is a binary rotation in D_n, or which bisects the angle between two adjacent binary axes u_k and u_{k+1} of D_n. For the former we have $\overline{c}' = \overline{u}_k$ so that $\overline{u}_k u_k = \overline{1}$ and for the latter $\overline{c}_2' u_k = c_{2n}$.

Table 5.7. *Improper point groups, $P_{\bar{z}} = P + \bar{z}P$*

Present notation	P	$\bar{z}$	Isomorphism	Schönflies notation	International notation	Generators
$SO(3)_i$	$SO(3)$	$\bar{1}$	$SO(3) \times C_i$	$SO(3)_{\mathrm{h}}$	$\infty\infty m$	$\infty\infty\bar{1}$
$C_{\infty i}$	C_∞	$\bar{1}$	$C_\infty \times C_i$	$C_{\infty\mathrm{h}}$	∞/m	$\infty\bar{1}$
$C_{\infty v}$	C_∞	$\bar{c}'_2$	D_∞	$C_{\infty\mathrm{v}}$	∞m	$\infty\bar{2}$
$D_{\infty i}$	D_∞	$\bar{1}$	$D_\infty \times C_i$	$D_{\infty\mathrm{h}}$	∞/mm	$\infty 2\bar{1}$
C_{ni}	C_n	$\bar{1}$	$C_n \times C_i$	$C_{n_{\mathrm{e}}\mathrm{h}}$, $S_{2n_{\mathrm{o}}}$	n_{e}/m, $\bar{n}_{\mathrm{o}}$	$n\bar{1}$
C_{np}	C_n	$\bar{c}_{2n}$	C_{2n}	$C_{n_{\mathrm{o}}\mathrm{h}}$, $S_{2n_{\mathrm{e}}}$	$\overline{2n}$	$\overline{2n}$
C_{nv} $(n>1)$	C_n	$\bar{c}'_2$	D_n	$C_{n\mathrm{v}}$	$n_{\mathrm{e}}mm$, $n_{\mathrm{o}}m$	$n\bar{2}$
D_{ni} $(n>1)$	D_n	$\bar{1}$	$D_n \times C_i$	$D_{n_{\mathrm{e}}\mathrm{h}}$, $D_{n_{\mathrm{o}}\mathrm{d}}$	n_{e}/mmm, $\bar{n}_{\mathrm{o}}m$	$n2\bar{1}$
D_{np} $(n>1)$	D_n	$\bar{c}_{2n}$	D_{2n}	$D_{n_{\mathrm{o}}\mathrm{h}}$, $D_{n_{\mathrm{e}}\mathrm{d}}$	$(\overline{2n_{\mathrm{o}}})m2$, $(\overline{2n_{\mathrm{e}}})2m$	$\overline{2n}2$
T_i	T	$\bar{1}$	$T \times C_i$	T_{h}	$m3$	$33\bar{1}$
T_p	T	$\bar{c}_4$	O	T_{d}	$\bar{4}3m$	$\bar{4}3$
O_i	O	$\bar{1}$	$O \times C_i$	O_{h}	$m3m$	$43\bar{1}$
Y_i	Y	$\bar{1}$	$Y \times C_i$	Y_{h}	$53m$	$53\bar{1}$

(1) $\bar{c}_{2n} = \bar{1}c_{2n}$, $\bar{c}'_2 = \bar{1}c'_2$; $\bar{1}$ = inversion.
(2) n_{e} (n_{o}) is an even (odd) integer.
(3) The two different notations on the fifth and sixth lines mean the correspondences; e.g. $C_{n_{\mathrm{e}}i} \leftrightarrow C_{n_{\mathrm{e}}\mathrm{h}} \leftrightarrow n_{\mathrm{e}}/m$ and $C_{n_{\mathrm{o}}i} \leftrightarrow S_{2n_{\mathrm{o}}} \leftrightarrow \bar{n}_{\mathrm{o}}$.

reflection is given by the product of the inversion $\bar{1}$ and a binary rotation u, i.e. $m = \bar{1}u$ so that

$$\bar{c}'_2 = m_{\mathrm{v}}, \qquad \bar{c}_2 = (\bar{c}_{2n_{\mathrm{o}}})^{n_{\mathrm{o}}} = m_{\mathrm{h}}; \qquad c'_2 \| c_n \perp c_2 \tag{5.5.7}$$

where m_{v} and m_{h} are vertical and horizontal reflection planes, respectively, with respect to the principal axis of rotation c_n.

5.5.2 *Presentations of improper point groups*

According to the general definition (5.5.1), an improper point group $P_{\bar{z}}$ is isomorphic to a proper point group defined by $P_z = P + zP$ $(z \notin P)$:

$$P_{\bar{z}} \simeq P_z \tag{5.5.8a}$$

via the correspondence $\bar{z} \leftrightarrow z$. The isomorphism holds because inversion $\bar{1}$ commutes with any point operations. Specifically,

$$C_{np} \simeq C_{2n}, \qquad C_{nv} \simeq D_n, \qquad D_{np} \simeq D_{2n}, \qquad T_p \simeq O; \qquad n>1 \tag{5.5.8b}$$

via the correspondence $\bar{c}_{2n} \leftrightarrow c_{2n}$ or $\bar{c}' \leftrightarrow c'_2$. When $z = e$, we have P_i, which is equal to the direct product of P and the group of inversion C_i:

$$P_i = P \times C_i; \qquad C_i = \{e, \bar{1}\} \tag{5.5.9}$$

Hence, the presentation of $P_{\bar{z}}$ follows from that of the proper point group P_z by replacing z in P_z with $\bar{z}$ if $z \neq e$, whereas the presentation of P_i follows from those of

P and C_i. Instead of giving the presentations of the improper point groups $P_{\bar{z}}$, we simply write down their axis order notations analogous to those of the corresponding proper point groups given by (5.3.8e):

$$C_{np} = \{\overline{2n}, 0, 0\}, \qquad C_{nv} = \{n, \bar{2}, \bar{2}\}, \qquad D_{np} = \{\overline{2n}, 2, \bar{2}\}$$
$$T_p = \{\bar{4}, 3, \bar{2}\}, \qquad P_i = \{n, m, l; \bar{1}\}; \qquad P = \{n, m, l\} \qquad (5.5.10)$$

where $\overline{2n}$ denotes the rotation–inversion, $2n$-fold rotation followed by the inversion, and the inversion $\bar{1}$ in the group P_i is in the center of the group. The corresponding short notations by the generators in the standard presentations are given in Table 5.7. Note that every improper rotation group isomorphic to a proper point group P is obtained by replacing an even-order generator of P with the improper one in the *standard presentation* of P. Thus, there can be no improper point group isomorphic to a uniaxial group C_n of an odd order, or to the tetrahedral group T or to the icosahedral group Y, because they have no even-order generator in their standard presentations.

Exercise 1. Write down the elements of an improper point group from the corresponding proper point group, using the isomorphisms (5.5.8b).

1. $C_{2n} = \{e, c_{2n}, c_{2n}^2, \ldots, c_{2n}^{2n-1}\}$
   $C_{np} = \{e, \bar{c}_{2n}, c_{2n}^2, \ldots, \bar{c}_{2n}^{2n-1}\}$
2. $D_n = \{C_n; \quad u_0, u_1, \ldots, u_{n-1}\}$
   $C_{nv} = \{C_n; \quad \bar{u}_0, \bar{u}_1, \ldots, \bar{u}_{n-1}\}; \qquad u_k = c_n^k u_0$
3. $D_{2n} = \{e, c_{2n}, c_{2n}^2, \ldots, c_{2n}^{2n-1}, u_0, u_1, u_2, \ldots, u_{2n-1}\}$
   $D_{np} = \{e, \bar{c}_{2n}, c_{2n}^2, \ldots, \bar{c}_{2n}^{2n-1}, u_0, \bar{u}_1, u_2, \ldots, \bar{u}_{2n-1}\}$
   $u_k = c_{2n}^k u_0$
4. $O = \{e, 6c_4, 8c_3, 3c_2, 6c_2'\}$
   $T_p = \{e, 6\bar{c}_4, 8c_3, 3c_2, 6\bar{c}_2'\}$
5. $P = \{R\}, \qquad P_i = \{R, \bar{R}\}$

Exercise 2. Write down the classes of $D_{np} = \{\overline{2n}, 2, \bar{2}\}$ from those of $D_{2n} = \{2n, 2, 2\}$.

$$C(e) = \{e\}, \qquad C(\bar{a}^m) = \{\bar{a}^m, \bar{a}^{-m}\}; \qquad m = 1, 2, \ldots, n$$
$$C(b) = \{u_0, u_2, \ldots, u_{2n-2}\}$$
$$C(\bar{a}b) = \{\bar{u}_1, \bar{u}_3, \ldots, \bar{u}_{2n-1}\} \qquad (5.5.11)$$

Remark 1. There exist also less obvious isomorphisms for P_i,

$$C_{n_o i} \simeq C_{2n_o}, \qquad D_{n_o i} \simeq D_{2n_o}, \qquad C_{2i} \simeq D_2 \qquad (5.5.12)$$

via the correspondence $\bar{1} \leftrightarrow c_2 \in P$. Here n_o is an odd integer, and c_2 is along the principal axis so that c_2 is in the center of each proper point group in (5.5.12), just as $\bar{1}$ is in the center of the corresponding improper point group. The isomorphisms (5.5.12) are less important than (5.5.8) in the sense that the former cannot be extended to their so-called double groups (see Chapter 11).

Remark 2. There exist alternative presentations for $C_{n_o i}$ and T_i with smaller numbers of generators. In fact, the group $C_{n_o i}$ is a cyclic group of order $2n$ with the single

Table 5.8. *The maximal subgroups of point groups of finite order*

$Y > T, D_5, D_3$
$Y_i > T_i, D_{5i}, D_{3i}, T$
$O > \underline{T}, D_4, D_3$
$O_i > \underline{T}_i, D_{4i}, D_{3i}, \underline{O}, \underline{T}_p$
$T_p > \underline{T}, D_{2p}, C_{3v}$
$T > \underline{D}_2, C_3$
$T_i > \underline{D}_{2i}, C_{3i}, \underline{T}$
$D_n > \underline{C}_n, C_2, D_{n/v}$ ($D_{n/2}$ is a normal subgroup, if n is even)
$D_{ni} > \underline{C}_{ni}, C_{2i}, D_{n/v,i}, \underline{D}_n, \underline{C}_{nv}, \underline{D}_{n/2,p}$
$D_{np} > \underline{C}_{np}, D_{n/v,p}, \underline{D}_n, C_{nv}, C_{2v}$
$C_n > \underline{C}_{n/v}$
$C_{ni} > \underline{C}_{n/v,i}, C_n, \underline{C}_{n/2,p}$
$C_{np} > \underline{C}_{n/v,p}, \underline{C}_n$
$C_{nv} > C_n, C_{n/v,v}$ ($\underline{C}_{n/2}$ is a normal subgroup, if n is even)

generator $s = \overline{c}_{n_o}$ which satisfies

$$s^{2n_o} = e \tag{5.5.13}$$

For T_i, we set $a = c_2$ and $s = \overline{c}_3^2$, then the defining relations are given by

$$a^2 = s^6 = (as^2)^3 = (as^3)^2 = e \tag{5.5.14}$$

where $(as^3)^2 = e$ means that $as^3 = s^3 a$, which is required because s^3 $(= \overline{1})$ is in the center of the group T_i. These presentations are, however, not very convenient for classifying their irreducible representations by the *gerade* and *ungerade* characteristics with respect to the inversion operator $\overline{1}$.

Remark 3. The isomorphism given in (5.5.8b) may obviously be described in terms of the other notations. For example, the isomorphism $D_{np} \simeq D_{2n}$ is expressed in terms of Schönflies notation as follows:

$$D_{n_o h} \simeq D_{2n_o}, \qquad D_{n_e d} \simeq D_{2n_e} \tag{5.5.15}$$

where n_o (n_e) is an odd (even) integer. The dependence of the isomorphism on the oddness or evenness of the order of the principal axis is a common characteristic of the Schönflies notation. It is emphasized that the present notation is free from this dependence. This is one of the reasons that the present notation is suitable for describing the general structures of the matrix representations of the point groups, as will be shown explicitly, later, in Chapter 11. The present system of notation will be extended also to describe the magnetic point groups in Section 16.3.

5.5.3 *Subgroups of point groups of finite order*

Previously in (5.3.10), we presented the maximal subgroups of proper point groups of finite order determined by their defining relations. This is easily extended to improper point groups because the inversion commutes with any point operation. Note also that one- or two-sidedness of an improper axis of rotation $\overline{R}(\boldsymbol{\theta}) = \overline{1}R(\boldsymbol{\theta})$ is determined by that of the proper part $R(\boldsymbol{\theta})$, as was discussed in Section 5.1.2. For example, analogous to $O = \{4, 3, 2\}$, all axes of $T_p = \{\overline{4}, 3, \overline{2}\}$ are two-sided and its maximal subgroups are given by $D_{2p} = \{\overline{4}, 2, \overline{2}\}$, $C_{3v} = \{3, \overline{2}, \overline{2}\}$, $T = \{3, 3, 2\}$. Moreover, three axes of $D_{np} = \{\overline{2n}, 2, \overline{2}\}$ are also two-sided and its maximal subgroups are given by $C_{np} = \{\overline{2n}, 0, 0\}$, $D_n = \{n, 2, 2\}$, $C_{nv} = \{n, \overline{2}, \overline{2}\}$, $C_{2v} = \{2, \overline{2}, \overline{2}\}$ and $D_{n/\nu,p}$, where ν is an integral divisor of n. For convenience, we have presented the maximal subgroups of every point group of finite order in Table 5.8, where those underlined are normal subgroups.

5.6 The angular distribution of the axis-vectors of rotation for regular polyhedral groups

5.6.1 General discussion

For the tetrahedral group T and the octahedral group O, we have expressed their symmetry elements by the axis-vector representations $\boldsymbol{n}_s$ of rotation. To extend this notation to the icosahedral group Y, it is necessary to determine the angular distribution of the axis-vectors of rotation. We can achieve this via the relative angular distribution of the characteristic symmetry points of the regular polyhedron. We shall first calculate the angular distances for every pair of the nearest characteristic points for a regular polyhedron in general; then, specializing them to the dodecahedron, we determine the polar coordinates of the axis-vectors for the icosahedral group Y.

Let us consider a regular polyhedron characterized by a set of n-fold vertices, m-fold faces and two-fold edges. Let $\boldsymbol{s}_n$, $\boldsymbol{s}_m$ and $\boldsymbol{s}_2$ be the axis-vectors of rotation passing through an n-fold vertex, the mid-point of an m-fold face and the mid-point of a two-fold edge, respectively, from the center O of the polyhedron. If ⓝ, ⓜ and ② denote the characteristic points at which these axis-vectors meet with the surface of the unit sphere centered at O, then a set of the closest three points ⓝ, ⓜ and ② forms a right spherical triangle on the sphere with the spherical angles π/n, π/m and $\pi/2$ (angles formed by the lines tangential to the sides, see Figure 5.6). Let θ_{PQ} be the angular distance (the arc) between a pair of closest symmetry points P and Q on the unit sphere of the polyhedron, then we shall show that

$$\cos\theta_{n2} = \cos(\pi/m)/\sin(\pi/n)$$

$$\cos\theta_{m2} = \cos(\pi/n)/\sin(\pi/m)$$

$$\cos\theta_{nm} = \cos\theta_{n2}\cos\theta_{m2} = \cot(\pi/n)\cot(\pi/m) \tag{5.6.1a}$$

Knowledge of only two of these angular distances is sufficient to determine the complete distribution of the characteristic symmetry points of the regular polyhedron. However, the following additional information helps one to understand the symmetry structure of the polyhedron:

$$\theta_{nn} = 2\theta_{n2}, \qquad \theta_{mm} = 2\theta_{m2}$$

$$\cos\theta_{22} = 1 - 2\sin^2(\pi/n) + 2\cos^2(\pi/m) \tag{5.6.1b}$$

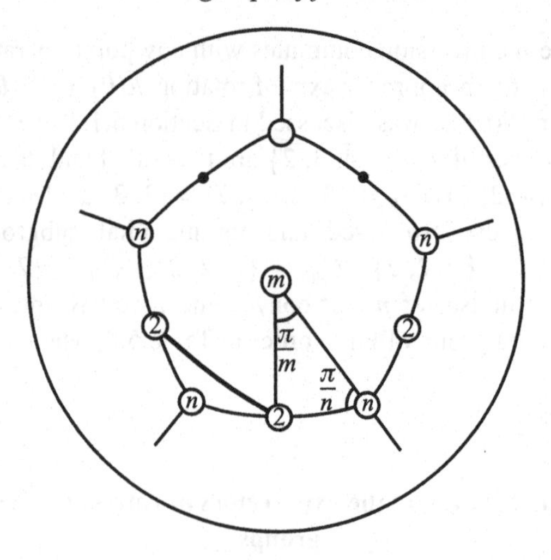

Figure 5.6. The characteristic symmetry points ⓝ, ⓜ and ② on the unit molecular sphere of the dodecahedron.

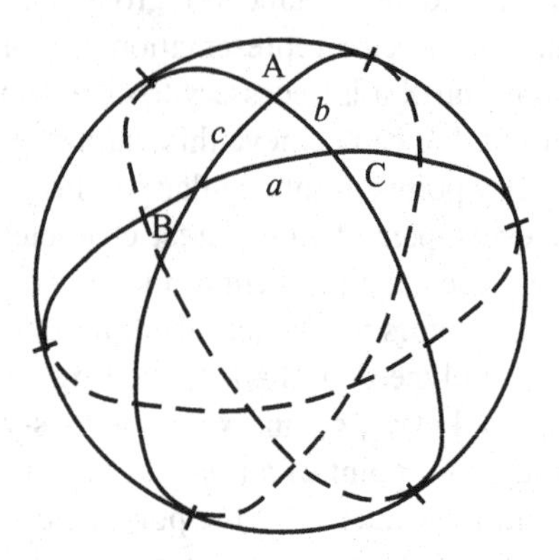

Figure 5.7. A spherical triangle ABC.

where θ_{pp} is the angular distance between a pair of two closest points of the same kind.

The proof of (5.6.1) will be based on the spherical trigonometry. Let ABC be a spherical triangle bounded by three arcs a, b and c of great circles of the radius unity (Figure 5.7). Let A, B and C be the spherical angles formed by lines tangential to the arcs a, b and c and intersecting at the points A, B and C on the sphere. Then, the following laws hold:

$$\frac{\sin a}{\sin A} = \frac{\sin b}{\sin B} = \frac{\sin c}{\sin C} \qquad \text{(the law of sines)} \qquad (5.6.2a)$$

$$\cos a = \cos b \cos c + \sin b \sin c \cos A \qquad \text{(the law of cosines)} \qquad (5.6.2b)$$

and the simultaneous permutations of (a, b, c) and (A, B, C) for the last relation. In the special case of a right spherical triangle, say $B = \pi/2$, the above laws are greatly simplified to express the three arcs a, b and c in terms of the spherical angles A and C as follows:

Table 5.9. *Angular distances (arcs) of nearest pairs of axis-vectors of rotation for regular polyhedral groups*

Groups	(n, m)	θ_{n2}	θ_{m2}	θ_{nm}	θ_{nn}	θ_{mm}	θ_{22}
T	(3, 3)	54.736°	54.736°	70.529°	109.471°	109.471°	90°
O	(3, 4)	35.264°	45°	54.736°	70.529°	90°	60°
Y	(3, 5)	20.905°	31.718°	37.377°	41.812°	63.435°	36°

$$\cos a = \cos A/\sin C, \qquad \cos c = \cos C/\sin A$$
$$\cos b = \cot A \cot C \tag{5.6.2c}$$

From these follow the three relations in (5.6.1a) via the correspondence of the spherical triangles $\Delta(m)(2)(n) \leftrightarrow \Delta ABC$ with $A = \pi/m$, $B = \pi/2$ and $C = \pi/n$ and $a = \theta_{2n}$, $b = \theta_{nm}$ and $c = \theta_{m2}$. The last relation in (5.6.1b) is obtained by applying the law of cosines (5.6.2b) for the isosceles spherical triangle $\Delta(2)(n)(2)$ defined in Figure 5.6 with $a = \theta_{22}$, $b = c = \theta_{n2}$ and $A = 2\pi/n$.

The angular distances among the nearest axis-vectors s_n, s_m and s_2 for the regular polyhedral groups T, O and Y are calculated via (5.6.1) and presented in Table 5.9. From these we can calculate all the polar coordinates of the axis-vectors for any polyhedral group, which will be demonstrated for Y.

Exercise. Verify that $\theta_{22} = 90°$, $60°$ and $36°$ for T, O and Y, respectively, from their subgroups, $D_2 < T$, $D_3 < O$, $D_5 < Y$, given in (5.3.10).

5.6.2 The icosahedral group Y

It is the proper symmetry group of a regular dodecahedron (or icosahedron), see Figure 5.3. The set of axis orders of Y is given by $\{5, 3, 2\}$, so that its order $|Y|$ equals 60 from the Wyle relation (5.3.9). The characteristics of the dodecahedron have been described in Table 5.5; i.e. it has 12 five-fold faces, 20 three-fold vertices and 30 two-fold edges. Correspondingly, from (5.3.15e), it contains six c_5-axes joining the mid-points of opposite faces, ten c_3-axes joining opposite vertices and 15 c_2-axes joining the mid-points of the opposite edges. These are all two-sided so that the icosahedral group Y contains the dihedral groups D_5, D_3 and D_2 as its subgroups in addition to the tetrahedral group given in (5.3.10). It has a total of five classes, all of which are ambivalent, as given in (5.3.15e). Since the regular dodecahedron has the inversion symmetry, its symmetry group is Y_i so that there exist 15 great planes of reflection passing through the center O of the body and perpendicular to 15 binary axes.

The icosahedral group Y is also defined by its coset decomposition with respect to the subgroup D_5:

$$Y = D_5 + v_1 D_5 + \cdots + v_5 D_5 \tag{5.6.3a}$$ [6]

[6] Obviously, Y may be expressed by the coset decomposition with respect to the subgroup T:

$$Y = T + aT + a^2T + a^3T + a^4T \tag{5.6.3b}$$

where $a = c_5 \in D_5$.

where the coset representatives v_ν ($\nu \neq 0$) are the binary axes of rotation not contained in D_5 and are mutually equivalent with respect to $c_5 \in D_5$, i.e.

$$v_{\nu+1} = c_5^\nu v_1 c_5^{-\nu}; \qquad \nu = 1, 2, 3, 4, 5$$

If we take $c_5 \in D_5$ in the z-direction, then there exist five $c_2 \in D_5$ on the horizontal plane but the coset representatives v_ν introduced above are not on the horizontal plane. Accordingly any one of their product $v_\nu v_\mu$ represents a rotation that is not contained in D_5, in view of (4.3.15b). One may take these v_ν to be the closest binary axes to the $c_5 \in D_5$ with the angular distance $\theta_{52} = 31.718°$ given in Table 5.9, then all five of them are on one five-fold face of the dodecahedron (see Figure 5.9). The above coset decomposition is very effective when we construct the unirreps of Y by induction from the unirreps of D_5. Note, however, the D_5 is not an invariant subgroup of Y, as was mentioned before.

Now, guided by the symmetry of the dodecahedron, we shall determine the polar

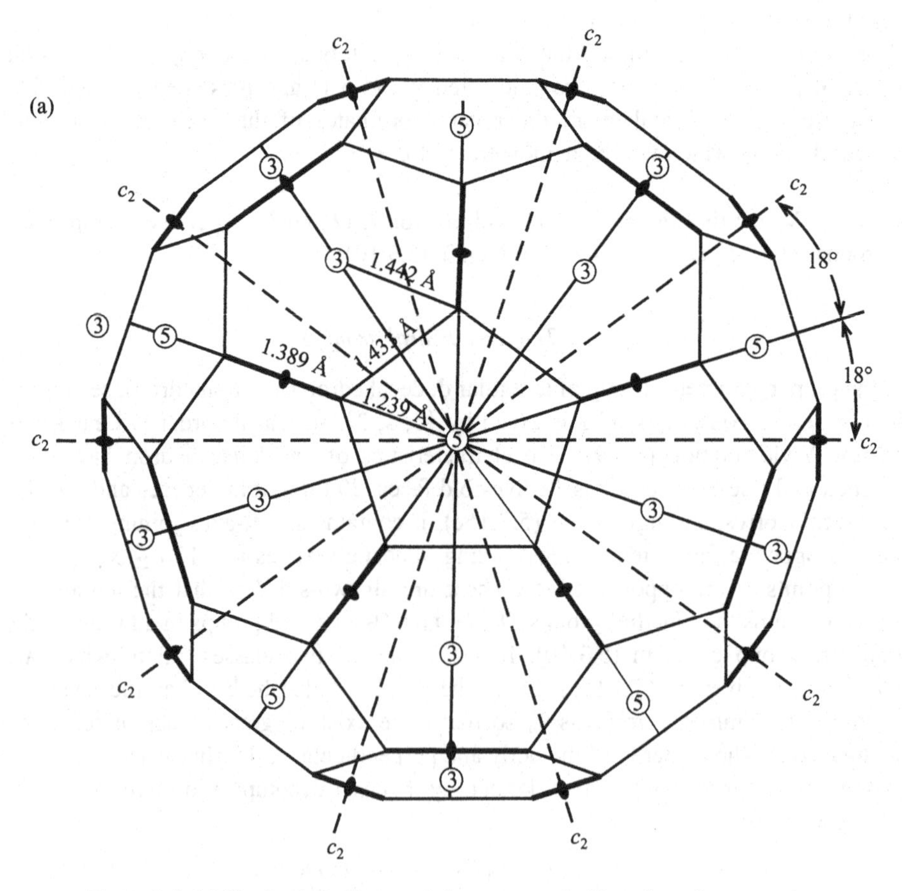

Figure 5.8. (a) The buckyball viewed along a c_5-axis. Note the subgroup symmetry $D_{5i} < Y_i$. (b) The buckyball viewed along a c_3-axis; note the subgroup symmetry $D_{3i} < Y_i$. (c) The buckyball viewed along a c_2-axis; note the subgroup symmetry $T_i < Y_i$. The polar coordinates of a $s_3 \in T$ denoted as ③ is given by $\vartheta = 54.736°$ and $\varphi = 45°$. Note that the horizontal plane is a reflection plane.

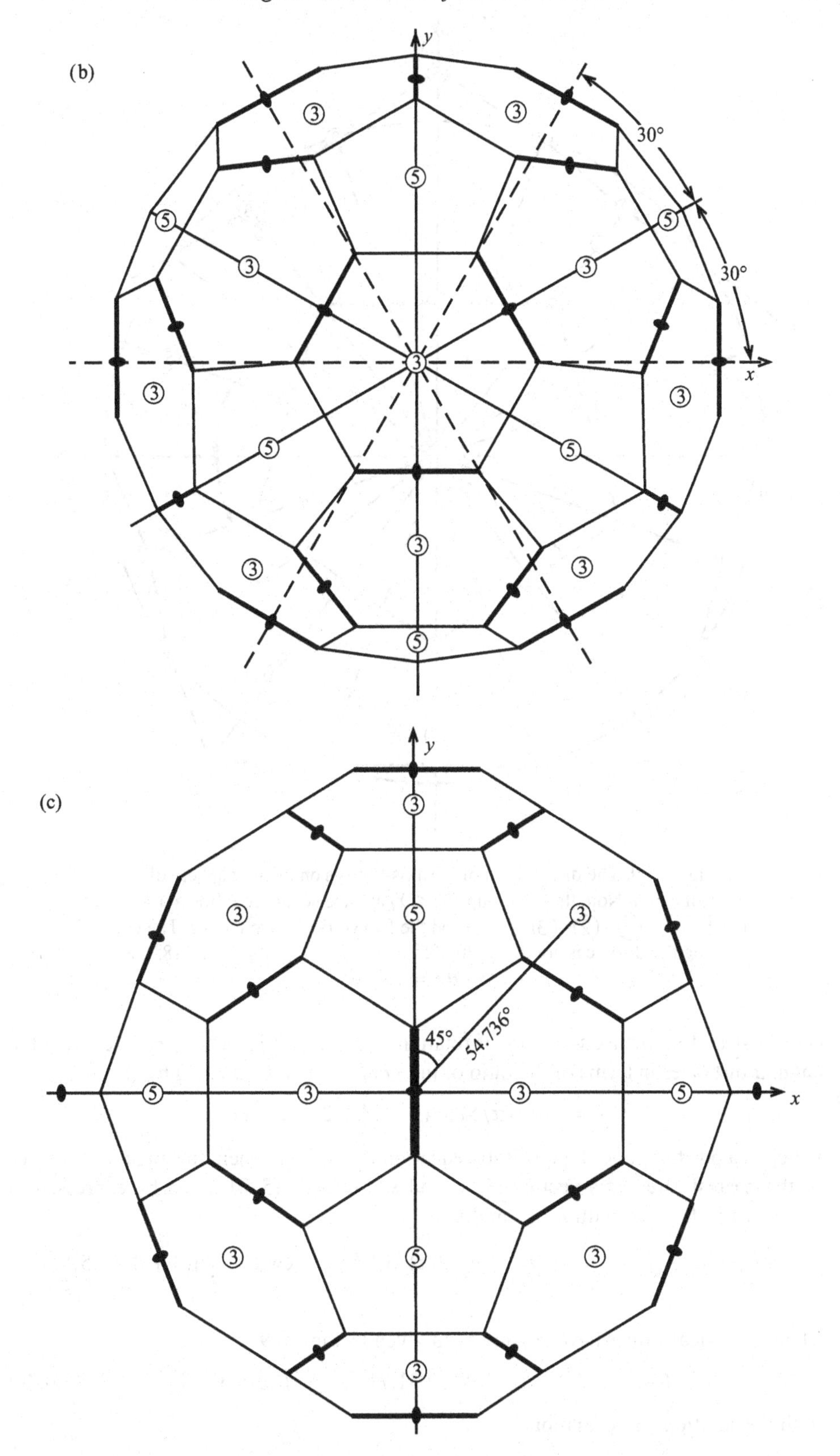
(b)
y
x
30°
30°
③
⑤
(c)
y
x
45°
54.736°

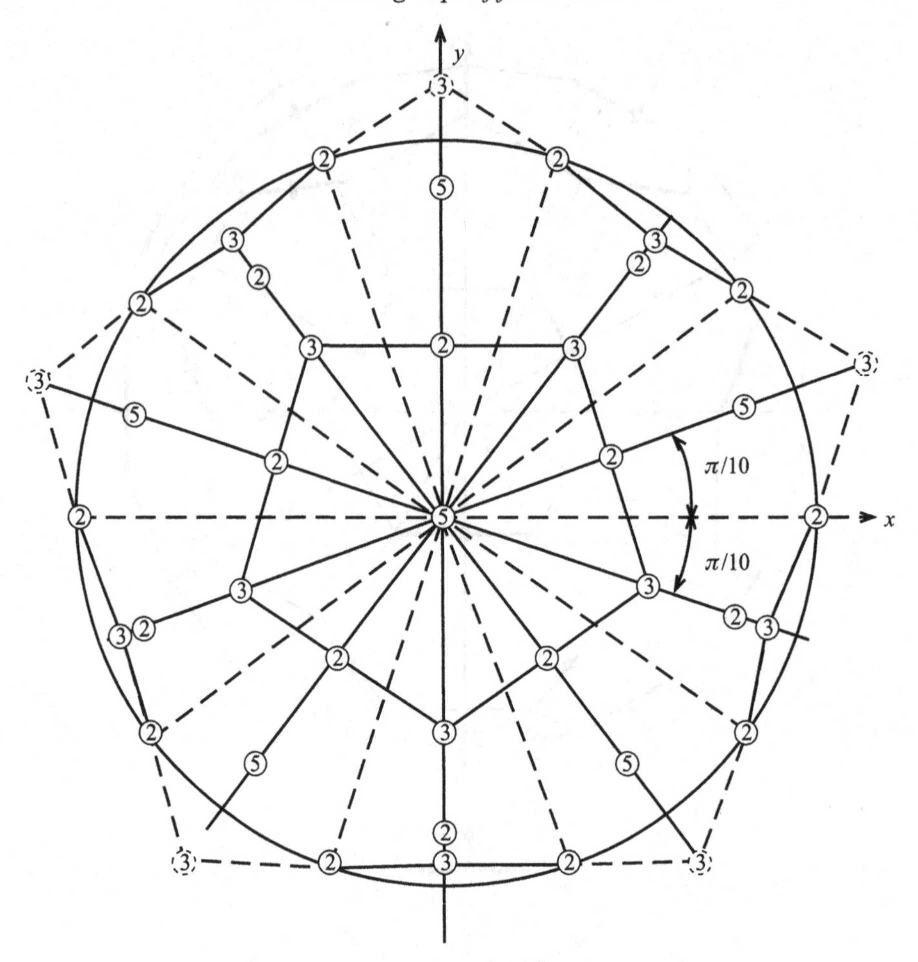

Figure 5.9. The distribution of the axis-vectors on the unit sphere of the dodecahedron. Note the subgroup $D_{5i} < Y_i$ and the sequence of the axis-vectors [-②-③-⑤-②-③-⑤-]$_2$ on the five vertical great circles. The closest angular distances are $\theta_{23} = 20.905°$, $\theta_{35} = 37.377°$, $\theta_{52} = 31.718°$ and $\theta_{22} = 36°$.

coordinates of all the axis-vectors of rotation in Y. Here, it is customary to express the angular distances in terms of the ratio of the *golden section* τ defined by

$$\tau = 2\cos(\pi/5) = (1 + \sqrt{5})/2 = 1.6180$$

which is a positive root of a quadratic equation $\tau^2 = \tau + 1$. Then, the angular distances of the nearest three axis-vectors s_2, s_3 and s_5 given by (5.6.1a) can be expressed in terms of τ as follows, with $n = 3$ and $m = 5$:

$$\cos\theta_{32} = \tau/\sqrt{3}, \qquad \cos\theta_{52} = [(2+\tau)/5]^{1/2}, \qquad \cos\theta_{35} = [(3+4\tau)/15]^{1/2} \tag{5.6.4a}$$

Their numerical values have already been given in Table 5.9:

$$\theta_{32} = 20.905°, \qquad \theta_{52} = 31.718°, \qquad \theta_{35} = 37.377° \tag{5.6.4b}$$

with the additional information

$$\theta_{32} + \theta_{35} + \theta_{52} = 90^\circ, \qquad \theta_{22} = 36^\circ \tag{5.6.4c}$$

The first relation in (5.6.4c) means that the end-to-end angular distance spanned by a sequence of axis-vectors ②–③–⑤–② on a great circle of the unit sphere of the dodecahedron equals 90° (see Figure 5.8(c)). This is in accordance with the fact that the group Y has the dihedral group D_2 as a subgroup. Moreover, $\theta_{22} = 36^\circ$ is the nearest-neighbor angle between the binary axes of rotations belonging to D_5 (see Figure 5.9). These simple relations greatly simplify the determination of the polar coordinates of the axis-vectors, as will be explained below.

To determine the polar coordinates of the axis-vectors of rotation in the icosahedral group Y, it is most convenient to take the z-axis along a highest symmetry axis-vector s_5 as in Figure 5.9. On account of the subgroup symmetry D_{5i} $(= D_{5d}) < Y_i$, there exist the horizontal great plane which contains five c_2-axes equally distributed with the nearest-neighbor angle $\theta_{22} = 36^\circ$ and five vertical great planes of reflection perpendicular to these five c_2-axes, each of which contains the whole sequence of axis-vectors [-②-③-⑤-②-⑤-③-]$_2$ in accordance with (5.6.4c). See also Figure 5.8(c). Let us take the x-axis along an axis-vector $s_2 \in D_5$ on the horizontal plane. Then the polar coordinates (ϑ, φ) of the ten $s_2 \in D_5$ on the horizontal x, y plane are given by

$$\vartheta = \pi/2, \qquad \varphi_\nu(s_2) = \nu\pi/5; \qquad \nu = 0, 1, \ldots, 9 \tag{5.6.5a}$$

The coordinates of the axis-vectors on the vertical reflection planes are given by

$$\vartheta = \sum \theta_{PQ}, \qquad \varphi_\nu = \nu\pi/5 \pm \pi/10; \qquad \nu = 0, 1, \ldots, 9 \tag{5.6.5b}$$

where θ_{PQ} with P, $Q = 2, 3, 5$ are given in Table 5.9 and $\varphi_0 = \pi/10$ or $-\pi/10$ is for either of the partial sequences ②–⑤ and ③–②–③ in Figure 5.9. These give the polar coordinate of all axis-vectors $\boldsymbol{s}$ on the positive hemisphere of the unit sphere of the dodecahedron. Those on the negative hemisphere are given by $-\boldsymbol{s}$ on account of the inversion symmetry $C_i < Y_i$.

5.6.3 Buckminsterfullerene C_{60} (buckyball)

This celebrated molecule of 60 carbon atoms was discovered by Kroto and Smalley in 1985 (Kroto *et al.* 1985). It belongs to the icosahedral-inversion group $Y_i = Y \times C_i$. The structure may be regarded as a truncated dodecahedron, which may be viewed along an axis-vector s_5, s_3 or s_2 of Y passing through the mid-points of a five-fold face, a six-ring face or a two-fold edge as shown in Figure 5.8(a), (b) or (c). All carbon atoms lie on the vertices, each of which is on a symmetry great plane of reflection but not along any symmetry axis of rotation. It has 60 vertices (one-fold), 12 five-fold faces and 20 six-ring faces with three-fold symmetry, 30 edges with two-fold symmetry and 60 edges with no symmetry axis of rotation (one-fold). These equivalence characteristics are in accordance with (5.3.12).

All carbon atoms of the molecule are equivalent but there exist two kinds of C—C bonds: one of which may be considered as double-bonded and the other as single-bonded. Only the double bonds have two-fold symmetry about the mid-points. The average bond lengths are 1.389 Å for the double bonds formed by six–six ring fusions and 1.433 Å for the single bonds formed by six–five ring fusions. The radius of the cavity is 3.512 Å (the average distance of a tricoordinate carbon from the center of the molecule). The polar coordinates of the carbon atoms may be determined by the following additional angular distances with obvious notations (see Figure 5.8(a)):

$$\theta_{C=C} = 22.811°, \qquad \theta_{C-C} = 23.544°$$
$$\theta_{5,C} = 20.313°, \qquad \theta_{3,C} = 23.691°$$

where, for example, $\theta_{5,C}$ is the angular distance between s_5 and one of the nearest carbon atoms.

5.7 Coset enumeration

There exists a mechanical technique by which one can construct the multiplication table of a finite group G starting from the defining relations of the group G without using the geometric interpretation like (5.3.2a). The algorithm is a useful tool with a wide range of applications (Coxeter and Moser 1984). Let H be a subgroup of G, then the primary object of coset enumeration is to determine the complete set of distinct cosets of H in G in terms of the generators of G and thereby identify every element of G. Let the *right coset decomposition* of G by H be written as follows:

$$G = H_1 + H_2 + H_3 + \cdots$$

where H_1 is the subgroup H. Denote these cosets H_1, H_2, ... by numerals 1, 2, ... Then coset 1 is the subgroup H itself. The algorithm is such that, as the computation proceeds, each right coset will be defined as a product $H_k g_i$, where H_k is a previously defined coset and g_i is an appropriately chosen generator of G. We shall explain the further steps of the algorithm through an example.

Consider *the tetrahedral group T* defined by the presentation

$$a^2 = b^3 = (ab)^3 = e \tag{5.7.1}$$

Let $H = \{e, b, b^2\}$ be the subgroup of T by which the right coset decomposition of T will be constructed. For this purpose, we form tables that express the fact that $Hb = bH = H$, $H_k bbb = H_k$, $H_k aa = H_k$ and $H_k ababab = H_k$ for any k on account of the presentation (5.7.1). For the first few k, the table takes the form

b b b	a a	a b a b a b
1 1 1 1	1 1	1 1 1
2 2	2 2	2 2

where the first line, for example, expresses the fact that $b1 = 1b = 1$, $1a = ?$, $1aa = 1$ and $1ababab = 1$; here, 1 denotes the subgroup H defined above and 2 is a right coset of H to be defined. The vacant spots in the table indicate that the right cosets of H are not yet all defined. To fill the vacant spots in the first line we may define the right cosets of H by

$$1, \qquad 1a = 2, \qquad 2b = 3, \qquad 3a = 4 \tag{5.7.2}$$

and insert them into the first line to obtain

b b b	a a	a b a b a b
1 1 1 1	1 2 1	1 2 3 4 2 1 1

which rewards us with $2a = 1$ and $4b = 2$. By extending the table to $k = 2, 3, 4$ and inserting all available information we obtain the following table:

b b b	a a	a b a b a b
1 1 1 1	1 2 1	1 2 3 4 2 1 1
2 3 4 2	2 1 2	2 1 1 2 3 4 2
3 4 2 3	3 4 3	3 4 2 1 1 2 3
4 2 3 4	4 3 4	4 3 4 3 4 3 4

which closes up completely, rewarding us with the additional information $3b = 4$ and $4a = 3$. This means that the coset enumeration of H in the group T is completed by the four right cosets of H given by (5.7.2). This also confirms that the group T is well defined by the presentation (5.7.1).

Next, we shall construct the multiplication table of the group T using the coset enumeration given by (5.7.2) and the presentation (5.7.1). For this purpose we shall denote the coset representative of the coset k by $\underline{k}$ and choose the coset representatives as follows: set $\underline{1} = e$ and $\underline{j} = \underline{k}g_i$ if $j = kg_i$. Then from (5.7.2) we obtain the following coset representatives:

$$\underline{1} = e, \qquad \underline{2} = a, \qquad \underline{3} = \underline{2}b = ab, \qquad \underline{4} = \underline{3}a = aba \tag{5.7.3}$$

Now, we can identify any element of T as a member of a right coset $H_k = (\underline{k}, b\underline{k}, b^{-1}\underline{k})$ of H, where $k = 1, 2, 3, 4$. With this identification of the elements, the multiplication table of T is given below in a reduced form,[7] using the defining relations (5.7.1) for T: it is the multiplication table of the coset representatives and the generators of T or their inverses. The inverses of the coset representatives are also included in the table, for convenience.

	a	b	b^{-1}	
$e = \underline{1}$	$\underline{2}$	b	b^{-1}	$\underline{1}^{-1} = \underline{1}$
$a = \underline{2}$	$\underline{1}$	$\underline{3}$	$b\underline{4}$	$\underline{2}^{-1} = \underline{2}$
$ab = \underline{3}$	$\underline{4}$	$b\underline{4}$	$\underline{2}$	$\underline{3}^{-1} = b^{-1}\underline{2}$
$aba = \underline{4}$	$\underline{3}$	$b^{-1}\underline{2}$	$b^{-1}\underline{3}$	$\underline{4}^{-1} = b\underline{3}$

The table reads, for example, $\underline{2}b = \underline{3}$, $\underline{2}b^{-1} = b\underline{4}$.

As an application of the multiplication table, we shall reconstruct the conjugate classes of the group T. We may identify the conjugates of any element of T by a right

[7] Using the defining relations (5.7.1) for T we have

$$\underline{2}b^{-1} = \underline{2}(ab)^2a = \underline{2}ababa = baba = b\underline{4}$$
$$\underline{4}b = abab = (ab)^{-1} = b^{-1}a^{-1} = b^{-1}a = b^{-1}\underline{2}$$
$$\underline{3}^{-1} = (ab)^{-1} = b^{-1}a^{-1} = b^{-1}a = b^{-1}\underline{2}$$
$$\underline{4}^{-1} = a^{-1}b^{-1}a^{-1} = ab^{-1}a = \underline{2}b^{-1}a = b\underline{4}a = b\underline{3}$$

etc.

coset element of H. Then distinct conjugates of the generator a are computed as follows:

$$bab^{-1} = b^{-1}\underline{4}, \qquad b^{-1}ab = b^{-1}\underline{3}$$

The conjugates of a by the remaining elements are all redundant. Thus the class of a is given by

$$C(a) = \{a,\ b^{-1}\underline{4},\ b^{-1}\underline{3}\} = \{a,\ bab^{-1},\ b^{-1}ab\}$$

Analogously, for the class of b we compute

$$\underline{2}b\underline{2}^{-1} = \underline{3}a = \underline{4}, \qquad b\underline{4}b^{-1} = bb^{-1}\underline{3} = \underline{3}, \qquad b\underline{3}b^{-1} = b\underline{2}$$

and obtain

$$C(b) = \{b,\ \underline{4},\ \underline{3},\ b\underline{2}\} = \{b,\ aba,\ ab,\ ba\}$$

The class of b^{-1} is given, taking the inverse of $C(b)$, by

$$C(b^{-1}) = \{b^{-1},\ b\underline{3},\ b^{-1}\underline{2},\ b\underline{4}\} = \{b^{-1},\ bab,\ b^{-1}a,\ ab^{-1}\}$$

The classes of the group T thus obtained are in complete agreement with the classes already given in (5.3.4).

In an analogous manner one can perform the coset enumeration of *the octahedral group O* on the basis of the presentation $a^4 = b^3 = (ab)^2 = e$. We shall give here only the multiplication table of the group O in the reduced form, choosing the period $\langle a \rangle$ as the subgroup H:

	a	a^{-1}	b	b^{-1}	
$e = \underline{1}$	a	a^{-1}	$\underline{2}$	$\underline{3}$	$\underline{1}^{-1} = \underline{1}$
$b = \underline{2}$	$a^{-1}\underline{3}$	$\underline{5}$	$\underline{3}$	$\underline{1}$	$\underline{2}^{-1} = \underline{3}$
$b^{-1} = \underline{3}$	$\underline{4}$	$a\underline{2}$	$\underline{1}$	$\underline{2}$	$\underline{3}^{-1} = \underline{2}$
$b^{-1}a = \underline{4}$	$a\underline{5}$	$\underline{3}$	$\underline{5}$	$\underline{6}$	$\underline{4}^{-1} = a^{-1}\underline{2}$
$ba^{-1} = \underline{5}$	$\underline{2}$	$a^{-1}\underline{4}$	$\underline{6}$	$\underline{4}$	$\underline{5}^{-1} = a\underline{3}$
$ba^{-1}b = \underline{6}$	$a^{-1}\underline{6}$	$a^{-1}\underline{6}$	$\underline{4}$	$\underline{5}$	$\underline{6}^{-1} = \underline{6}$

One can show that the classes of the group O constructed by the multiplication table coincide with those given by (5.1.3a).

Finally, for the *icosahedral group Y* defined by $a^5 = b^3 = (ab)^2 = e$ we choose $H = \langle a \rangle$. Then the reduced multiplication table is given by

	a	a^{-1}	b	b^{-1}	
$e=\underline{1}$	a	a^{-1}	$\underline{2}$	$a\underline{3}$	$\underline{1}$
$b=\underline{2}$	$\underline{3}$	$a\underline{6}$	$a\underline{3}$	$\underline{1}$	$\underline{2}^{-1}=a\underline{3}$
$ba=\underline{3}$	$\underline{4}$	$\underline{2}$	a^{-1}	$a^{-1}\underline{2}$	$\underline{3}^{-1}=\underline{3}$
$ba^2=\underline{4}$	$\underline{5}$	$\underline{3}$	$\underline{6}$	$\underline{8}$	$\underline{4}^{-1}=a^{-1}\underline{3}$
$ba^3=\underline{5}$	$a\underline{6}$	$\underline{4}$	$\underline{7}$	$\underline{9}$	$\underline{5}^{-1}=a^{-2}\underline{3}$
$ba^2b=\underline{6}$	$a^{-1}\underline{2}$	$a^{-1}\underline{5}$	$\underline{8}$	$\underline{4}$	$\underline{6}^{-1}=a^2\underline{4}$
$ba^3b=\underline{7}$	$\underline{8}$	$a^{-1}\underline{11}$	$\underline{9}$	$\underline{5}$	$\underline{7}^{-1}=a\underline{9}$
$ba^2b^2=\underline{8}$	$a^{-1}\underline{9}$	$\underline{7}$	$\underline{4}$	$\underline{6}$	$\underline{8}^{-1}=\underline{9}$
$ba^3b^2=\underline{9}$	$\underline{10}$	$a\underline{8}$	$\underline{5}$	$\underline{7}$	$\underline{9}^{-1}=\underline{8}$
$ba^3b^2a=\underline{10}$	$\underline{11}$	$\underline{9}$	$a^{-1}\underline{11}$	$a^{-1}\underline{12}$	$\underline{10}^{-1}=a^{-1}\underline{8}$
$ba^3b^2a^2=\underline{11}$	$a\underline{7}$	$\underline{10}$	$\underline{12}$	$a\underline{10}$	$\underline{11}^{-1}=a^{-2}\underline{8}$
$ba^3b^2a^2b=\underline{12}$	$a^{-1}\underline{12}$	$a\underline{12}$	$a\underline{10}$	$\underline{11}$	$\underline{12}^{-1}=\underline{12}$

From the multiplication table, we have constructed the classes of the icosahedral group Y, given as follows, by successive conjugations with the generators and taking the inverses:

$$C(e)=e$$

$$C(a)=\{a,\ a^{-1}\underline{2},\ a\underline{6},\ a\underline{5},\ a^2\underline{4},\ a^3\underline{3},\ a^{-1},\ a\underline{4},\ a^2\underline{3},\ a^3\underline{2},\ \underline{6},\ \underline{5}\}$$

$$C(a^2)=\{a^2,\ \underline{8},\ a\underline{7},\ a\underline{11},\ a^3\underline{11},\ a^3\underline{9},\ a^3,\ \underline{9},\ a^2\underline{8},\ a^3\underline{7},\ a^2\underline{11},\ a^{-1}\underline{11}\}$$

$$C(b)=\{b,\ a^2\underline{6},\ a^2\underline{5},\ a^3\underline{4},\ a^{-1}\underline{3},\ \underline{7},\ \underline{11},\ a\underline{10},\ a^2\underline{9},\ a^{-1}\underline{8},\ a\underline{3},\ a^2\underline{2},\ a^{-1}\underline{6},\ a^{-1}\underline{5},\ \underline{4},$$
$$a\underline{9},\ a^{-2}\underline{8},\ a^{-1}\underline{7},\ a^{-1}\underline{11},\ \underline{10}\}$$

$$C(ab)=\{a\underline{2},\ a^3\underline{6},\ a^3\underline{5},\ a^{-1}\underline{4},\ \underline{3},\ a\underline{8},\ a^2\underline{7},\ a^2\underline{11},\ a^3\underline{10},\ a^{-1}\underline{9},\ \underline{12},\ a^2\underline{12},$$
$$a^{-1}\underline{12},\ a\underline{12},\ a^3\underline{12}\}$$

Their orders, $|C(a)|=|C(a^2)|=12$, $|C(b)|=20$ and $|C(ab)|=15$, are in complete agreement with the results (5.3.15e) obtained by the class order theorem. The construction of the multiplication table and the classes of a point group through the method of coset enumeration may be somewhat tedious but is never difficult.

6

Theory of group representations

6.1 Hilbert spaces and linear operators

6.1.1 Hilbert spaces

The concept of 'Hilbert space' may be defined in several ways. For our purpose, it is the space of functions $f(x)$ of a real variable x (or a set of real variables $x = (x_1, x_2, \ldots)$) that are integrable in the sense that the integral of $f^*(x)f(x)$ over the whole range of x is finite, i.e.

$$\langle f, f \rangle = \int f^*(x)f(x)\,\mathrm{d}x < \infty \tag{6.1.1}$$

which is zero only if $f(x) = 0$. When the integral is not zero, it is always possible to make $\langle f, f \rangle = 1$, in which case $f(x)$ is said to be *normalized to unity.* A function $f(x)$ is also called a *vector* in the Hilbert space, without making any distinction between functions and vectors in the space.

The (Hermitian) scalar product of two vectors in the Hilbert space is defined by

$$\langle f, g \rangle = \int f^*(x)g(x)\,\mathrm{d}x \tag{6.1.2}$$

It is finite because of the Schwartz inequality

$$|\langle f, g \rangle|^2 \leqslant \langle f, f \rangle \langle g, g \rangle \tag{6.1.3}$$

which follows from the inequality

$$\langle (f + \lambda g), (f + \lambda g) \rangle = \langle f, f \rangle + \lambda \langle f, g \rangle + \lambda^* \langle g, f \rangle + |\lambda|^2 \langle g, g \rangle \geqslant 0$$

with $\lambda = -\langle f, g \rangle^* / \langle g, g \rangle$. When the scalar product is zero, i.e. $\langle f, g \rangle = 0$, the two vectors or functions are said to be *orthogonal.*

The number of linearly independent vectors in a given Hilbert space is called *the dimensionality of the space*: it is infinite in general. We are, however, frequently interested in a subspace of a finite dimension, which is also called a Hilbert space. Let us assume that there exists a set of linearly independent vectors $\{\psi_\nu(x)\}$ in a Hilbert space such that any vector $f(x)$ in the space can be expressed by a linear combination of the set $\{\psi_\nu(x)\}$:

$$f(x) = \sum_\nu \psi_\nu(x) f_\nu \tag{6.1.4}$$

Such a set $\{\psi_\nu(x)\}$ is called a *complete set of vectors* (or simply a *basis*) in the Hilbert space. The expansion coefficients $\{f_\nu\}$ are called the *coordinates* of $f(x)$ with respect to the basis $\{\psi_\nu(x)\}$. It is assumed further that a Hilbert space is compact, i.e. the limit of any sequence of vectors in the space, if one exists, belongs to the space.

6.1.1.1 Orthogonalization

If the complete set of vectors $\{\psi_\nu(x)\}$ is not orthogonal, it may be replaced by an orthogonal set $\{\phi_\nu\}$: let

$$\psi_1 = \phi_1$$
$$\psi_2 = a\phi_1 + \phi_2$$
$$\psi_3 = b\phi_1 + c\phi_2 + \phi_3$$

etc. Then, the linear coefficients, a, b, c, ..., are given, from the orthogonality of the set $\{\phi_\nu\}$, by

$$a = \langle \phi_1, \psi_2 \rangle / \langle \phi_1, \phi_1 \rangle$$
$$b = \langle \phi_1, \psi_3 \rangle / \langle \phi_1, \phi_1 \rangle$$
$$c = \langle \phi_2, \psi_3 \rangle / \langle \phi_2, \phi_2 \rangle$$

etc. In a Hilbert space of a finite dimension n, the above orthogonalization process comes to an end after n steps and we obtain a complete orthogonal set of vectors in the space. This process is called the *Schmidt orthogonalization*. The process may be extended to the infinite dimension through an appropriate limiting process. Hereafter, we shall assume that there exists a complete set of orthogonal vectors in a Hilbert space even if the dimensionality is infinite. If one takes *a complete orthogonal and normalized set* $\{\psi_\nu(x)\}$ as a basis of the space, then the scalar product (6.1.2) is given by

$$\langle f, g \rangle = \sum_\nu f_\nu^* g_\nu \tag{6.1.5}$$

where f_ν and g_ν are the νth coordinates of $f(x)$ and $g(x)$ with respect to the basis $\{\psi_\nu(x)\}$, respectively.

6.1.2 Linear operators

Let $f(x)$ be an arbitrary vector in a Hilbert space. Then, an operator A in the space is defined to act on the function $f(x)$ and bring it into another function $g(x)$ in the space such that

$$g(x) = Af(x) \tag{6.1.6}$$

Examples of an operator are a constant, variables, differential operators, square root ($\surd$) and quantum mechanical operators such as Hamiltonians and angular momenta. Hereafter we shall limit our discussion to linear operators: a linear operator A is defined to satisfy

$$A(af(x) + bg(x)) = aAf(x) + bAg(x)$$

where a and b are arbitrary constants. The product of two operators A and B is defined by

$$ABf = A(Bf)$$

Thus a product of operators satisfies the distribution law $A(BC) = (AB)C$. The inverse A^{-1}, if it exists, is defined to satisfy

$$A^{-1}A = AA^{-1} = 1 \tag{6.1.7}$$

Through repeated multiplications of A and/or A^{-1} one defines a power of A^n of A with an integer n that is positive, negative or zero, with $A^0 = 1$. Thus one can define a function of A by $F(A)$, if $F(x)$ can be expanded in powers of x.

The above definition of the function of an operator may be extended as follows. Let us assume that there exists a set of the eigenfunctions $\{\phi_\nu\}$ of an operator A such that

$$A\phi_\nu(x) = a_\nu\phi_\nu(x); \qquad \nu = 1, 2, \ldots \tag{6.1.8a}$$

where $\phi_\nu(x)$ is an eigenfunction of A belonging to an eigenvalue a_ν. If the set of eigenfunctions provides a *complete orthonormalized set* in the Hilbert space, then any function $f(x)$ in the Hilbert space is expanded by the set $\{\phi_\nu(x)\}$:

$$f(x) = \sum_\nu \phi_\nu(x) f_\nu \tag{6.1.8b}$$

In terms of the set $\{\phi_\nu(x)\}$, a function $F(A)$ of an operator A may be redefined to satisfy

$$F(A)\phi_\nu(x) = F(a_\nu)\phi_\nu(x) \qquad \text{for all } \nu \tag{6.1.8c}$$

then the action of $F(A)$ on an arbitrary function $f(x)$ in the Hilbert space is given, from (6.1.8b), by

$$F(A)f(x) = \sum_\nu F(a_\nu)\phi_\nu(x) f_\nu$$

This definition of $F(A)$ does not require that $F(x)$ be expansible in powers of x; instead, it requires the completeness of the eigenfunctions $\{\phi_\nu\}$ of the operator A. In quantum mechanics, one of the basic assumptions is that the set of the eigenfunctions of an operator which represents a physical observable provides a complete orthonormalized set in the Hilbert space. Note that the above definition of $F(A)$ defined by (6.1.8c) is very much parallel to that of the function of a matrix introduced in Section 1.3.

6.1.2.1 Special operators

Let A be an operator in the Hilbert space. The *adjoint operator* $A^\dagger$ is defined, in terms of two arbitrary vectors in the Hilbert space, as follows:

$$\langle g, A^\dagger f\rangle = \langle Ag, f\rangle \tag{6.1.9}$$

By definition, we have

$$(AB)^\dagger = B^\dagger A^\dagger$$

In terms of the adjoint operator, a *Hermitian operator* H is defined to satisfy

$$H^\dagger = H$$

i.e. a Hermitian operator is *self-adjoint*. A unitary operator U is defined to satisfy

$$U^\dagger U = 1 \qquad \text{or} \qquad U^\dagger = U^{-1}$$

It leaves a scalar product invariant:

$$\langle Ug, Uf\rangle = \langle g, U^\dagger Uf\rangle = \langle g, f\rangle$$

These concepts are very much parallel to those of the corresponding matrices discussed in Chapter 1.

An arbitrary operator A is always expressed in terms of two Hermitian operators H_1 and H_2 as follows:

$$A = H_1 + \mathrm{i}H_2$$

where $H_1 = (A + A^\dagger)/2$ and $H_2 = (A - A^\dagger)/(2\mathrm{i})$. Accordingly, the adjoint operator $A^\dagger$ of A is given by

$$A^\dagger = H_1 - \mathrm{i}H_2$$

Moreover, a unitary operator U may be expressed, in terms of a Hermitian operator H, by

$$U = \exp(\mathrm{i}H)$$

6.1.3 The matrix representative of an operator

The properties of operators are very much parallel to those of the matrices discussed previously in Chapter 1. Here we shall discuss the close correlation between two concepts by introducing the matrix representative of a given operator. Let $\{\psi_\nu(x)\}$ be a basis in a Hilbert space and let A be an operator in the Hilbert space. Then, the transformed basis $\{A\psi_\nu\}$ also belongs to the space,

$$A\psi_\nu(x) = \sum_\mu \psi_\mu(x)M(A)_{\mu\nu} \tag{6.1.10}$$

where the matrix $M(A) = \|M(A)_{\mu\nu}\|$ defined by the linear coefficients is called the *matrix representative* of the operator A with respect to the basis $\{\psi_\nu(x)\}$. The basis may be expressed by a row vector $\Psi(x)$:

$$\Psi(x) = [\psi_1(x), \psi_2(x), \ldots] \tag{6.1.11}$$

the components of which are the basis vectors of the space. Then (6.1.10) is rewritten formally as follows:

$$A\Psi(x) = \Psi(x)M(A) \tag{6.1.12a}$$

where the operator A acts on Ψ from the left while the matrix representative $M(A)$ comes in on the right of Ψ: explicitly,

$$\begin{aligned} A[\psi_1(x), \psi_2(x), \ldots] &= [A\psi_1(x), A\psi_2(x), \ldots] \\ &= [\psi_1(x), \psi_2(x), \ldots]\begin{bmatrix} M(A)_{11} & M(A)_{12} & \cdots \\ M(A)_{21} & M(A)_{22} & \cdots \\ \cdot & \cdot & \cdots \\ \cdot & \cdot & \cdots \end{bmatrix} \end{aligned} \tag{6.1.12b}$$

This form is very convenient for the actual construction of the matrix representative $M(A)$ from (6.1.10), which will be used frequently in the future.

A simple example. Let P be a permutation operator which interchanges two basis functions ψ_1 and ψ_2 such that

$$P\psi_1 = \psi_2, \qquad P\psi_2 = \psi_1$$

The matrix representative of P based on $[\psi_1, \psi_2]$ is determined via (6.1.12b) as follows:

$$P[\psi_1, \psi_2] = [\psi_2, \psi_1] = [\psi_1, \psi_2]\begin{bmatrix} 0 & 1 \\ 1 & 0 \end{bmatrix}$$

From (6.1.12a) it follows that the matrix representative of a product of two operators A and B in the space is given by the product of the corresponding matrix representatives, i.e.

$$M(AB) = M(A)M(B) \tag{6.1.13a}$$

which follows from

$$AB\Psi = \Psi M(AB)$$

$$AB\Psi = A(B\Psi) = A\Psi M(B) = \Psi M(A)M(B) \tag{6.1.13b}$$

Obviously the identity operator 1 is represented by the unit matrix $\mathbf{1}$. Thus, if there exists the inverse operator A^{-1} such that

$$AA^{-1} = A^{-1}A = 1$$

then the matrix representations of these equations yield

$$M(A)M(A^{-1}) = M(A^{-1})M(A) = M(1) = \mathbf{1}$$

so that

$$M(A^{-1}) = M(A)^{-1} \tag{6.1.13c}$$

If the basis $\Psi = [\psi_1, \psi_2, \ldots]$ is orthonormalized, i.e.

$$\langle \Psi, \Psi \rangle = \mathbf{1} \qquad \text{or} \qquad \langle \psi_\mu, \psi_\nu \rangle = \delta_{\mu\nu}; \qquad \forall \mu, \nu \tag{6.1.14a}$$

where $\forall$ means for all, then, from (6.1.10), the matrix representative $M(A)$ of A is given directly by the scalar product

$$\langle \Psi, A\Psi \rangle = M(A) \qquad \text{or} \qquad \langle \psi_\mu, A\psi_\nu \rangle = M(A)_{\mu\nu} \tag{6.1.14b}$$

Accordingly, from the definition (6.1.9) of the adjoint operator $A^\dagger$, its matrix representative $M(A^\dagger)$ is given by the adjoint matrix $M(A^\dagger)$:

$$M(A^\dagger) = M(A)^\dagger \qquad \text{or} \qquad \langle \psi_\mu, A^\dagger\psi_\nu \rangle = M(A)^*_{\nu\mu} \tag{6.1.15}$$

Accordingly, if A is Hermitian, then so is the matrix representative:

$$\text{if } A^\dagger = A, \text{ then } M(A)^\dagger = M(A) \tag{6.1.16a}$$

Moreover, if A is unitary then so is $M(A)$:

$$\text{if } A^{-1} = A^\dagger, \text{ then } M(A)^{-1} = M(A)^\dagger \tag{6.1.16b}$$

because $M(A)^{-1} = M(A^{-1}) = M(A^\dagger) = M(A)^\dagger$.

Next, we discuss the transformation properties of the matrix representatives. Under a transformation of the basis by a non-singular matrix S

$$\Psi' = \Psi S \qquad \text{or} \qquad \psi'_\nu = \sum \psi_\mu S_{\mu\nu} \tag{6.1.17a}$$

the matrix representative of A transforms according to

$$\overline{M}(A) = S^{-1}M(A)S \tag{6.1.17b}$$

which follows from

$$A\Psi' = A\Psi S = \Psi M(A)S = \Psi' S^{-1}M(A)S = \Psi'\overline{M}(A)$$

Since Ψ is an orthonormalized basis, we have

$$\langle \Psi', \Psi' \rangle = \langle \Psi S, \Psi S \rangle = S^{\dagger}S \tag{6.1.18}$$

so that the transformed basis is also orthonormalized if the transformation matrix S is unitary, and vice versa. The present section provides a general preparation for the matrix representations of a group which will be discussed next.

6.2 Matrix representations of a group

6.2.1 Homomorphism conditions

Let $G = \{g_i\}$ be a group and let $D(G) = \{D(g_i); \forall\, g_i \in G\}$ be a set of non-singular matrices in a certain dimension n. If the correspondence $g_i \to D(g_i)$ is such that (cf. (6.1.13a))

$$D(g_i)D(g_j) = D(g_i g_j) \tag{6.2.1}$$

for all elements of G, then the set $D(G) = \{D(g_i)\}$ is called a *matrix representation* of G. The above condition (6.2.1) is called the *homomorphism condition*, from which the set $D(G)$ is also a group: the unit element e of G is represented by the unit matrix $D(e) = \mathbf{1}$, for $D(g_i)D(e) = D(e)D(g_i) = D(g_i)$ for all $g_i \in G$; moreover,

$$D(g_i^{-1})D(g_i) = D(g_i)D(g_i^{-1}) = D(e) = \mathbf{1}$$

so that $D(g_i^{-1}) = D(g_i)^{-1}$.

The group $G = \{g_i\}$ is homomorphic to one of its representations $D(G) = \{D(g_i)\}$ via the correspondence $g_i \to D(g_i)$ for all $g_i \in G$. If the elements of $D(G)$ are all different, then the group G is isomorphic to $D(G)$ via one-to-one correspondence $g_i \leftrightarrow D(g_i)$ for all $g_i \in G$. In such a case the representation is said to be *faithful*. As discussed in Section 3.5, the elements of G represented by the unit matrix form an invariant subgroup N of G which is called the *kernel of the homomorphism*: $G \to D(G)$, and there exists an isomorphism between $D(G)$ and the factor group G/N.

The simplest representation of G is the identity representation which assigns 1 (the unit matrix in one dimension) to all elements of G. For an Abelian group, one-dimensional representations are sufficient to describe all the representations, because all the elements of the group commute with each other (see (6.6.18) for a rigorous proof). Before presenting general methods of constructing matrix representations we shall consider some simple examples.

Example 1. The dihedral group D_2. The defining relations are

$$a^2 = b^2 = (ab)^2 = e \tag{6.2.2}$$

Since it is Abelian we shall seek one-dimensional representations. Let $D(a)$ and $D(b)$

Table 6.1. *The irreducible representations of D_2*

D_2	e	2_z	2_y	2_x	Bases
A	1	1	1	1	$1, x^2, x^2, y^2, z^2, xyz$
B_1	1	1	-1	-1	z, xy
B_2	1	-1	1	-1	y, xz
B_3	1	-1	-1	1	x, yz

The notations A, B_1, B_2 and B_3 (Mulliken's symbols) are used here because two-fold axes are all equivalent.

be the representatives of the generators a and b, respectively. Then the matrix representation of (6.2.2) yields, assigning $D(e) = 1$,

$$D(a)^2 = D(b)^2 = D(a)^2 D(b)^2 = 1 \tag{6.2.3}$$

The solutions are

$$D(a) = \pm 1, \qquad D(b) = \pm 1$$

Thus, we obtain four one-dimensional representations A, B_1, B_2 and B_3 of D_2, as given in Table 6.1, with the realization $a = 2_x$, $b = 2_y$, $ab = 2_z$. There is no other one-dimensional representation for D_2, because there is no other one-dimensional solution for (6.2.3). The trivial representation A is called *the identity representation*. These are all unfaithful representations because the homomorphism $G \to D(G)$ is either four-to-one or two-to-one. For B_1, for example, the kernel of the homomorphism is $N = \{e, 2_z\}$.

Example 2. The group D_{2i}. It is a direct product group $D_2 \times C_i$, where C_i is the group of inversion with elements $\{e, \bar{1}\}$. The representations of C_i are one-dimensional and given by

	e	$\bar{1}$
Γ_g	1	1
Γ_u	1	-1

Accordingly, for each unirrep $\Gamma(D_2)$, there exist two unirepps of D_{2i} defined through the direct product representations

$$\Gamma_g(D_{2i}) = \Gamma(D_2) \times \Gamma_g(C_i)$$
$$\Gamma_u(D_{2i}) = \Gamma(D_2) \times \Gamma_u(C_i) \tag{6.2.4}$$

Thus, we have the eight one-dimensional representations presented in Table 6.2.

Example 3. The quaternion group Q. This group has been defined by (3.2.8) with a set of eight elements $\{\pm\mathbf{1}, \pm\mathbf{i}, \pm\mathbf{j}, \pm\mathbf{k}\}$. It has been shown also in (3.5.2) that Q is homomorphic to group D_2 with the correspondence

$$(\pm\mathbf{1}) \to e, \qquad \{\pm\mathbf{i}\} \to 2_x, \qquad \{\pm\mathbf{j}\} \to 2_y, \qquad \{\pm\mathbf{k}\} \to 2_z \tag{6.2.5}$$

Table 6.2. *The irreducible representations of D_{2i}*

D_{2i}	e	2_z	2_y	2_x	$\bar{1}$	$\bar{2}_z$	$\bar{2}_y$	$\bar{2}_x$	Bases
A_g	1	1	1	1	1	1	1	1	1
B_{1g}	1	1	−1	−1	1	1	−1	−1	xy
B_{2g}	1	−1	1	−1	1	−1	1	−1	zx
B_{3g}	1	−1	−1	1	1	−1	−1	1	yz
A_u	1	1	1	1	−1	−1	−1	−1	xyz
B_{1u}	1	1	−1	−1	−1	−1	1	1	z
B_{2u}	1	−1	1	−1	−1	1	−1	1	y
B_{3u}	1	−1	−1	1	−1	1	1	−1	x

Table 6.3. *The irreducible representations of the quaternion group Q*

Q	e	i	j	k	e'	i'	j'	k
Γ_1	1	1	1	1	1	1	1	1
Γ_2	1	1	−1	−1	1	1	−1	−1
Γ_3	1	−1	1	−1	1	−1	1	−1
Γ_4	1	−1	−1	1	1	−1	−1	1
E	$\begin{bmatrix}1 & 0\\0 & 1\end{bmatrix}$	$\begin{bmatrix}0 & -\mathrm{i}\\-\mathrm{i} & 0\end{bmatrix}$	$\begin{bmatrix}0 & -1\\1 & 0\end{bmatrix}$	$\begin{bmatrix}-\mathrm{i} & 0\\0 & \mathrm{i}\end{bmatrix}$	$\begin{bmatrix}-1 & 0\\0 & -1\end{bmatrix}$	$\begin{bmatrix}0 & \mathrm{i}\\\mathrm{i} & 0\end{bmatrix}$	$\begin{bmatrix}0 & 1\\-1 & 0\end{bmatrix}$	$\begin{bmatrix}\mathrm{i} & 0\\0 & -\mathrm{i}\end{bmatrix}$

$e = \mathit{1}$, $e' = -\mathit{1}$; $i' = -i$ and so on.

Accordingly Q is also homomorphic to a representation of D_2. Thus, we obtain four one-dimensional representations of Q through those of D_2 given in Table 6.3. Table 6.3 also contains a two-dimensional spinor representation introduced by (3.2.10), which is the only faithful representation of Q.

6.2.2 *The regular representation*

It is one of the most basic representations of any finite group G. It is based on the group property that a product of two elements of G is also an element of G. Let us order the elements of G by the subscripts such that $G = \{g_i;\ i = 1, 2, \ldots, |G|\}$ and let g be any arbitrary element of G. Then gg_i must be an element, say g_j, of G so that we have

$$gg_i = \sum_{j=1}^{|G|} g_j\delta(g_j,\ gg_i); \qquad i = 1, 2, \ldots, |G| \tag{6.2.6a}$$

where $\delta(g_j,\ gg_i)$ is the Kronecker delta defined below. It means that the set of group elements $\{g_i\}$ acts as a basis of the $|G| \times |G|$ matrix representation $D^{(\mathrm{R})}$ of G defined by the coefficients in the expansion (6.2.6a):

$$D^{(R)}(g)_{ji} = \delta(g_j, gg_i) = \delta(g_j g_i^{-1}, g)$$
$$= \begin{cases} 1, & \text{if } g = g_j g_i^{-1} \\ 0, & \text{otherwise} \end{cases} \qquad (6.2.6b)$$

It is called *the regular representation* of G, of which every matrix entry is 1 or 0: for a given g, $D(g)_{ji} = 1$ occurs only once, for the jth row and ith column at which $g = g_j g_i^{-1}$. Accordingly, from the group table for multiplication between $\{g_j\}$ and $\{g_i^{-1}\}$ one can immediately write down $D^{(R)}(G)$. It is a real orthogonal representation of G because $D^{(R)}(g)D^{(R)}(g)^{\sim} = \mathbf{1}$.

Example 4. The regular representation $D^{(R)}$ of the dihedral group D_2. Let $e = g_1$, $2_z = g_2$, $2_y = g_3$ and $2_x = g_4$. By definition, $D^{(R)}(g)_{ji} = 1$ when and only when $g = g_j g_i^{-1}$. However, since D_2 is Abelian and $g_j = g_j^{-1}$ for all j, we have $D^{(R)}(g)_{ji} = 1$, if and only if $g = g_j g_i = g_i g_j$. This means that the representation is symmetric with respect to (i, j). From the group multiplication table of D_2

D_2	e	2_z	2_y	2_x
e	e	2_z	2_y	2_x
2_z	2_z	e	2_x	2_y
2_y	2_y	2_x	e	2_z
2_x	2_x	2_y	2_z	e

the regular representation of D_2 is given by

$$D^{(R)}(e) = \begin{bmatrix} 1 & 0 & 0 & 0 \\ 0 & 1 & 0 & 0 \\ 0 & 0 & 1 & 0 \\ 0 & 0 & 0 & 1 \end{bmatrix}, \qquad D^{(R)}(2_z) = \begin{bmatrix} 0 & 1 & 0 & 0 \\ 1 & 0 & 0 & 0 \\ 0 & 0 & 0 & 1 \\ 0 & 0 & 1 & 0 \end{bmatrix}$$

$$D^{(R)}(2_y) = \begin{bmatrix} 0 & 0 & 1 & 0 \\ 0 & 0 & 0 & 1 \\ 1 & 0 & 0 & 0 \\ 0 & 1 & 0 & 0 \end{bmatrix}, \qquad D^{(R)}(2_x) = \begin{bmatrix} 0 & 0 & 0 & 1 \\ 0 & 0 & 1 & 0 \\ 0 & 1 & 0 & 0 \\ 1 & 0 & 0 & 0 \end{bmatrix} \qquad (6.2.7)$$

The regular representation $D^{(R)}(G)$ of a group G plays a crucial role in the representation theory because it contains every irreducible representation of G once and only once, as will be shown later.

6.2.3 Irreducible representations

Let $D(G) = \{D(g); \forall g \in G\}$ be a representation of a group G. Then, from the given representation, new ones can be produced by the similarity transformation with a given matrix S:

$$D(g)' = S^{-1}D(g)S, \qquad \forall g \in G \qquad (6.2.8)$$

where $\forall$ means for all. Since a similarity transformation does not affect the multiplication properties of the matrices, two representations connected by a similarity transformation are said to be *equivalent*. Equivalent representations are regarded as essentially the same.

If a representation $D(G)$ can be brought into a direct sum of two or more representations by a similarity transformation S such that

$$D'(g) = S^{-1}D(g)S = \begin{bmatrix} D^{(1)}(g) & 0 & \dots & 0 \\ 0 & D^{(2)}(g) & \dots & 0 \\ \vdots & \vdots & & \vdots \\ 0 & 0 & \dots & D^{(w)}(g) \end{bmatrix}; \qquad \forall g \in G \quad (6.2.9a)$$

then the representation is said to be reducible; otherwise, it is said to be irreducible. A one-dimensional representation is irreducible by definition. Equation (6.2.9a) may be expressed in terms of the direct sum notation as follows:

$$D'(g) = D^{(1)}(g) \oplus D^{(2)}(g) \oplus \cdots \oplus D^{(w)}(g), \qquad \forall g \in G \qquad (6.2.9b)$$

Then, from the multiplication law of the direct sums, we see that each matrix in the sum provides a representation of G. A very simple theorem on the irreducibility condition for a representation will be developed later in Section 6.6, but here some examples are in order.

Example 5. The representations of D_2 given by Table 6.1 are all irreducible because they are one-dimensional.

Example 6. Reduce the following representation of D_2 given by

$$D(e) = \begin{bmatrix} 1 & 0 \\ 0 & 1 \end{bmatrix}, \quad D(2_z) = \begin{bmatrix} 1 & 0 \\ 0 & 1 \end{bmatrix}, \quad D(2_y) = \begin{bmatrix} 0 & 1 \\ 1 & 0 \end{bmatrix}, \quad D(2_x) = \begin{bmatrix} 0 & 1 \\ 1 & 0 \end{bmatrix} \qquad (6.2.10a)$$

utilizing the fact that the representation involves only two matrices:

$$\mathbf{1} = \begin{bmatrix} 1 & 0 \\ 0 & 1 \end{bmatrix} \quad \text{and} \quad \sigma_x = \begin{bmatrix} 0 & 1 \\ 1 & 0 \end{bmatrix}$$

Since the unit matrix **1** commutes with any matrix, it is only necessary to reduce σ_x to

$$\sigma_z = \begin{bmatrix} 1 & 0 \\ 0 & -1 \end{bmatrix}$$

This is achieved by an involutional transformation introduced in Section 2.1, i.e.

$$Y\sigma_x Y = \sigma_z; \qquad Y = (\sigma_x + \sigma_z)/\sqrt{2} = \frac{1}{\sqrt{2}}\begin{bmatrix} 1 & 1 \\ 1 & -1 \end{bmatrix} \qquad (6.2.10b)$$

Thus, the representation defined by (6.2.10a) is reduced, by the involutional transformation, to the following form

$$D'(e) = \begin{bmatrix} 1 & 0 \\ 0 & 1 \end{bmatrix}, \quad D'(2_z) = \begin{bmatrix} 1 & 0 \\ 0 & 1 \end{bmatrix}, \quad D'(2_y) = \begin{bmatrix} 1 & 0 \\ 0 & -1 \end{bmatrix},$$

$$D'(2_x) = \begin{bmatrix} 1 & 0 \\ 0 & -1 \end{bmatrix}$$

That is, $D' = A \oplus B_1$, where A and B_1 are the irreducible representations of D_2 given in Table 6.1.

Example 7. Reduce the regular representation $D^{(R)}$ of D_2 given by (6.2.7) using the fact that $D^{(R)}$ can be rewritten in the following direct product form:

$$D^{(R)}(e) = \mathbf{1} \times \mathbf{1}, \qquad D^{(R)}(2_z) = \mathbf{1} \times \sigma_x, \qquad D^{(R)}(2_y) = \sigma_x \times \mathbf{1},$$
$$D^{(R)}(2_x) = \sigma_x \times \sigma_x$$

where $\mathbf{1}$ is the 2×2 unit matrix.

Since the involutional matrix $Y$of (6.2.10b) transforms σ_x to σ_z we can diagonalize $D^{(R)}$ of D_2 by the direct product $Y \times Y$ (which is also involutional) as follows:

$$D'(g) = (Y \times Y)D^{(R)}(g)(Y \times Y); \qquad \forall g \in G \qquad (6.2.11)$$

where

$$D'(e) = \mathbf{1} \times \mathbf{1}, \qquad D'(2_z) = \mathbf{1} \times \sigma_z, \qquad D'(2_y) = \sigma_z \times \mathbf{1}, \qquad D'(2_x) = \sigma_z \times \sigma_z$$

which are all diagonal matrices because σ_z is diagonal. Explicitly,

$$D'(e) = \begin{bmatrix} 1 & 0 & 0 & 0 \\ 0 & 1 & 0 & 0 \\ 0 & 0 & 1 & 0 \\ 0 & 0 & 0 & 1 \end{bmatrix}, \qquad D'(2_z) = \begin{bmatrix} 1 & 0 & 0 & 0 \\ 0 & -1 & 0 & 0 \\ 0 & 0 & 1 & 0 \\ 0 & 0 & 0 & -1 \end{bmatrix}$$

$$D'(2_y) = \begin{bmatrix} 1 & 0 & 0 & 0 \\ 0 & 1 & 0 & 0 \\ 0 & 0 & -1 & 0 \\ 0 & 0 & 0 & -1 \end{bmatrix}, \qquad D'(2_x) = \begin{bmatrix} 1 & 0 & 0 & 0 \\ 0 & -1 & 0 & 0 \\ 0 & 0 & -1 & 0 \\ 0 & 0 & 0 & 1 \end{bmatrix}$$

By comparison with the irreducible representations of D_2 given by Table 6.1, the regular representation $D^{(R)}$ of D_2 is reduced to the following direct sum:

$$D' = A \oplus B_2 \oplus B_1 \oplus B_3$$

which contains every irreducible representation of D_2 given in Table 6.1 once and only once.

6.3 The basis of a group representation

6.3.1 The carrier space of a representation

Let us consider a linear vector space $V^{(n)}$ spanned by a set of n linear independent vectors $\{\psi_1(x), \psi_2(x), \ldots, \psi_n(x)\}$ in a Hilbert space. If the space $V^{(n)}$ is invariant with respect to a group of operators $G = \{A\}$, then the transform $A\psi_\nu(x)$ also belongs to $V^{(n)}$ such that

$$A\psi_\nu(x) = \sum_{\sigma=1}^{n} \psi_\sigma(x) D_{\sigma\nu}(A); \qquad \nu = 1, 2, \ldots, n \qquad (6.3.1)$$

for all $A \in G$. It means that the set $\Psi = [\psi_1(x), \psi_2(x), \ldots, \psi_n(x)]$ provides a row vector basis of a matrix representation $D(G) = \{D(A); \forall A \in G\}$ of the group $G = \{A\}$ analogous to (6.1.12a) and thus leads to the homomorphism relations:

$$D(BA) = D(B)D(A) \qquad \forall B, A \in G \qquad (6.3.2)$$

analogous to (6.1.13a). The linear vector space $V^{(n)}$ spanned by the basis Ψ of $D(G)$ is called the *basis space* or the *carrier space* of the representation $D(G)$.

Since (6.3.1) is a special case of (6.1.10), the properties of the matrix representation $D(G) = \{D(A)\}$ follow from those of the matrix representative $M(A)$ of an operator A discussed in Section 6.1. In particular, if the basis is orthonormal, we have from (6.3.1), analogous to (6.1.14b),

$$\langle \Psi, A\Psi \rangle = D(A) \tag{6.3.3}$$

which is unitary if the group G is unitary from (6.1.16b). Moreover, under a transformation of the basis by $\Psi' = \Psi S$ with a unitary matrix S, the matrix representative $D(A)$ transforms, analogous to (6.1.17b), according to

$$\overline{D}(A) = S^{\dagger} D(A) S; \qquad S^{\dagger} S = 1 \tag{6.3.4}$$

which is again unitary.

Any basis space $V^{(n)}$ of a unitary representation $D(G)$ of a group G contains two trivial subspaces, the null space consisting of only the vector $[0, 0, \ldots, 0]$ and the entire basis space $V^{(n)}$. If there exists a proper subspace $V^{(m)}$ $(0 < m < n)$ that is invariant with respect to G, the representation $D(G)$ can be reduced to the following form by a unitary transformation S:

$$\overline{D}(A) = S^{\dagger} D(A) S = \begin{bmatrix} D^{(1)}(A) & 0 \\ 0 & D^{(2)}(A) \end{bmatrix}; \qquad \forall R \in G \tag{6.3.5}$$

where $\{D^{(1)}(A)\}$ is an m-dimensional unitary representation of G and $D^{(2)}(R)$ is an $(n - m)$-dimensional unitary representation of G. This is shown as follows. We choose an orthonormal basis Ψ' such that

$$\Psi' = [\psi_1^{(1)}, \psi_2^{(1)}, \ldots, \psi_m^{(1)}, \psi_1^{(2)}, \ldots, \psi_{n-m}^{(2)}]$$

where $\Psi^{(1)} = [\psi_1^{(1)}, \ldots, \psi_m^{(1)}]$ spans the invariant subspace $V^{(m)}$ and $\Psi^{(2)} = [\psi_1^{(2)}, \ldots, \psi_{n-m}^{(2)}]$ spans the complement of $V^{(m)}$ in $V^{(n)}$, which may but need not be an invariant subspace of $V^{(m)}$. Then, the representation $\overline{D}(A)$ based on Ψ' may take the following form:

$$\overline{D}(A) = \begin{bmatrix} D^{(1)}(A) & M(A) \\ 0 & D^{(2)}(A) \end{bmatrix}; \qquad \forall A \in G \tag{6.3.6}$$

Since $\{\overline{D}(A)\}$ is a unitary representation of the group G, we have $\overline{D}(A^{-1}) = \overline{D}(A)^{\dagger}$ for all $A \in G$. This means that the above form (6.3.6) should hold both for $\overline{D}(A)$ and for its adjoint $\overline{D}(A)^{\dagger}$. This is possible if and only if $M(A) = 0$ for all $A \in G$. Q.E.D.

According to (6.3.5), if $V^{(m)}$ $(0 < m < n)$ is an invariant subspace of $V^{(n)}$ with respect to G, then so is its complement $V^{(m-n)}$ of dimensions $m - n$. If the basis space $V^{(n)}$ of $D(G)$ does not contain any proper invariant subspace, then $D(G)$ is irreducible and the basis space $V^{(n)}$ is said to be irreducible. Two irreducible basis spaces are said to be inequivalent, if the corresponding irreducible representations are inequivalent. The basis functions belonging to two inequivalent basis spaces are linearly independent (actually they are orthogonal to each other according to Schur's lemma, which will be discussed in the next but one section) and cannot be combined to form an irreducible basis of G. Thus the whole basis space of a unitary group G of a finite order can be classified into a certain number of irreducible basis spaces of G: the number equals the number of the conjugacy classes of G, as will be shown in the next section.

6.3.2 *The natural basis of a matrix group*

More frequently than not, it is profitable to introduce a formal basis for a matrix group G to allow one to understand it as a group of linear transformations. The well-known examples are the basis $[\boldsymbol{i}, \boldsymbol{j}, \boldsymbol{k}]$ of the rotation group $O^{(3)}$ $(= O(3, r))$ formed by the three mutually orthogonal unit vectors in the Cartesian coordinate system and the formal basis $\{\xi_1, \xi_2]$ for the special unitary group $SU(2)$.

Consider the group $G = GL(n)$ of linear transformations in $V^{(n)}$ and let $M = \|M_{ij}\|$ be an $n \times n$ matrix belonging to G and observe the identity

$$M\mathbf{1} = \mathbf{1}\|M_{ij}\| \tag{6.3.7}$$

where $\mathbf{1}$ is the $n \times n$ unit matrix. Let us express the unit matrix in the form

$$\mathbf{1} = \begin{bmatrix} 1 & 0 & \dots & 0 \\ 0 & 1 & \dots & 0 \\ \vdots & \vdots & \dots & \vdots \\ 0 & 0 & & 1 \end{bmatrix} = [e_1, e_2, \dots, e_n] \equiv e \tag{6.3.8}$$

where $e_1, e_2, \dots, e_n$ are the unit column vectors defined by

$$e_1 = \begin{bmatrix} 1 \\ 0 \\ \vdots \\ 0 \end{bmatrix}, \qquad e_2 = \begin{bmatrix} 0 \\ 1 \\ \vdots \\ 0 \end{bmatrix}, \qquad \dots, e_n = \begin{bmatrix} 0 \\ 0 \\ \vdots \\ 1 \end{bmatrix}$$

Then, the identity (6.3.7) is rewritten in the form

$$\mathring{M}[e_1, e_2, \dots, e_n] = [e_1, e_2, \dots, e_n]M \tag{6.3.9a}$$

or

$$\mathring{M}e_j = \sum_i e_i M_{ij}; \qquad \forall\, M \in GL(n) \tag{6.3.9b}$$

where the operator $\mathring{M}$ is introduced to stress the fact that matrix M is acting on each basis vector e_j as an operator. The basis $e = [e_1, e_2, \dots, e_n]$ is called *the natural basis* of the matrix group G. Any (column) vector $x = (x_1, x_2, \dots, x_n)$ in $V^{(n)}$ is expressed by the natural basis as follows:

$$x = e_1 x_1 + e_2 x_2 + \dots + e_n x_n = \sum_i e_i x_i$$

Under the matrix transformation $M \in G$, the vector x transforms according to

$$x' = \mathring{M}x = \sum_{ij} e_i M_{ij} x_j = \sum_i e_i x'_i$$

where

$$x'_i = \sum_j M_{ij} x_j; \qquad i = 1, 2, \dots, n$$

which is the coordinate transformation of x induced by the transformation of the natural basis e.

6.3.2.1 Examples

1. The three orthogonal unit vectors $[\boldsymbol{i}, \boldsymbol{j}, \boldsymbol{k}]$ in the Cartesian coordinate system provide the natural basis for the three-dimensional rotation group $O^{(3)} = \{R\}$, i.e.

$$\mathring{R}[\boldsymbol{i}, \boldsymbol{j}, \boldsymbol{k}] = [\boldsymbol{i}', \boldsymbol{j}', \boldsymbol{k}'] = [\boldsymbol{i}, \boldsymbol{j}, \boldsymbol{k}]R \tag{6.3.10}$$

   From this relation we have shown in Lemma 5.1.1 that the matrix representation of R is given by $R = [\boldsymbol{i}', \boldsymbol{j}', \boldsymbol{k}']$. Moreover, by the multiplication law given by (5.1.8), we have formed the multiplication table for the octahedral group, Table 5.6.

2. The natural basis in $V^{(2)}$ defined by two column vectors

$$\xi_1 = \begin{bmatrix} 1 \\ 0 \end{bmatrix}, \qquad \xi_2 = \begin{bmatrix} 0 \\ 1 \end{bmatrix} \tag{6.3.11}$$

   provides a basis for the spinor transformation $S \in SU(2)$. The basis $\xi = [\xi_1, \xi_2]$ is referred to as *the elementary spinor basis.* Let $S = \|S_{ij}\|$ be a 2×2 matrix belonging to $SU(2)$, then $\mathring{S}\xi = \xi S$ or

$$\mathring{S}[\xi_1, \xi_2] = [\mathring{S}\xi_1, \mathring{S}\xi_2] = [\xi_1, \xi_2]\begin{bmatrix} S_{11} & S_{12} \\ S_{21} & S_{22} \end{bmatrix} \tag{6.3.12}$$

   Note that the elementary spinors ξ_1 and ξ_2 are the eigenvectors of the z-component $\sigma_z \in SU(2)$ of the Pauli spin matrix belonging to the eigenvalues 1 and -1, respectively, i.e.

$$\mathring{\sigma}_z\xi_1 = \xi_1, \qquad \mathring{\sigma}_z\xi_2 = -\xi_2 \tag{6.3.13}$$

   It will be shown in Section 10.3 that a set of homogeneous monomials of the elementary spinors ξ_1 and ξ_2 provides a basis of a representation of $SU(2)$.

6.4 Transformation of functions and operators

6.4.1 General discussion

Let $f(x) = f(x_1, x_2, \ldots, x_n)$ be a function of a point $x = (x_1, x_2, \ldots, x_n)$ in an n-dimensional configuration space $V^{(n)}$ and consider the transformation of $f(x)$ under *a real orthogonal transformation* (or a rotation) $R \in O(n, r)$ of the point x:

$$x' = Rx \qquad \text{or} \qquad x'_i = \sum_j R_{ij}x_j; \qquad i = 1, 2, \ldots, n \tag{6.4.1}$$

Substitution of $x = R^{-1}x'$ into $f(x)$ yields a new function f' of the new point $x' \in V^{(n)}$

$$f(x) = f(R^{-1}x') = f'(x') \tag{6.4.2a}$$

This transformation may also be expressed in terms of a transformation operator $\mathring{R}$ that acts on the function such that

$$f'(x') = f(R^{-1}x') \equiv \mathring{R}f(x') \tag{6.4.2b}$$

Here $\mathring{R}f = f'$ is regarded as the symbol for a function just as f or g is a symbol. Analogously, if we start from $f(x')$ we obtain

$$f(x') = f(Rx) = \mathring{R}^{-1}f(x) \tag{6.4.2c}$$

Since the variable x' in the functional relation (6.4.2b) is a dummy common variable, it may be replaced by x as follows:

$$f'(x) = f(R^{-1}x) = \mathring{R}f(x) \tag{6.4.2d}$$

However, when it is necessary to distinguish the transformed point x' from the original point x, the relation (6.4.2b) or (6.4.2c) is more convenient than (6.4.2d). The geometric interpretation of (6.4.2d) is interesting: the rotation of a function $f(x)$ of a point x is brought about by the inverse rotation of the point x. Before going any further, we give some simple examples of basis transformations.

Example 1. Let $\boldsymbol{r} = (x, y, z)$ be a (column) vector of a point in $V^{(3)}$. Then its transpose provides an elementary basis $\boldsymbol{r}^\sim = [x, y, z]$, which is a special case of the row vector basis defined by (6.1.11) with $\psi_1 = x$, $\psi_2 = y$ and $\psi_3 = z$. Under a rotation R: $\boldsymbol{r} \to \boldsymbol{r}' = R\boldsymbol{r}$, we have from (6.4.2d)

$$\mathring{R}\boldsymbol{r}^\sim = [R^{-1}\boldsymbol{r}]^\sim = \boldsymbol{r}^\sim R \tag{6.4.3a}$$

for $R^\sim = R^{-1}$. Accordingly, in view of (6.1.12b),

$$\mathring{R}[x, y, z] = [\mathring{R}x, \mathring{R}y, \mathring{R}z] = [x, y, z]R \tag{6.4.3b}$$

Comparing this with (6.3.10), we conclude that the elementary basis $[x, y, z]$ transforms like the natural basis $[\boldsymbol{i}, \boldsymbol{j}, \boldsymbol{k}]$ under a rotation. Thus, for example, analogous to (5.1.7a) and (5.1.7c), we have

$$\mathring{3}_{xyz}x = y, \qquad \mathring{3}_{xyz}y = z, \qquad \mathring{3}_{xyz}z = x \tag{6.4.3c}$$

$$\mathring{4}_z x = y, \qquad \mathring{4}_x y = -x, \qquad \mathring{4}_z z = z \tag{6.4.3d}$$

More generally, from (6.4.3b), a function $f(x, y, z)$ of $\boldsymbol{r} = (x, y, z)$ transforms according to

$$\begin{aligned}\mathring{R}f(x, y, z) &= f(\mathring{R}x, \mathring{R}y, \mathring{R}z)\\ &= f(xR_{11} + yR_{21} + zR_{31}, xR_{12} + yR_{22} + zR_{32}, xR_{13} + yR_{23} + zR_{33})\end{aligned} \tag{6.4.3e}$$

which is, obviously, in agreement with $\mathring{R}f(\boldsymbol{r}) = f(R^{-1}\boldsymbol{r})$, if it is expressed by the coordinates.

Example 2. Consider a function $\phi(\boldsymbol{r}, \boldsymbol{r}')$ of two points $\boldsymbol{r}$ and $\boldsymbol{r}'$ in the three-dimensional space that is invariant under a rotation R such that

$$\mathring{R}\phi(\boldsymbol{r}, \boldsymbol{r}') = \phi(R^{-1}\boldsymbol{r}, R^{-1}\boldsymbol{r}') = \phi(\boldsymbol{r}, \boldsymbol{r}') \tag{6.4.4a}$$

A simple example of such a function is given by $\phi(|\boldsymbol{r} - \boldsymbol{r}'|)$ or $\phi(\boldsymbol{r} \cdot \boldsymbol{r}')$. If we let $\boldsymbol{r}' = \boldsymbol{r}_\nu^{\text{o}}$ be a fixed point in space, then we obtain an invariant (or a scalar) function $\phi_\nu(\boldsymbol{r}) \equiv \phi(\boldsymbol{r}, \boldsymbol{r}_\nu^{\text{o}})$ located at $\boldsymbol{r}_\nu^{\text{o}}$, which transforms under a rotation R according to

$$\mathring{R}\phi_\nu(\boldsymbol{r}) = \phi_\nu(R^{-1}\boldsymbol{r}) = \phi(R^{-1}\boldsymbol{r}, \boldsymbol{r}_\nu^{\text{o}}) = \phi(\boldsymbol{r}, R\boldsymbol{r}_\nu^{\text{o}}) \tag{6.4.4b}$$

i.e. the function $\phi_\nu(\boldsymbol{r})$ at $\boldsymbol{r}_\nu^{\text{o}}$ is transformed into the function at $R\boldsymbol{r}_\nu^{\text{o}}$. More specifically, let $\{\boldsymbol{r}_1^{\text{o}}, \boldsymbol{r}_2^{\text{o}}, \boldsymbol{r}_3^{\text{o}}\}$ be a set of three equivalent points under three-fold rotation c_3 such that

$$c_3\boldsymbol{r}_1^{\text{o}} = \boldsymbol{r}_2^{\text{o}}, \qquad c_3\boldsymbol{r}_2^{\text{o}} = \boldsymbol{r}_3^{\text{o}}, \qquad c_3\boldsymbol{r}_3^{\text{o}} = \boldsymbol{r}_1^{\text{o}}$$

then the set of scalar functions $\{\phi_1(\boldsymbol{r}), \phi_2(\boldsymbol{r}), \phi_3(\boldsymbol{r})\}$ transforms exactly like the set of equivalent points:

$$\overset{\circ}{c}_3\phi_1(\boldsymbol{r}) = \phi_2(\boldsymbol{r}), \qquad \overset{\circ}{c}_3\phi_2(\boldsymbol{r}) = \phi_3(\boldsymbol{r}), \qquad \overset{\circ}{c}_3\phi_3(\boldsymbol{r}) = \phi_1(\boldsymbol{r}) \tag{6.4.4c}$$

See Figure 6.1.

6.4.2 The group of transformation operators

If a set of transformations $\{R\}$ forms a group G, then the set of corresponding transformation operators $\{\overset{\circ}{R}\}$ also forms an operator group that is isomorphic to $G = \{R\}$ via the correspondence $\overset{\circ}{R} \leftrightarrow R$. To show this we shall first prove the homomorphism relations:

$$\overset{\circ}{S}\overset{\circ}{R} = (SR)^{\circ}; \qquad \forall\, S, R \in G \tag{6.4.5}$$

By applying $\overset{\circ}{S}$ on (6.4.2d) we obtain

$$\overset{\circ}{S}\overset{\circ}{R}f(x) = \overset{\circ}{S}(\overset{\circ}{R}f)(x) = (\overset{\circ}{R}f)(S^{-1}x) = f(R^{-1}S^{-1}x) = f((SR)^{-1}x) = (SR)^{\circ}f(x)$$

which yields (6.4.5) because the function $f(x)$ is an arbitrary function of x. Accordingly, the set $\{\overset{\circ}{R}\}$ forms a group isomorphic to $G = \{R\}$ via the correspondence $\overset{\circ}{R} \leftrightarrow R$. On account of the isomorphism, the operator group may also be expressed by the same notation $G = \{\overset{\circ}{R}\}$ as $G = \{R\}$ unless confusion arises.

Remark 1. Let $f' = \overset{\circ}{R}f$, then it seems natural to write

$$\overset{\circ}{S}\overset{\circ}{R}f(x) = \overset{\circ}{S}f'(x) = f'(S^{-1}x) = \overset{\circ}{R}f(S^{-1}x) \tag{6.4.6}$$

Now, if we substitute $f(S^{-1}x) = \overset{\circ}{S}f(x)$ into the last term in (6.4.6), we would obtain $\overset{\circ}{S}\overset{\circ}{R}f(x) = \overset{\circ}{R}\overset{\circ}{S}f(x)$, which is contradictory unless $\overset{\circ}{S}$ commutes with $\overset{\circ}{R}$. The contradiction is avoided by writing $f'(S^{-1}x) = (\overset{\circ}{R}f)(S^{-1}x)$, which clearly indicates where R^{-1} should act and also does not allow $\overset{\circ}{S}$ to come in between $\overset{\circ}{R}$ and f.

Remark 2. Let $f(x) = f(x_1, x_2, \ldots, x_n)$ be a function of a point $x = (x_1, x_2, \ldots, x_n)$ in $V^{(n)}$ as before. In (6.4.1), we have regarded x as a column vector in describing its transformation $x \rightarrow x' = Rx$ under an orthogonal transformation R. Alternatively, we can equally well regard $f(x_1, x_2, \ldots, x_n)$ as a function $f(x^{\sim})$ of a row vector $x^{\sim} = [x_1, x_2, \ldots, x_n]$. Then we have, analogous to (6.4.3a),

$$\overset{\circ}{R}x^{\sim} = (R^{-1}x)^{\sim} = x^{\sim}R \tag{6.4.7}$$

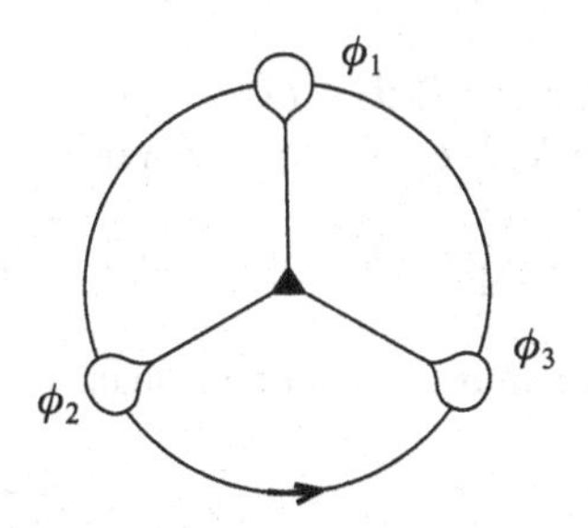

Figure 6.1. Rotation of a set of equivalent scalar functions under a three-fold rotation c_3.

which is quite parallel to (6.3.9b). Thus the transformation of $f(x^\sim)$ is given by

$$\mathring{S}\mathring{R}f(x^\sim) = \mathring{S}f(x^\sim R) = f(x^\sim SR) = (SR)^\circ f(x^\sim)$$

This formalism is quite convenient because one may by-pass the inverse operation R^{-1}. Hereafter, we may use either definition of $\mathring{R}$ freely, whichever is more convenient for the situation.

Further properties of an operator group $G = \{\mathring{R}\}$ corresponding to a real orthogonal group $G = \{R)$ are the following.

(i) Given two functions $f(x)$ and $g(x)$ and two constants a and b, then we have the linearity

$$\mathring{R}(af(x) + bg(x)) = a\mathring{R}f(x) + b\mathring{R}g(x)$$

because the point transformation $x \to Rx$ must apply to every x. For the same reason

$$\mathring{R}[f(x)g(x)] = \mathring{R}f(x)\mathring{R}g(x) \tag{6.4.8a}$$

If $F(f(x), g(x), \ldots)$ is a function of functions $f(x), g(x), \ldots$, then

$$\mathring{R}F(f(x), g(x), \ldots) = F(\mathring{R}f(x), \mathring{R}g(x), \ldots) \tag{6.4.8b}$$

which plays an important role in describing the transformation of a composite function of functions whose transformation properties are known.

(ii) $G = \{\mathring{R}\}$ is unitary. To see this, introduce a real orthogonal transformation R: $x \to x' = Rx$ for the integral variable x of a scalar product (6.1.2). Then from (6.4.2a)

$$\langle f, g\rangle = \int f(x)^* g(x)\,\mathrm{d}x = \int f'(x')^* g'(x')\,\mathrm{d}x'$$

using $\mathrm{d}x = \mathrm{d}x'$ under an orthogonal transformation R. From (6.4.2b), this means that

$$\langle f, g\rangle = \langle \mathring{R}f, \mathring{R}g\rangle = \langle f, \mathring{R}^\dagger\mathring{R}g\rangle$$

which yields $\mathring{R}^\dagger R = 1$ because f and g are arbitrary functions. This can be written as follows:

$$\mathring{R}\langle f, g\rangle = \langle f, g\rangle \tag{6.4.9}$$

assuming that $\mathring{R}$ acts on the integrand.

Example 5. The rotation operator $\mathring{R}(\boldsymbol{\theta}) \in O^{(+3)}$. The transformation operator corresponding to a proper rotation $\mathring{R}(\boldsymbol{\theta}) \in O^{(+3)}$ is given by

$$\mathring{R}(\boldsymbol{\theta}) = \exp(-\mathrm{i}\boldsymbol{\theta}\cdot\boldsymbol{L}), \qquad \boldsymbol{L} = \frac{1}{\mathrm{i}}[\boldsymbol{r}\times\boldsymbol{\nabla}] \tag{6.4.10}$$

where $\boldsymbol{L}$ is the angular momentum operator in quantum mechanics and $\boldsymbol{\nabla} = (\partial/\partial x, \partial/\partial y, \partial/\partial z)$.

Proof. We shall determine $\mathring{R}(\boldsymbol{\theta})$ through infinitesimal rotation. On expanding $R(\theta\boldsymbol{n}) = \exp[\theta\omega(\boldsymbol{n})]$ given in (4.3.2a) for an infinitesimal rotation angle θ, we obtain

$$r' = e^{\theta\omega(n)} r = r + \theta\omega(n) r + \cdots = r + \theta[n \times r] + \cdots$$

to the first order in θ. Thus, under the infinitesimal transformation, a function $f(r)$ transforms according to

$$f(r') = f(r) + \theta[n \times r] \cdot \nabla f(r) + \cdots$$

Using the identity

$$[n \times r] \cdot \nabla = n \cdot [r \times \nabla] = \mathrm{i}(n \cdot L) \equiv \mathrm{i}L_n$$

we obtain

$$\left.\frac{\partial f}{\partial \theta}\right|_{\theta=0} = \mathrm{i}L_n f; \qquad L_n = n \cdot L$$

which yields, upon integration with respect to θ,

$$f(r') = \exp(\mathrm{i}\theta L_n) f(r) = \mathring{R}^{-1} f(r)$$

in view of (6.4.2c). Since $f(r)$ is an arbitrary function we arrive at (6.4.10). The operator is unitary because the angular momentum operator L is Hermitian.

6.4.3 *Transformation of operators under* $G = \{\mathring{R}\}$

Let $Q(x)$ be an operator that is a function of x in a configuration space $V^{(n)}$ that brings an arbitrary function $f(x)$ in a Hilbert space to another function $F(x)$ in the space such that

$$F(x) = Q(x) f(x) \tag{6.4.11}$$

Under a real orthogonal transformation R: $x \to x' = Rx$ we have, from (6.4.2c),

$$F(x') = F(Rx) = \mathring{R}^{-1}[Q(x) f(x)] = \mathring{R}^{-1} Q(x) \mathring{R} f(x')$$

On the other hand, $F(x') = Q(x') f(x')$ so that

$$Q(x') = \mathring{R}^{-1} Q(x) \mathring{R} \tag{6.4.12}$$

which describes the transformation law of the operator $Q(x)$ under a rotation R of the point x in the configuration space $V^{(n)}$.

Example 1. If the operator $Q(x)$ is a scalar operator, then it is invariant under the group $G = \{\mathring{R}\}$ of transformations of x such that $Q(x') = Q(x)$. Accordingly, from (6.4.12),

$$Q(x) = \mathring{R}^{-1} Q(x) \mathring{R} \quad \text{or} \quad \mathring{R} Q(x) = Q(x) \mathring{R} \tag{6.4.13}$$

for all $\mathring{R} \in G$. Conversely, a group of transformations that leaves an operator $Q(x)$ invariant is called the *symmetry group of the operator* $Q(x)$. In quantum mechanics, the symmetry group G of the Hamiltonian of a system plays an essential role in classifying the energy eigenstates by the complete set of irreducible representations of G; in particular, because the Schrödinger equations are seldom solvable exactly. Moreover, the selection rule for the transitions between the eigenstates is most effectively discussed in terms of the irreducible representations of the symmetry group of the Hamiltonian (see Section 6.8).

Example 2. Let $V(\boldsymbol{r})$ be a vector operator in three dimensions $V^{(3)}$ such that, under a rotation R: $\boldsymbol{r} \to \boldsymbol{r}' = R\boldsymbol{r}$, it transforms according to

$$V(\boldsymbol{r}') = RV(\boldsymbol{r})$$

On the other hand, $V(\boldsymbol{r})$ is an operator, so it must obey the operator transformation law (6.4.12). Accordingly,

$$\mathring{R}^{-1}V(\boldsymbol{r})\mathring{R} = RV(\boldsymbol{r})$$

Substitutions of (6.4.10) and $R = \exp[\theta\omega(\boldsymbol{n})]$ into this yield the transformation law of a vector operator

$$\mathrm{e}^{-\mathrm{i}\theta L_n} V \mathrm{e}^{\mathrm{i}\theta L_n} = \mathrm{e}^{\theta\omega(\boldsymbol{n})} V \tag{6.4.14}$$

If we apply this for an infinitesimal rotation, we obtain the commutation relations between a vector operator V and the angular momentum operator $\boldsymbol{L}$. To see this we expand both sides of this equation with respect to an infinitesimal angle θ up to the order θ^2 to obtain

$$(1 + \mathrm{i}\theta L_n)V(1 - \mathrm{i}\theta L_n) = (1 + \theta\omega(\boldsymbol{n}))V + \cdots$$

which, using $\omega(\boldsymbol{n})V = \boldsymbol{n} \times V$, leads to

$$\mathrm{i}[L_n, V] = \boldsymbol{n} \times V \tag{6.4.15a}$$

If we write this out in terms of the vector components of the operators V and $\boldsymbol{L}$, we obtain the desired commutation relations

$$\begin{array}{lll} [L_x, V_x] = 0, & [L_y, V_x] = -\mathrm{i}V_z, & [L_z, V_x] = \mathrm{i}V_y \\ [L_x, V_y] = \mathrm{i}V_z, & [L_y, V_y] = 0, & [L_z, V_y] = -\mathrm{i}V_x \\ [L_x, V_z] = -\mathrm{i}V_y, & [L_y, V_z] = \mathrm{i}V_x, & [L_z, V_z] = 0 \end{array}$$

These may be rewritten as

$$[L_\alpha, V_\beta] = \mathrm{i}\epsilon_{\alpha\beta\gamma}V_\gamma; \qquad \alpha, \beta, \gamma = 1, 2, 3 \tag{6.4.15b}$$

where $\epsilon_{\alpha\beta\gamma}$ is the Levi Civita tensor whose non-null elements are limited to

$$\epsilon_{123} = \epsilon_{231} = \epsilon_{312} = 1, \qquad \epsilon_{231} = \epsilon_{321} = \epsilon_{123} = -1$$

In the special case in which $V = \boldsymbol{L}$, the above commutation relations reduce to the commutation relations for the angular momentum operators (L_x, L_y, L_z).

6.5 Schur's lemma and the orthogonality theorem on irreducible representations

Theorem 6.5.1. Any matrix representation of a finite group with non-vanishing determinants can be transformed into a unitary representation through a similarity transformation.

Proof. Following Wigner (1962), let $D(G) = \{D(g)\}$ be a matrix representation of a finite group $G = \{g\}$, and let $D^\dagger(g)$ be the adjoint of $D(g)$. Then a matrix defined by

$$P = \sum_{g \in G} D(g)D^\dagger(g) \tag{6.5.1}$$

is a positive definite Hermitian matrix since its quadratic form with any vector ψ in the Hilbert space is positive definite:

$$\langle \psi, P\psi \rangle = \sum_g \langle D^\dagger(g)\psi, D^\dagger(g)\psi \rangle > 0$$

Accordingly, one can define its positive square root $P^{1/2}$, which has a positive eigenvalue and which commutes with P, being a function of P. A further important property of P that is essential for the proof is that, for an element s of G, we have

$$D(s)PD^\dagger(s) = \sum_g D(sg)D^\dagger(sg) = P \tag{6.5.2}$$

which follows from the group property that $G = \{g\} = \{sg\}$ for a fixed element s of G. Now we shall demonstrate that the representation defined by

$$U(s) = P^{-1/2} D(s) P^{1/2}; \qquad \forall\, s \in G$$

is unitary because

$$\begin{aligned} U(s)U^\dagger(s) &= P^{-1/2} D(s) P^{1/2} P^{1/2} D^\dagger(s) P^{-1/2} \\ &= P^{-1/2} D(s) P D^\dagger(s) P^{-1/2} \\ &= P^{-1/2} P P^{-1/2} \\ &= \mathbf{1} \end{aligned}$$

where (6.5.2) is used for the third equality.

For most cases of practical interest, we are involved with the representations of a group of unitary transformations $G = \{\mathring{R}\}$. In such a case, Theorem 6.5.1 is easily achieved by choosing an orthonormal basis for the representation in view of (6.3.3).

Theorem 6.5.1′. If two unitary representations $D^{(1)}(G) = \{D^{(1)}(g)\}$ and $D^{(2)}(G) = \{D^{(2)}(g)\}$ of the same group $G = \{g\}$ are equivalent (but not necessarily irreducible), then they can be transformed to one another by a unitary transformation.

Proof. By assumption, there exists a non-singular matrix M such that

$$MD^{(1)}(g)M^{-1} = D^{(2)}(g); \qquad \forall\, g \in G \tag{6.5.3a}$$

then

$$MD^{(1)}(g) = D^{(2)}(g)M, \qquad D^{(1)}(g)M^\dagger = M^\dagger D^{(2)}(g) \tag{6.5.3b}$$

where the second equation is the adjoint of the first. Combining these two, we have

$$MD^{(1)}M^\dagger = D^{(2)}MM^\dagger = MM^\dagger D^{(2)}(g)$$

which means that $MM^\dagger$ commutes with $D^{(2)}(g)$ for all $g \in G$. Since $MM^\dagger$ is a positive definite Hermitian matrix, let $Q = (MM^\dagger)^{1/2}$ be its positive square root; then it is also Hermitian and commutes with $D^{(2)}(g)$ because Q is a function of $MM^\dagger$. Now, the required unitary matrix is given by $U = Q^{-1}M$. Firstly, it is unitary because

$$UU^\dagger = Q^{-1}MM^\dagger Q^{-1} = Q^{-1}Q^2Q^{-1} = 1$$

Secondly, it transforms $D^{(1)}(g)$ into $D^{(2)}(g)$ for all $g \in G$;

$$UD^{(1)}(g)U^{-1} = Q^{-1}MD^{(1)}(g)M^{-1}Q = Q^{-1}D^{(2)}(g)Q = D^{(2)}(g)$$

because Q commutes with $D^{(2)}(g)$. (Q.E.D.)

The importance of these two theorems lies in the fact that they allow us to restrict both the representations of a finite group G and their transformations to be unitary. Thus, if a unitary representation is reducible, it is reducible by a unitary transformation to a direct sum of unitary irreducible representations (unirreps in short).

Theorem 6.5.2. Schur's lemma. It is the fundamental theorem on an irreducible collection of matrices.

Let $D^{(\alpha)}(G) = \{D^{(\alpha)}(g)\}$ and $D^{(\beta)}(G) = \{D^{(\beta)}(g)\}$ be two irreducible representations of the same group $G = \{g\}$ and let their dimensions be n $(= d_\alpha)$ and m $(= d_\beta)$ respectively. If there exists an $n \times m$ intertwining matrix M such that

$$D^{(\alpha)}(g)M = MD^{(\beta)}(g); \qquad \forall\, g \in G \tag{6.5.4}$$

then,

(a) M is either a null matrix or a square non-singular matrix, and
(b) M is a constant matrix when $\alpha = \beta$.

If we assume further that $D^{(\alpha)}(G)$ and $D^{(\beta)}(G)$ are inequivalent unless $\alpha = \beta$, we can combine (a) and (b) to express the matrix M in the following form:

$$M = \delta_{\alpha\beta}\lambda\mathbf{1} \tag{6.5.5}$$

where $\delta_{\alpha\beta}$ is Kronecker's delta which equals zero for $\alpha \neq \beta$ and unity for $\alpha = \beta$, λ is a constant that can be any complex number and $\mathbf{1}$ is the unit matrix of order $n \times n$.

The proof of this celebrated theorem is straightforward. From the outset we may assume that $n \geqslant m$; if $n < m$ we merely take the transpose of (6.5.4) and what follows applies without change because the group property is not used. For the same reason, the theorem also holds for any irreducible collection of matrices, which need not constitute a group.

Let M be expressed as a row vector $M = [\boldsymbol{v}_1, \boldsymbol{v}_2, \ldots, \boldsymbol{v}_m]$, where the components are the n-dimensional column vectors $\boldsymbol{v}_i$ of M. Then (6.5.4) is rewritten as

$$D^{(\alpha)}(g)\boldsymbol{v}_i = \sum_{j=1}^{m} \boldsymbol{v}_j D^{(\beta)}_{ji}(g); \qquad \forall\, g \in G \tag{6.5.6}$$

where $i = 1, 2, \ldots, m$. This means that the linear space $V^{(m)}$ spanned by the m column vectors $[\boldsymbol{v}_1, \boldsymbol{v}_2, \ldots, \boldsymbol{v}_m]$ is an invariant subspace of the n-dimensional basis space $V^{(n)}$ of the irreducible representation $D^{(\alpha)}(G)$. Since $D^{(\alpha)}(G)$ is irreducible, by assumption, the linear space $V^{(m)}$ is either the null space or the entire basis space $V^{(n)}$ of $D^{(\alpha)}(G)$. In the first case, M is the null $n \times m$ matrix. In the second case, M is square $(n = m)$ and non-singular because the set of n vectors $[\boldsymbol{v}_1, \boldsymbol{v}_2, \ldots, \boldsymbol{v}_n]$ spans the basis space of the irreducible representation $D^{(\alpha)}(G)$. Thus we have proven the first part (a) of the lemma.

To prove the second part (b), let $\alpha = \beta$ in (6.5.4), then

$$D^{(\alpha)}(g)M = MD^{(\alpha)}(g); \qquad \forall\, g \in G \tag{6.5.7a}$$

which may be rewritten in the form

$$D^{(\alpha)}(g)[M - \lambda\mathbf{1}] = [M - \lambda\mathbf{1}]D^{(\alpha)}(g); \qquad \forall\, g \in G \tag{6.5.7b}$$

where $\lambda\mathbf{1}$ is a constant matrix that commutes with $D^{(\alpha)}(g)$. We choose λ as one of the characteristic roots of M, then $\det[M - \lambda\mathbf{1}] = 0$, i.e. the matrix $[M - \lambda\mathbf{1}]$ is singular. This leads to the conclusion that $[M - \lambda\mathbf{1}]$ is a null matrix because the intertwining matrix can only be either a null matrix or a square non-singular one. Thus we have $M = \lambda\mathbf{1}$ for $\alpha = \beta$. (Q.E.D.)

From the second half (b) of Theorem 6.5.2 we deduce the following corollary.

Corollary 6.5.2. If there exists a non-constant matrix that commutes with all matrices of a representation of a group G, then the representation is reducible; if there exists none, it is irreducible.

This corollary is also called Schur's lemma. It is frequently used to prove the irreducibility of a given representation.

Theorem 6.5.3. This theorem is the most important of all for practical purposes; it describes *the orthogonality relations* of the matrix elements of irreducible representations in the space of the group elements of a finite group $G = \{g\}$.

Let $D^{(\alpha)}(G) = \{D^{(\alpha)}(g)\}$ and $D^{(\beta)}(G) = \{D^{(\beta)}(g)\}$ be two irreducible representations of the same group G, which are inequivalent unless $\alpha = \beta$. Then they satisfy the orthogonality relations in the group space

$$\sum_{g\in G} D^{(\alpha)}_{ij}(g)D^{(\beta)}_{j'i'}(g^{-1}) = (|G|/d_\alpha)\delta_{\alpha\beta}\delta_{ii'}\delta_{jj'} \tag{6.5.8}$$

where $|G|$ is the order of the group G and d_α is the dimension of the irreducible representation $D^{(\alpha)}(G)$.

The proof proceeds as follows. Using the group property of the representations, we first introduce an intertwining matrix M that satisfies (6.5.4) via a bilinear form of $D^{(\alpha)}(G)$ and $D^{(\beta)}(G)$. Then the orthogonality relations (6.5.8) follow from Schur's lemma expressed by (6.5.5).

From the group property, any matrix of the form

$$M = \sum_{s\in G} D^{(\alpha)}(s)XD^{(\beta)}(s^{-1}) \tag{6.5.9}$$

satisfies (6.5.4), where X is a $d_\alpha \times d_\beta$ rectangular matrix; in fact, for an element g of G,

$$\begin{aligned} D^{(\alpha)}(g)M &= \sum_{s\in G} D^{(\alpha)}(gs)XD^{(\beta)}(s)^{-1} \\ &= \sum_{s} D^{(\alpha)}(gs)XD^{(\beta)}(gs)^{-1}D^{(\beta)}(g) \\ &= MD^{(\beta)}(g) \end{aligned}$$

which is (6.5.4). Here, we have used $\{gs\} = \{s\} = G$ for any fixed element $g \in G$. Now, from Schur's lemma expressed with (6.5.5), we have

$$M_{ii'} = \sum_{s,t} \sum_{g} D_{is}^{(\alpha)}(g) X_{st} D_{ti'}^{(\beta)}(g^{-1}) = \delta_{\alpha\beta} \delta_{ii'} \lambda(X)$$

where λ is a constant that may depend on the matrix X. Set all the matrix elements $X_{st} = 0$ except for one element $X_{jj'} = 1$ in the above equation, then

$$M_{ii'} = \sum_{g} D_{ij}^{(\alpha)}(g) D_{j'i'}^{(\beta)}(g^{-1}) = \delta_{\alpha\beta} \delta_{ii'} \lambda(j, j') \tag{6.5.10}$$

where $\lambda(j, j')$ is the constant $\lambda(X)$ for the above particular choice of X.

To determine the constant $\lambda(j, j')$ we set $\alpha = \beta$ and $i = i'$ in (6.5.10) and sum over i to obtain

$$\sum_{i=1}^{d_\alpha} \sum_{g} D_{ij}^{(\alpha)}(g) D_{j'i}^{(\alpha)}(g^{-1}) = \sum_{g} D_{j'j}^{(\alpha)}(e) = |G| \delta_{j'j} = d_\alpha \lambda(j, j')$$

which yields $\lambda(j, j') = (|G|/d_\alpha)\delta_{jj'}$. Substitution of this into (6.5.10) gives the desired orthogonality theorem (6.5.8). If it is assumed further that the two representations are unitary, then (6.5.8) can be rewritten in the form

$$\sum_{g} D_{ij}^{(\alpha)}(g) D_{i'j'}^{(\beta)}(g)^* = (|G|/d_\alpha) \delta_{\alpha\beta} \delta_{ii'} \delta_{jj'} \tag{6.5.11a}$$

which will be referred to as the orthogonality relations of the *unitary irreducible representations (unirreps)* in the group space $G = \{g\}$. Since the right-hand side of (6.5.11a) is real, the asterisk $*$ on $D^{(\beta)}$ can be transferred to $D^{(\alpha)}$. Hereafter, every representation will be assumed to be unitary unless otherwise specified.

If one regards each matrix element $D_{ij}^{(\alpha)}(g)$ of the unirrep as a vector in the $|G|$-dimensional space of the group elements of G, then the orthogonality relations (6.5.11a) can be expressed in terms of a Hermitian scalar product in the group space as follows:

$$\langle D_{ij}^{(\alpha)}, D_{i'j'}^{(\beta)} \rangle_G = (|G|/d_\alpha) \delta_{\alpha\beta} \delta_{ii'} \delta_{jj'}; \tag{6.5.11b}$$

$$i, j = 1, 2, \ldots, d_\alpha; \qquad i', j' = 1, 2, \ldots, d_\beta; \qquad \alpha, \beta = 1, 2, \ldots, c$$

where c is the number of the inequivalent unirreps of G. Thus, the number of the orthogonal vectors in the group space is given by the sum of the squares of the dimensions of the unirreps, $d_1^2 + d_2^2 + \cdots + d_c^2$. This sum is at most equal to the order $|G|$ of G, because the number of orthogonal vectors cannot exceed the dimensionality of the space. As will be shown later in (6.6.13), the sum is exactly equal to the order $|G|$ of G

$$d_1^2 + d_2^2 + \cdots + d_c^2 = |G| \tag{6.5.12}$$

which provides *the completeness condition* for the irreducible representations of G. From this it follows that the total number of the vectors $\{D_{ij}^{(\alpha)}(g)/(|G|/d_\alpha)^{1/2}$; $i, j = 1, 2, \ldots, d_\alpha$; $\alpha = 1, 2, \ldots, c\}$ in the group space of G equals the order $|G|$ of the group G so that the set forms an orthonormal complete set in the $|G|$-dimensional group space of G. As a result, they satisfy the completeness relations written as

$$\sum_{\alpha=1}^{c} \sum_{i,j=1}^{d_\alpha} d_\alpha D_{ij}^{(\alpha)}(g)^* D_{ij}^{(\alpha)}(g') = |G| \delta_{gg'}; \qquad \forall\, g, g' \in G \tag{6.5.13}$$

This is simply due to the fact that the unitary conditions $UU^\dagger = U^\dagger U = 1$ are expressed by the elements in the following two ways:

$$\sum_m U_{nm} U^*_{n'm} = \delta_{nn'}, \qquad \sum_n U^*_{nm} U_{nm'} = \delta_{mm'}$$

one of which is regarded as the set of orthogonality relations for the row (or column) vectors of U, whereas the other is regarded as the set of their completeness relations.

Exercise 1. For the unirreps of D_2, D_{2i} and Q given in Tables 6.1–6.3, verify the orthogonality relations (6.5.11a), the completeness relations (6.5.13) and the completeness condition (6.5.12). Note that the orthogonality relations are over any two rows of a table of irreducible representations whereas the completeness relations are over all the corresponding matrix elements belonging to two elements g and g' of G.

6.6 The theory of characters

In group theory, the trace of a representative $D(g)$ of $g \in G$

$$\chi(g) = \sum_i D(g)_{ii} \tag{6.6.1}$$

is called the *character* of g in the representation $D^{(\alpha)}(G)$. The set of all characters $\chi(G) = \{\chi(g); \forall\, g \in G\}$ over all elements of G is called *the character of the representation* $D(G)$. The specification of a representation by means of the character has the advantage that it remains *invariant under a similarity transformation.*

6.6.1 Orthogonality relations

From the orthogonality relations of unirreps follow the orthogonality relations of their characters; in fact, we set $i = j$ and $i' = j'$ in (6.5.11a) and sum over j and j' to obtain

$$\sum_g \chi^{(\alpha)}(g)^* \chi^{(\beta)}(g) = |G|\delta_{\alpha\beta} \tag{6.6.2a}$$

From this it follows that two inequivalent unirreps cannot have the same character; hence, *two irreducible representations with the same character are equivalent.* The above equation is again expressed as a scalar product in the group space as follows:

$$\langle \chi^{(\alpha)}, \chi^{(\beta)} \rangle_G = |G|\delta_{\alpha\beta} \tag{6.6.2b}$$

Now, the elements of a conjugate class have the same character since their representations are connected by similarity transformations. Thus the character is a class function so that we may denote the character of an element belonging to a class C_ρ in a representation $D^{(\alpha)}(G)$ by $\chi^{(\alpha)}(C_\rho)$. Suppose that there exist k classes $C_1, C_2, \ldots, C_k$ in a group G. Then the character of a representation $D^{(\alpha)}(G)$ is completely specified by the characters $\chi^{(\alpha)}(C_1), \chi^{(\alpha)}(C_2), \ldots, \chi^{(\alpha)}(C_k)$ of the k classes. Let $|C_\rho|$ be the order of a class C_ρ (i.e. the number of elements in C_ρ). Then

$$\sum_{\rho=1}^{k} |C_\rho| = |C_1| + |C_2| + \cdots + |C_k| = |G|$$

and the orthogonality relations (6.6.2a) are rewritten as

$$\sum_{\rho=1}^{k} \chi^{(\alpha)}(C_\rho)^* \chi^{(\beta)}(C_\rho)^* |C_\rho| = |G|\delta_{\alpha\beta} \tag{6.6.3}$$

This equation means that the set $\{\chi^{(\alpha)}(C_\rho)|C_\rho|^{1/2}; \alpha = 1, 2, \ldots, c\}$ forms an orthogonal vector system in the k-dimensional space of the classes (the class space). Since the number of the orthogonal vectors c in the class space cannot exceed the dimensionality of the space k, one obtains $c \leqslant k$. It will be shown later in (6.6.17) that c equals exactly k:

$$c = k \tag{6.6.4}$$

i.e. *the total number c of the unirreps of a group G equals the number k of the classes in G.* Thus from (6.6.3) the set of vectors $\{\chi^{(\alpha)}(|C_\rho|/|G|)^{1/2}\}$ forms an orthonormalized complete set in the k-dimensional class space of G, so that there follow the completeness relations of the characters

$$\sum_{\alpha=1}^{c} \chi^{(\alpha)}(C_\rho)^* \chi^{(\alpha)}(C_{\rho'}) = (|G|/|C_\rho|)\delta_{\rho\rho'} \tag{6.6.5}$$

where $\rho, \rho' = 1, 2, \ldots, k$.

6.6.2 *Frequencies and irreducibility criteria*

If a unitary representation $D(G) = \{D(g)\}$ of a finite group $G = \{g\}$ is reducible, it can be reduced to a direct sum of unirreps by a unitary transformation:

$$S^\dagger D(g) S = D^{(1)}(g) \oplus D^{(2)}(g) \oplus \ldots; \qquad \forall\, g \in G \tag{6.6.6}$$

where some of the unirreps may be equivalent. Let us assume that the equivalent unirreps contained in (6.6.6) are the same and let the number of times that a unirrep $D^{(\alpha)}$ is contained in a representation $D(G)$ be f_α, then the character $\chi(g)$ of $g \in G$ for the representation $D(G)$ is given by

$$\chi(g) = \sum_{\alpha=1}^{c} f_\alpha \chi^{(\alpha)}(g) \tag{6.6.7}$$

where f_α is called *the frequency of the unirrep $D^{(\alpha)}$* in a representation D. From the orthogonality theorem (6.6.2), the frequency is given by

$$f_\alpha = |G|^{-1} \sum_{g} \chi(g)^* \chi^{(\alpha)}(g) \equiv |G|^{-1} \langle \chi, \chi^{(\alpha)} \rangle_G \tag{6.6.8}$$

which is a non-negative integer by definition. Thus, the scalar product of any two characters $\chi(G)$ and $\chi'(G)$ in the group space is also a non-negative integer given by

$$\langle \chi, \chi' \rangle_G = \sum_{g} \chi(g)^* \chi'(g) = |G| \sum_{\alpha=1}^{c} f_\alpha f'_\alpha \geqslant 0 \tag{6.6.9a}$$

This equation reduces to the following form for the same two characters:

$$\sum_{g} |\chi(g)|^2 = |G| \sum_{\alpha=1}^{c} f_\alpha^2 \tag{6.6.9b}$$

If the representation $D(G)$ is irreducible, then only one of f_α equals unity while the remaining f_α terms should be zero. Accordingly, the irreducibility condition for $D(G)$ is given by

$$\sum_g |\chi(g)|^2 = |G| \tag{6.6.10}$$

6.6.2.1 The completeness condition for unirreps

Another important application of (6.6.9b) is to deduce the completeness condition (6.5.12) for the unirreps of G, which has been introduced without proof. This will be shown via the regular representation $D^{(\mathrm{R})}(G)$ of a group G defined by (6.2.6b). Since its character $\chi^{(\mathrm{R})}(G)$ is given by

$$\chi^{(\mathrm{R})}(g) = \sum_i \delta(g_i g_i^{-1}, g) = |G|\delta(e, g) = \begin{cases} |G|, & \text{if } g = e \\ 0 & \text{otherwise} \end{cases} \tag{6.6.11}$$

the frequency $f_\alpha^{(\mathrm{R})}$ of an irreducible representation $D^{(\alpha)}(G)$ contained in $D^{(\mathrm{R})}(G)$ is given, using (6.6.8), by

$$f_\alpha^{(\mathrm{R})} = d_\alpha \tag{6.6.12}$$

i.e. the *frequency of an irreducible representation $D^{(\alpha)}(G)$ contained in $D^{(\mathrm{R})}(G)$ equals the dimensionality d_α of the irreducible representation.* Since $d_\alpha \geqslant 1$, every irreducible representation $D^{(\alpha)}(G)$ of G is contained in $D^{(\mathrm{R})}(G)$ at least once. Applying (6.6.9b) for $\chi^{(\mathrm{R})}(G)$, we obtain

$$|G| = \sum_{\alpha=1}^{c} d_\alpha^2 \tag{6.6.13}$$

which is the completeness condition for the irreducible representations of a group G (stated previously by (6.5.12) without proof), because c is the total number of the irreducible representations of G. It will be shown that c equals the number k of the classes in G.

6.6.2.2 Exercises

1. Show that the irreducible representations of the dihedral group D_2 given by Table 6.1 are complete.
2. Show that the representation E of the quaternion group Q given in Table 6.3 is irreducible.
3. Show that the irreducible representations of Q given in Table 6.3 are complete.

6.6.3 Group functions

Let $G = \{g\}$ be a group, then *a group function* is defined on G such that a certain real or complex number $\psi(g)$ is assigned to each element g of G. The matrix element $D_{ij}^{(\alpha)}(g)$ for a given set of indices $\{\alpha, i, j\}$ is an example of a group function. The character $\chi^{(\alpha)}(g)$ is another example. Since the latter depends only on the conjugate class of G, it is more proper to call it *a class function*. A class function $\phi(g)$ of $g \in G$ is a group function that satisfies

$$\phi(g) = \phi(s^{-1}gs); \qquad \forall\, g, s \in G \tag{6.6.14}$$

By definition, a group function $\psi(g)$ of $g \in G$ is a vector in the $|G|$-dimensional space of the group elements of G. Since the complete set of the unirreps $\{D_{ij}^{(\alpha)}(g);\ \alpha = 1, 2, \ldots, c\}$ of G spans the complete set of vectors in the $|G|$-dimensional group space, any group function $\psi(g)$ must be expanded by the set as follows (Lyubarskii 1960):

$$\psi(g) = \sum_{\alpha=1}^{c} \sum_{i,j=1}^{d_\alpha} D_{ij}^{(\alpha)}(g) a_{ij}^{\alpha}; \qquad \forall\, g \in G \tag{6.6.15}$$

where the expansion coefficients can be calculated using the orthogonality relations of the unirreps.

Analogous expansion holds for a class function $\phi(g)$ with respect to the complete set of the characters $\{\chi^{(\alpha)}(g);\ \alpha = 1, 2, \ldots, c\}$. Since a class function is also a group function with the condition (6.6.14), we have, from the expansion (6.6.15),

$$\begin{aligned}\phi(g) &= \sum_{\alpha,i,j} D_{ij}^{(\alpha)}(s^{-1}gs) a_{ij}^{\alpha} \\ &= \sum_{\alpha,i,j} \sum_{i',j'} D_{ii'}^{(\alpha)}(s^{-1}) D_{i'j'}^{\alpha}(g) D_{j'j}^{(\alpha)}(s) a_{ij}^{\alpha}\end{aligned}$$

for all $g, s \in G$. Summation of both sides of this equation over s followed by the use of the orthogonality relations (6.5.8) leads to the desired expansion of an arbitrary class function $\phi(g)$

$$\phi(g) = \sum_{\alpha,i} \chi^{(\alpha)}(g) a_{ii}^{\alpha}/d_\alpha \equiv \sum_{\alpha=1}^{c} \chi^{(\alpha)}(g) b_\alpha \tag{6.6.16}$$

where the b_α terms are the expansion coefficients. This expansion means that an arbitrary class function $\phi(g)$ of G in the k-dimensional class space is expressed by a linear combination of c inequivalent vectors $\{\chi^{(\alpha)}(g);\ \alpha = 1, 2, \ldots, c\}$ in the space. Accordingly, the set of c vectors must be a complete set in the k-dimensional space so that

$$c = k \tag{6.6.17}$$

i.e. *the number of inequivalent irreducible representations of a group G equals the number of classes in G.* This has been stated in (6.6.4) without proof.

For an Abelian group G, the number of the classes equals the order $|G|$. Moreover, *all irreducible representations of an Abelian group are one-dimensional,* because the completeness condition (6.6.13) with $c = |G|$,

$$d_1^2 + d_2^2 + \cdots + d_c^2 = c \tag{6.6.18}$$

has only one solution, $d_1 = d_2 = \cdots = d_c = 1$.

6.7 Irreducible representations of point groups

There are many ways of forming the unitary irreducible representations (unirreps) of a point group. In the following, one-dimensional unirreps will be formed by the defining relations of the group whereas the higher dimensional unirreps will be formed by the method of induction from the unirreps of the subgroups. The general theory of induced representations will be developed later in Chapter 8 in full detail. Here we shall describe its simple application.

6.7.1 The group C_n

Since $C_n = \{c_n^k;\ k = 0, 1, 2, \ldots, n-1\}$ is Abelian, all its irreducible representations are one-dimensional and their number is equal to the order of the group. It has one and only one generator, $a = c_n$, and the defining relation is given by

$$a^n = e$$

Let D be a unirrep of C_n, then it is one-dimensional. Then, from $[D(a)]^n = 1$, we obtain n and only n solutions given by

$$D^{(m)}(c_n) = \exp(-\mathrm{i}2\pi m/n)$$

which defines a total of n one-dimensional unirreps of C_n

$$M_m = \{D^{(m)}(c_n^k) = \exp(-\mathrm{i}2\pi mk/n); \qquad k = 0, 1, \ldots, n-1\}$$

$$m = \begin{cases} 0, \pm 1, \ldots, \pm(n-1)/2 & \text{for an odd } n \\ 0, \pm 1, \ldots, \pm\dfrac{n}{2} - 1, \dfrac{n}{2} & \text{for an even } n \end{cases} \tag{6.7.1}$$

for $M_m = M_{m\pm n}$. These constitute the complete set of the unirreps of C_n. To introduce the bases of the representations, let $c_n = n_z$ (the n-fold rotation axis in the z-direction) and let θ be the polar angle in the x, y plane measured from an appropriate axis (see Figure 6.2). Then the basis of M_m is given by

$$\phi_m(\theta) = \exp(\mathrm{i}m\theta) \in M_m \tag{6.7.2}$$

where $\phi_m(\theta)$ and $\phi_{m\pm n}(\theta)$ belong to the same unirrep M_m.

6.7.2 The group D_n

The defining relations are

$$a^n = b^2 = (ab)^2 = e$$

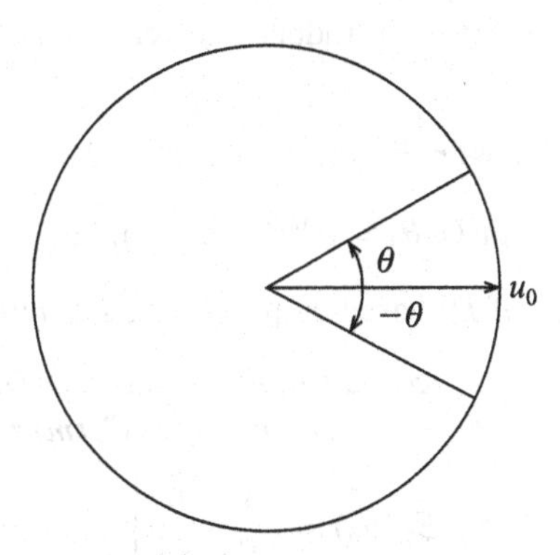

Figure 6.2. The polar angle θ in the x, y plane measured from an appropriately chosen axis u_0.

The direct matrix solutions of these equations are not very convenient in general, because D_n is not Abelian for $n \geqslant 3$. For one-dimensional representations, however, the direct matrix solutions are still effective, because the matrix representatives commute with each other; in fact, one-dimensional representations satisfy

$$D(a)^n = D(b)^2 = D(a)^2 D(b)^2 = 1$$

which yields

$$D(a) = \begin{cases} 1, & \text{if } n \text{ is odd} \\ \pm 1, & \text{if } n \text{ is even} \end{cases} \qquad D(b) = \pm 1$$

With the realizations $a = c_n = n_z$ and $b = u_0$, where u_0 is a binary axis in the x, y plane, these one-dimensional representations are denoted A_1, A_2 and B_1, B_2 and defined by

$$\begin{aligned} &A_1: \quad D(n_z) = 1, \quad D(u_0) = 1; \qquad A_2: \quad D(n_z) = 1, \quad D(u_0) = -1 \\ &B_1: \quad D(n_z) = -1, \quad D(u_0) = 1; \qquad B_2: \quad D(n_z) = -1, \quad D(u_0) = -1 \end{aligned} \tag{6.7.3}$$

where B_1 and B_2 occur only for an even n.

To determine the higher dimensional unirreps of D_n, we shall induce them from those of its subgroup C_n, using the left coset decomposition of D_n with respect to C_n

$$D_n = C_n + u_0 C_n; \qquad u_0 \perp n_z \in C_n$$

We apply the augmentor u_0 to the basis $\phi_m(\theta) = \mathrm{e}^{\mathrm{i}m\theta}$ of the unirrep $M_m(C_n)$ and obtain

$$\mathring{u}_0 \phi_m(\theta) = \phi_m(-\theta) = \mathrm{e}^{-\mathrm{i}m\theta} = \phi_{-m}(\theta)$$

measuring the angle θ from the u_0-axis (see Figure 6.2). This means that the basis $\phi_m(\theta)$ of $M_m(C_n)$ is connected to the basis $\phi_{-m}(\theta)$ of $M_{-m}(C_n)$ by the augmentor u_0. Accordingly, the combined set $[\phi_m(\theta), \phi_{-m}(\theta)]$ provides a basis of a two-dimensional representation E_m of D_n, which is irreducible if $M_m \neq M_{-m}$ for the obvious reason that two bases belonging to inequivalent unirreps of the subgroup C_n cannot be combined to form a basis of a unirrep of C_n. If $M_m = M_{-m}$, on the other hand, E_m becomes reducible because the augmentor u_0 simply transforms a basis function $\phi_m(\theta)$ to another basis function $\phi_{-m}(\theta)$ belonging to the same unirrep of C_n.

In any case, it is convenient to introduce the real bases of representation by the linear combinations

$$\begin{aligned} \cos(m\theta) &= (\mathrm{e}^{\mathrm{i}m\theta} + \mathrm{e}^{-\mathrm{i}m\theta})/2 \\ \sin(m\theta) &= (\mathrm{e}^{\mathrm{i}m\theta} - \mathrm{e}^{-\mathrm{i}m\theta})/(2\mathrm{i}) \end{aligned} \tag{6.7.4a}$$

Then the representation E_m of D_n based on $[\cos(m\theta), \sin(m\theta)]$ is given by

$$\begin{aligned} E_m(n_z) &= \begin{bmatrix} \cos(2\pi m/n) & -\sin(2\pi m/n) \\ \sin(2\pi m/n) & \cos(2\pi m/n) \end{bmatrix} \\ E_m(u_0) &= \begin{bmatrix} 1 & 0 \\ 0 & -1 \end{bmatrix} \end{aligned} \tag{6.7.4b}$$

which is irreducible if $M_m(C_n) \neq M_{-m}(C_n)$ as discussed above. If $M_m(C_n) =$

$M_{-m}(C_n)$ we have $\sin(2\pi m/n) = 0$ from (6.7.1) so that $m = 0$ or $n/2$, of which the latter is possible only if n is an even integer, for m being an integer; in either case, the two-dimensional representation becomes a direct sum of two one-dimensional unirreps given by (6.7.3) as follows:

$$E_0 = A_1 \oplus A_2, \qquad E_{n/2} = B_1 \oplus B_2 \tag{6.7.4c}$$

Now, for the general two-dimensional irreducible representation E_m, the basis space spanned by $[\cos(m\theta), \sin(m\theta)]$ is equivalent under $m \to -m$ so that E_m and E_{-m} are equivalent and $E_m = E_{m\pm n}$. Therefore, the two-dimensional unirreps E_m of D_n are limited, in view of (6.7.1), to

$$m = \begin{cases} 1, 2, \ldots, (n-1)/2 & \text{for an odd } n \\ 1, 2, \ldots, \dfrac{n}{2} - 1, & \text{for an even } n \end{cases} \tag{6.7.4d}$$

Once the representations of the generators n_z and u_0 have been determined, those of the remaining elements $c_n^k = n_z^k$ and $u_q = c_n^q u_0$ of D_n are obtained by direct matrix multiplications. These are given in Table 6.4. The completeness of these unirreps follows from the sums of the squares of the dimensions:

$$1^2 + 1^2 + 2^2(n-1)/2 = 2n = |D_n|; \qquad \text{for an odd } n$$

$$1^2 + 1^2 + 1^2 + 1^2 + 2^2\left(\frac{n}{2} - 1\right) = 2n = |D_n|; \qquad \text{for an even } n$$

Note also that the number of the unirreps equals $(n+3)/2$ for an odd n and $\frac{1}{2}n + 3$ for an even n. These are precisely the numbers of the classes of D_n given in Section 5.2, in accordance with (6.6.17).

6.7.2.1 Exercises

1. From the characters of the representations E_m of D_n given in Table 6.4, show that these are irreducible, satisfying (6.6.10).
2. From Table 6.4 write down the unirreps of the groups D_3 and D_4 explicitly and obtain Tables 6.5 and 6.6. Compare these with the character tables given in the Appendix.

6.7.3 The group T

From the defining relations

$$a^2 = b^3 = (ab)^3 = e$$

the one-dimensional representations satisfy

$$D(a)^2 = 1, \qquad D(a)^3 = 1, \qquad D(b)^3 = 1$$

which yields

$$D(a) = 1, \qquad D(b) = 1, \omega, \omega^* \tag{6.7.5}$$

where $\omega = \exp(-2\pi i/3)$ is a cubic root of unity. The three one-dimensional representations A, A' and A'' thus obtained are given in Table 6.7 with the realization $a = 2_z$, $b = 3_{xyz}$. To determine the higher dimensional irreducible representations of T, introduce the left coset decomposition of T by the subgroup D_2

$$T = D_2 + bD_2 + b^2 D_2; \qquad b = 3_{xyz}$$

Table 6.4. *The irreducible representations of D_n*

D_n	c_n^k	$u_q = c_n^q u_0$	Bases
A_1	1	1	$1, \cos(n\theta), z^2, x^2+y^2$
A_2	1	-1	$\sin(n\theta)$, z
B_1	$(-1)^k$	$(-1)^q$	$\cos(n_e\theta/2)$
B_2	$(-1)^k$	$(-1)^{q+1}$	$\sin(n_e\theta/2)$
E_m	$\begin{bmatrix} \cos(2\pi mk/n) & -\sin(2\pi mk/n) \\ \sin(2\pi mk/n) & \cos(2\pi mk/n) \end{bmatrix}$	$\begin{bmatrix} \cos(2\pi mq/n) & \sin(2\pi mq/n) \\ \sin(2\pi mq/n) & -\cos(2\pi mq/n) \end{bmatrix}$	$[\cos(m\theta), \sin(m\theta)]$

1. $k, q = 0, 1, \ldots, (n-1)$ and

$$m = \begin{cases} 1, 2, \ldots, (n-1)/2, & \text{when } n \text{ is odd} \\ 1, 2, \ldots, \dfrac{n}{2} - 1, & \text{when } n \text{ is even} \end{cases}$$

2. B_1 and B_2 occur only for an even integer $n = n_e$.
3. The angle θ is measured from the u_0-axis, which may be chosen parallel to the x- or y-axis of the coordinate system; for the u_q-axis see Figure 6.3.

Table 6.5. *The irreducible representations of D_3*

D_3	e	3_z	3_z^{-1}	u_0	u_1	u_{-1}	Bases
A_1	1	1	1	1	1	1	1
A_2	1	1	1	-1	-1	-1	z
E	$\begin{bmatrix} 1 & 0 \\ 0 & 1 \end{bmatrix}$	$\begin{bmatrix} -\frac{1}{2} & -\frac{\sqrt{3}}{2} \\ \frac{\sqrt{3}}{2} & -\frac{1}{2} \end{bmatrix}$	$\begin{bmatrix} -\frac{1}{2} & \frac{\sqrt{3}}{2} \\ -\frac{\sqrt{3}}{2} & -\frac{1}{2} \end{bmatrix}$	$\begin{bmatrix} 1 & 0 \\ 0 & -1 \end{bmatrix}$	$\begin{bmatrix} -\frac{1}{2} & \frac{\sqrt{3}}{2} \\ \frac{\sqrt{3}}{2} & \frac{1}{2} \end{bmatrix}$	$\begin{bmatrix} -\frac{1}{2} & -\frac{\sqrt{3}}{2} \\ -\frac{\sqrt{3}}{2} & \frac{1}{2} \end{bmatrix}$	$[\cos\theta, \sin\theta]$

Table 6.6. *The irreducible representations of D_4*

D_4	e	4_z	4_z^{-1}	2_z	u_0	u_2	u_1	u_{-1}	Bases
A_1	1	1	1	1	1	1	1	1	1
A_2	1	1	1	1	-1	-1	-1	-1	z
B_1	1	-1	-1	1	1	1	-1	-1	$\cos(2\theta)$
B_2	1	-1	-1	1	-1	-1	1	1	$\sin(2\theta)$
E	$\begin{bmatrix} 1 & 0 \\ 0 & 1 \end{bmatrix}$	$\begin{bmatrix} 0 & -1 \\ 1 & 0 \end{bmatrix}$	$\begin{bmatrix} 0 & 1 \\ -1 & 0 \end{bmatrix}$	$\begin{bmatrix} -1 & 0 \\ 0 & -1 \end{bmatrix}$	$\begin{bmatrix} 1 & 0 \\ 0 & 1 \end{bmatrix}$	$\begin{bmatrix} -1 & 0 \\ 0 & 1 \end{bmatrix}$	$\begin{bmatrix} 0 & 1 \\ 1 & 0 \end{bmatrix}$	$\begin{bmatrix} 0 & -1 \\ -1 & 0 \end{bmatrix}$	$[\cos\theta, \sin\theta]$

Then the augmentor 3_{xyz} connects the bases x, y and z of the three one-dimensional representations B_1, B_2 and B_3 of D_2 given in Table 6.1, as was shown by (6.4.3c). Thus we obtain a three-dimensional representation of T based on $[x, y, z]$, which must be irreducible because x, y and z are the bases of inequivalent unirreps of the subgroup D_2. Consequently, we obtain a complete set of four unirreps of T as given by Table 6.7,

Table 6.7. *The irreducible representations of T*

T	e	2_z	3_{xyz}	3_{xyz}^{-1}	Bases
A	1	1	1	1	$1, x^2+y^2+z^2, xyz$
A'	1	1	ω	ω^*	$u+\mathrm{i}v$
A''	1	1	ω^*	ω	$u-\mathrm{i}v$
T	$\begin{bmatrix}1&0&0\\0&1&0\\0&0&1\end{bmatrix}$	$\begin{bmatrix}-1&0&0\\0&-1&0\\0&0&1\end{bmatrix}$	$\begin{bmatrix}0&0&1\\1&0&0\\0&1&0\end{bmatrix}$	$\begin{bmatrix}0&1&0\\0&0&1\\1&0&0\end{bmatrix}$	$[x, y, z], [yz, zx, xy]$

1. $u = 3z^2 - r^2$, $v = 3^{1/2}(x^2 - y^2)$; $\omega = \exp(-2\pi\mathrm{i}/3)$.
2. See Table 11.7 for the general representations of the group T.

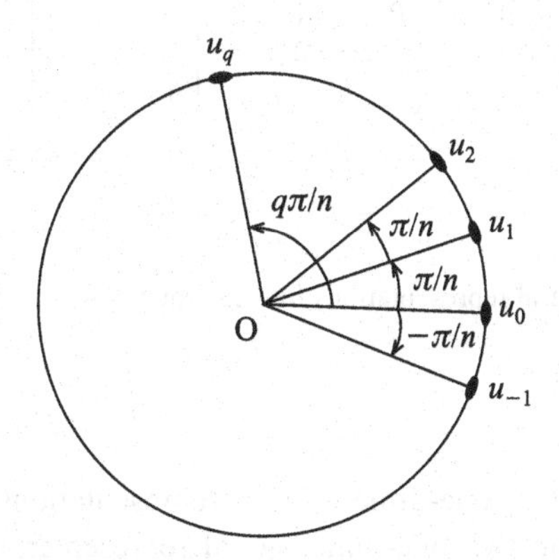

Figure 6.3. The binary axes of rotation u_q in D_n.

corresponding to the four classes of the group T. These also satisfy the completeness condition

$$1^2 + 1^2 + 1^2 + 3^2 = 12 = |T|$$

Table 6.7 contains the representatives of the elements which are required in order for one to write down the character table of the group T. The representatives of the remaining elements of T are obtained by using the multiplication table, Table 5.3, or by the general representations of T given by Table 11.7 later. Note also that the character of the three-dimensional representation denoted by T (the same symbol as that of the group T) satisfies the irreducibility condition

$$3^2 + 3(-1)^2 = 12 = |T|$$

The bases $u \pm \mathrm{i}v$ (where $u = 3z^2 - r^2$ and $v = 3^{1/2}(x^2 - y^2)$) of A' and A'' in Table 6.7 will be derived later in (8.4.16).

6.7.4 *The group O*

From the defining relations

$$a^4 = b^3 = (ab)^2 = e$$

Table 6.8. *The irreducible representations of O*

O	e	4_z	3_{xyz}	2_z	2_{yz}	Bases
A_1	1	1	1	1	1	$1, x^2 + y^2 + z^2$
A_2	1	-1	1	1	-1	xyz
E	$\begin{bmatrix} 1 & 0 \\ 0 & 1 \end{bmatrix}$	$\begin{bmatrix} 1 & 0 \\ 0 & -1 \end{bmatrix}$	$\begin{bmatrix} -\frac{1}{2} & -\frac{\sqrt{3}}{2} \\ \frac{\sqrt{3}}{2} & -\frac{1}{2} \end{bmatrix}$	$\begin{bmatrix} 1 & 0 \\ 0 & 1 \end{bmatrix}$	$\begin{bmatrix} -\frac{1}{2} & -\frac{\sqrt{3}}{2} \\ -\frac{\sqrt{3}}{2} & \frac{1}{2} \end{bmatrix}$	$[u, v]$
T_1	$\begin{bmatrix} 1 & 0 & 0 \\ 0 & 1 & 0 \\ 0 & 0 & 1 \end{bmatrix}$	$\begin{bmatrix} 0 & -1 & 0 \\ 1 & 0 & 0 \\ 0 & 0 & 1 \end{bmatrix}$	$\begin{bmatrix} 0 & 0 & 1 \\ 1 & 0 & 0 \\ 0 & 1 & 0 \end{bmatrix}$	$\begin{bmatrix} 0 & -1 & 0 \\ 1 & 0 & 0 \\ 0 & 0 & 1 \end{bmatrix}$	$\begin{bmatrix} -1 & 0 & 0 \\ 0 & 0 & 1 \\ 0 & 1 & 0 \end{bmatrix}$	$[x, y, z]$
T_2	$\begin{bmatrix} 1 & 0 & 0 \\ 0 & 1 & 0 \\ 0 & 0 & 1 \end{bmatrix}$	$\begin{bmatrix} 0 & 1 & 0 \\ -1 & 0 & 0 \\ 0 & 0 & -1 \end{bmatrix}$	$\begin{bmatrix} 0 & 0 & 1 \\ 1 & 0 & 0 \\ 0 & 1 & 0 \end{bmatrix}$	$\begin{bmatrix} 0 & -1 & 0 \\ 1 & 0 & 0 \\ 0 & 0 & 1 \end{bmatrix}$	$\begin{bmatrix} 1 & 0 & 0 \\ 0 & 0 & -1 \\ 0 & -1 & 0 \end{bmatrix}$	$[yz, zx, xy]$

1. $2_{yz} = 4_z 3_{xyz}$; $u = 2z^2 - r^2$, $v = 3^{1/2}(x^2 - y^2)$.
2. $T_2 = A_2 \times T_1$
3. See Table 11.6 for the general representations for the group O.

the allowed solutions for one-dimensional representations are $D(a) = \pm 1$ and $D(b) = 1$. Thus we obtain two one-dimensional representations A_1 and A_2 given in Table 6.8 with the realization $a = 4_z$, $b = 3_{xyz}$. The higher dimensional unirreps of O will be induced from the unirreps of the subgroup T using the left coset decomposition

$$O = T + 4_z T$$

We may apply the generators 4_z and 3_{xyz} of O to the basis $[u, v]$ obtained from the linear combinations of the bases $u \pm iv$ of the subgroup T and obtain the two-dimensional representation E of O given in Table 6.8. The three-dimensional unirrep T_1 of O is generated by the irreducible basis $[x, y, z]$ of the group T. There exists one more three-dimensional unirrep defined by the direct product $T_2 = A_2 \times T_1$, which is irreducible because A_2 is one-dimensional and T_1 is irreducible. Thus we obtain the five unirreps of O given by Table 6.8. These form a complete set, because the number of the classes of O is also equal to five. Table 6.8 also contains the representatives of 2_z and $2_{yz} = 4_z \times 3_{xyz}$ besides those of the generators because the characters of the five classes of O may be determined by those of e, 4_z, 2_z, 3_{xyz} and 2_{yz}. The completeness of those unirreps of O may also be seen via their dimensions by

$$1^2 + 1^2 + 2^2 + 3^2 + 3^2 = 24 = |O|$$

The irreducibility condition for T_1 is satisfied because the character satisfies

$$3^2 + 6 \times 1^2 + 3 \times 1^2 + 6 \times 1^2 = 24 = |O|$$

6.7.5 The improper point groups

The irreducible representations of C_{np}, D_{np}, C_{nv} and T_p are determined from those of the proper point groups through the isomorphism described in Section 5.5.2:

$$C_{np} \simeq C_{2n}, \qquad D_{np} \simeq D_{2n} \qquad \text{via } \overline{c}_{2n} \leftrightarrow c_{2n}$$
$$C_{nv} \simeq D_n \qquad \text{via } \overline{u}_q \leftrightarrow u_q$$
$$T_p \simeq O \qquad \text{via } \overline{4}_z \leftrightarrow 4_z$$

Moreover, the irreducible representations of the rotation–inversion groups $P_{\mathrm{i}} = (C_{n\mathrm{i}}, D_{n\mathrm{i}}, T_{\mathrm{i}}, O_{\mathrm{i}})$ are determined by the direct product representations of the proper point groups P and the group of inversion C_{i}. See (6.2.4), for example. The character tables of the crystallographic point group based on the unirreps derived in this section are given in the Appendix.

6.8 Properties of irreducible bases

Let $D^{(\alpha)}(G) = \{D^{(\alpha)}(g);\ g \in G\}$ be a d_α-dimensional unirrep of a transformation group $G = \{R\}$. Then a basis $\{\psi_i^\alpha;\ i = 1, 2, \ldots, d_\alpha\}$ of the unirrep $D^{(\alpha)}(G)$ transforms according to

$$\mathring{R}\psi_i^\alpha = \sum_{j=1}^{d_\alpha} \psi_j^\alpha D_{ji}^{(\alpha)}(R); \qquad i = 1, 2, \ldots, d_\alpha \tag{6.8.1a}$$

for all $R \in G$. Then a basis function ψ_i^α is said to belong to the ith row of the unirrep $D^{(\alpha)}(G)$ and the remaining functions in the basis are called the *partner functions* of ψ_i^α. This statement is well defined only if $D^{(\alpha)}(G)$ is specified completely (not just up to a similarity transformation) because once one of its basis functions has been given, all its partners are completely determined by the unirrep $D^{(\alpha)}(G)$. To see this, we solve (6.8.1a) for ψ_j^α, using the orthogonality theorem on the unirreps (6.5.11a) in the group space, and obtain *the fundamental equation for the partners* of any given basis function ψ_m^α:

$$\psi_j^\alpha = (d_\alpha/|G|) \sum_{R \in G} D_{jm}^{(\alpha)}(R)^* \mathring{R}\psi_m^\alpha; \qquad j = 1, 2, \ldots, d_\alpha \tag{6.8.1b}$$

It should be noted, however, that infinitely many functions may belong to the same row of the same unirrep. This fact should be kept in mind whenever we talk about the classification of the basis functions. Frequently, a set of functions that satisfies (6.8.1a) or (6.8.1b) is called a set of *symmetry-adapted functions* belonging to the unirrep $D^{(\alpha)}(G)$.

We shall first discuss the orthogonality of the basis functions using (6.8.1a), and then discuss how to generate a basis of $D^{(\alpha)}(G)$ from any given function F in the basis space of G under consideration with a slight extension of (6.8.1b).

6.8.1 The orthogonality of basis functions

Two functions ψ_i^α and ψ_j^β are orthogonal if they belong to different unirreps or to different rows of the same unirrep of a group G. From the unitary nature of the operators $\mathring{R}$ of the group, we have

$$\langle \psi_i^\alpha, \phi_j^\beta \rangle = \langle \mathring{R}\psi_i^\alpha, \mathring{R}\phi_j^\beta \rangle$$
$$= \sum_{i',j'} D_{i'i}^{(\alpha)}(R)^* D_{j'j}^{(\beta)}(R) \langle \psi_{i'}^\alpha, \phi_{j'}^\beta \rangle$$

then, averaging these over all $R \in G$ (summing over R and dividing by $|G|$) we obtain, using the orthogonal theorem (6.5.11b),

$$\langle \psi_i^\alpha, \phi_j^\beta \rangle = (1/d_\alpha)\delta_{\alpha\beta}\delta_{ij} \sum_{i'} \langle \psi_{i'}^\alpha, \phi_{i'}^\alpha \rangle \tag{6.8.2a}$$

This implies firstly that $\langle \psi_i^\alpha, \phi_j^\beta \rangle$ vanishes for $\alpha \neq \beta$ or $i \neq j$ and secondly that the diagonal elements $\langle \psi_i^\alpha, \phi_i^\alpha \rangle$ are all equal and hence independent of i. Thus, if one of the basis functions ψ_i^α is not null, then each of its partners is not null.

Suppose further that there exists a symmetry operator S that is invariant with respect to G such that $\mathring{R}S = S\mathring{R}$ for all $R \in G$. Then from the unitarity of $\mathring{R}$ applied to the matrix representative $M(S) = \|\langle \psi_i^\alpha | S | \phi_j^\beta \rangle\|$ of S, we have

$$\langle \psi_i^\alpha | S | \phi_j^\beta \rangle = \sum_{R \in G} \langle \mathring{R}\psi_i^\alpha | S | \mathring{R}\phi_j^\beta \rangle / |G|$$
$$= \delta_{\alpha\beta}\delta_{ij} d_\alpha^{-1} \sum_k \langle \psi_k^\alpha | S | \phi_k^\alpha \rangle \tag{6.8.2b}$$

analogous to the scalar product (6.8.2a). Thus the matrix representative $M(S)$ for $\alpha = \beta$ is a constant matrix. This is consistent with Schur's lemma.

6.8.2 *Application to perturbation theory*

Consider the symmetry group of G of a Hamiltonian H_0 for a quantum mechanical system. Then, H_0 is invariant with respect to the group of transformations $G = \{\mathring{R}\}$ such that

$$\mathring{R}H_0\mathring{R}^{-1} = H_0 \quad \text{or } \mathring{R}H_0 = H_0\mathring{R}; \quad \forall R \in G \tag{6.8.3a}$$

Let ψ_k^Δ $(k = 1, 2, \ldots, d_\Delta)$ be a set of degenerate eigenfunctions of H_0 belonging to an energy E_Δ,

$$H_0\psi_k^\Delta = E_\Delta\psi_k^\Delta; \qquad k = 1, 2, \ldots, d_\Delta \tag{6.8.3b}$$

Then $\mathring{R}\psi_k^\Delta$ are also eigenfunctions of H_0 belonging to the same energy E_Δ, because

$$H_0(\mathring{R}\psi_k^\Delta) = \mathring{R}H_0\psi_k^\Delta = E_\alpha(\mathring{R}\psi_k^\Delta)$$

Accordingly, $\mathring{R}\psi_k^\Delta$ can be expanded by the degenerate set $\{\psi_k^\Delta\}$:

$$\mathring{R}\psi_k^\Delta = \sum_{k'} \psi_{k'}^\Delta \Delta_{k'k}(R); \qquad k = 1, 2, \ldots, d_\Delta$$

for all $\mathring{R} \in G$. Since G is the symmetry group of the Hamiltonian H_0, there is no symmetry reason that the representation $\Delta(G)$ should be reducible; in fact, $\Delta(G)$ is irreducible for almost all cases. If this is not the case we say that it is *accidentally degenerate* owing to some hitherto unrecognized symmetries; it is rare but it happens.[1]

[1] A well-known example is the case of the hydrogen atom, in which all the excited energy levels are accidentally degenerate with respect to the rotation group $O(3, r)$. This degeneracy has been removed by recognizing that the symmetry group of the hydrogen atom is $O(4, r)$ $(> O(3, r))$.

Thus, the degeneracy of an energy level E_Δ is determined by the dimensions of the irreducible representations of the symmetry group G (unless it is accidentally degenerate). Thus, for example, if the symmetry group H_0 is the dihedral group D_2, there would be no degeneracy; if the symmetry group G is D_n $(n \geqslant 3)$, the degeneracy is at most 2 from Table 6.4 of the irreducible representations of D_n.

Now we introduce a perturbation V to the system; then, the Hamiltonian is given by $H = H_0 + V$. Suppose that the perturbation V is invariant with respect to the symmetry group G of H_0, and also that there exists no accidental degeneracy for the representation $\Delta(G)$, then the matrix representation of the perturbation V based on $\{\psi_k^\Delta\}$ is a constant matrix according to (6.8.2b). Thus there will be a shift in energy given by $\langle\psi_k^\Delta|V|\psi_k^\Delta\rangle$ but no splitting. Following Wigner (1962), we state that, 'under a symmetric pertubation, an eigenvalue with an irreducible representation retains its representation and cannot split.' In this case, the basis $\{\psi_k^\Delta\}$ is already a set of correct linear combinations for the zeroth-order eigenfunctions.

On the other hand, if there exists accidental degeneracy for the energy E_Δ, then the $d_\Delta \times d_\Delta$ representation $\Delta(G)$ is reducible such that

$$\Delta(G) \sim \sum_\alpha \oplus\, n_\alpha D^{(\alpha)}(g); \qquad \forall\, g \in G \tag{6.8.4}$$

where n_α is the number of times a unirrep $D^{(\alpha)}(G)$ is contained in $\Delta(G)$ so that

$$d_\Delta = \sum_\alpha n_\alpha d_\alpha$$

Thus, the symmetry perturbation V may lift the accidental degeneracy of the eigenvalue E_Δ such that there will be n_α eigenvalues corresponding to n_α unirreps $D^{(\alpha)}(G)$ and these eigenvalues may be different. Accordingly, the zeroth-order energy E_Δ may split into a total of N_Δ different energies such that

$$N_\Delta = \sum_\alpha n_\alpha \tag{6.8.5}$$

To calculate the splittings of the energy due to the perturbation V, let us classify the zeroth-order eigenfunctions of the energy E_Δ in terms of the basis sets of the unirreps contained in $\Delta(G)$. Postponing how to construct these bases from the zeroth-order eigenfunctions $\{\psi_k^\Delta\}$ to the next section, let $\{\psi_{j\rho}^\alpha;\ \rho = 1, 2, \ldots, n_\alpha\}$ be the n_α bases belonging to the jth row of $D^{(\alpha)}(G)$ constructed from the set $\{\psi_k^\Delta\}$. Then, in view of (6.8.2b), the matrix elements of the perturbation V are given by

$$\langle\psi_{i\rho}^\alpha|V|\psi_{j\rho'}^\beta\rangle = \delta_{\alpha\beta}\delta_{ij}V_{\rho\rho'}^{(\alpha)}; \qquad i, j = 1, 2, \ldots, d_\alpha; \qquad \rho, \rho' = 1, 2, \ldots, n_\alpha \tag{6.8.6}$$

Accordingly the secular determinant is factorized into a product of lower order secular determinants, of which a factor belonging to the unirrep $D^{(\alpha)}(G)$ is given by an $n_\alpha \times n_\alpha$ secular determinant

$$\det|V_{\rho\rho'}^{(\alpha)} - \Delta E^{(\alpha)}\mathbf{1}| = 0; \qquad \rho, \rho' = 1, 2, \ldots, n_\alpha \tag{6.8.7}$$

The n_α roots of this equation determine the splitting $\Delta E^{(\alpha)}$ of the eigenvalue E_Δ. The correct zeroth-order linear combinations of $\{\psi_k^\Delta\}$ are given by the linear combinations of the subset $\{\psi_{j\rho}^\alpha;\ \rho = 1, 2, \ldots, n_\alpha\}$ for each α and j and the linear coefficients are independent of j.

If the perturbation V has less symmetry than H_0, the energy eigenvalues are classified by the symmetry group of the total Hamiltonian $H = H_0 + V$, which must be a subgroup of the symmetry group G of H_0.

6.9 Symmetry-adapted functions

6.9.1 Generating operators

The problem is to construct a basis $\{\psi_i^\alpha\}$ of a unirrep $D^{(\alpha)}(G)$ of a transformation group $G = \{\mathring{R}\}$ from a given set of functions $\{F(x)\}$ in the Hilbert space to which the symmetry operators $\mathring{R}$ of G can be applied. Such a set of functions may be referred to as a set of functions in a basis space of the group G. In the perturbation theory in quantum mechanics, such a set is provided by the zeroth-order wave functions of the unperturbed Hamiltonian H_0 belonging to an eigenvalue E_Δ. In constructing approximate molecular orbitals (MOs) of a molecule from linear combinations of the atomic orbitals (LCAOs), such a set is provided by a set of equivalent atomic orbits of the molecule with respect to the symmetry group G of the molecule. There exist two approaches to the problem: one is based on *the generating operators* (or projection operators); the other is based on *the correspondence theorem* in the basis space of a point group. We shall discuss the former in this section, postponing the latter to the next chapter.

To begin with, we simply rewrite the fundamental set of equations (6.8.1b) for a partner in the following suggestive form:

$$\psi_j^\alpha = \mathring{P}_{jm}^{(\alpha)} \psi_m^\alpha; \qquad j = 1, 2, \ldots, d_\alpha \tag{6.9.1a}$$

where the operator set $\{\mathring{P}_{jm}^{(\alpha)}\}$ for a given m is defined by

$$\mathring{P}_{jm}^{(\alpha)} = (d_\alpha/|G|) \sum_{R \in G} D_{jm}^{(\alpha)}(R)^* \mathring{R}; \qquad j = 1, 2, \ldots, d_\alpha \tag{6.9.1b}$$

Application of this set on a basis function $\psi_i^\beta \in D^{(\beta)}(G)$ yields, via the orthogonality relations of the unirreps in the group space,

$$\mathring{P}_{jm}^{(\alpha)} \psi_i^\beta = \delta_{\alpha\beta} \delta_{mi} \psi_j^\beta; \qquad j = 1, 2, \ldots, d_\alpha \tag{6.9.1c}$$

that is, the oeprator set $\{\mathring{P}_{jm}^{(\alpha)}\}$ for a given m transforms a basis function ψ_m^α belonging to the mth row of $D^{(\alpha)}(G)$ into its partner set, but annihilates all other basis functions of all unirreps of G.

Now, any given function F in the basis space of G must be expressed by a linear combination of all basis functions of a complete set of the unirreps of G:

$$F = \sum_{\beta=1}^{c} \sum_{i=1}^{d_\beta} \psi_i^\beta \tag{6.9.2}$$

where c is the total number of all unirreps of G. Here the linear coefficients are included in the basis functions and some of the basis functions could be null. It is stressed here that two different basis functions ψ_m^α and $\psi_{m'}^\alpha$ $(m' \neq m)$ in the expansion need not belong to the same partner set of $D^{(\alpha)}(G)$, because it is quite possible that the partner set of ψ_m^α is linearly independent from the partner set of $\psi_{m'}^\alpha$; for example, they may differ by a scalar factor.

On applying the operator set $\{\mathring{P}^{(\alpha)}_{im}\}$ for a given m on the function F, we obtain *a set of partners* $\{\psi^\alpha_j\}$ of ψ^α_m, in view of (6.9.1c) and (6.9.2),

$$\begin{aligned} \psi^\alpha_j &= \mathring{P}^{(\alpha)}_{jm} F \\ &= (d_\alpha/|G|) \sum_{R \in G} D^{(\alpha)}_{jm}(R)^* \mathring{R} F; \qquad j = 1, 2, \ldots, d_\alpha \end{aligned} \tag{6.9.3}$$

The set would be null, if ψ^α_m were not contained in F, in view of (6.9.1c). For this reason, a proper choice of m has to be made for a given F. Obviously, a different choice of m, say m', for a given F may give different bases of $D^{(\alpha)}(G)$ because ψ^α_m and $\psi^\alpha_{m'}$ in the sum (6.9.2) need not belong to the same partner set. With different choices of F and different m for a given F we may obtain a variety of bases belonging to a unirrep $D^{(\alpha)}$ of G. The operator set $\{\mathring{P}^{(\alpha)}_{jm}\}$ may be called the *generator set for the partners of a basis function* ψ^α_m.

The structure of the above set of functions given by (6.9.3) is interesting; it is *a set of symmetry-adapted linear combinations (SALCs)* of the equivalent functions $\{\mathring{R}F;\ \forall R \in G\}$. In general, the equivalent set $\{\mathring{R}F\}$ need not be linearly independent. Let $\{F_s;\ s = 1, 2, \ldots, n\}$ be a set of n linearly independent functions contained in the set $\{\mathring{R}F\}$ and let $\Delta(G)$ be the $n \times n$ matrix representation of G based on the set $\{F_s\}$. Then the frequency n_α of $D^{(\alpha)}(G)$ contained in the representation $\Delta(G)$ is given by the frequency rule

$$n_\alpha = \frac{1}{|G|} \sum_{R \in G} \chi^{(\alpha)}(R)^* \chi^{(\Delta)}(R) \tag{6.9.4a}$$

where $\chi^{(\alpha)}(R)$ and $\chi^{(\Delta)}(R)$ are the characters of $D^{(\alpha)}(R)$ and $\Delta(R)$, respectively. The frequencies satisfy

$$\sum_\alpha n_\alpha d_\alpha = n, \qquad n_\alpha \leqslant d_\alpha \tag{6.9.4b}$$

The frequency n_α gives the number of linearly independent basis sets of $D^{(\alpha)}(G)$ contained in F.

Equation (6.9.3) gives a general method of constructing a basis of any given irreducible representation $D^{(\alpha)}(G)$ starting from a function F suitably chosen in the basis space of G under consideration. For example, in the perturbation theory in quantum mechanics, F is chosen from the zeroth-order eigenfunctions. With particular choices of the initial function F, many kinds of irreducible bases will be formed with proper choices of m. The method is known as *the generating operator method.* The method is particularly useful when the equivalent set of functions $\{\mathring{R}F\}$ is given to begin with. This is the case, for example, when we construct the molecular orbitals (MOs) from linear combinations of atomic orbitals (LCAOs).

Example 1. Consider a set of free-electron wave functions

$$F_1 = e^{i2\pi x}, \qquad F_2 = e^{i2\pi x}, \qquad F_3 = e^{i2\pi y}, \qquad F_4 = e^{-i2\pi y} \tag{6.9.5}$$

belonging to an energy $E = h^2/(2m)$ of the free electron, where h is Planck's constant and m is the mass of the electron. The above set of functions $\{F_s\}$ is equivalent with respect to the point group C_{4v}. The problem is to determine the correct zeroth-order wave functions by using the SALCs for the point group C_{4v}. This kind of problem arises in the theory of band energies for an electron in a metal (see Section 15.2).

Table 6.9. *The irreducible representations of C_{4v}*

C_{4v}	e	4_z	4_z^{-1}	2_z	$\overline{2}_x$	$\overline{2}_y$	$\overline{2}_{xy}$	$\overline{2}_{\overline{xy}}$	Bases
A_1	1	1	1	1	1	1	1	1	$1, z$
A_2	1	1	1	1	−1	−1	−1	−1	$xy \times (x^2 - y^2)$
B_1	1	−1	−1	1	1	1	−1	−1	$x^2 - y^2$
B_2	1	−1	−1	1	−1	−1	1	1	xy
E	$\begin{bmatrix} 1 & 0 \\ 0 & 1 \end{bmatrix}$	$\begin{bmatrix} 0 & -1 \\ 1 & 0 \end{bmatrix}$	$\begin{bmatrix} 0 & 1 \\ -1 & 0 \end{bmatrix}$	$\begin{bmatrix} -1 & 0 \\ 0 & -1 \end{bmatrix}$	$\begin{bmatrix} -1 & 0 \\ 0 & 1 \end{bmatrix}$	$\begin{bmatrix} 1 & 0 \\ 0 & -1 \end{bmatrix}$	$\begin{bmatrix} 0 & -1 \\ -1 & 0 \end{bmatrix}$	$\begin{bmatrix} 0 & 1 \\ 0 & 0 \end{bmatrix}$	$[x, y]$
$\mathring{R}F_1$	F_1	F_3	F_4	F_2	F_2	F_1	F_4	F_3	
$\chi^{(\Delta)}$	4	0	0	0	2	2	0	0	

From Table 6.9 for the unirreps of C_{4v} and the character of the representation Δ based on $\{F_s\}$, the unirreps of C_{4v} contained in Δ are given by

$$\Delta = \underset{1}{A_1} + \underset{x^2-y^2}{B_1} + \underset{[x,\, y]}{E} \tag{6.9.6}$$

where we have characterized each unirrep contained in Δ by their elementary bases.

Let us choose F_1 in (6.9.5) for F in (6.9.3). Then, using the equivalent set $\{\mathring{R}F_1; \forall R \in C_{4v}\}$ given in Table 6.9, the symmetry-adapted linear combinations for C_{4v} are given by, with $m = 1$ in (6.9.3),

$$A_1: \quad \psi_1 = \mathring{P}_{11}^{A_1} F_1 = \tfrac{1}{4}(F_1 + F_2 + F_3 + F_4) = \tfrac{1}{2}[\cos(2\pi x) + \cos(2\pi y)]$$

$$B_1: \quad \psi_{x^2-y^2} = \mathring{P}_{11}^{B_1} F_1 = \tfrac{1}{4}(F_1 + F_2 - F_3 - F_4) = \tfrac{1}{2}[\cos(2\pi x) - \cos(2\pi y)]$$

$$E: \quad \psi_1^E = \psi_x = \mathring{P}_{11}^{E} F_1 = \tfrac{1}{2}(F_1 - F_2) = \mathrm{i}\sin(2\pi x)$$

$$\psi_2^E = \psi_y = \mathring{P}_{21}^{E} F_1 = \tfrac{1}{2}(F_3 - F_4) = \mathrm{i}\sin(2\pi y) \tag{6.9.7a}$$

Note that the choice $m = 2$ in (6.9.3) gives the null result for the unirrep E:

$$\mathring{P}_{12}^{E} F_1 = 0; \qquad \mathring{P}_{22}^{E} F_1 = 0 \tag{6.9.7b}$$

This is because $F_1 = \exp(\mathrm{i}2\pi x)$ does not contain $\psi_2^E \propto \sin(2\pi y)$.

Note also that ψ_1 given in (6.9.7a) is an invariant function of C_{4v} belonging to the identity representation A_1, whereas $\psi_{x^2-y^2} \in B_1$ and $(\psi_x, \psi_y) \in E$ transform like the elementary bases $x^2 - y^2$ and $[x, y]$, respectively. If we use the fact that the set of differential opertors $(\partial/\partial x, \partial/\partial y, \partial/\partial z)$ transforms like the set (x, y, z) under an orthogonal transformation, then the last two bases may be obtained simply, from the invariant basis ψ_1, by

$$B_1: \quad \psi_{x^2-y^2} \sim (\partial_x^2 - \partial_y^2)\psi_1 = 2\pi^2[\cos(2\pi x) + \cos(2\pi y)] \tag{6.9.8a}$$

$$E: \quad \psi_x \sim \partial_x \psi_1 = -\pi \sin(2\pi x)$$

$$\psi_y \sim \partial_y \psi_1 = -\pi \sin(2\pi y) \tag{6.9.8b}$$

where $(\partial_x, \partial_y) = (\partial/\partial x, \partial/\partial y)$.

The correspondence between a SALC and the corresponding elementary basis belonging to the same irreducible representation used above will be discussed further in the correspondence theorem developed later in Section 7.2.

6.9.2 The projection operators

In the special case in which $m = j$, Equation (6.9.3) takes the form

$$\psi_m^\alpha = \mathring{P}_{mm}^{(\alpha)} F \tag{6.9.9}$$

which also can be used to calculate the basis functions taking $m = 1, 2, \ldots, d_\alpha$, provided that none of them is null. Substitution of this and (6.9.3) into (6.9.1a) leads to

$$\mathring{P}_{jm}^{(\alpha)} \mathring{P}_{mm}^{(\alpha)} = \mathring{P}_{jm}^{(\alpha)} \tag{6.9.10a}$$

which reduces to the idempotent relation, for $j = m$ (cf. (1.4.12a)),

$$[\mathring{P}_{mm}^{(\alpha)}]^2 = \mathring{P}_{mm}^{(\alpha)} \qquad (6.9.10\text{b})^2$$

Accordingly, the operator $\mathring{P}_{mm}^{(\alpha)}$ is a projection operator that projects out the basis function ψ_m^α belonging to the mth row of $D^{(\alpha)}(G)$ from an arbitrary function F in the basis space of G. Its repeated applications do not bring out anything new since it satisfies the idempotent relation $x^2 = x$. Its eigenvalues are given by the roots of $x^2 = x$, which are 1 and zero. The corresponding eigenvectors are also easily determined: we simply specialize (6.9.1c) for $j = m$ to obtain

$$\mathring{P}_{mm}^{(\alpha)} \psi_i^\beta = \delta_{\alpha\beta} \delta_{mi} \psi_i^\beta; \qquad m = 1, 2, \ldots, d_\alpha; \qquad i = 1, 2, \ldots, d_\beta;$$
$$\alpha, \beta = 1, 2, \ldots, c \tag{6.9.11}$$

where the eigenvalue $\delta_{\alpha\beta}\delta_{mi}$ equals unity if and only if $\alpha = \beta$ and $m = i$. Thus we have that ψ_m^α is the only eigenvector of $\mathring{P}_{mm}^{(\alpha)}$ belonging to the eigenvalue 1 and all the remaining basis functions in the basis space of G belong to the zero eigenvalue.

Next, substitution of (6.9.9) into (6.9.11) leads to the orthogonality relations

$$\mathring{P}_{mm}^{(\alpha)} \mathring{P}_{ii}^{(\beta)} = \delta_{\alpha\beta} \delta_{mi} \mathring{R}_{ii}^{(\beta)}; \qquad m = 1, 2, \ldots, d_\alpha; \qquad i = 1, 2, \ldots, d_\beta;$$
$$\alpha, \beta = 1, 2, \ldots, c \tag{6.9.12}$$

whereas its substitution into (6.9.2) leads to the completeness relation

$$\sum_{\beta=1}^{c} \sum_{i=1}^{d_\beta} \mathring{P}_{ii}^{(\alpha)} = \mathring{e} \tag{6.9.13}$$

where $\mathring{e}$ is the unit operator of G. These are expected relations for a set of projection operators. The completeness relation is often called the *spectral decomposition* of the unit element. Application of this relation to a given function F in the basis space brings back the expansion of F given by (6.9.2).

6.9.2.1 Concluding remarks

We have given above two alternative ways of forming a basis $\{\psi_j^\alpha\}$ of a unirrep $D^{(\alpha)}(G)$ from a given function F in the basis space of G: either by using a set of

[2] The generating operator $P_{jm}^{(\alpha)}$ is not a projection operator since $[P_{jm}^{(\alpha)}]^2 = \delta_{jm} P_{jm}^{(\alpha)}$ unless $j = m$.

generating operators based on (6.9.3) with *a proper choice* of m; or by using the projection operators based on (6.9.9) for all $m = 1, 2, \ldots, d_\alpha$, where each projection operator $\mathring{R}^{(\alpha)}_{mm}$ of different m acts on F quite independently from each other. In the former, it follows from (6.9.1a) that the set $\{\psi^\alpha_j\}$ thus formed is a set of partners belonging to the unirrep $D^{(\alpha)}(G)$, unless it is null for the choice of m. In the latter, however, the set $\{\psi^\alpha_m\}$ requires that $\mathring{P}^{(\alpha)}_{mm}F \neq 0$ for every m; even then, the set thus formed need not be a set of partners belonging to the unirreps $D^{(\alpha)}(G)$. We may state that the projection operator method based on (6.9.9) is less effective than the generating operator method based on (6.9.3). This point is stressed here, because frequently people seem to prefer the projection operator method over the generating operator method, probably owing to the beauty of the spectral theorem (6.9.13). See the following examples.

Example 2. Recalculate the basis $[\psi^E_1, \psi^E_2]$ belonging to the unirrep E of $C_{4\mathrm{v}}$ introduced in Example 1 using the projection operator method.

We choose F_1 in (6.9.5) for F in (6.9.9) just like in Example 1. Then, from (6.9.9) we have, for the unirrep E

$$\mathring{P}^E_{11}F_1 = \tfrac{1}{2}(F_1 - F_2) = \mathrm{i}\sin(2\pi x)$$
$$\mathring{P}^E_{22}F_1 = \tfrac{2}{8}(F_1 - F_2 + F_2 - F_1) = 0 \qquad (6.9.14a)$$

The null result is due to the fact that ψ^E_2 is not contained in F_1. To obtain a non-null result for E, we may choose $F = F_1 + cF_3$ where c is a scalar factor that is invariant under rotation. Then

$$\mathring{P}^E_{11}[F_1 - cF_3] = \tfrac{1}{2}(F_1 - F_2) = \mathrm{i}\sin(2\pi x)$$
$$\mathring{P}^E_{22}[F_1 + cF_3] = \frac{c}{2}(F_3 - F_4) = c\,\mathrm{i}\sin(2\pi y) \qquad (6.9.14b)$$

These form a partner set of the unirrep E if $c = 1$, in accordance with (6.9.7a).

Example 3. Consider the group of inversion $C_\mathrm{i} = \{e, I\}$, where I is the inversion. The projection operators are

$$P^{(1)} = \tfrac{1}{2}(e + I), \qquad P^{(2)} = \tfrac{1}{2}(e - I)$$

and the symmetry-adapted functions are given by

$$\mathring{P}^{(1)}F(x) = \tfrac{1}{2}(F(x) + F(-x)) = F_\mathrm{e}(x)$$
$$\mathring{P}^{(2)}F(x) = \tfrac{1}{2}(F(x) - F(-x)) = F_\mathrm{o}(x) \qquad (6.9.15a)$$

where $F_\mathrm{e}(x)$ ($F_\mathrm{o}(x)$) is an even (odd) function of x. A function $F(x)$ is expanded, in accordance with (6.9.2), to

$$F(x) = F_\mathrm{e}(x) + F_\mathrm{o}(x) \qquad (6.9.15b)$$

Exercise. Let $\mathring{P}^{(\alpha)\dagger}_{jm}$ be the adjoint operator of $\mathring{P}^{(\alpha)}_{jm}$. Show via the unitarity of the operator $\mathring{R}$ that

$$\mathring{P}^{(\alpha)\dagger}_{jm} = \mathring{P}^{(\alpha)}_{jm}; \qquad \mathring{P}^{(\alpha)\dagger}_{jm}\mathring{P}^{(\beta)}_{j'm'} = \delta_{\alpha\beta}\delta_{jj'}\mathring{P}^{(\alpha)}_{mm'}$$

Using these relations one can also show the orthogonality relations (6.8.2a) for the irreducible basis functions.

6.9.2.2 The projection operators based on the characters

The method of constructing the irreducible bases via (6.9.3) is useless unless the representation is known. In the following we shall discuss the method of construction based on the characters of G. We set $j = m$ in (6.9.1b) and sum over m to obtain another projection operator:

$$\mathring{P}^{(\alpha)} = \sum_m \mathring{P}^{(\alpha)}_{mm} = (d_\alpha/|G|)\sum_R \chi^{(\alpha)}(R)^* \mathring{R} \tag{6.9.16a}$$

which depends only on the character $\chi^{(\alpha)}(G)$ of the unirrep $D^{(\alpha)}(G)$. From (6.9.12) and (6.9.13) these satisfy the projective relations and the completeness relation

$$\mathring{P}^{(\alpha)}\mathring{P}^{(\beta)} = \delta_{\alpha\beta}\mathring{P}^{(\beta)}, \qquad \sum_\alpha \mathring{P}^{(\alpha)} = 1 \tag{6.9.16b}$$

These operators are not affected by the similarity transformation of the unirreps, differently from the previous operators $\mathring{P}^{(\alpha)}_{mm}$.

The eigenvalue problem (6.9.11) is reduced to

$$\mathring{P}^{(\alpha)}\psi_i^\beta = \delta_{\alpha\beta}\psi_i^\beta$$

which holds for any allowed i. Hence, any linear combination $\psi^\beta = \sum_i \psi_i^\beta$ satisfies

$$\mathring{P}^{(\alpha)}\psi^\beta = \delta_{\alpha\beta}\psi^\beta \tag{6.9.17}$$

where $\beta = 1, 2, \ldots, c$. Here again ψ^α is only one eigenvector that belongs to the eigenvalue unity of $\mathring{P}^{(\alpha)}$ whereas all the remaining ψ^β $(\beta \neq \alpha)$ belong to the zero eigenvalue. The function ψ^α which satisfies $\mathring{P}^{(\alpha)}\psi^\alpha = \psi^\alpha$ is said to belong to the unirrep $D^{(\alpha)}(G)$. This fact, like the characters, is independent of the specific form of the representation. Moreover, by summing (6.9.9) over m, we obtain

$$\psi^\alpha = \mathring{P}^{(\alpha)}F \tag{6.9.18}$$

This is also useful for obtaining the function ψ^α belonging to the unirrep $D^{(\alpha)}(G)$ from a given function F with knowledge merely of the character $\chi^{(\alpha)}(G)$. A shortcoming of (6.9.18) is that one has to form at least d_α linear independent F in order to obtain a set of functions that spans the d_α-dimensional basis space of $D^{(\alpha)}$, which is a laborious procedure unless $d_\alpha = 1$. In the special case of the identity representation we have

$$\psi^1 = \mathring{P}^{(1)}F = (1/|G|)\sum_R \mathring{R}F \tag{6.9.19}$$

which is called *an invariant function* with respect to the group G. It plays the crucial role in the theory of 'selection rules' which will be discussed in the next section. It also provides an invariant basis for a point group G, from which the remaining SALCs are constructed via the correspondence theorem, as was exemplified by (6.9.8a) and (6.9.8b).

From the completeness relation of $\{P^{(\alpha)}\}$ given in (6.9.16b) and (6.9.18), we have the expansion of any function F in the basis space of G:

$$F = \sum_{\alpha} \mathring{P}^{(\alpha)} F = \sum_{\alpha} \psi^{\alpha} \tag{6.9.20}$$

corresponding to the expansion (6.9.2).

6.10 Selection rules

In quantum mechanics, we are concerned with the matrix elements of a set of operators $\{V_l^{\lambda}\}$ defined by

$$V_{ijl}^{\alpha\beta\lambda} = \langle \psi_i^{\alpha}, V_l^{\lambda} \psi_j^{\beta} \rangle = \int \psi_i^{\alpha *} V_l^{\lambda} \psi_j^{\beta} \, dx \tag{6.10.1}$$

where $\{\psi_i^{\alpha}\}$ and $\{\psi_j^{\beta}\}$ are the eigenfunctions of the Hamiltonian H_0 of a quantum mechanical system. For example, the square of the absolute value of a matrix element $|V_{ijl}^{\alpha\beta\lambda}|^2$ determines the probability of transition between two states ψ_i^{α} and ψ_j^{β} due to the perturbation V_l^{λ}. In many problems, however, there is no need to calculate these matrix elements completely: it is often sufficient to establish *selection rules* that state which matrix elements are non-zero and what the linear relations between non-zero matrix elements are. Here, the projection operator $\mathring{P}^{(1)}$ which projects out the invariant part of a basis function with respect to the symmetry group $G = \{g\}$ of the system defined in (6.9.19), i.e.

$$\mathring{P}^{(1)} = (1/|G|) \sum_{g \in G} \mathring{g} \tag{6.10.2}$$

plays the crucial role for this problem.

As a preparation we shall look into the invariance property of an integral defined over the whole space of an n-dimensional configuration space of $x = (x_1, x_2, \ldots, x_n)$

$$J = \int F(x) \, dx; \qquad dx = dx_1 \, dx_2 \ldots dx_n \tag{6.10.3a}$$

Let $G = \{g\}$ be a group of transformations of the variable x that preserves the volume element; for example, G is a group of orthogonal transformations. Then, under a transformation $x \to x' = gx$, we have

$$F(x) = F(g^{-1}x') = \mathring{g} F(x')$$

with $dx = dx'$ so that[3] we obtain the invariance of the integeral J:

$$J = \int \mathring{g} F(x') \, dx' \equiv \mathring{g} J, \qquad \forall \, g \in G \tag{6.10.3b}$$

under the transformation of the integrand $F(x)$ with respect to $g \in G$. Since the integral J is simply a number, $\mathring{g} J$ is meaningful only through the action $\mathring{g}$ on the integrand $F(x)$ of J via the transformation of the integral variable (cf. (6.4.8)).

[3] In the case in which the integral J is given by a Hermitian scalar product of two functions $\Psi(x)$ and $\Phi(x)$,

$$J = \int \Psi^*(x) \Phi(x) \, dx = \langle \Psi, \Phi \rangle$$

(6.10.3b) means that

$$J = \mathring{g} J \equiv \langle \mathring{g} \Psi, \mathring{g} \Phi \rangle$$

Averaging (6.10.3b) over the group $G = \{g\}$ we obtain

$$J = \frac{1}{|G|}\sum_{g\in G} \mathring{g}J = \int F^1(x)\,\mathrm{d}x \tag{6.10.4a}$$

where $F^1(x)$ is the invariant part of $F(x)$ with respect to G defined by the projection operator $\mathring{P}^{(1)}$ corresponding to the identity representation $D^{(1)}(G)$:

$$F^1(x) = \mathring{P}^{(1)}F(x) = \frac{1}{|G|}\sum_{g} \mathring{g}F(x) \tag{6.10.4b}$$

Thus we arrive at the following rule.

Rule 1. Let J be an integral defined over the whole configuration space of x:

$$J = \int F(x)\,\mathrm{d}x$$

and let $G = \{g\}$ be a volume-preserving group of transformations of x. Then, the integral J is given by the invariant part $F^1(x)$ of the integrand $F(x)$ with respect to G:

$$J = \int F^1(x)\,\mathrm{d}x$$

Thus J is zero, if the invariant part $F^1(x)$ of the integrand is zero.

Example 1. Consider an integral that is invariant under the group of inversion $C_\mathrm{i} = \{e, I\}$, where I is the inversion which brings x to $-x$. For this group the invariant part of $F(x)$ is given by

$$F^1(x) = \mathring{P}^{(1)}F(x) = \tfrac{1}{2}[F(x) + F(-x)] = F_\mathrm{e}(x)$$

which is an even function of x. Thus, if $F(x)$ is an odd function of x, i.e. $F(-x) = -F(x)$, then the integral vanishes.

The selection rule 1 for the group of inversion leads to Laporte's rule in molecular spectroscopy: In the dipole approximation, the probability of transition between two states ψ_i^α and ψ_j^β is determined by the absolute square of the following integral:

$$\langle\psi_i^\alpha|\boldsymbol{\mu}|\psi_j^\beta\rangle = \int \psi_i^{\alpha*}\boldsymbol{\mu}\psi_j^\beta\,\mathrm{d}x$$

where $\boldsymbol{\mu}$ is the dipole moment of the system which changes the sign under inversion. Accordingly the transitions between even and odd states, i.e. the states with different parities, are allowed.

Let us extend the above rule to a case in which J is a set of integrals $\boldsymbol{J} = \{J_s\}$ defined by

$$J_s = \int F_s(x)\,\mathrm{d}x; \qquad s = 1, 2, \ldots, n$$

which may be expressed formally as

$$\boldsymbol{J} = \int \boldsymbol{F}(x)\,\mathrm{d}x \tag{6.10.5}$$

where $\boldsymbol{J} = [J_1, J_2, \ldots, J_n]$ and $\boldsymbol{F}(x) = [F_1(x), F_2(x), \ldots, F_n(x)]$ are regarded as row vectors. It is assumed that the integrand $\boldsymbol{F}(x)$ transforms according to a representation $\Delta(G)$ of a volume-preserving group G

$$\mathring{g}\boldsymbol{F}(x) = \boldsymbol{F}(x)\Delta(g); \qquad \forall\, g \in G$$

Then the invariant part of $\boldsymbol{F}(x)$ satisfies

$$\boldsymbol{F}^1(x) = \mathring{P}^{(1)}\boldsymbol{F}(x) = \sum_g \boldsymbol{F}(x)\Delta(g)/|G| = \sum_g \boldsymbol{F}^1(x)\Delta(g)/|G| \tag{6.10.6a}$$

where in the last expression use of the projective relation $\mathring{P}^{(1)}\boldsymbol{F}^1(x) = \boldsymbol{F}^1(x)$ has been made. Thus we obtain, from (6.10.5),

$$\boldsymbol{J} = \boldsymbol{J}\sum_g \Delta(g)/|G| \qquad \text{or} \qquad J_s = \sum_t J_t \sum_g \Delta_{ts}(g)/|G| \tag{6.10.6b}$$

which means that the set of integrals $\boldsymbol{J} = \{J_s\}$ is not linearly independent unless $\Delta(G)$ is the identity representation.

To determine the linearly independent elements in $\boldsymbol{J}$ we introduce a similarity transformation with a transformation matrix T that reduces $\Delta(G)$ to the irreducible components:

$$T^{-1}\Delta(g)T = \sum_\gamma \oplus\, n_\gamma D^{(\gamma)}(g); \qquad \forall\, g \in G$$

where n_γ is the frequency of the irreducible representaion $D^{(\gamma)}(G)$ contained in $\Delta(G)$. If we take the average of the above equation over $g \in G$, all $D^{(\gamma)}(g)$ with $\gamma \neq 1$ vanish on account of the orthogonality relation, leaving only the n_1 identity representation $\{D^{(1)}(g) = 1;\ \forall\, g \in G\}$, so that

$$\frac{1}{|G|}\sum_g T^{-1}\Delta(g)T = \operatorname{diag}[\overbrace{1, \ldots 1}^{n_1}, 0, \ldots 0] \tag{6.10.7}$$

Accordingly, the linear transform of the integral set defined by $\boldsymbol{K} = \boldsymbol{J}T = [K_1, K_2, \ldots]$ takes the following form, in view of (6.10.6b) and (6.10.7):

$$\begin{aligned} \boldsymbol{K} = \boldsymbol{J}T &= \boldsymbol{K}\sum_g T^{-1}\Delta(g)T/|G| \\ &= [K_1, K_2, \ldots, K_{n_1}, 0, \ldots 0] \end{aligned} \tag{6.10.8}$$

i.e. the non-zero components of $\boldsymbol{K}$ are limited to the following n_1 integrals:

$$\boldsymbol{K}^1 = [K_1, K_2, \ldots, K_{n_1}]$$

corresponding to the n_1 identity representations contained in $\Delta(G)$. Thus the original integrals $\boldsymbol{J} = \{J_s\}$ are given by the linear combinations of the linearly independent non-zero components $\boldsymbol{J} = \boldsymbol{K}T^{-1}$, i.e.

$$J_s = \sum_{s'=1}^{n_1} K_{s'}(T^{-1})_{s's}; \qquad s = 1, 2, \ldots, n \tag{6.10.9}$$

Here the frequency n_1 of the identity representation of G contained in $\Delta(G)$ is given by

$$n_1 = \frac{1}{|G|} \sum_g \chi^{(\Delta)}(g) \tag{6.10.10}$$

where $\chi^{(\Delta)}(g) = \operatorname{tr} \Delta(g)$. Therefore we arrive at the following rule.

Rule 2. If a set of integrals $\boldsymbol{J} = \{J_s\}$ transforms according to a representation $\Delta(G) = \{\Delta(g); \forall\, g \in G\}$, the maximum number of linearly independent elements in $\boldsymbol{J}$ equals n_1, which is the number of identity representations contained in $\Delta(G)$.

Example 2. Let $\boldsymbol{J} = \{J_1, J_2, J_3\}$ be a set of integrals defined by

$$J_s = \int_{-\infty}^{+\infty} \mathrm{e}^{-r} x_s^2 \,\mathrm{d}x = \int_{-\infty}^{+\infty} F_s \,\mathrm{d}x; \qquad s = 1, 2, 3$$

where $r = (x_1^2 + x_2^2 + x_3^2)^{1/2}$ and $\mathrm{d}x = \mathrm{d}x_1\,\mathrm{d}x_2\,\mathrm{d}x_3$. The set of integrands $\{F_1, F_2, F_3\}$ is invariant under the cyclic permutation group C_3. The invariant parts of F_1, F_2 and F_3 are all equal and given by

$$F_1^1 = F_2^1 = F_3^1 = \tfrac{1}{3}\mathrm{e}^{-r} r^2$$

Accordingly, the integrals J_s are all equal and given by

$$J_s = \tfrac{1}{3} \int_{-\infty}^{+\infty} \mathrm{e}^{-r} r^2 4\pi r^2 \,\mathrm{d}r = 64\pi$$

for any s.

Let us now return to the selection rule on the set of matrix elements $V = \{V_{ijl}^{\alpha\beta\gamma}\}$ defined by (6.10.1). The invariance of the set under a group $G = \{g\}$ is expressed by

$$V_{ijl}^{\alpha\beta\gamma} = \mathring{g} V_{ijl}^{\alpha\beta\gamma} = \langle \mathring{g}\psi_i^\alpha, \mathring{g} V_l^\lambda \mathring{g}^{-1} \mathring{g}\psi_j^\beta \rangle, \qquad \forall\, g \in G$$

Let us assume that the sets $\{\psi_i^\alpha\}$ and $\{\psi_j^\beta\}$ transform according to the representations $D^{(\alpha)}$ and $D^{(\beta)}$ of G, respectively, and the set $\{V_l^\lambda\}$ regarded as a tensor obeys the transformation law

$$\mathring{g} \mathring{V}_l^\lambda \mathring{g}^{-1} = \sum_{l'} \mathring{V}_{l'}^\lambda D_{l'l}^{(\lambda)}(g); \qquad l = 1, 2, \ldots, d_\lambda \tag{6.10.11}$$

Then the integral set $\boldsymbol{J} = \{V_{ijl}^{\alpha\beta\lambda}\}$ transforms according to the direct product representation $D^{(\alpha)}(g)^* \times D^{(\beta)}(g) \times D^{(\lambda)}(g)$ of G. Thus from (6.10.10) the number n_1 of linear independent matrix elements contained in a total of $d_\alpha d_\beta d_\lambda$ integrals is given by

$$n_1 = \frac{1}{|G|} \sum_g \chi^{(\alpha)}(g)^* \chi^{(\beta)}(g) \chi^{(\lambda)}(g) \tag{6.10.12}$$

where $\chi^{(\alpha)}$, $\chi^{(\beta)}$ and $\chi^{(\lambda)}$ are the characters of $D^{(\alpha)}$, $D^{(\beta)}$ and $D^{(\lambda)}$, respectively.

For most cases of practical interest, the representations $D^{(\alpha)}$ and $D^{(\beta)}$ are the unirreps of G, in which case n_1 given by (6.10.12) can be interpreted as the number of the unirreps, $D^{(\alpha)}$, contained in the direct product representation $D^{(\alpha)} \times D^{(\lambda)}$ of G. Thus we have the following.

Rule 3. If a set of integrals $\{V_{ijl}^{\alpha\beta\lambda}\}$ transforms according to a direct product representation $D^{(\alpha)}(g)^* \times D^{(\beta)}(g) \times D^{(\lambda)}(g)$ of a group G, then the number of linearly independent integrals in the set equals the number n_1 of the identity irreducible representations of G contained in the direct product. If $D^{(\alpha)}(G)$ is a unirrep of G, then the frequency n_1 equals the number of times $D^{(\alpha)}(G)$ is contained in the direct product representation $\{D^{(\beta)}(g) \times D^{(\lambda)}(g)\}$ of G.

Let us apply the above rule to the selection rules on the electric-dipole transitions for an atom. An electric dipole $\boldsymbol{\mu}$ transforms according to the three-dimensional representation $D^{(1)}$ of the full rotation group $O^{(3)} = O(3, r)$, whereas the electronic states of an atom are classified in terms of the $(2j+1)$-dimensional unirreps $D^{(j)}$ of $O^{(3)}$. Accordingly, the transition moments for transitions between two sets of states $\{\psi_{m_1}^{j_1}\} \in D^{(j_1)}$ and $\{\psi_{m_2}^{j_2}\} \in D^{(j_2)}$ are defined by

$$\langle j_1 m_1 | \boldsymbol{\mu} | j_2 m_2 \rangle = \int \psi_{m_1}^{j_1} * \boldsymbol{\mu} \psi_{m_2}^{j_2} \, \mathrm{d}\tau$$

and these transform according to the direct product representation

$$D(g) = D^{(j_1)^*}(g) \times D^{(j_2)}(g) \times D^{(1)}(g); \qquad \forall\, g \in O_3$$

Now, according to the vector addition model which will be discussed in Section 10.4.3, we have

$$D^{(j_2)} \times D^{(1)} = D^{(j_2+1)} \oplus D^{(j_2)} \oplus D^{(j_2-1)} \qquad \text{for } j_2 \neq 0$$

$$D^{(0)} \times D^{(1)} = D^{(1)} \qquad \text{for } j_2 = 0$$

so that, for $j_2 \neq 0$,

$$\langle j_1 m_1 | \boldsymbol{\mu} | j_2 m_2 \rangle = 0 \qquad \text{unless } j_1 = j_2 + 1,\ j_2 \text{ or } j_2 - 1$$

and, for $j_2 = 0$,

$$\langle j_1 | \boldsymbol{\mu} | 0 \rangle = 0 \qquad \text{unless } j_1 = 1$$

Here, the transition between $j_1 = 0$ and $j_2 = 0$ is forbidden.

7

Construction of symmetry-adapted linear combinations based on the correspondence theorem

7.1 Introduction

In the eigenvalue problem of a Hamiltonian in quantum mechanics, the eigenfunctions of the Hamiltonian are classified in terms of the unitary irreducible representations (unirreps) of the symmetry group G of the Hamiltonian. In constructing approximate eigenfunctions by LCAO-MOs of a molecule belonging to a certain symmetry group G, the corresponding problem is to find the irreducible basis sets of G constructed by the linear combinations of the atomic orbitals belonging to the equivalent atoms of the molecule. Such a set is called *a set of symmetry-adapted linear combinations (SALC)* of the equivalent basis functions or equivalent orbitals. A standard method for such a problem is the generating operator method introduced in Section 6.9: it generates the desired basis set from an appropriate basis function. This method is very general and powerful but it is often extremely laborious to use; Cotton (1990). It is so very formal that one has little feeling until one arrives at the final result, which often could simply be obtained by inspection.

For point groups and their extensions, there exists a simple direct method of constructing the SALC belonging to a unirrep of a symmetry group G. The method requires knowledge of the basis functions of a space vector $\boldsymbol{r} = (x, y, z)$ in three dimensions belonging to the unirrep. The basis sets are well known for all point groups; e.g., those for the point groups T_p and D_{3p} are reproduced in Tables 7.1 and 7.2, respectively, from the Appendix. Note, here, that each basis set is given by a set of homogeneous polynomials of a certain degree with respect to the Cartesian coordinates x, y and z. Hereafter, such a basis set is called an *elementary basis set*. As we shall see, the elementary basis sets defined on a set of symmetrically equivalent points with respect to the group G play the fundamental role in the present work.

For example, let us consider the SALCs of *equivalent scalar functions* $\{(\phi_\nu(\boldsymbol{r}) = \phi(\boldsymbol{r}, \boldsymbol{r}_\nu^0)\}$ defined on a set of n equivalent points $S^{(n)} = \{\boldsymbol{r}_\nu^0\}$ with respect to G: such a set of equivalent functions has been introduced in (6.4.4b) by an invariant (or scalar) function $\phi(\boldsymbol{r}, \boldsymbol{r}')$ of two points $\boldsymbol{r}$ and $\boldsymbol{r}'$ in space with respect to G. Let $\Delta^{(n)}(G)$ be the representation based on the set $\{\phi_\nu(\boldsymbol{r})\}$ and let $\{u_i^\alpha(\boldsymbol{r});\ i = 1, 2, \ldots, d_\alpha\}$ be an elementary basis belonging to a unirrep $D^{(\alpha)}(G)$ contained in $\Delta^{(n)}$, then the SALCs of $\{\phi_\nu(\boldsymbol{r})\}$ belonging to $D^{(\alpha)}(G)$ are given, as will be shown in the next section, by

$$\psi_i^\alpha(\boldsymbol{r}) = \sum_\nu u_i^\alpha(\boldsymbol{r}_\nu^0)\phi_\nu(\boldsymbol{r}); \qquad i = 1, 2, \ldots, d_\alpha \tag{7.1.1}$$

unless the set is null. The correspondence between the SALC $\{\psi_i^\alpha(\boldsymbol{r})\}$ and the set of the linear coefficients $\{u_i^\alpha(\boldsymbol{r}_\nu^0)\}$ via (7.1.1) is a consequence of *the correspondence theorem* on basis function introduced by Kim (1981a). According to this theorem, a $D^{(\alpha)}(G)$ SALC of any given equivalent orbitals can be formed via an

Table 7.1. *The elementary bases of the unirreps of T_p (= T_d)*

A_1:	$1, x^2+y^2+z^2$ or xyz
A_2:	$(x^2-y^2)(y^2-z^2)(z^2-x^2)$ or $\tilde{x}\tilde{y}\tilde{z}$
E:	$[u, v]$
T_1:	$[\tilde{x}, \tilde{y}, \tilde{z}]$ or $[x(y^2-z^2), y(z^2-x^2), z(x^2-y^2)]$
T_2:	$[x, y, z], [yz, zx, xy]$ or $[x(3x^2-r^2), y(3y^2-r^2), z(3z^2-r^2)]$

$u = 2z^2 - x^2 - y^2$, $v = 3^{1/2}(x^2 - y^2)$ and $\tilde{\boldsymbol{r}} = [\boldsymbol{r} \times \boldsymbol{r}'] = [\tilde{x}, \tilde{y}, \tilde{z}]$; i.e. $\tilde{x} = yz' - zy'$, $\tilde{y} = zx' - xz'$ and $\tilde{z} = xy' - yx'$.

Table 7.2. *The elementary bases of the unirreps of D_{3p} (= D_{3h})*

A_1:	$1, x^2+y^2, z^2$ or y^3-3yx^2
A_2:	x^3-3xy^2 or $\tilde{z}$
B_1:	$(x^3-3xy^2)z$ or $\tilde{z}z$
B_2:	z
E_1:	$[\tilde{x}, \tilde{y}]$ or $[yz, -xz]$
E_2:	$[x, y], [2xy, x^2-y^2]$ or $[y\tilde{z}, -x\tilde{z}]$

appropriate two-point basis $\{f_i^\alpha(\boldsymbol{r}, \boldsymbol{r}')\}$ of $D^{(\alpha)}(G)$. In fact, from this theorem, we shall derive a general expression for the SALC of any given equivalent orbitals in Section 7.3. It is effective for degenerate as well as for non-degenerate irreducible representations. It requires neither additional symmetry consideration of the equivalent basis functions nor the actual matrix representations of the irreducible representations. It simply requires knowledge of the elementary basis sets. This is quite a contrast to the conventional method of using the generating operators or projection operators.

The correspondence theorem will also be applied for construction of the symmetry coordinates of vibration for a molecule or a crystal; cf. Kim (1981b). The correspondence theorem is also very effective in constructing the energy band eigenfunctions of solids, as will be shown in Chapter 14.

7.2 The basic development

7.2.1 *Equivalent point space $S^{(n)}$*

Let us consider a finite group G of symmetry operations R in the three-dimensional configuration space. Let $S^{(n)} = \{\boldsymbol{r}_1^0, \boldsymbol{r}_2^0, \ldots, \boldsymbol{r}_n^0\}$ be a set of n points in the space generated by the symmetry operations R of G from a single given point in space. The set $S^{(n)}$ spans an invariant space of G such that any two points $\boldsymbol{r}_\mu^0$ and $\boldsymbol{r}_\nu^0$ in $S^{(n)}$ are connected by at least one symmetry operation $R \in G$:

$$\boldsymbol{r}_\mu^0 = R\boldsymbol{r}_\nu^0 \quad \text{or} \quad \mu = \nu(R) \tag{7.2.1a}$$

Since the members of the set $S^{(n)}$ permute among themselves under R of G, the set provides a representation of G defined by

$$R\boldsymbol{r}_\nu^0 = \sum_\sigma \boldsymbol{r}_\sigma^0 \delta(\boldsymbol{r}_\sigma^0, R\boldsymbol{r}_\nu^0) \equiv \sum_{\sigma=1}^{n} \boldsymbol{r}_\sigma^0 \Delta_{\sigma\nu}^{(n)}(R); \qquad \nu = 1, \ldots, n$$

where $\{\Delta^{(n)}(R); \quad \forall R \in G\} = \Delta^{(n)}(G)$ is an $n \times n$ matrix representation of G defined by Kronecker's delta,

$$\Delta_{\sigma\nu}^{(n)}(R) = \delta(\boldsymbol{r}_\sigma^0, R\boldsymbol{r}_\nu^0) = \begin{cases} 1 & \text{if } \boldsymbol{r}_\sigma^0 = R\boldsymbol{r}_\nu^0 \\ 0 & \text{if } \boldsymbol{r}_\sigma^0 \neq R\boldsymbol{r}_\nu^0 \end{cases} \tag{7.2.1b}$$

An example of such a set $S^{(n)}$ of n equivalent points is provided by the positions of the symmetrically equivalent atoms in a polyatomic molecule or a crystal.

The set of operations $\{R\}$ of G that leaves a point of $S^{(n)}$, say $\boldsymbol{r}_1^0$, invariant forms a subgroup H of G with the index $n = |G|/|H|$. Let the coset decomposition of G with respect to H be

$$G = R_1^0 H + R_2^0 H + \cdots + R_n^0 H$$

where $R_1^0 = E$, the identity element of G. Then the set of the coset representatives of H in G may be correlated to the set of n equivalent points $S^{(n)}$ by

$$\boldsymbol{r}_\nu^0 = R_\nu^0 \boldsymbol{r}_1^0; \qquad \nu = 1, 2, \ldots, n$$

Thus, the representation $\Delta^{(n)}(G)$ is rewritten as

$$\Delta_{\sigma\nu}^{(n)}(R) = \delta(\boldsymbol{r}_\sigma^0, R\boldsymbol{r}_\nu^0) = \sum_{h \in H} \delta(h, R_\sigma^{0-1} R R_\nu^0)$$

so that the character is given by

$$\chi^{(\Delta)}(R) = \sum_\nu \sum_{h \in H} \delta(h, R_\nu^{0-1} R R_\nu^0) \tag{7.2.2}$$

where h are elements of H. The representation $\Delta^{(n)}(G)$ is called *the permutation representation* of G based on $S^{(n)}$. It is also called the permutation representation of G furnished by the subgroup H. In a special case, in which H is the trivial identity group, it becomes *the regular representation* of G.

The permutation representation $\Delta^{(n)}$ of G is a real orthogonal representation of G, for $\Delta^{(n)}(R)^{-1} = \Delta^{(n)}(R)^{\sim}$ (the transpose); hence it is reducible by a unitary transformation. Let n_α be the number of times a unirrep $D^{(\alpha)}(G)$ is contained in $\Delta^{(n)}(G)$ and let $\chi^{(\alpha)}(G)$ and $\chi^{(\Delta)}(G)$ be the respective characters. Then n_α is given by

$$\begin{aligned} n_\alpha &= \frac{1}{|G|} \sum_{R \in G} \chi^{(\alpha)}(R) \chi^{(\Delta)}(R) \\ &= \frac{1}{|G|} \sum_\nu \sum_{h \in H} \chi^{(\alpha)}(R_\nu^{0-1} h R_\nu^0) \\ &= \frac{1}{|H|} \sum_{h \in H} \chi^{(\alpha)}(h) \end{aligned} \tag{7.2.3}$$

where (7.2.2) is used. In a special case, in which $D^{(\alpha)}$ is the identity representation of

G, we have $\chi^{(\alpha)}(h) = 1$ for all $h \in H$ so that $n_\alpha = 1$, i.e. $\Delta^{(n)}(G)$ always contains the identity representation once and only once. In general,

$$\sum_\alpha n_\alpha d_\alpha = n; \qquad n_\alpha \leqslant d_\alpha \tag{7.2.4}$$

where d_α is the dimensionality of $D^{(\alpha)}(G)$ and gives the possible maximum of n_α, as will be shown shortly. In cases in which $n_\alpha > 1$, it is often profitable to redefine the set $S^{(n)}$ by means of an intermediate subgroup K ($H < K < G$), which ensures that $n_\alpha = 1$ for all $D^{(\alpha)}(G)$ contained in $\Delta^{(n)}(G)$, see Kim (1986a).

7.2.2 *The correspondence theorem on basis functions*

Let $D^{(\gamma)}(G)$ be a unirrep of a symmetry group G and let $\{f_k^\gamma(\boldsymbol{r}, \boldsymbol{r}'); k = 1, 2, \ldots, d_\gamma\}$ be a set of basis functions of two points $\boldsymbol{r}$ and $\boldsymbol{r}'$ in space belonging to $D^{(\gamma)}(G)$ such that

$$f_k^\gamma(R^{-1}\boldsymbol{r}, R^{-1}\boldsymbol{r}') = \sum_j f_j^\gamma(\boldsymbol{r}, \boldsymbol{r}')D_{jk}^{(\gamma)}(R); \qquad \forall R \in G \tag{7.2.5}$$

Hereafter, such a basis is simply called *a two-point basis* of a unirrep $D^{(\gamma)}(G)$. A simple but important example of a two-point basis is provided by $\{f_k^\gamma(\boldsymbol{r} - \boldsymbol{r}')\}$, where $\{f_k^\gamma(\boldsymbol{r})\}$ is an ordinary one-point basis of $D^{(\gamma)}(G)$. Another example is the set $\{u_i^\alpha(\boldsymbol{r}')\phi(\boldsymbol{r}, \boldsymbol{r}')\}$ used in (7.1.1), where $\phi(\boldsymbol{r}, \boldsymbol{r}')$ is a scalar function. If the set $\{f_k^\gamma(\boldsymbol{r}, \boldsymbol{r}_\nu^0)\}$ defined on the set of equivalent points $S^{(n)} = \{\boldsymbol{r}_\nu^0\}$ is linearly independent, then it provides a set of *equivalent basis functions* (or *equivalent orbitals* in short) on $S^{(n)}$ belonging to the direct product representation $\Delta^{(n)} \times D^{(\gamma)}$ of G:

$$\begin{aligned} f_k^\gamma(R^{-1}\boldsymbol{r}, \boldsymbol{r}_\nu^0) &= \sum_j f_j^\gamma(\boldsymbol{r}, R\boldsymbol{r}_\nu^0)D_{jk}^{(\gamma)}(R) \\ &= \sum_{j,\mu} f_j^\gamma(\boldsymbol{r}, \boldsymbol{r}_\mu^0)\Delta_{\mu\nu}^{(n)}(R)D_{jk}^\gamma(R) \end{aligned} \tag{7.2.6}$$

Thus, e.g., the set of equivalent scalar orbitals $\{\phi_\nu(\boldsymbol{r}) = \phi(\boldsymbol{r}, \boldsymbol{r}_\nu^0)\}$ in (7.1.1) belongs to the permutation representation $\Delta^{(n)}(G)$. The following simple theorem is basic to the present method of constructing SALCs.

Theorem 7.2.1 (the correspondence theorem). Let $\{f_k^\gamma(\boldsymbol{r}, \boldsymbol{r}')\}$ be a two-point basis of a unirrep $D^{(\gamma)}(G)$. If the corresponding set $\{f_k^\gamma(\boldsymbol{r}, \boldsymbol{r}_\nu^0)\}$ on $S^{(n)}$ is linearly independent, then its sum over $S^{(n)}$

$$\psi_k^\gamma(\boldsymbol{r}) = \sum_{\nu=1}^n f_k^\gamma(\boldsymbol{r}, \boldsymbol{r}_\nu^0); \qquad k = 1, 2, \ldots, d_\gamma \tag{7.2.7}$$

provides a basis of the unirrep $D^{(\gamma)}(G)$.

The proof is trivial. If we sum both sides of (7.2.6) over ν and map off $\Delta^{(n)}(R)$ on the right-hand side, using $\sum_\nu \Delta_{\mu\nu}^{(n)}(R) = 1$, then we obtain the required result

$$\psi_k^\gamma(R^{-1}\boldsymbol{r}) = \sum_{j=1}^n \psi_j^\gamma(\boldsymbol{r})D_{jk}^{(\gamma)}(R)$$

Note that $D^{(\gamma)}$ is contained in $\Delta^{(n)} \times D^{(\gamma)}$ at least once because the identity representation is always contained in $\Delta^{(n)}$. The basis of $D^{(\gamma)}(G)$ defined by (7.2.7) may be called *the characteristic SALC* corresponding to the two-point basis $\{f_k^\gamma(\boldsymbol{r}, \boldsymbol{r}')\}$ of $D^{(\gamma)}(G)$.

7.2.2.1 The SALC of equivalent scalar orbitals

According to the correspondence theorem, the SALC belonging to $D^{(\alpha)}$ given by (7.1.1) is trivially true; because of its basic importance, however, we shall look into it in some detail. Let $\phi(\boldsymbol{r}, \boldsymbol{r}')$ be a two-point invariant function with respect to G, then the equivalent scalar orbitals $\{\phi_\nu(\boldsymbol{r}) = \phi(\boldsymbol{r}, \boldsymbol{r}_\nu^0)\}$ defined on $S^{(n)}$ belong to the permutation representation $\Delta^{(n)}(G)$, provided that the set is linearly independent. This set may be called a set of *equivalent scalar orbitals*. Let $D^{(\alpha)}(G)$ be a unirrep of G contained in $\Delta^{(n)}(G)$ and let $\{u_i^\alpha(\boldsymbol{r}); i = 1, 2, \ldots, d_\alpha\}$ be a basis of $D^{(\alpha)}(G)$. Then the SALC of $\{\phi_\nu(\boldsymbol{r})\}$ belonging to $D^{(\alpha)}(G)$ is given by the characteristic set of the two-point basis $\{u_i^\alpha(\boldsymbol{r}')\phi(\boldsymbol{r}, \boldsymbol{r}')\}$ of $D^{(\alpha)}(G)$

$$\psi_i^\alpha(\boldsymbol{r}) = \sum_{\nu=1}^{n} u_i^\alpha(\boldsymbol{r}_\nu^0)\phi_\nu(\boldsymbol{r}); \qquad i = 1, 1, \ldots, d_\alpha \tag{7.2.8}$$

unless the coefficient set is null. This proves (7.1.1).

It is to be noted that the sets $\{f_k^\gamma(\boldsymbol{r}, \boldsymbol{r}')\}$ and $\{f_k^\gamma(\boldsymbol{r} - \boldsymbol{r}', \boldsymbol{r}')\}$ belong to the same unirrep, $D^{(\gamma)}(G)$. Thus frequently the characteristic set of the former is defined by the characteristic set of the latter:

$$\psi_k^\gamma(\boldsymbol{r}) = \sum_{\nu} f_k^\gamma(\boldsymbol{r} - \boldsymbol{r}_\nu^0, \boldsymbol{r}_\nu^0) \in D^{(\gamma)}(G) \tag{7.2.9}$$

The further significance of the correspondence theorem will be shown from the fact that a SALC of any given equivalent orbitals can be given by the characteristic set of an appropriate two-point basis. This will be achieved by straightforward extension of (7.2.8), for which, however, it is necessary to discuss the mathematical properties of the coefficient matrix in the transformation (7.2.8), in particular, to understand the non-null condition imposed on the coefficient set $\{u_i^\alpha(\boldsymbol{r}_\nu^0)\}$.

7.2.3 Mathematical properties of bases on $S^{(n)}$

Let $\{u_i^\alpha(\boldsymbol{r})\}$ be a basis of a unirrep $D^{(\alpha)}(G)$ contained in $\Delta^{(n)}(G)$, then at the equivalent points $\boldsymbol{r} = \boldsymbol{r}_\nu^0 \in S^{(n)}$, we have

$$u_i^\alpha(R^{-1}\boldsymbol{r}_\nu^0) = \sum_{j=1}^{d_\alpha} u_j^\alpha(\boldsymbol{r}_\nu^0)D_{ji}^{(\alpha)}(R); \qquad \forall R \in G \tag{7.2.10}$$

This means that, if the coefficient matrix $\|u_{j\nu}^\alpha\| = \|u_j^\alpha(\boldsymbol{r}_\nu^0)\|$ of order $d_\alpha \times n$ is not null, its row vectors denoted $\boldsymbol{u}_j^\alpha = [u_j^\alpha(\boldsymbol{r}_1^0), \ldots, u_j^\alpha(\boldsymbol{r}_n^0)]$ provide a basis set in the n-dimensional space $S^{(n)}$ belonging to the jth row of $D^{(\alpha)}(G)$. Accordingly, if $\{\boldsymbol{v}_j^\beta\}$ is another basis set belonging to a unirrep $D^{(\beta)}(G)$ contained in $\Delta^{(n)}(G)$, then, from the orthogonality relations on the unirreps of G, we obtain

$$\sum_{R\in G} u_i^\alpha(R^{-1}\boldsymbol{r}_\nu^0)^* v_j^\beta(R^{-1}\boldsymbol{r}_\nu^0) = (|G|/d_\alpha)\delta_{\alpha\beta}\delta_{ij}\sum_{k}^{d_\alpha} u_k^\alpha(\boldsymbol{r}_\nu^0)^* v_k^\alpha(\boldsymbol{r}_\nu^0); \qquad \nu = 1, 2, \ldots, n$$

Let H be the subgroup of G which leaves $\boldsymbol{r}_1^0$ invariant, then the set $\{R^{-1}\boldsymbol{r}_\nu^0; \forall R \in G\}$ for a given $\boldsymbol{r}_\nu^0$ reproduces the set $S^{(n)}$ $|H|$ times when R sweeps through all elements of G. Thus, we obtain the following orthogonality relations, using $|G|/|H| = n$:

$$\sum_{\sigma=1}^{n} u_i^\alpha(\boldsymbol{r}_\sigma^0)^* v_j^\beta(\boldsymbol{r}_\sigma^0) = (n/d_\alpha)\delta_{\alpha\beta}\delta_{ij}\sum_{k=1}^{d_\alpha} u_k^\alpha(\boldsymbol{r}_\nu^0)^* v_k^\alpha(\boldsymbol{r}_\nu^0); \qquad i, j = 1, 2, \ldots, d_\alpha \tag{7.2.11}$$

where * denotes the complex conjugate and the right-hand side (rhs) is independent of ν. The left-hand side (lhs) defines a scalar product $\langle \boldsymbol{u}_i^\alpha, \boldsymbol{v}_j^\beta \rangle_n$ on $S^{(n)}$, while the sum on the rhs defines a scalar product $\langle \boldsymbol{u}^\alpha(\boldsymbol{r}_\nu^0), \boldsymbol{v}^\alpha(\boldsymbol{r}_\nu^0)\rangle_c$ in the d_α-dimensional carrier space of $D^{(\alpha)}(R)$. In the special case in which $\alpha = \beta$ and $i = j$, these two kinds of scalar products are proportional to each other and hence one can define a single scalar product by

$$\langle \boldsymbol{u}^\alpha, \boldsymbol{v}^\alpha \rangle \equiv \langle \boldsymbol{u}_i^\alpha, \boldsymbol{v}_i^\alpha \rangle_n = (n/d_\alpha)\langle \boldsymbol{u}^\alpha(\boldsymbol{r}_\nu^0), \boldsymbol{v}^\alpha(\boldsymbol{r}_\nu^0)\rangle_c \tag{7.2.12}$$

which is independent of i and ν. This has the very significant consequence that the number of linearly independent basis sets of $D^{(\alpha)}(R)$ in the n-dimensional space $S^{(n)}$ cannot exceed d_α. This then proves the statement that $n_\alpha \leqslant d_\alpha$ given in (7.2.4).

Since the $d_\alpha \times n$ matrix $\|u_i^\alpha(\boldsymbol{r}_\nu^0)\|$ transforms a basis $\{\phi_\nu(\boldsymbol{r})\}$ of the permutation representation $\Delta^{(n)}(G)$ to a basis $\{\psi_i^\alpha(\boldsymbol{r})\}$ of $D^{(\alpha)}(G)$ according to (7.2.8), it must transform the $n \times n$ matrix representation $\Delta^{(n)}(G)$ to the $d_\alpha \times d_\alpha$ matrix representation $D^{(\alpha)}(G)$. To see this we rewrite (7.2.10), using $\Delta_{\mu\nu}^{(n)}(R^{-1}) = \Delta_{\nu\mu}^{(n)}(R)$, in the form

$$\sum_\mu \Delta_{\nu\mu}^{(n)}(R) u_i^\alpha(\boldsymbol{r}_\mu^0) = \sum_j u_j^\alpha(\boldsymbol{r}_\nu^0) D_{ji}^{(\alpha)}(R)$$

then the orthogonality relations (7.2.11) lead to the expected transformation

$$\sum_{\nu\mu} U_j^\alpha(\boldsymbol{r}_\nu^0)^* \Delta_{\nu\mu}^{(n)}(R) U_i^\alpha(\boldsymbol{r}_\mu^0) = D_{ji}^{(\alpha)}(R) \tag{7.2.13}$$

where

$$U_j^\alpha(\boldsymbol{r}_\nu^0) = u_j^\alpha(\boldsymbol{r}_\nu^0)\langle \boldsymbol{u}^\alpha, \boldsymbol{u}^\alpha \rangle^{-1/2}$$

If $D^{(\alpha)}(G)$ appears n_α times in $\Delta^{(n)}(G)$, then there exist n_α and only n_α linearly independent basis sets of $D^{(\alpha)}(G)$ on $S^{(n)}$, each of which brings $\Delta^{(n)}(G)$ into $D^{(\alpha)}(G)$ as given by (7.2.13). It can be shown that it is always possible to construct a basis set of functions of $\boldsymbol{r}$ that reduces to a given basis set on $S^{(n)}$. Thus, one can state that, from all the basis sets of functions $\{u_i^\alpha(\boldsymbol{r})\}$ belonging to $D^{(\alpha)}(G)$, one can obtain n_α and only n_α linear independent basis sets of vectors on $S^{(n)}$. Thus, if $D^{(\alpha)}(G)$ is not contained in $\Delta^{(n)}(G)$, i.e. $n_\alpha = 0$, then all basis sets of $D^{(\alpha)}(G)$ become null on $S^{(n)}$. In the most important special case in which $n_\alpha = 1$, all basis sets belonging to $D^{(\alpha)}(G)$ are reduced to a single basis set of vectors on $S^{(n)}$ (apart from a constant factor), unless it is null. For convenience, we shall call a basis set $\{u_i^\alpha(\boldsymbol{r})\}$ *proper* on $S^{(n)}$ if the set is not null on $S^{(n)}$ and *improper* if the set is null on $S^{(n)}$. Then, all the bases belonging to $D^{(\alpha)}(G) \notin \Delta^{(n)}(G)$ are improper whereas a basis belonging to $D^{(\alpha)}(G) \in \Delta^{(n)}(G)$ may but need not be proper since there exist only n_α linearly independent basis sets on $S^{(n)}$ belonging to $D^{(\alpha)}(G)$. Let us summarize the results obtained thus far in the form of a theorem.

Theorem 7.2.2. Let n_α be the number of times a unirrep $D^{(\alpha)}(G)$ is contained in the permutation representation $\Delta^{(n)}(G)$, then from all bases of $D^{(\alpha)}(G)$ one can construct n_α and only n_α linearly independent bases $\{u_i^{\alpha\rho}(\boldsymbol{r}_\nu^0);\ \rho = 1, 2, \ldots, n_\alpha\}$ of $D^{(\alpha)}(G)$ on $S^{(n)} \in \Delta^{(n)}(G)$.

By combining this theorem with (7.2.8), we may state that, if $D^{(\alpha)}(G)$ is contained in $\Delta^{(n)}(G)$ n_α times, then there exist n_α and only n_α linearly independent SALCs of equivalent scalars $\{\phi_\nu(\boldsymbol{r})\}$ on $S^{(n)}$, which can be expressed by

$$\psi_i^{\alpha\rho}(\boldsymbol{r}) = \sum_\nu u_i^{\alpha\rho}(\boldsymbol{r}_\nu^0)\phi_\nu(\boldsymbol{r}); \qquad \rho = 1, 2, \ldots, n_\alpha \tag{7.2.14}$$

where the coefficient sets are linearly independent basis vectors on $S^{(n)}$ belonging to $D^{(\alpha)}(G)$.

7.2.4 Illustrative examples of the SALCs of equivalent scalars

Example 7.2.1. Construct the SALCs of the s-orbitals (s_1, s_2, s_3 and s_4) of the four equivalent H atoms in the methane molecule $CH_4 \in T_p$ $(= T_d)$.

Place the carbon atom at the center of a cube and the equivalent H atoms on the four vertices of the cube as shown in Figure 7.1. Introduce a Cartesian coordinate system with the origin at the central C atom and let each axis be parallel to an edge of the cube. Then the coordinates of H atoms are given, on a relative scale, by

$$(x_\nu^0, y_\nu^0, z_\nu^0):\ (1, 1, 1), (-1, -1, 1), (1, -1, -1), (-1, 1, -1) \tag{7.2.15a}$$

These four points define a set of four equivalent points $S^{(4)} = \{\boldsymbol{r}_\nu^0\}$ with respect to the point group T_p. Note that the relative coordinates given in (7.2.15a) are sufficient to determine the SALCs, because an elementary basis that describes the linear coefficients consists of homogeneous polynomials of a certain degree such that they affect only the normalization constants of the basis vectors on $S^{(n)}$.

From the elementary bases of T_p given in Table 7.1, we see that the non-null bases on $S^{(4)}$ are $1 \in A_1$ and $[x, y, z] \in T_2$. No other irreducible representations are

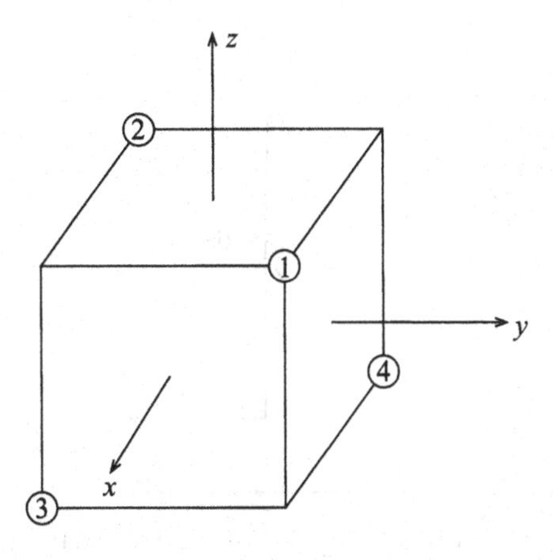

Figure 7.1. The coordinates of equivalent H atoms in the methane molecule $CH_4 \in T_p$.

contained in $\Delta^{(4)}$ since (7.2.4) is satisfied by $1 + 3 = 4$. Thus one may conclude that

$$\Delta^{(4)} = \underset{1}{A_1} + \underset{[x,\, y,\, z]}{T_2} \tag{7.2.15b}$$

Accordingly, from (7.2.8) the SALCs for A_1 and T_2 are given by

$$A_1: \quad \psi = \frac{1}{2}\sum_{\nu} s_\nu = (s_1 + s_2 + s_3 + s_4)/2$$

$$T_2: \quad \psi_x = \frac{1}{2}\sum_{\nu} x_\nu^0 s_\nu = (s_1 - s_2 + s_3 - s_4)/2$$

$$\psi_y = \frac{1}{2}\sum_{\nu} y_\nu^0 s_\nu = (s_1 - s_2 - s_3 + s_4)/2$$

$$\psi_z = \frac{1}{2}\sum_{\nu} z_\nu^0 s_\nu = (s_1 + s_2 - s_3 - s_4)/2 \tag{7.2.15c}$$

where the linear coefficients are normalized by setting the sums of their squares equal to unity. With the normalization, the transformation (7.2.15c) is described by a real orthogonal matrix given by

$$L = \frac{1}{2}\begin{bmatrix} 1 & 1 & 1 & 1 \\ 1 & -1 & 1 & -1 \\ 1 & -1 & -1 & 1 \\ 1 & 1 & -1 & -1 \end{bmatrix} \tag{7.2.15d}$$

Example 7.2.2. The SALCs of the s-orbitals of three equivalent Y atoms surrounding the central X atom in a planar molecule $XY_3 \in D_{3p}$ $(= D_{3h})$.

Take the coordinate origin at the central X atom and let the z-axis be perpendicular to the molecular plane. Then the coordinates (x_ν^0, y_ν^0) of Y atoms on the molecular plane define a set of three equivalent points $S^{(3)}$ as follows (see Figure 7.2):

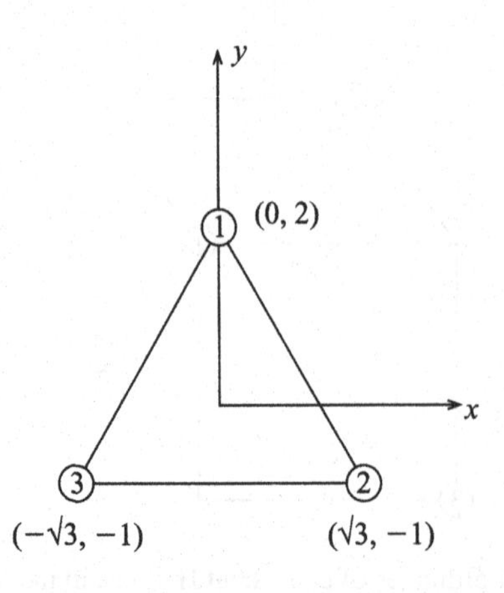

Figure 7.2. The coordinates of equivalent Y atoms in $XY_3 \in D_{3p}$.

$$\nu = 1, \quad 2, \quad 3$$
$$x_\nu^0 = 0, \sqrt{3}, -\sqrt{3}$$
$$y_\nu^0 = 2, -1, -1 \tag{7.2.16a}$$

Inspection of Table 7.2 leads to

$$\underset{}{\Delta^{(3)}} = \underset{1}{A_1} + \underset{[x,\, y]}{E_2} \tag{7.2.16b}$$

i.e. the non-null bases on $S^{(3)}$ are $1 \in A_1$ and $[x, y] \in E_2$; therefore, the required SALCs of the s-orbitals of the equivalent Y atoms are given, from (7.2.8), by

$$A_1: \qquad \psi_1(\boldsymbol{r}) = \frac{1}{\sqrt{3}} \sum_\nu s_\nu = \frac{1}{\sqrt{3}}(s_1 + s_2 + s_3)$$

$$E_2: \qquad \psi_x(\boldsymbol{r}) = \frac{1}{\sqrt{6}} \sum_\nu x_\nu^0 s_\nu = \frac{1}{\sqrt{2}}(s_2 - s_3)$$

$$\psi_y(\boldsymbol{r}) = \frac{1}{\sqrt{6}} \sum_\nu y_\nu^0 s_\nu = \frac{1}{\sqrt{6}}(2s_1 - s_2 - s_3) \tag{7.2.16c}$$

where each set of the linear coefficients is normalized by setting the sum of their squares equal to unity.

Note that three elementary bases are given for the unirrep E_2 of D_{3p} in Table 7.2. Instead of $[x, y]$ we could have used $[2xy, x^2 - y^2]$ to form $[\psi_x, \psi_y]$ in (7.2.16c); the result is the same because these two bases are mutually proportional on $S^{(3)}$:

$$[2xy, x^2 - y^2]_\nu^0 = -2[x, y]_\nu^0 \tag{7.2.16d}$$

in accordance with Theorem 7.2.2. Here $[x, y]_\nu^0$ means $[x_\nu^0, y_\nu^0]$ for example. This confirms also that both bases belong to the same matrix representation E_2 precisely (not up to a similarity transformation). The third basis of E_2, $[y\tilde{z}, -x\tilde{z}]$ does not apply for the present problem; it will be applied for the π-orbital system (see Example 7.3.2).

Example 7.2.3. The Bloch function and the Wannier function.

As a further application of (7.2.8) we shall discuss the Bloch function and the Wannier function in solid state physics. Consider a solid composed of N^3 lattice points located at $\{\boldsymbol{r}_\nu^0\}$ with the periodic boundary condition. The *Bloch function* $b_{\boldsymbol{k}}(\boldsymbol{r})$ may be defined as a basis function belonging to the irreducible representation $\Gamma_{\boldsymbol{k}}(T) = \{\exp(-\mathrm{i}\boldsymbol{k} \cdot \boldsymbol{t})\}$ of the translation group $T = \{\boldsymbol{t}\}$ of the solid, where the $\boldsymbol{t}$ are the lattice vectors and $\boldsymbol{k}$ is a wave vector in the first Brillouin zone of the reciprocal lattice (see Chapter 14). Obviously, an elementary basis function belonging to $\Gamma_{\boldsymbol{k}}(T)$ is given by $\exp(\mathrm{i}\boldsymbol{k} \cdot \boldsymbol{r})$. Now, let $\phi(\boldsymbol{r} - \boldsymbol{r}_\nu^0)$ be an atomic s-orbital located at $\boldsymbol{r}_\nu^0$. Then, from (7.2.8), a Bloch function belonging to $\Gamma_{\boldsymbol{k}}(T)$ is given by

$$b_{\boldsymbol{k}}(\boldsymbol{r}) = N^{-3/2} \sum_\nu \exp(\mathrm{i}\boldsymbol{k} \cdot \boldsymbol{r}_\nu^0)\, \phi(\boldsymbol{r} - \boldsymbol{r}_\nu^0) \tag{7.2.17a}$$

The above basis function is called the tight-binding approximation for the Bloch function. Now, the inverse transformation of (7.2.17a) gives the equivalent set of s-orbitals with respect to the translation group,

$$\phi(\boldsymbol{r} - \boldsymbol{r}_\nu^0) = N^{-3/2} \sum_{\boldsymbol{k}} \exp(-i\boldsymbol{k} \cdot \boldsymbol{r}_\nu^0)\, b_{\boldsymbol{k}}(\boldsymbol{r}) \tag{7.2.17b}$$

known as the *Wannier function.*

7.3 SALCs of equivalent orbitals in general

7.3.1 The general expression of SALCs

Let $D^{(A)}(G)$ be a unitary representation of a symmetry group G that may but need not be irreducible. The problem is to construct the SALC of any given equivalent orbitals $\{f_s^A(\boldsymbol{r}, \boldsymbol{r}_\nu^0);\ s = 1, 2, \ldots, d_A\}$ belonging to a direct product representation $\Delta^{(n)} \times D^{(A)}$ of G. We shall achieve this by extending the SALC of equivalent scalar orbitals given by (7.2.8).

We shall begin with an invariant function $\phi(\boldsymbol{r}, \boldsymbol{r}')$ with respect to G expressed by 'a scalar product' of any two basis sets $\{g_s^A(\boldsymbol{r}, \boldsymbol{r}')\}$ and $\{f_s^A(\boldsymbol{r}, \boldsymbol{r}')\}$ in the carrier space of the unitary representation $D^{(A)}(G)$:

$$\phi(\boldsymbol{r}, \boldsymbol{r}') = \sum_s g_s^A(\boldsymbol{r}, \boldsymbol{r}')^* f_s^A(\boldsymbol{r}, \boldsymbol{r}') \tag{7.3.1}$$

This is the only invariant function in the bilinear forms of the two basis sets. It is emphasized here that each basis set may depend only on one vector $\boldsymbol{r}$ or $\boldsymbol{r}'$. Thus, in a special case, we have

$$\phi(\boldsymbol{r}, \boldsymbol{r}') = \sum_s g_s^A(\boldsymbol{r}')^* f_s^A(\boldsymbol{r}) \tag{7.3.2}$$

The well-known addition theorem of the Legendre polynomials is a further special case of this result. For the theory of SALCs, it suffices to introduce a two-point invariant function defined by

$$\phi(\boldsymbol{r}, \boldsymbol{r}') = \sum_s v_s^A(\boldsymbol{r}')^* f_s^A(\boldsymbol{r}, \boldsymbol{r}') \tag{7.3.3}$$

where $\{v_s^A(\boldsymbol{r}')\}$ is an elementary basis of $D^{(A)}(G)$. Then, the corresponding set of equivalent scalar functions (or ϕ-functions) on $S^{(n)} = \{\boldsymbol{r}_\nu^0\}$ is given by

$$\phi(\boldsymbol{r}, \boldsymbol{r}_\nu^0) = \sum_s v_s^A(\boldsymbol{r}_\nu^0)^* f_s^A(\boldsymbol{r}, \boldsymbol{r}_\nu^0); \qquad \nu = 1, 2, \ldots, n \tag{7.3.4}$$

unless the set is null. This provides the general expression for a set of equivalent scalars with respect to G formed by the equivalent orbitals $\{f_s^A(\boldsymbol{r}, \boldsymbol{r}_\nu^0)\}$. Later, in Section 7.5, the set (7.3.4) will be applied to the theory of hybrid atomic orbitals.

Now, let $D^{(\gamma)}(G)$ be a unirrep contained in the direct product representation $\Delta^{(n)} \times D^{(A)}$ of G, and let $\{u_k^\gamma(\boldsymbol{r})\}$ be an elementary basis of $D^{(\gamma)}(G)$ and $\phi(\boldsymbol{r}, \boldsymbol{r}')$ be the invariant function defined by (7.3.3). Then, a two-point basis of $D^{(\gamma)}(G)$ is defined by

$$f_k^\gamma(\boldsymbol{r}, \boldsymbol{r}') = u_k^\gamma(\boldsymbol{r})\phi(\boldsymbol{r}, \boldsymbol{r}') = \sum_s u_k^\gamma(\boldsymbol{r}') v_s^A(\boldsymbol{r}')^* f_s^A(\boldsymbol{r}, \boldsymbol{r}') \tag{7.3.5a}$$

and its characteristic SALC defined by the correspondence theorem (7.2.7) provides the $D^{(\gamma)}$ SALC of the equivalent orbitals $\{f_s^A(\boldsymbol{r}, \boldsymbol{r}_\nu^0)\}$

$$\psi_k^\gamma(\boldsymbol{r}) = \sum_{\nu,s} u_k^\gamma(\boldsymbol{r}_\nu^0) v_s^A(\boldsymbol{r}_\nu^0)^* f_s^A(\boldsymbol{r}, \boldsymbol{r}_\nu^0); \qquad k = 1, 2, \ldots, d_\alpha \tag{7.3.5b}$$

unless the set is null. This set, however, has a very limited applicability as it stands, because the set is null unless both $D^{(\gamma)}(G)$ and $D^{(A)}(G)$ are contained in $\Delta^{(n)}(G)$ according to Theorem 7.2.2 (even though $D^{(\gamma)} \in \Delta^{(n)} \times D^{(A)}$). In the following, we shall remove this difficulty by introducing an operator basis, which transforms exactly like an ordinary basis except for the fact that it contains the differential operator $\boldsymbol{\nabla} = (\partial_x, \partial_y, \partial_z)$, which transforms like the Cartesian coordinates (x, y, z) under an orthogonal transformation. Here $\partial_x = \partial/\partial x$, $\partial_y = \partial/\partial y$ and $\partial_z = \partial/\partial z$.

Let $\{\mathring{T}_k^\gamma(\boldsymbol{r})\}$ be a basis set of linear operators belonging to a unirrep $D^{(\gamma)}(G)$ such that

$$\mathring{T}_k^\gamma(R^{-1}\boldsymbol{r}) = \sum_s \mathring{T}_s^\gamma(\boldsymbol{r}) D_{sk}^{(\gamma)}(R)$$

for all $R \in G$. Then a two-point basis belonging to $D^{(\gamma)}(G)$ is defined by

$$\text{(I)} \qquad f_k^\gamma(\boldsymbol{r}, \boldsymbol{r}') = \sum_{s=1}^{d_A} [\mathring{T}_k^\gamma(\boldsymbol{r}') v_s^A(\boldsymbol{r}')^*] f_s^A(\boldsymbol{r}, \boldsymbol{r}') \tag{7.3.6a}$$

where the operators act only on $v_s^A(\boldsymbol{r}')^*$, and the corresponding characteristic set

$$\psi_k^\gamma(\boldsymbol{r}) = \sum_{\nu=1}^{n} \sum_{s=1}^{d_A} [\mathring{T}_k^\gamma(\boldsymbol{r}') v_s^A(\boldsymbol{r}')^*]_{\boldsymbol{r}'=\boldsymbol{r}_\nu^0} f_s^A(\boldsymbol{r}, \boldsymbol{r}_\nu^0) \tag{7.3.6b}$$

provides a general expression for the $D^{(\gamma)}$ SALC of the equivalent orbitals $\{f_s^A(\boldsymbol{r}, \boldsymbol{r}_\nu^0)\}$, unless it is null.

An operator basis $\{\mathring{T}_k^\gamma(\boldsymbol{r})\} \in D^{(\gamma)}(G)$ is easily constructed through an appropriate two-point elementary basis $T_k^\gamma(\boldsymbol{r}, \boldsymbol{r}') \in D^{(\gamma)}(G)$ by replacing $\boldsymbol{r}'$ with $\boldsymbol{\nabla} = (\partial_x, \partial_y, \partial_z)$,

$$\mathring{T}_k^\gamma(\boldsymbol{r}) \equiv T_k^\gamma(\boldsymbol{r}, \boldsymbol{\nabla}) \tag{7.3.7}$$

In (7.3.6a), the polynomial degree of $T_k^\gamma(\boldsymbol{r}, \boldsymbol{\nabla})$ with respect to $\boldsymbol{\nabla}$ should be no higher than the degree of the homogeneous polynomial of the elementary basis $\{v_s^A(\boldsymbol{r})\}$. For most cases, it suffices to choose the operator basis linear in $\boldsymbol{\nabla}$ (see (7.3.11)).

To understand the non-null condition imposed on (7.3.6b), let us consider the case in which $D^{(A)}(G)$ is not contained in $\Delta^{(n)}(G)$. Then an elementary basis $\{v_s^A(\boldsymbol{r})\}$ of $D^{(A)}(G)$ is null for all $\boldsymbol{r} = \boldsymbol{r}_\nu^0 \in S^{(n)}$, i.e. the $\boldsymbol{r}_\nu^0$ are the roots of the set of polynomial equations $\{v_s^A(\boldsymbol{r}) = 0\}$ of $\boldsymbol{r}$. This does not necessarily mean that the $\boldsymbol{r}_\nu^0$ are the roots of all the derivatives of the polynomials. Thus, if the order of the operator $\boldsymbol{\nabla}$ in $\mathring{T}_k^\gamma$ is less than or equal to the degree of the polynomials $\{v_s^A(\boldsymbol{r})\}$, the coefficient matrix in (7.3.6b) need not be null. The crucial point is that we can easily construct such a set of basis operators that yields a non-null SALC, as will be seen through examples. There exists also a formal proof for the existence of such an operator, see Kim (1981a). There are infinitely many elementary bases for $D^{(\gamma)}(G)$ and $D^{(A)}(G)$, but there exist only n_γ^A linear independent coefficient matrices in (7.3.6b), where n_γ^A is the number of times $D^{(\gamma)}(G)$ is contained in $D^{(A)} \times \Delta^{(n)}$ of G.

An alternative general expression for a two-point basis may be defined by

$$\text{(II)} \qquad \overline{f}^{\gamma}_{k}(\boldsymbol{r}, \boldsymbol{r}') = \sum_{s=1}^{d_A} f^{A}_{s}(\boldsymbol{r}, \boldsymbol{r}')\mathring{T}^{A}_{s}(\boldsymbol{r}')^{*} v^{\gamma}_{k}(\boldsymbol{r}') \tag{7.3.8a}$$

which may be more convenient than (7.3.6a), if the polynomial degree of $\mathring{T}^{A}_{s}(\boldsymbol{r}')$ is lower than that of $\mathring{T}^{\gamma}_{k}(\boldsymbol{r}')$. The corresponding characteristic set is given by

$$\overline{\psi}^{\gamma}_{k}(\boldsymbol{r}) = \sum_{s,\nu} f^{A}_{s}(\boldsymbol{r}, \boldsymbol{r}^{0}_{\nu})[\mathring{T}^{A}_{s}(\boldsymbol{r}')^{*} v^{\gamma}_{k}(\boldsymbol{r}')]_{\boldsymbol{r}'=\boldsymbol{r}^{0}_{\nu}} \tag{7.3.8b}$$

One may use either the expression I or II, whichever is more convenient. See the illustrative examples given at the end of this section.

7.3.2 *Two-point bases and operator bases*

To facilitate the application of the correspondence theorem, we shall give some examples of two-point bases and the corresponding operator bases frequently encountered in point groups. The elementary bases to be considered are

$$\begin{aligned}
&1; \qquad x, y, z; \qquad x^2 + y^2 + z^2 \\
&u = 2z^2 - x^2 - y^2, \qquad v = \sqrt{3}(x^2 - y^2) \\
&\tilde{x} = yz' - zy', \qquad \tilde{y} = zy' - xz', \qquad \tilde{z} = xy' - yx' \\
&yz, \qquad zx, \qquad xy \\
&xyz
\end{aligned} \tag{7.3.9}$$

where $[\tilde{x}, \tilde{y}, \tilde{z}]$ is already a two-point basis. Let $\{v^{\gamma}_{k}(\boldsymbol{r})\}$ be an elementary basis, then, via the general expression II given by (7.3.8a), a two-point basis corresponding to a one-point basis $\{v^{\gamma}_{k}(\boldsymbol{r})\}$ is formed by

$$v^{\gamma}_{k}(\boldsymbol{r}, \boldsymbol{r}') = (x\partial_{x'} + y\partial_{y'} + z\partial_{z'})v^{\gamma}_{k}(\boldsymbol{r}') \tag{7.3.10a}$$

where we have used the fact that $xx' + yy' + zz'$ is a scalar. Thus, the two-point bases corresponding to the one-point bases given in (7.3.9) are

$$\begin{aligned}
&[x, y, z] \\
&x^2 + y^2 + z^2 \sim xx' + yy' + zz' \\
&[u, v] \sim [2zz' - xx' - yy', \sqrt{3}(xx' - yy')] \\
&[\tilde{x}, \tilde{y}, \tilde{z}] \sim [yz' - zy', zx' - xz', xy' - yx'] \\
&[yz, zx, xy] \sim [yz' + zy', zx' + xz', xy' + yx'] \\
&xyz \sim x(yz)' + y(zx)' + z(xy)'
\end{aligned} \tag{7.3.10b}$$

where $(yz)' \equiv y'z'$ for example. The bases $[x, y, z]$ and $[\tilde{x}, \tilde{y}, \tilde{z}]$ are included here again for convenience of later use. These two-point bases given in (7.3.10b) could have been written down by inspection from the one-point bases given in (7.3.9) with proper symmetrizations.

Now, from (7.3.7), the operator bases corresponding to the two point bases in (7.3.10b) are obtained by replacing $\boldsymbol{r}'$ with $\boldsymbol{\nabla}$. These are linear in $[\partial_x, \partial_y, \partial_z]$ and given by

$$
\begin{aligned}
&(x, y, z) \sim (\partial_x, \partial_y, \partial_z) \\
&x^2 + y^2 + z^2 \sim x\partial_x + y\partial_y + z\partial_z \\
&[u, v] \sim [2z\partial_z - x\partial_x - y\partial_y, \sqrt{3}(x\partial_x - y\partial_y)] \\
&[\tilde{x}, \tilde{y}, \tilde{z}] \sim [y\partial_z - z\partial_y, z\partial_x - x\partial_z, x\partial_y - y\partial_x] \\
&[yz, zx, xy] \sim [y\partial_z + z\partial_y, z\partial_x + x\partial_z, x\partial_y + y\partial_x] \\
&xyz \sim yz\partial_x + zx\partial_y + xy\partial_z
\end{aligned} \tag{7.3.11}
$$

7.3.3 *Notations for equivalent orbitals*

Let $[f_\xi(\boldsymbol{r}), f_\eta(\boldsymbol{r}), \ldots]$ be a basis set of functions which transforms like an elementary basis $[\xi(\boldsymbol{r}), \eta(\boldsymbol{r}), \ldots]$ under the symmetry operations of a given group G. Then we may call $f_\xi(\boldsymbol{r})$ a ξ-like orbital, $f_\eta(\boldsymbol{r})$ a η-like orbital etc., and write

$$\xi_\nu = f_\xi(\boldsymbol{r} - \boldsymbol{r}_\nu^0), \qquad \eta_\nu = f_\eta(\boldsymbol{r} - \boldsymbol{r}_\nu^0), \ldots \tag{7.3.12a}$$

For example, two basis sets $[x, y, z]$ and $[yz, zx, xy]$ belong to the same unirrep T_2 of T_p so that x_ν may mean

$$x_\nu = (x - x_\nu^0)\phi(|\boldsymbol{r} - \boldsymbol{r}_\nu^0|) \qquad \text{or} \qquad (y - y_\nu^0)(z - z_\nu^0)\phi(|\boldsymbol{r} - \boldsymbol{r}_\nu^0|) \tag{7.3.12b}$$

where $\boldsymbol{r}_\nu^0 = (x_\nu^0, y_\nu^0, z_\nu^0)$ and $\phi(|\boldsymbol{r} - \boldsymbol{r}_\nu^0|)$ is an appropriate scalar factor. The equivalent basis $\{[x_\nu, y_\nu, z_\nu]\}$ thus defined may be called a set of *p-like equivalent orbitals*.

To show the effectiveness of the notation introduced above we apply the correspondence theorem on the two-point bases $[2zz' - xx' - yy', 3^{1/2}(xx' - yy')]$ given in (7.3.10b) and obtain the SALC of p-like orbitals $\{[x_\nu, y_\nu, z_\nu]\}$

$$\psi_u(\boldsymbol{r}) = \sum_\nu (2z_\nu^0 z_\nu - x_\nu^0 x_\nu - y_\nu^0 y_\nu), \qquad \psi_v(\boldsymbol{r}) = \sum_\nu \sqrt{3}(x_\nu^0 x_\nu - y_\nu^0 y_\nu) \tag{7.3.13a}$$

which is a special case of (7.2.9). Note *the formal resemblance* between the elementary basis $[u, v]$ and its characteristic SALC $[\psi_u, \psi_v]$. Such a correspondence between the bases holds irrespective of the group G to which $[u, v]$ belongs. Analogously, for the basis of xyz, the corresponding SALC of the p-like orbitals is given, via (7.3.10b), by

$$\psi_{xyz}(\boldsymbol{r}) = \sum_\nu (yz)_\nu^0 x_\nu + (zx)_\nu^0 y_\nu + (xy)_\nu^0 z_\nu \tag{7.3.13b}$$

where $(yz)_\nu^0 = y_\nu^0 z_\nu^0$ etc.

7.3.4 *Alternative elementary bases*

If we use the general expression (7.3.6a), we can form an alternative basis from a given elementary basis for a unirrep. We note that a direct product of a unirrep Γ_α with the identity representation Γ_1 equals Γ_α; i.e. $\Gamma_\alpha \times \Gamma_1 = \Gamma_\alpha$. Thus their direct product basis also belongs to the same unirrep because the basis of Γ_1 is an invariant function of the group. For example, from the elementary bases of T_p given by Table 7.1, we know that

$$[\boldsymbol{r}' \times \boldsymbol{\nabla}] = [y'\partial_z - z'\partial_y, z'\partial_x - x'\partial_z, x'\partial_y - y'\partial_x] \in T_1, \qquad xyz \in A_1$$

Thus, applying the operator basis $\boldsymbol{r}' \times \boldsymbol{\nabla}$ on xyz we obtain an alternative basis of T_1 given by

$$[x'(y^2 - z^2),\ y'(z^2 - x^2),\ z'(x^2 - y^2)] \in T_1$$

which reduces to a one-point basis by $\boldsymbol{r}' \to \boldsymbol{r}$. Moreover, applying $[\partial_x, \partial_y, \partial_z] \in T_2$ on $xyz \in A_1$, we obtain $[yz, zx, xy] \in T_2$. Also, applying $[\partial'_x, \partial'_y, \partial'_z] \in T_2$ on $[u(\boldsymbol{r}')u(\boldsymbol{r}) + v(\boldsymbol{r}')v(\boldsymbol{r})] \in A_1$, where $[u, v] \in E$, we obtain

$$[x'(3x^2 - r^2), \qquad y'(3y^2 - r^2), \qquad z'(3z^2 - r^2)] \in T_2$$

Accordingly, three bases of T_2 given in Table 7.1 belong to exactly the same unirrep T_2 of T_p (not up to a similarity transformation).

Analogously, from the elementary bases of D_{3p} given in Table 7.2, we apply $[\partial'_x, \partial'_y] \in E_2$ on $\tilde{z}\tilde{z} \in A_1$, where $\tilde{z} = xy' - yx' \in A_2$, to obtain an alternative basis of E_2 given by

$$[\partial_{x'}, \partial_{y'}]\tilde{z}\tilde{z} \propto [y\tilde{z}, -x\tilde{z}] \in E_2 \tag{7.3.14a}$$

Moreover, since $y^3 - 3yx^2$ belongs to the identity representation A_1 of D_{3p} according to Table 7.2, we obtain another basis for E_2 given by

$$[\partial_x, \partial_y](y^3 - 3yx^2) \propto [2xy, x^2 - y^2] \in E_2 \tag{7.3.14b}$$

Thus, we may conclude that the three bases $[x, y]$, $[z\tilde{y}, -x\tilde{z}]$ and $[2xy, x^2 - y^2]$ belong to the same unirrep E_2 of D_{3p} as given in Table 7.2. Furthermore, applying the operator basis $[y\partial_z - z\partial_y, z\partial_x - x\partial_z] \in E_1$ of D_{3p} on $x^2 + y^2 \in A_1$ we obtain $[yz, -zx] \in E_1$. Thus $[\tilde{x}, \tilde{y}]$ and $[yz, -xz]$ belong to the same unirrep E_1 of D_{3p}.

The above method can be extended to a case in which a direct product of two unirreps Γ_α and Γ_β is equivalent to a unirrep Γ_γ, i.e. $\Gamma_\alpha \times \Gamma_\beta \simeq \Gamma_\gamma$. In this case, the direct product basis of $\Gamma_\alpha \times \Gamma_\beta$ is equivalent to a basis of Γ_γ but need not be proportional to the latter. For example, from the character table of D_{3p}, we know that $E_2 \times A_2 \simeq E_2$. Now, from $[x, y] \in E_2$ and $x^3 - 3xy^2 \in A_2$, a direct product basis of $E_2 \times A_2$ is given by

$$[\partial_x, \partial_y](x^3 - 3xy^2) = 3[(x^2 - y^2, -2xy)]$$

which is equivalent to the basis $[2xy, x^2 - y^2]$ of E_2 given by (7.3.14b) but not proportional.

7.3.5 Illustrative examples

Using the correspondence theorem we shall work out the SALCs of equivalent orbitals for typical rigid molecules.

Example 7.3.1. Construct the SALCs of the d-orbitals $\{[u_\nu, v_\nu]\}$ of four equivalent Y atoms surrounding the central X atom in a tetrahedral molecule $XY_4 \in T_p$. Here, the basis $[u, v]$ is defined in Table 7.1 and $u_\nu = 2z_\nu^2 - x_\nu^2 - y_\nu^2$ and $v_\nu = 3^{1/2}(x_\nu^2 - y_\nu^2)$ following the notation (7.3.12a) with a p-like orbital $[x_\nu, y_\nu, z_\nu]$.

According to Table 7.1, the elementary basis $[u, v]$ belongs to E of T_p. Hence, the equivalent orbitals $\{[u_\nu, v_\nu]\}$ on $S^{(4)}$ belong to $E \times \Delta^{(4)}$, where $\Delta^{(4)} = A_1 + T_2$ from (7.2.15b). The irreducible representations contained in the direct product are

$$E \times \Delta^{(4)} = E(A_1 + T_2) = \underset{[u,\, v]}{E} + \underset{[\tilde{x},\, \tilde{y},\, \tilde{z}]}{T_1} + \underset{[x,\, y,\, z]}{T_2} \tag{7.3.15a}$$

with use of the character table of T_p given in the Appendix. Note here that the unirreps E and T_1 are not contained in $\Delta^{(4)}$. Now, the E SALC is given directly by the characteristic SALC of $[u, v]$

$$\psi_u(\boldsymbol{r}) = \frac{1}{2}\sum_\nu u_\nu, \qquad \psi_v(\boldsymbol{r}) = \frac{1}{2}\sum_\nu v_\nu \tag{7.3.15b}$$

To form T_1 and T_2 SALCs of $\{[u_\nu, v_\nu]\}$, we use the general expression (7.3.6b) written in the form

$$\psi_k^\gamma(\boldsymbol{r}) = \sum_\nu [\mathring{T}_k^\gamma(\boldsymbol{r}')(u(\boldsymbol{r}')u_\nu(\boldsymbol{r}) + v(\boldsymbol{r}')v_\nu(\boldsymbol{r}))]_{\boldsymbol{r}'=\boldsymbol{r}_\nu^0}$$

then, substituting $[\boldsymbol{r} \times \boldsymbol{\nabla}] \in T_1$ and $\boldsymbol{\nabla} \in T_2$ for $[\mathring{T}_k^\gamma]$, we obtain, with use of $x_\nu^0 y_\nu^0 = z_\nu^0$ from (7.2.15a),

$$T_1: \qquad \psi_{\tilde{x}}(\boldsymbol{r}) = -\frac{1}{4}\sum_\nu x_\nu^0(\sqrt{3}u_\nu + v_\nu), \qquad \psi_{\tilde{y}}(\boldsymbol{r}) = \frac{1}{4}\sum_\nu y_\nu^0(\sqrt{3}u_\nu - v_\nu)$$

$$\psi_{\tilde{z}}(\boldsymbol{r}) = \frac{1}{2}\sum_\nu z_\nu^0 v_\nu$$

$$T_2: \qquad \psi_x(\boldsymbol{r}) = \frac{1}{4}\sum_\nu x_\nu^0(-u_\nu + \sqrt{3}v_\nu), \qquad \psi_y(\boldsymbol{r}) = \frac{1}{4}\sum_\nu y_\nu^0(u_\nu + \sqrt{3}v_\nu)$$

$$\psi_z(\boldsymbol{r}) = \frac{1}{2}\sum_\nu z_\nu^0 u_\nu \tag{7.3.15c}$$

It is interesting to note that, if we use the identities

$$-\tfrac{1}{2}(\sqrt{3}u + v) = \sqrt{3}(y^2 - x^2), \qquad \tfrac{1}{2}(\sqrt{3}u - v) = \sqrt{3}(z^2 - x^2), \qquad v = \sqrt{3}(x^2 - y^2)$$

$$\tfrac{1}{2}(-u + \sqrt{3}v) = 3x^2 - r^2, \qquad -\tfrac{1}{2}(u + \sqrt{3}v) = 3y^2 - r^2, \qquad u = 3z^2 - r^2 \tag{7.3.15d}$$

then the above SALCs can be rewritten as

$$T_1: \qquad \psi_{\tilde{x}} = \frac{\sqrt{3}}{2}\sum_\nu x_\nu^0(y^2 - z^2)_\nu, \qquad \psi_{\tilde{y}} = \frac{\sqrt{3}}{2}\sum_\nu y_\nu^0(x^2 - x^2)_\nu,$$

$$\psi_{\tilde{z}} = \frac{\sqrt{3}}{2}\sum_\nu z_\nu^0(x^2 - x^2)_\nu$$

$$T_2: \qquad \psi_x = \frac{1}{2}\sum_\nu x_\nu^0(3x^2 - r^2)_\nu, \qquad \psi_y = \frac{1}{2}\sum_\nu y_\nu^0(3y^2 - r^2)_\nu,$$

$$\psi_z = \frac{1}{2}\sum_\nu z_\nu^0(3z^2 - r^2)_\nu$$

which can be understood from the alternative bases of T_1 and T_2 given in Table 7.1. Note here that $(y^2 - z^2)_\nu = y_\nu^2 - z_\nu^2$ etc.

Example 7.3.2. Construct the SALCs of p-like orbitals $\{x_\nu, y_\nu, z_\nu\}$ belonging to three equivalent Y atoms surrounding the central X atom in a planar molecule $XY_3 \in D_{3p}$.

According to Table 7.2, $z \in B_2$ and $[x, y] \in E_2$, so that $\{z_\nu\} \in B_2 \times \Delta^{(3)}$ and $\{[x_\nu, y_\nu]\} \in E_2 \times \Delta^{(3)}$. The set of three equivalent points $S^{(3)}$ was defined by (7.2.16a) and $\Delta^{(3)} = A_1 + E_2$ from (7.2.16b). The decompositions of the direct products yield, with use of the character table of D_{3p} given in the Appendix,

$$B_2 \times \Delta^{(3)} = B_2(A_1 + E_2) = \underset{z}{B_2} + \underset{[yz,\ -xz]}{E_1}$$

$$E_2 \times \Delta^{(3)} = E_2(A_1 + E_2) = \underset{1}{A_1} + \underset{\tilde{z}}{A_2} + \underset{[x,\ y],\ [y\tilde{z},\ -x\tilde{z}]}{2E_2} \tag{7.3.16a}$$

Let us first construct the equivalent orbitals corresponding to one-dimensional unirreps A_1, A_2 and B_2 of D_{3p}. From their elementary bases

$$x'x + y'y \in A_1; \qquad \tilde{z} = x'y - y'x \in A_2; \qquad z \in B_2$$

the equivalent orbitals on $S^{(3)}$ are given by

$$\sigma_\nu = (x_\nu^0 x_\nu + y_\nu^0 y_\nu)/2 \in A_1 \times \Delta^{(3)}$$

$$\tilde{z}_\nu = (x_\nu^0 y_\nu - y_\nu^0 x_\nu)/2 \in A_2 \times \Delta^{(3)}$$

$$z_\nu \in B_2 \times \Delta^{(3)} \tag{7.3.16b}$$

where $\{\sigma_\nu\}$ is a set of σ-orbitals (cylindrical with respect to each chemical bond) while $\{\tilde{z}_\nu\}$ and $\{z_\nu\}$ are sets of π-orbitals (perpendicular with respect to the chemical bond). See Figure 7.3. Since the three orbitals σ_ν, $\tilde{z}_\nu$ and z_ν for a given ν are mutually

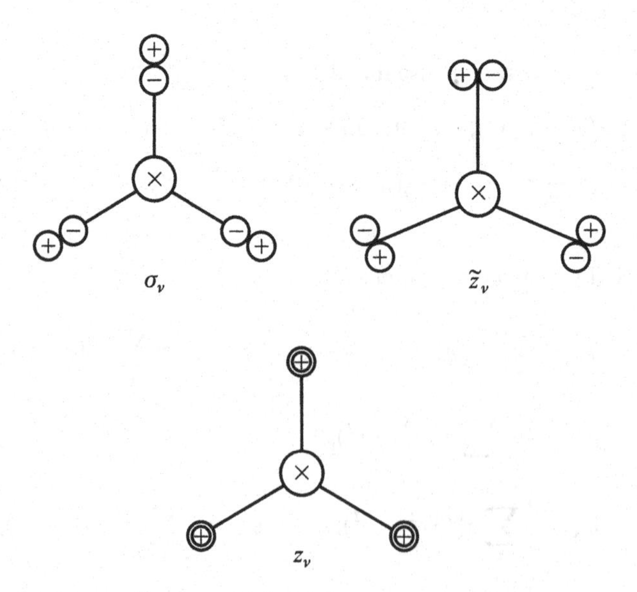

Figure 7.3. The σ, π' and π orbitals for the three equivalent Y atoms surrounding the central X atom in a planar molecule $XY_3 \in D_{3p}$. Note that $\sigma_\nu = \boldsymbol{r}_\nu^0 \cdot \boldsymbol{r}_\nu$ is cylindrical with respect to $\boldsymbol{r}_\nu^0$, whereas $\tilde{z}_\nu^0 = [\boldsymbol{r}_\nu^0 \times \boldsymbol{r}_\nu]_z = x_\nu^0 y_\nu - y_\nu^0 x_\nu$ is on the molecular x, y plane and perpendicular to $\boldsymbol{r}_\nu^0$, and obviously z_ν is perpendicular to the x, y plane.

orthogonal and span the complete space spanned by the p-orbitals x_ν, y_ν and z_ν, all of the SALCs of the equivalent orbitals $\{x_\nu, y_\nu, z_\nu\}$ are constructed by those of $\{\sigma_\nu, \tilde{z}_\nu, z_\nu\}$. Thus, from the correspondence theorem, the SALCs of p-orbitals for $XY_3 \in D_{3p}$ are given by

$$A_1: \quad \psi_1 = \sum_\nu \sigma_\nu/\sqrt{3}$$

$$A_2: \quad \psi_{\tilde{z}} = \sum_\nu \tilde{z}_\nu/\sqrt{3}$$

$$B_2: \quad \psi_z = \sum_\nu z_\nu/\sqrt{3}$$

$$E_1: \quad \psi_{yz} = \sum_\nu y_\nu^0 z_\nu/\sqrt{6}, \qquad \psi_{-xz} = -\sum_\nu x_\nu^0 z_\nu/\sqrt{6}$$

$$2E_2: \quad \psi_x = \sum_\nu x_\nu^0 \sigma_\nu/\sqrt{6}, \qquad \psi_y = \sum_\nu y_\nu^0 \sigma_\nu/\sqrt{6}$$

$$\psi_{y\tilde{z}} = \sum_\nu y_\nu^0 \tilde{z}_\nu/\sqrt{6}, \qquad \psi_{-x\tilde{z}} = -\sum_\nu x_\nu^0 \tilde{z}_\nu/\sqrt{6} \qquad (7.3.16c)$$

where the linear coefficients are normalized by setting the sums of their squares equal to unity.

An alternative choice for the two bases of E_2 would be those corresponding to the elementary bases $[x, y]$ and $[2xy, x^2 - y^2]$ from Table 7.2:

$$E_2: \quad \psi_x(\boldsymbol{r}) = \sum_\nu x_\nu/\sqrt{3}, \qquad \psi_y(\boldsymbol{r}) = \sum_\nu y_\nu/\sqrt{3}$$

$$\psi_{2xy}(\boldsymbol{r}) = \sum_\nu (x_\nu^0 x_\nu + y_\nu^0 y_\nu)/\sqrt{6}, \qquad \psi_{x^2-y^2}(\boldsymbol{r}) = \sum_\nu (x_\nu^0 y_\nu - y_\nu^0 x_\nu)/\sqrt{6}$$

$$(7.3.16d)$$

The previous bases given for E_2 in (7.3.16c) are useful for the theory of chemical bonding whereas the alternative bases given above are useful in the theory of molecular vibration: regard (x_ν, y_ν) as the displacement vector of the νth atom in the molecular symmetry plane, then the set $[\psi_x, \psi_y]$ represents the pure translational mode in the x- and y-directions, whereas $(\psi_{x^2-y^2}, \psi_{-2xy})$ represents the internal mode of vibration (see Section 7.6).

Example 7.3.3. Construct the SALCs of the p-orbitals $\{x_\nu, y_\nu, z_\nu\}$ belonging to n equivalent Y atoms surrounding the X atom in a planar molecule $XY_n \in D_{n\mathrm{h}}$.

This problem is a straightforward extension of Example 7.3.2. Take the coordinate origin on the central atom X and let the z-axis be perpendicular to the molecular plane. Then the σ-orbitals and the two kinds of π-orbitals are

$$\sigma_\nu = x_\nu^0 x_\nu + y_\nu^0 y_\nu, \qquad \tilde{z}_\nu = x_\nu^0 y_\nu - y_\nu^0 x_\nu, \qquad z_\nu$$

where $\nu = 1, 2, \ldots, n$. Now let $D^{(\alpha)}(G)$ be a unirrep contained in $\Delta^{(n)}$ and let $\{u_i^\alpha(\boldsymbol{r})\}$ be a proper basis of $D^{(\alpha)}(G)$ on $S^{(n)}$, then the unirreps of $D_{n\mathrm{h}}$ are characterized by the direct product bases $u_i^a(\boldsymbol{r})\sigma$, $u_i^a(\boldsymbol{r})\tilde{z}$ and $u_i^a(\boldsymbol{r})z$ because σ, $\tilde{z}$ and z are bases of one-dimensional unirreps of D_{nh}. Thus the required SALCs are given by

$$\psi_i^{a\sigma}(\boldsymbol{r}) = \sum_\nu u_i^a(\boldsymbol{r}_\nu^0)\sigma_\nu$$

$$\psi_i^{a\tilde{z}}(\boldsymbol{r}) = \sum_\nu u_i^a(\boldsymbol{r}_\nu^0)\tilde{z}_\nu$$

$$\psi_i^{az}(\boldsymbol{r}) = \sum_\nu u_i^a(\boldsymbol{r}_\nu^0)z_\nu \tag{7.3.17}$$

for every unirrep $D^{(a)}$ contained in $\Delta^{(n)}$. There exists a total of $3n$ basis functions in (7.3.17), which exhausts all the SALCs that can be constructed from $3n$ p-orbitals of n equivalent Y atoms in XY_n.

Note that the SALCs given by $\{\psi_i^{az}(\boldsymbol{r})\}$ in (7.3.17) describe the SALCs of the conjugated π-bond system of a plane molecule like the benzene molecule.

Exercise. Construct the SALCs of the p-orbitals $\{z_\nu\}$ of the six carbon atoms in the benzene molecule $C_6H_6 \in D_{6i}$ using $\{\psi_i^{az}(\boldsymbol{r})\}$ given in (7.3.17).

Hint. The coordinates of the six C atoms on the molecular plane may be given by

$$\begin{array}{lcccccc} \nu = 1 & 2 & 3 & 4 & 5 & 6 \\ x_\nu^0 = 1 & \frac{1}{2} & -\frac{1}{2} & -1 & -\frac{1}{2} & \frac{1}{2} \\ y_\nu^0 = 0 & \dfrac{\sqrt{3}}{2} & \dfrac{\sqrt{3}}{2} & 0 & -\dfrac{\sqrt{3}}{2} & -\dfrac{\sqrt{3}}{2} \end{array}$$

Note that $z \in A_{2u}$, and that $A_{2u} \times \Delta^{(6)}$ is reduced, via the elementary bases of the unirreps of D_{6i} given in the Appendix, to

$$\begin{aligned} A_{2u} \times \Delta^{(6)} &= A_{2u}(A_{1g} + B_{2u} + E_{1u} + E_{2g}) \\ &= \underset{z}{A_{2u}} + \underset{z(x^3 - 3xy^2)}{B_{1g}} + \underset{z[x,\, y]}{E_{1g}} + \underset{z[2xy,\, x^2 - y^2]}{E_{2u}} \end{aligned} \tag{7.3.18}$$

where each basis function is proportional to the z-orbital, as it should be. The rest follows from $\psi_i^{az}(\boldsymbol{r})$ given in (7.3.17).

7.4 The general classification of SALCs

Before actual construction of the SALCs from any given set of equivalent orbitals $\{f_s^A(\boldsymbol{r} - \boldsymbol{r}_\nu^0)\} \in D^{(A)} \times \Delta^{(n)}$ on $S^{(n)} = \{\boldsymbol{r}_\nu^0\}$, it is always profitable to decompose the direct product representation into two parts:

$$D^{(A)} \times \Delta^{(n)} = \Delta_A^{(\sigma)} + \Delta_A^{(\pi)}; \qquad \Delta_A^{(\sigma)} = n_A\Delta^{(n)} \tag{7.4.1}$$

where n_A is the number of times a unirrep $D^{(A)}(G)$ is contained in $\Delta^{(n)}(G)$. The first part $\Delta_A^{(\sigma)}$ contains n_A linearly independent scalar bases $\{\phi_\nu^\rho(\boldsymbol{r});\ \rho = 1, 2, \ldots, n_A\}$ of $\Delta^{(n)}(G)$ formed by (7.3.4) with n_A linearly independent bases $\{v_s^{A\rho}(\boldsymbol{r}_\nu^0);\ \rho = 1, 2, \ldots, n_A\}$ of $D^{(A)}$ on $S^{(n)}$ from Theorem 7.2.2. The second part $\Delta_A^{(\pi)}$ contains the rest of SALCs orthogonal to the first part. In the important special case of equivalent p-orbitals, the above classification (7.4.1) reduces to the SALCs of σ- and π-orbitals; e.g. see (7.3.17).

Now, returning to (7.4.1), let d_A be the dimensionality of $D^{(A)}$, then the dimensionality of $D^{(A)} \times \Delta^{(n)}$ is $d_A \times n$. Thus, if $n_A = d_A$, then $\Delta_A^{(\pi)}$ is null; on the other hand, if $D^{(A)}(R)$ is not contained in $\Delta^{(n)}(R)$, i.e. $n_A = 0$, then $\Delta_A^{(\sigma)}$ is null. Accordingly, we are left with the case in which $d_A > n_A \geqslant 1$, in which case $D^{(A)}$ is also contained in $\Delta_A^{(\pi)}$ so that $D^{(A)} \times \Delta^{(n)}$ is not simply reducible, even if $\Delta^{(n)}$ is simply reducible. To see this, let n_A^A be the frequency of $D^{(A)}$ contained in $D^{(A)} \times \Delta^{(n)}$. Then, using (7.2.2) we have, analogously to (7.2.3),

$$\begin{aligned} n_A^A &= \frac{1}{|G|} \sum_{R \in G} \chi^{(A)}(R)^* \chi^{(A)}(R) \chi^{(\Delta)}(R) \\ &= \frac{1}{|H|} \sum_{h \in H} |\chi^{(A)}(h)|^2 \\ &\geqslant \frac{1}{|H|} \sum_{h \in H} \chi^{(A)}(h) = n_A \end{aligned} \tag{7.4.2}$$

Accordingly, $n_A^A \geqslant n_A$, where the equality holds if and only if the subduced representation $\{D^{(A)}(h);\ h \in H\}$ onto H ($\leqslant G$) is the identity representation. Thus, for the case in which $d_A > n_A \geqslant 1$, we have $n_A^A > n_A$, i.e. the unirrep $D^{(A)}$ is contained in $D^{(A)} \times \Delta^{(n)}$ more than once even if $\Delta^{(n)}$ is simply reducible; in fact, it will be shown below by construction that $n_A^A = n_A + 1$.

For example, consider the SALCs of the p-orbitals of four equivalent Y atoms in $XY_4 \in T_p$. From Table 7.1, the basis $[x, y, z]$ belongs to T_2 of T_p while $\Delta^{(4)} = A_1 + T_2$ from (7.2.15b), i.e. $\Delta^{(4)}$ is simply reducible. Decomposition of $T_2 \times \Delta^{(4)}$ yields, however,

$$T_2 \times \Delta^{(4)} = A_1 + E + T_1 + 2T_2 \tag{7.4.3}$$

which is not simply reducible. From (7.4.1), these irreducible components are classified into

$$\Delta^{(\sigma)} = \Delta^{(4)} = A_1 + T_2, \qquad \Delta^{(\pi)} = E + T_1 + T_2 \tag{7.4.4}$$

where both subspaces are simply reducible.

7.4.1 $D^{(A)}$ SALCs from the equivalent orbitals $\in D^{(A)} \times \Delta^{(n)}$

Since the unirrep $D^{(A)}$ is contained in $D^{(A)} \times \Delta^{(n)}$ more than once when $d_A > n_A \geqslant 1$, it is desirable to give the general expression for the $D^{(A)}$ SALCs of the equivalent orbitals $\{f_s^A(\boldsymbol{r}, \boldsymbol{r}_\nu^0)\} \in D^{(A)} \times \Delta^{(n)}$. We shall find it convenient to use the vector notation for the bases of $D^{(A)}(G)$. Set

$$\begin{aligned} \boldsymbol{f}^A(\boldsymbol{r}, \boldsymbol{r}') &= [f_1^A(\boldsymbol{r}, \boldsymbol{r}'), \ldots, f_{d_A}^A(\boldsymbol{r}, \boldsymbol{r}')] \\ \boldsymbol{A}^\rho(\boldsymbol{r}) &= [A_1^\rho(\boldsymbol{r}), \ldots, A_{d_A}^\rho(\boldsymbol{r})]; \quad \rho = 1, 2, \ldots, n_A \end{aligned} \tag{7.4.5}$$

where n_A sets $\{\boldsymbol{A}^\rho(\boldsymbol{r})\}$ are proper elementary bases on $S^{(n)}$ belonging to $D^{(A)}(G)$ that are orthonormalized by

$$\delta_{\rho\rho'} = \langle (A^\rho(r_\nu^0), A^{\rho'}(r_\nu^0) \rangle = \sum_s A_s^\rho(r_\nu^0)^* A_s^{\rho'}(r_\nu^0)$$

$$= (d_A/n) \sum_\mu A_j^\rho(r_\mu^0)^* A_j^{\rho'}(r_\mu^0)$$

Here the third equality follows from (7.2.11). These relations are independent of ν and j according to (7.2.12). The parallel and perpendicular components of $f^A(r, r')$ with respect to $A^\rho(r')$ are defined by

$$f_\sigma^{A\rho}(r, r') = A^\rho(r')\langle A^\rho(r'), f^A(r, r')\rangle; \quad \rho = 1, \ldots, n_A$$

$$f_\pi^A(r, r') = f^A(r, r') - \sum_{\rho=1}^{n_A} f_\sigma^{A\rho}(r, r') \tag{7.4.6a}$$

The required SALCs belonging to $D^{(A)}(G)$ are given, from the correspondence theorem, by

$$\psi_\sigma^{A\rho}(r) = \sum_\nu f_\sigma^{A\rho}(r, r_\nu^0); \qquad \rho = 1, \ldots, n_A \in \Delta_A^{(\sigma)}(G)$$

$$\psi_\pi^A(r) = \sum_\nu f_\pi^A(r, r_\nu^0) \in \Delta_A^{(\pi)}(G) \tag{7.4.6b}$$

These $n_A + 1$ SALCs are mutually orthogonal in the Hilbert space if the overlap integrals are neglected. It is evident that the second set becomes null when $n_A = d_A$ owing to the closure of the orthonormalized set $\{A^\rho(r_\nu^0)\}$.

There exists an alternative choice for $D^{(A)}$ SALCs, which disregards the classification (7.4.1). They are given by

$$\psi_A(r) = \sum_\nu f^A(r, r_\nu^0)$$

$$\psi_A^\rho(r) = \psi_A(r) - d_A \psi_\sigma^{A\rho}(r); \qquad \rho = 1, \ldots, n_A \tag{7.4.7}$$

where $\psi_\sigma^{A\rho}(r)$ are defined in (7.4.6b). The second SALCs ψ_A^ρ are obtained from $\psi_A(r)$ and $\psi_\sigma^{A\rho}(r)$ by the Schmidt orthogonalization method. This choice is important in the theory of molecular vibration: Suppose that $A_\nu(r)$ represents a displacement vector δr_ν of the νth equivalent atom of a symmetrical molecule, then $\psi_A(r)$ represents the pure translational modes while $\psi_A^\rho(r)$ represents the internal modes of the vibration belonging to $D^{(A)}$ and orthogonal to $\psi_A(r)$ (see Section 7.6).

We shall specialize (7.4.6b) and (7.4.7) for the important special case in which $1 = n_A < d_A$. In this case there exists only one set of proper bases $\{A(r)\} \in D^{(A)}(G)$ on $S^{(n)}$ (apart from the phase factor). Thus the parallel and perpendicular components of $f^A(r, r')$ with respect to $A(r')$ are defined by

$$f_\sigma^A(r, r') = A(r')\langle A(r'), f^A(r, r')\rangle$$

$$f_\pi^A(r, r') = f^A(r, r') - f_\sigma^A(r, r')$$

$$= -A(r') \times [A(r') \times f^A(r, r')] \tag{7.4.8a}$$

where in the last equality we have used the vector identity $-a \times [a \times c] = c - a(a, c)$ with $(a, a) = 1$. This is possible because $d_A = 2$ or 3 for a crystallographic point

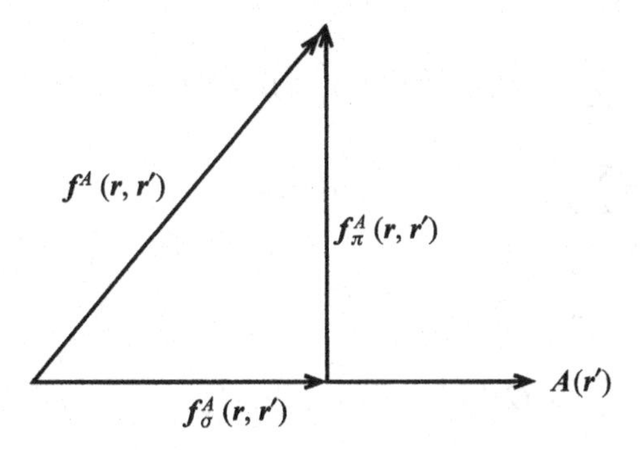

Figure 7.4. $f^A(r, r') = f^A_\sigma(r, r') + f^A_\pi(r, r')$.

group (see Figure 7.4). Thus (7.4.6b) are simplified to two mutually orthogonal SALCs belonging to $D^{(A)}(G)$:

$$\psi^A_\sigma(r) = \sum_\nu f^A_\sigma(r, r^0_\nu) \in \Delta^{(\sigma)}_A$$

$$\psi^A_\pi(r) = \sum_\nu f^A_\pi(r, r^0_\nu) \in \Delta^{(\pi)}_A \tag{7.4.8b}$$

Analogously, (7.4.5) is simplified to two SALCs. These results will be used in the examples given below.

Example 7.4.1. Construct the SALCs of the p-orbitals $\{[x_\nu, y_\nu, z_\nu]\}$ on four equivalent Y atoms of a molecule $XY_4 \in T_p$.

The set of four equivalent points $S^{(4)}$ defined by the coordinates of Y atoms is already given in (7.2.15a) and the elementary bases of the irreducible representations of T_p are given in Table 7.1. Since $[x, y, z] \in T_2$, we have $\{[x_\nu, y_\nu, z_\nu]\} \in T_2 \times \Delta^{(4)}$. The unirreps contained in $T_2 \times \Delta^{(4)} = \Delta^{(\sigma)} + \Delta^{(\pi)}$ are classified, via (7.4.4), by

$$\Delta^{(\sigma)} = \underset{1}{A_1} + \underset{[x,\ y,\ z]}{T_2}$$

$$\Delta^{(\pi)} = \underset{[u,\ v]}{E} + \underset{[\tilde{x},\ \tilde{y},\ \tilde{z}]}{T_1} + \underset{[x,\ y,\ z]}{T_2} \tag{7.4.9a}$$

(i) The SALCs $\in \Delta^{(\sigma)}$. In terms of the σ-orbitals defined by

$$\sigma_\nu = (x^0_\nu x_\nu + y^0_\nu y_\nu + z^0_\nu z_\nu)/\sqrt{3}$$

where x^0_ν y^0_ν and z^0_ν are defined by (7.2.15a), the SALCs belonging to A_1 and T_2 are given by

$$A_1: \quad \psi^\sigma = \sum_\nu \sigma_\nu/2$$

$$T_2: \quad \psi^\sigma_x = \sum_\nu x^0_\nu \sigma_\nu/2, \qquad \psi^\sigma_y = \sum_\nu y^0_\nu \sigma_\nu/2, \qquad \psi^\sigma_z = \sum_\nu z^0_\nu \sigma_\nu/2 \tag{7.4.9b}$$

(ii) The SALCs $\in \Delta^{(\pi)}$:

$$E: \quad \psi_u^\pi = \sum_\nu (2z_\nu^0 z_\nu - x_\nu^0 x_\nu - y_\nu^0 y_\nu)/\sqrt{24}$$

$$\psi_v^\pi = \sum_\nu \sqrt{3}(x_\nu^0 x_\nu - y_\nu^0 y_\nu)/\sqrt{24}$$

$$T_1: \quad \psi_x^\pi = \sum_\nu (y_\nu^0 z_\nu - z_\nu^0 y_\nu)/\sqrt{8}, \qquad \psi_y^\pi, \qquad \psi_z^\pi$$

$$T_2: \quad \psi_x^\pi = \sum_\nu x_\nu^0(2x_\nu^0 x_\nu - y_\nu^0 y_\nu - z_\nu^0 z_\nu)/\sqrt{24}, \qquad \psi_y^\pi, \qquad \psi_z^\pi \tag{7.4.9c}$$

where the SALCs for E and T_1 are formed directly by the correspondence theorem whereas those for T_2 are formed by (7.4.8b); those SALCs not given above explicitly are easily obtained by the cyclic permutation of (x, y, z) from the first member of the partners.

The alternative bases of T_2 based on (7.4.7) are given by

$$T_2: \quad \psi_x = \sum_\nu x_\nu/2, \qquad \psi_y, \qquad \psi_z$$

$$\psi'_x = \sum_\nu (y_\nu^0 z_\nu + z_\nu^0 y_\nu)/\sqrt{8}, \qquad \psi'_y, \qquad \psi'_z \tag{7.4.9d}$$

It should be noted that the bases $[\psi_x^\pi, \psi_y^\pi, \psi_z^\pi]$ and $[\psi'_x, \psi'_y, \psi'_z]$ of T_2 given in (7.4.9c) and (7.4.9d) respectively are understood from the correspondence theorem with the alternative bases of T_2 given in Table 7.1:

$$[\psi_x^\pi, \psi_y^\pi, \psi_z^\pi] \leftrightarrow [x(2x^2 - y^2 - z^2), y(2y^2 - x^2 - z^2), z(2z^2 - x^2 - y^2)]$$

$$[\psi'_x, \psi'_y, \psi'_x] \leftrightarrow [yz, zx, xy]$$

7.5 Hybrid atomic orbitals

Let XY_n be a molecule belonging to a symmetry group G with a central atom X surrounded by n equivalent Y atoms located on $S^{(n)} = \{\boldsymbol{r}_\nu^0\}$, where the origin of the coordinate system is at X. It is assumed that the atomic orbitals (AOs) of every atom in the molecule are classified by the unirreps of the group G. In the LCAO-MO theory, a given set of equivalent AOs of the Y atoms are linearly combined into the SALCs belonging to the unirreps of the group G; each irreducible basis thus formed is linearly combined with the AOs on the central atom X (belonging to the same unirrep) to describe the first-order approximation to the molecular orbital of the molecule XY_n. In the theory of directed valence, on the other hand, the AOs on the central atom X are linearly combined to form the equivalent orbitals in the directions of n equivalent Y atoms. These are called the hybrid AOs of the central atom X, and are to match with the equivalent orbitals on the surrounding Y atoms to form chemical bonding. Thus, the present problem of forming hybrid AOs from the AOs on X is inverse to the former problem of forming LCAO MOs via the SALCs of the equivalent AOs of the surrounding Y atoms. We shall begin with the following simple case.

7.5.1 The σ-bonding hybrid AOs

Let $XY_n \in G$ and the AOs on the surrounding Y atoms be equivalent scalar orbitals (or σ-orbitals) $\{\phi_\nu(\boldsymbol{r}) = \phi(|\boldsymbol{r} - \boldsymbol{r}_\nu^0|)\}$ belonging to the permutation representation $\Delta^{(n)}(G)$. Let $\{u_i^\alpha(\boldsymbol{r})\}$ be a set of atomic orbitals of the central atom X belonging to a unirrep $D^{(\alpha)}$ of G, which is contained in $\Delta^{(n)}(G)$. Then, via the two-point invariant function introduced by (7.3.4), we obtain a set of equivalent orbitals belonging to $\Delta^{(n)}$ for the central atom

$$\varphi_\nu^\alpha(\boldsymbol{r}) \equiv \varphi^\alpha(\boldsymbol{r}, \boldsymbol{r}_\nu^0) = \sum_{i=1}^{d_\alpha} v_i^\alpha(\boldsymbol{r}_\nu^0)^* u_i^\alpha(\boldsymbol{r}); \qquad \nu = 1, 2, \ldots, n \tag{7.5.1}$$

where $\{v_i^\alpha(\boldsymbol{r})\}$ is a basis of $D^{(\alpha)}(G)$ that is proper on $S^{(n)}$. The νth equivalent φ-orbital $\varphi_\nu^\alpha(\boldsymbol{r})$ belonging to the central X atom is to match with the equivalent orbital $\phi_\nu(\boldsymbol{r})$ on the νth Y atom in the molecule. The n equivalent φ-orbitals defined by (7.5.1) are, however, not linearly independent in general because it is a set of n linear combinations of d_α linearly independent basis functions $\{u_i^\alpha(\boldsymbol{r})\}$ and $d_\alpha \leqslant n$ from (7.2.4). To obtain a linearly independent set, it is necessary to introduce further linear combinations over all the irreducible representations contained in $\Delta^{(n)}(G)$. Thus, a linearly independent set may be written in the form

$$\varphi_\nu(\boldsymbol{r}) = \sum_\alpha \sum_{\rho=1}^{n_\alpha} \sum_{i=1}^{d_\alpha} c_{\alpha\rho} v_i^{\alpha,\rho}(\boldsymbol{r}_\nu^0)^* u_i^{\alpha,\rho}(\boldsymbol{r}); \qquad \nu = 1, 2, \ldots, n \tag{7.5.2}$$

where the superscript ρ denotes the different bases belonging to the same unitary representation and $\sum_\alpha n_\alpha d_\alpha = n$; the coefficients $c_{\alpha\rho}$ are arbitrary constants as long as they ensure the linear independence of the set. This set is called a set of *σ-bonding hybrid AOs* of the central X atom.

If one requires that the transformation (7.5.2) is unitary, then the constants $c_{\alpha\rho}$ are determined up to the phase factors. To see this, let us assume that both $\{u_i^{\alpha,\rho}(\boldsymbol{r})\}$ and $\{v_i^{\alpha,\rho}(\boldsymbol{r})\}$ are n orthonormalized basis sets of functions in the Hilbert space. It is also assumed that each set is proper and provides an orthogonal set on $S^{(n)}$. Then, we have the following set of orthonormalized equivalent φ-orbitals:

$$\varphi_\nu(\boldsymbol{r}) = \sum_{\alpha,\rho,i} (1/N_v^{\alpha,\rho}) v_i^{\alpha\rho}(\boldsymbol{r}_\nu^0)^* u_i^{\alpha,\rho}(\boldsymbol{r})$$

$$(N_v^{\alpha,\rho})^2 = \sum_\nu |v_i^{\alpha\rho}(\boldsymbol{r}_\nu^0)|^2 = \frac{n}{d_\alpha} \sum_i |v_i^{\alpha\rho}(\boldsymbol{r}_\sigma^0)|^2 \tag{7.5.3}$$

where we have used (7.2.12). Here there still remains a certain degree of arbitrariness in choosing $v_i^{\alpha,\rho}(\boldsymbol{r}_\nu^0)$ for a given set of $u_i^{\alpha,\rho}(\boldsymbol{r})$ and also the arbitrariness in the phases of the coefficients. In the theory of directed valence, this arbitrariness is removed by *Pauling's criterion on the maximum bonding strength.* The bonding strength of the νth equivalent orbital $\varphi_\nu(\boldsymbol{r})$ of X in the direction of $\boldsymbol{r}$ may be defined by $\mathrm{Re}\,\varphi_\nu(\boldsymbol{r})$, where Re denotes the real part, extending Pauling's definition to a complex orbital. Then, the equivalent orbital with the maximum bonding strength in the direction of $\boldsymbol{r}_\nu^0$ is given by

$$\psi_\nu(\boldsymbol{r}) = \sum_{\alpha,\rho,i} (1/N_u^{\alpha,\rho}) u_i^{\alpha,\rho}(\boldsymbol{r}_\nu^0)^* u_i^{\alpha,\rho}(\boldsymbol{r}) \tag{7.5.4}$$

where $N_u^{\alpha,\rho}$ is defined analogous to $N_v^{\alpha,\rho}$ of (7.5.3) assuming $N_u^{\alpha,\rho} > 0$.

The proof is simple: it follows from the inequality

$$\psi_\nu(\boldsymbol{r}_\nu^0) - \operatorname{Re} \varphi_\nu(\boldsymbol{r}_\nu^0) = \frac{1}{2} \sum_{\alpha,\rho,i} (1/N_u^{\alpha,\rho}) |u_i^{\alpha,\rho}(\boldsymbol{r}_\nu^0) - (N_u^{\alpha,\rho}/N_v^{\alpha,\rho}) v_i^{\alpha,\rho}(\boldsymbol{r}_\nu^0)|^2 \geqslant 0 \tag{7.5.5}$$

The maximum bonding strength of $\psi_\nu(\boldsymbol{r})$ in the direction of $\boldsymbol{r}_\nu^0$ is given by

$$\psi_\nu(\boldsymbol{r}_\nu^0) = \frac{1}{n} \sum_{\alpha,\rho} d_\alpha N_u^{\alpha,\rho} \tag{7.5.6}$$

Example 7.5.1. The σ-bonding hybrid AOs of the central atom X for a molecule $XY_4 \in T_p$.

From (7.2.15b), we have $\Delta^{(4)} = A_1 + T_2$, where the unirreps are characterized by the elementary bases as follows:

A_1: 1 or xyz

T_2: $[x, y, z]$ or $[yz, zx, xy]$ or $[x(3x^2 - r^2), y(3y^2 - r^2), z(3z^2 - r^2)]$ (7.5.7a)

From (7.5.4), the required hybrid AOs may be written as

$$\psi_\nu(\boldsymbol{r}) = \tfrac{1}{2}(u^1 + x_\nu^0 u_x + y_\nu^0 u_y + z_\nu^0 u_z), \qquad \nu = 1, 2, 3, 4 \tag{7.5.7b}$$

where u^1 and (u_x, u_y, u_z) are properly normalized bases of the atomic orbitals of X belonging to A_1 and T_2, respectively, and $S^{(4)} = \{\boldsymbol{r}_\nu^0\}$ is defined by (7.2.15a). For example, for the s, p_x, p_y and p_z orbitals of X we have

$$u^1 = \phi, \qquad u_x = \sqrt{3}x\phi, \qquad u_y = \sqrt{3}y\phi, \qquad u_z = \sqrt{3}z\phi \tag{7.5.7c}$$

where $\phi = \phi(|\boldsymbol{r}|)$ is the radial part. To achieve the maximum bonding strength, the phase of ϕ is chosen such that $\phi(|\boldsymbol{r}_\nu^0|) > 0$. The maximum bonding strength for this case is given by

$$\psi_\nu(\boldsymbol{r}_\nu^0) = \tfrac{1}{2}(1 + 3\sqrt{3})\phi(|\boldsymbol{r}_\nu^0|) \tag{7.5.7d}$$

The transformation matrix defined by (7.5.7b) is given by

$$L^{\sim} = \frac{1}{2} \begin{bmatrix} 1 & 1 & 1 & 1 \\ 1 & -1 & -1 & 1 \\ 1 & 1 & -1 & -1 \\ 1 & -1 & 1 & -1 \end{bmatrix} \tag{7.5.7e}$$

which is the inverse of L given by (7.2.15d), as expected.

In view of the degrees of the polynomial bases for A_1 and T_2 given in (7.5.7a), we may conclude that the set of equivalent orbitals given by (7.5.7b) can be applied for the sp^3-, sd^3-, sf^3-, fp^3- and f^4-hybrid orbitals. Here s-, p-, d- and f-orbitals refer to the elementary bases with the polynomial degrees 0, 1, 2 and 3, respectively.

Analogously to (7.5.7b), the sp^2-hybridized orbitals of the central atom X for a planar molecule $XY_3 \in D_{3p}$ are given by

$$\varphi'_\nu(\boldsymbol{r}) = \tfrac{1}{2}(u^1 + x^0_\nu u_x + y^0_\nu u_y); \qquad \nu = 1, 2, 3 \tag{7.5.8}$$

where $S^{(3)} = \{x^0_\nu, y^0_\nu\}$ is defined by (7.2.16a). Moreover, the sp-hybridized orbitals of X for a linear molecule $YXY \in D_{\infty i}$ are given by

$$\varphi''_\pm(\boldsymbol{r}) = \frac{1}{\sqrt{2}}(u_1 \pm u_x) \tag{7.5.9}$$

7.5.2 General hybrid AOs

Consider the general case of $XY_n \in G$. Let $D^{(\gamma)}(G)$ be a unirrep of G and let $\{f^\gamma_k(\boldsymbol{r} - \boldsymbol{r}^0_\nu)\}$ be the equivalent orbitals of the surrounding Y atoms belonging to a direct product representation $D^{(\gamma)} \times \Delta^{(n)}$ of G and let $D^{(A)}$ be a unirrep of G contained in $D^{(\gamma)} \times \Delta^{(n)}$ with the frequency n^γ_A. Let $\{u^{A\rho}_s(\boldsymbol{r}); \rho = 1, 2, \ldots, n^\gamma_A\}$ be the n^γ_A sets of the atomic orbitals of the central X atom, each of which transforms according to the unirrep $D^{(A)}$ of G for a given ρ. Then the hybrid AOs of the central atom $X \in D^{(\gamma)} \times \Delta^{(n)}$ matching with the equivalent orbitals $\{f^\gamma_k(\boldsymbol{r} - \boldsymbol{r}^0_\nu)\}$ of the surrounding Y atoms may be expressed by

$$\varphi^\gamma_{k\nu}(\boldsymbol{r}) = \sum_A \sum_{\rho=1}^{n^\gamma_A} c_{A\rho} \sum_{s=1}^{d_A} L^{\gamma A\rho}_{ks}(\boldsymbol{r}^0_\nu) u^{A\rho}_s(\boldsymbol{r}); \qquad L^{\gamma A\rho}_{ks}(\boldsymbol{r}^0_\nu) = \mathring{T}^{\gamma\rho}_k(\boldsymbol{r}') v^{A\rho}_s(\boldsymbol{r}')^*|_{r=r^0_\nu} \tag{7.5.10a}$$

where $k = 1, 2, \ldots, d_\gamma$; $\nu = 1, 2, \ldots, n$; $\rho = 1, 2, \ldots, n^\gamma_A$ and

$$\sum_A n^\gamma_A d_A = n d_\gamma \tag{7.5.10b}$$

Previously in the calculation of SALCs by (7.3.6b), we have transformed $d_A \times n$ equivalent orbitals $\{f^A_s(\boldsymbol{r} - \boldsymbol{r}^0_\nu)\}$ of the surrounding Y atoms belonging to $D^{(A)} \times \Delta^{(n)}$ of G to their SALCs belonging to $D^{(\gamma)}(G)$. Here in (7.5.10a) we named the representations such that we transform a set of the $D^{(A)}$-bases of the central atom X to the equivalent orbitals belonging to $D^{(\gamma)} \times \Delta^{(n)}$ of G. Accordingly, the transformation matrices L for both cases are essentially similar because in both cases we transform '$D^{(A)}$-bases' to '$D^{(\gamma)}$-bases'. Thus, the transformation matrix calculated previously for the SALCs can be used to obtain the corresponding hybrid AOs (see (7.5.13)).

The coefficients $c_{A\rho}$ in (7.5.10a) are arbitrary constants insofar as they leave the hybrid AOs linearly independent. If we require the transformation to be unitary, then the constants are determined up to the phase factors by

$$(1/c_{A\rho})^2 = \sum_{\nu=1}^{n} \sum_{k=1}^{d_\gamma} |L^{\gamma A\rho}_{ks}(\boldsymbol{r}^0_\nu)|^2 \tag{7.5.11}$$

Obviously, if the set $\{u^{\gamma\rho}_k(\boldsymbol{r})\}$ is chosen orthonormalized in the Hilbert space, then so is the set of hybrid AOs $\{\varphi^A_{s\nu}(\boldsymbol{r})\}$, provided that the transformation is unitary.

In the theory of valence, frequently some of the atomic orbitals $\{u^{\gamma\rho}_k(\boldsymbol{r})\}$ of X are not available from energy considerations. In such a case, the hybrid AOs given above are not linearly independent.

Example 7.5.2. The hybrid AOs of the central X atom for $XY_4 \in T_p$ matching with the p-like orbitals of the equivalent Y atoms.

From Table 7.1, $[x, y, z] \in T_2$ of T_p and also from (7.4.3) we have

$$T_2 \times \Delta^{(4)} = \underset{1}{A_1} + \underset{[u,\, v]}{E} + \underset{[\tilde{x},\, \tilde{y},\, \tilde{z}]}{T_1} + \underset{[x,\, y,\, z]}{2T_2} \tag{7.5.12}$$

Thus, the required hybrid AOs $\in T_2 \times \Delta^{(4)}$ are formed from the atomic orbitals of X belonging to A_1, E, T_1 and $2T_2$. From (7.5.10a) and (7.5.11), we give here only the x-components:

$$\varphi_{x\nu}^{(5)}(r) = \frac{1}{\sqrt{12}} x_\nu^0 f_1 + \frac{1}{\sqrt{48}} x_\nu^0(-f_u + \sqrt{3} f_v) + \frac{1}{\sqrt{8}}(y_\nu^0 f_{\tilde{z}} - z_\nu^0 f_{\tilde{y}})$$
$$+ \frac{1}{\sqrt{12}} x_\nu^0(x_\nu^0 f_x + y_\nu^0 f_y + z_\nu^0 f_z) + \frac{1}{\sqrt{24}} x_\nu^0(2x_\nu^0 h_x - y_\nu^0 h_y - z_\nu^0 h_z) \tag{7.5.13}$$

where $\{x_\nu^0, y_\nu^0, z_\nu^0;\ \nu = 1, 2, 3, 4\}$ has been given by (7.2.15a). The y- and z-components not given above are obtained by the permutations of x, y and z from the x-component. The notations of these bases are in accordance with the definition (7.3.12a); e.g. $[f_u, f_v]$ is a basis that transforms like the elementary basis $[u, v]$, whereas the basis $[f_{\tilde{x}}, f_{\tilde{y}}, f_{\tilde{z}}] \in T_1$ transforms like $[\tilde{x}, \tilde{y}, \tilde{z}]$ or $[x(y^2 - z^2), y(z^2 - x^2), z(x^2 - y^2)]$ given in Table 7.1. The two bases $[f_x, f_y, f_z]$ and $[h_x, h_y, h_z]$ of T_2 are mutually orthogonal p-like orbitals as defined by (7.3.12b) at $\boldsymbol{r}_\nu^0 = 0$. Moreover, each term on the rhs of (7.5.13) transforms like x_ν^0 as in the first term: the second term is obtained analogously to ψ_x of (7.3.15c), the third term is based on the fact that $[\boldsymbol{r} \times \tilde{\boldsymbol{r}}]$ transforms like a vector, the fourth term is analogous to ψ_x^σ of (7.4.9b) and the last term is analogous to ψ_x^π of (7.4.9c). The normalization constants may be checked by (7.5.11); here it is necessary to write down $\varphi_{y\nu}^5$ and $\varphi_{z\nu}^5$ in addition to $\varphi_{x\nu}^5$ given above to arrive at the correct normalization constants given in (7.5.13). Finally, we can show that the first and fourth terms are σ-orbitals whereas the remaining terms are π-orbitals according to the classification (7.4.4) and (7.4.8a).

7.6 Symmetry coordinates of molecular vibration based on the correspondence theorem

In the theory of molecular vibration of a polyatomic molecule belonging to a point group G, the primary step is to construct the symmetry coordinates of molecular vibration belonging to an irreducible representation of the group G. These may be classified by the SALCs of infinitesimal displacements of the atoms in the molecule. Since the infinitesimal displacements of equivalent atoms transform like equivalent p-orbitals, the problem of finding the symmetry coordinates of vibration is reduced to the problem of constructing the SALCs of equivalent p-orbitals, which has been discussed fully in the previous sections; cf. Kim (1981b, 1986a).

The symmetry coordinates formed by the SALCs of atomic displacements are called the *external vibrational coordinates* of a molecule. There exist also *internal vibrational coordinates* that describe the relative displacements of atoms in a molecule, such as a bond-stretching and a valence angle bending. These are free from the total displacement, such as translation or rotation of the molecule as a whole. Let N be the total number of atoms in a rigid molecule, then there exists a total of $3N - 6$ (5) internal vibrational coordinates because there exist three translational degrees of freedom and three (or two) rotational degrees of freedom for a non-linear (or linear) molecule. We shall begin with the discussion of the external coordinates.

7.6.1 External symmetry coordinates of vibration

The external coordinates are defined by the atomic displacements in the Cartesian coordinates fixed on the molecular frame at equilibrium. Let $\delta \boldsymbol{r}_\nu$ be the Cartesian displacement of the νth equivalent atom located at $\boldsymbol{r}_\nu^0$. Then the set $\{\delta \boldsymbol{r}_\nu\}$ transforms like an equivalent set of vectors or p-orbitals, i.e. it transforms according to the direct product representation $D^{(1)}(R) \times \Delta^{(n)}(R)$ for all $R \in G$, where $D^{(1)}(R)$ is the three-dimensional representation of orthogonal transformation $R \in G$ and $\Delta^{(n)}(R)$ is the permutation representation of $R \in G$ based on the set of n equivalent positions $S^{(n)} = \{\boldsymbol{r}_\nu^0\}$ of symmetrically equivalent atoms in the molecule. Let $D^{(\gamma)}(G)$ be a unirrep contained in $D^{(1)} \times \Delta^{(n)}$ of G, then a set of external symmetry coordinates belonging to $D^{(\gamma)}(G)$ is expressed by

$$q_k^\gamma = \sum_\nu^n \boldsymbol{u}_k^\gamma(\boldsymbol{r}_\nu^0) \cdot \delta \boldsymbol{r}_\nu \tag{7.6.1}$$

where $\boldsymbol{u}_k^\gamma(\boldsymbol{r}_\nu^0)$ is a vector function at $\boldsymbol{r} = \boldsymbol{r}_\nu^0$ defined, from (7.3.6b) or (7.3.8b), by

$$\boldsymbol{u}_k^\gamma(\boldsymbol{r}_\nu^0) = \mathring{T}_k^\gamma(\boldsymbol{r})\boldsymbol{r}|_{\boldsymbol{r}=\boldsymbol{r}_\nu^0} \qquad \text{or} \qquad \nabla v_k^\gamma(\boldsymbol{r})|_{\boldsymbol{r}=\boldsymbol{r}_\nu^0}$$

If $D^{(\gamma)}$ is contained in $\Delta^{(n)}$, the operator basis will obviously be an ordinary proper basis on $S^{(n)}$. These coordinates q_k^γ are nothing other than the general expressions of the SALCs of equivalent p-orbitals discussed fully in the previous sections; in particular, the second expression should be compared with (7.3.10a). As we shall see in the examples discussed below, for most cases, direct use of the correspondence theorem suffices to construct any required SALCs instead of these general expressions (7.6.1). In any case, the set of the coefficient vectors $\{\boldsymbol{u}_k^\gamma(\boldsymbol{r}_\nu^0)\}$ describes the mode of vibration corresponding to the symmetry coordinates q_k^γ of vibration.

Example 7.6.1. Determine the external vibrational symmetry coordinates of an equilateral triangular molecule $X_3 \in D_{3p}$ $(= D_{3h})$.

Take the coordinate origin at the center of the triangle as in Figure 7.5 and let the coordinates of three equivalent atoms be given by (7.2.16a). Let us use the following abbreviations for the atomic displacements:

$$x_\nu = \delta x_\nu, \qquad y_\nu = \delta y_\nu, \qquad z_\nu = \delta z_\nu; \qquad \nu = 1, 2, 3 \tag{7.6.2a}$$

Since the set $\{[x_\nu, y_\nu, z_\nu]\}$ transforms like an equivalent set of p-orbitals, the problem is reduced to the problem of constructing the SALCs of equivalent p-orbitals on $S^{(3)}$, which already has been discussed in Example 7.3.2. We discuss this problem once more, analogously to Example 7.3.2, but stressing the differences in the physical interpretation; in particular, in choosing the elementary bases.

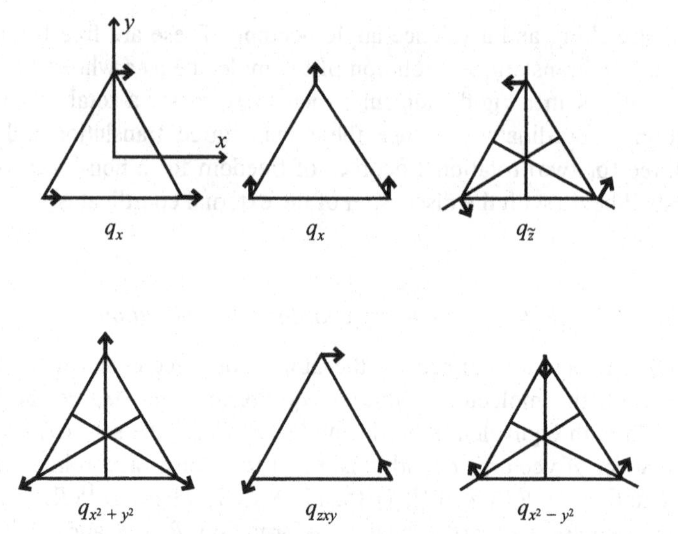

Figure 7.5. The external symmetry coordinates of an equilateral triangular molecule belonging to D_{3p}. Each vector at a νth atom is drawn proportional to the coefficient vector $\boldsymbol{u}_k^{\gamma}(\boldsymbol{r}_\nu^0)$ defined in (7.6.1).

From the elementary bases $[x, y] \in E_2$ and $z \in B_2$ given in Table 7.2, the in-plane modes of motion are classified, using $\Delta^{(3)} = A_1 + E_2$, by

$$E_2 \times \Delta^{(3)} = \underset{x^2+y^2}{A_1} + \underset{\tilde{z}}{A_2} + \underset{[x,\, y],\ [2xy,\, x^2-y^2]}{2E_2} \tag{7.6.2b}$$

while the out-of-plane modes are classified by

$$B_2 \times \Delta^{(3)} = \underset{z}{B_2} + \underset{[\tilde{x},\, \tilde{y}]}{E_1} \tag{7.6.2c}$$

Let us first construct the symmetry coordinates belonging to the one-dimensional unirreps A_1, A_2 and B_2 of D_{3p}. We obtain, via the correspondence theorem,

$$A_1: \quad q_{x^2+y^2} = \sum_\nu (x_\nu^0 x_\nu + y_\nu^0 y_\nu)/N$$

$$= (2y_1 + \sqrt{3}x_2 - y_2 - \sqrt{3}x_3 - y_3)/\sqrt{12}$$

$$A_2: \quad q_{\tilde{z}} = \sum_\nu (x_\nu^0 y_\nu - y_\nu^0 x_\nu)/N$$

$$= (-2x_1 + \sqrt{3}y_2 + x_2 - \sqrt{3}y_3 + x_3)/\sqrt{12}$$

$$B_2: \quad q_z = \sum_\nu z_\nu/N = (z_1 + z_2 + z_3)/\sqrt{3}$$

where we have used the equilibrium coordinates $(x_\nu^0, y_\nu^0, z_\nu^0)$ given in (7.2.16a). From Figure 7.5, we see that the coordinate $q_{x^2+y^2}$ describes *the symmetric stretching mode*, whereas $q_{\tilde{z}}$ describes the pure rotational mode about the z-axis, and the coordinate q_z describes the pure translational mode in the z-direction.

Next, the degenerate symmetry coordinates are

$$E_1: \quad q_{\tilde{x}} = \sum_{\nu} (y^0_\nu z_\nu - z^0_\nu y_\nu)/N = (2z_1 - z_2 - z_3)/\sqrt{6}$$

$$q_{\tilde{y}} = \sum_{\nu} (z^0_\nu x_\nu - x^0_\nu z_\nu)/N = (-z_2 + z_3)/\sqrt{2}$$

$$E_2: \quad q_x = \sum_{\nu} x_\nu/N = (x_1 + x_2 + x_3)/\sqrt{3}$$

$$q_y = \sum_{\nu} y_\nu/N = (y_1 + y_2 + y_3)/\sqrt{3}$$

$$E_2: \quad q_{2xy} = \sum_{\nu} (x^0_\nu y_\nu + y^0_\nu x_\nu)/N$$

$$= (2x_1 + \sqrt{3}y_2 - x_2 - \sqrt{3}y_3 - x_3)/\sqrt{12}$$

$$q_{x^2-y^2} = \sum_{\nu} (x^0_\nu x_\nu - y^0_\nu y_\nu)/N$$

$$= (-2y_1 + \sqrt{3}x_2 + y_2 - \sqrt{3}x_3 - y_3)/\sqrt{12}$$

Analogously to $q_{\tilde{z}}$, the basis $[q_{\tilde{x}}, q_{\tilde{y}}] \in E_1$ describes the pure rotational modes about the x- and y-axes, while $[q_x, q_y]$ describes the pure translational modes in the directions of the x- and y-axes, analogously to q_z, whereas the set $[q_{2xy}, q_{x^2-y^2}] \in E_2$ describes *the asymmetric stretching* (see Figure 7.5).

Thus, out of $9 = 3 \times 3$ degrees of freedom, there exist only three vibrational degrees of freedom described by $q_{x^2+y^2} \in A_1$ and $[q_{2xy}, q_{x^2-y^2}] \in E_2$. The remaining degrees of freedom are three pure translational and three pure rotational coordinates described by (q_x, q_y, q_z) and $(q_{\tilde{x}}, q_{\tilde{y}}, q_{\tilde{z}})$, respectively: these six modes belong to the eigenvalue zero of the Hamiltonian or the zero frequency modes.

Exercise. Construct the external vibrational coordinates of CH_4 from the SALCs of the p-orbitals. Use the SALCs of the $[p_x, p_y, p_z]$ orbitals of the four equivalent H atoms given in Example 7.4.1.

7.6.2 Internal vibrational coordinates

The types of internal vibrational coordinates of a molecule which will be considered here are

Bond stretchings,
Valence-angle bendings,
Bond-plane angle changes (the angle between a bond and a plane defined by two bonds) or vertex liftings, and
Bond twistings (or torsions).

The accurate definitions of these coordinates have been given in the classic work of Wilson *et al.* (1955).

The transformation properties of these internal coordinates are determined by the following rules of correspondence first introduced by Kim (1981b).

(1) *A bond stretching* transforms like a scalar function located at the mid-point or the free end point of the bond.

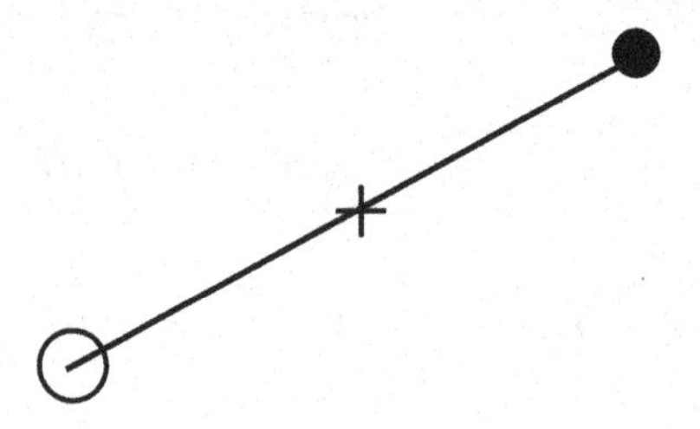

(2) *A valence-angle bending* transforms like a scalar function located at a point on the line which bisects the angle in the plane of two bonds.

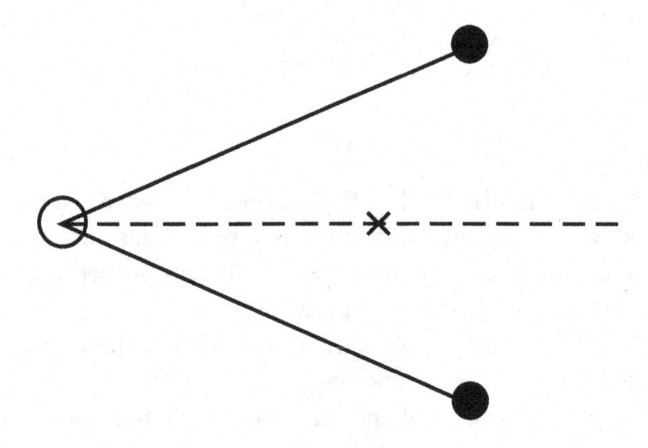

(3) *A bond-plane angle change* is defined by a set of three bonds initially in a plane meeting at the νth equivalent atom: if the plane is perpendicular to the z-axis, then the change of the angle α_ν between one bond and the plane of the remaining two bonds transforms like $z_\nu = z - z_\nu^0$. It can also be regarded as *lifting of the vertex atom.*

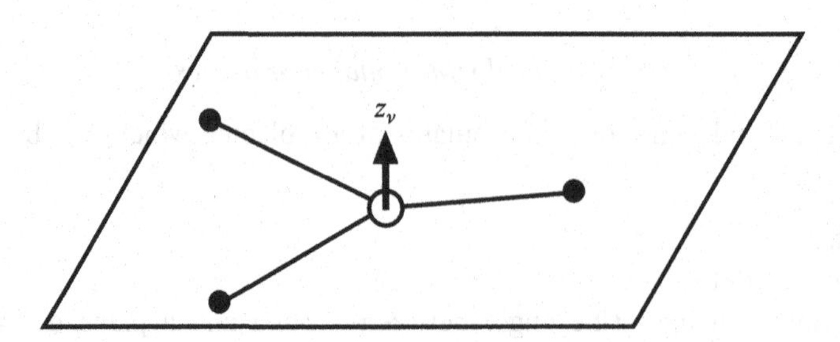

(4) *A bond twisting* (or *torsion*) transforms like a pseudo-scalar with respect to the symmetry group G located at the mid-point of the bond. Here a pseudo-scalar means a function of $\boldsymbol{r}$ that is invariant under all proper rotations of G but changes

its sign under all improper rotations of G, if they are contained in G. Examples of pseudo-scalar bases with respect to a point group are

$$xyz\tilde{x}\tilde{y}\tilde{z} \in O_i, \qquad \tilde{x}\tilde{y}\tilde{z} \in T_p, \qquad z\tilde{z} \in D_{n\mathrm{h}} \tag{7.6.3}$$

where $\tilde{x} = y'z - z'y$ etc. Note that $\tilde{x}\tilde{y}\tilde{z}$ changes its sign under $\overline{4}_z \in T_p$ even though it is an even function with respect to the inversion.

The above rules 1–4 may be referred to as *the correspondence rule* for internal vibrational coordinates. Each kind of the internal vibration coordinate, say ξ, introduced above transforms according to a one-dimensional representation, say $D^{(\xi)}$, of the symmetry group G of the molecule, being represented by a scalar or a pseudo-scalar basis, for example. Thus, the set of equivalent internal coordinates $\{\xi_\nu;\ \nu = 1, 2, \ldots, n\}$ defined on a set of equivalent points $S^{(n)} = \{\boldsymbol{r}_\nu^0\}$ of the molecule at equilibrium transforms according to a direct product representation $\Delta^{(n)} \times D^{(\xi)}$ of G, where $\Delta^{(n)}(G)$ is the permutation representation based on $S^{(n)}$. Since $D^{(\xi)}(G)$ is one-dimensional, a unirrep contained in $\Delta^{(n)} \times D^{(\xi)}$ may be characterized by a direct product $D^{(\alpha)} \times D^{(\xi)}$ of G, where $D^{(\alpha)}$ is a unirrep contained in $\Delta^{(n)}(G)$. Thus, a set of *internal symmetry coordinates (ISCs)* belonging to the unirrep $D^{(\alpha)} \times D^{(\xi)}$ of G is given by the SALC of $\{\xi_\nu\}$:

$$\xi_i^\alpha = \sum_{\nu=1}^{n} u_i^\alpha(\boldsymbol{r}_\nu^0)\xi_\nu; \qquad i = 1, 2, \ldots, d_\alpha \tag{7.6.4}$$

where $\{u_i^\alpha(\boldsymbol{r})\}$ is a proper basis of $D^{(\alpha)}(G)$ on $S^{(n)}$. As was shown in Theorem 7.2.2, there exist linear independent basis vectors $\{u_i^\alpha(\boldsymbol{r}_\nu^0)\}$ on $S^{(n)}$ as many times as the number of times $D^{(\alpha)}(G)$ is contained in $\Delta^{(n)}(G)$. Also, note that the ISC given by (7.6.4) is simpler than the external symmetry coordinates given by (7.6.1) because $D^{(\xi)}(G)$ is one-dimensional. From (7.6.4) we shall construct the ISCs of typical rigid molecules as prototype examples.

7.6.3 *Illustrative examples*

Example 7.6.2. The ISCs of $H_2O \in C_{2v}$. There exist two O—H stretches and one HOH angle bending. Let the coordinate origin be on the O atom and take the x-axis perpendicular to the molecular symmetry plane (see Figure 7.6(a)). Let the equilibrium coordinates of two H on the molecular plane be

$$(y, z) = (1, h), \qquad (-1, h) \in S^{(2)} \tag{7.6.5a}$$

The irreducible bases of C_{2v} are given by

$$A_1\colon\quad 1, z \qquad A_2\colon\quad xy, \tilde{z}$$
$$B_1\colon\quad x, \tilde{y} \qquad B_2\colon\quad y, \tilde{x}$$

from the Appendix, and the unirreps contained in $\Delta^{(2)}$ based on $S^{(2)}$ are

$$\Delta^{(2)} = \underset{1}{A_1} + \underset{y}{B_2} \tag{7.6.5b}$$

According to the rule 1 of correspondence, two O—H stretches s_1 and s_2 transform like scalars located at the free end point of the respective bond. Thus, from (7.6.4), the required ISCs are given by the SALCs of s_1 and s_2:

$$A_1: \qquad S_1 = \sum_{\nu} s_\nu / N = (s_1 + s_2)/\sqrt{2}$$

$$B_2: \qquad S_y = \sum_{\nu} y_\nu^0 s_\nu / N = (s_1 - s_2)/\sqrt{2}$$

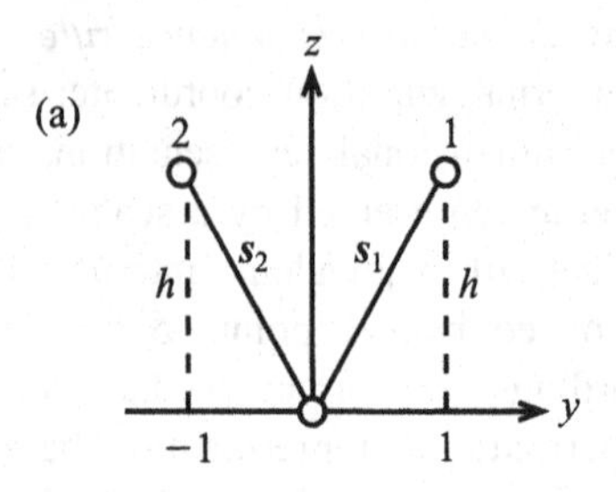

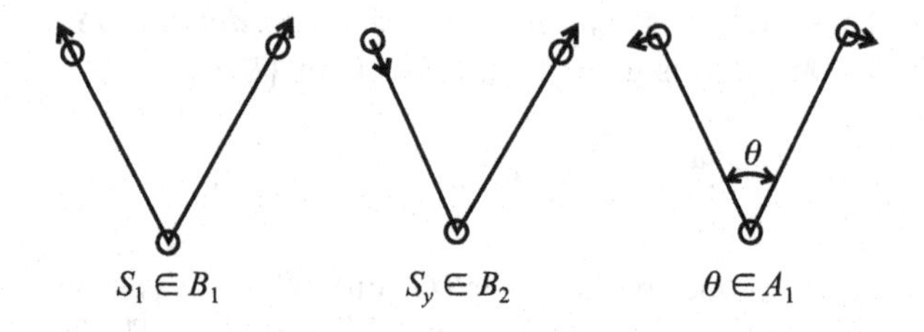

(b)

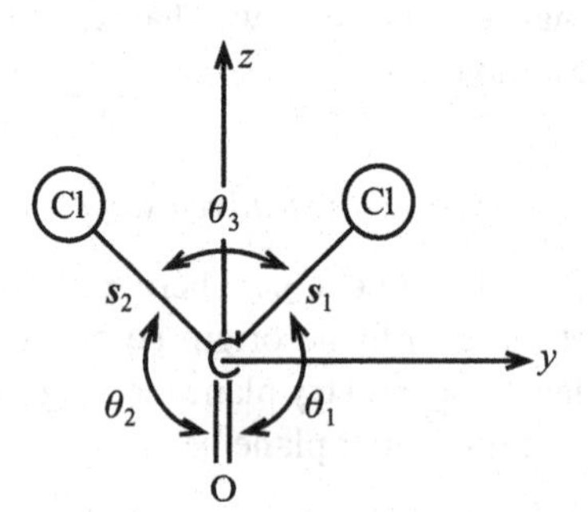

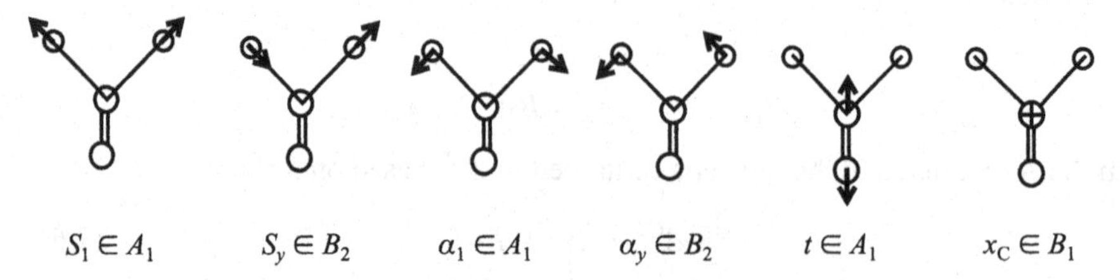

Figure 7.6. (a) The modes of vibration for the H_2O molecule belonging to C_{2v}: S_1 is symmetric stretching, S_y is asymmetric stretching and θ is valence-angle bending. (b) The vibrational coordinates and the modes of vibration for $COCl_2 \in C_{2v}$.

From rule 3, the HOH valence-angle change denoted by θ transforms like a scalar so that its ISC is θ itself belonging to A_1 (see Figure 7.6(a)):

$$A_1: \qquad \theta$$

Example 7.6.3. ISCs of

$$\mathrm{O{=}C}\begin{matrix}\diagup \mathrm{Cl}\\ \diagdown \mathrm{Cl}\end{matrix} \in C_{2v}$$

Let the coordinate origin be on the C atom and the x-axis be perpendicular to the molecular plane (see Figure 7.6(b)). There exist two C—Cl stretches (s_1 and s_2), one C═O stretch t, two O═C—Cl bendings (θ_1 and θ_2) and one lifting mode x_C for the central C atom. Proceeding analogously to the case of H_2O, we have the following internal symmetry vibrational coordinates of $COCl_2 \in C_{2v}$:

$$A_1: \qquad S_1 = (s_1 + s_2)/\sqrt{2}, \qquad \alpha_1 = (\theta_1 + \theta_2)/\sqrt{2}$$

$$B_2: \qquad S_y = (s_1 - s_2)/\sqrt{2}, \qquad \alpha_y = (\theta_1 - \theta_2)/\sqrt{2}$$

$$A: \qquad S_{CO} = t$$

$$B_1: \qquad x_C$$

where x_C is the lifting of the carbon atom C in the x-direction from the equilibrium position $x_C = 0$.

Example 7.6.4. The ISCs of $CO_3^{2-} \in D_{3p}$.

It is an equilateral triangular ion with a carbon atom in the center. Place the coordinate origin at the C atom and take the z-axis perpendicular to the molecular plane as in Figure 7.7. The coordinates of the three oxygen atoms are given by (7.2.16a). From (7.2.16b), the permutation representation $\Delta^{(3)}$ on $S^{(3)}$ is reduced to

$$\Delta^{(3)} = \underset{1}{A_1} + \underset{[x,\,y]}{E_2} \tag{7.6.6}$$

Thus from the rule 1 and (7.6.4), the ISCs of three CO stretching s_1, s_2 and s_3 are given by

$$A_1: \qquad S_1 = \sum s_\nu/N = (s_1 + s_2 + s_3)/\sqrt{3}$$

$$E_2: \qquad S_x = \sum x_\nu^0 s_\nu/N = (s_2 - s_3)/\sqrt{2}$$

$$S_y = \sum y_\nu^0 s_\nu/N = (2s_1 - s_2 - s_3)/\sqrt{6}$$

Likewise, from rule 2, the bending modes of the three OCO angles are given by the SALCs of the three bond angle changes θ_1, θ_2 and θ_3 as follows:

$$A_1: \qquad \alpha_1 = (\theta_1 + \theta_2 + \theta_3) = 0$$

$$E_2: \qquad \alpha_x = (\theta_2 - \theta_3)/\sqrt{2}$$

$$\alpha_y = (\theta_2 + \theta_3)/\sqrt{2}$$

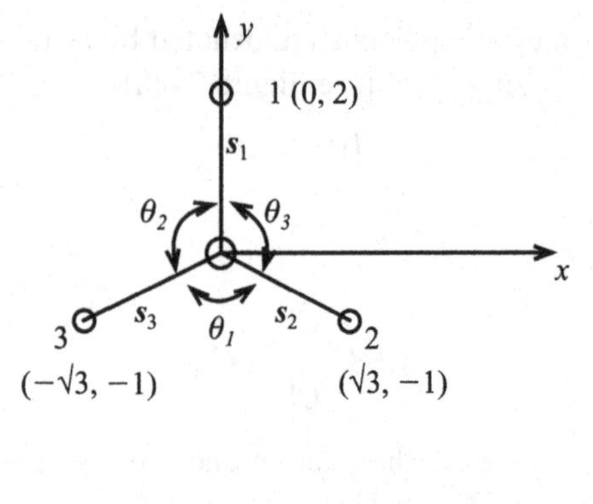

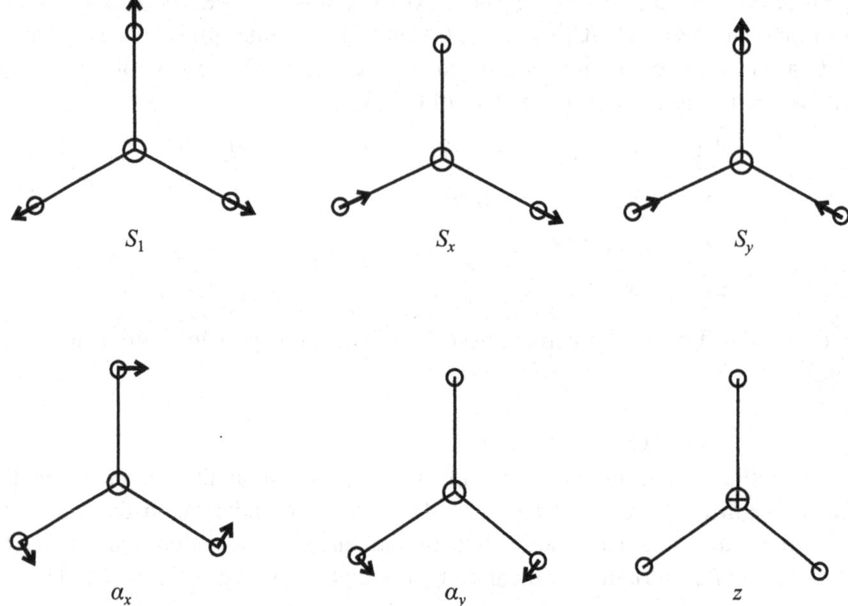

Figure 7.7. The vibrational modes of $CO_3^{2-} \in D_{3p}$.

Note that $\alpha_1 = 0$ is due to the fact that the sum of the three bond angles OCO equals 2π. There exists one more ISC, which describes the lifting of the C atom relative to the three O atoms: it is simply described by the z-coordinate of the C atom

$$B_2: \qquad z$$

Thus we have obtained a total of six ISCs for CO_3^{2-}, as expected from the number of internal degrees of freedom $3 \times 4 - 6 = 6$. See Figure 7.7 for the vibrational modes.

Example 7.6.5. The ISCs of the ammonia molecule $NH_3 \in C_{3v}$.

It is a pyramidal molecule with an equal lateral triangular base. Thus it can be treated analogously to the case of CO_3^{2-}. We place three equivalent H atoms on the x, y coordinate plane in the places of the three O atoms at CO_3^{2-}. The irreducible bases of C_{3v} are given by

$$\begin{array}{ccc} A_1 & A_2 & E \\ 1 & \tilde{z} & [x, y] \end{array}$$

and the permutation representation $\Delta^{(3)}$ based on the three equivalent H atoms is decomposed into the unirreps, as in (7.6.6):

$$\Delta^{(3)} = \underset{1}{A_1} + \underset{[x,\, y]}{E}$$

Thus, the ISCs of the three stretches (s_1, s_2 and s_3) of N—H bonds and three bond-angle changes (θ_1, θ_2 and θ_3) are given by

$$A_1: \quad S_1 = (s_1 + s_2 + s_3)/\sqrt{3}; \qquad \alpha_1 = (\theta_1 + \theta_2 + \theta_3)/\sqrt{3}$$

$$E: \quad S_x = (s_2 - s_3)/\sqrt{2}; \qquad \alpha_x = (\theta_2 - \theta_3)/\sqrt{2}$$

$$S_y = (2s_1 - s_2 - s_3)/\sqrt{6}; \qquad \alpha_y = (2\theta_1 - \theta_2 - \theta_3)/\sqrt{6}$$

Note here that α_1 is not null for NH_3 because the three N—H bonds are not on a plane (cf. Example 7.6.4). Note also that the lifting mode of CO_3^{2-} is replaced by the symmetric bending mode α_1 of NH_3.

Example 7.6.6. The ISCs of the ethylene molecule $CH_2{=}CH_2 \in D_{2i}$.

Let the coordinate origin be at the center of the molecule and let the z-axis be perpendicular to the molecular plane, the x-axis parallel to the C=C bond and the y-axis perpendicular to the C=C bond in the molecular plane, as given in Figure 7.8. The irreducible bases for D_{2i} are given by

$$\begin{array}{llll} A_g: & 1 & A_u: & xyz,\ z\tilde{z} \\ B_{1g}: & \tilde{z},\ xy & B_{1u}: & z \\ B_{2g}: & \tilde{y},\ zx & B_{2u}: & y \\ B_{3g}: & \tilde{x},\ yz & B_{3u}: & x \end{array} \tag{7.6.7}$$

Let the equilibrium coordinates (x, y) of the four H atoms be

$$(a, b), \qquad (a, -b), \qquad (-a, b), \qquad (-a, -b) \tag{7.6.8a}$$

and the coordinates of the two C atoms be

$$(d, 0), \qquad (-d, 0) \tag{7.6.8b}$$

Then, it will be shown that the final results of ISCs are independent of these parameters a, b and d upon normalization.

(i) The four C—H stretches (s_1, s_2, s_3 and s_4). The decomposition of the permutation representation $\Delta^{(4)}$ based on the set of four equivalent points $S^{(4)}$ defined by (7.6.8a) is given by

$$\Delta^{(4)} = \underset{1}{A_g} + \underset{y}{B_{2u}} + \underset{xy}{B_{1g}} + \underset{x}{B_{3u}}$$

Thus, from (7.6.4), the required ISCs are given by (see Figure 7.8)

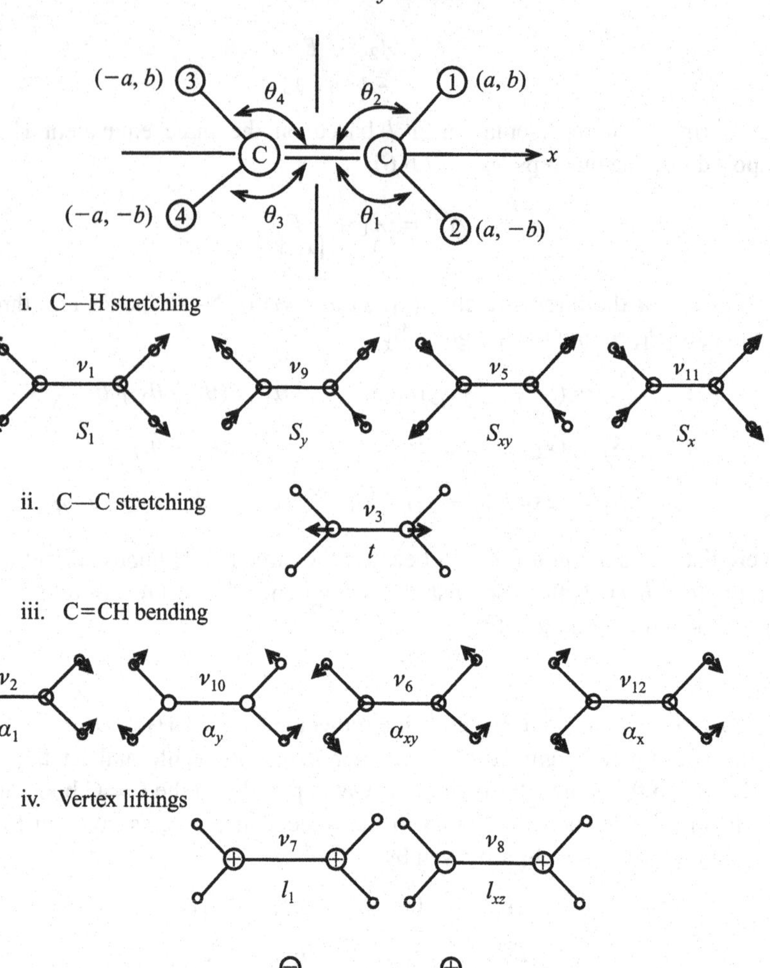

Figure 7.8. Internal vibrational modes of $C_2H_4 \in D_{2i}$.

$$A_g: \quad S_1 = \sum_\nu s_\nu/2 = (s_1 + s_2 + s_3 + s_4)/2$$

$$B_{2u}: \quad S_y = \sum_\nu y_\nu^0 s_\nu/N = (s_1 - s_2 + s_3 - s_4)/2$$

$$B_{1g}: \quad S_{xy} = \sum_\nu (xy)_\nu^0 s_\nu/N = (s_1 - s_2 - s_3 + s_4)/2$$

$$B_{3u}: \quad S_x = \sum_\nu x_\nu^0 s_\nu/N = (s_1 + s_2 - s_3 - s_4)/2$$

(ii) One C=C stretching:

$$A_g\text{: } t$$

(iii) Four C=C—H bendings: analogously to (i),

$$\begin{aligned} A_g&: & \alpha_1 &= (\theta_1 + \theta_2 + \theta_3 + \theta_4)/2 \\ B_{2u}&: & \alpha_y &= (\theta_1 - \theta_2 + \theta_3 - \theta_4)/2 \\ B_{1g}&: & \alpha_{xy} &= (\theta_1 - \theta_2 - \theta_3 + \theta_4)/2 \\ B_{3u}&: & \alpha_x &= (\theta_1 + \theta_2 - \theta_3 - \theta_4)/2 \end{aligned}$$

(iv) Two vertex liftings. Let $\Delta^{(2)}$ be the permutation representation based on the set $S^{(2)}$ of two equivalent carbon atoms in $H_2C{=}CH_2$, then

$$\Delta^{(2)} = \underset{1}{A_g} + \underset{x}{B_{3u}}$$

A lifting mode transforms like $z \in B_{1u}$ according to rule 3 so that two equivalent lifting modes are classified by the unirreps of G as follows:

$$B_{3u} \times \Delta^{(2)} = \underset{z}{B_{1u}} + \underset{zx}{B_{2g}}$$

Accordingly, the ISCs of these two lifting modes are given, using (7.6.4), by

$$\begin{aligned} B_{1u}&: & l_z &= (z_1 + z_2)/\sqrt{2} \\ B_{2g}&: & l_{xz} &= (z_1 - z_2)/\sqrt{2} \end{aligned}$$

where z_1 and z_2 are the z-coordinates of the two carbon atoms relative to the molecular plane in equilibrium.

(v) The C=C torsion. According to rule 4 and the bases of D_{2i} given by (7.6.7), this mode transforms like a pseudo-scalar $xyz \in A_u$. Consequently the torsion is described by the ISC given by

$$A_u\text{:} \qquad \tau_{xyz} = \sum_{\nu} (xy)^0_{\nu} z_{\nu}/N = (z_1 - z_2 - z_3 + z_4)/2$$

Note that these z-coordinates are those of four equivalent H atoms. The motion of the two C atoms does not contribute because $y^0_{\nu} = 0$ for them. This is expected because this mode is the torsion about the C=C bond.

Thus, we have obtained a total of 12 ($6 \times 3 - 6$) ISCs for the ethylene molecule, as shown in Figure 7.8. This example is interesting because it contains all of the kinds of internal symmetry coordinates which have been considered in the rules for ISCs. The frequencies $\nu_1 - \nu_{12}$ denoted for the modes of vibration for each ISC are those given by Herzberg (1951). The ISCs of $H_2C{=}CH_2$ belonging to each unirrep of the symmetry group D_{2i} in summary are

$$\begin{aligned} A_g&: & S_1, t, \alpha_1 & \qquad A_u: & \tau_{xyz} \\ B_{2u}&: & S_y, \alpha_y & \qquad B_{2g}: & l_{xz} \\ B_{1g}&: & S_{xy}, \alpha_{xy} & \qquad B_{1u}: & l_z \\ B_{3u}&: & S_x, \alpha_x & & \end{aligned}$$

It is noted here that the torsion and the two lifting modes are not coupled with any other modes of vibration. The three ISCs belonging to the unirrep A_g are linearly combined to give a set of normal coordinates that diagonalizes the Hamiltonian and likewise with the ISCs belonging to B_{2u}, to B_{1g} and to B_{3u}. Since these couplings are rather small, the ISCs given above are very good approximations to the true normal modes.

Example 7.6.7. The ISCs of the CH_4 molecule $\in T_p$.

There exist four C—H stretches and the six HCH angle bendings. The position coordinates of the four H atoms with respect to the central C atom have been given by (7.2.15a) and the permutation representation $\Delta^{(4)}$ based on these four equivalent points $S^{(4)}$ is decomposed into two unirreps A_1 and T_2 of the symmetry group T_p, as was shown by (7.2.15b).

(i) ISCs of the four C—H stretches s_ν ($\nu = 1-4$). According to the rule 1 of correspondence, s_ν transforms like a scalar function located at $r_\nu^0 \in S^{(4)}$, so that, from the general expression (7.6.4), we have

$$A_1: \quad S_1 = \sum_\nu s_\nu$$

$$T_2: \quad S_x = \sum_\nu x_\nu^0 s_\nu, \qquad S_y = \sum_\nu y_\nu^0 s_\nu, \qquad S_z = \sum_\nu z_\nu^0 s_\nu$$

(ii) The ISCs of the six valence-angle bendings $\alpha_{\nu\mu}$ ($\nu > \mu$; 1, 2, 3, 4) of HCH angles. These transform like a set of scalar functions located at the six equivalent points defined, according to rule 1, by

$$\boldsymbol{r}_{\nu\mu}^0 = (\boldsymbol{r}_\nu^0 + \boldsymbol{r}_\mu^0)/2; \qquad \nu > \mu;\ 1, 2, 3, 4$$

Their explicit values are, from (7.2.15a),

$$\begin{aligned} &\boldsymbol{r}_{12}^0 = (0, 0, 1), \qquad &&\boldsymbol{r}_{13}^0 = (0, 1, 0), \qquad &&\boldsymbol{r}_{14}^0 = (1, 0, 0) \\ &\boldsymbol{r}_{23}^0 = (-1, 0, 0), &&\boldsymbol{r}_{24}^0 = (0, -1, 0), &&\boldsymbol{r}_{34}^0 = (0, 0, -1) \end{aligned} \tag{7.6.9}$$

The decomposition of $\Delta^{(6)}$ based on the set $S^{(6)}$ of these six points yields

$$\Delta^{(6)} = \underset{1}{A_1} + \underset{[u,\, v]}{E} + \underset{[x,\, y,\, z]}{T_2}$$

Thus, from (7.6.4), we have

$$A_1: \quad \alpha_1 = \sum_{\nu>\mu} \alpha_{\nu\mu}$$

$$E: \quad \alpha_u = \sum_{\nu>\mu} u(\boldsymbol{r}_{\nu\mu}^0)\alpha_{\nu\mu}, \qquad \alpha_v = \sum_{\nu>\mu} v(\boldsymbol{r}_{\nu\mu}^0)\alpha_{\nu\mu}$$

$$T_2: \quad \alpha_x = \sum_{\nu>\mu} x_{\nu\mu}^0 \alpha_{\nu\mu}, \qquad \alpha_y = \sum_{\nu>\mu} y_{\nu\mu}^0 \alpha_{\nu\mu}, \qquad \alpha_z = \sum_{\nu>\mu} z_{\nu\mu}^0 \alpha_{\nu\mu}$$

or explicitly, with use of (7.6.9),

$$A_1: \quad \alpha_1 = (\alpha_{12} + \alpha_{13} + \alpha_{14} + \alpha_{23} + \alpha_{24} + \alpha_{34})/\sqrt{6} = 0$$

$$E: \quad \alpha_u = (-2\alpha_{12} + \alpha_{13} + \alpha_{14} + \alpha_{23} + \alpha_{24} - 2\alpha_{34})/\sqrt{8}$$

$$\alpha_v = (-\alpha_{13} + \alpha_{14} + \alpha_{23} - \alpha_{24})/\sqrt{2}$$

$$T_2: \quad \alpha_x = (\alpha_{14} - \alpha_{23})/\sqrt{2}, \qquad \alpha_y = (\alpha_{13} - \alpha_{24})/\sqrt{2},$$

$$\alpha_z = (\alpha_{12} - \alpha_{34})/\sqrt{2}$$

where α_1 is a redundant coordinate equal to zero due to the fact that the sum of the six bond angles HCH is a constant.

Thus we have obtained a total of nine ISCs for CH_4 molecule, as expected from the number of internal degrees of freedom $3 \times 5 - 6 = 9$.

8

Subduced and induced representations

Let G be a finite group and H be a subgroup of G. Then a representation of the group G automatically describes a representation of the subgroup H of G. Such a representation is called a representation of H *subduced* by a representation of G. Conversely, from a given representation of a subgroup H of G we can form a representation of the group G. Such a representation is called a representation of G *induced* by a representation of its subgroup H. The problem is to form the irreducible representations (irreps in short) of G from the irreps of its subgroup H. If the group G is finite and *solvable* (see Section 8.4.1), the problem of forming the irreps of G may be solved by a step-by-step procedure from the trivial irrep of the trivial identity subgroup. This method is possible, for example, for a crystallographic point group. An alternative approach is via the induced irreps of G from the so-called *small representations* of *the little groups* of the irreps of H. As a preparation, we shall discuss subduced representations first.

8.1 Subduced representations

Let $G = \{g\}$ be a group and $H = \{h\}$ be a subgroup of G. Let $\Gamma(G) = \{\Gamma(g);\ g \in G\}$ be a representation of G, then it provides a representation of H by $\{\Gamma(h);\ h \in H\}$. This representation is called *the subduced representation of* $\Gamma(G)$ *onto* H or the representation of H subduced by the representation $\Gamma(G)$. It is often expressed by

$$\Gamma^{\downarrow}(H) = [\Gamma(G) \downarrow H] \tag{8.1.1}$$

Even if $\Gamma(G)$ is irreducible, the subduced representation $\Gamma^{\downarrow}(H)$ need not necessarily be irreducible. However, if $\Gamma^{\downarrow}(H)$ is irreducible then so is $\Gamma(G)$, because, if $\Gamma(G)$ is reducible, then $\Gamma^{\downarrow}(H)$ must also be reducible. Thus we arrive at the following simple theorem.

Theorem 8.1.1. Let H be a subgroup of a group G. Then a representation $\Gamma(G)$ of G is irreducible, if its subduced representation $\Gamma^{\downarrow}(H)$ onto H is irreducible.

This simple theorem frequently provides an easy proof of the irreducibility of $\Gamma(G)$. Moreover, the following theorem also provides an easy proof for the orthogonalities of the representations of G.

Theorem 8.1.2. Let $\Gamma(G)$ and $\Gamma'(G)$ be two representations of G. If there exists a subgroup H of G such that their subduced representations onto H are orthogonal (i.e. they contain no irreps of H in common), then the two representations $\Gamma(G)$ and $\Gamma'(G)$ are orthogonal.

Proof. Let the left coset decomposition of G by H be $G = \sum_\nu s_\nu H$, then a general element g of G can be expressed as $g = s_\nu h$ with $h \in H$ for some ν, so that

$$\Gamma(g)_{ij} = \Gamma(s_\nu h)_{ij} = \sum_k \Gamma(s_\nu)_{ik}\Gamma(h)_{kj}$$

and we have an analogous expression for $\Gamma'(g)$. Thus

$$\sum_{g\in G} \Gamma(g)^*_{ij}\Gamma'(g)_{i'j'} = \sum_{\nu,k,k'} \Gamma(s_\nu)^*_{ik}\Gamma'(s_\nu)_{i'k'} \sum_{h\in H} \Gamma(h)^*_{kj}\Gamma'(h)_{k'j'}$$
$$= 0$$

because the sum over $h \in H$ vanishes on making the assumption that $\Gamma^\downarrow(H)$ and $\Gamma'^\downarrow(H)$ are orthogonal. Note that the converse of Theorem 8.1.2 need not be true, i.e. $\Gamma(G)$ and $\Gamma'(G)$ can be orthogonal even if $\Gamma^\downarrow(H)$ and $\Gamma'^\downarrow(H)$ are not.

8.2 Induced representations

The problem is to construct a representation of a group from a known representation of its subgroup. It is the inverse of the process of subduction. Let $G = \{g\}$ be a finite group and $H = \{h\}$ be a subgroup of G with the index $r = |G|/|H|$. Then the left coset decomposition of G by H may be written as

$$G = \sum_{i=1}^{r} s_i H, \qquad r = |G|/|H| \tag{8.2.1}$$

where $s_1 = e$ is the identity. This means that any element of G belongs to one and only one coset of H in G. Thus a product $gs_j \in G$ must also belong to some coset of H in G. This simple fact may be expressed in a formal way as follows (Kim 1986a):

$$gs_j = \sum_{i=1}^{r}\sum_{h\in H} s_i h\delta(h, s_i^{-1}gs_j); \qquad j = 1, 2, \ldots, r \tag{8.2.2}$$

where $\delta(h, s_i^{-1}gs_j)$ is Kronecker's delta defined by

$$\delta(h, s_i^{-1}gs_j) = \begin{cases} 1, & \text{if } gs_j = s_i h \\ 0, & \text{otherwise} \end{cases}$$

The above relation (8.2.2) is basic to the induced representation of G from a representation of its subgroup H. To see this, let $\gamma(H) = \{\gamma(h);\ h \in H\}$ be a $d_\gamma \times d_\gamma$ matrix representation of H and let $\psi = [\psi_1, \psi_2, \ldots, \psi_{d_\gamma}]$ be a basis of $\gamma(H)$ such that

$$\hat{h}\psi = \psi\gamma(h) \qquad \text{or} \qquad \hat{h}\psi_i = \sum_j \psi_j\gamma_{ji}(h); \qquad \forall\, h \in H$$

Then, operating both sides of (8.2.2) on ψ, we arrive at the defining relations of the induced representation of G from $\gamma(H)$:

$$\hat{g}(\hat{s}_j\psi) = \sum_{i,h}(\hat{s}_i\psi)\gamma(h)\delta(h, s_i^{-1}gs_j); \qquad \forall g \in G \tag{8.2.3}$$

Here it is safe to assume that the set $\{\mathring{s}_i\psi\}$ is linearly independent, then it defines a basis called *the induced basis* $\psi^{\uparrow}$

$$\psi^{\uparrow} = [\psi, \mathring{s}_2\psi, \ldots, \mathring{s}_r\psi], \qquad r = |G|/|H| \tag{8.2.4a}$$

which spans an invariant vector space of the dimensionality rd_γ with respect to G such that

$$\mathring{g}\psi^{\uparrow} = \psi^{\uparrow}\gamma^{\uparrow}(g), \qquad \forall g \in G \tag{8.2.4b}$$

Here, the set $\{\gamma^{\uparrow}(g)\} = \gamma^{\uparrow}(G)$ is an $rd_\gamma \times rd_\gamma$ matrix representation of G called *the induced representation* of G from a representation $\gamma(H)$ of a subgroup H of G: the matrix elements are

$$\begin{aligned}\gamma^{\uparrow}(g)_{ij} &= \sum_{h\in H} \gamma(h)\delta(h, s_i^{-1}gs_j)\\ &\equiv \dot{\gamma}(s_i^{-1}gs_j); \qquad i, j = 1, 2, \ldots, r = |G|/|H|\end{aligned} \tag{8.2.5}$$

with

$$\dot{\gamma}(s_i^{-1}gs_j) = \begin{cases}\gamma(s_i^{-1}gs_j), & \text{if } s_i^{-1}gs_j \in H\\ 0, & \text{otherwise}\end{cases}$$

According to this definition, if the (i, j)th entry $\gamma^{\uparrow}(g)_{ij}$ is not null, then the remaining entries in the ith column and jth row are all null. A matrix having only one non-zero entry in each row and each column is known as *a monomial matrix*. Then the induced representation $\gamma^{\uparrow}(G)$ is *a super monomial matrix* because each non-zero entry is a matrix $\gamma(s_i^{-1}gs_j)$ instead of a pure number.

The induced representation $\gamma^{\uparrow}(G)$ defined by (8.2.5) is denoted by

$$\gamma^{\uparrow}(G) = [\gamma(H) \uparrow G] \tag{8.2.6}$$

It is completely defined by (8.2.5) with the coset decomposition (8.2.1) quite independently from the induced basis $\psi^{\uparrow}$ (Lomont 1959). However, it is undoubtedly easier to understand the structure of the induced representation with use of the induced basis.

In the special case in which $\gamma(H)$ is the trivial identity representation, i.e. $\gamma_1(h) = 1$ for all $h \in H$, we have, from (8.2.5),

$$\gamma_1^{\uparrow}(g)_{ij} = \sum_{h\in H} \delta(h, s_i^{-1}gs_j) = \begin{cases}1, & \text{if } s_i^{-1}gs_j \in H\\ 0, & \text{otherwise}\end{cases} \tag{8.2.7}$$

which is called *the principal induced representation* of G relative to H. This representation is monomial and it is particularly important to understand the structure of the induced representation $\gamma_1^{\uparrow}(G)$, because, if we replace each unit element $\gamma_1^{\uparrow}(g)_{ij} = 1$ by the matrix $\gamma(s_i^{-1}gs_j)$, we obtain the general element $\gamma^{\uparrow}(g)_{ij}$.

For a further special case, in which H is the trivial subgroup of the identity element e, the principal induced representation $\gamma_1^{\uparrow}(G)$ becomes *the regular representation* of G

$$D^{(R)}(g)_{ij} = \delta(g, s_is_j^{-1}) = \begin{cases}1, & \text{if } g = s_is_i^{-1}\\ 0, & \text{otherwise}\end{cases} \tag{8.2.8}$$

Here the s_i terms are simply the elements of G.

8.2.1 Transitivity of induction

Let F, G and H be groups such that $F > G > H$. Let the left coset decompositions of G by H and of F by G be

$$G = \sum_i s_i H, \qquad F = \sum_j t_j G \tag{8.2.9a}$$

Then, combining these, we obtain the left coset decomposition of F by H:

$$F = \sum_{j,i} t_j s_i H \tag{8.2.9b}$$

Since an induced representation of a group from a representation of its subgroup is completely defined by the left coset decomposition of the group via the subgroup, we have

$$[\gamma(H)\uparrow G]\uparrow F = [\gamma(H)\uparrow F] \tag{8.2.10}$$

This proves the transitivity of induction.

8.2.2 Characters of induced representations

Let $\chi(H)$ be the character of $\gamma(H)$, then the character $\chi^{\uparrow}(G)$ of the induced representation $\gamma^{\uparrow}(G)$ is given by

$$\chi^{\uparrow}(g) = \sum_i \sum_h \chi(h)\delta(h, s_i^{-1} g s_i) \tag{8.2.11a}$$

Let $H^i = \{s_i h s_i^{-1};\ h \in H\}$ be *the conjugate group* of H with respect to a coset representative s_i of H in G, then, H^i is isomorphic to H via the correspondence $s_i h s_i^{-1} \leftrightarrow h$. Let an element of H^i be $\xi = s_i h s_i^{-1}$, then, one can define a representation of H^i by $\{\gamma^i(\xi) \equiv \gamma(s_i^{-1}\xi s_i) = \gamma(h)\}$ from the representation $\gamma(H)$. Let $\chi^i(\xi)$ and $\chi(h)$ be the traces of $\gamma^i(\xi)$ and $\gamma(h)$, respectively, then, $\chi^i(\xi) = \chi(s_i^{-1}\xi s_i) = \chi(h)$ so that, from (8.2.11a), the character of the induced representation $\chi^{\uparrow}(g)$ can be rewritten, in terms of $\chi^i(\xi)$, as follows:

$$\chi^{\uparrow}(g) = \sum_i \sum_{\xi \in H^i} \chi^i(\xi)\delta(\xi, g) \tag{8.2.11b}$$

This form will be used in formulating the irreducibility condition of $\gamma^{\uparrow}(g)$ in the next sub-section.

8.2.3 The irreducibility condition for induced representations

Let us discuss the irreducibility condition for the representation $\gamma^{\uparrow}(G)$ of a group G induced by a representation $\gamma(H)$ of a subgroup H of G. First of all, the representation $\gamma(H)$ itself should be irreducible because $\gamma^{\uparrow}(G)$ is linear in $\gamma(H)$. Now, according to the general theorem on the irreducibility condition, the induced representation $\gamma^{\uparrow}(G)$ is irreducible if and only if its character $\chi^{\uparrow}(G)$ satisfies the following condition:

$$\sum_{g \in G} |\chi^{\uparrow}(g)|^2 = |G| \tag{8.2.12}$$

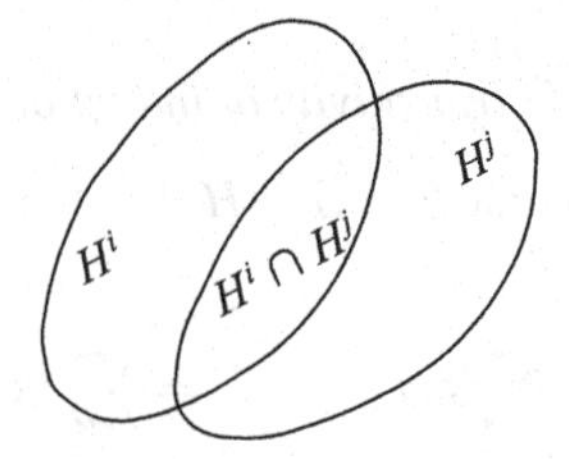

Figure 8.1. The intersection $H^i \cap H^j$.

Substitution of (8.2.11b) into this condition yields

$$\sum_{i,j=1}^{r} \sum_{\xi \in H^i \cap H^j} \chi^i(\xi)^* \chi^j(\xi) = |G| \tag{8.2.13}$$

where the sum on ξ is over the intersection $H^i \cap H^j$ of every two conjugate groups H^i and H^j of H in G (see Figure 8.1). An intersection $H^i \cap H^j$ is a subgroup of G: firstly, it is not empty because it contains the identity; secondly, if ξ_1 and ξ_2 are contained in $H^i \cap H^j$, then so is the product, because ξ_1 and ξ_2 are contained in each of the groups H^i and H^j. In special cases in which $i = j$, we have $H^i \cap H^j = H^i$, which is isomorphic to H.

To simplify the irreducibility condition (8.2.13), we observe that the sums defined by

$$Q_{ij} = \sum_{\xi \in H^i \cap H^j} \chi^i(\xi)^* \chi^j(\xi) \geqslant 0; \qquad i, j = 1, 2 \ldots, r \tag{8.2.14}$$

are non-negative quantities because Q_{ij} is a scalar product of two characters in the group space of $H^i \cap H^j$, as was shown in (6.6.9a). Let us calculate the diagonal elements Q_{ii}, where $H^i \cap H^i = H^i$. Since H^i is isomorphic to H via $\xi = s_i h s_i^{-1} \leftrightarrow h$, we may replace the sum over ξ in Q_{ii} by the sum over h, using $\chi^i(\xi) = \chi(h)$, and obtain

$$Q_{ii} = \sum_{h \in H} \chi(h)^* \chi(h) = |H|$$

because $\gamma(H)$ is assumed to be irreducible. This means that the sum of the diagonal terms Q_{ii} over i in (8.2.13) equals $|G|$ $(= r|H|)$ so that every off-diagonal term Q_{ij} $(i \neq j)$ must vanish, since otherwise each non-zero off-diagonal element Q_{ij} $(i \neq j)$ would contribute a positive number to (8.2.13). Therefore, the irreducibility condition (8.2.13) for $\gamma^{\uparrow}(G) = [\gamma(H) \uparrow G]$ is reduced to

$$Q_{ij} = \sum_{\xi} \chi^i(\xi)^* \chi^j(\xi) = 0, \qquad \text{for all } i \neq j \tag{8.2.15a}$$

where $\xi \in H^i \cap H^j$. That is, the induced representation $\gamma^{\uparrow}(G)$ is irreducible if its subduced representations $\{\gamma^{(i)}(\xi)\}$ and $\{\gamma^{(j)}(\xi)\}$ onto the intersections $H^i \cap H^j$ are mutually orthogonal (contain no irreps in common) for all pairs $(i \neq j)$. Moreover, since the conjugacy relation is transitive, any conjugate group H^j can be regarded as a conjugate group of H^i (because $s_j h s_j^{-1} = s_j s_i^{-1} s_i h s_i^{-1} s_i s_j^{-1}$) and, since H^i is isomorphic to H, the irreducibility condition is further reduced to

$$\sum_{\xi \in H \cap H^j} \chi(\xi)^* \chi^j(\xi) = 0; \qquad j = 2, \ldots, r = |G|/|H| \tag{8.2.15b}$$

Thus we arrive at the following theorem; cf. Jansen and Boon (1967).

Theorem 8.2.1. Let $H < G$ and $G = \sum_j s_j H$. The representation $\gamma^{\uparrow}(G)$ of G induced by an irrep $\gamma(H)$ of H is irreducible, if and only if the subduced representations $\{\gamma(\xi)\}$ and $\{\gamma^{(j)}(\xi)\}$ onto the intersection $\{\xi \in H \cap H^j\}$ are orthogonal (contain no irreps in common) for all $j = 2, 3, \ldots, r = |G|/|H|$.

The irreducibility criterion given above is quite general but rather complicated. If H is a normal subgroup of G, however, the criterion is considerably simplified because $H^j = H$. This will be discussed in the next section.

8.3 Induced representations from the irreps of a normal subgroup

8.3.1 *Conjugate representations*

If H is a normal subgroup of a group G, then any conjugate group $H^j = s_j H s_j^{-1}$ of H in G coincides with H, possibly with a different order. Thus one can define a conjugate representation of a given representation $\gamma(H)$ with respect to a coset representative s_j of H in G by $\gamma^j(H) = \{\gamma(s_j^{-1} h s_j)\}$, which is irreducible if $\gamma(H)$ is irreducible. Then, from (8.2.5), we have

$$\begin{aligned} &\gamma^{\uparrow}(h)_{ij} = \delta_{ij}\gamma(s_j^{-1} h s_j), \qquad \text{for } g = h \in H \\ &\gamma^{\uparrow}(g)_{jj} = 0, \qquad \text{for } g \notin H \end{aligned} \tag{8.3.1a}$$

That is, the induced representation $\gamma^{\uparrow}(g)$ from $\gamma(H)$ becomes a direct sum of the conjugate representations $\{\gamma^j(h)\}$ for $g = h \in H$ and off diagonal (i.e. the diagonal elements are zero) for $g \notin H$. A conjugate representation $\gamma^j(H)$ may but need not be equivalent to $\gamma(H)$, because we should not write $\gamma(s_j^{-1} h s_j) = \gamma(s_j)^{-1}\gamma(h)\gamma(s_j)$ unless s_j belongs to H. However, the conjugacy relation is transitive like the equivalence relation. Let ψ be a basis of $\gamma(H)$, then the basis of the conjugate representation $\gamma^j(H)$ is given by $\mathring{s}_j\psi$ from the fundamental relation (8.2.3) or directly from

$$\mathring{h}(\mathring{s}_j\psi) = \mathring{s}_j(\mathring{s}_j^{-1} h \mathring{s}_j)\psi = (\mathring{s}_j\psi)\gamma(s_j^{-1} h s_j), \qquad \forall h \in H \triangleleft G \tag{8.3.1b}$$

The basis $\mathring{s}_j\psi$ may be called a conjugate basis of $\psi \in \gamma(H)$.

Now, let $\chi^j(h) = \operatorname{tr}\gamma^j(h)$, then the character $\chi^{\uparrow}(g)$ of $\gamma^{\uparrow}(g)$ is rewritten from (8.3.1a) as

$$\chi^{\uparrow}(g) = \begin{cases} \sum_{j=1}^{r} \chi^j(h), & \text{if } g = h \in H \\ 0, & \text{otherwise} \end{cases} \tag{8.3.1c}$$

Moreover, since $H \cap H^j = H$, the irreducibility condition (8.2.15b) is reduced to the following simple form:

$$\langle \chi, \chi^j \rangle_H \equiv \sum_{h \in H} \chi(h)^* \chi^j(h) = 0; \qquad j = 2, \ldots, r \tag{8.3.2}$$

which means that all the conjugate irreps $\gamma^j(H)$ of H in G are inequivalent to $\gamma(H)$. Thus, we have the following theorem.

Theorem 8.3.1. Let $H \lhd G$ and $G = \sum_j s_j H$, then the representation of G induced by an irrep $\gamma(H)$ is irreducible, if and only if all the conjugate irreps $\gamma^j(H)$ of $\gamma(H)$ with respect to the coset representatives s_j $(\neq e)$ of H in G are inequivalent to $\gamma(H)$.

The above theorem also implies that all the conjugate irreps $\gamma^j(H)$ of H in G are mutually inequivalent, because any pair of conjugate representations is reduced to a pair $\{\gamma(H), \gamma^i(H)\}$ by a further conjugation.

8.3.2 Little groups and orbits

To understand Theorem 8.3.1 given above, a systematic characterization of the conjugacy relation for the irreps of H seems to be in order. Let $\gamma(H)$ and $\gamma'(H)$ be two irreps of H and let $H \lhd G$. If there exists an element s in G such that $\gamma'(H) \sim \gamma^s(H) = \{\gamma(s^{-1}hs); \forall h \in H\}$ ($\sim$ denotes equivalence), then $\gamma(H)$ and $\gamma'(H)$ are said to be mutually conjugate relative to G. The conjugacy relation of irreps is clearly different from their equivalence relation unless $s \in H$; however, it is transitive (like the equivalence relations) so that we can classify all the irreps of H by their conjugacy relation.

Let us introduce the concept of *the little group* G_γ of an irrep $\gamma(H)$ of a normal subgroup H of G. It is defined by the subset of G that leaves $\gamma(H)$ equivalent under conjugation; here, the subset is a subgroup of G because the conjugacy relation is transitive. Since any element of H leaves $\gamma(H)$ equivalent under conjugation, H is a subgroup of G_γ; in fact, H is an invariant subgroup of G_γ because H is an invariant subgroup of G. Thus,

$$H \trianglelefteq G_\gamma \leqslant G \tag{8.3.3}$$

That is, for a given pair $\{H \trianglelefteq G\}$, the possible maximum of G_γ equals G while its possible minimum equals H. At the maximum, all the conjugate irreps of $\gamma(H)$ relative to G are equivalent. For example, the identity representation $\gamma_1(H) = 1$ is invariant under any conjugation so that its little group is always G. Also, if H is in the center of G, then the little group of every irrep of H is G. At the minimum G_γ $(= H)$, the conjugate representation of $\gamma(H)$ with respect to any coset representative $(\neq e)$ of H in G is inequivalent to $\gamma(H)$. Accordingly, from Theorem 8.3.1, the irreducibility condition for $\gamma^\uparrow(G) = [\gamma(H) \uparrow G]$ is given by

$$G_\gamma = H \tag{8.3.4}$$

Next, we introduce the concept of *an orbit* which is closely related to the little group G_γ. An orbit of a normal subgroup H of G is a set of *mutually conjugate irreps of H relative to G which are inequivalent*. Then the complete set of irreps of H can be classified into the orbits of H relative to G. These are disjoint because the conjugacy relation is transitive. An orbit of H that contains an irrep $\gamma(H)$ is called *the orbit* of $\gamma(H)$. Obviously, any member of an orbit can be used to identify the orbit. Now, since the little group G_γ is the subgroup of G which leaves $\gamma(H)$ equivalent under conjugation, the orbit O_γ of $\gamma(H)$ is formed by the conjugate representations of $\gamma(H)$ with respect to the coset representatives of G_γ, i.e. let $G = \sum_m p_m G_\gamma$, then the orbit O_γ is given by

$$O_\gamma = \{\gamma^m(h) = \gamma(p_m^{-1} h p_m);\ h \in H;\ m = 1, 2, \ldots, |O_\gamma| = |G|/|G_\gamma|\} \tag{8.3.5}$$

Table 8.1. *The unirreps of D_2 with bases*

D_2	E	2_z	2_y	2_x	Bases
A	1	1	1	1	$1, x^2, y^2, z^2, xyz$
B_1	1	1	-1	-1	z, xy
B_2	1	-1	1	-1	y, zx
B_3	1	-1	-1	1	x, yz

where $|O_\gamma|$ is the order of the orbit O_γ given by the index of G_γ in G. Thus, the inequality (8.3.3) is rewritten with $|O_\gamma|$ as follows:

$$1 \leqslant |O_\gamma| = |G|/|G_\gamma| \leqslant |G|/|H| \tag{8.3.6}$$

Accordingly, the irreducibility condition for the induced representation $\gamma^\uparrow(G)$ given by (8.3.4) is also characterized by

$$|O_\gamma| = |G|/|H| \tag{8.3.7}$$

which is the possible maximum order of the orbit $|O_\gamma|$ of H in G.

Remark 1. All members of an orbit have the same little group up to isomorphism because the conjugacy relation is transitive; in fact, the little group of a conjugate irrep $\gamma^m(H)$ in O_γ of (8.3.5) is simply given by $G_\gamma^m = p_m G_\gamma p_m^{-1}$. In particular, when G_γ is an invariant subgroup of G, we have $G_\gamma^m = G_\gamma$. This is the case, for example, when $G_\gamma = H$ or G (see Section 8.4).

Remark 2. There is no need to construct the induced representations of G for more than one member (arbitrarily chosen) per orbit of H in G. The reason for this is that the representations of G induced by any two members of an orbit O_γ of $H \triangleleft G$ are equivalent:

$$[\gamma^i(h) \uparrow G] \sim [\gamma(h) \uparrow G], \qquad \forall\, h \in H \tag{8.3.8}$$

because from (8.3.1c) it follows that the characters of both sides of (8.3.8) are equal.

We see that one can therefore speak of a representation of G induced by an orbit of H. This property of an orbit is easy to understand because all conjugate irreps of H relative to G participate in forming $\gamma^\uparrow(G)$ according to (8.3.1a).

8.3.3 Examples

Example 1. The orbits of the dihedral group D_2 relative to the tetrahedral group T.

The group T is formed by augmenting the group D_2 with the augmentor a $(= 3_{xyz})$ and the index of D_2 in T equals 3. Thus, from (8.3.6), the order of an orbit of D_2 satisfies $1 \leqslant |O_\gamma| \leqslant 3$. The unirreps of D_2 are given in Table 8.1 with bases.

To determine the orbits of a group, only outer automorphism of D_2 need be considered. The conjugate elements of D_2 for the augmentor $a = 3_{xyz}$ are

$$a^{-1}2_z a = 2_y, \qquad a^{-1}2_y a = 2_x, \qquad a^{-1}2_x a = 2_z \tag{8.3.9}$$

so that from Table 8.1 the conjugate unirreps of D_2 by a are given by

$$A^a = A, \qquad B_1^a = B_3, \qquad B_2^a = B_1, \qquad B_3^a = B_2 \tag{8.3.10}$$

Therefore, the orbits of D_2 relative to the group T are

$$O_A = \{A\}, \qquad O_B = \{B_1, B_2, B_3\} \tag{8.3.11a}$$

Let the little group of a unirrep $\gamma(D_2)$ with respect to the group T be expressed by $T(\gamma)$, then we have

$$T(A) = T, \qquad T(B_1) = T(B_2) = T(B_3) = D_2 \tag{8.3.11b}$$

From the irreducibility criterion (8.3.7), the induced representation $A^\uparrow$ is reducible because $|O_A| = 1\ (<3)$, whereas $B_1^\uparrow(T)$ is irreducible because $|O_B| = 3$, the maximum. For the orbit O_B, it is necessary to construct only $B_1^\uparrow(T)$, $B_2^\uparrow(T)$ or $B_3^\uparrow(T)$ because they are mutually equivalent from (8.3.8).

An easy way to find the conjugacy relation of the irreps of $H \lhd G$ is to use their bases of representation: since the basis of a conjugate irrep $\gamma^j(H)$ is given by the *conjugate basis* $\mathring{s}\psi$ according to (8.3.1b), the bases of inequivalent irreps of H in an orbit are connected by a coset representative of H in G. For the above example $D_2 \lhd T$, the bases of the irreps of D_2 are

$$1 \in A,\ z \in B_1,\ Y \in B_2,\ x \in B_3$$

Since these are connected by the augmentor $a = 3_{xyz}$ via

$$\mathring{a}1 = 1, \qquad \mathring{a}z = x, \qquad \mathring{a}x = y, \qquad \mathring{a}y = z$$

the orbits of D_2 with respect to T are those given by (8.3.11a).

Example 2. The orbits of the cyclic group C_n relative to the dihedral group D_n.

The group D_n is formed by augmenting C_n with 2_x ($\perp n_z \in C_n$) and the index of C_n in D_n equals 2. The unirreps of C_n and their bases may, from (6.7.1) and (6.7.2), be expressed by

$$\phi_m(\theta) = \mathrm{e}^{\mathrm{i}m\theta} \in M_m(C_n); \qquad m = 0, 1, 2, \ldots, n-1$$

Since the augmentor 2_x connects the bases by $\mathring{2}_x\phi_m(\theta) = \phi_{-m}(\theta)$, the orbits of C_n relative to D_n are given by

$$O_0 = \{M_0\}, \qquad O_m = \{M_m, M_{-m}\}, \qquad O_{n_e/2} = \{M_{n_e/2}\} \tag{8.3.12a}$$

where, from the condition $M_m \neq M_{-m}$, we have

$$m = \begin{cases} 1, 2, \ldots, (n-1)/2, & \text{for an odd } n \\ 1, 2, \ldots, \dfrac{n}{2} - 1, & \text{for an even } n \end{cases}$$

and $M_{n_e/2}$ occurs only for an even n denoted by n_e. From the irreducibility criterion (8.3.7), the induced representation $M_m^\uparrow$ for every m in (8.3.12a) is irreducible because $|O_m| = 2$, whereas $M_0^\uparrow$ and $M_{n_e/2}^\uparrow$ are reducible. Let the little group of M_ν with respect to D_n be expressed by $D_n(M_\nu)$, then

$$D_n(M_0) = D_n, \qquad D_n(M_m) = D_n(M_{-m}) = C_n, \qquad D_n(M_{n_e/2}) = D_n \tag{8.3.12b}$$

8.4 Irreps of a solvable group by induction

From the irreducibility criterion for an induced representation introduced in the previous section, we shall develop the methods of constructing the complete set of the irreps of a group G by induction. At present two general methods are available:

(i) When G is solvable (as will be defined below), we can construct all irreps of G via induction by a step by step procedure starting from the one-dimensional representations of the Abelian subgroup.
(ii) The complete set of the irreps of G can be constructed using the theorem that *the representation of G induced by an irrep called a small representation of the little group G_γ of an irrep $\gamma(H)$ is irreducible.* Here the difficulty is to construct the small representations of G_γ, which will be discussed in Section 8.5. However, for an important special case in which the subgroup H is Abelian, the small representations can be formed by the projective representations of the factor group G_γ/H (see Theorem 12.4.1 given in Chapter 12).

In the following, we shall describe the method (i) first, then the method (ii), after introducing some general theorems on the induced and subduced representations.

8.4.1 Solvable groups

As a preparation we state some basic theorems on solvable groups without proofs; see Lomont (1959).

Theorem. Every finite group G of order greater than 1 has a finite series of subgroups, $G, H_1, \ldots, H_s, E$ such that

$$G \triangleright H_1 \triangleright H_2 \triangleright \cdots \triangleright H_s \triangleright E \tag{8.4.1}$$

where H_{i+1} is a maximal (proper) normal subgroup of H_i.

A series given by (8.4.1) is called *a composition series of G* and the quotient groups of $G/H_1, H_1/H_2, \ldots, H_{s-1}/H_s, H_s/E$ are called the *composition quotient groups of* G. The orders of the composition quotient groups $|H_i|/|H_{i+1}|$ are called *the composition indices,* and the integer $s+1$ is called *the length* of the composition series.

Example

$$O \triangleright T \triangleright D_2 \triangleright C_2 \triangleright E \tag{8.4.2a}$$

The composition quotient groups are

$$O/T \simeq C_2, \qquad T/D_2 \simeq C_3, \qquad D_2/C_2 \simeq C_2, \qquad C_2/E = C_2 \tag{8.4.2b}$$

Here $\simeq$ means isomorphism and C_n denotes the cyclic group of order n. The composition indices are 2, 3, 2 and 2 (all prime) and the length of the composition series is 4.

Definition. A solvable (or integrable) group is a group whose composition indices are all prime numbers (a prime number being a positive integer that is divisible only by itself and by unity). Since a group of a prime order is cyclic, the composition quotient groups of a solvable group must be cyclic.

(1) Every group of order < 60 is solvable.
(2) Every Abelian group is solvable.
(3) Every subgroup of a solvable group is solvable.
(4) Every crystallographic point group is solvable because their orders are less than 60.
(5) The order of the icosahedral group Y equals 60 and hence it is not solvable.

8.4.2 Induced representations for a solvable group

Let G be a group and H be a normal subgroup of G with a prime index $p = |G|/|H|$. If we can construct all irreps of G by induction from the irreps of H, then by a step by step procedure we can construct all irreps of any solvable group starting from the one-dimensional irreps of an Abelian subgroup (Raghavacharyulu 1961, Miller and Love 1967).

By assumption, the factor group G/H is isomorphic to a cyclic group of a prime order p $(= |G|/|H|)$ and hence there exists a single augmentor a such that the left coset decomposition of G by H can be expressed as follows:

$$G = H + aH + \cdots + a^{p-1}H, \qquad a^p \in H \tag{8.4.3a}$$

Let ψ be the basis of a unirrep $\gamma(H)$ of H, then from (8.2.4a), the induced basis $\psi^{\uparrow}$ is given by

$$\psi^{\uparrow} = [\psi, \mathring{a}\psi, \ldots, \mathring{a}^{p-1}\psi] \tag{8.4.3b}$$

Thus, from (8.2.5) and (8.3.1a) the induced representation $\gamma^{\uparrow}(G) = [\gamma(H) \uparrow G]$ is given by

$$\gamma^{\uparrow}(h) = \begin{bmatrix} \gamma(h) & 0 & \ldots & 0 \\ 0 & \gamma(a^{-1}ha) & \ldots & 0 \\ \vdots & \vdots & & \vdots \\ 0 & 0 & \ldots & \gamma(a^{-p+1}ha^{p-1}) \end{bmatrix} \tag{8.4.4a}$$

$$\gamma^{\uparrow}(a) = \begin{bmatrix} 0 & 0 & \ldots & 0 & \gamma(a^p) \\ \mathbf{1} & 0 & \ldots & 0 & 0 \\ 0 & \mathbf{1} & \ldots & 0 & 0 \\ \vdots & \vdots & & \vdots & \vdots \\ 0 & 0 & \ldots & \mathbf{1} & 0 \end{bmatrix} \tag{8.4.4b}$$

where $\mathbf{1} = \gamma(e)$ is the unit matrix of order d_γ. Here $\gamma^{\uparrow}(a)$ follows from (8.2.5) through

$$\gamma^{\uparrow}(a)_{ij} = \sum_h \gamma(h)\delta(h, a^{-i+1+j}); \qquad i, j = 0, 1, 2, \ldots, p-1$$

whose non-zero elements occur when $-i + 1 + j = 0$ or p, i.e.

$$\gamma^{\uparrow}(h)_{i,i-1} = \gamma(e), \qquad \gamma^{\uparrow}(a)_{0,p-1} = \gamma(a^p)$$

where $i = 1, 2, \ldots, p-1$. The remaining representatives $\gamma(a^\nu h)$ are not given, because they are easily calculated by matrix multiplication of $\gamma^{\uparrow}(a)$ and $\gamma^{\uparrow}(h)$. Note, however, that $\gamma^{\uparrow}(a^\nu h)_{ii} = 0$ unless $a^\nu h \in H$, from (8.3.1a).

Now, according to (8.4.4a), every conjugate irrep $\gamma^i(H)$ with respect to a^i is on the diagonal entries of $\gamma^{\uparrow}(h)$: they are either all equivalent (*case I*) or all inequivalent (*case II*). The reason is that, if $\gamma^i(H) \sim \gamma(H)$ for any one i, then it is true for all i because p is prime.[1] Accordingly, the order $|O_\gamma|$ of the orbit O_γ equals one or the prime index p $(= |G|/|H|)$:

$$|O_\gamma| = \begin{cases} 1, & \text{for case I} \\ p = |G|/|H|, & \text{for case II} \end{cases} \qquad (8.4.5)$$

The previous examples for the orbits given by (8.3.11a) and (8.3.12a) are in accordance with this result. From the irreducibility criterion (8.3.7), the induced representation $\gamma^\uparrow(G)$ is reducible for case I and irreducible for case II. For the former, it will be shown that $\gamma^\uparrow(G)$ is *reducible to p inequivalent irreps* $D^{(m)}(G)$ $(m = 0, 1, \ldots, p-1)$ with the dimensionality d_γ, each of which subduces $\gamma(H)$ onto H. Assuming this to be true and for the moment postponing its proof, we shall show that the complete set of the irreps of G can be formed by the induced representations of G from the irreps of H, one per orbit of H relative to G.

Let $\gamma_\alpha(H)$ be an irrep belonging to the αth orbit of the order unity (case I), let $\gamma_\beta(H)$ be an irrep belonging to the βth orbit of order p (case II) and let their dimensionalities be d_α and d_β, respectively. Then we form p inequivalent irreps of G with dimensionality d_α by reducing the induced representation $\gamma_\alpha^\uparrow(G)$ and one induced irrep $\gamma_\beta^\uparrow(G)$ with the dimensionality pd_β from $\gamma_\beta(H)$. Now, on summing the squares of the irreps' dimensions over all orbits of G we obtain

$$p\sum_\alpha d_\alpha^2 + \sum_\beta (pd_\beta)^2 = p\left(\sum_\alpha d_\alpha^2 + p\sum_\beta d_\beta^2\right) = p|H| = |G|$$

where the second equality follows from the completeness relation of the irreps of H and from the fact that each irrep of H is contained in one and only one orbit of H. The above equation is nothing but the completeness relation for the irreps of G. In the following we shall construct explicitly the irreducible components of $\gamma^\uparrow(G)$ for case I and modify the representation (8.4.4) for case II into a more convenient form.

8.4.3 Case I (reducible)

$|O_\gamma| = 1$ and $G_\gamma = G$. All conjugate irreps of $\gamma(H)$ relative to G are equivalent so that the induced representation $\gamma^\uparrow(G)$ given by (8.4.4) is reducible. The irreducible components contained in $\gamma^\uparrow(G)$ may be called *the irreps of G associated with* $\gamma(H)$. There exist p inequivalent associated irreps of G with the dimensionality equal to that of $\gamma(H)$; these will be determined by the equivalence of the conjugate irreps of $\gamma(H)$ and the fact that the factor group G/H is a cyclic group of order p.

Let us assume that $\gamma(H)$ is a unirrep of H, then, from the assumed equivalence $\gamma(a^{-1}ha) \sim \gamma(h)$ for all $h \in H$, there exists a unitary transformation $M(a)$ such that

$$\gamma(a^{-1}ha) = M(a)^{-1}\gamma(h)M(a), \qquad \forall\, h \in H \qquad (8.4.6a)$$

Then, the transformation matrix $M(a)$ serves as a representative of the augmentor a provided that it satisfies the boundary condition

$$M(a)^p = \gamma(a^p); \qquad a^p \in H \qquad (8.4.6b)$$

[1] Suppose that $\gamma^1(H) \sim \gamma(H)$, then $\gamma^2(H) \sim \gamma^1(H) \sim \gamma(H)$ and so on, so that $\gamma^i(H) \sim \gamma(H)$ for all i. Suppose that $\gamma^i(H) \sim \gamma(H)$, then one can find integers n and m such that $mi = np + 1$, which leads to $\gamma^1(H) \sim \gamma(H)$.

To show that this boundary condition holds, we observe, with repeated use of (8.4.6a), that

$$\gamma(a^p)^{-1}\gamma(h)\gamma(a^p) = \gamma(a^{-p}ha^p) = M(a)^{-p}\gamma(h)M(a)^p$$

for all $h \in H$. Thus, $M(a)^p\gamma(a^p)^{-1}$ commutes with the unirrep $\gamma(H)$ and hence is a constant, according to Schur's lemma. By including the constant factor into $M(a)$ we arrive at (8.4.6b). Now, we see that the matrix $M(a)$ is defined by (8.4.6a) and (8.4.6b) up to a phase factor which is a pth root of unity. Thus if $M(a)$ is a matrix solution, then we obtain a total of p associate representations $D^{(m)}(G)$ of $\gamma(H)$ defined by

$$D^{(m)}(h) = \gamma(h), \qquad D^{(m)}(a) = \omega^m M(a); \qquad m = 0, 1, 2, \ldots, p-1 \tag{8.4.7}$$

where $\omega = \exp(-2\pi i/p)$. These are irreducible from Theorem 8.1.1 because their subduced representation $[D^{(m)}(G) \downarrow H] = \gamma(H)$ is irreducible; moreover, they are all inequivalent since their characters are all different and hence they are orthogonal with respect to each other.

Next, let ψ be a basis of $\gamma(H)$, then, from (8.4.6a) and $\mathring{h}\mathring{a}\psi = \mathring{a}\psi\gamma(a^{-1}ha)$, it follows that $\mathring{a}\psi M(a)^{-1} = \phi$ is also a basis of $\gamma(H)$, so that

$$\mathring{a}\psi = \phi M(a) \tag{8.4.8a}$$

where ϕ may but need not coincide with ψ. In the case in which $\mathring{a}\psi$ is linearly dependent on ψ we can set $\phi = \psi$ to obtain

$$\mathring{a}\psi = \psi M(a) \tag{8.4.8b}$$

i.e. a basis of $\gamma(H)$ also serves as a basis of the associate irrep $D^{(0)}$ (with $m = 0$) defined in (8.4.7) for this case. It turns out that this is frequently the case for a higher dimensional representation $\gamma(H)$, i.e. $d_\gamma > 1$.

According to (8.4.7), the p associate unirreps $\{D^{(m)}(G)\}$ of $\gamma(H)$ are explicit except for $M(a)$, which has to be determined from (8.4.6) or (8.4.8); see Miller and Love (1967). For a one-dimensional representation, however, $M(a)$ is given simply by a pth root of $\gamma(a^p)$ from the boundary condition (8.4.6b) because (8.4.6a) holds independently from $M(a)$. For a higher dimensional case it is determined from (8.4.8b) for most cases of practical importance; see examples given at the end of this section.

8.4.3.1 The induced representation from the identity representation

The identity representation $\{\gamma_1(h) = 1\}$ of H always forms an orbit of its own and thus belongs to case I for any proper super-group G of H. Since $\gamma_1(a^p) = 1$, we may take $M(a) = 1$ from (8.4.6b). Then, from (8.4.7), the p associate unirreps $D_1^{(m)}(G)$ of the identity unirrep $\gamma_1(H)$ are given by

$$D_1^{(m)}(h) = 1, \qquad D_1^{(m)}(a) = \omega^m; \qquad m = 0, 1, \ldots, p-1 \tag{8.4.9a}$$

where $\omega = \exp(-2\pi i/p)$. Let ψ_1 be a basis of the identity representation $\gamma_1(H)$, then a basis of $D_1^{(m)}(G)$ is given, using the projection operator method introduced in Section 6.9, by

$$\begin{aligned}\psi_1^{(m)} &= \sum_{j=0}^{p-1}(\omega^{-m}\mathring{a})^j\psi_1 \\ &= \psi_1 + \omega^{-m}\psi_2 + \cdots + \omega^{-m(p-1)}\psi_p\end{aligned} \tag{8.4.9b}$$

where $\psi_{j+1} = \mathring{a}^j \psi_1$, $j = 0, 1, \ldots, p-1$, are bases of the identity representation $\gamma_1(H)$ according to (8.4.8a) with $M(a) = 1$. The basis ψ_1 in (8.4.9b) has to be chosen appropriately in order to avoid the null result; for example, if $\mathring{a}\psi_1 = \psi_1$ then (8.4.9b) yields the null result, for ω being a pth root of unity.

In view of (8.4.9a), the general associate representations $D^{(m)}(G)$ given by (8.4.7) can be written as the direct product representations of G as follows:

$$D^{(m)}(G) = D_1^{(m)}(G) \times D^{(0)}(G); \qquad m = 0, 1, \ldots, p-1 \tag{8.4.10}$$

where $D_1^{(m)}(G)$ are the associate unirreps of the identity representation of G and $D^{(0)}(G)$ is the unirrep defined by (8.4.7) with $m = 0$.

8.4.4 Case II (irreducible)

$|O_\gamma| = p$ and $G_\gamma = H$. The induced representation $\gamma^\uparrow(G)$ from the orbit O_γ is irreducible and given by (8.4.4a) and (8.4.4b). One unsatisfactory feature of (8.4.4a) is that every one of the conjugate irreps $\gamma^j(H)$ for all j has to be calculated, even though each of them must be equivalent to one or other irrep of H that is assumed to be known. To remedy this, we shall transform $\gamma^\uparrow(g)$ such that $\gamma^\uparrow(h)$ given by (8.4.4a) becomes a direct sum of the known set of irreps of H. Then it becomes necessary to calculate only the representative of the augmenting operator a.

Let $\{\gamma(H), \gamma_1(H), \ldots, \gamma_{p-1}(H)\}$ be an orbit of H in G for which the members are the unirreps of H ordered such that their conjugacy relations are expressed by

$$\gamma_{n-1}(a^{-1}ha) \sim \gamma_n(h); \qquad n = 1, 2, \ldots, p$$

where $\gamma_0(h) = \gamma_p(h) = \gamma(h)$. This means that there exists a set of unitary matrices $N_n(a)$ such that

$$\gamma_{n-1}(a^{-1}ha) = N_n^{-1}(a)\gamma_n(h)N_n(a) \tag{8.4.11}$$

Correspondingly, the basis set $\{\psi^{(n)}\}$ of $\{\gamma_n(H)\}$ satisfies

$$\mathring{a}\psi^{(n-1)} = \psi^{(n)}N_n(a); \qquad n = 1, 2, \ldots, p \tag{8.4.12}$$

with $\psi^{(0)} = \psi^{(p)} = \psi$. Now we introduce a new basis of representation for G by

$$\Psi = [\psi^{(0)}, \psi^{(1)}, \ldots, \psi^{(p-1)}] \tag{8.4.13}$$

then it provides a representation $\Gamma(G)$ defined by

$$\Gamma(h) = \begin{bmatrix} \gamma(h) & 0 & \ldots & 0 \\ 0 & \gamma_1(h) & \ldots & 0 \\ \cdot & \cdot & & 0 \\ \cdot & \cdot & \cdot & \cdot \\ \cdot & \cdot & \cdot & \cdot \\ 0 & 0 & \ldots & \gamma_{p-1}(h) \end{bmatrix} \tag{8.4.14a}$$

$$\Gamma(a) = \begin{bmatrix} 0 & 0 & \ldots & \cdot & N_p(a) \\ N_1(a) & 0 & \ldots & \cdot & 0 \\ 0 & N_2(a) & \ldots & \cdot & 0 \\ \vdots & \vdots & & & \vdots \\ 0 & 0 & \ldots & N_{p-1}(a) & 0 \end{bmatrix} \tag{8.4.14b}$$

where $N_p(a)$ is determined by

$$N_p(a)N_{p-1}(a)\ldots N_1(a) = \gamma(a^p) \tag{8.4.15}$$

because repeated use of (8.4.12) leads to

$$\begin{aligned}\mathring{a}^p\psi^{(0)} &= \psi^{(p)}N_p(a)N_{p-1}(a)\ldots N_1(a)\\ &= \psi^{(0)}\gamma(a^p)\end{aligned}$$

with $\psi^{(0)} = \psi^{(p)} = \psi$ and $a^p \in H$.

In the new representation $\Gamma(G)$ given by (8.4.14) which is equivalent to $\gamma^{\uparrow}(G)$ of (8.4.4), we see that the subduced representation $\Gamma^{\downarrow}(H)$ is given by a direct sum of the original irreps $\gamma(h), \gamma_1(H), \ldots, \gamma_{p-1}(H)$ belonging to the orbit of $\gamma(H)$. To calculate $\Gamma(a)$, however, one has to determine the transformation matrices $N_n(a)$ from (8.4.11) or (8.4.12). In the special case in which one can take $N_n(a) = \mathbf{1}$ for n ($\neq p$), from (8.4.15) we have $N_p(a) = \gamma(a^p)$ so that (8.4.11) reduces to $\gamma_{n-1}(a^{-1}ha) = \gamma_n(h)$. Thus, for this case, $\Gamma(G)$ given by (8.4.14) coincides completely with $\gamma^{\uparrow}(G)$ given by (8.4.4). This is the case, for example, when $\gamma(H)$ is one-dimensional.

The following two examples have been discussed already in Section 6.7 via the defining relations of the point groups and inductions from the unirreps of appropriate subgroups. Here, we shall discuss them again in terms of the general expressions of induced representations given by (8.4.10) and (8.4.14).

8.4.5 Examples

Example 1. The unirreps of the tetrahedral group T by induction from the unirreps of the dihedral group D_2 ($\triangleleft$ T) (cf. Section 6.7.3).

The orbits of D_2 relative to T are given by $O_A = \{A\}$ and $O_B = \{B_1, B_2, B_3\}$ from (8.3.11a). From (8.4.5), the orbit O_A belongs to case I whereas O_B belongs to case II.

For case I, $O_A = \{A\}$, there exist three one-dimensional unirreps of T associated with the identity representation A of D_2. These are determined by (8.4.9a) with $\omega = \exp(-2\pi i/3)$. The unirreps thus obtained coincide with A, A' and A'' given in Table 6.7. Here, we shall determine their bases by (8.4.9b). Choosing $\psi_1 = z^2 \in A$ of D_2 given in Table 8.1 and using $\mathring{a}z^2 = x^2$ and $\mathring{a}x^2 = y^2$ with $a = 3_{xyz}$, we obtain

$$\begin{aligned} A\!: &\quad \psi_1^{(0)} = z^2 + x^2 + y^2\\ A'\!: &\quad \psi_1^{(1)} = z^2 + \omega^* x^2 + \omega y^2 = (u + iv)/2\\ A''\!: &\quad \psi_1^{(2)} = z^2 + \omega x^2 + \omega^* y^2 = (u - iv)/2 \end{aligned} \tag{8.4.16}$$

where $u = 3z^2 - r^2$ and $v = \sqrt{3}(x^2 - y^2)$. Note that, if we had chosen $\psi_1 = 1$ for the basis of $A(D_2)$, we would have obtained $\psi_1^{(1)} = \psi_1^{(2)} = 0$, because $\mathring{a}\psi_1 = \psi_1$ and $1 + \omega + \omega^2 = 0$. Previously, in Table 6.7, we had written down the bases of A' and A'' without derivation.

For case II, $O_B = \{B_1, B_2, B_3\}$, since the unirreps of D_2 are all one-dimensional, we may take $N_n(a) = 1$ in (8.4.11). Then the induced representation is determined by the general expression (8.4.14), i.e. $\Gamma(h)$ for $h \in D_2$ is given by the direct sum of B_1, B_2, $B_3 \in D_2$ via (8.4.14a) while $\Gamma(3_{xyz})$ is determined by (8.4.14b) with $N_n(a) = 1$. The unirrep thus obtained is the representation T given in Table 6.7. The corresponding basis is given by $[x, y, z]$, from Table 8.1 and (8.4.13).

Example 2. The unirreps of the octahedral group O by induction from the unirreps of the tetrahedral group T ($\triangleleft$ O) (cf. Section 6.7.4).

The group O is defined by augmenting the group T with $a = 4_z$. From the unirreps of the group T given by Table 6.7, the orbits of T relative to the group O are

$$O_1 = \{A\}, \qquad O_2 = \{A', A''\}, \qquad O_3 = \{T\} \tag{8.4.17}$$

because $\dot{4}_z[x, y, z] = [y, -x, z]$ and $\dot{4}_z(u + iv) = (u - iv)$. From (8.4.5) the orbits O_1 and O_3 belong to case I, whereas the orbit O_2 belongs to case II.

For case I, $O_1 = \{A\}$ and $O_3 = \{T\}$, from the orbit O_1, using (8.4.9a) with $\omega = \pm 1$, we obtain two unirreps A_1 and A_2 of the group O as given in Table 6.8. For the orbit O_3, the basis $\psi = [x, y, z]$ of the representation T satisfies (8.4.8b) so that we obtain the unirreps T_1 given in Table 6.8 with use of the basis. Next, from (8.4.10) we obtain one more unirrep T_2 given by $A_2 \times T_1$.

For case II, $O_2 = \{A', A''\}$, where the members of the orbit are one-dimensional unirreps of the subgroup T. Let E be the two-dimensional induced unirrep of the group O, then it is determined by the general expression (8.4.14): $E(h)$ for $h \in T$ is given by the direct sum of A' and A'' of T from (8.4.14a) while $E(4_z)$ is given by (8.4.14b) with $N_n(a) = 1$. The corresponding basis is given by $[u + iv, u - iv]$ from (8.4.12). Previously, in Table 6.8 we have presented the unirrep E based on $[u, v]$.

Exercise. Write down the two-dimensional unirreps E of the group O using (8.4.14) based on $[u + iv, u - iv]$.

8.5 General theorems on induced and subduced representations and construction of unirreps via small representations

In the previous sections we have described how to construct the irreps of a solvable group through induction from the irreps of a subgroup with a prime index and applied it for the construction of the irreps of the point groups. Here in this section we shall introduce the concept of '*the small representations of the little groups*' from which irreps of the group can be induced. We shall first introduce some of the basic theorems which describe the relations between induced representations and subduced representations in general.

8.5.1 Induction and subduction

We begin with the following basic theorem which connects induced and subduced representations.

Theorem 8.5.1. The Frobenius reciprocity theorem. Let H be a subgroup of a group G and let $\gamma(h)$ and $\Gamma(G)$ be their respective irreps. Then the frequency $F_{\gamma\uparrow\Gamma}$ of $\Gamma(G)$ contained in the induced representation $\gamma^{\uparrow}(G)$ from $\gamma(H)$ equals the frequency $f_{\Gamma\downarrow\gamma}$ of $\gamma(H)$ contained in the subduced representation $\Gamma^{\downarrow}(H)$ from $\Gamma(G)$, i.e. if

$$\gamma^{\uparrow}(G) \sim \sum_{\Gamma} \oplus\, F_{\gamma\uparrow\Gamma}\Gamma(G), \qquad \Gamma^{\downarrow}(H) \sim \sum_{\gamma} \oplus\, f_{\Gamma\downarrow\gamma}\gamma(H)$$

then

$$F_{\gamma\uparrow\Gamma} = f_{\Gamma\downarrow\gamma} \tag{8.5.1}$$

Proof. Let $H = \{h\}$, $G = \{g\}$ and $G = \sum_i s_i H$ and let $\chi_\gamma(h)$ and $\chi_\Gamma(g)$ be the characters of $\gamma(h)$ and $\Gamma(g)$, respectively. Then $F_{\gamma\uparrow\Gamma}$ is given, using (8.2.11a), by

$$F_{\gamma\uparrow\Gamma} = |G|^{-1} \sum_{g,i} \chi_\Gamma(g)^* \chi_\gamma(s_i^{-1} g s_i) \tag{8.5.2}$$

where g is limited by $s_i^{-1} g s_i = h \in H$ so that

$$\chi_\Gamma(g) = \chi_\Gamma(s_i h s_i^{-1}) = \chi_\Gamma(h) \qquad (\because s_i \in G) \tag{8.5.3}$$

Substitution of this into (8.5.2) yields

$$f_{\gamma\uparrow\Gamma} = |G|^{-1} \sum_{h,i} \chi_\Gamma(h)^* \chi_\gamma(h) = |H|^{-1} \sum_h \chi_\Gamma(h)^* \chi_\gamma(h) = f_{\Gamma\downarrow\gamma} \tag{8.5.4}$$

Note that $\chi_\Gamma(h) = \operatorname{tr} \Gamma^\downarrow(h)$. Q.E.D.

Example 1. For $D_2 \lhd T$, from (8.4.16) and Table 6.7,

$$A^\uparrow(T) \sim A(T) \oplus A'(T) \oplus A''(T)$$

while

$$A^\downarrow(D_2) = A'^\downarrow(D_2) = A''^\downarrow(D_2) = A(D_2)$$

Thus, for example, $A^\uparrow(T)$ induced by $A(D_2)$ contains $A(T)$ exactly as often as $A^\downarrow(D_2)$ subduced by $A(T)$ contains $A(D_2)$; namely once for this case.

Example 2. For $T \lhd O$, from (8.4.17) and Table 6.8,

$$A'^\uparrow(O) = A''^\uparrow(O) = E(O); \qquad E^\downarrow(T) = A'(T) \oplus A''(T)$$

We shall next consider induction of $\gamma(H)$ followed by subduction, i.e. $[\gamma^\uparrow(G)]^\downarrow(H)$. It does not return to the original representation $\gamma(H)$ because induction of a representation $\gamma(H)$ increases the original dimensionality to the index $(= |G|/|H|)$ times, whereas subduction does not change the dimensionality. Hereafter, we assume that H is a normal subgroup of G and fully utilize the concepts of the little groups and the orbits introduced in Section 8.3.

Theorem 8.5.2. Let $H \lhd G$ and $\gamma(H)$ be an irrep of H and let G_γ be the little group of $\gamma(H)$ relative to G. Let $\boldsymbol{O}_\gamma(H)$ be the direct sum of the conjugate irreps of $\gamma(H)$ in the orbit O_γ. Then the subduced induced representation $[\gamma^\uparrow(G)]^\downarrow(H)$ is equivalent to a multiple of $\boldsymbol{O}_\gamma(H)$:

$$[\gamma^\uparrow(G)]^\downarrow(H) \sim f_\gamma \boldsymbol{O}_\gamma(H); \qquad f_\gamma = |G_\gamma|/|H| \tag{8.5.5}$$

Here $f_\gamma \boldsymbol{O}_\gamma(H)$ should be regarded as a direct sum of $\boldsymbol{O}_\gamma$ matrices.

Proof. It is only necessary to show that the characters of both sides of (8.5.5) are equal. From $H \trianglelefteq G_\gamma \leqslant G$ and $H \lhd G$ we express the left coset decomposition of G by H, via $G = \sum_m p_m G_\gamma$ and $G_\gamma = \sum_n q_n H$, in the form

$$G = \sum_m \sum_n p_m q_n H$$

Let $\chi_\gamma(h) = \mathrm{tr}\, \gamma(h)$. Then the character of the left-hand side (lhs) of (8.5.5) is given, from (8.3.1c), by

$$\mathrm{tr}\, [\gamma^{\uparrow}(G)]^{\downarrow}(h) = \sum_{nm} \chi_\gamma(q_n^{-1} p_m^{-1} h p_m q_n)$$

$$= (|G_\gamma|/|H|) \sum_m \chi_\gamma(p_m^{-1} h p_m)$$

because q_n ($\in G_\gamma$) leaves $\gamma(H)$ equivalent under conjugation. The last expression of the above equation is nothing but the character of the right-hand side (rhs) of (8.5.5). Q.E.D.

In the special case of the induced representation $\gamma^{\uparrow}(G_\gamma)$ of the little group G_γ from $\gamma(H)$, we have $\boldsymbol{O}_\gamma = \gamma(H)$ so that we have the following corollary.

Corollary 8.5.2

$$[\gamma^{\uparrow}(G_\gamma)]^{\downarrow}(H) \sim f_\gamma \gamma(H); \qquad f_\gamma = |G_\gamma|/|H| \tag{8.5.6}$$

The corollary is interesting because an induced representation of the little group G_γ from an irrep $\gamma(H)$ subduces only a multiple of $\gamma(H)$ itself onto H. It will be seen that this simple result leads to a general method of inducing the irreps of G through the small representations of G_γ, which will be introduced below.

8.5.2 Small representations of a little group

Let G_γ be the little group of an irrep $\gamma(H)$ relative to G. Then *any irrep of G_γ which subduces a multiple of $\gamma(H)$ is called a small representation of the little group G_γ*. It will be shown that the induced representation of G from a small representation of G_γ is irreducible and that the complete set of the irreps of G can be constructed by induction from the small representations of the G_γ. From the definition of the small representation[2] and Corollary 8.5.2, one can expect that any irrep of G_γ contained in the induced representation $\gamma^{\uparrow}(G_\gamma)$ is a small representation of G_γ. The amazing thing is that the inverse of this statement is also true due to the Frobenius reciprocity theorem; in fact, we have the following theorem.

Theorem 8.5.3. Any irrep of the little group G_γ contained in the induced representation $\gamma^{\uparrow}(G_\gamma)$ is a small representation of G_γ. Conversely, any small representation of G_γ is always contained in the induced representation $\gamma^{\uparrow}(G_\gamma)$.

Proof. Let Λ_γ^i be an irrep of the little group G_γ contained in the induced $\gamma^{\uparrow}(G_\gamma)$. Then

$$\gamma^{\uparrow}(G_\gamma) \sim \sum_i \oplus\, F_\gamma^i \Lambda_\gamma^i(G_\gamma) \tag{8.5.7}$$

[2] Some authors call a small representation of G_γ defined above 'an allowed or permitted small representation' of G_γ, reserving 'a small representation of G_γ' for any irrep of G_γ. We feel that any irrep of G_γ can simply be referred to as 'an irrep' of G_γ without causing any confusion.

where F^i_γ is the frequency of Λ^i_γ contained in $\gamma^\uparrow(G_\gamma)$. Since $\gamma^\uparrow(G_\gamma)$ subduces only a multiple of $\gamma(H)$, as shown by (8.5.6), an irrep $\Lambda^i_\gamma(G_\gamma)$ contained in $\gamma^\uparrow(G_\gamma)$ must also subduce some multiple of $\gamma(H)$, because all submatrices $\Lambda^i_\gamma(G_\gamma)$ contained in $\gamma^\uparrow(G_\gamma)$ are on the diagonal of the rhs of (8.5.7) and thus each submatrix subduces independently of each other. This proves the first half of the theorem. Therefore, we may set

$$\Lambda^{i\downarrow}_\gamma(H) \sim f^i_\gamma \gamma(H), \qquad f^i_\gamma = |\Lambda^i_\gamma(G_\gamma)|/|\gamma(H)| \tag{8.5.8}$$

where $|M|$ denotes the dimensionality of a matrix M. The crucial part of the second half of the proof is based on the Frobenius reciprocity theorem, i.e. any small representation $\Lambda^i_\gamma(G_\gamma)$ that subduces a multiple of $\gamma(H)$ must be contained in $\gamma^\uparrow(G_\gamma)$, because $f^i_\gamma = F^i_\gamma$. Q.E.D.

The dimensionalities of the two sides of (8.5.7) must be equal, so that we have the following relations:

$$|\gamma^\uparrow(G_\gamma)| = |\gamma(H)|\,|G_\gamma|/|H| = \sum_i |\Lambda^i_\gamma(G_\gamma)|^2/|\gamma(H)| \tag{8.5.9}$$

where we have used $F^i_\gamma = f^i_\gamma = |\Lambda^i_\gamma(G_\gamma)|/|\gamma(H)|$. The above relation may be rewritten as

$$\sum_i |\Lambda^i_\gamma(G_\gamma)|^2/|G_\gamma| = |\gamma(H)|^2/|H| \tag{8.5.10a}$$

or

$$\sum_i |f^i_\gamma|^2 = f_\gamma; \qquad f_\gamma = |G_\gamma|/|H| \tag{8.5.10b}$$

where f_γ is the order of the factor group G_γ/H and the summation over i is for all the small representations of G_γ.

The above relation may be called *the completeness condition for the small representations of* G_γ. Note that the rhs of (8.5.10a) is smaller than 1 because

$$\sum_\gamma |\gamma(H)|^2 = |H| \tag{8.5.11}$$

so that the lhs of (8.5.10a) must also be smaller than 1. This means that only a small number of the irreps of G_γ are the small representations of G_γ, because $|G_\gamma|$ equals the sum of the squares of the dimensions of all irreps of G_γ. The following simple example may illustrate the theorems introduced above.

Example. When $\gamma(H)$ is a one-dimensional representation, (8.5.10a) reduces to the form

$$\sum_i |\Lambda^i_\gamma(G_\gamma)|^2 = |G_\gamma|/|H|$$

8.5.3 Induced representations from small representations

The following theorem is basic to the method of constructing the irreps of G by induction from the small representations of the little group.

Theorem 8.5.4. Let $H \lhd G$ and let $\Lambda^i_\gamma(G_\gamma)$ be a small representation of the little group G_γ of an irrep $\gamma(H)$ relative to G. Then, we have the following.

(i) The induced representation $\Lambda^{i\uparrow}_\gamma(G) = [\Lambda^i_\gamma(G_\gamma) \uparrow G]$ of G from a small representation $\Lambda^i_\gamma(G_\gamma)$ is irreducible.
(ii) The set of all induced representations $\{\Lambda^{i\uparrow}_\gamma(G)\}$ from the irreps of H (one per orbit of H relative to G) provides the complete set of the irreps of G.

Proof. Since G_γ need not be a normal subgroup of G, the irreducibility of the induced representations $\Lambda^{i\uparrow}_\gamma(G)$ should be proven by application of the general irreducibility criterion given by Theorem 8.2.3. Let the left coset decomposition of G with respect to G_γ be $\sum_m p_m G_\gamma$ with $p_1 = 1$ and let ζ be a general element of the intersection $G_\gamma \cap G^m_\gamma$, where $G^m_\gamma = p_m G_\gamma p_m^{-1}$ is a conjugate group of G_γ. Then the irreducibility condition for the induced representation $\Lambda^{i\uparrow}_\gamma(G)$ is that the two representations $\{\Lambda^i_\gamma(\zeta)\}$ and $\{\Lambda^i_\gamma(p_m^{-1}\zeta p_m)\}$ subduced onto the intersection $G_\gamma \cap G^m_\gamma$ are orthogonal (contain no irreps in common) for all m ($\neq 1$). To show this we note first that H is a common normal subgroup of all the intersections, i.e. $H \trianglelefteq (G_\gamma \cap G^m_\gamma)$ for all m (see Figure 8.2), because $H \trianglelefteq G^m_\gamma$ for any m is obtained by the conjugations of the two sides of $H \trianglelefteq G_\gamma$ with respect to p_m and $H^m = H$. Therefore, from Theorem 8.1.2 the proof will be given by the orthogonality of the further subduced representations $\{\Lambda^i_\gamma(p_m^{-1} h p_m)\}$ via $\zeta \to h$ onto H. Now the conjugate representations satisfy, from (8.5.8),

$$\Lambda^i_\gamma(p_m^{-1} h p_m) \sim f^i_\gamma \gamma(p_m^{-1} h p_m); \qquad m = 1, 2, \ldots, |O_\gamma|; \qquad \forall h \in H \tag{8.5.12}$$

where $p_m^{-1} h p_m \in H$. Here the set of the conjugate irreps on the rhs of (8.5.12) forms the orbit O_γ of $\gamma(H)$ in G according to (8.3.5) and hence they are inequivalent and orthogonal. Therefore, the conjugate representations $\{\Lambda^i_\gamma(p_m^{-1} h p_m)\}$ on the lhs of (8.5.12) must also be mutually orthogonal. This proves the first half of the theorem.

Next, for the proof of the completeness of the irreps $\{\Lambda^{i\uparrow}_\gamma(G)\}$, it is necessary only to show that the squares of their dimensions satisfy the general completeness condition of the irreps of the group G:

$$|G|^{-1} \sum_\gamma{}' \sum_i |\Lambda^{i\uparrow}_\gamma(G)|^2 = 1 \tag{8.5.13}$$

where the sum over i is for all the small representations of G_γ whereas the sum over γ is for the irreps of H, one per orbit O_γ of H relative to G. By the definition of an induced representation, the dimensionality $|\Lambda^{i\uparrow}_\gamma(G)|$ is given by

$$|\Lambda^{i\uparrow}_\gamma(G)| = |\Lambda^i_\gamma(G_\gamma)||O_\gamma|, \qquad |O_\gamma| = |G|/|G_\gamma| \tag{8.5.14a}$$

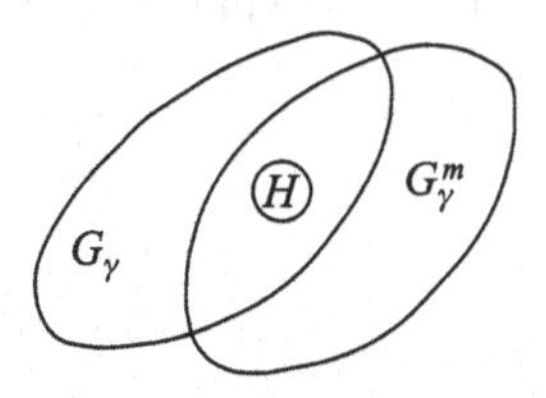

Figure 8.2. The intersection $G_\gamma \cap G^m_\gamma$.

Substitution of this into the lhs of (8.5.13) and use of the completeness condition of the small representations, (8.5.10a), lead to

$$\sum_{\gamma}{}' \frac{1}{|G_\gamma|} \sum_i |\Lambda^i_\gamma(G_\gamma)|^2 |O_\gamma| = \frac{1}{|H|} \sum_{\gamma}{}' |\gamma(H)|^2 |O_\gamma| = 1 \tag{8.5.14b}$$

where the last equality follows because the sum over γ, one each orbit O_γ of H, multiplied by $|O_\gamma|$ describes the completeness relation of the irreps of H. Q.E.D.

Since the totality of the induced representations $\{\Lambda^{i\uparrow}_\gamma(G)\}$ forms the complete set of the irreps of G we have the following corollary.

Corollary 8.5.4. Any irrep of G is equivalent to one of the induced irreps $\Lambda^{i\uparrow}_\gamma(G)$ from a small representation $\Lambda^i_\gamma(G_\gamma)$.

The following theorem is an extension of Theorem 8.5.2.

Theorem 8.5.5 (Clifford's theorem). Let $H \lhd G$ and $\Gamma(G)$ be an irrep of G, then its subduced representation $\Gamma^{\downarrow}(H)$ onto H is equivalent to a multiple of the direct sum $\boldsymbol{O}_\gamma(H)$ of the irreps in a certain orbit $O_\gamma(H)$ of H in G.

Summary. In this section a method of constructing all the irrep of a group G from those of a normal subgroup H of G through the small representations of the little groups has been discussed. It proceeds as follows.

1) Construct all the irreps $\gamma(H)$ of H.
2) Group the irreps into different orbits of the $\gamma(H)$ relative to G and select arbitrarily one member from each orbit.
3) Determine the respective little group G_γ for each selected irrep $\gamma(H)$ of H.
4) Find the small representations $\{\Lambda^i_\gamma(G_\gamma)\}$ of the little group G_γ by reducing the induced representation $\gamma^{\uparrow}(G_\gamma)$.
5) The irreps of G are constructed by the induced irreps $[\Lambda^i_\gamma(G_\gamma) \uparrow G]$ from the small representations $\Lambda^i_\gamma(G_\gamma)$ of G_γ.
6) The set of all irreps of G thus formed is complete.

It is noted here that the most difficult step of the above procedure is step 4 involving the reduction of $\gamma^{\uparrow}(G_\gamma)$. Since G_γ is in between H and G, i.e. $H \trianglelefteq G_\gamma \leqslant G$, the most favorable case occurs when $H = G_\gamma$, in which case $\gamma(H) \uparrow G$ is already irreducible. In the other extreme case in which $G_\gamma = G$ (which occurs, for example, for the identity representation of H), the concept of the little group does not help at all. In the special case in which H is in the center of G, however, the difficulty involved in step 4 is removed by application of the theory of projective (or ray) representations: the small representations of the little group G_γ are constructed by the projective irreps of a group (called the little co-group) that is isomorphic to the factor group G_γ/H. This case will be discussed in Chapter 14 and applied for constructing the unirreps of the wave vector space groups.

9

Elements of continuous groups

9.1 Introduction

An infinite group is a group that contains an infinite number of elements. The group axioms still hold for infinite groups. Among infinite groups, there are two categories: discrete and continuous ones. If the number of elements of a group is denumerably infinite, the group is said to be discrete, whereas if the number of elements is non-denumerably infinite, it is called a continuous group. For example, the whole set of rational numbers forms an infinite group that is discrete, whereas the whole set of real positive numbers is a continuous group. A continuous group G is a set of group elements that can be characterized by a set of continuous real parameters in a certain region called the parameter domain (or space) Ω such that *there exists a one-to-one correspondence between group elements in G and points (the parameter sets) in the parameter domain Ω*. For example, an element of the rotation group $SO(3, r) = \{R(\boldsymbol{\theta})\}$ is characterized by a set of three real parameters $\boldsymbol{\theta} = (\theta_1, \theta_2, \theta_3)$ in the parameter sphere Ω of the radius π, i.e. $0 \leqslant |\boldsymbol{\theta}| \leqslant \pi$ with the cyclic boundary condition (see Equation (4.3.6)). In a continuous group G, the nearness of group elements is characterized by the nearness of their parameters in Ω. Thus the neighborhood of a group element is characterized by the neighborhood of the corresponding parameter set. The parameter domain may be finite or infinite. A continuous group G is said to be *compact* if the parameter space is closed, i.e. the limiting point of any sequence of points in the space is contained in the space; in this case, the parameter space should be finite.

A group whose elements can be expressed by a finite number of continuous parameters is a finite continuous group. In an h-parameter continuous group G, any group element T is expressed by a function of h real parameters $a = (a_1, a_2, \ldots, a_h)$:

$$T = g(a) = g(a_1, a_2, \ldots, a_h) \tag{9.1.1}$$

The set of parameters should be necessary and sufficient to characterize all elements of the group. The parameter set of the identity may be denoted by $e = (e_1, e_2, \ldots, e_h)$. Then

$$g(e)g(a) = g(a)g(e) = g(a)$$

Let the parameter sets of the group elements T, S and their product ST be a, b and c, respectively, i.e. $T = g(a)$, $S = g(b)$ and $ST = g(c)$, then

$$g(b)g(a) = g(c) \tag{9.1.2a}$$

so that the parameter set c is given by a function of the parameter sets a and b:

$$c = \phi(b, a) \tag{9.1.2b}$$

Likewise, let $T^{-1} = g(\overline{a})$, then $\overline{a}$ is given by a function of a:

$$\overline{a} = \overline{\phi}(a) \tag{9.1.3}$$

If these functions ϕ and $\overline{\phi}$ are piecewise continuously differentiable with respect to the parameters, then the continuous group G is called a *Lie group*. For most cases, however, it is sufficient to assume that the group elements are continuous functions of the group parameters, although we occasionally make use of the concept of continuous differentiability of functions of the group parameters.

Example 1. The special real orthogonal group in two dimensions $SO(2, r)$ is a one-parameter compact Lie group defined by

$$R(\theta) = \begin{bmatrix} \cos\theta & -\sin\theta \\ \sin\theta & \cos\theta \end{bmatrix}, \qquad 0 \leqslant \theta \leqslant 2\pi \tag{9.1.4a}$$

with a real parameter θ, where two boundary points $\theta = 0$ and 2π are regarded as one point in the parameter domain Ω, i.e. $\theta + 2\pi \equiv \theta \pmod{2\pi}$, corresponding to the periodicity $R(\theta + 2\pi) = R(\theta)$. With this convention, we can define any function (even an aperiodic one) of θ in the range $-\infty < \theta < \infty$ without θ leaving the parameter space Ω for the group.

Example 2. The group of one-dimensional transformation defined by

$$x \to x' = ax, \qquad 0 < a \leqslant 1 \tag{9.1.4b}$$

is a non-compact Lie group because the range of the parameter a is finite but open at $a = 0$.

9.1.1 Mixed continuous groups

A group is said to be connected if an arbitrary element of the group can be obtained continuously from the identity E. This means that there exists a continuous manifold $T(t)$ of elements via a continuous path variable t $(0 \leqslant t \leqslant 1)$ such that the manifold begins at $T(0) = E$ and ends at $T(1) = T$. For example, the rotation group $SO(3, r) = \{R(\boldsymbol{\theta})\}$ is clearly connected, i.e. every element $R(\boldsymbol{\theta})$ is connected to the unit element E through a path variable t such that $T(t) \equiv R(t\boldsymbol{\theta})$.

The full rotation group $O(3, r)$ is not connected, because it is not possible to pass continuously from the orthogonal matrices of determinant $+1$ to those of determinant -1; therefore, the group $O(3, r)$ has two disjoint pieces. Such a group G is called a *mixed continuous group*. Here, the piece that is connected to the identity element forms a subgroup H of G. Let T and S be two elements of this subspace, then the product ST can be obtained from the identity in a continuous manner through a path $T(t)S(t)$. The same holds for the inverse T^{-1} through $T(t)^{-1}$. Moreover, the subgroup H with elements connected to the identity forms an invariant subgroup of the mixed continuous group G. If T can be reached continuously from the identity, then so can $X^{-1}TX$ for any element X of G along the path $X^{-1}T(t)X$ because $X^{-1}T(0)X = E$. The cosets of this invariant subgroup H are the other disjoint pieces. Let the coset decomposition be

$$G = H + X_1H + X_2H + \cdots \tag{9.1.5}$$

where a coset representative X_ν is simply any member of the corresponding disjoint piece $X_\nu H$. Let $\{a_i\}$ be the set of continuous real parameters of H, then each disjoint piece $X_\nu H$ is described by the same set of continuous parameters $\{a_i\}$ but with an additional parameter corresponding to X_ν. The order of the factor group G/H equals the number of disjoint pieces of the parameter space.

Example 3. For $O(3, r)$, the coset decomposition by the invariant subgroup $SO(3, r)$ is given by

$$O(3, r) = SO(3, r) + I\, SO(3, r) \tag{9.1.6}$$

where I is the inversion defined as the unit matrix multiplied by -1. The order of the factor group $O(3, r)/SO(3, r)$ equals 2. The two disjoint pieces $SO(3, r)$ and $I\, SO(3, r)$ are described by the same continuous parameter set $\{\theta_1, \theta_2, \theta_3\}$ but are characterized by the different signs of the determinants of the group elements.

9.2 The Hurwitz integral

In the representation theory of a finite group $G = \{T\}$, the following rearrangement theorem has been essential (Wigner 1962):

$$\sum_{T \in G} F(T) = \sum_{T \in G} F(ST) \tag{9.2.1}$$

where $F(T)$ is any function of the group elements $T \in G$ and S is a fixed element of G. For example, the orthogonality theorem of the irreducible representations of the group G is based on the sum

$$\sum_{T} D^{(\alpha)}(T) X D^{(\beta)}(T^{-1})$$

For a Lie group such a sum must be replaced by an integral; this requires one to introduce a measure of integration that is common to both sides of (9.2.1). Through a measure for the number of group elements in an extension of the parameter space, we may introduce the density distribution of the group elements in the parameter space. Obviously, the concept of the 'absolute number of group elements' in an extension is meaningless on account of the continuous distribution of group elements. However, their ratio is meaningful through the one-to-one correspondence such as that between the left-hand side and the right-hand side of (9.2.1). Hereafter, the number of group elements should be understood in this relative sense.

Suppose that the parameter a of the element T draws an extension V_a in the parameter space, then the parameters c of the product ST also draws an extension V_c via $c = \phi(b, a)$ for a fixed b (see Figure 9.1). Since there exists a one-to-one correspondence between the group elements T in the extension V_a and ST in the extension V_c for a given S, we may assign an equal measure to 'the numbers' of the group elements in both extensions such that, for all T and a given S,

$$N_T = N_{ST} \tag{9.2.2a}$$

where N_T and N_{ST} are the assigned measures for the numbers of the elements in V_a and V_c, respectively. This equation determines only the ratio of the measures but is sufficient to determine the relative density distribution of group elements in the

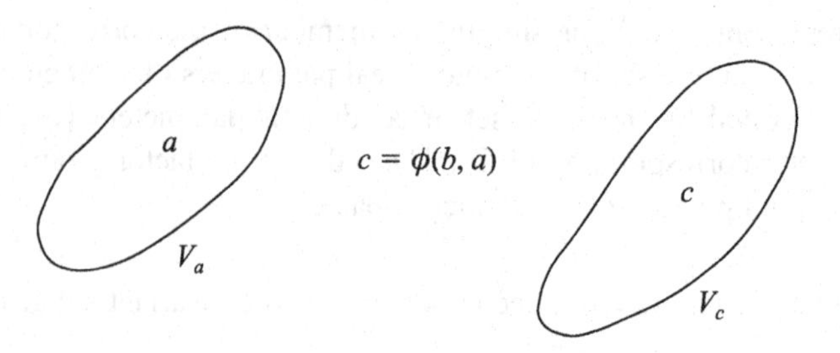

Figure 9.1. The parameter spaces V_a and V_c connected by $c = \phi(b, a)$ for a fixed b.

parameter space. Let the infinitesimal measures of both sides of (9.2.2a) be expressed in the differential forms

$$\mathrm{d}N_T = \mathrm{d}N_{ST} \tag{9.2.2b}$$

and the corresponding infinitesimal extensions located at a and c be $\mathrm{d}a = \mathrm{d}a_1\,\mathrm{d}a_2 \ldots$ and $\mathrm{d}c = \mathrm{d}c_1\,\mathrm{d}c_2 \ldots$, respectively. Then we set, for a given b,

$$\rho(a)\,\mathrm{d}a = \rho(c)\,\mathrm{d}c, \qquad c = \phi(b, a) \tag{9.2.3}$$

where $\rho(a)$ and $\rho(c)$ are the relative densities of the group elements in infinitesimal extensions $\mathrm{d}a$ at a and $\mathrm{d}c$ at c, respectively. Via the infinitesimal measure, the sum on the left-hand side of (9.2.1) may be expressed by the integral

$$\int_G F(T)\,\mathrm{d}N_T \equiv \int_{[G]} F[T(a)]\rho(a)\,\mathrm{d}a \tag{9.2.4}$$

where the left-hand side is simply an abbreviation of the right-hand side and the integration is extended over the whole parameter space $[G]$ of G. It is called *the Hurwitz integral* or the invariant integral in the parameter space. In terms of the Hurwitz integral, the rearrangement theorem (9.2.1) is given, for a fixed S, by

$$\int F(T)\,\mathrm{d}T = \int F(ST)\,\mathrm{d}T \tag{9.2.5}$$

where $\mathrm{d}T$ stands for the infinitesimal measure $\mathrm{d}N_T$ following the notation introduced by Wigner (1962).

Next, we shall determine the relative density function ρ in the parameter space from (9.2.3). Since the infinitesimal extensions $\mathrm{d}c = \mathrm{d}c_1\,\mathrm{d}c_2 \ldots \mathrm{d}c_h$ and $\mathrm{d}a = \mathrm{d}a_1\,\mathrm{d}a_2 \ldots \mathrm{d}a_h$ are related by

$$\mathrm{d}c = J\left[\frac{\partial c}{\partial a}\right]\mathrm{d}a$$

where $J[\partial c/\partial a]$ is *the Jacobian* defined by

$$J\left[\frac{\partial c}{\partial a}\right] = \frac{\partial(c_1, c_2, \ldots, c_h)}{\partial(a_1, a_2, \ldots, a_h)} \equiv \det\left\|\frac{\partial c_i}{\partial a_j}\right\|$$

we obtain, from (9.2.3),

$$\rho(c) = \rho(a)/J[\partial c/\partial a] \tag{9.2.6}$$

which determines only the ratio of densities as expected. Thus, one may assign an arbitrary constant ρ_0 for the density in the neighborhood of the identity element $E = g(e)$ and obtain

$$\rho(c) = \rho_0/J[\partial c/\partial e], \qquad c = \phi(b, e) = b \tag{9.2.7a}$$

where ρ_0 may be determined by the normalization

$$\int_G \mathrm{d}N_T = \int_{[G]} \rho(c)\,\mathrm{d}c = 1 \tag{9.2.7b}$$

provided that the integral converges.

Now, according to (9.2.7a) the density function $\rho(c)$ is calculated via the Jacobian $J[\partial c/\partial e]$, for which it is necessary only to consider the product relation $g(c) = g(b)g(a)$ with the parameter a in the vicinity of the identity parameter e; therefore, we set $a = e + \epsilon$ and then

$$g(c) = g(b)g(e + \epsilon) \tag{9.2.8a}$$

where ϵ is an infinitesimal deviation from the identity parameter e. This equation may be solved to the first order in ϵ to obtain the corresponding functional relation for the parameters

$$c = \phi(b, e + \epsilon) \tag{9.2.8b}$$

The required Jacobian is given by $\partial c/\partial e = \partial c/\partial\epsilon|_{\epsilon=0}$ for the given b, so that we have, from (9.2.7a),

$$\rho(b) = \rho_0/J[\partial c/\partial e] \tag{9.2.8c}$$

Example 1. The density function for the group $SO(2, r) = \{R(\theta)\}$ defined by (9.1.4a). This is a one-parameter group with an additive parameter θ such that $R(\theta_1)R(\theta_2) = R(\theta_1 + \theta_2)$. Thus, from $R(\theta') = R(\theta)R(\epsilon) = R(\theta + \epsilon)$, where $R(\epsilon)$ is the infinitesimal transformation, we have

$$\theta' = \theta + \epsilon \tag{9.2.9}$$

corresponding to (9.2.8b). Accordingly $\mathrm{d}\theta'/\mathrm{d}\epsilon = 1$, so that from (9.2.8c) we arrive at a constant result: $\rho(\theta) = \rho_0$. Through the normalization

$$\int_{-\pi}^{\pi} \rho(\theta)\,\mathrm{d}\theta = 2\pi\rho_0 = 1$$

we obtain $\rho(\theta) = 1/(2\pi)$.

Example 2. The density function for the one-dimensional transformation defined by

$$x' = ax, \qquad 0 < a \leqslant 1$$

Here the group elements are the parameter a itself, i.e. $g(a) = a$, and the identity element is given by $a = 1$. Thus the product relation (9.2.8a) takes $c = b(1 + \epsilon)$ so that $\mathrm{d}c/\mathrm{d}\epsilon = b \doteq c(\epsilon = 0)$. Accordingly, from (9.2.7a), it follows that $\rho(c) = \rho_0/c$, which cannot be normalized because its integration over c diverges logarithmically at $c \to 0$.

Example 3. The density function for $SO(3, r)$. From the group element $R(\boldsymbol{\theta}) = \exp(\theta\omega)$ given by (4.3.2a), the infinitesimal rotation $R(\epsilon)$ is given by

$$R(\epsilon) = 1 + \epsilon\omega = \begin{bmatrix} 1 & -\epsilon_3 & \epsilon_2 \\ \epsilon_3 & 1 & -\epsilon_1 \\ -\epsilon_2 & \epsilon_1 & 1 \end{bmatrix}$$

which is correct to the first order in the magnitude of the infinitesimal parameter $\epsilon = |\epsilon|$. Since all the rotation axes are equivalent, the density function $\rho(\boldsymbol{\theta})$ should depend only on the rotational angle $\theta = |\boldsymbol{\theta}|$, so that calculation of $R(\boldsymbol{\theta}') = R(\boldsymbol{\theta})R(\epsilon)$ may be carried out with $\boldsymbol{\theta} = (0, 0, \theta)$ for $R(\boldsymbol{\theta})$. Since

$$R(00\theta) = \begin{bmatrix} \cos\theta & -\sin\theta & 0 \\ \sin\theta & \cos\theta & 0 \\ 0 & 0 & 1 \end{bmatrix}$$

we obtain, from $R(\boldsymbol{\theta}') = R(\boldsymbol{\theta})R(\epsilon)$,

$$R(\boldsymbol{\theta}') = \begin{bmatrix} \cos\theta - \epsilon_3 \sin\theta & -\sin\theta - \epsilon_3\cos\theta & \epsilon_2\cos\theta + \epsilon_1\sin\theta \\ \sin\theta + \epsilon_3\cos\theta & \cos\theta - \epsilon_3\sin\theta & \epsilon_2\sin\theta - \epsilon_1\cos\theta \\ -\epsilon_2 & \epsilon_1 & 1 \end{bmatrix} \quad (9.2.10a)$$

By equating the traces of the two sides, we obtain $\cos\theta' = \cos\theta - \epsilon_3\sin\theta$ which yields, to the first order in ϵ,

$$\theta' = \theta + \epsilon_3 \quad (9.2.10b)$$

The axis vector $\boldsymbol{n}'$ of $\boldsymbol{\theta}'$ is determined from (4.3.9b), i.e.

$$\omega(\boldsymbol{n}') = (R(\boldsymbol{\theta}') - R(\boldsymbol{\theta}')^{\sim})/(2\sin\theta')$$

which yields, to the first order in ϵ,

$$n_1' = \tfrac{1}{2}\epsilon_2 + \epsilon_1(1 + \cos\theta)/(2\sin\theta)$$
$$n_2' = \tfrac{1}{2}\epsilon_1 + \epsilon_2(1 + \cos\theta)/(2\sin\theta)$$
$$n_3' = 1$$

On substituting these into $\theta_i' = (\theta + \epsilon_3)n_i'$; $i = 1, 2, 3$, we obtain the required Jacobian:

$$J = \frac{\partial(\theta_1', \theta_2', \theta_3')}{\partial(\epsilon_1, \epsilon_2, \epsilon_3)} = \begin{vmatrix} \dfrac{\theta(1+\cos\theta)}{2\sin\theta} & -\dfrac{1}{2}\theta & 0 \\ \dfrac{1}{2}\theta & \dfrac{\theta(1+\cos\theta)}{2\sin\theta} & 0 \\ 0 & 0 & 1 \end{vmatrix}$$
$$= \theta^2/[4\sin^2(\theta/2)]$$

Therefore, from (9.2.8c),

$$\rho(\theta) = 4\rho_0 \sin^2(\theta/2)/\theta^2 = \sin^2(\theta/2)/(2\pi^2\theta^2) \quad (9.2.11)$$

where $\rho_0 = 1/(8\pi^2)$ is obtained via the normalization

$$\int dR = \int_0^{\pi} \rho(\theta)4\pi\theta^2\, d\theta = 1$$

9.2.1 Orthogonality relations

The actual calculation of the density function $\rho(c)$ for the Hurwitz integral is often quite laborious if it is done directly from (9.2.7a). For many purposes, in particular, for the derivation of the orthogonality relations for Lie groups, the knowledge of the existence of the invariant integral is all that is needed.

Proceeding in exactly the same way as for finite groups, it follows from (9.2.5) that every representation can be transformed into a unitary representation, provided that

$$\int D(T)_{\mu\sigma} D(T)^*_{\nu\tau} \, \mathrm{d}T$$

converges where $\mathrm{d}T = \mathrm{d}N_T$. This is always the case for a compact Lie group. We may state without proof that the integral $\int \rho(a)\,\mathrm{d}a$ always converges for a compact Lie group (see Examples 1 and 2). Then, the orthogonal relations of the unirreps for a compact Lie group may be expressed in the form

$$\int D^{(\alpha)}(T)^*_{\nu\mu} D^{(\beta)}(T)_{\nu'\mu'} \, \mathrm{d}T = (1/d_\alpha)\delta_{\alpha\beta}\delta_{\nu\nu'}\delta_{\mu\mu'} \int \mathrm{d}T \tag{9.2.12}$$

and correspondingly

$$\int \chi^{(\alpha)}(T)^* \chi^{(\beta)}(T) \, \mathrm{d}T = \delta_{\alpha\beta} \int \mathrm{d}T \tag{9.2.13}$$

for the characters. For explicit examples, see the orthogonality relations for the unirreps of the special unitary group $SU(2)$ and the rotation group $SO(3, r)$ given by (10.3.12a) and (10.3.13), respectively.

9.3 Group generators and Lie algebra

It is convenient to choose the continuous parameters of a Lie group G such that the parameters of the identity element E are at the origin of the parameter space, i.e. $E = g(0, 0, \ldots, 0)$. Then an element near the identity may be written, due to the analytic properties of the Lie group, as

$$g(\epsilon_1, \epsilon_2, \ldots, \epsilon_h) = g(0, \ldots, 0) + \sum_j \epsilon_j I_j \tag{9.3.1}$$

to the first order in the infinitesimal parameters ϵ_j. The operator I_j is called an infinitesimal generator of the Lie group and determined from (9.3.1) by

$$I_j = \lim_{\epsilon_j \to 0} [g(0, \ldots, \epsilon_j, \ldots, 0) - g(0, \ldots, 0)]/\epsilon_j$$

All the properties of the Lie group can be derived from the infinitesimal generators $I_1, I_2, \ldots, I_h$ of the group G which are defined near the identity element of the group.

By the successive application of the infinitesimal transformations, one can arrive at an element of the group that is a finite distance away from the identity. Let us express the parameters by $a_j = N\epsilon_j$, where N is a large positive integer so that a_j is a finite quantity. Then we may obtain the element corresponding to the set of finite parameters $\{a_j\}$ as follows:

$$g(a_1, a_2, \ldots, a_h) = [g(\epsilon_1, \epsilon_2, \ldots, \epsilon_h)]^N$$
$$= \left(E + \sum_j \epsilon_j I_j\right)^N$$
$$= \left[E + \sum_j (a_j I_j)/N\right]^N$$

Allowing N to tend to infinity and using

$$\lim_{N\to\infty} (1 + x/N)^N = e^x$$

we obtain, in the vicinity of the identity E,

$$g(a_1, a_2, \ldots, a_h) = E \exp\left(\sum_j a_j I_j\right) = E \exp(a \cdot I) \tag{9.3.2}$$

where $a = \{a_i\}$ and $I = \{I_j\}$. This is an exact result. It means that all the elements of the Lie group belonging to a closed parameter space containing the identity can be obtained by giving various values to the parameters a_j of the parameter space; at least, in the vicinity of the identity. The following basic theorem holds.

Theorem 9.3.1. Let $\{I_1, I_2, \ldots, I_h\}$ be the set of generators of an h-parameter Lie group G, then the generator set satisfies

$$[I_k, I_l] = \sum_{m=1}^{h} C_{kl}^m I_m; \qquad k, l = 1, 2, \ldots, h \tag{9.3.3}$$

where $[I_k, I_l] = I_k I_l - I_l I_k$ is the commutator of the generators I_k and I_l while C_{kl}^m are constants called *the structure constants* of the Lie group G.

According to the theorem, the generator set $\{I_k\}$ forms a vector space such that the commutator of each pair of vectors belongs to the space. Such a vector space is called a *Lie algebra*. Thus, we can state that the set of the generators $\{I_k\}$ of a Lie group forms *the Lie algebra* of the Lie group. The study of Lie algebras is of fundamental importance in the study of Lie groups. Indeed, in many respects, physicists are more interested in the algebraic structure than they are in the group structure.

The proof of the theorem is straightforward; see Wybourne (1974). For sufficiently small values of the parameters $a = (a_1, a_2, \ldots, a_h)$, we may represent an element $g(a)$ of G lying close to the identity by the Taylor expansion (9.3.2) as follows:

$$g(a) = g(0) \exp(a \cdot I)$$
$$= g(0) + (a \cdot I) + \tfrac{1}{2}(a \cdot I)^2 + \cdots \tag{9.3.4a}$$

Then the inverse $g(a)^{-1}$ is given by

$$g(a)^{-1} = g(0) - (a \cdot I) + \tfrac{1}{2}(a \cdot I)^2 + \cdots \tag{9.3.4b}$$

Define the commutator of two group elements $g(b)$ and $g(c)$ by $g(b)^{-1} g(c)^{-1} g(b) g(c)$. Then, from (9.3.4a) and (9.3.4b), the commutator is given, to the second order in b and c, by

$$g(b)^{-1}g(c)^{-1}g(b)g(c) = g(0) + [(b \cdot I), (c \cdot I)] + \cdots \tag{9.3.5}$$

where

$$[(b \cdot I), (c \cdot I)] = \sum b_k c_l [I_k, I_l]$$

Now, the commutator (9.3.5) itself must be a group element, say $g(a)$, lying close to $g(0)$; hence, we have

$$g(a) = g(0) + \sum b_k c_l [I_k, I_l] + \cdots$$

Comparing this with (9.3.4a), we conclude that the parameter set $a = \{a_m\}$ must be bi-linear with respect to the parameter sets b and c

$$a_m = \sum_{k,l} c_{kl}^m b_k c_l; \qquad m = 1, 2, \ldots, h \tag{9.3.6}$$

so that

$$[I_k, I_l] = \sum_m c_{kl}^m I_m$$

which is the required result (9.3.3).

The important properties of the structure constants are the following.

1. They are antisymmetric with respect to their lower indices

$$c_{kl}^m = -c_{lk}^m \tag{9.3.7}$$

2. From *the Jacobi identity* defined by the infinitesimal generators

$$[[I_k, I_l], I_m] + [[I_l, I_m], I_k] + [[I_m, I_k], I_l] = 0 \tag{9.3.8}$$

the structure constants must satisfy

$$c_{kl}^n c_{mn}^p + c_{lm}^n c_{kn}^p + c_{mk}^n c_{ln}^p = 0 \tag{9.3.9}$$

Example 1. The Euclidean group in one dimension $E(1)$ is defined by the transformation

$$x \to x' = ax + b$$

where a and b are real parameters. The matrix representation of this group is expressed by the transformation T:

$$\begin{bmatrix} x' \\ 1 \end{bmatrix} = \begin{bmatrix} a & b \\ 0 & 1 \end{bmatrix} \begin{bmatrix} x \\ 1 \end{bmatrix} \equiv T \begin{bmatrix} x \\ 1 \end{bmatrix}$$

So that the infinitesimal generators of the Euclidean group $E(1)$ are defined by

$$I_a = \left.\frac{\partial T}{\partial a}\right|_{(00)} = \begin{bmatrix} 1 & 0 \\ 0 & 0 \end{bmatrix}$$

$$I_b = \left.\frac{\partial T}{\partial b}\right|_{(00)} = \begin{bmatrix} 0 & 1 \\ 0 & 0 \end{bmatrix}$$

These satisfy the commutation relations

$$[I_a, I_b] = I_b$$

which defines the Lie algebra $e(1)$ corresponding to the Euclidean group $E(1)$. It is customary to denote the Lie algebra corresponding to a Lie group by the lower case of the lie group symbol.

Example 2. The Lie algebra $so(3, r)$ of the proper rotation group $SO(3, r)$. The general element of the proper rotation group is given by (4.3.2a), i.e.

$$R(\boldsymbol{\theta}) = \mathrm{e}^{\theta\omega}; \qquad \theta\omega = \begin{bmatrix} 0 & -\theta_3 & \theta_2 \\ \theta_3 & 0 & -\theta_1 \\ -\theta_2 & \theta_1 & 0 \end{bmatrix} \tag{9.3.10}$$

The infinitesimal generators of $R(\boldsymbol{\theta})$ with respect to the parameters $\{\theta_j\}$ are defined by

$$I_j = \left.\frac{\partial R(\boldsymbol{\theta})}{\partial \theta_j}\right|_{(000)}; \qquad j = 1, 2, 3$$

which yield

$$I_1 = \begin{bmatrix} 0 & 0 & 0 \\ 0 & 0 & -1 \\ 0 & 1 & 0 \end{bmatrix}, \qquad I_2 = \begin{bmatrix} 0 & 0 & 1 \\ 0 & 0 & 0 \\ -1 & 0 & 0 \end{bmatrix}, \qquad I_3 = \begin{bmatrix} 0 & -1 & 0 \\ 1 & 0 & 0 \\ 0 & 0 & 0 \end{bmatrix} \tag{9.3.11}$$

These satisfy the commutation relations

$$[I_1, I_2] = I_3, \qquad [I_2, I_3] = I_1, \qquad [I_3, I_1] = I_2 \tag{9.3.12a}$$

and form the Lie algebra $so(3, r)$ of the rotation group $SO(3, r)$. From (9.3.10) and (9.3.11) we have $\theta\omega = \boldsymbol{\theta} \cdot \boldsymbol{I}$ so that $R(\boldsymbol{\theta})$ is rewritten in the form

$$R(\boldsymbol{\theta}) = \exp(\boldsymbol{\theta} \cdot I) \tag{9.3.12b}$$

as expected from the general expression (9.3.2).

Exercise. Prove, from (9.3.10) and (9.3.12b), that the infinitesimal rotation about a unit vector $\boldsymbol{n}$ given by $\omega(\boldsymbol{n}) = \boldsymbol{n} \cdot \boldsymbol{I}$ satisfies the following commutation relation:

$$[\omega(\boldsymbol{n}), \omega(\boldsymbol{n}')] = (\boldsymbol{n} \times \boldsymbol{n}') \cdot \boldsymbol{I}$$

Example 3. The $su(3)$ algebra of the group $SU(3)$. From (4.1.3), the element of $SU(3)$ is given by

$$U = \exp(-\mathrm{i}H); \qquad H = \begin{bmatrix} a_7 & a_1 - \mathrm{i}a_2 & a_3 - \mathrm{i}a_4 \\ a_1 + \mathrm{i}a_2 & -a_7 + a_8 & a_5 - \mathrm{i}a_6 \\ a_3 + \mathrm{i}a_4 & a_5 + \mathrm{i}a_6 & -a_8 \end{bmatrix}$$

where H is a traceless Hermitian matrix expressed by the eight real parameters $a_1, \ldots, a_8$. The generators of $SU(3)$ are defined by

$$I_i = \left.\frac{\partial H}{\partial a_i}\right|_0; \qquad i = 1, 2, \ldots, 8$$

which yield

$$I_1 = \begin{bmatrix} 0 & 1 & 0 \\ 1 & 0 & 0 \\ 0 & 0 & 0 \end{bmatrix}, \quad I_2 = \begin{bmatrix} 0 & -\mathrm{i} & 0 \\ \mathrm{i} & 0 & 0 \\ 0 & 0 & 0 \end{bmatrix}, \quad I_3 = \begin{bmatrix} 0 & 0 & 1 \\ 0 & 0 & 0 \\ 1 & 0 & 0 \end{bmatrix}$$

$$I_4 = \begin{bmatrix} 0 & 0 & -\mathrm{i} \\ 0 & 0 & 0 \\ \mathrm{i} & 0 & 0 \end{bmatrix}, \quad I_5 = \begin{bmatrix} 0 & 0 & 0 \\ 0 & 0 & 1 \\ 0 & 1 & 0 \end{bmatrix}, \quad I_6 = \begin{bmatrix} 0 & 0 & 0 \\ 0 & 0 & -\mathrm{i} \\ 0 & \mathrm{i} & 0 \end{bmatrix}$$

$$I_7 = \begin{bmatrix} 1 & 0 & 0 \\ 0 & -1 & 0 \\ 0 & 0 & 0 \end{bmatrix}, \quad I_8 = \begin{bmatrix} 0 & 0 & 0 \\ 0 & 1 & 0 \\ 0 & 0 & -1 \end{bmatrix}$$

These form the Lie algebra $su(3)$ of the Lie group $SU(3)$. The only generators of $SU(3)$ which commute with each other are the diagonal matrices I_7 and I_8. The minimum number of mutually commuting generators of a Lie group is called its *rank*. Then, the rank of $SU(3)$ equals 2. Determine the structure constants of $SU(3)$ as an exercise.

9.4 The connectedness of a continuous group and the multivalued representations

A continuous connected group G may further be simply connected or multiply connected. A parameter space Ω of G is said to be m-fold connected if there exist m distinct paths connecting any given two points of the space *that cannot be brought into each other by continuous deformation without going out of the parameter space.* A group manifold G is said to be *m-fold connected* if its parameter space Ω is m-fold connected. If a group is m-fold connected, we may expect that some of the matrix representations will be multivalued (m-valued at most) because each distinct path may result in different value for the representation when a point of the parameter space Ω returns to the initial point after the completion of each distinct path. Thus, it is necessary to examine the connectivity of a continuous group in order to determine the multivaluedness of the representations.

To understand the relation between the multivalued representations of a group G and its connectedness we shall consider a special group manifold $G = \{g(\alpha, \phi)\}$, where ϕ is the only cyclic parameter and α is the set of the remaining parameters such that

$$g(\alpha, \phi + \tau) = g(\alpha, \phi)$$

where τ is the period of ϕ. Then the parameter space Ω of G is defined such that two points with the parameter sets $(\alpha, \phi + \tau)$ and (α, ϕ) are regarded as one point in the parameter space in order to ensure the one-to-one correspondence between the group manifold $G = \{g(\alpha, \phi)\}$ and its parameter space Ω. Thus infinitely many parameter sets $\{\alpha, \phi + l\tau\}$ with arbitrary integers l correspond to one point of the parameter space Ω and hence to one group element. Now, a representation of a continuous group is assumed to be a continuous function of the parameters of the group. Accordingly, for the group manifold $G = \{g(\alpha, \phi)\}$ introduced above, any one of its representations $D = D(\alpha, \phi)$ is also a continuous function of the parameters, in particular, of the cyclic parameter. This, however, does not necessarily mean that the representation $D(\alpha, \phi)$ has the same periodicity with respect to ϕ as the periodicity of $g(\alpha, \phi)$. Thus, if $D(\alpha, \phi + \tau) \neq D(\alpha, \phi)$, then the representation is not single-valued. Suppose that the periodicity of $D(\alpha, \phi)$ with respect to ϕ is such that $D(\alpha, \phi + n\tau) = D(\alpha, \phi)$ for which n is the smallest integer. Then the representation $D(\alpha + \phi)$ is n-valued, i.e. n different representatives

$$D(\alpha, \phi), D(\alpha, \phi + \tau), \ldots, D(\alpha, \phi + (n-1)\tau) \tag{9.4.1}$$

correspond to the same group element $g(\alpha, \phi)$. Here, the possible maximum number of n is determined by the connectedness of the group manifold; in fact, for an m-connected group G we have $1 \leqslant n \leqslant m$ because by assumption there exist only m distinct closed curves (passing through any given point in the parameter space Ω) which cannot be brought together by continuous variation of the parameters. For example, it will be shown in Example 2 that $SO(3, r)$ is doubly connected due to the cyclic boundary condition $R(\pi s) = R(-\pi s)$ given by (4.3.6). Accordingly, its representations will be either single-valued or double-valued.

These multivalued representations cannot simply be neglected since they are essential in describing the symmetry properties of many physical problems. One can state, however, that, for any m-connected group $G = \{g\}$, there exists a simply connected group $G' = \{g'\}$ that is homomorphic to G with m-to-one correspondence. The kernel of the homomorphism is a discrete invariant subgroup N ($\triangleleft G'$) of order m so that the factor group G'/N is isomorphic to G. The group G' is called '*the universal covering group*' of G. On account of the homomorphism, every representation of G (be it single-valued or multivalued) is a representation of G' that has to be a single-valued one because G' is simply connected. Accordingly, to find all the irreducible representations of G it is necessary only to study the vector representations of G'. Here a vector representation means a single-valued representation. This is truly a life-saving conclusion.

To understand the concept of *the universal covering group* more clearly, let us return to the group $G = \{g(\alpha, \phi)\}$ with one cyclic parameter ϕ and let $G' = \{g'(\alpha, \phi)\}$ be the universal covering group of G. Then the m-to-1 homomorphism between G' and G is described by the m-to-1 correspondence

$$g'(\alpha, \phi), g'(\alpha, \phi + \tau), \ldots, g'(\alpha, \phi + (m-1)\tau) \to g(\alpha, \phi) \tag{9.4.2}$$

Note that m different points $\phi, \phi + \tau, \ldots, \phi + (m-1)\tau$ in the parameter space Ω' of G' correspond to one point ϕ in Ω of G. Thus a single-valued representation $\{D(g')\}$ of G' becomes a multivalued representation of G; at most m-valued because some of the $D(g')$ of $g'(\alpha, \phi + l\tau) \in G'$ may coincide for different values of l. For example, in the next chapter, it will be shown that the universal covering group of the proper rotation group $SO(3, r)$ is the special unitary group $SU(2)$. Thus the vector irreps of $SU(2)$ exhaust all the possible irreps of $SO(3, r)$ which may be either single-valued or double-valued.

It should be noted here that the difficulty of describing the orthogonality theorems on the multivalued irreducible representations of a group G is easily removed in terms of the corresponding single-valued irreducible representations of the covering group G' of G (see Chapter 10). This difficulty is also removed when a multivalued representation of G is regarded as a *projective representation* of G with the factor system which depends on the distinctive paths, because the projective unirreps obey the same orthogonality relations as do the ordinary unirreps (see Chapter 12).

Example 1. The one-dimensional unitary group $U(1) = \{g(\phi)\}$. The general element of the group $U(1)$ is given by

$$g(\phi) = \exp(\mathrm{i}\phi), \qquad 0 \leqslant \phi \leqslant 2\pi \ (\equiv 0) \tag{9.4.3a}$$

with one cyclic parameter, ϕ. It is isomorphic to the two-dimensional rotation group $SO(2, r)$ defined by (9.1.4a). The group manifold consists of points on the unit circle in the complex plane and the parameter space Ω is the circle itself described by the phase angle ϕ. On account of the periodicity $g(\phi + 2\pi) = g(\phi)$, the two points $\phi = 0$ and 2π are regarded as the same point in Ω; i.e. $\phi + 2\pi \equiv \phi \pmod{2\pi}$. Owing to this condition, the parameter space is infinitely connected. To see this we introduce a path variable t that varies from 0 to 1 and describes a closed curve $\phi = \phi(t)$ in the parameter space Ω such that $g[\phi(0)] = g[\phi(1)]$. Then the path $\phi = 2\pi t$ is a single loop around the unit circle, which cannot be deformed to a single point, say $\phi = 0$, by a continuous deformation because $\phi = 0$ and $\phi = 2\pi$ represent the same point in Ω which cannot be separated. Analogously, the path $\phi = 2\pi l t$ with an integer l is a closed curve of l loops in Ω which cannot be deformed to $\phi = 2\pi l' t$ if $l' \neq l$. Thus the one-dimensional unitary group $U(1)$ is infinitely connected.

A unitary representation of the unitary group $U(1)$ is given by

$$D^{(k)}(\phi) = \exp(\mathrm{i}k\phi) \tag{9.4.3b}$$

where k is a real number. When k is an integer, $D^{(k)}(\phi)$ is a single-valued representation of $U(1)$. When k is a rational number q/n, where there is no common denominator of the two integers q and n, $D^{(k)}(\phi)$ is n-valued because we have n different values for the identity element of $U(1)$:

$$D^{(k)}(2s\pi) = \exp(\mathrm{i}2\pi sq/n); \qquad s = 0, 1, \ldots, n-1$$

Finally, when k is irrational, $D^{(k)}(\phi)$ is infinitely many-valued. This is consistent with the infinite connectedness of $U(1)$.

The universal covering group $U'(1)$ of $U(1)$ is given by the group of real numbers x:

$$U'(1) = \{x; -\infty < x < \infty\} \tag{9.4.4}$$

with addition as the law of multiplication: the homomorphism $x \to \mathrm{e}^{\mathrm{i}\phi}$ is described by

$$x = \phi + 2l\pi, \qquad 0 \leqslant \phi \leqslant 2\pi \ (\equiv 0)$$

where l is any integer (positive, negative or zero); $-\infty < l < \infty$. The kernel of the homomorphism is $N = \{2l\pi; -\infty < l < \infty\}$. The unitary representation $D^{(k)}(x) = \exp(\mathrm{i}kx)$ of $U(1)'$ is a single-valued function of a point x in the parameter space Ω': $-\infty < x < \infty$, whereas $D^{(k)}(\phi) = \exp(\mathrm{i}k\phi)$ is a single-valued or multivalued function of a point ϕ in the parameter space Ω: $\phi + 2\pi \equiv \phi \pmod{2\pi}$, depending on the value of k, as discussed above (Hamermesh 1962).

Example 2. The connectedness of $SO(3, r)$. Previously, in Section 4.3, we have discussed the one-to-one correspondence between a proper rotation $R(\boldsymbol{\theta})$ in $V^{(3)}$ and a rotation vector $\boldsymbol{\theta}$ in the parameter sphere Ω defined by

$$\Omega: \quad 0 \leqslant |\boldsymbol{\theta}| \leqslant \pi, \qquad \boldsymbol{\theta} = \theta \boldsymbol{n} \tag{9.4.5}$$

where two opposite poles $\pi\boldsymbol{n}$ and $-\pi\boldsymbol{n}$ are regarded as one point in Ω corresponding to the cyclic boundary condition $R(\pi\boldsymbol{n}) = R(-\pi\boldsymbol{n})$. From this condition, we shall show that the parameter space Ω of $SO(3, r)$ is doubly connected.

In the parameter space Ω defined by (9.4.5), we may consider two kinds of paths that connect two points 1 and 2 in Ω (see Figure 9.3). The path (a) connects 1 and 2 directly, whereas the path (b) connects them after one antipodal jump from x to x'. The points x and x' are the same point in Ω so that the path $1 \to x = x' \to 2$ is a continuous one. It can be seen that the path (b) cannot be made to coincide with the path (a) by a continuous distortion, because, as we move the point x on the surface, its antipodal point x' also moves and remains always diametrically opposite to x. Note further that a path (c) that has two antipodal jumps ($1 \to x = x' \to y = y' \to 2$ in Figure 9.3) can be brought into coincidence with the path (a) by a continuous distortion, which moves y' ($= y$) to x ($= x'$) on the surface and removes them from the surface completely. In such a case, paths (a) and (c) are said to be *homotopic* or to belong to the same *homotopy class*. In general, all paths that connect 1 to 2 with an even number of antipodal jumps belong to one homotopy class whereas those with an odd number of jumps belong to the other homotopy class.

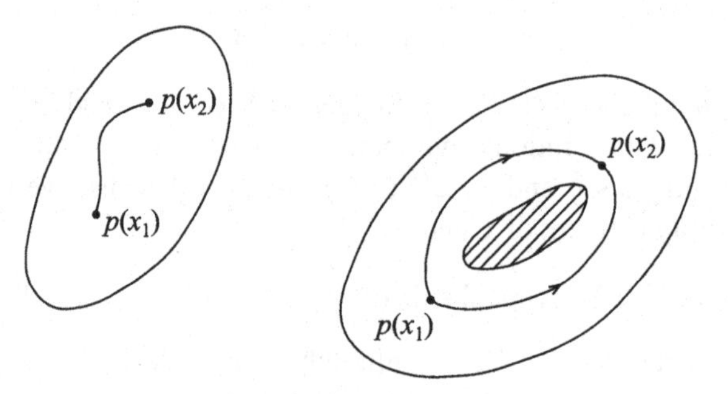

Figure 9.2. (a) A simply connected space. (b) A multiply connected space (schematic).

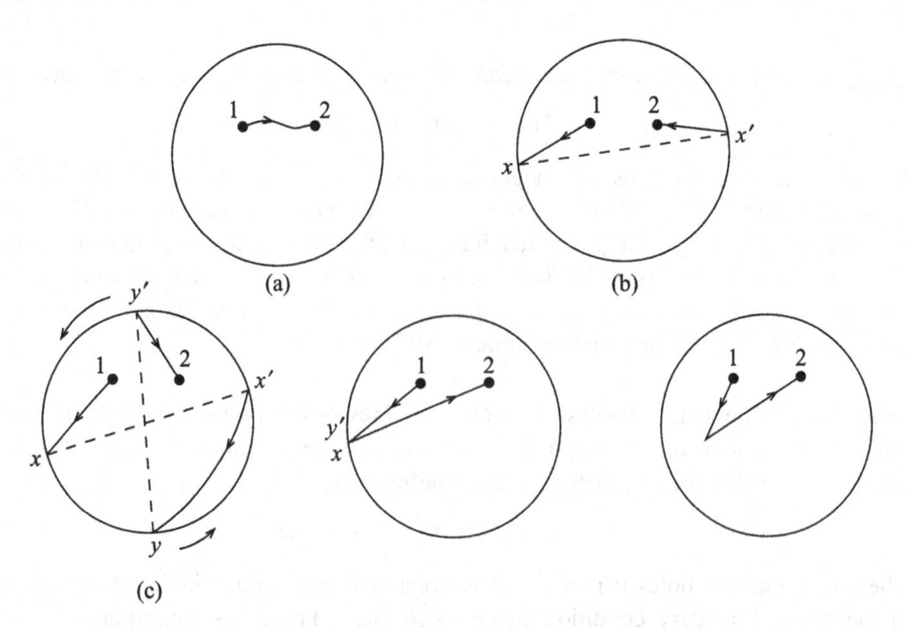

Figure 9.3. The double-connectedness of the parameter space of $SO(3, r)$.

Thus one may conclude that the parameter space Ω of $SO(3, r)$ is doubly connected. Accordingly, from the general discussion given in this section, we may expect that there exist single-valued as well as double-valued irreps for $SO(3, r)$. These will be determined from the vector irreps of the universal covering group, which will be shown to be $SU(2)$ in the next chapter.

10
The representations of the rotation group

As has been shown in Section 9.4, the parameter space of the proper rotation group $SO(3, r)$ is doubly connected so that there exist single-valued as well as double-valued representations for $SO(3, r)$. In this chapter, it will be shown that the special unitary group $SU(2)$ is simply connected and homomorphic to $SO(3, r)$ with two-to-one correspondence. Accordingly, the representations of $SU(2)$ are single-valued, but these provide all the single- and double-valued representations of $SO(3, r)$. In particular, $SU(2)$ itself provides a 2×2 matrix representation of $SO(3, r)$ that is double-valued.[1] As was mentioned at the end of Section 4.1, an element of $SU(2)$ is called *a spinor transformation*, because of the role it plays in the theory of the spinning electron. Moreover, the double-valued unirreps of $SO(3, r)$ given by the unirreps of $SU(2)$ are called *the spinor representations* of $SO(3, r)$. We shall begin with a discussion on the special unitary group $SU(2)$.

10.1 The structure of *SU(2)*

10.1.1 The generators of SU(2)

Previously, in Section 4.1, we have shown that an element of the special unitary group in n dimensions $SU(n)$ is expressed by a matrix of the form $U_0 = \exp K_0$, where K_0 is an anti-Hermitian traceless matrix. Thus, we may write a general element of S of $SU(2)$ in the form

$$S(\boldsymbol{\theta}) = \exp\,[-\mathrm{i}H(\boldsymbol{\theta})] \tag{10.1.1a}$$

where $H(\boldsymbol{\theta})$ is a 2×2 traceless Hermitian matrix that can be expressed by

$$H(\boldsymbol{\theta}) = \frac{1}{2}\begin{bmatrix} \theta_3 & \theta_1 - \mathrm{i}\theta_2 \\ \theta_1 + \mathrm{i}\theta_2 & -\theta_3 \end{bmatrix} \tag{10.1.1b}$$

with three real parameters θ_1, θ_2 and θ_3. These may be regarded as three components of a vector $\boldsymbol{\theta} = (\theta_1, \theta_2, \theta_3)$ that will be called *the rotation vector* of the spinor transformation S analogous to the case of a rotation $R(\boldsymbol{\theta}) \in SO(3, r)$. The numerical factor $\frac{1}{2}$ in (10.1.1b) is introduced to correlate $\boldsymbol{\theta}$ to the rotation vector of a rotation $R(\boldsymbol{\theta}) \in SO(3, r)$.

In terms of the Pauli spin vector $\boldsymbol{\sigma}$ with the components

$$\sigma_1 = \begin{bmatrix} 0 & 1 \\ 1 & 0 \end{bmatrix}, \qquad \sigma_2 = \begin{bmatrix} 0 & -\mathrm{i} \\ \mathrm{i} & 0 \end{bmatrix}, \qquad \sigma_3 = \begin{bmatrix} 1 & 0 \\ 0 & -1 \end{bmatrix}$$

[1] If a group A is homomorphic to a group B with m-to-one correspondence, the group A itself provides an m-valued representation of the group B while the group B itself provides a single-valued representation of the group A.

the matrix $H(\boldsymbol{\theta})$ of (10.1.1b) can be expressed by the scalar product $(\boldsymbol{\theta}\cdot\boldsymbol{\sigma})$ as follows:

$$\begin{aligned} \mathrm{H}(\boldsymbol{\theta}) &= \tfrac{1}{2}(\theta_1\sigma_1+\theta_2\sigma_2+\theta_3\sigma_3) \\ &= \tfrac{1}{2}(\boldsymbol{\theta}\cdot\boldsymbol{\sigma}) \\ &= \tfrac{1}{2}\theta\sigma_{\boldsymbol{n}}; \qquad \sigma_{\boldsymbol{n}} = (\boldsymbol{\sigma}\cdot\boldsymbol{n}) \end{aligned} \tag{10.1.2}$$

where $\theta = |\boldsymbol{\theta}|$ is the magnitude of $\boldsymbol{\theta}$ and $\boldsymbol{n} = \boldsymbol{\theta}/\theta$, the axis-vector of $\boldsymbol{\theta}$. Since the Pauli spin matrices σ_υ are unit involutional satisfying $\sigma_\nu^2 = 1$ $(\nu = 1, 2, 3)$ and also anti-commute with each other, the spin component $\sigma_{\boldsymbol{n}}$ in the direction of a unit vector $\boldsymbol{n}$ is also unit involutional, satisfying $\sigma_{\boldsymbol{n}}^2 = 1$. From (10.1.1a) and (10.1.2) the general element of $SU(2)$ is written in the form

$$S(\boldsymbol{\theta}) = \exp\left(-\frac{1}{2}(\boldsymbol{\theta}\cdot\boldsymbol{\sigma})\right) = \exp\left(-\frac{1}{2}\theta\sigma_{\boldsymbol{n}}\right) \tag{10.1.3a}$$

It follows then that two successive transformations $S(\theta_1\boldsymbol{n})$ and $S(\theta_2\boldsymbol{n})$ about a given axis-vector $\boldsymbol{n}$ are additive for the rotation angles:

$$S(\theta_1\boldsymbol{n})S(\theta_2\boldsymbol{n}) = S((\theta_1+\theta_2)\boldsymbol{n}) \tag{10.1.3b}$$

which is quite analogous to the rotations $R(\boldsymbol{\theta}) \in SO(3, r)$.

The infinitesimal generator of $S(\boldsymbol{\theta})$, with respect to each component of the rotation vector $\boldsymbol{\theta}$, is defined by

$$\tau_\nu = \left.\frac{\partial S(\boldsymbol{\theta})}{\partial\theta_\nu}\right|_{(0.0.0)} = -\frac{1}{2}\sigma_\nu; \qquad \nu = 1, 2, 3 \tag{10.1.4a}$$

These anticommute with each other analogously to the Pauli spin matrices and satisfy the commutation relations

$$[\tau_1, \tau_2] = \tau_3, \qquad [\tau_2, \tau_3] = \tau_1, \qquad [\tau_3, \tau_1] = \tau_2 \qquad (10.1.4\mathrm{b})^2$$

The set (τ_1, τ_2, τ_3) defines the Lie algebra *su*(2) which is identical to the *so*(3, *r*) algebra defined by (9.3.12a), i.e.

$$[I_1, I_2] = I_3, \qquad [I_2, I_3] = I_1, \qquad [I_3, I_1] = I_2 \tag{10.1.4c}$$

In terms of $\boldsymbol{\tau} = (\tau_1, \tau_2, \tau_3)$, the spinor transformation $S(\boldsymbol{\theta})$ is rewritten in the exponential form

$$S(\boldsymbol{\theta}) = \exp(\boldsymbol{\theta}\cdot\boldsymbol{\tau}) = \exp(\theta\tau_{\boldsymbol{n}}) \tag{10.1.5}$$

which is quite analogous to (9.3.12b) for $R(\boldsymbol{\theta}) \in SO(3, r)$. From (10.1.4b) and (10.1.4c), there exists a one-to-one correspondence between the infinitesimal generators of $SU(2)$ and those of $SO(3, r)$ under the commutation relations via

$$\tau_\nu \leftrightarrow I_\nu \qquad (\nu = 1, 2, 3) \tag{10.1.6}$$

Such a correspondence is called *the local isomorphism* between two groups, since it does not necessarily mean that two groups themselves are isomorphic. Actually, it will be shown later that $SU(2)$ is homomorphic to $SO(3, r)$ via the two-to-one correspondence $\pm S(\boldsymbol{\theta}) \rightarrow R(\boldsymbol{\theta})$.

To proceed further we shall calculate the matrix elements of $S(\boldsymbol{\theta})$ explicitly. On expanding the last expression of $S(\boldsymbol{\theta})$ given by (10.1.3a) and using $\sigma_{\boldsymbol{n}}^2 = 1$, we find that

[2] In terms of the vector product notation, (10.1.4b) is expressed by $\boldsymbol{\tau}\times\boldsymbol{\tau} = \boldsymbol{\tau}$.

$$S(\boldsymbol{\theta}) = \sum_{N=0}^{\infty}\left[\frac{(-1)^N}{(2N)!}\left(\frac{1}{2}\theta\right)^{2N} + \frac{(-1)^N}{(2N+1)!}\left(\frac{1}{2}\theta\right)^{2N+1}(-\mathrm{i}\sigma_{\boldsymbol{n}})\right]$$

$$= \sigma_0 \cos\left(\frac{\theta}{2}\right) - \mathrm{i}\sigma_{\boldsymbol{n}} \sin\left(\frac{\theta}{2}\right) \tag{10.1.7a}$$

where σ_0 is the unit matrix in two dimensions. Explicitly,

$$S(\boldsymbol{\theta}) = \begin{bmatrix} \cos\left(\frac{\theta}{2}\right) - \mathrm{i}n_z \sin\left(\frac{\theta}{2}\right) & -(\mathrm{i}n_x + n_y)\sin\left(\frac{\theta}{2}\right) \\ (-\mathrm{i}n_x + n_y)\sin\left(\frac{\theta}{2}\right) & \cos\left(\frac{\theta}{2}\right) + \mathrm{i}n_z \sin\left(\frac{\theta}{2}\right) \end{bmatrix} \tag{10.1.7b}$$

which may be written in the form

$$S(\boldsymbol{\theta}) \equiv S(a, b) = \begin{bmatrix} a & b \\ -b^* & a^* \end{bmatrix}; \qquad |a|^2 + |b|^2 = 1 \tag{10.1.7c}$$

where

$$a = \cos\left(\frac{\theta}{2}\right) - \mathrm{i}n_z \sin\left(\frac{\theta}{2}\right), \qquad b = -(\mathrm{i}n_x + n_y)\sin\left(\frac{\theta}{2}\right) \tag{10.1.7d}$$

The expression (10.1.7b) is called *the Euler–Rodrigues parametrizations* of $S(\boldsymbol{\theta})$ whereas the expression (10.1.7c) is called *the Cayley–Kline parametrization* of $S(\boldsymbol{\theta})$. Note that $\boldsymbol{\theta}$ is a set of three real independent parameters whereas the set (a, b) is complex and is not independent because of the normalization condition $|a|^2 + |b|^2 = 1$.

Example. Let $c_z(\theta) = R(\theta \boldsymbol{e}_z)$ be a rotation about the z-axis through an angle θ and $u_0 = R(\pi \boldsymbol{e}_x)$ be the binary rotation about the x-axis, then a rotation u_β defined by

$$u_\beta = c_z(\beta)u_0 = R(\pi \boldsymbol{h}_\beta); \qquad \boldsymbol{h}_\beta = (\cos(\beta/2), \sin(\beta/2), 0)$$

is a binary rotation about the unit vector $\boldsymbol{h}_\beta$ in the x, y plane which makes an angle $\beta/2$ with the x-axis. The corresponding spinor transformations with the rotation vectors $\theta \boldsymbol{e}_z$ and $\pi \boldsymbol{h}_\beta$ are given, from (10.1.7b) and (10.1.7c) by

$$\check{c}_z(\theta) \equiv S(\theta \boldsymbol{e}_z) = \begin{bmatrix} \mathrm{e}^{-\mathrm{i}\theta/2} & 0 \\ 0 & \mathrm{e}^{\mathrm{i}\theta/2} \end{bmatrix} = S(\mathrm{e}^{-\mathrm{i}\theta/2}, 0) \tag{10.1.7e}$$

$$\check{u}_\beta \equiv S(\pi \boldsymbol{h}_\beta) = \begin{bmatrix} 0 & -\mathrm{i}\mathrm{e}^{-\mathrm{i}\beta/2} \\ -\mathrm{i}\mathrm{e}^{\mathrm{i}\beta/2} & 0 \end{bmatrix} = S(0, -\mathrm{i}\mathrm{e}^{-\mathrm{i}\beta/2}) \tag{10.1.7f}$$

Note that $\check{u}_\beta = \check{c}_z(\beta)\check{u}_0 = -\mathrm{i}\boldsymbol{\sigma} \cdot \boldsymbol{h}_\beta$, where $\check{u}_0 = -\mathrm{i}\sigma_x$. These two spinor transformations will describe all elements of the spinor groups corresponding to the point groups C_∞, C_n, D_∞ and D_n.

10.1.2 The parameter space Ω' of $SU(2)$

From (10.1.7a), we observe that

$$S(2\pi \boldsymbol{n}) = -\mathbf{1}, \qquad S(4\pi \boldsymbol{n}) = S(\theta = 0) = \mathbf{1} \tag{10.1.8a}$$

for any given axis-vector $\boldsymbol{n}$. Thus, $S(\boldsymbol{\theta})$ changes its sign when $\boldsymbol{\theta}$ makes a complete 2π

revolution about an axis-vector $\boldsymbol{n}$, quite on the contrary to $R(\boldsymbol{\theta})$, which is unchanged after a 2π rotation, i.e. $R(2\pi\boldsymbol{n}) = \mathbf{1}$. From the periodicity of $S(\boldsymbol{\theta})$, the parameter space Ω' of $SU(2)$ may be defined by

$$\Omega': \quad 0 \leqslant |\boldsymbol{\theta}| \leqslant 2\pi \tag{10.1.8b}$$

where whole surface points $2\pi\boldsymbol{n}$ are regarded as one point in Ω' because of the boundary condition $S(2\pi\boldsymbol{n}) = -1$. Then there exists one-to-one correspondence between $S(\boldsymbol{\theta}) \in SU(2)$ and $\boldsymbol{\theta} \in \Omega'$:

$$S(\boldsymbol{\theta}) \leftrightarrow \boldsymbol{\theta} \in \Omega' \tag{10.1.8c}$$

In fact, for a given spinor transformation $S(\boldsymbol{\theta})$, its rotation vector $\boldsymbol{\theta} \in \Omega'$ is uniquely determined, from (10.1.7a) and (10.1.7b), by

$$\operatorname{tr} S(\boldsymbol{\theta}) = 2\cos\left(\frac{\theta}{2}\right), \qquad (S^{\dagger} - S)\Big/\left[2\mathrm{i}\sin\left(\frac{\theta}{2}\right)\right] = \sigma_{\boldsymbol{n}} \tag{10.1.8d}$$

except on the surface of Ω', where $\theta = 2\pi$ and $\boldsymbol{n}$ is indefinite.

10.1.2.1 The connectedness of SU(2)

One may trivially establish that the parameter space Ω' of $SU(2)$ defined by (10.1.8b) is simply connected, quite on the contrary to the parameter space Ω of $SO(3, r)$, which is doubly connected, as was shown in Section 9.4. To see this, let us consider two kinds of paths that connect two points 1 and 2 in Ω' (see Figure 10.1) as in the case of $SO(3, r)$. The path (a) connects two points 1 and 2 directly, whereas the path (b) connects them after one jump from a surface point x of the parameter sphere to another surface point x'. Since all points of the surface are regarded as the same point, the path $1 \to x = x' \to 2$ is a continuous one in Ω'. Now, the path (b) can be made to coincide with the path (a) by a continuum distortion because one can move the point x to the point x' and remove them together as one point from the surface. Thus $SU(2)$ is simply connected, in contrast to the double-connectedness of $SO(3, r)$: the difference is due to the boundary condition that $S(2\pi\boldsymbol{n}) = -\mathbf{1}$ for $SU(2)$ whereas $R(-\pi\boldsymbol{n}) = R(\pi\boldsymbol{n})$ for $SO(3, r)$.

Since $SU(2)$ is simply connected, its representations are all single-valued ordinary representations or vector representations in short. Now, on account of the pseudo-periodicity $S(2\pi\boldsymbol{n}) = -\mathbf{1}$, one can express the whole set of $SU(2)$ by a set $\{\pm S(\boldsymbol{\theta})\}$, where $\boldsymbol{\theta}$ is the rotation vector in the parameter domain Ω $(0 \leqslant |\boldsymbol{\theta}| \leqslant \pi)$ of $SO(3, r)$.

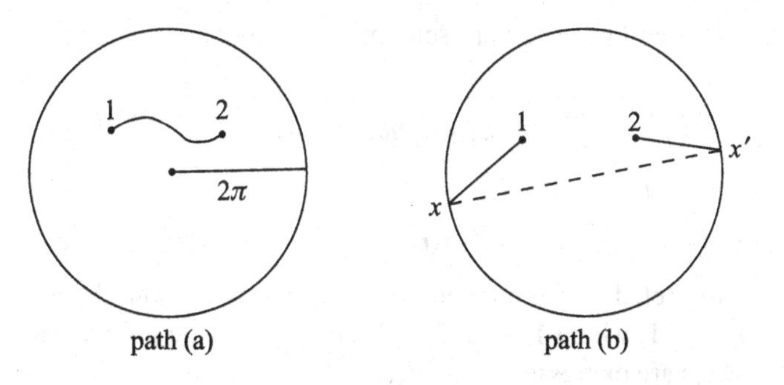

Figure 10.1. Simple connectedness of Ω': $0 \leqslant |\boldsymbol{\theta}| \leqslant 2\pi$.

Then there exists a two-to-one correspondence between $SU(2) = \{\pm S(\boldsymbol{\theta})\}$ and $SO(3, r) = \{R(\boldsymbol{\theta})\}$ via

$$\pm S(\boldsymbol{\theta}) \to R(\boldsymbol{\theta}); \qquad \boldsymbol{\theta} \in \Omega \tag{10.1.9}$$

Note that the two-to-one correspondence $\pm S(\pi\boldsymbol{n}) \to R(\pi\boldsymbol{n})$ holds on the surface of Ω as well. In the next section it will be shown that this correspondence leads to their two-to-one homomorphism.

10.1.3 Spinors

A spinor in $V^{(2)}$ is a two-component vector $\psi = (\psi_1, \psi_2)$ that transforms according to a spinor transformation

$$\psi' = \mathring{S}(\boldsymbol{\theta})\psi; \qquad S(\boldsymbol{\theta}) \in SU(2) \tag{10.1.10}$$

under a rotation characterized by the rotation vector $\boldsymbol{\theta} = \theta\boldsymbol{n}$. Since $S(2\pi\boldsymbol{n}) = -\mathbf{1}$, the spinor ψ does change sign when one applies a complete revolution 2π. The elementary spinors ξ_1 and ξ_2 introduced by (6.3.11) provide the natural basis $\xi = [\xi_1, \xi_2]$ of $SU(2)$, as was shown by (6.3.12). In terms of the two elementary spinors, the general spinor ψ in $V^{(2)}$ is expressed by $\psi = \psi_1\xi_1 + \psi_2\xi_2$: it may also depend on the spatial coordinates, as will be discussed in Section 10.4. In quantum mechanics, a spinor describes a spin state of the electron; e.g. ξ_1 and ξ_2 are the eigenvectors of the z-component of the spin $s_z = \sigma_z/2$ belonging to the eigenvalues $\frac{1}{2}$ and $-\frac{1}{2}$, respectively.

According to Dirac (1974), 'spinors, like tensors, are geometric objects embodied in a space and have components that transform linearly under transformations of the coordinates of the space. Spinors differ from tensors in that they change signs when one applies a complete revolution about an axis, while tensors are unchanged.'

Remark 1. For the sake of comparison with the existing literature, we shall give the parameterization of $S(\boldsymbol{\theta})$ by the Euler angles α, β and γ. Let $[\boldsymbol{e}_x, \boldsymbol{e}_y, \boldsymbol{e}_z]$ be a Cartesian basis, then, analogously to (4.3.18), we have

$$S(\boldsymbol{\theta}) \equiv S(\alpha, \beta, \gamma) \equiv S(\gamma\boldsymbol{e}_z)S(\beta\boldsymbol{e}_y)S(\alpha\boldsymbol{e}_z) \tag{10.1.11a}$$

Using (10.1.7b) for each factor on the right-hand side, the Cayley–Kline parameters are given by

$$a = \mathrm{e}^{-\mathrm{i}(\gamma+\alpha)/2}\cos\left(\frac{\beta}{2}\right), \qquad b = -\mathrm{e}^{-\mathrm{i}(\gamma-\alpha)/2}\sin\left(\frac{\beta}{2}\right) \tag{10.1.11b}$$

This parametrization will be used very seldom in this book.

10.1.4 Quaternions

When $\theta = \pi$ we have, from (10.1.7a),

$$S(\pi\boldsymbol{n}) = -\mathrm{i}\sigma_{\boldsymbol{n}} \tag{10.1.12a}$$

which may be interpreted as the p (pseudo)-binary rotation about the axis-vector $\boldsymbol{n}$: it satisfies $(-\mathrm{i}\sigma_{\boldsymbol{n}})^2 = -\mathbf{1}$, where $\mathbf{1}$ is the 2×2 unit matrix. The p-binary rotations about the x-, y- and z-axes are expressed by

$$\check{2}_x = -\mathrm{i}\sigma_x, \qquad \check{2}_y = -\mathrm{i}\sigma_y, \qquad \check{2}_z = -\mathrm{i}\sigma_z$$

Then, these satisfy the following algebraic relations:

$$\check{2}_x^2 = -\mathbf{1}, \qquad \check{2}_x\check{2}_y = \check{2}_z = -\check{2}_y\check{2}_x \qquad (x,\ y,\ z \text{ cyclic}) \tag{10.1.12b}$$

that is, $\check{2}_x$ and $\check{2}_y$ anticommute; whereas the ordinary binary rotations 2_x and 2_y commute. From (10.1.12b), it follows that the set of eight elements

$$D_2' = \{\pm\mathbf{1},\ \pm\check{2}_x,\ \pm\check{2}_y,\ \pm\check{2}_z\} \tag{10.1.13}$$

forms a group called *the quaternion group* or the double group of the four-group D_2, since D_2' is homomorphic to D_2.

Extending the axis-vector notation of the N-fold rotation $N_{\boldsymbol{n}} = R((2\pi/N)\boldsymbol{n})$ to the spinor transformation we write, from (10.1.7a),

$$\begin{aligned}\check{N}_{\boldsymbol{n}} = S\left(\frac{2\pi}{N}\boldsymbol{n}\right) &= \cos\left(\frac{\pi}{N}\right)\mathbf{1} + \sin\left(\frac{\pi}{N}\right)(n_x\check{2}_x + n_y\check{2}_y + n_z\check{2}_z)\\ &= q_0\mathbf{1} + q_x\check{2}_x + q_y\check{2}_y + q_z\check{2}_z\end{aligned} \tag{10.1.14}$$

where q_0, q_x, q_y and q_z are real coefficients normalized by

$$q_0^2 + q_x^2 + q_y^2 + q_z^2 = 1$$

Hereafter, the spinor transformation $\check{N}_{\boldsymbol{n}}$ may be conveniently called the $\check{N}$-fold rotation about the axis-vector $\boldsymbol{n}$; e.g.

$$\check{3}_{xyz} = (\mathbf{1} + \check{2}_x + \check{2}_y + \check{2}_z)/2, \qquad \check{4}_x = (\mathbf{1} + \check{2}_x)/\sqrt{2}, \qquad \check{2}_{xy} = (\check{2}_x + \check{2}_y)/\sqrt{2} \tag{10.1.15}$$

Note that there exists a simple correlation between an axis-vector symbol and the corresponding set of the quaternion coefficients $\{q_0, q_1, q_2, q_3\}$ for an element of a cubic group. From this correlation, one can easily write down all elements of the spinor group O' corresponding to the octahedral group O and construct the multiplication table of O'; see Table 11.2 later.

Historically, Hamilton is the one who introduced a quantity called a quaternion Q as an extension of a complex number. It is nothing but a spinor transformation multiplied by a non-zero real number q: $Q = qS(\boldsymbol{\theta})$. Then a product of two quaternions is also a quaternion.

10.2 The homomorphism between $SU(2)$ and $SO(3, r)$

We shall begin with the following lemma (Kim 1981c).

Lemma 10.2.1. Under a similarity transformation by $S(\boldsymbol{\alpha}) \in SU(2)$, the generator set $\boldsymbol{\tau} = [\tau_1, \tau_2, \tau_3]$ of $SU(2)$ defined by (10.1.4a) transforms like a basis of the rotation $R(\boldsymbol{\alpha}) \in SO(3, r)$:

$$S(\boldsymbol{\alpha})\tau_\mu S(\boldsymbol{\alpha})^{-1} = \sum_\sigma \tau_\sigma R_{\sigma\mu}(\boldsymbol{\alpha}); \qquad \mu = 1, 2, 3 \tag{10.2.1}$$

where $S(\boldsymbol{\alpha}) = \exp(\alpha\tau_{\boldsymbol{n}})$, $R(\boldsymbol{\alpha}) = \exp[\alpha\omega(\boldsymbol{n})]$ and $\omega(\boldsymbol{n})$ is the infinitesimal rotation about the unit vector $\boldsymbol{n} = \boldsymbol{\alpha}/\alpha$ defined by (4.3.2a).

The proof is based on the infinitesimal transformations. Let α be an infinitesimal angle, then

$$S(\boldsymbol{\alpha}) \approx (1 + \alpha\tau_{\boldsymbol{n}}), \qquad R(\boldsymbol{\alpha}) \approx (1 + \alpha\omega) \tag{10.2.2}$$

Thus, to the first order in α, we have for the left-hand side of (10.2.1)

$$(1 + \alpha\tau_{\boldsymbol{n}})\tau_\mu(1 - \alpha\tau_{\boldsymbol{n}}) \approx \tau_\mu + \alpha[\tau_{\boldsymbol{n}}, \tau_\mu] = \sum_\sigma \tau_\sigma(\mathbf{1} + \alpha\omega)_{\sigma\mu}$$

with use of

$$[\tau_{\boldsymbol{n}}, \tau_\mu] = [\boldsymbol{\tau} \times \boldsymbol{n}]_\mu = \sum_\sigma \tau_\sigma \omega_{\sigma\mu} \tag{10.2.3}$$

The case for a finite α given by (10.2.1) can be built up by successive infinitesimal transformations.

Exercise. The Dirac representation of the infinitesimal rotation $\hat{\omega} = \hat{\omega}(\boldsymbol{n})$ is defined by $\hat{\omega} = \sum_{\nu\mu}\tau_\nu\omega_{\nu\mu}\tau_\mu$. Then, $\hat{\omega} = \tau_{\boldsymbol{n}}$ so that $S(\boldsymbol{\theta}) = \exp(\theta\hat{\omega})$, which is analogous to $R(\boldsymbol{\theta}) = \exp(\theta\omega)$ (Dirac 1974).

Now we are ready to show two basic theorems on $SU(2)$ that lead to the homomorphism between $SU(2)$ and $SO(3, r)$. Each of these two theorems has an important significance in its own right.

Theorem 10.2.1. The conjugation of $S(\boldsymbol{\theta})$ by $S(\boldsymbol{\alpha}) \in SU(2)$ simply rotates the rotation vector $\boldsymbol{\theta}$ by $R(\boldsymbol{\alpha}) \in SO(3, r)$:

$$S(\boldsymbol{\alpha})S(\boldsymbol{\theta})S(\boldsymbol{\alpha})^{-1} = S[R(\boldsymbol{\alpha})\boldsymbol{\theta}] \tag{10.2.4}$$

which is quite analogous to the conjugate rotation of $R(\boldsymbol{\theta})$ by $R(\boldsymbol{\alpha}) \in SO(3, r)$:

$$R(\boldsymbol{\alpha})R(\boldsymbol{\theta})R(\boldsymbol{\alpha})^{-1} = R[R(\boldsymbol{\alpha})\boldsymbol{\theta}]$$

Proof. Since $S(\boldsymbol{\theta}) = \exp(\boldsymbol{\tau} \cdot \boldsymbol{\theta})$, the similarity transformation of the scalar product $(\boldsymbol{\tau} \cdot \boldsymbol{\theta})$ with respect to $S(\boldsymbol{\alpha})$ will provide the proof. Using Lemma 10.2.1, we have

$$\begin{aligned} S(\boldsymbol{\alpha})(\boldsymbol{\tau} \cdot \boldsymbol{\theta})S(\boldsymbol{\alpha})^{-1} &= \sum_{\sigma,\mu} \tau_\sigma R(\boldsymbol{\alpha})_{\sigma\mu}\theta_\mu \\ &= \boldsymbol{\tau} \cdot R(\boldsymbol{\alpha})\boldsymbol{\theta} \end{aligned}$$

from which follows the required result (10.2.4).

This theorem implies that two spinor transformations $S(\boldsymbol{\theta})$ and $S(\boldsymbol{\theta}')$ are equivalent with respect to $SU(2)$ if and only if their rotation vectors $\boldsymbol{\theta}$ and $\boldsymbol{\theta}'$ in the parameter domain Ω' are equivalent with respect to $SO(3, r)$. Thus spinor transformations with the same rotation angle are in the same class of $SU(2)$, quite analogously to $SO(3, r)$. Now, when we describe $SU(2)$ by $\{\pm S(\boldsymbol{\theta}); 0 \leqslant |\boldsymbol{\theta}| \leqslant \pi\}$, the classes of $S(\boldsymbol{\theta})$ and $-S(\boldsymbol{\theta})$ are different except when $|\boldsymbol{\theta}| = \pi$. This follows from their characters:

$$\mathrm{tr}\,[\pm S(\boldsymbol{\theta})] = \pm 2\cos\left(\frac{\theta}{2}\right), \qquad 0 \leqslant |\boldsymbol{\theta}| \leqslant \pi \tag{10.2.5}$$

which are different unless $\theta = \pi$. When $\theta = \pi$, we have $-S(\pi\boldsymbol{n}) = S(-\pi\boldsymbol{n})$ so that the classes of $S(\pi\boldsymbol{n})$ and $-S(\pi\boldsymbol{n})$ are combined into a single class of $SU(2)$, because

$\pi \boldsymbol{n}$ and $-\pi \boldsymbol{n}$ are equivalent with respect to $SO(3, r)$. (This is true for a subgroup of $SU(2)$, if and only if $\pi \boldsymbol{n}$ is two-sided.)

Exercise. Let $S(\pi \boldsymbol{v}) = \check{2}_{\boldsymbol{v}}$ be a $\check{2}$-fold rotation about a unit vector $\boldsymbol{v}$ and $S(\boldsymbol{\alpha})$ be a spinor transformation about a rotation vector $\boldsymbol{\alpha}$ perpendicular to $\boldsymbol{v}$. Show that

$$S(\boldsymbol{\alpha})\check{2}_{\boldsymbol{v}} = \check{2}_{\boldsymbol{v}} S(\boldsymbol{\alpha})^{\dagger} = \check{2}_{\boldsymbol{h}}, \qquad \boldsymbol{h} = R(\boldsymbol{\alpha}/2)\boldsymbol{v}$$

This is analogous to (4.3.15c):

$$R(\boldsymbol{\alpha})2_{\boldsymbol{v}} = 2_{\boldsymbol{v}} R(\boldsymbol{\alpha})^{\sim} = 2_{\boldsymbol{h}}$$

where $2_{\boldsymbol{v}}$ is an ordinary two-fold rotation about the unit vector $\boldsymbol{v}$ in three dimensions. This means that a simple multiplication of $\check{2}_{\boldsymbol{v}}$ by $S(\boldsymbol{\alpha})$ from the left rotates the axis vector $\boldsymbol{v}$ through an angle $\alpha/2$ that is half the angle rotated by the conjugation with $S(\boldsymbol{\alpha})$. For the proof show that $\check{2}_{\boldsymbol{h}}\check{2}_{\boldsymbol{v}}^{\dagger} = S(\boldsymbol{\alpha})$, analogously to (4.3.15b).

Theorem 10.2.2. The group $SU(2) = \{\pm S(\boldsymbol{\theta})\}$ is homomorphic to $SO(3, r) = \{R(\boldsymbol{\theta})\}$ via the two-to-one correspondence in the parameter space Ω of $SO(3, r)$:

$$\pm S(\boldsymbol{\theta}) \to R(\boldsymbol{\theta}), \qquad \boldsymbol{\theta} \in \Omega \tag{10.2.6a}$$

that is, it preserves multiplication such that for $\boldsymbol{\theta}_1, \boldsymbol{\theta}_2, \boldsymbol{\theta}_3 \in \Omega$

$$S(\boldsymbol{\theta}_1)S(\boldsymbol{\theta}_2) = \pm S(\boldsymbol{\theta}_3), \qquad \text{if } R(\boldsymbol{\theta}_1)R(\boldsymbol{\theta}_2) = R(\boldsymbol{\theta}_3) \tag{10.2.6b}$$

where the $\pm$ signs depend on $\boldsymbol{\theta}_1$ and $\boldsymbol{\theta}_2$.

Proof. Let $S_i = S(\boldsymbol{\theta}_i)$ and $R_i = R(\boldsymbol{\theta}_i)$. Then, substitution of $S_1 S_2 S_3^{-1}$ into $S(\boldsymbol{\alpha})$ of (10.2.1) yields

$$(S_1 S_2 S_3^{-1})\tau_{\mu}(S_1 S_2 S_3^{-1})^{-1} = \sum_{\sigma} \tau_{\sigma}(R_1 R_2 R_3^{-1})_{\sigma\mu}; \qquad \mu = 1, 2, 3 \tag{10.2.7}$$

Thus if $R_1 R_2 = R_3$, then $S_1 S_2 S_3^{-1}$ commutes with the complete set of the generators $\{\tau_{\mu}\}$ of $SU(2)$ and hence is a constant. Now, let $S_1 S_2 S_3^{-1} = k$, then, from $\det S_j = 1$, we obtain $k^2 = 1$, for S_j being a special unitary matrix in two dimensions. Accordingly $k = \pm 1$ so that $S_1 S_2 = \pm S_3$, which proves (10.2.6b).

To understand the homomorphism relation (10.2.6b) consider rotations about a given axis of rotation: then the rotation angles are additive for two successive rotations, i.e. $R(\theta_1)R(\theta_2) = R(\theta_1 + \theta_2)$ and $S(\theta_1)S(\theta_2) = S(\theta_1 + \theta_2)$. Now, let $R(\theta_1)R(\theta_2) = R(\theta_3)$, where $-\pi < \theta_1, \theta_2, \theta_3 \leqslant \pi$, then we have $\theta_1 + \theta_2 = \theta_3$ or $\theta_1 + \theta_2 = \theta_3 \pm 2\pi$, so that

$$S(\theta_1)S(\theta_2) = \begin{cases} S(\theta_3), & \text{if } \theta_1 + \theta_2 = \theta_3 \\ -S(\theta_3), & \text{if } \theta_1 + \theta_2 = \theta_3 \pm 2\pi \end{cases}$$

which explains the homomorphism relation for this simple case.

The homomorphism (10.2.6a) means that $SO(3, r)$ is a three-dimensional representation of $SU(2)$ whereas $SU(2)$ is a two-dimensional double-valued representation of $SO(3, r)$. Thus, $SU(2)$ may be called *the double rotation group* or *the basic spinor representation* of $SO(3, r)$. From the single connectedness of $SU(2)$ and the homomorphism (10.2.6a), we can state that $SU(2)$ is *a universal covering group* of $SO(3, r)$ in the sense that all representations (single- or double-valued) of $SO(3, r)$ are provided by the vector representations of $SU(2)$. $SU(2)$ is often called *a central extension* of $SO(3, r)$, since the kernel N of the homomorphism, $N = \{\pm S(\boldsymbol{\theta} = 0) = \pm 1\}$, is in the center of $SU(2)$. Further generalization of this concept will be discussed in the theory of projective representations in Chapter 12.

The concept of the double group can easily be extended to a proper point group P that is a subgroup of $SO(3, r)$. Let $P = \{R(\boldsymbol{\theta}_i); \boldsymbol{\theta}_i \in \Omega\}$ be a point group and let $P' = \{\pm S(\boldsymbol{\theta}_i)\}$ be the corresponding set of elements of the order $|P'| = 2|P|$ in $SU(2)$. Then, from Theorem 10.2.2, P' also forms a group homomorphic to P and is called the *double group* of P. Thus, the single- and double-valued representations of P are provided by the vector representations of P' (see Chapter 11 for further detail).

Remark. From (4.3.10) and (10.1.12a), a binary rotation $R(\pi \boldsymbol{n})$ and the corresponding p-binary rotation $S(\pi \boldsymbol{n})$ are given by

$$R(\pi \boldsymbol{n}) = 2\boldsymbol{nn} - \boldsymbol{1}, \qquad S(\pi \boldsymbol{n}) = -\mathrm{i}\sigma_{\boldsymbol{n}} \tag{10.2.8}$$

These show explicitly that $R(\pi \boldsymbol{n}) = R(-\pi \boldsymbol{n})$ whereas $S(-\pi \boldsymbol{n}) = -S(\pi \boldsymbol{n}) \neq S(\pi \boldsymbol{n})$, exhibiting the two-to-one correspondence between $\{\pm S(\pi \boldsymbol{n})\}$ and $\{R(\pi \boldsymbol{n})\}$ on the surface of Ω.

10.3 The unirreps $D^{(j)}(\boldsymbol{\theta})$ of the rotation group

As discussed in Section 10.1, the parameter space Ω': $0 \leqslant |\boldsymbol{\theta}| \leqslant 2\pi$ of $SU(2)$ is singly connected and thus its representations are vector representations. In this section, we shall first construct the unirreps of $SU(2)$ from the monomials of the two elementary spinors ξ_1 and ξ_2 and discuss their irreducibilities and completeness. Then we classify these unirreps as the single- and double-valued unirreps of the rotation group $SO(3, r)$, and discuss their orthogonality relations.

According to (6.3.12), the natural basis $[\xi_1, \xi_2]$ of the spinor transformation $S(a, b)$ is described by the two elementary spinors:

$$\xi_1 = \begin{bmatrix} 1 \\ 0 \end{bmatrix}, \qquad \xi_2 = \begin{bmatrix} 0 \\ 1 \end{bmatrix}$$

Under an element $S(a, b) \in SU(2)$ given by (10.1.7c), the basis transforms according to

$$[\xi_1', \xi_2'] = \mathring{S}[\xi_1, \xi_2] = [\xi_1, \xi_2]\begin{bmatrix} a & b \\ -b^* & a^* \end{bmatrix}$$

i.e.

$$\begin{aligned} \xi_1' &= \mathring{S}\xi_1 = a\xi_1 - b^*\xi_2 \\ \xi_2' &= \mathring{S}\xi_2 = b\xi_1 + a^*\xi_2 \end{aligned} \tag{10.3.1a}$$

where $\mathring{S} = S(a, b)$ is regarded as an operator acting on the basis vectors.

To construct the irreducible representations of $SU(2)$, we introduce a complete set of homogeneous monomials of degree $2j$ defined by the direct products of the elementary spinors ξ_1 and ξ_2 as follows:

$$\phi^j_m(\xi_1, \xi_2) = \xi_1^{j+m}\xi_2^{j-m}/[(j+m)!(j-m)!]^{1/2}; \qquad m = j, j-1, \ldots, -j;$$
$$j = 0, \tfrac{1}{2}, 1, \ldots \tag{10.3.1b}$$

where j is an integer or half integer but $j \pm m$ are integers. The numerical factors in (10.3.1b) are introduced in order to achieve the greatest symmetry for the resulting representations. Since the monomial set for a given j is closed under the linear transformation (10.3.1a), it will provide a $(2j+1) \times (2j+1)$ matrix representation $D^{(j)}(\boldsymbol{\theta}) \equiv D^{(j)}(S(\boldsymbol{\theta}))$ of $SU(2)$ defined by

$$\begin{aligned}\mathring{S}\phi^{(j)}_m(\xi_1, \xi_2) &\equiv \phi^{(j)}_m(\mathring{S}\xi_1, \mathring{S}\xi_2)\\ &= (a\xi_1 - b^*\xi_2)^{j+m}(b\xi_1 + a^*\xi_2)^{j-m}/[(j+m)!(j-m)!]^{1/2}\\ &= \sum_{m'=j}^{-j} \phi^{(j)}_{m'}(\xi_1, \xi_2) D^{(j)}[S(\boldsymbol{\theta})]_{m'm}\end{aligned} \tag{10.3.2}$$

To calculate the explicit form of the representation $D^{(j)}$, it is most convenient to express the binomial expansion in (10.3.2) in the following form:

$$(a\xi_1 - b^*\xi_2)^{j+m}(b\xi_1 + a^*\xi_2)^{j-m} =$$
$$\sum_t \binom{j+m}{t}(a\xi_1)^t(-b^*\xi_2)^{j+m-t}\sum_k \binom{j-m}{k}(b\xi_1)^{j-m-k}(a^*\xi_2)^k$$

Then, comparing this with (10.3.2), we have $m' = t - m - k$ and obtain

$$D^{(j)}(\boldsymbol{\theta})_{m'm} \equiv D^{(j)}[S(\boldsymbol{\theta})]_{m'm} = D^{(j)}[S(a, b)]_{m'm}$$

$$= a^{m'+m}(-b^*)^{j-m'}b^{j-m}\sum_k \frac{[(j+m')!(j-m')!(j+m)!(j-m)!]^{1/2}}{(m'+m+k)!(j-m'-k)!(j-m-k)!k!}\left(-\left|\frac{a}{b}\right|^2\right)^k;$$
$$m', m = j, j-1, \ldots, -j \quad (10.3.3)^3$$

where the limit of summation over k is automatically determined by the condition $1/(-N)! = 0$, which holds for any positive integer N. The representation $D^{(j)}$ is called an *integral* (*half integral*) representation of $SU(2)$ when j is an integer (half integer). Note that $D^{(1/2)}[S(\boldsymbol{\theta})]$ equals the spinor transformation $S(\boldsymbol{\theta})$ itself, for $\phi^{1/2} = [\xi_1, \xi_2]$. The symmetry properties of the representation $D^{(j)}(S) = D^{(j)}(\boldsymbol{\theta})$ are the following.

10.3.1 The homogeneity of $D^{(j)}(S)$

Since the basis of the representation is a set of $2j$-degree homogeneous monomials, $D^{(j)}(S)$ must be a $2j$-degree homogeneous function of S such that $D^{(j)}(\lambda S) = \lambda^{2j}D^{(j)}(S)$ for an arbitrary constant λ: this can be seen by replacing $\mathring{S}$ with $\lambda\mathring{S}$ in (10.3.2). In the special case in which $\lambda = -1$, we have

$$D^{(j)}(-S) = (-1)^{2j}D^{(j)}(S) \tag{10.3.4a}$$

[3] The present representation defined by (10.3.2)–(10.3.3) is different from the conventional representation based on the point transformation introduced by (6.3.2) (cf. Wigner (1962)). Two representations are mutually complex conjugate (inverse transpose); see also Kim (1969).

Now, through the homomorphism $\pm S(\boldsymbol{\theta}) \rightarrow R(\boldsymbol{\theta})$ given in (10.2.6a), a representation of $SO(3, r)$ is defined by

$$D^{(j)}[R(\boldsymbol{\theta})] \equiv D^{(j)}[\pm S(\boldsymbol{\theta})] = (\pm)^{2j} D^{(j)}[S(\boldsymbol{\theta})] \equiv (\pm)^{2j} D^{(j)}(\boldsymbol{\theta}) \tag{10.3.4b}$$

where $\boldsymbol{\theta} \in \Omega$ of $SO(3, r)$. It is a single-valued (vector) representation of $SO(3, r)$ for an integral j but a double-valued (spinor) representation of $SO(3, r)$ for a half integral j. Moreover, from the representation of the homomorphism relation (10.2.6b), we have

$$D^{(j)}(\boldsymbol{\theta}_1) D^{(j)}(\boldsymbol{\theta}_2) = (\pm 1)^{2j} D^{(j)}(\boldsymbol{\theta}_3), \qquad \text{if } R(\boldsymbol{\theta}_1) R(\boldsymbol{\theta}_2) = R(\boldsymbol{\theta}_3) \tag{10.3.4c}$$

where the $\pm$ signs depend on $\boldsymbol{\theta}_1$ and $\boldsymbol{\theta}_2$. This inconvenient feature of a double-valued factor ($\pm$) in the representations will be effectively controlled if we regard $D^{(j)}(\boldsymbol{\theta})$ as a projective unirrep of $SO(3, r)$ belonging to a factor system, as will be discussed in the next chapter.

10.3.2 *The unitarity of $D^{(j)}(S)$*

It is to be noted from (10.3.3) that the matrix $D^{(j)}(S)$ is transposed by the interchange $m \rightleftarrows m'$ or by $b \rightleftarrows -b^*$, which transposes the matrix S. Thus,

$$D^{(j)}(S)^{\sim} = D^{(j)}(S^{\sim}) \tag{10.3.5a}$$

where $\sim$ denotes the transpose. Combining this with an obvious complex conjugate symmetry $D^j(S)^* = D^{(j)}(S^*)$, we obtain

$$D^{(j)}(S)^{\dagger} = D^{(j)}(S^{\dagger}) = D^{(j)}(S^{-1}) = D^{(j)}(S)^{-1} \tag{10.3.5b}$$ [4]

which means that $D^{(j)}(S)$ is a unitary representation of $SU(2)$. Here the last equality follows from the fact that $D^{(j)}(S)$ is a representation of $SU(2)$.

10.3.3 *The irreducibility of $D^{(j)}(\boldsymbol{\theta})$*

To show this, we shall first consider two special cases of $D^{(j)}(S)$. Firstly, let $\boldsymbol{\theta} = (0, 0, \theta)$, then $a = e^{-i\theta/2}$ and $b = 0$ from (10.1.7e) so that (10.3.2) leads to a diagonal matrix

$$D^{(j)}(0, 0, \theta)_{m'm} = \delta_{m'm} e^{-im\theta} \tag{10.3.6}$$

Secondly, on setting $m' = j$ in (10.3.3), the jth row of $D^{(j)}(\boldsymbol{\theta})$ is given by

$$D^{(j)}(\boldsymbol{\theta})_{jm} = [(2j)!/(j+m)!(j-m)!]^{1/2} a^{j+m} b^{j-m} \tag{10.3.7}$$

Now, according to Schur's lemma, the representation $\{D^{(j)}(\boldsymbol{\theta})\}$ is irreducible, if a matrix that commutes with all elements of the representation is a constant matrix. To show this, let W be a matrix that commutes with $D^{(j)}(\boldsymbol{\theta})$, i.e.

$$W D^{(j)}(\boldsymbol{\theta}) - D^{(j)}(\boldsymbol{\theta}) W = 0 \tag{10.3.8a}$$

[4] An alternative proof for the unitarity of the transformation $D^{(j)}(S)$ follows from the invariance of the bilinear form (Unsoeld's theorem)

$$\sum_{m=-j}^{j} \phi_m^{(j)\dagger} \phi_m^{(j)} = (\xi_1^{\dagger}\xi_1 + \xi_2^{\dagger}\xi_2)^{2j}/(2j)! = 2^{2j}/(2j)! \tag{10.3.5'}$$

For further symmetry properties of $D^{(j)}(S)$ see the original work of Kim (1969).

For the special case in which $\boldsymbol{\theta} = (0, 0, \theta)$, $D^{(j)}$ is diagonal, as is given by (10.3.6), so that the (m', m)th element of the above equation reduces to

$$(\mathrm{e}^{-\mathrm{i}m\theta} - \mathrm{e}^{\mathrm{i}m'\theta})W_{m'm} = 0$$

which means that W is diagonal, i.e. $W_{m'm} = 0$ for $m' \neq m$. Let $W = \|W_{m'}\delta_{m'm}\|$, then (10.3.8a) takes the form

$$(w_{m'} - w_m)D^{(j)}(\boldsymbol{\theta})_{m'm} = 0 \tag{10.3.8b}$$

Here, we set $m' = j$ and use (10.3.7) to obtain, for a fixed j,

$$(w_j - w_m)(a/b)^m = 0, \qquad \text{for all } m$$

which gives $w_m = w_j$ for all m and the given j. Thus, any matrix W that commutes with the representation $\{D^{(j)}(\boldsymbol{\theta})\}$ is a constant matrix; therefore, the representation is irreducible.

10.3.4 The completeness of the unirreps $\{D^{(j)}(\boldsymbol{\theta}); j = 0, \frac{1}{2}, 1, \ldots\}$

This will be based on the characters of the unirreps. Since the characters depend only on the magnitude θ of the rotation vector, we may set $\boldsymbol{\theta} = (0, 0, \theta)$ in $D^{(j)}(\boldsymbol{\theta})$ and obtain a diagonal matrix as given by (10.3.6). Accordingly, the character $\chi^{(j)}(\theta)$ of $D^{(j)}(\theta)$ is given by

$$\chi^{(j)}(\theta) = \operatorname{tr} D^{(j)}(\theta) = \sum_{m=-j}^{j} \mathrm{e}^{-\mathrm{i}m\theta} = \sin\left[\left(j + \frac{1}{2}\right)\theta\right] \Big/ \sin\left(\frac{\theta}{2}\right) \tag{10.3.9}$$

where $0 \leqslant \theta \leqslant 2\pi$. From this it follows that the group $SU(2)$ has no irreducible representation other than $\{D^{(j)}; j = 0, \frac{1}{2}, 1, \frac{3}{2}, \ldots\}$. Suppose that there exists such a representation and let $f(\theta)$ be its character, then, after multiplication by a density function, it must be orthogonal to all $\chi^{(j)}(\theta)$ and therefore to

$$\chi^{(0)}(\theta) = 1, \qquad \chi^{1/2}(\theta) = 2\cos\left(\tfrac{1}{2}\theta\right)$$

$$\chi^{(j)}(\theta) - \chi^{(j-1)}(\theta) = 2\cos(j\theta), \qquad \text{with } j = 1, \tfrac{3}{2}, \ldots$$

However, such a function $f(\theta)$ must vanish in the region $0 \leqslant \theta \leqslant 2\pi$ according to Fourier's theorem, because if we let $\theta/2 = \phi$ and $l = 2j$ then $\{\cos(j\theta)\} = \cos(l\phi)$; $l = 0, 1, 2, \ldots\}$ forms a complete set of functions in the interval $0 \leqslant \phi \leqslant \pi$.

10.3.5 Orthogonality relations of $D^{(j)}(\boldsymbol{\theta})$

Before formulation of the orthogonality relations of the unirreps $D^{(j)}(\boldsymbol{\theta})$ of $SU(2)$, it may be worthwhile to verify the orthogonality relations of the characters $\chi^{(j)}(\theta)$ given by (10.3.9) via direct integration. As will be shown shortly in (10.3.14), the Hurwitz density function for $SU(2)$ is given by

$$\rho'(\theta) = \sin^2(\theta/2)/(4\pi^2\theta^2), \qquad 0 \leqslant \theta \leqslant 2\pi \tag{10.3.10}$$

which is half of $\rho(\theta)$ of $SO(3, r)$ given by (9.2.11). From this and (10.3.9), the orthogonality relations for the characters $\chi^{(j)}(\theta)$ are shown explicitly as follows:

$$\int_0^{2\pi} \chi^{(j)}(\theta)^* \chi^{(j')}(\theta)\rho'(\theta) 4\pi\theta^2 \, d\theta = \frac{1}{\pi}\int_0^{2\pi} \sin\left[\left(j+\frac{1}{2}\right)\theta\right] \sin\left[\left(j'+\frac{1}{2}\right)\theta\right] d\theta = \delta_{jj'} \tag{10.3.11}$$

Now, let $\boldsymbol{\theta} = (\theta, \vartheta, \varphi)$, where ϑ and φ are the polar angles of $\boldsymbol{\theta}$, and let $D^{(j)}(\boldsymbol{\theta}) = D^{(j)}(\theta, \vartheta, \varphi)$. Then, with use of the density function $\rho'(\theta)$ given by (10.3.10), the orthogonality relations of the unirreps $D^{(j)}(\boldsymbol{\theta})$ of $SU(2)$ are expressed as follows:

$$\frac{1}{4\pi^2}\int_{\theta=0}^{2\pi} \sin^2\left(\frac{\theta}{2}\right) d\theta \int_{\vartheta=0}^{\pi} \sin\vartheta \, d\vartheta \int_{\varphi=0}^{2\pi} D^{(j)}_{\nu\mu}(\theta, \vartheta, \varphi)^* D^{(j')}_{\nu'\mu'}(\theta, \vartheta, \varphi) \, d\varphi = (2j+1)^{-1}\delta_{jj'}\delta_{\nu\nu'}\delta_{\mu\mu'} \tag{10.3.12a}$$

These integrals are over the parameter domain $\Omega' = (0 \leqslant |\boldsymbol{\theta}| \leqslant 2\pi)$ of $SU(2)$. If we use the two-to-one correspondence $\pm S(\boldsymbol{\theta}) \rightarrow R(\boldsymbol{\theta})$ for $\boldsymbol{\theta} \in \Omega = (0 \leqslant |\boldsymbol{\theta}| \leqslant \pi)$, the integration domain Ω' can be reduced to the parameter domain Ω of $SO(3, r)$; in fact, using the homogeneity relation $D^{(j)}(-S(\boldsymbol{\theta})) = (-1)^{2j} D^{(j)}(S(\boldsymbol{\theta}))$, the integral (10.3.12a) is reduced to

$$\text{lhs of (10.3.12a)} = [1 + (-1)^{2j+2j'}] \int_{\theta=0}^{\pi} I^{jj'}_{\nu\nu',\mu\mu'}(\theta) \, d\theta \tag{10.3.12b}$$

where $I(\theta)$ is defined by the integrand of (10.3.12a). Note that the numerical factor $f(j, j') = [1 + (-1)^{2j+2j'}]$ is due to the representation of the kernel $N = \{E, E'\}$ of the two-to-one homomorphism $SU(2) \rightarrow SO(3, r)$. If one of j and j' is an integer and the other is a half integer, then $f(j, j') = 0$ so that both sides of (10.3.12a) are identically zero: $0 = 0$. On the other hand, if both j and j' are either integers or half integers, we have $f(j, j') = 2$, so that the orthogonality relations (10.3.12a) reduce to the integrals over the parameter domain Ω of $SO(3, r)$:

$$\frac{1}{2\pi^2}\int_0^{\pi} \sin^2\left(\frac{\theta}{2}\right) d\theta \int_0^{\pi} \sin\vartheta \, d\vartheta \int_0^{2\pi} D^{(j)}(\theta, \vartheta, \varphi)^*_{\nu\mu} D^{(j')}(\theta, \vartheta, \varphi)_{\nu'\mu'} \, d\varphi = (2j+1)^{-1}\delta_{jj'}\delta_{\nu\mu'}\delta_{\mu\mu'} \tag{10.3.13}$$

Here, the Hurwitz density function is that of $SO(3, r)$ given by (9.2.11). Accordingly, these may be regarded as the orthogonality relations for the unirreps of $SO(3, r) = \{R(\boldsymbol{\theta})\}$ in the parameter space Ω: $0 \leqslant |\boldsymbol{\theta}| \leqslant \pi$. They are identical in form for the vector unirreps and for the double-valued spinor unirreps of $SO(3, r)$. However, they do not apply for relations between vector and spinor unirreps.

10.3.6 The Hurwitz density function for SU(2)

According to (9.2.8a), the density function $\rho'(\boldsymbol{\theta})$ for $SU(2)$ is calculated from a product $S(\boldsymbol{\theta}') = S(\boldsymbol{\theta})S(\epsilon)$ in $SU(2)$, where $\epsilon = (\epsilon_1, \epsilon_2, \epsilon_3)$ is an infinitesimal rotation vector. Since all rotation axes in the parameter domain Ω' are equivalent, $\rho'(\boldsymbol{\theta})$ should depend only on the angle of rotation θ. Thus the product $S(\boldsymbol{\theta}')$ may be calculated through $S(\boldsymbol{\theta}') = S(0, 0, \theta)S(\epsilon)$. Using (10.1.7b) we obtain

$$S(\boldsymbol{\theta}') = \begin{bmatrix} (1 - i\epsilon_3/2)e^{-i\theta/2} & -\frac{1}{2}(i\epsilon_1 + \epsilon_2)e^{-i\theta/2} \\ \frac{1}{2}(-i\epsilon_1 + \epsilon_2)e^{i\theta/2} & (1 + i\epsilon_3/2)e^{i\theta/2} \end{bmatrix}$$

which is correct to the first order in $\epsilon = |\boldsymbol{\epsilon}|$. From the traces of both sides, we have

$$2\cos(\theta'/2) = 2\cos(\theta/2) - \epsilon_3 \sin(\theta/2)$$

with $\theta' = |\boldsymbol{\theta}'|$. This gives, to the first order in ϵ,

$$\theta' = \theta + \epsilon_3$$

The axis-vector $\boldsymbol{n}'$ of the new rotation vector $\boldsymbol{\theta}'$ is determined by applying (10.1.8d) on $S(\boldsymbol{\theta}')$, i.e.

$$(\boldsymbol{\sigma}, \boldsymbol{n}') = (S(\boldsymbol{\theta}')^\dagger - S(\boldsymbol{\theta}'))/[2\mathrm{i}\sin(\theta'/2)]$$

From θ' and $\boldsymbol{n}'$ thus determined, the three components of the rotation vector $\boldsymbol{\theta}' = \theta'\boldsymbol{n}'$ are given by

$$\begin{aligned}
\theta_1' &= (\theta/2)[\epsilon_1 \cot(\theta/2) - \epsilon_2] \\
\theta_2' &= (\theta/2)[\epsilon_2 \cot(\theta/2) + \epsilon_1] \\
\theta_3' &= \theta + \epsilon_3
\end{aligned}$$

Thus, the required Jacobian is given by

$$\begin{aligned}
\frac{\partial(\theta_1', \theta_2', \theta_3')}{\partial(\epsilon_1, \epsilon_2, \epsilon_3)} &= \begin{vmatrix} (\theta/2)\cot(\theta/2) & -\theta/2 & 0 \\ \theta/2 & (\theta/2)\cot(\theta/2) & 0 \\ 0 & 0 & 1 \end{vmatrix} \\
&= \theta^2/[4\sin^2(\theta/2)]
\end{aligned}$$

which is the same as the Jacobian for $SO(3, r)$ given by (9.2.11). Finally, from (9.2.8c), the density function of $SU(2)$ is given by

$$\rho'(\theta) = 4\rho_0' \sin^2(\theta/2)/\theta^2$$

Since the parameter space $\Omega' = (0 \leqslant |\boldsymbol{\theta}| \leqslant 2\pi)$ for $SU(2)$ is given by a sphere of radius 2π, upon normalization via

$$\int \mathrm{d}R = \int_0^{2\pi} \rho'(\theta) 4\pi\theta^2 \,\mathrm{d}\theta = 16\pi^2 \rho_0' = 1$$

we obtain

$$\rho'(\theta) = \sin^2(\theta/2)/[4\pi^2\theta^2]; \qquad 0 \leqslant \theta \leqslant 2\pi \tag{10.3.14}$$

which is half of the density function $\rho(\theta)$ of $SO(3, r)$ given by (9.2.11). This is understood since the radius of the parameter sphere Ω' of $SU(2)$ is twice that of the parameter sphere Ω of $SO(3, r)$.

10.4 The generalized spinors and the angular momentum eigenfunctions

10.4.1 The generalized spinors

A two-component spinor $\psi(\boldsymbol{r})$ that depends on the space coordinate $\boldsymbol{r}$ can be expressed by a linear combination of the elementary spinors ξ_1 and ξ_2 as follows:

$$\psi(\boldsymbol{r}) = \begin{bmatrix} \psi_1(\boldsymbol{r}) \\ \psi_2(\boldsymbol{r}) \end{bmatrix} = \xi_1\psi_1(\boldsymbol{r}) + \xi_2\psi_2(\boldsymbol{r}) \tag{10.4.1}$$

where $\psi_1(\boldsymbol{r})$ and $\psi_2(\boldsymbol{r})$ are simply functions of the space point $\boldsymbol{r}$. A general rotation characterized by a rotation vector $\boldsymbol{\theta}$ affects the position vector $\boldsymbol{r}$ as well as the spinor basis $[\xi_1, \xi_2]$. Therefore, the general spinor transformation may be expressed by

$$\psi(\boldsymbol{r}) \to \psi'(\boldsymbol{r}') = \mathring{S}(\boldsymbol{\theta})\psi(R^{-1}\boldsymbol{r}') = \mathring{S}(\boldsymbol{\theta})\mathring{R}(\boldsymbol{\theta})\psi(\boldsymbol{r}') \tag{10.4.2}$$

where $\boldsymbol{r}' = R(\boldsymbol{\theta})\boldsymbol{r}$. Here the product of two operations $\mathring{S}$ and $\mathring{R}$ is an element of the direct product group $SU(2) \times SO(3, r)$ and their order is irrelevant: they commute with each other because $\mathring{S}$ acts on the spinor basis $[\xi_1, \xi_2]$ whereas $\mathring{R}$ acts on the coefficients $\psi_1(\boldsymbol{r})$ and $\psi_2(\boldsymbol{r})$.

The general spinor transformation given above may be expressed in the form

$$\psi'(\boldsymbol{r}) = \mathring{U}_J(\boldsymbol{\theta})\psi(\boldsymbol{r}) \tag{10.4.3a}$$

with the total rotation operator $\mathring{U}_J(\boldsymbol{\theta})$ defined by

$$\mathring{U}_J(\boldsymbol{\theta}) = \mathring{S}(\boldsymbol{\theta})\mathring{R}(\boldsymbol{\theta}) = \exp(-\mathrm{i}\boldsymbol{J}\cdot\boldsymbol{\theta}); \qquad \boldsymbol{J} = \boldsymbol{L} + \boldsymbol{S} \tag{10.4.3b}$$

where $\boldsymbol{J}$ is the total angular momentum operator in quantum mechanics defined by the orbital angular momentum $\boldsymbol{L} = (1/\mathrm{i})\boldsymbol{r} \times \nabla$ and the spin angular momentum $\boldsymbol{s} = 1/2\boldsymbol{\sigma}$, using $\mathring{R}(\boldsymbol{\theta}) = \exp(-\mathrm{i}\boldsymbol{L}\cdot\boldsymbol{\theta})$ from (6.4.10) and $\mathring{S}(\boldsymbol{\theta}) = \exp(-\mathrm{i}\boldsymbol{s}\cdot\boldsymbol{\theta})$.

The infinitesimal generators of $\mathring{U}_J(\boldsymbol{\theta}) \in SU(2) \times SO(3, r)$ with respect to each component of the rotation vector $\boldsymbol{\theta}$ are given by $-\mathrm{i}\boldsymbol{J} = (-\mathrm{i}J_1, -\mathrm{i}J_2, -\mathrm{i}J_3)$ and their commutation relations are

$$[-\mathrm{i}J_1, -\mathrm{i}J_2] = -\mathrm{i}J_3 \qquad (1, 2, 3 \text{ cyclic}) \tag{10.4.4}$$

which follow from those of $\boldsymbol{L}$ and $\boldsymbol{S}$.

Since there exists the one-to-one correspondence

$$\mathring{U}_J(\boldsymbol{\theta}) = \exp(-\mathrm{i}\boldsymbol{J}\cdot\boldsymbol{\theta}) \leftrightarrow \mathring{S}(\boldsymbol{\theta}) = \exp(-\mathrm{i}\boldsymbol{\sigma}\cdot\boldsymbol{\theta}/2) \tag{10.4.5}$$

we can construct all unirreps of $SU(2)$ either through $\{\mathring{S}(\boldsymbol{\theta})\}$ based on pure spinor bases or through $\{\mathring{U}_J(\boldsymbol{\theta})\}$ based on angular momentum eigenfunctions that diagonalize $\boldsymbol{J}^2$ and J_z. These two approaches will provide identical results if we use the phase convention due to Condon and Shortley (1935), which will be defined in the next section. This means also that we can describe the transformation of the general angular momentum eigenfunctions under a rotation by the unirreps of $SU(2)$ using the pure spinor basis. This is one more reason that $SU(2)$ is referred to simply as the rotation group.

10.4.2 The transformation of the total angular momentum eigenfunctions under the general rotation $\mathring{U}_J$

On account of the commutation relations for the total angular momentum operator $\boldsymbol{J}$ given by (10.4.4), there exists a set of simultaneous eigenfunctions that diagonalizes $\boldsymbol{J}^2$ and J_z:

$$\boldsymbol{J}^2\psi(j, m) = j(j+1)\psi(j, m)$$

$$J_z\psi(j, m) = m\psi(j, m) \tag{10.4.6a}$$

where $m = j, j-1, \ldots, -j$ for a given j that is an integer or a half integer. It is well established that the above set of eigenfunctions $\{\psi(j, m)\}$ is determined by the set of equations

$$J_z\psi(j, m) = m\psi(j, m)$$
$$J_\pm\psi(j, m) = e^{i\delta}[(j \mp m)(j \pm m + 1)]^{1/2}\psi(j, m \pm 1);$$
$$m = j, j-1, \ldots, -j; \qquad j = 0, \tfrac{1}{2}, 1, \tfrac{3}{2}, \ldots \tag{10.4.6b}$$

where $J_\pm = J_x \pm iJ_y$ are the ladder operators and δ is an arbitrary real number. We set $\delta = 0$ following the Condon–Shortley convention. From (10.4.6b), the matrix representation of $\boldsymbol{J} = (J_x, J_y, J_z)$ is completely determined by the set of the eigenfunctions $\{\psi(j, m)\}$ with the convention that $\delta = 0$. Accordingly, we can introduce a representation of the general rotation $\mathring{U}_J = \exp(-i\boldsymbol{J}\cdot\boldsymbol{\theta}) \in SU(2) \times SO(3, r)$ based on the set $\{\psi(j, m)\}$ as follows:

$$\begin{aligned}\mathring{U}_J\psi(j, m) &= \exp(-i\boldsymbol{J}\cdot\boldsymbol{\theta})\psi(j, m)\\ &= \sum_{m'}\psi(j, m')\overline{D}^{(j)}(\boldsymbol{\theta})_{m'm}\end{aligned} \tag{10.4.7}$$

On account of the one-to-one correspondence (10.4.5), the representation $\overline{D}^{(j)}$ also provides the representation of $SU(2)$. In the following, we shall show that this representation $\overline{D}^{(j)}(\boldsymbol{\theta})$ is identical to the unirrep $D^{(j)}(\boldsymbol{\theta})$ defined by (10.3.3) from the pure spinor basis $\phi(j, m) = \phi^j_m(\xi_1, \xi_2)$. If this is the case, we can describe the transformation of angular momentum eigenvectors in quantum mechanics by the representation theory of $SU(2)$. The following theorem holds (Kim 1983b).

Theorem 10.4.1. Under the general rotation $U_J(\boldsymbol{\theta}) = \exp(-i\boldsymbol{J}\cdot\boldsymbol{\theta})$, a set of $2j+1$ total angular momentum eigenfunctions $\{\psi(j, m)\}$ for a system of any number of electrons transforms according to the unirrep $D^{(j)}(\boldsymbol{\theta})$ of $SU(2)$ defined by (10.3.2) provided that the set $\{\psi(j, m)\}$ satisfies the phase convention due to Condon and Shortley.

Proof. Since any set of angular momentum eigenfunctions that satisfies (10.4.6a) or (10.4.6b) with the Condon–Shortley convention defines the same representation $\overline{D}^{(j)}$ defined by (10.4.7), all that is necessary is to construct a special set of angular momentum eigenfunctions that can easily be shown to transform according to the unirrep $D^{(j)}$ of $SU(2)$. For this purpose, we introduce a set of pure spin angular momentum eigenfunctions $\{\Phi(j, m)\}$ that satisfies (10.4.6b): for a system of $2j$ electrons, it is given by a set of symmetrized spinors

$$\Phi(j, m) =$$
$$[(2j)!(j+m)!(j-m)!]^{-1/2}\sum_P P\xi_1(1)\xi_1(2)\ldots\xi_1(j+m)\xi_2(j+m+1)\ldots\xi_2(2j) \tag{10.4.8}$$

where ξ_1 and ξ_2 are the elementary spinors which transform according to (10.3.1a) and the summation is over $(2j)!$ permutations P of the $2j$ electrons $1, 2, \ldots, 2j$. Here the Condon–Shortley convention amounts to the positive sign convention for the square root in (10.4.8). Now, let $s(n)$ $(= \sigma(n)/2)$ be the spin of the nth electron and S be the total spin angular momentum of $2j$ electrons. Then direct application of $S_z = \sum_n s_z(n)$ and of the ladder operators $S_\pm = \sum_n s_\pm(n)$ to $\Phi(j, m)$ with use of $s_+(n)\xi_1(n) = 0$, $s_+(n)\xi_2(n) = \xi_1(n)$, $s_-(n)\xi_1(n) = \xi_2(n)$ and $s_-(n)\xi_2(n) = 0$, leads to

$$\mathring{S}_z \Phi(j, m) = m\Phi(j, m)$$

$$\mathring{S}_\pm \Phi(j, m) = [(j \mp m)(j \pm m + 1)]^{1/2} \Phi(j, m \pm 1)$$

$$m = j, j-1, \ldots, -j; \qquad j = 0, \tfrac{1}{2}, 1, \ldots \tag{10.4.9}$$

which verifies that the set $\{\Phi(j, m)\}$ indeed satisfies (10.4.6b) and hence (10.4.6a) with $J = S$, in the pure spinor space. Comparison of the basis $\{\Phi(j, m)\}$ with the basis $\{\phi(j, m)\}$ of the unirrep $D^{(j)}(\theta)$ of $SU(2)$ defined by (10.3.1b) shows that the transformation of the set $\{\Phi(j, m)\}$ under

$$\mathring{U}_s = \exp(-\mathrm{i}S \cdot \theta) = \mathrm{e}^{-\mathrm{i}s(1)\cdot\theta} \mathrm{e}^{-\mathrm{i}s(2)\cdot\theta} \ldots \mathrm{e}^{-\mathrm{i}s(2j)\cdot\theta}$$

is exactly the same as the transformation of $\{\phi(j, m)\}$ under $\mathring{S}(\theta) = \exp(-\mathrm{i}s \cdot \theta)$ since each factor both in $\Phi(j, m)$ and in $\Phi(j, m)$ transforms according to (10.3.1a). Q.E.D.

10.4.3 The vector addition model

Consider two electronic systems characterized by the angular momentum quantum numbers (j_1, m_1) and (j_2, m_2), respectively. Then the allowed set of the quantum numbers (j, m) for the resultant angular momentum of the coupled system is given by

$$m = m_1 + m_2 = j, j-1, \ldots, -j$$

$$j = j_1 + j_2, j_1 + j_2 - 1, \ldots, |j_1 - j_2| \tag{10.4.10}$$

This rule is called *the vector addition model* or *the triangular rule*: it can be understood by regarding a quantum number as a vector; e.g. $\boldsymbol{j} = (j, m)$. It is of basic importance for spectroscopy. The two systems to be combined need not consist of single electrons, but could themselves be composite systems. It applies even to the composition of the total electronic quantum number and the nuclear spin, etc. In the following, we shall give the group theoretical proof of the vector addition model.

Theorem 10.4.2. Let $D^{(j_1)}$ and $D^{(j_2)}$ be two unirreps of the rotation group $SU(2)$, then the direct product representation $D^{(j_1)} \times D^{(j_2)}$ is simply reducible as follows:

$$D^{(j_1)} \times D^{(j_2)} = D^{(j_1+j_2)} \oplus D^{(j_1+j_2-1)} \oplus \cdots \oplus D^{(|j_1-j_2|)} \tag{10.4.11a}$$

Proof. From their characters $\chi^{(j_1)}$ and $\chi^{(j_2)}$ given by (10.3.9), we have

$$\chi^{(j_1)}\chi^{(j_2)} = \sum_{m_2=-j_2}^{j_2} \mathrm{e}^{-\mathrm{i}m_2\theta} \sum_{m_1=-j_1}^{j_1} \mathrm{e}^{-\mathrm{i}m_1\theta}$$

Set $m = m_1 + m_2$, and assume $j_1 \geqslant j_2$ without loss of generality. Then

$$\chi^{(j_1)}\chi^{(j_2)} = \sum_{m_2=-j_2}^{j_2} \sum_{m=-j_1+m_2}^{j_1+m_2} e^{-im\theta}$$

$$= \sum_{m_2=-j_2}^{j_2} \sum_{m=-j_1-m_2}^{j_1+m_2} e^{-im\theta} = \sum_{m_2=-j_2}^{j_2} \chi^{(j_1+m_2)}$$

$$= \chi^{(j_1+j_2)} + \chi^{(j_1+j_2-1)} + \cdots + \chi^{(j_1-j_2)} \tag{10.4.11b}$$

Here, in the second step, the lower limit (only the lower limit) of the sum over m is changed from $-j_1 + m_2$ to $-j_1 - m_2$: this is allowed because each $\exp(-im\theta)$ term in the sum depends only on m while m_2 takes positive as well as negative values with equal frequencies. It changes the summation domain for (m_2, m) from a parallelogram to a trapezoid (see Figure 10.2) in such a way that it simply rearranges the terms $\exp(-im\theta)$ in the sum.

10.4.4 The Clebsch–Gordan coefficients

Let $\{\psi^{j_1}_{m_1}\}$ and $\{\psi^{j_2}_{m_2}\}$ be the bases (or the angular momentum eigenfunctions) belonging to the unirreps $D^{(j_1)}$ and $D^{(j_2)}$, respectively. From Theorem 10.4.2, it follows that the set of resultant angular momentum eigenfunctions $\{\psi^{j}_{m}\} \in D^{(j)}$ formed by the linear combinations of the direct product basis functions $\{\psi^{j_1}_{m_1}\psi^{j_2}_{m_2}\}$ for a given pair $\{j_1, j_2\}$ is expressed by

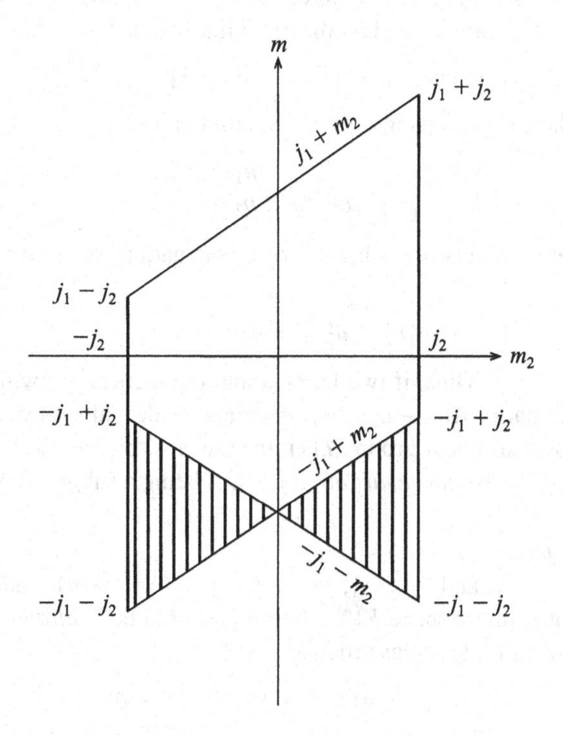

Figure 10.2. Change of the summation domain of (m_2, m) from a parallelogram to a trapezoid.

$$\psi_m^j = \sum_{m_1+m_2=m} C^j_{m_1 m_2} \psi^{j_1}_{m_1} \psi^{j_2}_{m_2} \tag{10.4.12a}$$

where the allowed values of j and m are

$$j = j_1 + j_2, j_1 + j_2 - 1, \ldots, |j_1 - j_2|; \qquad m = j, j-1, \ldots, -j$$

This is called *the Clebsch–Gordan (CG) series.* The CG *coefficients* $\{C^j_{m_1,m_2}\}$ for a given m can be chosen to be an orthogonal matrix in which j occurs as the column index and m_1 or m_2 as the row index for a given m. Thus the series is converted into

$$\psi^{j_1}_{m_1} \psi^{j_2}_{m_2} = \sum_j C^j_{m_1,m_2} \psi^j_m; \qquad (m_1 + m_2 = m) \tag{10.4.12b}$$

In the following, we shall discuss how to calculate CG coefficients.

The problem of constructing the eigenfunctions of the resulting angular momentum by coupling two angular momentum eigenfunctions is equivalent to the problem of constructing the *spinor invariant* with respect to $SU(2)$, which corresponds to the eigenfunction with zero resulting angular momentum formed by the linear combinations of the spinor basis functions belonging to the direct product representation $D^{(j_1)} \times D^{(j_2)} \times D^{(j_3)}$ of $SU(2)$. As a preparation, we shall discuss the spinor invariants, for which the notions of *covariance* and *contravariance* play the fundamental roles (van der Waerden 1974).

10.4.4.1 Covariance

Let $u = [u_1, u_2]$ and $v = [v_1, v_2]$ be two different bases in a two-dimensional vector space $V^{(2)}$ (e.g. belonging to two different systems). Suppose that these vectors are *covariant*, i.e. they are transformed by the same linear transformation T such that

$$[u'_1, u'_2] = [u_1, u_2]T, \qquad [v'_1, v'_2] = [v_1, v_2]T$$

then, by uniting these equations into a single matrix equation

$$\begin{bmatrix} u'_1 & u'_2 \\ v'_1 & v'_2 \end{bmatrix} = \begin{bmatrix} u_1 & u_2 \\ v_2 & v_2 \end{bmatrix} T$$

and taking the determinants of both sides of this equation, we arrive at an invariance under T:

$$u'_1 v'_2 - u'_2 v'_1 = u_1 v_2 - u_2 v_1 \tag{10.4.13}$$

provided that $\det T = 1$. Thus, if two bases u and v are covariant with respect to each other, the determinant $u_1 v_2 - u_2 v_1$ is invariant under all linear transformations belonging to the special linear group $SL(2)$. In cases in which u and v are spinor bases, the invariant is called a *spinor invariant*, for $SU(2)$ being a subgroup of $SL(2)$.

10.4.4.2 Contravariance

Let $x = [x_1, x_2, \ldots, x_n]$ and $y = [y_1, y_2, \ldots, y_n]$ be two linearly independent vectors in the n-dimensional linear space $V^{(n)}$. These are said to be *contravariant* with respect to each other if their simple scalar product

$$(x, y) = x_1 y_1 + x_2 y_2 + \cdots + x_n y_n$$

is invariant under the linear transformations $x' = xA$ and $y' = yB$ in $V^{(n)}$. The invariance requires

$$(x', y') = (xA, yB) = (x, AB^{\sim} y) = (x, y)$$

where $B^{\sim}$ is the transpose of B. Hence

$$AB^{\sim} = 1 \qquad \text{or} \qquad B = A^{\sim -1} = A^{\#} \tag{10.4.14}$$

The *transpose–inverse* $A^{\sim -1}$ of a matrix A is called the *contravariant* matrix of A and denoted by $A^{\#}$.

For example, for an orthogonal matrix R, the contravariant matrix is R itself, i.e. $R^{\#} = R$. For a unitary matrix U, the contravariant matrix is the complex conjugate, i.e. $U^{\#} = U^*$. This is consistent with the fact that a simple scalar product (x, y) is invariant under an orthogonal transformation whereas a Hermitian scalar product $\langle x, y\rangle = (x^*, y)$ is invariant under a unitary transformation.

Following van der Waerden (1974), we shall first discuss the spinor invariants of a two-spin system, for its simple extension leads to the invariants of the three-spin system. Let $u = [u_1, u_2]$ and $v = [v_1, v_2]$ be two elementary spinor bases (e.g. belonging to different systems) which transform according to $SU(2)$. Then, from (10.4.13), we have the monomial invariants under $SU(2)$

$$G_j = (u_1 v_2 - u_2 v_1)^{2j}/(2j)!; \qquad j = 0, \tfrac{1}{2}, 1, \ldots$$

Expansion of this yields

$$\begin{aligned} G_j &= \sum_m \frac{u_1^{j+m} u_2^{j-m}}{[(j+m)!(j-m)!]^{1/2}} \frac{(-1)^{j-m} v_1^{j-m} v_2^{j+m}}{[(j-m)!(j+m)!]^{1/2}} \\ &= \sum_m \phi^j_m(u) \tilde{\phi}^j_m(v) \end{aligned} \tag{10.4.15}$$

where $\phi^j_m(u)$ is the monomial basis of the unirrep $D^{(j)}(S)$ of $SU(2)$ defined by (10.3.1b), i.e.

$$\phi^j_m(u) = u_1^{j+m} u_2^{j-m}/[(j+m)!(j-m)!]^{1/2} \tag{10.4.16a}$$

while $\tilde{\phi}^j_m(v)$ is a basis of the contravariant representation $D^{(j)}(S)^*$ defined by the monomial basis $\phi^j_m(u) \in D^{(j)}(S)$ as follows:

$$\tilde{\phi}^j_m = (-1)^{j-m} \phi^j_{-m} \in D^{(j)}(S)^* \tag{10.4.16b}$$

This is so because the invariant G_j is a simple scalar product of two vectors $\phi^j_m(u)$ and $\tilde{\phi}^j_m(v)$ in the $(2j+1)$-dimensional space. It follows from (10.4.16b) that

$$D^{(j)}_{m'm}(S)^* = (-1)^{m'-m} D^{(j)}_{-m',-m}(S) \tag{10.4.16c}$$

which shows the well-known fact that the complex conjugate representation $D^{(j)}(S)^*$ is equivalent to the original representation $D^{(j)}(S)$ (cf. Equation (16.5.5c)).

For a three-spin system, let $u = [u_1, u_2]$, $v = [v_1, v_2]$ and $w = [w_1, w_2]$ be their elementary spinor bases, then the three determinants

$$\delta_1 = v_1 w_2 - v_2 w_1, \qquad \delta_2 = w_1 u_2 - w_2 u_1, \qquad \delta_3 = u_1 v_2 - u_2 v_1 \tag{10.4.17a}$$

are invariant under $S \in SU(2)$ and so are their monomials:

$$F_k = \delta_2^{k_1} \delta_2^{k_2} \delta_3^{k_3}/(k_1! k_2! k_2!) \tag{10.4.17b}$$

where $k = \{k_1, k_2, k_3\}$ is a set of any non-negative integers. Next, we shall convert this invariant into a linear combination of the products of three monomial bases $\phi^{j_1}_{m_1}(u)$, $\phi^{j_2}_{m_2}(v)$ and $\phi^{j_3}_{m_3}(w)$. For this purpose, we expand F_k via the binomial theorem

$$F_k = \sum_{\nu_1,\nu_2,\nu_3} (-1)^{\nu_1+\nu_2+\nu_3} \binom{k_1}{\nu_1}\binom{k_2}{\nu_2}\binom{k_3}{\nu_3} \times u_1^{k_3-\nu_3+\nu_2} u_2^{k_2-\nu_2+\nu_3} v_1^{k_1-\nu_1+\nu_3} v_2^{k_3-\nu_3+\nu_1} w_1^{k_2-\nu_2+\nu_1} w_2^{k_1-\nu_1+\nu_2} \tag{10.4.18}$$

and set

$$\begin{aligned} &k_3-\nu_3+\nu_2 = j_1+m_1, \qquad k_1-\nu_1+\nu_3 = j_2+m_2, \qquad k_2-\nu_2+\nu_1 = j_3+m_3 \\ &k_2-\nu_2+\nu_3 = j_1-m_1, \qquad k_3-\nu_3+\nu_1 = j_2-m_2, \qquad k_1-\nu_1+\nu_2 = j_3-m_3 \end{aligned} \tag{10.4.19a}$$

then $k_1+k_2+k_3 = j_1+j_2+j_3$, so that

$$\begin{aligned} &k_1 = j_2+j_3-j_1 \geqslant 0, \qquad k_2 = j_1+j_3-j_2 \geqslant 0, \qquad k_3 = j_1+j_2-j_3 \geqslant 0; \\ &m_1+m_2+m_3 = 0 \end{aligned} \tag{10.4.19b}$$

These reconfirm the triangular rule or the vector addition model for the quantum numbers $\{j_i, m_i\}$. Now, we change the summation variables (ν_1, ν_2, ν_3) of (10.4.18) to $(m_1, m_2, \nu = \nu_3\}$, expressing ν_1 and ν_2 by

$$\nu_1 = j_3 - j_1 - m_2 + \nu, \qquad \nu_2 = j_3 - j_2 + m_1 + \nu \tag{10.4.20}$$

with use of the first two equations of (10.4.19a) and (10.4.19b). Then F_k is rewritten in the desired form

$$F_k = \sum_{m_1,m_2} B^{j_3}_{m_1,m_2} \phi^{j_1}_{m_1}(u)\phi^{j_2}_{m_2}(v)\phi^{j_3}_{m_3}(w) \tag{10.4.21}$$

where $m_3 = -m_1 - m_2$ and the summation over ν is absorbed in the linear coefficients B, as will be shown in (10.4.23b).

Now, since F_k is a spinor invariant, the coefficients of $\phi^{j_3}_{m_3}(w)$ in (10.4.21) must form a basis $\tilde{\phi}^{j_3}_{m_3}(u, v)$ of the complex conjugate representation $D^{(i)*}$. Thus, from (10.4.16b), we obtain

$$\phi^{j_3}_{-m_3}(u, v) = \alpha(j_3)(-1)^{j_3-m_3} \sum_{m_1+m_2=-m_3} B^{j_3}_{m_1\,m_2} \phi^{j_1}_{m_1}(u)\phi^{j_2}_{m_2}(v) \tag{10.4.22}$$

where we have introduced a numerical factor $\alpha(j_3)$ for the given j_3 which will be chosen to make the transformation orthogonal. Now we rewrite the above in the form of (10.4.12a), identifying $j = j_3$ and $m = -m_3$:

$$\phi^{j}_{m}(u, v) = \sum_{m_1+m_2=m} C^{j}_{m_1,m_2} \phi^{j_1}_{m_1}(u)\phi^{j_2}_{m_2}(v) \tag{10.4.23a}$$

Since the basis vectors are all orthonormalized, the CG coefficients for a given m form a unitary matrix in which j occurs as the column index and m_1 or m_2 as the row index. Let $C^{j}_{m_1,m_2} \equiv \alpha_j c^{j}_{m_1,m_2}$ with $\alpha_j = \alpha(j)$, then $c^{j}_{m_1,m_2}$ are given by

Table 10.1. *The Clebsch–Gordan coefficients for* $j_2 = \frac{1}{2}$; $C^j_{m-m_2,m_2} = \alpha_j c^j_{m-m_2,m_2}$

$j_2 = \frac{1}{2}$	$m_2 = \frac{1}{2}$	$m_2 = -\frac{1}{2}$
$j = j_1 + \frac{1}{2}$	$\left(\dfrac{j_1 + m + \frac{1}{2}}{2j_1 + 1}\right)^{\frac{1}{2}}$	$\left(\dfrac{j_1 - m + \frac{1}{2}}{2j_1 + 1}\right)^{\frac{1}{2}}$
$j = j_1 - \frac{1}{2}$	$-\left(\dfrac{j_1 - m + \frac{1}{2}}{2j_1 + 1}\right)^{\frac{1}{2}}$	$\left(\dfrac{j_1 + m + \frac{1}{2}}{2j_1 + 1}\right)^{\frac{1}{2}}$

$$\begin{aligned} c^j_{m_1,m_2} &= (-1)^{j_3+m}(-1)^{k_1} B^j_{m_1,m_2} \\ &= \sum_\nu (-1)^\nu \frac{[(j_1+m_2)!(j_1-m_1)!(j_2+m_2)!(j_2-m_2)!]^{1/2}}{(j_2+m_2-\nu)!(j-j_1-m_2+\nu)!(j_1-m_1-\nu)!} \\ &\quad \times \frac{[(j+m)!(j-m)!]^{1/2}}{(j-j_2+m_1+\nu)!\nu!(j_1+j_2-j-\nu)!}; \qquad (m = m_1 + m_2) \end{aligned} \tag{10.4.23b}$$

where the limits of summation are determined by the relation that $1/(-N)! = 0$ for a positive integer N. Thus, for example, if $j_1 + j_2 = j$ then there remains only one term with $\nu = 0$ in the sum. Since $c^j_{m_1,m_2}$ are real, we choose α_j real. Then the unitary matrix $\|\alpha_j c^j_{m-m_2,m_2}\|$ becomes an orthogonal matrix if α_j is determined by the condition

$$\alpha_j^2 \sum_{m_2} (c^j_{m-m_2,m_2})^2 = 1$$

The calculation of α_j from this relation is elementary but rather involved (see Hamermesch (1962)). We simply write down the result

$$\alpha_j = \left(\frac{(2j+1)k_1!k_2!k_3!}{(j+j_1+j_2+1)!}\right)^{1/2} > 0 \tag{10.4.23c}$$

where $k_i = j_1 + j_2 + j_3 - 2j_i$; $i = 1, 2, 3$ as defined by (10.4.19b). This result can be obtained painlessly if we use the Schwinger representation or the Bargmann representation of the spin states (see Biedenharn and Van Dam (1965)). The transformation matrix is related to Dirac's bracket symbol by

$$\langle j_1 m_1 j_2 m_2 | jm \rangle = \alpha_j c^j_{m-m_2,m_2}$$

In Table 10.1 we have given the Clebsch–Gordan coefficients $C^j_{m-m_2,m_2} = \alpha_j c^j_{m-m_2,m_2}$ for $j_2 = 1/2$; cf. Wigner (1962).

10.4.5 The angular momentum eigenfunctions for one electron

When $\boldsymbol{J}$ equals the orbital angular momentum $\boldsymbol{L}$, the eigenfunctions defined by (10.4.6a) are described by a set of lth-order spherical harmonics $\{Y_{l,m}(\vartheta, \varphi); m = 0, \pm 1, \ldots, \pm l\}$, where ϑ and φ are the polar angles of the spatial coordinates $\boldsymbol{r}$

of the electron (see Wigner (1962)). Their parities are determined by l, i.e. under the inversion I which brings $\boldsymbol{r}$ to $-\boldsymbol{r}$ (or ϑ and φ to $\pi-\vartheta$ and $\pi+\varphi$, respectively) we have

$$IY_{l,m}(\vartheta, \varphi) = (-1)^l Y_{l,m}(\vartheta, \varphi); \qquad m = 0, \pm 1, \ldots, \pm l \tag{10.4.24}$$

Using Table 10.1, we may write down the eigenfunction $\psi(j, m; l, \frac{1}{2})$ by coupling a spherical harmonic $Y_1(\vartheta, \varphi)$ and the elementary spinor $[\xi_1, \xi_2]$ as follows:

$$\begin{aligned}\psi\left(l+\frac{1}{2}, m; l, \frac{1}{2}\right) &= \left(\frac{1}{2}+\frac{m}{2l+1}\right)^{1/2} Y_{l,m-1/2}\xi_1 + \left(\frac{1}{2}-\frac{m}{2l+1}\right)^{1/2} Y_{l,m+1/2}\xi_2\\ \psi\left(l-\frac{1}{2}, m; l, \frac{1}{2}\right) &= -\left(\frac{1}{2}-\frac{m}{2l+1}\right)^{1/2} Y_{l,m-1/2}\xi_1 + \left(\frac{1}{2}+\frac{m}{2l+1}\right)^{1/2} Y_{l,m+1/2}\xi_2\end{aligned} \tag{10.4.25}$$

Since these are linear with respect to ξ_1 and ξ_2, they change the sign after a 2π rotation; however, their parities are determined solely by l:

$$I\psi\left(l \pm \tfrac{1}{2}, m; l, \tfrac{1}{2}\right) = (-1)^l \psi\left(l \pm \tfrac{1}{2}, m; l, \tfrac{1}{2}\right) \tag{10.4.26a}$$

Note that I does not act on ξ_1 and ξ_2 simply because these are independent of the spatial coordinates.

Now, for a given half integral j, we have $l = j \pm 1/2$, which is either even or odd, so that the eigenfunction $\psi(j, m)$ may be further classified by its parity, i.e. $\psi_{\mathrm{g}}(j, m)$ and $\psi_{\mathrm{u}}(j, m)$, where g (u) stands for *gerade* (*ungerade*) such that

$$I\psi_{\mathrm{g}}(j, m) = \psi_{\mathrm{g}}(j, m), \qquad I\psi_{\mathrm{u}}(j, m) = -\psi_{\mathrm{u}}(j, m) \tag{10.4.26b}$$

that is, $\psi_{\mathrm{g}}(j, m)$ stands for a wave vector with an even spatial part whereas $\psi_{\mathrm{u}}(j, m)$ stands for a wave vector with an odd spatial part. These spinors $\psi_{\mathrm{g}}(j, m)$ and $\psi_{\mathrm{u}}(j, m)$ provide the basis functions for the direct product group $SU(2) \times O(3, r)$, where $O(3, r)$ is the full rotation group defined by $SO(3, r) \times C_{\mathrm{i}}$, where C_{i} is the group of inversion. The *gerade* and *ungerade* eigenfunctions for the first few l are easily written down from (10.4.25): for $l = 0$ and $j = \frac{1}{2}$

$$\psi_{\mathrm{g}}\left(\tfrac{1}{2}, \tfrac{1}{2}\right) = \xi_1, \qquad \psi_{\mathrm{g}}\left(\tfrac{1}{2}, -\tfrac{1}{2}\right) = \xi_2 \tag{10.4.27a}$$

for $l = 1$ and $j = \frac{1}{2}$

$$\begin{aligned}\psi_{\mathrm{u}}\left(\tfrac{1}{2}, \tfrac{1}{2}\right) &= \left(-Y_{1,0}\xi_1 + \sqrt{2}Y_{1,1}\xi_2\right)/\sqrt{3}\\ \psi_{\mathrm{u}}\left(\tfrac{1}{2}, -\tfrac{1}{2}\right) &= \left(-\sqrt{2}Y_{1,-1}\xi_1 + Y_{1,0}\xi_2\right)/\sqrt{3}\end{aligned} \tag{10.4.27b}$$

11

Single- and double-valued representations of point groups

On account of the homomorphism between $SU(2) = \{\pm S(\boldsymbol{\theta})\}$ and $SO(3, r) = \{R(\boldsymbol{\theta})\}$, there exist homomorphisms between their subgroups. In fact, one can define a proper double point group $P' \leqslant SU(2)$ that is homomorphic to an ordinary proper point group $P \leqslant SO(3, r)$ via a two-to-one correspondence. Since P' is a subgroup of $SU(2)$, its unirreps of P' may be obtained by reducing the subduced representations of the unirreps $\{D^{(j)}\}$ of $SU(2)$ onto P' (Kim 1981d). Then a unirrep of P' provides a single-valued (vector) unirrep or a double-valued (spinor) unirrep of P, analogously to (10.3.4b). The complication involved with the double-valued representations of the point groups shall be controlled by regarding them as 'projective representations' of P (Brown 1970, Altmann 1979, Kim 1981b). The discussion developed here may serve as a prototype example for more general discussions on the projective representations of a group in general, which will be discussed in Chapter 12.

11.1 The double-valued representations of point groups expressed by the projective representations

11.1.1 The projective set of a point group

Let $P = \{R(\boldsymbol{\theta}_i)\} \leqslant SO(3, r)$ be a proper point group, then its *double group* is defined by $P' = \{\pm S(\boldsymbol{\theta}_i)\} \leqslant SU(2)$ with the same set of rotation vectors $\{\boldsymbol{\theta}_i\}$ as those in the parameter space Ω of $SO(3, r)$ $(0 \leqslant |\boldsymbol{\theta}_i| \leqslant \pi)$. According to Theorem 10.2.2, the double group P' thus defined is homomorphic to P through the homomorphism relations (10.2.6b), i.e.

$$S(\boldsymbol{\theta}_1)S(\boldsymbol{\theta}_2) = \pm S(\boldsymbol{\theta}_3), \qquad \text{if } R(\boldsymbol{\theta}_1)R(\boldsymbol{\theta}_2) = R(\boldsymbol{\theta}_3) \tag{11.1.1}$$

where the $\pm$ signs depend on $S(\boldsymbol{\theta}_1)$ and $S(\boldsymbol{\theta}_2)$. The question of when to take the $+$ and when the $-$ sign arises. To answer this question, we form a subset $\check{P} < P'$ of the order $|\check{P}| = |P|$ whose elements are chosen from P' by arbitrarily assigning one of $\pm S(\boldsymbol{\theta}_i)$ to each $R(\boldsymbol{\theta}_i)$ of P. Then the set $\check{P}$ is in one-to-one correspondence with P. Let S_i be an element of $\check{P}$ corresponding to an element $R_i \in P$, then the homomorphism relation is described by

$$S_1 S_2 = f(S_1, S_2) S_3, \qquad \text{if } R_1 R_2 = R_3 \tag{11.1.2}$$

where $f(S_1, S_2)$ is a numerical factor defined by

$$f(S_1, S_2) = \begin{cases} 1, & \text{if } S_1 S_2 \in \check{P} \\ -1, & \text{if } S_1 S_2 \notin \check{P} \end{cases}$$

The set of $|P|^2$ numerical factors $\{f(S_1, S_2)\}$ is called the *factor system* and the set $\check{P}$ is called a *projective set* of P belonging to the factor system. By the one-to-one correspondence $\check{P} \leftrightarrow P$, the double group P' becomes a projective representation of P belonging to the factor system $\{f(S_1, S_2)\}$. For simplicity, we assume that $\check{P}$ contains the unit element $E = S(\theta = 0)$ of P', then $f(E, E) = 1$; in which case, the system is called the *standard factor system*.

Obviously there are many ways of choosing a projective set $\check{P}$ of P from the double group P'. However, it is most desirable to choose the set $\check{P}$ which preserves the class structure of P, i.e. if $R_1 \sim R_2$ in P, then $S_1 \sim S_2$ in $\check{P}$. (Here $\sim$ denotes the equivalence relative to the respective group.) On account of Theorem 10.2.1, one of the simplest projective sets which preserves the class structure of a given point group $P = \{R(\boldsymbol{\theta}_i)\}$, namely $\check{P} = \{S(\boldsymbol{\theta}_i)\}$, may be given by the following one-to-one correspondence:

$$R(\boldsymbol{\theta}_i) \leftrightarrow S(\boldsymbol{\theta}_i); \qquad 0 \leqslant |\boldsymbol{\theta}_i| \leqslant \pi \tag{11.1.3}$$

through the common rotation vectors with due caution when choosing equivalent axis-vectors for equivalent binary rotations if they are one-sided.[1] For a two-sided binary rotation $R(\pi\boldsymbol{n})$, utilizing the freedom of choosing either $+\boldsymbol{n}$ or $-\boldsymbol{n}$ as the axis-vector, we may choose the rotation vectors on the positive hemisphere; a domain defined by

$$\pi n_z > 0 \quad \text{or} \quad \pi n_z = 0, \quad \pi n_x \geqslant 0 \quad \text{or} \quad \pi n_z = \pi n_x = 0, \quad \pi n_y = \pi \tag{11.1.4}$$

This convention is due to Altmann (1979). It can be applied to the point groups $SO(3, r)$, T, O, Y and D_n with an even n, but not with an odd n, for which any binary axis is one-sided. The conventions (11.1.3) and (11.1.4) for choosing the projective set $\check{P}$ for a point group P can easily be extended to an improper point group because the inversion operator I commutes with any point operation so that, if $\boldsymbol{\theta}_1 \sim \boldsymbol{\theta}_2$, then $IR(\boldsymbol{\theta}_1) \sim IR(\boldsymbol{\theta}_2)$.

11.1.2 The orthogonality relations for projective unirreps

Any unirrep of a double point group $P' = \{\pm S(\theta_i)\}$ can be obtained by reducing the subduced representation from a unirrep $D^{(j)}$ of $SU(2)$ defined by (10.3.2), since P' is a subgroup of $SU(2)$. Let a unirrep of P' thus obtained be denoted $D^{(\alpha,j)}$, where we have introduced an additional superscript j to describe the homogeneity property

$$D^{(\alpha,j)}(-S) = (-1)^{2j} D^{(\alpha,j)}(S) \tag{11.1.5}$$

in accordance with the homogeneity relation (10.3.4a) of $D^{(j)}$. Then, the homomorphism relation (11.1.2) is expressed in terms of $D^{(\alpha,j)}$ as follows:

$$D^{(\alpha,j)}(S_1)D^{(\alpha,j)}(S_2) = [f(S_1, S_2)]^{2j} D^{(\alpha,j)}(S_3) \tag{11.1.6}$$

where $S_i = S(\boldsymbol{\theta}_i)$. Since $f(S_1, S_2) = \pm 1$ from (11.1.2), we have

$$[f(S_1, S_2)]^{2j} = \begin{cases} 1; & j = \text{integer} \\ f(S_1, S_2); & j = \text{half integer} \end{cases}$$

Thus, through the one-to-one correspondence $S(\boldsymbol{\theta}_i) \leftrightarrow R(\boldsymbol{\theta}_i)$, the set $\{D^{(\alpha,j)}(S_i) = P^{(\alpha,j)}(S(\boldsymbol{\theta}_i))\}$ provides a projective representation of $P = \{R(\boldsymbol{\theta}_i)\} \leqslant SO(3, r)$ belong-

[1] For one-sided binary rotations, the equivalence $R(\pi\boldsymbol{n}_1) \sim R(\pi\boldsymbol{n}_2)$ leads to $S(\pi\boldsymbol{n}_1) \sim S(\pi\boldsymbol{n}_2)$ by the choice $\pi\boldsymbol{n}_1 \sim \pi\boldsymbol{n}_2$, which, however, is automatically satisfied if the binary rotations are two-sided.

ing to a factor system $\{[f(S_1, S_2)]^{2j}\}$: the set becomes a vector representation of P for an integral j, whereas for a half-integral j, it is a projective representation of P belonging to the factor system $\{f(S_1, S_2)\}$ defined by (11.1.2). For simplicity the former case is referred to as *an integral representation* of P whereas the latter case is referred to as *a half-integral representation*, analogously to (10.3.4b). Needless to say, the set $\{D^{(\alpha,j)}(S_i), D^{(\alpha,j)}(-S_i)\}$ provides a vector representation of the double group $P' = \{\pm S_i\}$ for any value of j.

The orthogonality relations of the projective unirreps of a point group $P = \{R_i\} < SO(3, r)$ follow from those of the vector unirreps of its double group $P' = \{\pm S_i\}$. Using $D^{(\alpha,j)}(-S_i) = (-1)^{2j} D^{(\alpha,j)}(S_i)$, they may be expressed as

$$[1 + (-1)^{(2j+2j')}] \sum_i D^{(\alpha,j)}_{\nu\mu}(S(\boldsymbol{\theta}_i))^* D^{(\alpha',j')}_{\nu'\mu'}(S(\boldsymbol{\theta}_i)) = (|P'|/d_\alpha)\delta_{jj'}\delta_{\alpha\alpha'}\delta_{\nu\nu'}\delta_{\mu\mu'} \qquad (11.1.7)$$

where $\boldsymbol{\theta}_i \in \Omega$; $|P'| = 2|P|$ is the order of P' and d_α is the dimensionality of $D^{(\alpha,j)}$. If one of j and j' is an integer and the other is a half integer, then $[1 + (-1)^{2j+2j'}] = 0$ and $\delta_{jj'} = 0$ so that both sides of (11.1.7) are identically zero: $0 = 0$. This corresponds to the case in which two projective unirreps of $P = \{R_i\}$ belong to the different factor systems. On the other hand, if both unirreps belong to the same factor system (i.e. both j and j' are integers or half integers), we have $[1 + (-1)^{2j+2j'}] = 2$ so that the orthogonality relations (11.1.7) take the form

$$\sum_i D^{(\alpha,j)}_{\nu\mu}(\boldsymbol{\theta}_i)^* D^{(\alpha',j')}_{\nu'\mu'}(\boldsymbol{\theta}_i) = (|P|/d_\alpha)\delta_{\alpha\alpha'}\delta_{\nu\nu'}\delta_{\mu\mu'} \qquad (11.1.8)$$

which are identical in form to the case of the vector unirreps of the point group $P = \{R(\boldsymbol{\theta}_i)\} \leqslant SO(3, r)$ given by (6.5.11a). These correspond to the orthogonality relations for the single- and double-valued representations of $SO(3, r)$ given by (10.3.13).

Remark 1. For later use, we shall introduce the concept of the one- or two-sidedness of a spinor transformation. This is possible because of Theorem 10.2.1. A spinor transformation $S(\boldsymbol{\theta}) \in P'$ is said to be two-sided (one-sided) with respect to P' if the corresponding axis of rotation $R(\boldsymbol{\theta})$ is two-sided (one-sided) with respect to the homomorph P. Thus, if $S(\boldsymbol{\theta})$ is two-sided with respect to P', then there exists a 'binary' element $S(\pi\boldsymbol{n})$ in P' such that

$$S(\pi\boldsymbol{n})S(\boldsymbol{\theta})S(\pi\boldsymbol{n})^{-1} = S(-\boldsymbol{\theta}) = S(\boldsymbol{\theta})^{-1} \qquad (11.1.9)$$

where $\boldsymbol{n} \perp \boldsymbol{\theta}$. This means that the class of $S(\boldsymbol{\theta})$ in P' is ambivalent if and only if $S(\boldsymbol{\theta})$ is two-sided with respect to P'. An analogous statement holds for the point group P with the exception that the class of a binary rotation $R(\pi\boldsymbol{h})$ is always ambivalent in P, even if $\boldsymbol{h}$ is one-sided.

Remark 2. Before ending this section, we shall draw an interesting conclusion for the projective unirreps of a point group P that contains binary rotations that are two-sided. Let $S(\pi\boldsymbol{h})$ be the spinor representation of a two-sided binary rotation $R(\pi\boldsymbol{h})$, then, from (11.1.9), there exists another binary axis-vector $\boldsymbol{n}$ perpendicular to $\boldsymbol{h}$ such that

$$S(\pi\boldsymbol{n})S(\pi\boldsymbol{h})S(\pi\boldsymbol{n})^{-1} = -S(\pi\boldsymbol{h}) \qquad (11.1.10)$$

Let $\{D^{(\alpha,j)}(S)\}$ be a unirrep of P'. Then, from the representation of the above equation we have, using (11.1.5),

$$\operatorname{tr} D^{(\alpha,j)}[S(\pi \boldsymbol{h})] = (-1)^{2j} \operatorname{tr} D^{(\alpha,j)}[S(\pi \boldsymbol{h})]$$

Accordingly, if j is a half integer we have

$$\operatorname{tr} D^{(\alpha,j)}[S(\pi \boldsymbol{h})] = 0 \tag{11.1.11}$$

This holds for any half-integral representation in any dimensionality. Now, a one-dimensional representative of any element of $P = \{R(\boldsymbol{\theta}_i)\}$ cannot be zero, contradicting (11.1.11). This apparent contradiction is resolved if and only if a point group with two-sided binary rotations does not have any one-dimensional half-integral representation. This explains the fact that there exists no one-dimensional half-integral representation for the groups $SO(3, r)$, Y, T, O and D_n with an even n. It will be shown, however, that there exist one-dimensional half-integral irreps for D_n with an odd n. See Table 11.5 later.

11.2 The structures of double point groups

11.2.1 *Defining relations of double point groups*

In terms of the projective set $\check{P} = \{S(\boldsymbol{\theta}_i)\}$ of a point group $P = \{R(\boldsymbol{\theta}_i)\}$, its double group $P' = \{\pm S(\boldsymbol{\theta}_i)\}$ is expressed by

$$P' = \check{P} + E'\check{P}, \qquad E'^2 = E \tag{11.2.1}$$

where $E' = S(2\pi \boldsymbol{n}) = -\mathbf{1}$ is the spinor transformation for the 2π rotation and E is the unit element. The kernel $N = \{E, E'\}$ of the homomorphism between P' and P is in the center of P' so that P' is called the *central extension* of P by N. In view of (11.2.1), the set of defining relations (or presentation) of the double group P' simply follow from those of P. All that is necessary is to introduce 2π rotation E' as a second-order operator in the center of P'. Thus we have the following presentations of double point groups:

$$
\begin{aligned}
&C'_n: && A^n = E', && E'^2 = E \\
&D'_n: && A^n = B^2 = (AB)^2 = E', && E'^2 = E \\
&T': && A^3 = B^3 = (AB)^2 = E', && E'^2 = E \\
&O': && A^4 = B^3 = (AB)^2 = E', && E'^2 = E \\
&Y': && A^5 = B^3 = (AB)^2 = E', && E'^2 = E
\end{aligned} \tag{11.2.2}
$$

where A and B are generators of the respective double group P', E' is the 2π rotation and E is the identity. From the defining relations (11.2.2), the 2π rotation E' commutes with any element of the respective group, because it equals some power of each generator. From $E'^2 = E$, it follows that the possible matrix representatives of E' are $D(E') = \pm\mathbf{1}$ in an appropriate dimensionality, where $D(E') = \mathbf{1}$ identifies a vector representation of P whereas $D(E') = -\mathbf{1}$ identifies a projective or double-valued (spinor) representation of P. For example, the three-dimensional representation $\{R(\boldsymbol{\theta}_i) = D^{(1)}(\boldsymbol{\theta}_i)\}$ of P belongs to the former whereas the two-dimensional spinor representation $P' = \{\pm S(\boldsymbol{\theta}_i) = \pm D^{(\frac{1}{2})}(\theta_i)\}$ of $P = \{R(\boldsymbol{\theta}_i)\}$ belongs to the latter.

Table 11.1. *The multiplication table of the double group D'_n*

	A^l	$A^l B$
A^k	A^{k+l}	$A^{k+l}B$
$A^k B$	$A^{k-l}B$	$A^{k-l}E'$

An improper double point group is isomorphic either to a proper double point group P' or to a direct product group $P' \times C_i$, where C_i is the ordinary group of inversion defined by $\{E, I\}$, in which I is the inversion. The presentations for those belonging to the former case are

$$\begin{aligned} C'_{np}\ (\simeq C'_{2n}):&\quad (\overline{A})^{2n} = E, \qquad E'^2 = E \\ C'_{nv}\ (\simeq D'_{2n}):&\quad A^n = (\overline{B})^2 = (A\overline{B})^2 = E', \qquad E'^2 = E \\ D'_{np}\ (\simeq D'_{2n}):&\quad (\overline{A})^{2n} = B^2 = (\overline{A}B)^2 = E', \qquad E'^2 = E \\ T'_p\ (\simeq O'):&\quad (\overline{A})^4 = B^3 = (\overline{A}B)^2 = E', \qquad E'^2 = E \end{aligned} \tag{11.2.3}$$

where $\overline{A} = IA$ and $\overline{B} = IB$ are improper elements. There can be no improper point group isomorphic to C_n with an odd n, or to T and Y, because their canonical generators are all of odd orders whereas the inversion I is of even order (cf. Equation (5.5.10)). The double group of a direct product group $P_i = P \times C_i$ is defined by

$$P'_i; \qquad P',\ I^2 = E \tag{11.2.4}$$

where P' stands for the defining relations for a proper double point group and I is in the center of P'_i. For example, O'_i is defined by

$$A^4 = B^3 = (AB)^2 = E', \qquad I^2 = [I, A] = [I, B] = E'^2 = E \tag{11.2.5}$$

where $[I, A] = IAIA^{-1}$ is the commutator. These defining relations are quite effective for finding all projective representations of P' or all space groups with the point co-group P'.

11.2.2 The structure of the double dihedral group D'_n

From the defining relations of D'_n given in (11.2.2), we can form the multiplication table of D'_n. Proceeding in a way analogous to that for the multiplication table of D_n (Table 5.3), we have Table 11.1 for the double group D'_n.

Let us first summarize the structure of the group D_n described in Section 5.2 for comparison. Let $c_n = R((2\pi/n)\boldsymbol{e}_z)$ be the n-fold axis of rotation about the z-axis and let $u_0 = R(\pi \boldsymbol{e}_x)$ be the binary rotation about the x-axis. Then the elements of the group D_n are given by

$$c_n^k = R\left(\frac{2\pi k}{n}\boldsymbol{e}_z\right), \qquad u_q = c_n^q u_0 = R(\pi \boldsymbol{h}_q); \qquad k, q = 0, 1, \ldots, n-1 \tag{11.2.6}$$

where u_q is a binary rotation about the unit vector $\boldsymbol{h}_q = (\cos(\pi/n), \sin(\pi q/n), 0)$ on the x, y plane that makes an angle $(\pi q/n)$ with the x-axis.

Now for the double group D'_n, there exists a total of $4n$ elements in the double group. These are given by, using (10.1.7e) and (10.1.7f),

$$\check{c}_n^k = S\left(\frac{2\pi k}{n}\boldsymbol{e}_z\right) = \begin{bmatrix} \mathrm{e}^{-\mathrm{i}\pi k/n} & 0 \\ 0 & \mathrm{e}^{\mathrm{i}\pi k/n} \end{bmatrix} = S(\mathrm{e}^{-\mathrm{i}\pi k/n}, 0)$$

$$\check{u}_q = S(\pi\boldsymbol{h}_q) = \begin{bmatrix} 0 & -\mathrm{i}\mathrm{e}^{-\mathrm{i}\pi q/n} \\ -\mathrm{i}\mathrm{e}^{\mathrm{i}\pi q/n} & 0 \end{bmatrix} = S(0, -\mathrm{i}\mathrm{e}^{-\mathrm{i}\pi q/n}) \tag{11.2.7}$$

where $k, q = 0, 1, \ldots, 2n-1$. Here, for convenience, the rotation vectors are expressed in the parameter domain Ω' $(0 \leqslant |\boldsymbol{\theta}| \leqslant 2\pi)$ of $SU(2)$. Note that $\check{u}_q$ given above follows also from $\check{u}_q = \check{c}_n^q \check{u}_0$.

The symmetry properties of the elements of D'_n, from $\check{c}_n^n = -1$, are

$$\check{c}_n^{k+n} = -\check{c}_n^k, \qquad \check{u}_{q+n} = -\check{u}_q = \check{u}_q^{-1} \tag{11.2.8a}$$

The conjugacy relations are

$$\check{u}_q \check{c}_n^k \check{u}_q^{-1} = \check{c}_n^{-k}, \qquad \check{c}_n^k \check{u}_q \check{c}_n^{-k} = \check{u}_{q+2k} \tag{11.2.8b}$$

Thus, $\check{c}_n^k$ is always two-sided whereas $\check{u}_q$ is two-sided only if n is even (consistently with the case of D_n). Therefore, from (11.1.9), the class of $\check{c}_n^k$ is always ambivalent whereas the class of $\check{u}_q$ is ambivalent if and only if n is even. The class structure of D'_n is given by

$$\{E\}, \{E'\}, \{\check{c}_n^k, \check{c}_n^{-k}\}; \qquad k = 1, 2, \ldots, n-1$$

$$\{\check{u}_q;\ q = \text{even}\}, \{\check{u}_q;\ q = \text{odd}\}; \qquad q = 0, 1, 2, \ldots, 2n-1 \tag{11.2.8c}$$

There exist altogether $n + 3$ classes for D'_n. An analogous class structure holds for the dihedral group D_n with an even n but not with an odd n, as was shown in Section 5.2. From (11.2.8a), the class structure of D'_n may be expressed in terms of $\pm S(\boldsymbol{\theta})$ with $\boldsymbol{\theta}$ in the parameter domain $\Omega(0 \leqslant |\boldsymbol{\theta}| \leqslant \pi)$. For example, the classes of D'_3 and D'_4 are given by

$$D'_3\colon\ \{E\}, \{E'\}, \{\check{3}_z, \check{3}_z^{-1}\}, \{-\check{3}_z, -\check{3}_z^{-1}\}, \{\check{u}_0, \check{u}_2, \check{u}_{-2}\}, \{-\check{u}_0, -\check{u}_2, -\check{u}_{-2}\}$$

$$D'_4\colon\ \{E\}, \{E'\}, \{\check{4}_z, \check{4}_z^{-1}\}, \{-\check{4}_z, -\check{4}_z^{-1}\}, \{\check{2}_z, -\check{2}_z\}, \{\check{u}_0, \check{u}_2, -\check{u}_0, -\check{u}_2\},$$
$$\{\check{u}_1, \check{u}_{-1}, -\check{u}_1, -\check{u}_{-1}\}$$

As has been discussed in (10.2.5), $S(\boldsymbol{\theta})$ and $-S(\boldsymbol{\theta})$ are not equivalent unless they correspond to two-sided binary rotations.

Now, a projective set $\check{D}_n$ of D_n that preserves the class structure of D_n may be defined by

$$\{\check{c}_n^k = S(\mathrm{e}^{-\mathrm{i}\pi k/n}, 0), \check{u}_q = S(0, -\mathrm{i}\mathrm{e}^{-\mathrm{i}\pi q/n})\} \tag{11.2.9}$$

with

$$k, q = 0, \pm 1, \ldots, \pm\left(\frac{n}{2} - 1\right), \frac{n}{2} \qquad \text{for an even } n$$

$$2k, q = 0, \pm 2, \ldots, \pm(n-1) \qquad \text{for an odd } n$$

where we have used the positive hemisphere convention (11.1.4) for an even n and the simple equivalence convention (11.1.3) for an odd n.

Exercise. Write down the projective sets of D_3 and D_4 based on (11.2.9), following (11.1.3) and (11.1.4), respectively.

11.2.3 The structure of the double octahedral group O'

From the elements of the octahedral group O given in (5.1.3a), the projective set $\check{O}$ of O which preserves the class structure is given by

$$\begin{gathered}\{E\},\ \{\check{4}_x, \check{4}_{\bar{x}}, \check{4}_y, \check{4}_{\bar{y}}, \check{4}_z, \check{4}_{\bar{z}}\}\\ \{\check{3}_{xyz}, \check{3}_{\bar{x}yz}, \check{3}_{x\bar{y}z}, \check{3}_{xy\bar{z}}, \check{3}_{\bar{x}\,\bar{y}\,\bar{z}}, \check{3}_{x\bar{y}\,\bar{z}}, \check{3}_{\bar{x}y\bar{z}}, \check{3}_{\bar{x}\,\bar{y}z}\}\\ \{\check{2}_x, \check{2}_y, \check{2}_z\},\ \{\check{2}_{xy}, \check{2}_{xz}, \check{2}_{yz}, \check{2}_{x\bar{y}}, \check{2}_{\bar{x}z}, \check{2}_{\bar{y}z}\}\end{gathered} \tag{11.2.10}$$

where use of the positive hemisphere convention (11.1.4) for the binary rotations has been made because all axis-vectors of O are two-sided. These axis-vectors have been graphically presented in Figure 5.1. The 2×2 matrix representations of all elements are easily written down in terms of the Euler–Rodrigues parametrization using (10.1.7b) or the Cayley–Klein parameters given in Table 11.3 later.

The multiplication table for the projective set $\check{O}$ can be easily constructed by the direct matrix multiplication of the group elements of O' because they are 2×2 matrices. The multiplication table thus obtained is given in Table 11.2, which is a generalization of the multiplication table, Table 5.6, for the group O. There exists, however, an alternative method of constructing the multiplication table that is based on the quaternion expression (10.1.14) of the axis-vector symbol $\check{N}_{\boldsymbol{n}}$, i.e.

$$\check{N}_{\boldsymbol{n}} = S\left(\frac{2\pi}{N}\boldsymbol{n}\right) = \cos\left(\frac{\pi}{N}\right)\mathbf{1} + \sin\left(\frac{\pi}{N}\right)(n_x\check{2}_x + n_y\check{2}_y + n_z\check{2}_z) \tag{11.2.11a}$$

where the elementary quaternions satisfy the algebraic relations

$$\check{2}_x^2 = -\mathbf{1}, \qquad \check{2}_x\check{2}_y = -\check{2}_y\check{2}_x = \check{2}_z \qquad (x,\ y,\ z \text{ cyclic}) \tag{11.2.11b}$$

The typical elements of the double group O' are

$$\begin{gathered}\check{3}_{xyz} = (\mathbf{1} + \check{2}_x + \check{2}_y + \check{2}_z)/2\\ \check{4}_x = (\mathbf{1} + \check{2}_x)/\sqrt{2}, \qquad \check{2}_{xy} = (\check{2}_x + \check{2}_y)/\sqrt{2}\end{gathered} \tag{11.2.12a}$$

Observe the simple correlation between an axis-vector symbol and the coefficients of the unit quaternions. Analogously to these three expressions, we can write down the quaternion expressions for all elements of the double group O'; for example,

$$\begin{gathered}\check{3}_{xy\bar{z}} = (\mathbf{1} + \check{2}_x + \check{2}_y - \check{2}_z)/2\\ \check{4}_{\bar{x}} = (\mathbf{1} - \check{2}_x)/\sqrt{2}, \qquad \check{2}_{x\bar{y}} = (\check{2}_x - \check{2}_y)/\sqrt{2}\end{gathered}$$

etc. From these expressions, we can calculate the product of two elements using the algebraic relations (11.2.11b), e.g.

$$\begin{aligned}\check{3}_{xyz}\check{2}_{xy} &= (\mathbf{1} + \check{2}_x + \check{2}_y + \check{2}_z')(\check{2}_x + \check{2}_y)/2\sqrt{2}\\ &= -(\mathbf{1} - \check{2}_y)/\sqrt{2} = -\check{4}_{\bar{y}}\end{aligned} \tag{11.2.12b}$$

The same result is obtained from the multiplication table, Table 11.2. For later use, in

Table 11.2. *The multiplication table for the projective set $\check{O}$ of the octahedral group O*

Equivalent positions

Jones	x	x	$\bar{x}$	$\bar{x}$	z	$\bar{z}$	z	$\bar{z}$	y	$\bar{y}$	$\bar{y}$	y	x	x	$\bar{z}$	z	y	$\bar{y}$	$\bar{x}$	$\bar{x}$	$\bar{z}$	z	$\bar{y}$	y
	y	$\bar{y}$	y	$\bar{y}$	x	x	$\bar{x}$	x	z	z	$\bar{z}$	$\bar{z}$	z	$\bar{z}$	y	y	$\bar{x}$	x	$\bar{z}$	z	$\bar{y}$	$\bar{y}$	$\bar{x}$	x
	z	$\bar{z}$	$\bar{z}$	z	y	y	$\bar{y}$	$\bar{y}$	x	$\bar{x}$	x	$\bar{x}$	$\bar{y}$	y	x	$\bar{x}$	z	z	$\bar{y}$	y	$\bar{x}$	x	$\bar{z}$	$\bar{z}$
$i \setminus j$	e	2_x	2_y	2_z	3_{xyz}	$3_{x\bar{y}\bar{z}}$	$3_{\bar{x}y\bar{z}}$	$3_{\bar{x}\bar{y}z}$	$3_{\bar{x}\bar{y}\bar{z}}$	$3_{\bar{x}yz}$	$3_{x\bar{y}z}$	$3_{xy\bar{z}}$	$4_{\bar{x}}$	4_x	$4_{\bar{y}}$	4_y	$4_{\bar{z}}$	4_z	$2_{z\bar{y}}$	2_{zy}	$2_{z\bar{x}}$	2_{zx}	$2_{x\bar{y}}$	2_{xy}
e	1	2	3	4	5	6	7	8	9	10	11	12	13	14	15	16	17	18	19	20	21	22	23	24
2_x	2	1′	4	3′	7′	8′	5	6	12	11	10′	9′	14	13′	21′	22	24	23	20′	19	15	16′	18′	17′
2_y	3	4′	1′	2	8′	7	6′	5	10	9′	12	11′	20	19′	16	15′	23′	24	14	13′	22	21′	17	18′
2_z	4	3	2′	1′	6′	5	8	7′	11	12′	9′	10	19	20	22	21	18	17′	13′	14′	16′	15′	24	23′
3_{xyz}	5	8′	6′	7′	9′	12	10	11	1	4	2	3	18	24	14	20	16	22	23	17′	19	13′	21′	15′
$3_{x\bar{y}\bar{z}}$	6	7′	5	8	11	10′	12	9	4′	1	3′	2	17	23	20′	14	21′	15	24′	18	13	19	16′	22
$3_{\bar{x}y\bar{z}}$	7	6	8′	5	12	9	11′	10	2′	3	1	4′	23′	17	13	19′	22′	16	18	24	20	14	15	21′
$3_{\bar{x}\bar{y}z}$	8	5	7	6′	10	11	9	12′	3′	2′	4	1	24′	18	19	13	15	21	17′	23′	14′	20	22	16
$3_{\bar{x}\bar{y}\bar{z}}$	9	11	12	10	1	3′	4′	2′	5′	7	8	6	22′	15	24′	17	20′	13	21	16	23′	18	19	14
$3_{\bar{x}yz}$	10	12	11′	9′	3	1	2′	4	8	6′	5	7	21	16	18	23′	13	20	22	15′	17′	24	14	19′
$3_{x\bar{y}z}$	11	9′	10	12′	4	2	1	3′	6	8	7′	5	15	22	23	18	14	19	16′	21	24′	17′	13′	20
$3_{xy\bar{z}}$	12	10′	9′	11	2	4′	3	1	7	5	6	8′	16	21′	17	24	19′	14	15	22	18	23	20′	13′
$4_{\bar{x}}$	13	14	19′	20	16	15	22′	21	24′	23′	18	17	2′	1	8	7	9	10	4	3	6′	5	11	12
4_x	14	13′	20	19	22	21′	16	15	17	18	23	24	1	2	6	5	12	11	3′	4	8	7′	10′	9′
$4_{\bar{y}}$	15	22	16	21	18	23	17	24′	20′	13	19	14	9	11	3′	1	6	8	12′	10	2′	4	7′	5
4_y	16	21′	15′	22	24	17	23′	18	13	20	14	19′	10	12	1	3	7	5	11	9′	4	2	6	8′
$4_{\bar{z}}$	17	23	24	18	14	20′	19′	13	22′	16	15	21′	7	6	9	12	4′	1	8	5	10	11	3′	2
4_z	18	24	23′	17′	20	14	13	19	15	21	22	16	8	5	11	10	1	4	7′	6′	12′	9′	2	3

$2_{z\bar{y}}$	19	20	13	14′	21	22	15	16′	23	24′	17′	18	3′	4	7′	8	11	12′	1′	2′	5′	6′	9′	10
2_{zy}	20	19′	14′	13′	15′	16	21	22	18	17′	24	23′	4	3	5	6′	10	9′	2	1′	7′	8′	12	11′
$2_{z\bar{x}}$	21	16	22′	15′	23′	18	24′	17′	19	14′	20	13	12′	10	4	2′	8	6′	9′	11′	1′	3	5	7
2_{zx}	22	15′	21	16′	17′	24	18	23	14	19	13′	20	11	9′	2	4	5	7′	10′	12′	3′	1′	8′	6′
$2_{x\bar{y}}$	23	17′	18	24′	19	13′	14	20′	21′	15	16′	22	6	7′	10′	11	2	3′	5′	8	9	12′	1′	4
2_{xy}	24	18′	17′	23	13′	19′	20	14	16	22	21′	15′	5	8′	12	9′	3	2	6	7′	11	10′	4′	1′

1. The coordinates of equivalent positions (Jones notations) given in the first line are obtained by transforming a general column vector (x, y, z) by a rotation $R(\boldsymbol{\theta})$.
2. The entries give the subscript k such that $a_i a_j = a_k$.
3. In this table, the axis-vector rotation N_n stands for $\check{N}_n$ for a spinor transformation; e.g. 2_x means $\check{2}_x$ for a spinor group. The Cayley–Klein parameters of each element are given in Table 11.3.
4. The axis-vector of each element is given graphically in Figure 5.1.
5. $n' = 1'n$ where $1'$ is the 2π rotation E'; $1' = 1$ for a point group $< SO(3, r)$ and $1' = -1 = S(2\pi n)$ for a spinor point group $< SU(2)$.

Table 11.3. *The Cayley–Klein parameter sets $\{(a_\nu, b_\nu)\}$ for the elements $\{S(a_\nu, b_\nu)\}$ belonging to the projective set $\check{O}$ of the octahedral group O*

1. $\check{1}$: $(1, 0)$	2. $\check{2}_x$: $(0, -\mathrm{i})$
3. $\check{2}_y$: $(0, -1)$	4. $\check{2}_z$: $(-\mathrm{i}, 0)$
5. $\check{3}_{xyz}$: $(1-\mathrm{i}, -1-\mathrm{i})/2$	6. $\check{3}_{x\bar{y}\bar{z}}$: $(1+\mathrm{i}, 1-\mathrm{i})/2$
7. $\check{3}_{\bar{x}y\bar{z}}$: $(1+\mathrm{i}, -1+\mathrm{i})/2$	8. $\check{3}_{\bar{x}\bar{y}z}$: $(1-\mathrm{i}, 1+\mathrm{i})/2$
9. $\check{3}_{\bar{x}\bar{y}\bar{z}}$: $(1+\mathrm{i}, 1+\mathrm{i})/2$	10. $\check{3}_{\bar{x}yz}$: $(1-\mathrm{i}, -1+\mathrm{i})/2$
11. $\check{3}_{x\bar{y}z}$: $(1-\mathrm{i}, 1-\mathrm{i})/2$	12. $\check{3}_{xy\bar{z}}$: $(1+\mathrm{i}, -1-\mathrm{i})/2$
13. $\check{4}_{\bar{x}}$: $(1, \mathrm{i})/\sqrt{2}$	14. $\check{4}_x$: $(1, -\mathrm{i})/\sqrt{2}$
15. $\check{4}_{\bar{y}}$: $(1, 1)/\sqrt{2}$	16. $\bar{4}_y$: $(1, -1)/\sqrt{2}$
17. $\bar{4}_{\bar{z}}$: $(1+\mathrm{i}, 0)/\sqrt{2}$	18. $\check{4}_z$: $(1-\mathrm{i}, 0)/\sqrt{2}$
19. $\check{2}_{z\bar{y}}$: $(-\mathrm{i}, 1)/\sqrt{2}$	20. $\check{2}_{zy}$: $(-\mathrm{i}, -1)/\sqrt{2}$
21. $\check{2}_{z\bar{x}}$: $(-\mathrm{i}, \mathrm{i})/\sqrt{2}$	22. $\check{2}_{zx}$: $(-\mathrm{i}, -\mathrm{i})/\sqrt{2}$
23. $\check{2}_{x\bar{y}}$: $(0, 1-\mathrm{i})/\sqrt{2}$	24. $\check{2}_{xy}$: $(0, -1-\mathrm{i})/\sqrt{2}$

The number ν (= 1–24) assigned to each element $S(a_\nu, b_\nu)$ is the same as that given in Table 11.2. The Cayley–Klein parameters for the elements belonging to the projective set $\check{T}$ of the tetrahedral group T are given under $\nu = 1$–12.

Table 11.3, we shall give the Cayley–Klein (CK) parameter sets for all elements belonging to the projective set $\check{O}$ of the octahedral group O: these are obtained from the correlations given by (11.2.12a) or by using (10.1.7d) given in the form

$$a = \cos(\pi/N) - \mathrm{i}n_z \sin(\pi/N), \qquad b = (-n_y - \mathrm{i}n_x)\sin(\pi/N)$$

For example, the CK parameters (a, b) for typical elements are

$$\check{3}_{xyz}: \quad (1 - \overset{z}{\mathrm{i}}, - \overset{y}{1} - \overset{x}{\mathrm{i}})/2, \qquad \check{4}_x: \quad (1, - \overset{x}{\mathrm{i}})/\sqrt{2}, \qquad \check{2}_{xy}: \quad (0, - \overset{y}{1} - \overset{x}{\mathrm{i}})/\sqrt{2}$$

Here again observe the simple correlation between an axis-vector symbol and the corresponding parameter set via $z \to -\mathrm{i}$, $y \to -1$ and $x \to -\mathrm{i}$, analogously to (11.2.12a).

Finally, we observe that the 48 CK parameter sets $\{\pm(a_\nu, b_\nu)\}$ for O' obtained from Table 11.3 are the roots of the following polynomial equations, on account of the fixed condition $|a|^2 + |b|^2 = 1$:

$$a^8 = 1 \quad \text{or} \quad b^8 = 1 \quad \text{or} \quad a^4 = b^4 = \pm\tfrac{1}{4} \tag{11.2.13}$$

The set of these equations is called *the subgroup condition* for O' imposed on the CK parameters, because the 48 roots $\{\pm(a_\nu, b_\nu)\}$ of the set completely reproduce the CK parameters of all elements of the double point group O'. The 16 roots of $a^4 = b^4 = -\frac{1}{4}$ define the eight three-fold rotations in O', and those of the first two equations in (11.2.13) together with the fixed condition define the double group D'_4 whereas the 16 roots of the remaining equations $a^4 = b^4 = \frac{1}{4}$ provide the parameter sets for the remaining $\check{4}$-fold and $\check{2}$-fold rotations in O'. In the next section, using this subgroup condition, we shall form the *general unirreps* of O' expressed by the CK parameter set

(a, b), which will provide the explicit matrix representations of O' with the parameter sets $\{\pm(a_\nu, b_\nu)\}$ for O' given by Table 11.3.

11.2.3.1 The class structure of the double group O'

There are in total eight classes for the double group O' compared with five classes for the point group O given by (5.1.3a). They may be expressed as follows:

$$\{E\}, \{E'\}, \{6\check{c}_4\}, \{-6\check{c}_4\}, \{8\check{c}_3\}, \{-8\check{c}_3\}, \{3\check{c}_4^2, -3\check{c}_4^2\}, \{6\check{c}_2, -6\check{c}_2\} \quad (11.2.14)$$

with obvious abbreviations; cf. (11.2.10). These elements are all two-sided so that every class is ambivalent, according to (11.1.9).

11.3 The unirreps of double point groups expressed by the projective unirreps of point groups

Previously in Section 10.3 we have expressed the complete set of the unirreps $\{D^{(j)}[S(a, b)]\}$ of $SU(2)$ via the Cayley–Klein (CK) parametrization of the group elements. Since a proper double point group $P' = \{S(a_\nu, b_\nu)\}$ is a subgroup of $SU(2) = \{S(a, b)\}$, simple substitution of the CK parameters of $P' = \{S(a_\nu, b_\nu)\}$ into the defining equation (10.3.2) of $D^{(j)}[S(a, b)]$ may lead to the complete set of the unirreps of the double group P'. Following Section 11.1, these will be determined as the projective unirreps of its homomorph P through the projective set $\check{P}$ of P formed by the convention (11.1.3) or (11.1.4).

11.3.1 The uniaxial group C_∞

Let $c_z(\theta) = R(\theta \boldsymbol{e}_z)$ be a rotation about the z-axis through an angle θ, then the group C_∞ may be defined by $\{R(\theta\boldsymbol{e}_z);\ -\pi < \theta \leqslant \pi\}$. Via the convention (11.1.3), its projective set is defined by $\check{C}_\infty = \{S(\theta\boldsymbol{e}_z)\}$ with the elements

$$\check{c}_z(\boldsymbol{\theta}) = S(\theta\boldsymbol{e}_z) = \begin{bmatrix} \mathrm{e}^{-\mathrm{i}\theta/2} & 0 \\ 0 & \mathrm{e}^{\mathrm{i}\theta/2} \end{bmatrix} = S(\mathrm{e}^{-\mathrm{i}\theta/2}, 0)$$

so that the projective set is characterized by the Cayley–Klein parameters

$$(a = \mathrm{e}^{-\mathrm{i}\theta/2}, b = 0); \qquad -\pi < \theta \leqslant \pi \quad (11.3.1)$$

On substituting this into the defining equation (10.3.2) for the unirrep $D^{(j)}(a, b)$, we immediately obtain one-dimensional projective unirreps of C_∞; via $c_\infty(\theta) \leftrightarrow \check{c}_z(\theta)$:

$$M_m(\theta) = a^{2m} = \mathrm{e}^{-\mathrm{i}m\theta}; \qquad m = 0, \pm\tfrac{1}{2}, \pm 1, \ldots \quad (11.3.2)$$

with the bases $\phi(j, m)$ defined by (10.3.1b), i.e.

$$\phi(j, m) = \xi_1^{j+m}\xi_2^{j-m}[(j+m)!(j-m)!]^{-1/2}, \qquad j \geqslant |m|$$

From (11.1.6), the projective unirrep $M_m(\theta)$ belongs to the factor system $[f(S_1, S_2)]^{2m}$: it is an integral (or half-integral) representation of C_∞, if m is an integer (or half integer). It is needless to say that the set $\{M_m(\theta)\}$ given by (11.3.2) provides a complete set of the unirreps of the double group $C'_\infty = \{S(\theta\boldsymbol{e}_z);\ -2\pi < \theta \leqslant 2\pi\}$, because it is a complete set of functions of $\theta/2$, $\{\mathrm{e}^{-\mathrm{i}2m\theta/2};\ 2m = 0, 1, 2, \ldots\}$ in the domain $-\pi \leqslant \theta/2 \leqslant \pi$, according to Fourier's theorem.

11.3.2 The group C_n

The uniaxial group of the order n is a subgroup of C_∞ and is defined by $C_n = \{R((2\pi k/n)\boldsymbol{e}_z);\ k = 0, 1, \ldots, n-1\}$. The projective set of C_n is given by $\check{C}_n = \{S((2\pi k/n)\boldsymbol{e}_z)\}$ with the CK parameters

$$(a_k = \exp(-\mathrm{i}\pi k/n),\ b = 0); \qquad k = 0, 1, \ldots, n-1 \tag{11.3.3a}$$

Thus, substitution of $\theta_k = 2\pi k/n$ into θ of (11.3.2) provides a total of $2n$ projective unirreps of C_n:

$$M_m(\theta_k) = \exp(-2\mathrm{i}\pi mk/n); \qquad m = 0, \pm\frac{1}{2}, \pm 1, \ldots, \pm\frac{n-1}{2}, \frac{n}{2} \tag{11.3.3b}$$

Since $M_m = M_{m\pm n}$, the bases of the unirrep M_m are given by $\phi(j, m)$ and $\phi(j, m \pm n)$. The unirrep $M_m(\theta_k)$ belongs to the same factor system $[f(S_1, S_2)]^{2m}$ as that for C_∞. The set $\{M_m(\theta_k)\}$ given by (11.3.3b) also provides a complete set of unirreps of the double group

$$C'_n = \left\{S\left(\frac{2\pi k}{n}\boldsymbol{e}_z\right); \qquad k = 0, 1, \ldots, 2n-1\right\}$$

because it is one-dimensional and the total number of the unirreps equals the order $2n$ of the double group C'_n.

11.3.3 The group D_∞

This group is defined by adjoining the uniaxial group $C_\infty = \{c_z(\theta)\}$ with the binary rotation $u_0 = R(\pi\boldsymbol{e}_x)$ about the x-axis. Accordingly, the elements of the group D_∞ are given by

$$c_z(\theta) = R(\theta\boldsymbol{e}_z), \qquad u_\beta = c_z(\beta)u_0 = R(\pi\boldsymbol{h}_\beta); \qquad -\pi < \theta, \beta \leqslant \pi \tag{11.3.4}$$

where u_β is a binary rotation about the unit vector $\boldsymbol{h}_\beta = (\cos(\beta/2), \sin(\beta/2), 0)$ in the x, y plane which makes an angle $\beta/2$ with the x-axis. Since these binary rotations are all two-sided, their rotation vectors $\pi\boldsymbol{h}_\beta$ are placed on the positive hemisphere in accordance with the convention (11.1.4). The projective set $\check{D}_\infty$ of D_∞ is defined by the elements

$$\check{c}_z(\theta) = S(\theta\boldsymbol{e}_z) = S(a, 0), \qquad \check{u}_\beta = S(\pi\boldsymbol{h}_\beta) = S(0, b) \tag{11.3.5}$$

where the Cayley–Klein parameters are given, from (10.1.7e) and (10.1.7f), by

$$(a = \mathrm{e}^{-\mathrm{i}\theta/2}, 0), (0, b = -\mathrm{i}\mathrm{e}^{-\mathrm{i}\beta/2}); \qquad -\pi < \theta, \beta \leqslant \pi$$

If we substitute these CK parameter sets $(a, 0)$ and $(0, b)$ into the defining equation (10.3.2) of the unirrep $D^{(j)}$ of $SU(2)$ we see that the $(2j+1)$-dimensional representation $D^{(j)}$ spanned by the spinor basis $\{\phi(j, m)\}$ is reduced to a set of two-dimensional spaces spanned by $[\phi(j, m), \phi(j, -m)]$ with $m = j, j-1, \ldots, 1$ (or $\frac{1}{2}$) and a one-dimensional space spanned by $\phi(j, 0)$ when j is an integer.

Firstly, the one-dimensional representation based on $\phi(j, 0)$ with an integral j is given by

$$M_0^{(j)}[S(a, 0)] = 1, \qquad M_0^{(j)}[S(0, b)] = (-1)^j \tag{11.3.6a}$$

which gives two inequivalent unirreps: one with an even j and the other with an odd j. Secondly, the 2×2 representations are given by

$$\overline{M}_m^{(j)}[S(a, 0)] = \begin{bmatrix} a^{2m} & 0 \\ 0 & a^{*2m} \end{bmatrix}$$

$$\overline{M}_m^{(j)}[S(0, b)] = (-\mathrm{i})^{2j} \begin{bmatrix} 0 & (ib)^{2m} \\ (ib)^{*2m} & 0 \end{bmatrix} \tag{11.3.6b}$$

These are irreducible due to Schur's lemma because any 2×2 matrix W that commutes with both of these matrices is a constant matrix. Firstly, W is diagonal because it commutes with a diagonal matrix with all different diagonal elements. Secondly, a diagonal matrix that commutes with an off-diagonal matrix is a constant matrix. Now, the traces of these representatives are given by

$$\operatorname{tr} \overline{M}_m^{(j)}[S(a, 0)] = 2\cos(m\theta)$$

$$\operatorname{tr} \overline{M}_m^{(j)}[S(0, b)] = 0 \tag{11.3.6c}$$

Since these are independent both of j and of the sign of m, the inequivalent 2×2 representations may be classified solely by positive values of m, i.e.

$$m = \tfrac{1}{2}, 1, \tfrac{3}{2}, \ldots \tag{11.3.6d}$$

In fact, we have the following equivalence relations for the representations:

$$\sigma_x \overline{M}_m^{(j)} \sigma_x = \overline{M}_{-m}^{(j)}, \qquad \sigma_z \overline{M}_m^{(j)} \sigma_z = \overline{M}_m^{(j-1)} \tag{11.3.6e}$$

where σ_x and σ_z are the Pauli spin matrices. Thus the numerical factor $(-1)^{2j}$ in $\overline{M}_m^{(j)}[S(0, b)]$ can be reduced to $(-\mathrm{i})^{2m}$; however, we shall keep the factor $(-\mathrm{i})^{2j}$ as it is, because it identifies the basis $[\phi(j, m), \phi(j, -m)]$ of the representation $\overline{M}_m^{(j)}$.

For the 2×2 representations it is more convenient to introduce the new basis $[\phi_+(j, m), \phi_-(j, m)]$ defined by

$$\phi_+(j, m) = 2^{-1/2}(\phi(j, m) + \phi(j, -m))$$

$$\phi_-(j, m) = -\mathrm{i}2^{-1/2}(\phi(j, m) - \phi(j, -m)) \tag{11.3.7}$$

Then, the corresponding representation of D_∞ takes the form

$$E_m[c_z(\theta)] = \begin{bmatrix} \cos(m\theta) & -\sin(m\theta) \\ \sin(m\theta) & \cos(m\theta) \end{bmatrix}$$

$$E_m[u_\beta] = (-\mathrm{i})^{2j} \begin{bmatrix} \cos(m\beta) & \sin(m\beta) \\ \sin(m\beta) & -\cos(m\beta) \end{bmatrix}; \qquad -\pi < \theta, \beta \leqslant \pi \tag{11.3.8}$$

via $\check{D}_\infty \leftrightarrow D_\infty$, where $m = \tfrac{1}{2}, 1, \ldots$.

The final results for the integral and half-integral unirreps of D_∞ are given in Table 11.4. Note that, if the parameter domain is extended to $-2\pi < \theta, \beta \leqslant 2\pi$, then Table 11.4 becomes the unirreps for the double group D'_∞, as it should. According to (11.1.6), the one-dimensional unirreps given by (11.3.6a) are vector unirreps. When m is an integer, the two-dimensional unirrep E_m is a vector unirrep of D_∞, but when m is a half integer it is a projective unirrep belonging to the factor system $f(S_1, S_2)$ defined by (11.1.2).

Table 11.4. *The irreducible (vector and projective) representation of D_∞*

D_∞	$c_z(\theta)$	u_β	Bases
A_1	1	1	$\phi(j_e, 0)$
A_2	1	-1	$\phi(j_o, 0)$
E_m	$\begin{bmatrix} \cos(m\theta) & -\sin(m\theta) \\ \sin(m\theta) & \cos(m\theta) \end{bmatrix}$	$(-i)^{2j}\begin{bmatrix} \cos(m\beta) & \sin(m\beta) \\ \sin(m\beta) & -\cos(m\beta) \end{bmatrix}$	$[\phi_+(j, m), \phi_-(j, m)]$

1. $m = 1, 2, \ldots,$ for vector representations; $m = \frac{1}{2}, \frac{3}{2}, \ldots,$ for double-valued representations.
2. $-\pi < \theta, \beta < \pi$.
3. j_e (j_o) for A_1 (A_2) is an even (odd) integer.

11.3.4 The group D_n

If we compare the projective set $\check{D}_n$ defined by (11.2.9) with the projective set $\check{D}_\infty$ defined by (11.3.5), we see that $\check{D}_n$ is obtained from $\check{D}_\infty$ by the substitutions

$$\theta \rightarrow 2\pi k/n, \qquad \beta \rightarrow 2\pi q/n \tag{11.3.9}$$

with k and q given in (11.2.9). Thus, the same substitutions will provide the projective unirreps of D_n from those of D_∞ given by Table 11.4. Firstly, the one-dimensional unirreps A_1 and A_2 hold for D_n without modification. Secondly, the two-dimensional unirrep E_m of D_n takes the form

$$E_m(c_n^k) = \begin{bmatrix} \cos(2\pi mk/n) & -\sin(2\pi mk/n) \\ \sin(2\pi mk/n) & \cos(2\pi mk/n) \end{bmatrix}$$

$$E_m(u_q) = (-i)^{2j}\begin{bmatrix} \cos(2\pi mq/n) & \sin(2\pi mq/n) \\ \sin(2\pi mq/n) & -\cos(2\pi mq/n) \end{bmatrix} \tag{11.3.10a}$$

Since these satisfy the symmetry relations $E_m \sim E_{-m} = E_{n-m}$ for $\sigma_z E_m \sigma_z = E_{-m}$, the inequivalent unirreps E_m are limited to

$$m = \tfrac{1}{2}, 1, \ldots, \tfrac{1}{2}(n-1), n/2 \tag{11.3.10b}$$

which are irreducible except for $m = n/2$. When $m = n/2$, $E_{n/2}$ splits into two one-dimensional representations denoted B_1 and B_2, given in Table 11.5.

In Table 11.5, A_1 and A_2 are vector unirreps, whereas B_1 and B_2 are vector (projective) unirreps if n is even (odd). The two-dimensional unirrep E_m is a vector (or double-valued) unirrep if m is an integer (or a half integer) in accordance with the factor system $[f(S_1, S_2)]^{2m}$ from (11.1.6).

Note that the notations B_1, B_2 and E_m could be different from Mulliken's notation, which does not exist for a double group. The present notation E_m for a 2×2 representation seems most satisfactory because the suffix m of E_m distinguishes the integral and half-integral representations of D_n.

There exists a total of $n + 3$ unirreps for the double group D_n', of which four are one-dimensional and $n - 1$ are two-dimensional, satisfying the completeness condition

$$1^2 + 1^2 + 1^2 + 1^2 + (n-1)2^2 = 4n \tag{11.3.11}$$

Table 11.5. *The irreducible (vector and projective) representations of D_n*

D_n	c_n^k	u_q	Bases
A_1	1	1	$\phi(j_e, 0)$, $\phi_+(j_e, n)$, $\phi_-(j_o, n)$
A_2	1	-1	$\phi(j_o, 0)$, $\phi_+(j_o, n)$, $\phi_-(j_e, n)$
B_1	$(-1)^k$	$(-1)^{q+1}\mathrm{i}^{\theta(n)}$	$\phi_+(j_e, n/2)$, $\phi_-(j_o, n/2)$
B_2	$(-1)^k$	$(-1)^q\mathrm{i}^{\theta(n)}$	$\phi_+(j_o, n/2)$, $\phi_-(j_e, n/2)$
E_m	$\begin{bmatrix}\cos(2\pi mk/n) & -\sin(2\pi mk/n)\\ \sin(2\pi mk/n) & \cos(2\pi mk/n)\end{bmatrix}$	$(-\mathrm{i})^{2j}\begin{bmatrix}\cos(2\pi mq/n) & \sin(2\pi mq/n)\\ \sin(2\pi mq/n) & -\cos(2\pi mq/n)\end{bmatrix}$	$[\phi_+(j, m)$, $\phi_-(j, m)]$

1. $\theta(n) = 1$ for an odd n, $\theta(n) = 2$ for an even n; j_e (j_o) is a j with an even (odd) integral part.
2. $k, q = 0, \pm 1, \ldots, \pm(n/2 - 1),\ n/2$ for an even n; $2k, q = 0, \pm 2, \ldots, \pm(n-1)$ for an odd n.
3. $m = \frac{1}{2}, 1, \ldots, (n-1)/2$.
4. $\phi(j, m)$ and $\phi(j, m \pm n)$ belong to the same representation for $m \neq 0$.
5. Note that B_1 and B_2 given here are B_2 and B_1, respectively, defined originally in Table 11, *J. Math. Phys.* **22**, 2101 (1981), S. K. Kim; this change is to accommodate Mulliken's notation for the vector representations.

The completeness is also seen from the fact that the number $n + 3$ of inequivalent unirreps equals the number of the conjugate classes of the double group D'_n given in (11.2.8c).

11.3.5 The group O

We shall construct the unirreps of the double group O' directly subducing the unirreps $\{D^{(j)}\}$ of $SU(2)$ defined by (10.3.2), if necessary using the subgroup condition (11.2.13). The final results will be presented as the projective unirreps of the group O. Firstly, the identity unirrep $D^{(0)}$ of $SU(2)$ subduces the identity unirrep A_1 of O'. Another one-dimensional representation, A_2, is obtained by showing that the one-dimensional space spanned by $\phi_-(3, 2)$ is invariant with respect to O' via the subgroup condition. Next, the spinor transformation $S(a, b)$ of $SU(2)$ with the basis $[\phi(\frac{1}{2}, \frac{1}{2})$, $\phi(\frac{1}{2}, -\frac{1}{2})]$ simply subduces a 2×2 unirrep of O' denoted $E_{1/2}$ in Table 11.6; its explicit representation is given by $\{S(a_v, b_v)\}$ with CK parameter sets given by Table 11.3. The 3×3 representation $D^{(1)}$ of $SU(2)$ based on $[\phi(1, 1), \phi(1, 0), \phi(1, -1)]$ also subduces a unirrep of O'. This is then transformed into the unirrep denoted T_1 in Table 11.6 using the subgroup condition (11.2.13). The transformed basis is given by $[-\mathrm{i}\phi_-(1, 1), \mathrm{i}\phi_+(1, 1), \phi(1, 0)]$ for O' or by $[x, y, z]$ for O due to the correspondence theorem for the relationship between spinor bases and the angular momentum eigenfunctions given by Theorem 10.4.1. Next, the 2×2 unirrep E and 3×3 representation

Table 11.6. *The irreducible (vector and projective) representations of O*

O	E, $9S(\pi\boldsymbol{n})$, $8S((2\pi/3)\boldsymbol{n})$, $6S((\pi/2)\boldsymbol{n})$	Bases	Characters
Γ_1, A_1	1	1	1
Γ_2, A_2	$(-1)^\kappa \operatorname{sgn}(a^4+b^4)$	$\phi_-(3, 2)$ or xyz	$\chi_2 = \Gamma_2$
Γ_3, E	$\begin{bmatrix} (-1)^\kappa \lvert a^4+b^4\rvert & 2(3)^{1/2}a^2b^{*2} \\ 2(3)^{1/2}a^2b^2 & a^4+b^4 \end{bmatrix}$	$[\phi(2, 0), \phi_+(2, 2)]$, or $[3z^2-r^2, 3^{1/2}(x^2-y^2)]$	χ_3
Γ_4, T_1	$\begin{bmatrix} \mathrm{Re}\,(a^2-b^2) & \mathrm{Im}\,(a^2+b^2) & -2\,\mathrm{Re}\,(ab) \\ -\mathrm{Im}\,(a^2-b^2) & \mathrm{Re}\,(a^2+b^2) & 2\,\mathrm{Im}\,(ab) \\ 2\,\mathrm{Re}\,(ab^*) & 2\,\mathrm{Im}\,(ab^*) & aa^*-bb^* \end{bmatrix}$	$[-\mathrm{i}\phi_-(1, 1), \mathrm{i}\phi_+(1, 1), \phi(1, 0)]$ or $[x, y, z]$	$1+2\cos\theta$
Γ_5, T_2	$A_2 \times T_1$	$[\mathrm{i}\phi_+(2, 1), -\mathrm{i}\phi_-(2, 1), \phi(2, 2)]$ or $[yz, zx, xy]$	$\chi_2(1+2\cos\theta)$
$\Gamma_6, E_{1/2}$	$S(a, b)$	$\left[\phi\left(\frac{1}{2}, \frac{1}{2}\right), \phi\left(\frac{1}{2}, -\frac{1}{2}\right)\right]$	$2\cos(\theta/2)$
$\Gamma_7, E_{1/2}$	$A_2 \times S(a, b)$	$\phi_-(3, 2)\left[\phi\left(\frac{1}{2}, \frac{1}{2}\right), \phi\left(\frac{1}{2}, -\frac{1}{2}\right)\right]$	$2\chi_2\cos(\theta/2)$
Γ_8, Q	$E \times S(a, b) \cong D^{(3/2)}$	$\left[\phi\left(\frac{3}{2}, \frac{3}{2}\right), \phi\left(\frac{3}{2}, \frac{1}{2}\right), \phi\left(\frac{3}{2}, -\frac{1}{2}\right), \phi\left(\frac{3}{2}, -\frac{3}{2}\right)\right]$	$2\chi_3\cos(\theta/2)$

1. $\kappa = 2\lvert ab\rvert$; $\kappa = 0$ if $ab = 0$; $\kappa = 1$ if $ab \neq 0$.
2. $\Gamma_1 \sim \Gamma_8$, the notations used by Koster *et al.* (1963).
3. The parameter set (a_ν, b_ν) for each element in the projective set $\check{O}$ is given explicitly in Table 11.3.
4. The character χ_2 equals Γ_2, being one-dimensional, whereas $\chi_3 = (-1)^k\lvert a^4+b^4\rvert + a^4 + b^4$.
5. $D^{(3/2)}$ is a unirrep of the rotation group defined by (10.3.3).

T_2 in Table 11.6 are obtained by reducing the five-dimensional space spanned by the basis $[\phi(2, m);\ m = 2, 1, 0, -1, -2]$ into two- and three-dimensional invariant subspaces by means of the subgroup condition (11.2.13). The remaining 2×2 representation $E'_{1/2}$ and 4×4 representation Q are obtained by the representation of direct products:

$$E'_{1/2} = A_2 \times S(a, b)$$
$$Q = E \times S(a, b) \cong D^{(3/2)} \tag{11.3.12}$$

There are in total eight unirreps for O', corresponding to the eight classes of O' given by (11.2.14). The set also satisfies the completeness condition

$$1^2 + 1^2 + 2^2 + 3^2 + 3^2 + 2^2 + 2^2 + 4^2 = 48 = |O'| \tag{11.3.13}$$

Note that A_1, A_2, E, T_1 and T_2 are vector unirreps of O because the corresponding values of j are all integers whereas the remaining unirreps, $E_{1/2}$, $E'_{1/2}$ and Q, are projective unirreps of O belonging to half-integral j values. In Table 11.6, we have given only the most elementary basis sets for each irreducible representation. If necessary, additional basis sets can easily be obtained by the projection operator methods discussed in Chapter 6.

Remark. The unirreps of any double point group can be formed by reducing the unirreps $\{D^{(j)}\}$ of $SU(2)$ via its subgroup condition. For the group C'_∞, the subgroup condition is simply given, from (11.3.1), by

$$b = 0 \tag{11.3.14a}$$

which is the same as $|a| = 1$ because of the fixed condition $|a|^2 + |b|^2 = 1$. For the group C'_n the subgroup condition is given, from (11.3.3a), by

$$a^{2n} = 1 \tag{11.3.14b}$$

since its $2n$th roots $\{a_k = \mathrm{e}^{\mathrm{i}\pi k/n},\ k = 0, 1, \ldots, 2n-1\}$ determine the elements of $C'_n = \{(a_k, 0)\}$. From (11.3.5), the subgroup condition of D'_∞ is given by

$$ab = 0 \tag{11.3.15a}$$

whose solutions $(a, 0)$ and $(0, b)$ completely define all elements of D'_∞. Then from (11.2.9), the subgroup condition for D'_n is given

$$a^{2n} = 1 \qquad \text{or} \qquad (\mathrm{i}b)^{2n} = 1 \tag{11.3.15b}$$

The subgroup condition for O' has been given by (11.2.13) and the subgroup condition for T' will be discussed in the next sub-section. Originally, the general unirreps for the point groups given above were derived from these subgroup conditions (Kim 1981d).

11.3.6 The tetrahedral group T

This group is a subgroup of O and its double group T' is characterized by the subgroup condition

$$a^4 = 1 \qquad \text{or} \qquad b^4 = 1 \qquad \text{or} \qquad a^4 = b^4 = -\tfrac{1}{4} \tag{11.3.16}$$

imposed on the CK parameters of $S(a, b) \in SU(2)$. Note that the first two equations

Table 11.7. *The irreducible (vector and projective) representations of T*

T	$E, 3S(\pi\boldsymbol{n}), 8S((2\pi/3)\boldsymbol{n})$	Bases
Γ_1, A	1	1 or $\phi_-(3, 2)$ or xyz
Γ_2, A'	$a^4 + b^4 - \mathrm{i}2(3)^{1/2}a^2b^2$	$2^{-1/2}[\phi(2, 0) + \mathrm{i}\phi(2, 2)]$ or $2^{-1/2}(u + \mathrm{i}v)$
Γ_3, A''	$a^4 + b^4 + \mathrm{i}2(3)^{1/2}a^2b^2$	$2^{-1/2}[\phi(2, 0) - \mathrm{i}\phi_+(2, 2)]$ or $2^{-1/2}(u - \mathrm{i}v)$
Γ_4, T	T_1 of O	$[-\mathrm{i}\phi_-(1, 1), \mathrm{i}\phi_+(1, 1), \phi(1, 0)]$ or $[\mathrm{i}\phi_+(2, 1), -\mathrm{i}\phi_-(2, 1), \phi_-(2, 2)]$ or $[x, y, z]$ or $[yz, zx, xy]$
$\Gamma_5, E_{1/2}$	$S(a, b)$	$[\phi(\frac{1}{2}, \frac{1}{2}), \phi(\frac{1}{2}, -\frac{1}{2})]$
$\Gamma_6, E'_{1/2}$	$A' \times S(a, b)$	
$\Gamma_7, E''_{1/2}$	$A'' \times S(a, b)$	

1. $u = 3z^2 - r^2, v = 3^{1/2}(x^2 - y^2)$.
2. The first four unirreps are vector unirreps whereas the remaining unirreps are projective unirreps. $\Gamma_1 \sim \Gamma_7$ are the notations used by Koster *et al.* (1963).
3. The parameter sets for $\{S(a_\nu, b_\nu)\} \in \check{T}$ are given by $\nu = 1{-}12$ in Table 11.3.

define the group D'_2 whereas the remaining equations define the three-fold rotations just like in O'. These parameter sets $\{(a_\nu, b_\nu)\}$ are given in Table 11.3 with $\nu = 1{-}12$.

The irreducible representations of T given in Table 11.7 are obtained by reducing those of O given in Table 11.6. In fact, by using the subgroup conditions (11.3.16) one can easily show that

$$A_1, A_2 \to A, \qquad E \to A' + A'', \qquad T_1, T_2 \to T$$
$$Q = E \times E_{1/2} \to A' \times E_{1/2} + A'' \times E_{1/2} \tag{11.3.17}$$

Here we shall show only the reduction $E \to A' + A''$ since then the rest will be established accordingly. Firstly, the two-dimensional unirrep E of O' is simplified to E' given, using (11.3.16), by

$$E \to (a^4 + b^4)\mathbf{1} - 2\sqrt{3}\mathrm{i}a^2b^2\boldsymbol{\sigma}_y \equiv E'$$

Secondly, E' is diagonalized to the following form:

$$YE'Y = (a^4 + b^4)\mathbf{1} - 2\sqrt{3}\mathrm{i}a^2b^2\sigma_z$$

using (2.1.4) with an involutional transformation Y that diagonalizes σ_y to σ_z:

$$Y = (\sigma_y + \sigma_z)/\sqrt{2} = \frac{1}{\sqrt{2}}\begin{bmatrix} 1 & -\mathrm{i} \\ \mathrm{i} & -1 \end{bmatrix}; \qquad Y^2 = \mathbf{1}$$

Accordingly, the transformed basis is given by

$$[\phi(2, 0), \phi_+(2, 2)]Y = 2^{-1/2}[\phi(2, 0) + \mathrm{i}\phi_+(2, 2), -\mathrm{i}\phi(2, 0) - \phi_+(2, 2)] \tag{11.3.18}$$

which is equivalent to a basis $[u + \mathrm{i}v, u - \mathrm{i}v]$ for $T \in SO(3, r)$ where $u = 3z^2 - r^2$ and $v = 3^{1/2}(x^2 - y^2)$.

The unirreps of the double group T' given by Table 11.7 provide a complete set, because

$$1^2 + 1^2 + 1^2 + 3^2 + 2^2 + 2^2 + 2^2 = 24 = |T'|$$

Concluding remark. We have constructed the vector and projective unirreps of C_∞, C_n, D_∞, D_n, T and O by reducing the general unirreps $D^{(j)}$ of $SU(2)$ through the Cayley–Klein parametrization. The representations thus obtained are general unirreps in the sense that they hold for any element of the respective double group, being characterized by the general CK parameters of the group. The present results are easily extended to improper point groups because an improper point group is isomorphic to a proper point group or a rotation–inversion group; cf. Table 5.7 or the character tables for the crystallographic point groups given in the Appendix.

12
Projective representations

In the previous chapter, we have seen that the double-valued representations of a point group have been described very effectively as *the projective or ray representations* of the point group. In this chapter, we shall discuss the general structure of the projective unirreps of a finite group G and show that they can be constructed from the vector unirreps of the so-called *representation group* of G. Then, introducing the basic theorem that the representation group of a proper double point group P' is P' itself (Kim 1983a), we shall form all the projective unirreps of the double point groups through the vector representations of their representation groups. These will be applied to form the unirreps of the wave vector space groups in Chapter 14. For the projective representations of infinite point groups see Kim (1984c).

12.1 Basic concepts

Let $G = \{g\}$ be a finite group. Then, extending a vector representation, we define a *projective representation* of G by a set of non-singular square matrices $\check{D}(G) = \{\check{D}(g);\, g \in G\}$ defined on G that satisfies

$$\check{D}(g)\check{D}(s) = \lambda(g, s)\check{D}(gs); \qquad \forall\, g, s \in G \tag{12.1.1}$$

where $\lambda(g, s)$ is a non-zero complex number that depends on g and s. The set of $|G|^2$ numerical factors $\{\lambda(g, s)\}$ is called a *factor system* of the group G, whereas the set $\check{D}(G)$ is called a *projective representation* of G belonging to the factor system $\{\lambda(g, s)\}$. A vector representation of G is a projective representation belonging to the trivial *unit factor system*

$$\lambda(g, s) = 1, \qquad \text{for all } g, s \in G$$

Obviously, the identity representation of any group G necessarily belongs to the trivial unit factor system. In a case in which the projective representation is unitary, i.e. $\check{D}(g)^\dagger = \check{D}(g)^{-1}$ for all $g \in G$, the factor system is unimodular, $|\lambda(g, s)| = 1$, from the defining relations (12.1.1). Not every set of $|G|^2$ numbers is acceptable for a factor system of G; the necessary and sufficient condition for a factor system is given by the following theorem.

Theorem 12.1.1. Let G be a finite group, then a set of $|G|^2$ numbers $\{\lambda(g, s)\}$ provides a factor system of G, if and only if the set satisfies

$$\lambda(g, s)\lambda(gs, t) = \lambda(g, st)\lambda(s, t) \tag{12.1.2}$$

for all $g, s, t \in G$.

This condition will be referred to as *the associativity condition* for a factor system of the group of G. It is independent of any particular factor system, so it is a property of the group G.

Proof. Suppose that $\{\lambda(g, s)\}$ is a factor system of G and let $\check{D}(G) = \{\check{D}(g)\}$ be a projective representation belonging to the system. From the associative law of a matrix representation, we have

$$(\check{D}(g)\check{D}(s))\check{D}(t) = \check{D}(g)(\check{D}(s)\check{D}(t)); \qquad \forall\, g, s, t \in G$$

Then repeated applications of (12.1.1) to both sides of this equation lead to the condition (12.1.2): note that $\lambda(gs, t)$ is the projective factor for the product $D(gs)D(t)$.

Conversely, for every system of $|G|^2$ non-vanishing constants $\{\lambda(g, s)\}$ that satisfies the condition (12.1.2), there exists a projective representation of G with this system as the factor system. To prove this we order the elements of G by the subscripts such that $G = \{g_\nu;\ \nu = 1, 2, \ldots, |G|\}$ as for the regular representation $D^{(\mathrm{R})}(G)$ introduced in Section 6.2.2. Then a required representation is given by *the projective regular representation* of G defined by the $|G| \times |G|$ matrix system:

$$\check{D}^{(\mathrm{R})}(g)_{\nu\mu} = \lambda(g, g_\mu)\delta(g_\nu, gg_\mu); \qquad \forall\, g \in G; \qquad \nu, \mu = 1, 2, \ldots, |G| \quad (12.1.3)$$

where $\delta(g_\nu, gg_\mu)$ is Kronecker's delta; in fact, straightforward substitution of (12.1.3) into the left-hand side of (12.1.1) leads to the right-hand side of (12.1.1):

$$\begin{aligned}\sum_\mu \check{D}^{(\mathrm{R})}(g)_{\nu\mu}\check{D}^{(\mathrm{R})}(s)_{\mu\kappa} &= \sum_\mu \lambda(g, g_\mu)\delta(g_\nu, gg_\mu)\lambda(s, g_\kappa)\delta(g_\mu, sg_\kappa)\\ &= \lambda(g, sg_\kappa)\lambda(s, g_\kappa)\delta(g_\nu, gsg_\kappa)\\ &= \lambda(g, s)\lambda(gs, g_\kappa)\delta(g_\nu, gsg_\kappa)\\ &= \lambda(g, s)\check{D}^{(\mathrm{R})}(gs)_{\nu\kappa}\end{aligned}$$

where we have used, in the second step, the property of the delta function and in the third step the condition (12.1.2) backward with $t = g_\kappa$, and then the last step follows from the definition (12.1.3) of $\check{D}^{(\mathrm{R})}(gs)$. Q.E.D.

Remark 1. The projective regular representation $\check{D}^{(\mathrm{R})}(G)$ of G with a factor system defined by (12.1.3) plays the role that the ordinary regular representation $D^{(\mathrm{R})}(G)$ of G plays in the theory of vector representations (see (12.2.4)).

Remark 2. In general, a factor system $\{\lambda(g, s)\}$ is not symmetric with respect to the elements g and s. As a result, even if g and s commute, their projective representatives need not commute unless the factor system is symmetric, i.e. $\lambda(g, s) = \lambda(s, g)$, because $gs = sg$ leads to

$$\check{D}(g)\check{D}(s) - \check{D}(s)\check{D}(g) = [\lambda(g, s) - \lambda(s, g)]\check{D}(gs)$$

For example, consider the projective representations of the dihedral group D_2 (which is Abelian) based on the spinor representation defined by the Pauli matrices as follows:

$$\check{D}(e) = \sigma_0, \qquad \check{D}(2_x) = -\mathrm{i}\sigma_x, \qquad \check{D}(2_y) = -\mathrm{i}\sigma_y, \qquad \check{D}(2_z) = -\mathrm{i}\sigma_z \quad (12.1.4)$$

where σ_0 is the 2×2 unit matrix. From the properties of the Pauli matrices; e.g.

Table 12.1. *The factor system* $\{\lambda(g, s)\}$ *for* $\check{D}_2$ *based on the spinor representation*

g \ s	e	2_x	2_y	2_z
e	1	1	1	1
2_x	1	-1	1	-1
2_y	1	-1	-1	1
2_z	1	1	-1	-1

$\sigma_x\sigma_y = -\sigma_y\sigma_x = \sigma_z$, we obtain the factor system given by Table 12.1. From Table 12.1, we see that $\lambda(2_x, 2_y)$ is not symmetric, i.e. $\lambda(2_x, 2_y) = -\lambda(2_y, 2_x) = 1$. This is consistent with the fact that the projective representatives $\check{D}(2_x)$ and $\check{D}(2_y)$ do not commute even though 2_x and 2_y commute.

Remark 3. The following particular factors are symmetric:

$$\lambda(g, e) = \lambda(e, g) = \lambda(e, e)$$
$$\lambda(g, g^{-1}) = \lambda(g^{-1}, g) \tag{12.1.5}$$

where e is the identity element and g is any element of G. The relation $\lambda(g, e) = \lambda(e, e)$ follows if we set $s = t = e$ in the associative condition (12.1.2). Analogously, we have $\lambda(e, g) = \lambda(e, e)$. The last relation in (12.1.5) follows if we set $s = g^{-1}$ and $t = g$ in (12.1.2) and use $\lambda(e, g) = \lambda(g, e)$.

From (12.1.5), it follows that the representative $\check{D}(e)$ of the identity element of G commutes with any representative $\check{D}(g)$ and that $\check{D}(g)$ commutes with $\check{D}(g^{-1})$ for any $g \in G$. Here $\check{D}(g^{-1})$ should not be confused with the matrix $\check{D}(g)^{-1}$, which is the inverse of $\check{D}(g)$, because

$$\check{D}(g^{-1}) = \lambda(g^{-1}, g)\check{D}(e)\check{D}(g)^{-1}$$

where $\lambda(g^{-1}, g) = \lambda(g, g^{-1})$. This relation will become slightly simplified if we introduce the standard factor system for which $\check{D}(e) = 1$; see (12.2.3).

12.2 Projective equivalence

The terms equivalence and irreducibility have the same meanings for projective representations as they do for vector representations. Let $\check{D}(G)$ and $\check{D}'(G)$ be two projective representations of a group G. Then, if there exists a non-singular matrix T such that $\check{D}'(g) = T\check{D}(g)T^{-1}$ for all $g \in G$, then they are equivalent. Two equivalent projective representations belong to the same factor system, since (12.1.1) is invariant under a similarity transformation.

A projective representation $\check{D}(G)$ is reducible if it is equivalent to a direct sum of projective representations of lower dimensions. It is irreducible otherwise.

Next we shall introduce the concept of projective equivalence: two representations $\check{D}(G)$ and $\check{D}'(G)$ of a group G are *projective-equivalent (or p-equivalent)* if there exists a non-vanishing function $\mu(g)$ on $G = \{g\}$ and a non-singular matrix T such that

$$\check{D}'(g) = T\check{D}(g)T^{-1}/\mu(g), \qquad \forall\, g \in G \tag{12.2.1}$$

where $\mu(g)$ ($\neq 0$) is a single-valued function of G and is called *the gauge factor.* It should be noted that the set $\{\mu(g);\ g \in G\}$ is not required to form a representation of G so that $\mu(g)\mu(s)$ may not be equal to $\mu(gs)$. The special case of (12.2.1), $D'(g) = D(g)/\mu(g)$, is called the *gauge transformation.* The ordinary equivalence is p-equivalence with the trivial gauge factor $\mu(g) = 1$. To determine the factor system for the p-equivalent representation $\check{D}'(G)$, we divide both sides of (12.1.1) by $\mu(g)\mu(s)$ to obtain

$$\frac{\check{D}(g)}{\mu(g)}\frac{\check{D}(s)}{\mu(s)} = \lambda(g, s)\frac{\mu(gs)}{\mu(g)\mu(s)}\frac{\check{D}(gs)}{\mu(gs)} \tag{12.2.2a}$$

which implies that the factor system of $\check{D}'(g) = T\check{D}(g)T^{-1}/\mu(g)$ is given by

$$\lambda'(g, s) = \lambda(g, s)\mu(gs)/[\mu(g)\mu(s)] \tag{12.2.2b}$$

Two factor systems, $\{\lambda'(g, s)\}$ and $\{\lambda(g, s)\}$, related by (12.2.2b) are said to be *mutually p-equivalent* or *gauge equivalent.* In general, two p-equivalent factor systems are different unless $\{\mu(g)\}$ is a vector representation of G so that $\mu(gs) = \mu(g)\mu(s)$ for all $g, s \in G$.

By definition, p-equivalence is transitive, as is the ordinary equivalence. A set of all p-equivalent factor systems is called a *p-equivalence (gauge equivalence) class* of factor systems. By definition, factor systems belonging to different p-equivalence classes are not related by a gauge transformation: they are different types or p-inequivalent. The word 'class' becomes more meaningful when we establish that the whole set of factor systems of a group also forms a group.

It is to be noted that a one-dimensional projective representation of any group G is necessarily p-equivalent to the identity representation. To see this, let $\check{D}(G)$ be a one-dimensional projective representation satisfying (12.2.2a). If we let the representative $\check{D}(g)$ itself be the gauge factor $\mu(g)$ in (12.2.2a), then we arrive at the identity representation belonging to the unit factor system, $\lambda(g, s) = 1$.

12.2.1 Standard factor systems

In a projective representation, the representative $\check{D}(e)$ of the identity element $e \in G$ is not necessarily equal to the unit matrix $\mathbf{1}$, since by definition

$$\check{D}(e)\check{D}(e) = \lambda(e, e)\check{D}(e) \tag{12.2.3a}$$

so that

$$\check{D}(e) = \lambda(e, e)\mathbf{1}$$

This unsatisfactory feature is easily modified: divide both sides of the defining equation (12.1.1) by $[\lambda(e, e)]^2$ and introduce a gauge transform $\check{D}'(g) = \check{D}(g)/\lambda(e, e)$ for all $g \in G$, then we have

$$\check{D}'(g)\check{D}'(s) = \lambda'(g, s)\check{D}'(gs) \tag{12.2.3b}$$

where $\lambda'(g, s) = \lambda(g, s)/\lambda(e, e)$, which satisfies $\lambda'(e, e) = 1$. A factor system

$\{\lambda(g, s)\}$ is called *a standard factor system*, if $\lambda(e, e) = 1$, in which case $\check{D}(e) = \mathbf{1}$ from (12.2.3a). For a standard factor system we have from (12.1.5)

$$\lambda(g, e) = \lambda(e, g) = \lambda(e, e) = 1 \tag{12.2.3c}$$

Hereafter, every factor system is assumed to be standardized unless otherwise specified. A projective representation belonging to a standard factor system may be called a *standard projective representation.*

Exercise. Show that the character $\check{\chi}^{(R)}$ of *a standard projective regular representation* $\check{D}^{(R)}(G)$ of a group G is independent of the factor system and equal to the character $\chi^{(R)}$ of the ordinary regular representation of G given by (6.6.11).

Solution. From the trace of $\check{D}^{(R)}(G)$ defined by (12.1.3), its character is given by

$$\check{\chi}^{(R)}(g) = \sum_{\mu} \lambda(g, g_\mu)\delta(g_\mu, gg_\mu)$$

$$= \begin{cases} |G|, & \text{if } g = e \\ 0, & \text{if } g \neq e \end{cases} \tag{12.2.4}$$

where we have used $\sum_\mu \lambda(e, g_\mu) = \sum_\mu 1 = |G|$, because $\lambda(e, g_\mu) = 1$ from (12.2.3c). Q.E.D.

12.2.2 *Normalized factor systems*

Let $\{\lambda_0(g, s)\}$ be a factor system of a finite group G. If all members of the factor system are $|G|$th roots of unity, i.e.

$$[\lambda_0(g, s)]^{|G|} = 1; \qquad \forall\, g, s \in G \tag{12.2.5}$$

then the factor system is called a normalized factor system. By definition, every member of it is of modulus unity: $|\lambda_0(g, s)| = 1$.

Theorem 12.2.1. Every factor system of a finite group G is p-equivalent to a normalized factor system.

Proof. Let $\check{D}(G) = \{\check{D}(g)\}$ be a projective representation of G belonging to a given factor system $\{\lambda(g, s)\}$, then taking the determinants of both sides of the defining equation (12.1.1) and denoting $\Delta_g = \det \check{D}(g)$, we obtain

$$\Delta_g \Delta_s = [\lambda(g, s)]^d \Delta_{gs} \tag{12.2.6a}$$

where d is the dimensionality of $\check{D}(G)$. For the projective regular representation $\check{D}^{(R)}(G)$ belonging to the factor system $\{\lambda(g, s)\}$, we have $d = |G|$ so that (12.2.6a) becomes

$$\Delta_g^{(R)} \Delta_s^{(R)} = [\lambda(g, s)]^{|G|} \Delta_{gs}^{(R)}$$

Thus we arrive at *a normalized factor system*

$$\lambda_0(g, s) = \lambda(g, s)[\Delta_{gs}^{(R)}/(\Delta_g^{(R)}\Delta_s^{(R)})]^{1/|G|} \tag{12.2.6b}$$

which is p-equivalent to $\{\lambda(g, s)\}$. Q.E.D.

In principle, there exists an infinite number of factor systems $\{\lambda(g, s)\}$ which satisfy the associativity condition (12.1.2). From Theorem 12.2.1, however, these can be classified into a finite number of p-equivalence classes for a finite group G owing to the following corollary.

Corollary 12.2.1. There exists only a finite number of p-equivalence classes of factor systems for a finite group G.

Proof. From Theorem 12.2.1, the number of p-equivalence classes is given by the number of *p-equivalent normalized factor systems*, whose elements are $|G|$th roots of unity. Since the numbers of $|G|$th roots of unity equals $|G|$, the number of possible distributions of these $|G|$ roots on $|G|^2$ factors of a factor system equals

$$|G| \times |G| \times \cdots \times |G| = |G|^{|G|^2}$$

which must be an upper bound of the number of p-equivalence classes of the factor systems for the finite group G. Q.E.D.

12.2.3 Groups of factor systems and multiplicators

We shall begin with the following theorem.

Theorem 12.2.2. Let $\{\{\lambda(g, s)\}\}$ be the whole set of factor systems of a finite group G, then the set forms a group F under the multiplication law

$$\{\lambda(g, s)\}\{\lambda'(g, s)\} = \{\lambda(g, s)\lambda'(g, s)\} \tag{12.2.7}$$

for every pair of elements g and s in G. It is Abelian and called *the group of factor systems* of G.

Proof. If F is a group, then it is Abelian because the factor systems commute under the multiplication law (12.2.7) from the fact that each factor is simply a number. Now, the product of two factor systems defined by (12.2.7) also provides a factor system, since it satisfies the associativity condition (12.1.1) from the fact that each factor system satisfies the condition. The identity element of F is given by the unit factor system $\{\lambda(g, s) \equiv 1\}$ and the inverse $\{\lambda(g, s)\}^{-1}$ is given by $\{\lambda(g, s)^{-1}\}$. Accordingly, F forms a group which is Abelian. Q.E.D.

Let us look into the structure of the group of factor systems F of G. From Corollary 12.2.1, the number of p-equivalence classes of factor systems is finite for a finite group G. Suppose that there exist m distinctive p-equivalence classes of factor systems $\{K_\nu;\ \nu = 0, 1, \ldots, m-1\}$ for G. Then from (12.2.2b) and Theorem 12.2.1, the elements of K_ν are given by a p-equivalence set of factor systems

$$K_\nu = \{\lambda_0^{(\nu)}(g, s)\mu(gs)/\mu(g)\mu(s)\}; \qquad \forall\, g, s \in G \tag{12.2.8}$$

where $\{\lambda_0^{(\nu)}(g, s)\}$ is a normalized factor system belonging to K_ν. Let K_0 be the unit class containing the unit factor system $\{\lambda_0^{(0)}(g, s) = 1\}$, then its elements are given by the trivial factor systems

$$K_0 = \{\mu(gs)/\mu(g)\mu(s)\}$$

which forms an invariant subgroup of F because F is Abelian. Moreover, from

(12.2.8) each class of factor systems K_ν is a coset of K_0 in F with the coset representative $\{\lambda_0^{(\nu)}(g, s)\}$. Thus the set of p-equivalence classes $\{K_\nu\}$ forms the factor group of F by K_0:

$$M = F/K_0 \tag{12.2.9}$$

which is called *the multiplicator* of G. Since F is Abelian, the multiplicator M is also Abelian and the order $|M|$ equals the number m of the distinctive p-equivalence classes of factor systems in G. Thus, we obtain the following theorem.

Theorem 12.2.3. The set of all p-equivalence classes of factor systems of a finite group G forms an Abelian group of finite order called *the multiplicator* M of G.

In Section 12.4, we shall introduce the concept of a '*representation group*' G' of G, which is a group of minimum order whose vector irreps provide all the projective irreps of G. It will be shown that the order $|G'|$ of the representation group G' of G is given by

$$|G'| = |M| \times |G|$$

where $|M|$ is the order of the multiplicator M.

12.2.4 Examples of projective representations

Example 12.2.1. Every projective representation of a cyclic group C_n is p-equivalent to a vector representation. The proof will be based on the defining relation of the group C_n, since then we shall be concerned with only the representative of the group generator. Let a be the generator of C_n, then the defining relation of C_n is given by

$$a^n = e \tag{12.2.10a}$$

Let $\{\check{D}(a^k);\ k = 1, 2, \ldots, n\}$ be a standard projective representation of C_n, then $\check{D}(e) = \mathbf{1}$ so that

$$[\check{D}(a)]^n = \alpha\mathbf{1}$$

where α is a constant that depends on the factor system. If we introduce a gauge transformation

$$D(a) = \check{D}(a)/\alpha^{1/n} \tag{12.2.10b}$$

then $D(a)$ satisfies $[D(a)]^n = \mathbf{1}$, which is identical in form with the defining relation of the group C_n. Thus we can define a vector representation of C_n by, in terms of $D(a)$,

$$D(a^k) \equiv [D(a)]^k; \qquad k = 1, 2, \ldots, n \tag{12.2.10c}$$

Accordingly, it follows from (12.2.10b) that any projective representation $\check{D}(C_n)$ of C_n is p-equivalent to a vector representation $D(C_n)$. This means that there exists only the unit class of factor systems K_0 for a cyclic group and hence the multiplicator of C_n is K_0 itself.

Remark. Previously, in (11.3.3b), we have introduced the projective irreps of C_n by the double-valued (spinor) representations:

$$M_m(c_n^k) = \mathrm{e}^{-\mathrm{i}2\pi mk/n}; \qquad m = \tfrac{1}{2}, \tfrac{3}{2}, \ldots, n - \tfrac{1}{2}$$

where the m are all half integers. Let $m = m_0 + \frac{1}{2}$, where m_0 is an integer, then we have

$$M_{m_0+1/2}(c_n^k) = M_{m_0}(c_n^k)\,\mathrm{e}^{-\mathrm{i}\pi k/n} \tag{12.2.11}$$

i.e. the projective representation $M_{m_0+1/2}(C_n)$ is p-equivalent to the vector representation $M_{m_0}(C_n) = \{M_{m_0}(c_n^k)\}$. This does not mean that the double-valued representations are useless, because they are essential for describing the transformation of spinors. At the beginning of this section, we have shown also that a one-dimensional projective representation of a finite group is p-equivalent to the identity representation.

Example 12.2.2. The projective representations of the dihedral group D_n. This provides a prototype example of projective representations.

The defining relations of D_n are

$$a^n = b^2 = (ab)^2 = e \tag{12.2.12a}$$

Let $\check{D}(G)$ be a standard projective representation of $G = D_n$. Through simple gauge transformations, as introduced in Example 1, we may assume that $\check{D}(G)$ satisfies

$$\check{D}(a)^n = \check{D}(b)^2 = \mathbf{1}, \qquad [\check{D}(a)\check{D}(b)]^2 = \alpha\mathbf{1} \tag{12.2.12b}$$

where α is a constant that depends on the factor system of D_n. If we rewrite the last equation in the form $\check{D}(a) = \alpha\check{D}(b)\check{D}(a)^{-1}\check{D}(b)$, then its nth power yields $\alpha^n = 1$ so that α is given by the nth roots of unity:

$$\alpha = \exp(+\mathrm{i}2\pi m/n); \qquad m = 0, 1, \ldots, n-1 \tag{12.2.12c}$$

By a further gauge transformation, for a given m,

$$A = \check{D}(a)\,\mathrm{e}^{-\mathrm{i}\pi m/n}, \qquad B = \check{D}(b)\,\mathrm{e}^{-\mathrm{i}\pi m/2}$$

Equation (12.2.12b) is simplified to

$$A^n = B^2 = (AB)^2 = \gamma\mathbf{1}; \qquad \gamma = \mathrm{e}^{-\mathrm{i}\pi m} = \pm 1 \tag{12.2.13}$$

If we introduce $E' = \gamma\mathbf{1}$ as an abstract group element of order 2 satisfying $E'^2 = \mathbf{1} = E$, then the above set of equations defines the double group D'_n of D_n, where E' is in the center of D'_n. Thus, the projective irreps of D_n are provided by the vector irreps of the double group D'_n, as was discussed in Chapter 11.

To see the possibility of mapping off γ in (12.2.13), we introduce a further gauge transformation $x = A/\xi$, $y = B/\eta$ with gauge factors ξ and η. Then we obtain, for $\gamma = -1$,

$$x^n\xi^n = y^2\eta^2 = (xy)^2\xi^2\eta^2 = -\mathbf{1}$$

These are reduced to the form of the defining relations of D_n

$$x^n = y^2 = (xy)^2 = \mathbf{1} \tag{12.2.14a}$$

if and only if

$$\xi^n = -\xi^2 = -1, \qquad \eta^2 = -1 \tag{12.2.14b}$$

There are two cases.

(i) When n is odd, the set has solutions $\xi = -1$, $\eta = \pm\mathrm{i}$ so that $\gamma = -1$ is mapped off by the gauge transformation. This means that there exists only the unit class K_0 of p-equivalence factor systems for D_n with an odd n. Thus the projective representations for D_n given by the spinor representations in Section 11.3 are p-equivalent to vector representations for an odd n.

(ii) When n is even, it is impossible to map off $\gamma = -1$ by a gauge transformation. This means that there exist two p-equivalence classes of factor systems for D_n characterized by $\gamma = 1$ and -1, respectively. These classes may be denoted by $K(\gamma = 1)$ and $K(\gamma = -1)$, then the multiplicator M of D_n is defined by the set $\{K(1), K(-1)\}$ with the group property

$$K(1)^2 = K(1), \qquad K(1)K(-1) = K(-1), \qquad K(-1)^2 = K(1) \tag{12.2.15}$$

It is an Abelian group of order 2 with $K(1)$ as the identity. Moreover, the Abelian subgroup $H' = \{E, E'\}$ in the center of D'_n is isomorphic to the multiplicator $M = \{K(1), K(-1)\}$ of D_n via the correspondence

$$E \leftrightarrow K(1), \qquad E' \leftrightarrow K(-1) \tag{12.2.16}$$

This is a special case of the general relationship between a group G and the so-called *representation group* G' of G, as will be discussed later in Section 12.4. Also, note that the order of the group D'_n satisfies $|D'_n| = |M| \times |D_n|$, where $|M| = 2$ for this case.

12.3 The orthogonality theorem on projective irreps

Theorem 12.3.1. Every projective representation of a finite group G is equivalent to a unitary projective representation belonging to the same factor system.

Proof. Let $\check{D}(G)$ be a projective representation of G belonging to a normalized factor system $\{\lambda(g, s)\}$. Then, a positive definite Hermitian matrix P defined by

$$P = \sum_{S \in G} \check{D}(s)\check{D}(s)^\dagger$$

satisfies $\check{D}(g)P\check{D}(g)^\dagger = P$ for all $g \in G$. Thus it follows, as in Theorem 6.5.1 for a vector representation, that the similarity transform defined by

$$\check{U}(g) = P^{-1/2}\check{D}(g)P^{1/2}, \qquad \forall\, g \in G$$

satisfies the unitary condition $\check{U}(g)\check{U}(g)^\dagger = \mathbf{1}$ and hence provides the required unitary projective representation belonging to the factor system $\{\lambda(g, s)\}$.

Exercise. The projective regular representation $\check{D}^{(R)}(G)$ belonging to a normalized factor system is unitary. Use the complex conjugate of (12.1.3) with $|\lambda_0(g, g_\mu)| = 1$.

Theorem 12.3.2. (The orthogonality theorem.) Let $\check{D}^{(\alpha)}(G)$ and $\check{D}^{(\beta)}(G)$ be two projective unirreps of G belonging to a normalized factor system, then the following orthogonality relations hold:

$$\sum_g \check{D}^{(\alpha)}_{ij}(g)\check{D}^{(\beta)}_{i'j'}(g)^* = (|G|/d_\alpha)\delta_{\alpha\beta}\delta_{ii'}\delta_{jj'} \tag{12.3.1}$$

where d_α is the dimensionality of $\check{D}^{(\alpha)}(G)$. These are identical in form to the orthogonality relations for vector unirreps given by (6.5.11a).

Proof. Exactly like in Theorem 6.5.3 on vector unirreps, we define a rectangular matrix, analogous to (6.5.9),

$$M = \sum_{s \in G} \check{D}^{(\alpha)}(s) X \check{D}^{(\beta)}(s)^{\dagger} \tag{12.3.2a}$$

where X is a $d_\alpha \times d_\beta$ rectangular matrix. Then

$$\check{D}^{(\alpha)}(g) M = M \check{D}^{(\beta)}(g), \qquad \forall g \in G \tag{12.3.2b}$$

To show this, note first that

$$\check{D}^{(\alpha)}(g) M = \sum_{s \in G} \lambda(g, s) \check{D}^{(\alpha)}(gs) X \check{D}^{(\beta)}(s)^{\dagger} \tag{12.3.2c}$$

where $\check{D}^{(\beta)}(s)^{\dagger}$ is expressed, from the adjoint of (12.1.1) and the unitarity of $\check{D}^{(\beta)}(g)$, by

$$\check{D}^{(\beta)}(s)^{\dagger} = \lambda(g, s)^{*} \check{D}^{(\beta)}(gs)^{\dagger} \check{D}^{(\beta)}(g) \tag{12.3.2d}$$

Substitution of this into (12.3.2c) leads to (12.3.2b) with the use of $|\lambda(g, s)| = 1$. Then, by applying Schur's lemma (6.5.5) to (12.3.2b), we arrive at the orthogonality relations (12.3.1).

Since the orthogonality relations for projective unirreps of G are independent of the factor system as long as they belong to the same factor system, all algebraic properties derived from the orthogonality relations are the same for projective or vector unirreps:

(i) The orthogonality relations for the characters are

$$\sum_{g \in G} \check{\chi}^{(\alpha)}(g)^{*} \check{\chi}^{(\beta)}(g) = |G| \delta_{\alpha\beta} \tag{12.3.3}$$

(ii) The frequency f_α of a projective unirrep $\check{D}^{(\alpha)}(G)$ contained in a projective representation $\check{D}(G)$ is given by

$$f_\alpha = \sum_{g} \check{\chi}^{*}(g) \check{\chi}^{(\alpha)}(g) / |G| \tag{12.3.4}$$

where $\check{\chi}(g)$ is the trace of $\check{D}(g)$. Hence the sum of the absolute square $|\check{\chi}(g)|^2$ is given by

$$\sum_{g} |\check{\chi}(g)|^2 = |G| \sum_{\alpha} f_\alpha^2 \tag{12.3.5}$$

so that the irreducibility criterion for $\check{D}(G)$ is given by

$$\sum_{g} |\check{\chi}(g)|^2 = |G| \tag{12.3.6}$$

(iii) Finally, the completeness condition for a set of projective unirreps $\{\check{D}^{(\alpha)}(G)\}$ belonging to a factor system is given by

$$\sum_{\alpha} (d_\alpha)^2 = |G| \tag{12.3.7}$$

where d_α is the dimensionality of the projective unirrep $\check{D}^{(\alpha)}(G)$.

The completeness condition given in (iii) can be shown exactly in the same way as for vector unirreps, using the fact that the character of a projective regular representation $\breve{D}^{(R)}(G)$ equals the character of the ordinary regular representation $D^{(R)}(G)$ given by (12.2.4); assuming, of course, that these projective representations are standard.

Remark. For many aspects, the parallelism between projective and vector unirreps of a group G is remarkable, but some of their algebraic properties can be very different; for example, the elements of a group G belonging to the same conjugacy class of G need not have the same character in projective representations: suppose that $tst^{-1} = g$, where $g, s, t \in G$, then the characters for two equivalent elements s and g need not be the same, because they are related by

$$\breve{\chi}(s) = \lambda(t, s)\lambda(ts, t^{-1})\lambda(t^{-1}, t)^{-1}\breve{\chi}(g) \tag{12.3.8}$$

where the representation is assumed to be standard. Since equivalent elements of a group G need not have the same characters for a projective representation, the number of irreducible projective representations of G has nothing to do with the number of the *conjugacy classes* of G.

12.4 Covering groups and representation groups

12.4.1 Covering groups

Let G be a group and G' be another group that is homomorphic to G. If the kernel of the homomorphism H' is in the center of G', then G' is called *a covering group of G* extended by the subgroup H' in the center of G'. Obviously, H' is Abelian, being in the center of G', and there exists an isomorphism $G'/H' \simeq G$. A covering group G' of G is also called a *central extension* of G by a subgroup H' in the center of G'.

Example. The double group G' of a point group G is a covering group of G with the kernel $H' = \{E, E'\}$, where E is the identity and E' is the '2π rotation.' See also Example 12.2.2.

Theorem 12.4.1. Let G be a group and G' be a covering group of G extended by a subgroup H' in the center of G'. Then a vector irrep $D(G')$ of G' provides a projective irrep of G belonging to a factor system defined by the irrep $D(H')$ of H' subduced by $D(G')$. This theorem is the most basic for the projective representations.

Proof. Let $G = \{g_r\}$ and let the left coset decomposition of G' by $H' = \{h'\}$ be $G' = \sum_r g'_r H'$. Then, on account of the isomorphism $G'/H' \simeq G$, there exists the one-to-one correspondence

$$g'_r \leftrightarrow g_r; \qquad r = 1, 2, \ldots, |G| \tag{12.4.1}$$

and the homomorphism relation

$$g'_r g'_s = h'_{r,s} g'_{rs} \qquad \text{when } g_r g_s = g_{rs} \in G \tag{12.4.2}$$

between G' and G, where $h'_{r,s} \in H'$ depends on g_r and g_s. The set of coset representatives $\{g'_r\}$ is called a *projective set* of the group G. Its choice may affect the factor system through $h'_{r,s}$ in (12.4.2).

Let $D(G') = \{D(g'_r h');\ g'_r h' \in G'\}$ be a vector unirrep of G', then the representation of the homomorphism relation (12.4.2) by $D(G')$ gives

$$D(g'_r)D(g'_s) = D(h'_{r,s})D(g'_{rs}) \tag{12.4.3}$$

where $D^{\downarrow}(H') = \{D(h');\ h' \in H'\}$ is the subduced representation of $D(G')$ onto the subgroup H'. According to the Clifford theorem, Theorem 8.5.5, the subduced representation $D^{\downarrow}(H')$ is equivalent to a multiple of the direct sum of the irreps in a certain orbit of H' relative to G'. Since H' is Abelian, every one of its vector irreps is one-dimensional so that *every orbit of an irrep of H' is the irrep itself.* Thus $D(h')$ is expressed by

$$D(h') = \gamma(h')\mathbf{1}, \qquad \forall\, h' \in H' \tag{12.4.4}$$

where $\gamma(H')$ is a one-dimensional vector irrep of the Abelian subgroup H' and $\mathbf{1}$ is the unit matrix with the dimensionality equal to that of $D(G')$ (cf. Equation (8.5.6)). When (12.4.4) holds, the irrep $D(G')$ is said to be *associated with the irrep $\gamma(H')$*: apparently there may exist more than one associated irrep of G' for a given irrep $\gamma(H')$ in general (see Section 8.4.3, case I).

Now, in view of the one-to-one correspondence (12.4.1), Equations (12.4.3) and (12.4.4) imply that a matrix system defined by

$$\check{D}(g_r) \equiv D(g'_r), \qquad \forall\, g_r \in G \tag{12.4.5a}$$

is a projective representation of G satisfying

$$\check{D}(g_r)\check{D}(g_s) = \lambda^{(\gamma)}(g_r, g_s)\check{D}(g_r g_s) \tag{12.4.5b}$$

with the factor system

$$\lambda^{(\gamma)}(g_r, g_s) = \gamma(h'_{r,s}), \qquad \forall\, g_r, g_s \in G \tag{12.4.5c}$$

which is standard, if the identity element $g'_1 = e'$ of G' corresponds to $g_1 = e$ of G, since then $h'_{1,1} = e'$.

Finally, from (12.4.4) and (12.4.5a), we have

$$D(g'_r h') = \gamma(h')\check{D}(g_r), \qquad \forall\, g'_r h' \in G' \tag{12.4.6}$$

where $\gamma(h')$ is one-dimensional, so that $\check{D}(G) = \{\check{D}(g_r)\}$ is irreducible since $D(G') = \{D(g'_r h')\}$ has been assumed irreducible. Furthermore, the orthogonality relations of the projective unirreps $\check{D}(G)$ given by (12.3.1) follow also from those of the unirreps $D(G')$, since the one-dimensional unirrep $\gamma(H')$ is unimodular, i.e. $|\gamma(H')| = 1$.

Example. As was mentioned before, the double group G' of a point group G is a covering group of G with the kernel $H' = \{E, E'\}$. According to Theorem 12.4.1, the factor system for a double-valued representation of the point group G is determined by the representation of H'.

Exercise. Show directly from the defining relations (12.4.5c) for the factor system $\{\lambda^{(\gamma)}(g_r, g_s)\}$ that it satisfies the associativity condition (12.1.2).

Hint. From $(g'_r g'_s)g'_t = g'_r(g'_s g'_t)$ one obtains

$$h'_{r,s}h'_{rs,t} = h'_{r,st}h'_{s,t} \tag{12.4.7}$$

whose representation by $\gamma(H')$ provides the proof.

12.4.2 Representation groups

Let G be a group and G' be a covering group of G such that all projective irreps of G can be formed from the vector irreps of G'. Then such a covering group G' is called a *universal covering group* of G. A universal covering group of the minimal order is called a *representation group* of G.

Example 1. The representation group of the rotation group $SO(3, r)$ is the spinor group $SU(2)$.

Example 2. The double group D'_n of the dihedral group D_n is a universal covering group of D_n. It is also a representation group of D_n for an even n but not for an odd n. For the latter, the representation group is D_n itself, as was shown in Example 12.2.2.

In the following, we shall show that, for a representation group G' of G, the kernel H' of the homomorphism $G' \to G$ is isomorphic to the multiplicator M of G so that $|G'| = |M| \times |G|$. As a preparation we introduce the following lemma.

Lemma 12.4.2. Let $H = \{h_\nu; \nu = 1, 2, \ldots, |H|\}$ be an Abelian group of a finite order $|H|$. Then the set of vector irreps $\{\gamma^{(\nu)}(H); \nu = 1, 2, \ldots, |H|\}$ of H forms a group isomorphic to H.

Proof. First we note that every vector irrep of an Abelian group H is one-dimensional, so that the number of the inequivalent irreps of H equals the order $|H|$ of H. Moreover, the set of irreps of H forms a group because a product of one-dimensional irreps of H is also a one-dimensional irrep of H. To show the isomorphism between $\{\gamma^{(\nu)}(H)\}$ and $H = \{h_\nu\}$, note that every Abelian group H of a finite order can be expressed by a direct product of cyclic groups. Thus, it is sufficient for the proof for one to establish the isomorphism between a cyclic group C_n and the group of its irreps. Let $C_n = \{a^m; m = 1, 2, \ldots, n\}$. From $a^n = 1$, the generator a is represented by one of the nth roots of unity $\{\omega^\nu; \nu = 1, 2, \ldots, n\}$, where $\omega = \exp(-2\pi i/n)$. Thus, each root ω^ν characterizes a representation of the group C_n, so that the isomorphism between C_n and the group of its irreps is established via the one-to-one correspondence

$$a^\nu \leftrightarrow \omega^\nu; \qquad \nu = 1, 2, \ldots, n \tag{12.4.8}$$

because $\omega^\nu \omega^\mu = \omega^{(\nu+\mu)}$ corresponding to $a^\nu a^\mu = a^{\nu+\mu}$.

Theorem 12.4.3. Let G be a finite group and G' be a representation group of G extended by a subgroup H' in the center of G'. Then H' is isomorphic to the multiplicator M of G and hence $|G'| = |M| \times |G|$.

Proof. From Lemma 12.4.2, the set of irreps $\{\gamma^{(\nu)}(H'); \nu = 1, 2, \ldots, |H'|\}$ of H' forms a group isomorphic to the Abelian group H'. Accordingly, the set of factor systems $\{\lambda^{(\nu)}\}$ defined by the irreps $\{\gamma^{(\nu)}\}$ of H' via (12.4.5c) must also form a group isomorphic to H', i.e.

$$\{\lambda^{(\nu)}\} \simeq H' \tag{12.4.9}$$

Now, by definition a representation group G' is a universal covering group of minimal order, so that the factor systems defined by the inequivalent irreps of H' must be all p-inequivalent.[1] Thus, the group of the factor systems $\{\lambda^{(\nu)}\}$ is isomorphic to the group of p-equivalence classes of factor systems, that is, the multiplicator M of G: $\{\lambda^{(\nu)}\} \simeq M$. Accordingly, from (12.4.9) we have the desired isomorphism

$$M \simeq H' \tag{12.4.10}$$

If a covering group G' of G contains a representation group of G as a proper subgroup, some of the projective representations of G formed by the vector irreps of G' must be p-equivalent. Such a covering group is still a universal covering group of G. If a covering group of G does not contain a representation group of G as a subgroup, then its vector irreps do not provide all the projective irreps of G. Such a covering group is not a universal covering group. In the special case in which the representation group G' of G coincides with the group G itself, every projective irrep of G is p-equivalent to some vector irrep of G. This convenient situation actually occurs for all the proper double point groups, as will be shown in the next section; this is seldom the case for the ordinary point groups; see Bir and Pikus (1974).

12.5 Representation groups of double point groups

What are all the possible projective unirreps of a point group G? This question shall be answered by considering the projective representations of the double group G' of G, on account of the theorem that every projective representation of a proper double point group P' is p-equivalent to a vector representation of P', i.e. the representation group of P' is P' itself (Kim 1983a). Hence, it is only necessary to form the representation groups P_i'' of the rotation–inversion double point groups $P_i' = P' \times C_i$. Any other improper double point group C'_{nv}, C'_{np}, D'_{np} or T'_p is isomorphic to a proper double point group, so that the representation group is again the group itself. We shall first prove the basic theorem on the representation groups of proper double point groups and then construct the representation groups P_i'', postponing their representations to the next section.

12.5.1 Representation groups of double proper point groups P'

Theorem 12.5.1. Let P' be the double group of a proper point group P, then the representation group of P' is P' itself.

Proof. The theorem is true for a uniaxial double group C'_n since it can be shown for the uniaxial group C_n as in Example 12.2.1. Excluding this case, the defining relations of a proper double point group P' are given by

$$A^n = B^m = (AB)^2 = E', \qquad E'^2 = E \tag{12.5.1}$$

where A and B are the generators of P'; E' is the 2π rotation and E is the identity. The sets of the orders (n, m) of the generators are $(n, 2)$ for a dihedral double group D'_n; and (3, 3), (4, 3) and (5, 3) for the tetrahedral T', octahedral O' and icosahedral double

[1] Note here that two factor systems defined by two inequivalent irreps of H' are not necessarily p-inequivalent: we have seen in Example 12.2.2 that the factor systems of D_n defined by $\alpha = 1$ and $\alpha = -1$ are p-inequivalent for an even n but p-equivalent for an odd n. For the latter case, by a further gauge transformation, we have reduced H' to the trivial group of identity.

groups Y', respectively. Let $|P|$ be the order of a proper point group P, then it is determined by the orders n and m of the generators by the Wyle relation given by (5.4.4), i.e.

$$\frac{1}{n}+\frac{1}{m}-\frac{1}{2}=\frac{2}{|P|}(\equiv 1/\tau) \tag{12.5.2}$$

where $\tau = |P|/2$ is an integer that plays a crucial role in the proof.

Let $\check{D}(P') = \{\check{D}(g);\ g \in P'\}$ be a standard projective representation of P'. Then the projective representation of the defining relations (12.5.1) yields

$$\lambda_1 \check{D}(A)^n = \lambda_2 \check{D}(B)^m = \lambda_3[\check{D}(A)\check{D}(B)]^2 = \lambda_4 \check{D}(E'), \qquad \lambda_5 \check{D}(E')^2 = \check{D}(E) \tag{12.5.3a}$$

where λ_1 through λ_5 are numerical factors that depend on the assumed factor system. The theorem is proven if these factors are mapped off by a gauge transformation of $\check{D}(P)$. Firstly, by the gauge transformation

$$a = \check{D}(A)/\mu(A), \qquad b = \check{D}(B)/\mu(B), \qquad e' = \check{D}(E')/\mu(E'), \qquad e = \check{D}(E)$$

with appropriate gauge factors $\mu(A)$, $\mu(B)$ and $\mu(E')$, we reduce (12.5.3a) to the form

$$a^n = b^m = e', \qquad (ab)^2 = \alpha e', \qquad e'^2 = e \tag{12.5.3b}$$

where α is a constant parameter, which will be shown to be a τth root of unity given by

$$\alpha^\tau = 1; \qquad \alpha = \exp[-2\pi \mathrm{i} k/\tau]; \qquad k = 0, 2, \ldots, \tau - 1 \tag{12.5.4}$$

Proof of this relation is based on the self-consistency of (12.5.3b). It is elementary but somewhat involved. Let us postpone this proof and proceed with the rest of the proof assuming that $\alpha^\tau = 1$.

Introduce a further gauge transformation

$$x = \mathrm{e}^{\mathrm{i}\pi k/n} a, \qquad y = \mathrm{e}^{\mathrm{i}\pi k/m} b, \qquad \overline{e} = \mathrm{e}^{\mathrm{i}\pi k} e' \tag{12.5.5}$$

Then substitution of these into (12.5.3b) leads to

$$x^n = y^m = (xy)^2 = \overline{e}, \qquad \overline{e}^2 = e \tag{12.5.6}$$

with use of the Wyle relation (12.5.2). Since the defining relations (12.5.1) and (12.5.6) are identical in form, we may conclude that any projective representation of the group P' is p-equivalent to a vector representation of P'. This proves the basic theorem 12.5.1.

It should be noted that an analogous theorem need not hold for an ordinary proper point group P in general, because the proof requires a gauge transformation $\overline{e} = (-1)^k e'$ in (12.5.5). For example, the representation group of the dihedral group D_n is D_n itself when n is odd, but it is the double point group D'_n when n is even, as was shown in Example 12.2.2.

Proof of $\alpha^\tau = 1$. This relation has been given in (12.5.4) without proof. For convenience, it will be proven individually for the double dihedral groups D'_n and for double polyhedral groups T', O' and Y'.

1. For D'_n, the defining relations (12.5.3b) take the form

$$a^n = b^2 = e', \qquad (ab)^2 = \alpha e', \qquad e'^2 = e \tag{12.5.7a}$$

Rewrite the third equation in the form $a = \alpha e' b a^{-1} b$. Then its nth power yields

$$\alpha^n = 1$$

This proves $\alpha^\tau = 1$ for D_n because $\tau = n$ for this case.

2. For a polyhedral group, the defining relations (12.5.3) take the form

$$a^n = b^3 = e', \qquad (ab)^2 = \alpha e', \qquad e'^2 = e \tag{12.5.7b}$$

Rewrite the third equation in the following two forms:

$$aba = \alpha b^2, \qquad bab = \alpha a^{n-1} \tag{12.5.7c}$$

which are further rewritten as

$$b = \alpha a^{n-1} b^2 a^{n-1}, \qquad a = \alpha b^2 a^{n-1} b^2 \tag{12.5.7d}$$

Substituting the second equation of (12.5.7d) into the first equation and refactorizing it with use of $a^n = b^3 = e'$, we obtain

$$b = \alpha^{2n-1} e' b^2 (a^{n-1} b)^{n-2} a^{n-2} (b a^{n-1})^{n-2} b^2 \tag{12.5.7e}$$[2]

Then, by writing out this equation explicitly for each polyhedral group and using (12.5.7e), we arrive at $\alpha^\tau = 1$, i.e.

$$\alpha^6 = 1 \qquad \text{for } T' \text{ with } n = 3 \tag{12.5.7f}$$

$$\alpha^{12} = 1 \qquad \text{for } O' \text{ with } n = 4 \tag{12.5.7g}$$

$$\alpha^{30} = 1 \qquad \text{for } Y' \text{ with } n = 5 \tag{12.5.7h}$$

For example, for O' we have $n = 4$ so that (12.5.7e) takes the following explicit form:

$$b = \alpha^7 e' b^2 a^3 b a^3 b a^2 b a^3 b a^3 b^2$$

Then, using $aba = \alpha b^2$ and $a^4 = b^3 = e'$ we arrive at $b = \alpha^{12} b$, which yields (12.5.7g). The proof of (12.5.7f) is trivial. For the proof of (12.5.7h) we may use either $aba = \alpha b^2$ or $bab = \alpha a^4$.

12.5.2 Representation groups of double rotation–inverse groups P'_i

Let $P_i = P \times C_i$ be a rotation–inversion group, then its double group is defined by $P'_i = P' \times C_i$, so that its defining relations are given by

$$a^n = b^m = (ab)^2 = e', \qquad \bar{1}a = a\bar{1}, \qquad \bar{1}b = b\bar{1}, \qquad (\bar{1})^2 = e'^2 = e$$

where $\bar{1}$ is the inversion and the generator b is absent for C'_{ni}. Let $\check{D}(P'_i)$ be a projective representation of P'_i and set

$$A = \check{D}(a), \qquad B = \check{D}(b), \qquad I = \check{D}(\bar{1}), \qquad E' = \check{D}(e'), \qquad E = \check{D}(e)$$

Then, from Theorem 12.5.1, these may be assumed to satisfy

$$A^n = B^m = (AB)^2 = E', \qquad IA = \beta AI, \qquad IB = \gamma BI, \qquad I^2 = E'^2 = E \tag{12.5.8a}$$

[2] One may not rewrite (12.5.7e) in the form

$$b = \alpha^{2n-1} e' b^2 (a^{n-1} baba^{n-1})^{n-2} b^2$$

which may be a convenient form for the proof of (12.5.7f) but not for (12.5.7g) and (12.5.7h).

where β and γ are constant parameters to be determined. Express the inversion I in the forms $I = \beta AIA^{-1} = \gamma BIB^{-1}$ then from $I^2 = E$ we have $\beta^2 = \gamma^2 = 1$. Moreover, the similarity transformation of $A^n = B^m = (AB)^2 = E'$ by I leads, with use of $IAI = \beta A$ and $IBI = \gamma B$, to

$$\beta^n E' = \gamma^m E' = \beta^2\gamma^2 E' = IE'I$$

Since $\beta^2 = \gamma^2 = 1$, we have $\beta^n = \gamma^m = 1$ and $IE'I = E'$.

Accordingly, E' commutes with every element in (12.5.8a) so that it is a constant matrix. From $E'^2 = E$, we have $E' = \pm\mathbf{1}$ in an appropriate dimensionality: when $E' = \mathbf{1}$ the representation is integral and when $E' = -\mathbf{1}$ the representation is half integral, just like for an ordinary double point group. Furthermore, from $\beta^n = \gamma^m = 1$ and $\beta^2 = \gamma^2 = 1$, we conclude that

$$\beta = \begin{cases} \pm 1, & \text{if } n \text{ is even} \\ 1, & \text{if } n \text{ is odd} \end{cases}$$

$$\gamma = \begin{cases} \pm 1, & \text{if } m \text{ is even} \\ 1, & \text{if } m \text{ is odd} \end{cases} \qquad (12.5.8b)$$

Since these binary parameters cannot be mapped off by a further gauge transformation, the set of equations given by (12.5.8a) provides the general defining relations for the representation group of the double rotation–inversion group P'_i, regarding β and γ as abstract group elements in the center of the group.

For the special cases of the groups T'_i and Y'_i, the orders n and m in (12.5.8b) are odd so that $\beta = \gamma = 1$. This means that the representation groups of these two groups are the groups themselves. The representation groups of the remaining groups $C'_{2r,i}$, D'_{ni} and O'_i, where r is an integer, are given by

$$C''_{2r,i}: \quad A^{2r} = E', \qquad IAI = \beta A, \qquad I^2 = E'^2 = E \qquad (12.5.9a)^3$$

$$D''_{n,i}: \quad A^n = B^2 = (AB)^2 = E', \qquad IAI = \beta A, \qquad IBI = \gamma B,$$

$$I^2 = E'^2 = E \ (\beta = 1 \text{ for an odd } n) \qquad (12.5.9b)$$

$$O''_i: \quad A^4 = B^3 = (AB)^2 = E', \qquad IAI = \beta A, \qquad IBI = B, \qquad I^2 = E'^2 = E \qquad (12.5.9c)$$

Here, $\beta^2 = \gamma^2 = 1$. These binary parameters β and γ are regarded as abstract group elements of order 2 satisfying $\alpha^2 = \beta^2 = E$. Then these elements and the 2π rotation E' are in the centers of the respective representation groups.

It should be noted here that the representation group P''_i of each P'_i given in (12.5.9a)–(12.5.9c) is a central extension of each P'_i by an Abelian subgroup $H_\beta = \{E, \beta\}$, $H_\gamma = \{E, \gamma\}$ or $H_\beta \times H_\gamma$ in the center of P''_i. According to Theorem 12.4.1, every irrep of these subgroups in the center defines a factor system of P'_i which is characterized by $\beta = \pm 1$ and/or $\gamma = \pm 1$; hence there exist only two p-equivalence classes of factor systems denoted by $K(\beta = \pm 1)$ for $C'_{2r,i}$ and O'_i, and $K(\gamma = \pm 1)$ for $D'_{2r+1,i}$, whereas there are four classes $K(\beta = \pm 1, \gamma = \pm 1)$ of factor systems for $D'_{2r,i}$.

[3] Here we have excluded the possible representation group of $C'_{2r+1,i}$;

$$A^{2r+1} = E', \qquad IAI = \beta A, \qquad IE'I = \beta E', \qquad I^2 = E'^2 = E$$

with $\beta = \pm 1$, because it does not seem to have any useful application, and also because the representation group of the point group $C_{2r+1,i}$ is the group itself.

12.6 Projective unirreps of double rotation–inversion point groups P_i'

The unirreps of the proper double point groups P' have already been given in Section 11.3. Accordingly, it is necessary only to form the vector unirreps of the representation group P_i'' defined by (12.5.9). Let us characterize a representation group P_i'' of the double group P_i' in the form

$$G = H_z = H + zH \tag{12.6.1}$$

where z is the augmentor to the halving subgroup $H = \{h\}$ of G such that

$$z^2,\, z^{-1}hz \in H, \qquad \forall\, h \in H$$

If we compare (12.6.1) with (12.5.9), we see that the halving subgroup H stands for $P' \times H_\beta$, $P' \times H_\gamma$ or $P' \times H_\beta \times H_\gamma$ whereas the augmentor z stands for the inversion I. All the unirreps of H are given by the direct products of the unirreps of P', H_β and H_γ. For example, a unirrep of $P' \times H_\beta$ is given by $\Gamma(P') \times \Gamma_\beta$ where $\Gamma(P')$ is a unirrep of P' given in Section 11.3 and Γ_β is a one-dimensional unirrep of $H_\beta = \{E, \beta\}$ with $\beta = \pm 1$. Thus, explicitly we have

$$\Gamma(P') \times \Gamma_\beta(\beta) = \beta\Gamma(P') \tag{12.6.2}$$

Now, we can construct all the vector unirreps of H_z through induction from the unirreps of the halving subgroup H. For convenience, we shall directly construct the induced unirreps for this simple special case of (12.6.1), instead of applying the general results on the induced representation given in Section 8.4.

Let $\{\Gamma^{(\nu)}(H)\}$ be a complete set of the vector unirreps of the halving subgroup $H = \{h\}$ of H_z and let ψ^ν be a basis row vector of dimensionality d_ν belonging to the νth irrep $\Gamma^{(\nu)}(H)$ such that

$$\mathring{h}\psi^\nu = \psi^\nu\Gamma^{(\nu)}(h), \qquad \forall\, h \in H \tag{12.6.3a}$$

Then, $\mathring{z}\psi^\nu$ belongs to the conjugate unirrep $\Gamma^{(\nu)}(z^{-1}hz)$ because

$$\mathring{h}(\mathring{z}\psi^\nu) = \mathring{z}(z^{-1}hz)^0\psi^\nu = (\mathring{z}\psi^\nu)\Gamma^{(\nu)}(z^{-1}hz) \tag{12.6.3b}$$

Now the conjugate unirrep $\Gamma^{(\nu)}(z^{-1}hz)$ must be equivalent to one of the unirreps of H, say $\Gamma^{(\overline{\nu})}(H)$, so there exists a unitary transformation $N(z)$ such that

$$\Gamma^{(\nu)}(z^{-1}hz) = N(z)^{-1}\Gamma^{(\overline{\nu})}(h)N(z), \qquad \forall\, h \in H \tag{12.6.4a}$$

Let $\phi^{\overline{\nu}}$ be a basis of $\Gamma^{(\overline{\nu})}(H)$, then we have

$$\mathring{z}\psi^\nu = \phi^{\overline{\nu}}N(z) \tag{12.6.4b}$$

because substitution of (12.6.4a) into (12.6.3b) shows that $\mathring{z}\psi^\nu N(z)^{-1}$ belongs to $\Gamma^{(\overline{\nu})}(H)$. Moreover, on applying $\mathring{z}$ once more to (12.6.4b) and using $z^2 \in H$, we obtain

$$\mathring{z}\phi^{\overline{\nu}} = \psi^\nu\Gamma^{(\nu)}(z^2)N(z)^{-1} \tag{12.6.4c}$$

Thus the augmentor z connects two bases ψ^ν and $\phi^{\overline{\nu}}$ of H, so that the set $[\psi^\nu, \phi^{\overline{\nu}}]$ provides a basis of a representation $D^{(\nu,\overline{\nu})}(H_z)$ defined by

$$D^{(\nu,\overline{\nu})}(h) = \begin{bmatrix} \Gamma^{(\nu)}(h) & 0 \\ 0 & \Gamma^{(\overline{\nu})}(h) \end{bmatrix}; \qquad \forall\, h \in H$$

$$D^{(\nu,\overline{\nu})}(z) = \begin{bmatrix} 0 & \Gamma^{(\nu)}(z^2)N(z)^{-1} \\ N(z) & 0 \end{bmatrix} \tag{12.6.5}$$

This representation is irreducible if $\overline{\nu} \neq \nu$, because the augmentor z connects bases belonging to two different unirreps (case II in Section 8.4) and is reducible if $\overline{\nu} = \nu$ (case I in Section 8.4). For the latter case, we observe, by replacing h in (12.6.4a) with $z^{-1}hz$, using $\overline{\nu} = \nu$, that

$$\Gamma^{(\nu)}(z^{-2})\Gamma^{(\nu)}(h)\Gamma^{(\nu)}(z^2) = N(z)^{-2}\Gamma^{(\nu)}(h)N(z)^2$$

which means that $\Gamma^{(\nu)}(z^2)/N(z)^2 = \omega$ commutes with an irrep $\Gamma^{(\nu)}(H)$ of H so that it is a constant. The constant ω has modulus unity because the matrices involved are unitary. Thus, by adjusting the phase factor of $N(z)$ in (12.6.4b), we may set $\Gamma^{(\nu)}(z^2) = N(z)^2$ and see that the representation $D^{(\nu,\nu)}(H_z)$ defined by (12.6.5) is reducible; in fact, the two reduced unirreps of H_z are given by

$$D^{(\nu\pm)}(h) = \Gamma^{(\nu)}(h), \qquad D^{(\nu\pm)}(z) = \pm N(z) \tag{12.6.6}$$

via the bases defined by

$$\psi^{(\nu\pm)} = \psi^\nu \pm \phi^\nu = \psi^\nu \pm \mathring{z}\psi^\nu N(z)^{-1}$$

Here $\pm N(z)$ serve as the representatives of z. A proper choice of ψ^ν has to be made in order to avoid a null result for the bases.

The completeness of the induced unirreps for H_z thus obtained is easily established by calculating the sum of the squares of the dimensions of the induced unirreps provided that we use every unirrep of H once and once only in constructing the induced unirreps.

For the present case of the representation group P_i'', we have $z = I$ and $I^2 = E$ so that the unirreps of P_i'' given above are slightly simplified: for $\nu \neq \overline{\nu}$, (12.6.5) becomes

$$D^{(\nu,\overline{\nu})}(h) = \begin{bmatrix} \Gamma^{(\nu)}(h) & 0 \\ 0 & \Gamma^{(\overline{\nu})}(h) \end{bmatrix}, \qquad D^{(\nu,\overline{\nu})}(I) = \begin{bmatrix} 0 & N(I)^{-1} \\ N(I) & 0 \end{bmatrix} \tag{12.6.7}$$

with the basis $[\psi^\nu, \phi^{\overline{\nu}}]$, where $\phi^{\overline{\nu}} = \mathring{I}\psi^\nu N(I)^{-1}$; for $\nu = \overline{\nu}$, the two reduced representations (12.6.6) become

$$D^{(\nu\pm)}(h) = \Gamma^{(\nu)}(h), \qquad D^{(\nu\pm)}(I) = \pm N(I) \tag{12.6.8}$$

with the bases $\psi^{\nu\pm} = \psi^\nu \pm \phi^\nu$, where $\phi^\nu = \mathring{I}\psi^\nu N(I)$ for $N(I)^2 = \Gamma^{(\nu)}(I^2) = \mathbf{1}$.

These representations given by (12.6.7) and (12.6.8) are explicit in terms of the vector unirreps of P' except for the transformation matrix $N(I)$ which is to be determined from (12.6.4a). This equation takes the following forms for the generators A and B of P':

$$\Gamma^{(\nu)}(IAI) = \beta\Gamma^{(\nu)}(A) = N(I)^{-1}\Gamma^{(\overline{\nu})}(A)N(I)$$

$$\Gamma^{(\nu)}(IBI) = \gamma\Gamma^{(\nu)}(B) = N(I)^{-1}\Gamma^{(\overline{\nu})}(B)N(I) \tag{12.6.9}$$

using $IAI = \beta A$ and $IBI = \gamma B$ given in (12.5.9) and also (12.6.2). Since A or $\{A, B\}$ is the complete set of the generators of the respective double group P', this set of equations (12.6.9) is sufficient to determine the conjugate set $(\Gamma^{(\nu)}, \Gamma^{(\overline{\nu})})$ as well as the transformation matrix $N(I)$ for a set of given values of β and γ. Thus we can classify all unirreps of P' into the conjugate sets of orbits $\{(\Gamma^{(\nu)}, \Gamma^{(\overline{\nu})})\}$ of H with respect to the group H_i and determine the projective unirreps of P'_i for a given class of the factor systems $K(\alpha, \beta)$ defined by the values of α and β.

The following comments will be helpful for determining the transformation matrix $N(I)$ from (12.6.9) for a given $\Gamma^{(\nu)}(H)$. When the representation $\Gamma^{(\nu)}(H)$ is one-dimensional we may take $N(I) = 1$. When $\Gamma^{(\nu)}(H)$ is two-dimensional, it may be expressed in terms of the Pauli spin matrices σ_x, σ_y and σ_z in a form similar to the quaternion (10.1.14),

$$\Gamma^{(\nu)} = c_0\sigma_0 + c_x\sigma_x + c_y\sigma_y + c_z\sigma_z; \qquad c_0^2 - c_x^2 - c_y^2 - c_z^2 = \pm 1 \qquad (12.6.9')$$

where c_0 through c_z are constants and σ_0 is the unit matrix. Then, using the anticommutation relations $\sigma_i\sigma_j\sigma_i = -\sigma_j$ ($i \neq j$; 1, 2, 3), we can easily determine the conjugate set $(\Gamma^{(\phi)}, \Gamma^{(\overline{\nu})})$ as well as the transformation matrix $N(I)$ from (12.6.9) (see the examples for the projective unirreps of D_{ni} given below). The higher dimensional case (> 2) will be shown explicitly for the double group O'_i.

The projective representations of P'_i thus constructed for each class of factor systems $K(\beta, \gamma)$ will be denoted as

$$\check{D}(\Gamma^{(\nu)}, \Gamma^{(\overline{\nu})}; N(I)) \equiv D^{(\nu,\overline{\nu})} \text{ of } (12.6.7)$$

$$\check{D}(\Gamma^{(\nu)}; \pm N(I)) \equiv D^{(\nu\pm)} \text{ of } (12.6.8) \qquad (12.6.10a)$$

and $\Gamma^{(\nu\pm)}$ are the vector irreps of the double group P'_i belonging to the unit class $K_0 = K(1, 1)$ with $+$ $(-)$ denoting the *gerade* (*ungerade*) representation. In terms of the basis ψ^ν of $\Gamma^{(\nu)}(P')$, the bases of the projective representations are given by

$$[\psi^\nu, \mathring{I}\psi^\nu N(I)^{-1}] \in \check{D}(\Gamma^{(\nu)}, \Gamma^{(\overline{\nu})}; N(I))$$

$$\psi^\nu \pm \mathring{I}\psi^\nu N(I) \in \check{D}(\Gamma^{(\nu)}; \pm N(I)) \qquad (12.6.10b)$$

In the following, we shall construct all projective unirreps of the improper double point groups $C'_{2r,i}$, D'_{ni} and O'_i, using the complete sets of the vector unirreps of the double proper point groups given in Section 11.3.

12.6.1 The projective unirreps of $C'_{2r,i}$

According to the complete set of vector unirreps of C'_n given by (11.3.3b), there exist $2n$ $(= 4r)$ one-dimensional representations for C'_{2r}. The unitary representatives of the generator $A = c_{2r}$ of C'_{2r} are given by

$$M_m(c_{2r}) = e^{-i\pi m/r}; \qquad m = 0, \pm\tfrac{1}{2}, \pm 1, \ldots, \pm(r - \tfrac{1}{2}), r \qquad (12.6.11a)$$

According to the defining relation (12.5.9a), there exist only two classes of factor systems, $K(\beta = 1)$ and $K(\beta = -1)$. Since the class $K(\beta = 1)$ corresponds to the vector representation of C'_{2r}, it is necessary only to construct the projective representations belonging to the factor system $K(\beta = -1)$. From the fact that the representations M_m are one-dimensional, we may set $N(I) = 1$ in (12.6.9) and obtain, for $\beta = -1$,

$$-M_m(c_{2r}) = M_{m-r}(c_{2r}); \qquad m = \tfrac{1}{2}, 1, \ldots, r - \tfrac{1}{2}, r$$

which provide the conjugate pairs $\{(M_m, M_{m-r})\}$ that completely exhaust the whole

Table 12.2. *The vector unirreps of D'_n with respect to the generators*

$\Gamma^{(\nu)}(h)$	$\Gamma^{(\nu)}(c_n)$	$\Gamma^{(\nu)}(u_0)$
$A_1(h)$	1	1
$A_2(h)$	1	-1
$B_1(h)$	-1	1
$B_2(h)$	-1	-1
$E_m(h)$	$c_m\sigma_0 - \mathrm{i}s_m\sigma_y$	$(-\mathrm{i})^{2j}\sigma_z$

$c_m = \cos(2\pi m/n), \quad s_m = \sin(2\pi m/n);$
$m = \frac{1}{2}, 1, \ldots, (n-1)/2.$

Table 12.3. *The construction of projective unirreps of $D'_{2r,i} \in K(\beta = -1, \gamma = 1)$ based on Equation (12.6.9)*

$\Gamma^{(\nu)}(h^I)$	$-\Gamma^{(\nu)}(c_n)$	$\Gamma^{(\nu)}(u_0)$	$N(I)^{-1}\Gamma^{(\bar{\nu})}(h)N(I)$	$\check{D}^{(\nu,\mu)}$ or $\check{D}^{(\nu\pm)}$
$A_1(h^I)$	-1	1	$B_1(h)$	$\check{D}(A_1, B_1; 1)$
$A_2(h^I)$	-1	-1	$B_2(h)$	$\check{D}(A_2, B_2; 1)$
$E_m(h^I)$	$-c_m + \mathrm{i}s_m\sigma_y$	$(-\mathrm{i})^{2j}\sigma_z$	$\sigma_z E_{r-m}\sigma_z$	$\begin{cases}\check{D}(E_{r/2}; \pm\sigma_z)\\ \check{D}(E_m, E_{r-m}; \sigma_z)\end{cases}$

1. $c_m = \cos(\pi m/r), s_m = \sin(\pi m/r); m = \frac{1}{2}, 1, \ldots, \frac{1}{2}(r-1).$
2. $c_{r-m} = -c_m, s_{r-m} = s_m; c_{r/2} = 0, s_{r/2} = 1.$

set of unirreps given by (12.6.11a). Thus, from (12.6.7) we obtain $2r$ two-dimensional projective unirreps of $C'_{2r,i}$ belonging to the class $K(\beta = -1)$ expressed by

$$\check{D}(M_m, M_{m-r}; 1); \qquad m = \tfrac{1}{2}, 1, \ldots, r - \tfrac{1}{2}, r \tag{12.6.11b}$$

where m and $m - r$ are both integers (including zero) or half integers for r being an integer. The proof for completeness of this set will be given later in (12.6.13a).

12.6.2 *The projective unirreps of $D'_{n,i}$*

According to (12.5.9b), there exist two or four classes of factor systems for D'_{ni} depending on whether n is odd or even. From the complete set of vector irreps of D'_n, given in Table 11.5, the representatives of the generators $A = c_n$ and $B = u_0$ are as given in Table 12.2, in which two-dimensional representatives are expressed by the Pauli matrices following (12.6.9′).

We shall construct the projective unirreps of $D'_{2r,i}$ belonging to the class of factor systems $K(\beta = -1, \gamma = 1)$ as a prototype example. For this purpose, it is best to form a table of construction that contains $\Gamma^{(\nu)}(A^I) = \beta\Gamma^{(\nu)}(A)$ and $\Gamma^{(\nu)}(B^I) = \gamma\Gamma^{(\nu)}(B)$ as given by Table 12.3. Here $A^I = IAI$ and $B^I = IBI$. By comparing these with the vector unirreps of D'_{2r} given in Table 12.2 we determine $\Gamma^{(\bar{\nu})}(H)$ as well as $N(I)$ from

Table 12.4. *Projective unirreps $\check{D}(\Gamma^{(\nu)}, \Gamma^{(\overline{\nu})}; N(I))$ or $\check{D}(\Gamma^{(\nu)}; \pm N(I))$ or the double point groups $C'_{2r,i}$, D'_{ni} and O'_i via the unirreps of the respective proper double point groups*

$C'_{2r,i} \in K(\beta)$:
  $K(1)$; $M^{\pm}_m$; $m = 0, \pm\frac{1}{2}, \pm 1, \ldots, \pm(r - \frac{1}{2}), r$
  $K(-1)$; $\check{D}(M_m, M_{m-r}; 1)$; $m = \frac{1}{2} \ldots, r - \frac{1}{2}, r$

$D'_{n,i} \in K(\beta, \gamma)$:
  $K(1, 1)$; $A^{\pm}_1, A^{\pm}_2, B^{\pm}_1, B^{\pm}_2, E^{\pm}_m$; $m = \frac{1}{2}, 1, \ldots, (n-1)/2$
  $K(1, -1)$; $\check{D}(A_1, A_2; 1)$, $\check{D}(B_1, B_2; 1)$, $\check{D}(E_m; \pm\sigma_y)$; $m = \frac{1}{2}, 1, \ldots, (n-1)/2$
  $K(-1, 1)$; $\check{D}(A_1, B_1; 1)$, $\check{D}(A_2, B_2; 1)$, $\check{D}(E_{r/2}; \pm\sigma_z)$, $\check{D}(E_{m'}, E_{r-m'}; \sigma_z)$;
    $m' = \frac{1}{2}, 1, \ldots, \frac{1}{2}(r-1)$; $n = 2r$
  $K(-1, -1)$; $\check{D}(A_1, B_2; 1)$, $\check{D}(A_2, B_1; 1)$, $\check{D}(E_{r/2}; \pm\sigma_x)$, $\check{D}(E_{m'}, E_{r-m'}; \sigma_x)$;
    $m' = \frac{1}{2}, 1, \ldots, \frac{1}{2}(r-1)$; $n = 2r$

$O'_i \in K(\beta)$:
  $K(1)$; $A^{\pm}_1, A^{\pm}_2, E^{\pm}, T^{\pm}_1, T^{\pm}_2, E^{\pm}_{1/2}, E'^{\pm}_{1/2}, Q^{\pm}$
  $K(-1)$; $\check{D}(A_1, A_2; 1)$, $\check{D}(E; \pm\sigma_y)$, $\check{D}(T_1, T_2; 1)$, $\check{D}(E_{1/2}, E'_{1/2}; 1)$, $\check{D}(Q; \pm\sigma_0 \times \sigma_y)$

For the notations $\check{D}(\Gamma^{(\nu)}, \Gamma^{(\overline{\nu})}; N(I))$ and $\check{D}(\Gamma^{(\nu)}; \pm N(I))$ see Equation (12.6.10a).

(12.6.9). Analogously, the complete set of the projective unirreps of D'_{ni} belonging to the four classes of factor systems $K(1, 1)$, $K(1, -1)$, $K(-1, 1)$ and $K(-1, -1)$ is formed and summarized in Table 12.4.

Exercise. Construct the projective unirreps of D'_{ni} belonging to the class of factor systems $K(1, -1)$ through a table of construction similar to Table 12.3.

12.6.3 *The projective unirreps of O'_i*

From (12.5.9c) there exist two classes of factor systems $K(\beta = 1)$ and $K(\beta = -1)$ for this group. From the complete set of vector unirreps of O'_i given in Table 11.6, the representatives of the generators $A = 4_z$ and $B = 3_{xyz}$ are those given in Table 12.5. We shall construct the projective unirreps of O'_i again through the table of construction given in Table 12.6 for the class of factor systems $K(-1)$.

The projective unirreps of $C'_{2r,i}$, D'_{ni} and O'_i thus formed are summarized in Table 12.4, which was first obtained by Kim (1983a). From Table 12.4, we have the completeness conditions (12.3.7) for the projective unirreps belonging to each class of factor systems of $C'_{2r,i}$, D'_{ni} and O'_i given as follows. For $K(\beta = -1)$ of $C'_{2r,i}$

$$2^2 \times 2r = 8r = |C'_{2r,i}| \tag{12.6.12a}$$

For $K(\beta = 1, \gamma = -1)$ of D'_{ni}:

$$2^2 + 2^2 + 2(n-1) \times 2^2 = 8n = |D'_{n,i}| \tag{12.6.12b}$$

Table 12.5. *The vector unirreps of O'*

$\Gamma^{(v)}(h)$	$\Gamma^{(v)}(4_z)$	$\Gamma^{(v)}(3_{xyz})$
A_1	1	1
A_2	-1	1
E	σ_z	$-\frac{1}{2}(1+\mathrm{i}3^{1/2}\sigma_y)$
T_1	$\begin{bmatrix} 0 & -1 & 0 \\ 1 & 0 & 0 \\ 0 & 0 & 1 \end{bmatrix}$	$\begin{bmatrix} 0 & 0 & 1 \\ 1 & 0 & 0 \\ 0 & 1 & 0 \end{bmatrix}$
$T_2 = T_1 \times A_2$	$-T_1(4_z)$	$T_1(3_{xyz})$
$E_{1/2}$	$2^{-1/2}(1-\mathrm{i}\sigma_z)$	$\frac{1}{2}(1-\mathrm{i}\sigma_x-\mathrm{i}\sigma_y-\mathrm{i}\sigma_z) \equiv \sigma_{xyz}$
$E'_{1/2} = E_{1/2} \times A_2$	$-2^{-1/2}(1-\mathrm{i}\sigma_z)$	σ_{xyz}
$Q = E_{1/2} \times E$	$2^{-1/2}(1-\mathrm{i}\sigma_z)\otimes\sigma_z$	$-\sigma_{xyz}\otimes(1+\mathrm{i}3^{1/2}\sigma_y)/2$

Table 12.6. *Construction of the projective unirreps of O'_i* for $K(\beta=-1)$

$\Gamma^{(v)}(h^I)$	$-\Gamma^{(v)}(4_z)$	$\Gamma^{(v)}(3_{xyz})$	$N(I)^{-1}\Gamma^{(\bar{v})}(h)N(I)$	$D^{(v,\mu)}$ or $D^{(v\pm)}$
$A_1(h^I)$	-1	1	$A_2(h)$	$\check{D}(A_1, A_2; 1)$
$E(h^I)$	$-\sigma_z$	$-\frac{1}{2}(1+\mathrm{i}3^{1/2}\sigma_y)$	$\sigma_y E(h)\sigma_y$	$\check{D}(E; \pm\sigma_y)$
$T_1(h^I)$	$-T_1(4_z)$	$T_1(3_{xyz})$	$T_2(h)$	$\check{D}(T_1, T_2; 1)$
$E_{1/2}(h^I)$	$-2^{-1/2}(1-\mathrm{i}\sigma_z)$	σ_{xyz}	$E'_{1/2}(h)$	$\check{D}(E_{1/2}, E'_{1/2}; 1)$
$Q(h^I)$	$-2^{-1/2}(1-\mathrm{i}\sigma_z)\otimes\sigma_z$	$-\sigma_{xyz}\otimes(1+\mathrm{i}3^{1/2}\sigma_y)/2$	$\Sigma_y Q(h)\Sigma_y$	$\check{D}(Q; \pm\Sigma_y)$

Note that $\Sigma_y = \sigma_0 \otimes \sigma_y$.

For $K(\beta=-1, \gamma=\pm1)$ of $D'_{2r,i}$:

$$2^2 + 2^2 + 2\times 2^2 + (r-1)4^2 = 16r = |D'_{2r,i}| \qquad (12.6.12c)$$

For $K(\beta=-1)$ of O'_i:

$$2^2 + 2\times 2^2 + 6^2 + 4^2 + 2\times 4^2 = 96 = |O'_i| \qquad (12.6.12d)$$

Concluding remark. From Table 12.4 and the unirreps of the proper point groups given in Chapter 6, we can construct all the unirreps of any space group of wave vector $\hat{g}(\boldsymbol{k})$. This will be discussed, however, in Chapter 14 after the formation of the space groups in Chapter 13.

13
The 230 space groups

13.1 The Euclidean group in three dimensions[1] $E^{(3)}$

A linear transformation that leaves invariant the distance between two points $|\boldsymbol{x}_1 - \boldsymbol{x}_2|$ in the three-dimensional vector space $V^{(3)}$ is called a Euclidean transformation in three dimensions. It is described by an inhomogeneous orthogonal transformation of a vector $\boldsymbol{x}$ in $V^{(3)}$:

$$\boldsymbol{x}' = R\boldsymbol{x} + \boldsymbol{v} \tag{13.1.1}$$

where R is a 3×3 real orthogonal matrix and $\boldsymbol{v}$ is a vector in $V^{(3)}$. Following Seitz, we may denote the above transformation by

$$\boldsymbol{x}' = \{R|\boldsymbol{v}\}\boldsymbol{x} \tag{13.1.2}$$

The product of two Euclidean transformations and the inverse of a Euclidean transformation are also Euclidean transformations:

$$\begin{aligned} \{R'|\boldsymbol{v}'\}\{R|\boldsymbol{v}\} &= \{R'R|\boldsymbol{v}' + R'\boldsymbol{v}\} \\ \{R|\boldsymbol{v}\}^{-1} &= \{R^{-1}|-R^{-1}\boldsymbol{v}\} \end{aligned} \tag{13.1.3}$$

where the latter follows from $\{R'\boldsymbol{v}'\}$ with $R' = R^{-1}$ and $\boldsymbol{v}' + R^{-1}\boldsymbol{v} = 0$. Thus, the set of all Euclidean transformations in $V^{(3)}$ forms a group called *the Euclidean group* $E^{(3)}$ with the identity $\{E|0\}$. We shall study $E^{(3)}$ here to prepare for some of the general properties of space groups, which are subgroups of $E^{(3)}$ because any element of a space group of a crystal must leave the distance between two points in the crystal invariant.

The translation group T in $V^{(3)}$ is a subgroup of $E^{(3)}$ with the elements $\{E|\boldsymbol{t}\}$, where $\boldsymbol{t} \in V^{(3)}$: it is an Abelian group since

$$\{E|\boldsymbol{t}'\}\{E|\boldsymbol{t}\} = \{E|\boldsymbol{t} + \boldsymbol{t}'\} = \{E|\boldsymbol{t}\}\{E|\boldsymbol{t}'\}$$

Also, this means that a translation group T can be considered as a group of vectors $\{\boldsymbol{t}\}$ whose law of multiplication is the vector addition. It is an invariant subgroup of $E^{(3)}$ because

$$\{R|\boldsymbol{v}\}\{E|\boldsymbol{t}\}\{R|\boldsymbol{v}\}^{-1} = \{E|R\boldsymbol{t}\} \tag{13.1.4}$$

and $R\boldsymbol{t} \in V^{(3)}$. It is to be noted, however, that $\{E|\boldsymbol{t}\}$ does not commute with $\{R|\boldsymbol{v}\}$ unless R is an identity operation.

According to (13.1.3), the rotational part $\{R\}$ of $E^{(3)}$ also forms a group, which is the full rotation group $O^{(3)} = O(3, r)$ in three dimensions: it is a subgroup of $E^{(3)}$. Accordingly, $E^{(3)}$ may be regarded as the semi-direct product of T and $O^{(3)}$:

$$E^{(3)} = T \wedge O^{(3)} \tag{13.1.5}$$

[1] Some readers may prefer to read the next section 13.2 first, using this section as a reference.

We shall next consider a similarity transformation of an element $\{\{R|\boldsymbol{v}\}\}$ of $\boldsymbol{E}^{(3)}$ caused by an inhomogeneous coordinate transformation $\Lambda = [U|s]$ that describes a shift of the coordinate origin O to O′ by a vector $\boldsymbol{s}$ (see Figure 13.1) followed by a rotation U of the coordinate system. Under the coordinate transformation Λ, a vector $\boldsymbol{x}$ in $V^{(3)}$ transforms according to

$$\boldsymbol{x}' = U^{-1}(\boldsymbol{x} - \boldsymbol{s}) = \Lambda^{-1}\boldsymbol{x} \tag{13.1.6}$$

where $\Lambda^{-1} = [U^{-1}|-U^{-1}\boldsymbol{s}]$ is the inverse of the inhomogeneous linear transformation Λ. Here, the square bracket notation, $[U|\boldsymbol{s}]$, is introduced because the curly bracket notation, $\{U|\boldsymbol{s}\}$, is reserved for a group element[2] of $E^{(3)}$. Under the coordinate transformation Λ, an element $\{R|\boldsymbol{v}\} \in E^{(3)}$ transforms according to

$$\{R'|\boldsymbol{v}'\} = \Lambda^{-1}\{R|\boldsymbol{v}\}\Lambda \tag{13.1.7}$$

where

$$R' = U^{-1}RU$$
$$U\boldsymbol{v}' = \boldsymbol{v} - (E - R)\boldsymbol{s}$$

with use of (13.1.3). Here, R and R' are referred to the coordinate origins O and O′ respectively. In the special case of a pure shift $[e|\boldsymbol{s}]$ of the coordinate origin, we have

$$\boldsymbol{v}' = \boldsymbol{v} - [E - R]\boldsymbol{s} \tag{13.1.8}$$

where R' and R differ only in the locations of their operations. The similarity transformation (13.1.7) plays the essential role in establishing the equivalence of two space groups under a coordinate transformation.

As a simple application of the transformation (13.1.8), we shall consider the inverse problem of finding a shift $[e|\boldsymbol{s}]$ of the coordinate origin that brings a given element $\{R|\boldsymbol{v}\}$ to another element $\{R|\boldsymbol{v}'\}$. In the special case in which $\boldsymbol{v}' = 0$, it becomes the problem of reducing $\{R|\boldsymbol{v}\}$ to a pure rotation (proper or improper) $\{R|0\}$ by a shift of the coordinate origin. For this purpose we rewrite (13.1.8) in the form

$$[E - R]\boldsymbol{s} = \boldsymbol{v} - \boldsymbol{v}' \tag{13.1.9}$$

If this equation is soluble for $\boldsymbol{s}$, then the transformation of $\{R|\boldsymbol{v}\}$ to $\{R|\boldsymbol{v}'\}$ is possible by the shift $[E|\boldsymbol{s}]$. Obviously, the equation is soluble for any given $\boldsymbol{v} - \boldsymbol{v}'$ if the matrix $[E - R]$ is non-singular, i.e. if R does not have the eigenvalue unity. It is well known that, when R is a proper rotation p, the eigenvalues are 1, $e^{i\theta}$ and $e^{-i\theta}$, where θ is the

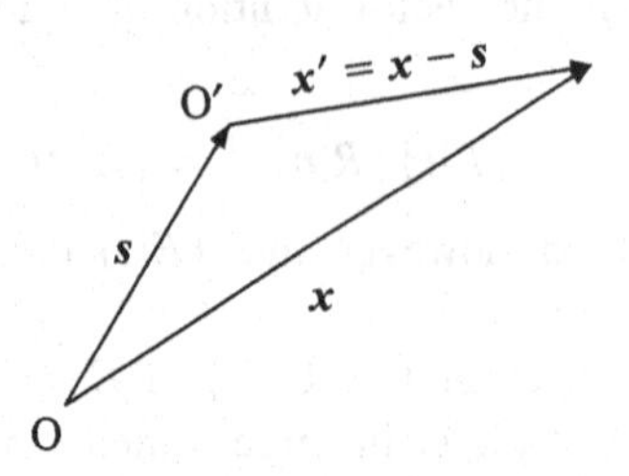

Figure 13.1. A shift $\boldsymbol{s}$ of the coordinate origin.

[2] Here $\Lambda = [U|s]$ could be a group element of $E^{(3)}$ simply being an inhomogeneous orthogonal transformation; however, it need not be an element of a subgroup (e.g. a space group) of $E^{(3)}$.

angle of rotation. Accordingly, when $R = p$, the matrix $[E - p]$ is always singular, whereas when R is a rotation–inversion $\overline{p} = -p$ that includes the pure inversion $\overline{1}$, the matrix $[E - \overline{p}]$ is always non-singular except when p is a binary rotation, for which $\theta = \pi$. In this case, $\overline{p} = m$ is a pure reflection, and its eigenvalues are -1, 1 and 1 so that $[E - m]$ is again singular. Accordingly, if $R = \overline{p}\ (\neq m)$, Equation (13.1.9) always has a general solution $s = [E - \overline{p}]^{-1}(\boldsymbol{v} - \boldsymbol{v}')$ for any given $\boldsymbol{v} - \boldsymbol{v}'$. This means that $\{\overline{p} \neq m|\boldsymbol{v}\}$ can be reduced to the pure rotation–inversion $\{\overline{p}|0\}$ at the coordinate origin O′, which is shifted from O by $\boldsymbol{s} = [E - \overline{p}]^{-1}\boldsymbol{v}$.

On the other hand, when $R = p\ (\neq E)$ or m, the equation (13.1.9) does not have a general solution but may have a special solution provided that $\boldsymbol{v} - \boldsymbol{v}'$ satisfies a certain condition. To see this, let ψ be a vector that is parallel to the axis of rotation p or the reflection plane m such that $R\psi = \psi$, where $R = p$ or m. Then ψ is orthogonal to the left-hand side of (13.1.9), i.e. the following scalar product is zero:

$$(\psi, [E - R]\boldsymbol{s}) = ([E - R^{\sim}]\psi, \boldsymbol{s}) = 0$$

because $\psi = R^{\sim}\psi$ from $R\psi = \psi$, where $R^{\sim}$ is the transpose of R. Accordingly, the right-hand side $\boldsymbol{v} - \boldsymbol{v}'$ of (13.1.9) must also be perpendicular to ψ:

$$[E - R]\boldsymbol{s} = (\boldsymbol{v} - \boldsymbol{v}') \perp \psi \qquad (13.1.10)$$

This means that (13.1.9) has a special solution for $\boldsymbol{s}$ if and only if $\boldsymbol{v} - \boldsymbol{v}'$ is perpendicular to ψ; see the algebraic solution (13.1.13). Conversely, under a given shift $[E|\boldsymbol{s}]$, only the perpendicular component $\boldsymbol{v}_\perp$ of $\boldsymbol{v}$ with respect to ψ can change while the parallel component remains unchanged, i.e. $\boldsymbol{v}'_\parallel = \boldsymbol{v}_\parallel$. For $R = p$, an invariant element $\{p|\boldsymbol{v}_\parallel\}$ with respect to a shift is called *a screw axis*, whereas for $R = m$, the invariant element $\{m|\boldsymbol{v}_\parallel\}$ is called *a glide reflection*. These may be denoted

$$S_p = \{p|\boldsymbol{v}_\parallel\}, \qquad g_m = \{m|\boldsymbol{v}_\parallel\} \qquad (13.1.11)$$

These cannot be reduced to the pure rotation $\{p|0\}$ or the pure reflection $\{m|0\}$ by a shift of the point operation unless $\boldsymbol{v}_\parallel = 0$ to begin with. See Figure 13.2.

Next, we shall determine the shift $\boldsymbol{s}$ which brings $\{R|\boldsymbol{v}_\perp\}$ for $R = p$ or m at the origin O to the pure rotation $\{R'|0\}$ at O′. Here R and R' differ only in their locations. The transformation $\{R|\boldsymbol{v}_\perp\}$ is a plane transformation because it leaves every point lying in the plane perpendicular to R ($R = p$ or m) in the same plane. The required shift $\boldsymbol{s}$ is obtained by solving Equation (13.1.10) with $\boldsymbol{v} - \boldsymbol{v}' = \boldsymbol{v}_\perp$, i.e.

$$[E - R]\boldsymbol{s} = \boldsymbol{v}_\perp \qquad (13.1.12)$$

For $R = m$, the solution is given by $\boldsymbol{s} = \boldsymbol{v}_\perp/2$, because $m\boldsymbol{v}_\perp = -\boldsymbol{v}_\perp$. For $R = p$, the algebraic solution is given as follows. Let us take a coordinate system such that the z-axis is parallel to the axis of p and the x-axis is parallel to $\boldsymbol{v}_\perp$. Then $p = R(\theta\boldsymbol{e}_z)$ and $\boldsymbol{v}_\perp = (v_0, 0, 0)$, where θ is the angle of rotation p and v_0 is the x-component of $\boldsymbol{v}_\perp$. Then, (13.1.12) takes the form

$$\begin{bmatrix} 1 - \cos\theta & \sin\theta \\ -\sin\theta & 1 - \cos\theta \end{bmatrix} \begin{bmatrix} s_x \\ s_y \end{bmatrix} = \begin{bmatrix} v_0 \\ 0 \end{bmatrix}; \qquad s_z = 0$$

which gives the following solution for the required shift $\boldsymbol{s}$ with components

$$s_x = \tfrac{1}{2}v_0, \qquad s_y = \tfrac{1}{2}v_0 \cot(\theta/2), \qquad s_z = 0 \qquad (13.1.13)$$

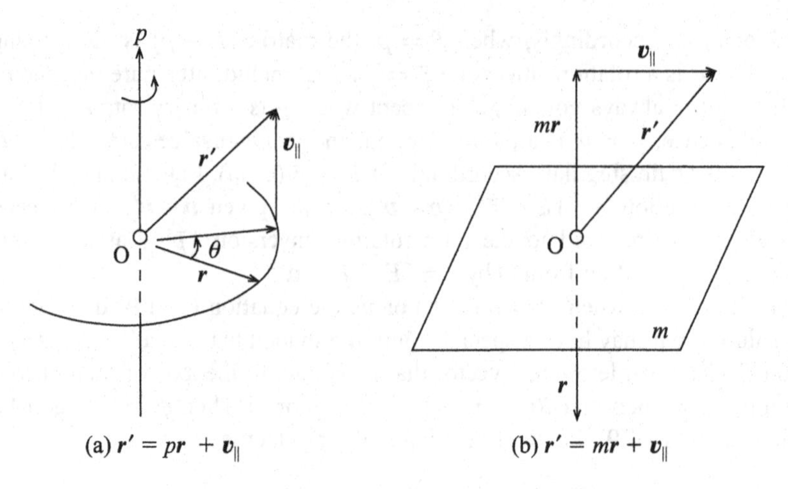

Figure 13.2. (a) The screw displacement $S_p = \{p|\boldsymbol{v}_\parallel\}$. (b) The glide reflection $g_m = \{m|v_\parallel\}$.

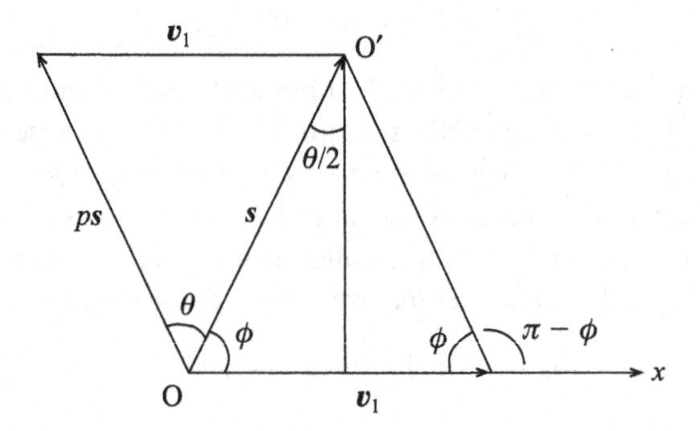

Figure 13.3. Graphical solution for the fixed point of the plane rotation $\{p|\boldsymbol{v}_\perp\}$: $\phi = (\pi - \theta)/2$.

The graphical solution for (13.1.12) is also given in Figure 13.3. Let O and O′ be the points at which the axes of p and p' cut the x, y plane. Draw the vector $\boldsymbol{v}_\perp$ from O along the x-axis and then draw two straight lines, one from O and the other from the end point of $\boldsymbol{v}_\perp$, making angles $\phi = (\pi - \theta)/2$ and $\pi - \phi$ relative to the x-axis, respectively, as in Figure 13.1. The point of intersection of these two lines is O′ and the shift is given by $\boldsymbol{s} = \vec{\text{OO}'}$. Note that O′ is the fixed point of the plane rotation $\{p|\boldsymbol{v}_\perp\}$. Note also that $s_y = \frac{1}{2}v_0 \tan\phi$.

From the above analysis, the elements of $E^{(3)}$ can be classified into pure proper rotations $(p|0)$ and pure rotation–inversions $\{\overline{p}|0\}$, screw displacements $S_p = \{p|\boldsymbol{v}_\perp\}$ and glide reflections $g_m = \{m|\boldsymbol{v}_\perp\}$ by an appropriate shift of the point of each operation. Even though these concepts may help one to understand the geometric structure of the group, we shall not be concerned with these classifications when we construct the space groups (each of which is a subgroup of $E^{(3)}$), because it is highly desirable to keep the point of every operation in the group at one fixed coordinate origin.

13.2 Introduction to space groups

The atomic structure of an ideal crystal in three dimensions may be considered as a periodic collection of sets of atoms (or ions or molecules) that is infinite in extent. Two points in such a crystal are said to be *equivalent* if all physical geometric properties are identical. Let us consider symmetry operations that take every point of the crystal into an equivalent point. Then the set of all these operations forms a group called *the space group of the crystal*, denoted by $\hat{G}$. It is a subgroup of the Euclidean group $E^{(3)}$, since the symmetry elements of $\hat{G}$ must leave the distance between two points in the crystal invariant. By purely mathematical reasoning based on the definition of space groups, a total of 230 space groups was constructed by Fedrov in 1895 and somewhat later by Schönflies.

We shall first discuss the translational periodicity which is the fundamental symmetry property of a crystal. It is characterized by a translation group $T = \{\{E|\boldsymbol{t}\}\}$ (or a group of vectors $T = \{\boldsymbol{t}\}$ under addition) that is a subgroup of the space group $\hat{G}$ of the crystal. It is discrete since the distance between atoms in a crystal cannot be arbitrarily small. That it is an invariant subgroup of $\hat{G}$ can be shown as in (13.1.4). By definition, an element of T brings an arbitrarily chosen point O of a crystal to an equivalent point in the crystal. The set of all equivalent points brought by the elements of T from O forms *a point lattice* called the *Bravais lattice* of the crystal, which may be regarded as a graphical representation of the group of discrete vectors $T = \{\boldsymbol{t}\}$. The end of a vector $\boldsymbol{t}$ is called *a lattice point* and $\boldsymbol{t}$ itself is called *a lattice vector.* Since the initial point O is chosen arbitrarily with respect to the crystal frame, it is meaningless to talk about the absolute position of the Bravais lattice relative to the given crystal (until the rotational symmetry of the crystal has been referred to a certain symmetry point of the crystal). In Figure 13.4, one finds a graphical representation of the Bravais lattice T of a two-dimensional crystal composed of two kinds of atoms, A and B. Any parallel shift of the point lattice T by an arbitrary vector $\boldsymbol{s}$ gives an equivalent representation of T; in fact,

$$[E|\boldsymbol{s}]^{-1}\{E|\boldsymbol{t}\}[E|\boldsymbol{s}] = \{E|\boldsymbol{t}\} \in T$$

Let us consider the symmetry point group $K = \{R\}$ of a Bravais lattice $T = \{\boldsymbol{t}\}$ with respect to a lattice point, then

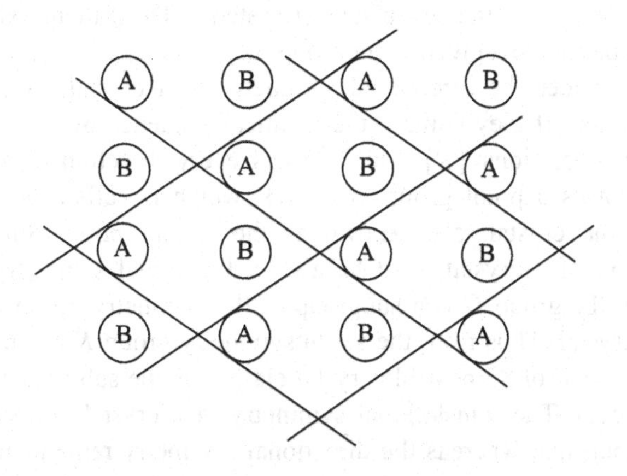

Figure 13.4. A Bravais lattice of a two-dimensional crystal.

$$Rt \in T; \qquad \forall R \in K, \forall t \in T \tag{13.2.1}$$

which is called *the compatibility condition* for compatibility between a Bravais lattice T and its point group K. In fact, from (13.2.1), it will be shown that the allowed rotation axes c_n for a Bravais lattice are limited to the following five kinds (see (13.3.3b) for the proof):

$$c_1, c_2, c_3, c_4, c_6 \tag{13.2.2}$$

Moreover, the inversion symmetry is always contained in K because if $\boldsymbol{t} \in \boldsymbol{T}$ then $-\boldsymbol{t} \in T$. Furthermore, if a Bravais lattice contains an n-fold axis of rotation c_n ($n > 2$), then it also contains a two-fold rotation c_2' perpendicular to c_n (see Lemma 3 in Section 13.3), i.e. c_n is two-sided. This means that a Bravais lattice with the principal axis c_n ($n > 2$) has the point symmetry D_{ni} ($n > 2$). When $n = 2$, there may but need not exist c_2' perpendicular to c_2: it turns out that both cases are possible, as will be shown by construction. Thus one may deduce that a possible point symmetry K of a Bravais lattice is one of the following seven groups:

$$C_i, C_{2i}, D_{2i}, D_{3i}, D_{4i}, D_{6i}, O_i \tag{13.2.3}$$

where the group O_i arises naturally when the three two-fold axes of D_{2i} become equivalent.

The Bravais lattices with the same point symmetry group K are said to belong to the same *crystal system*. From (13.2.3), there exists a total of seven crystal systems: their names and symbols are

1. triclinic, T (C_i),
2. monoclinic, M (C_{2i}),
3. orthorhombic, O (D_{2i}),
4. hexagonal, H (D_{6i}),
5. rhombohedral, RH (D_{3i}) (or trigonal),
6. tetragonal, Q (D_{4i}) and
7. cubic, C (O_i).

Bravais lattices belonging to a crystal system are further classified by the lattice types L. *Two lattices with the same point symmetry K are of the same type if they can be brought into each other by a continuous transformation of the lattices without lowering the point symmetry.* It will be shown later that there exists a total of 14 Bravais lattice types L for the seven crystal systems. Thus, there exists one or more lattice types compatible to a given point group K.

Let us next consider the directional symmetry of a crystal. Two directions in a crystal are equivalent if they contain the identical sequence of the equivalent points. The set of point operations $\{R\}$ which brings every direction of a crystal into an equivalent one forms a point group $G = \{R\}$, which is called the *crystallographic point group* of the crystal, also known as the *crystal class*. Since a directional symmetry element of a crystal must be also a directional symmetry element of its Bravais lattice T, the group G is a subgroup of the symmetry group K of the Bravais lattice T of the crystal. Thus from the seven symmetry group K given by (13.2.3), one can write down a total of 32 possible crystal classes by the subgroups of all K, as will be given in (13.5.1). The translational symmetry of a crystal is averaged out in the macroscopic dimension, whereas the directional symmetry remains unchanged in the macro-dimension.

The crystallographic point group G of a crystal may but need not be a subgroup of the space group $\hat{G}$ of the crystal, since it is not required that G should transform every point of the crystal into an equivalent one: it may need a further adjustment by a parallel displacement $\boldsymbol{t}_R$ to bring the crystal into coincidence with itself (like in the case of a screw displacement or a glide reflection). Thus a general element of a space group $\hat{G}$ may be expressed, using the Seitz notation, by

$$\hat{G} = \{\{R|\boldsymbol{t}_R\}\}, \qquad \forall R \in G \tag{13.2.4}$$

where the rotational part $\{R\}$ of $\hat{G}$ forms the crystal class G and the translational part $\boldsymbol{t}_R$ is given by

$$\boldsymbol{t}_R = \boldsymbol{v}_R + \boldsymbol{t}, \qquad \forall \boldsymbol{t} \in T$$

Here, $\boldsymbol{v}_R$ is the minimum translation characteristic to R, and is not a lattice translation unless it is null. Note, however, that $\boldsymbol{v}_E = 0$ for the identity element E.

According to (13.2.4) there exists a homomorphism between $\hat{G}$ and G via the correspondence $\{R|\boldsymbol{v}_R + \boldsymbol{t}\} \rightarrow R$. Since the kernel of the homomorphism is the translation group $T = \{\boldsymbol{t}\}$, the factor group $\hat{G}/T$ is isomorphic to G, i.e.

$$\hat{G}/T \simeq G \tag{13.2.5}$$

This isomorphism plays the essential role for constructing space groups belonging to a crystal class G.

In the actual description of a space group $\hat{G}$, a suitable choice of the coordinate origin (which shall be also the lattice origin) should be made on an appropriate symmetry point in the given crystal in order to minimize the non-lattice translations in the description of the space group $\hat{G}$. In the special case in which a space group $\hat{G}$ does not contain a non-lattice translation, an element of $\hat{G}$ is expressed in the form $\{R|\boldsymbol{t}\}$, where $\boldsymbol{t} \in T$. Since T is an invariant subgroup of $\hat{G}$, one may express $\hat{G}$ in the form of a semi-direct product:

$$\hat{G} = T \wedge G \tag{13.2.6}$$

Such a space group is called a *symmorphic space group*. The majority of the space groups are, however, non-symmorphic: this is different from the Euclidean group $E^{(3)}$, for which (13.1.5) always holds because there exists no minimum translation.

As a summary, we may state that there exist altogether seven crystal systems K, 14 lattice types L, 32 classes G, and 230 space groups $\hat{G}$ for crystals. These will be determined explicitly in the following sections in the order given. In particular, the space groups will be constructed through the isomorphism $\hat{G}/T \simeq G$ with use of their algebraic defining relations following the method developed by Kim (1986b).

13.3 The general structure of Bravais lattices

13.3.1 Primitive bases

For a given Bravais lattice or a translation group $T = \{\boldsymbol{t}\}$, there exists a set of three linearly independent basis vectors $[\boldsymbol{a}_1, \boldsymbol{a}_2, \boldsymbol{a}_3]$ such that every lattice vector $\boldsymbol{t}$ can be defined by

$$\boldsymbol{t} = n_1\boldsymbol{a}_1 + n_2\boldsymbol{a}_2 + n_3\boldsymbol{a}_3 \tag{13.3.1a}$$

where n_1, n_2 and n_3 are integers (positive, negative or zero). Such a set of basis vectors $[\boldsymbol{a}_1, \boldsymbol{a}_2, \boldsymbol{a}_3]$ is called a *primitive basis* of the Bravais lattice T. The numerical specification of a primitive basis is given by the lengths of the basis vectors $|\boldsymbol{a}_1|$, $|\boldsymbol{a}_2|$ and $|\boldsymbol{a}_3|$, and their mutual angles $\boldsymbol{a}_1$^$\boldsymbol{a}_2$, $\boldsymbol{a}_2$^$\boldsymbol{a}_3$ and $\boldsymbol{a}_3$^$\boldsymbol{a}_1$; it is called the set of *lattice parameters*. The spatial orientation of the lattice T as a whole is immaterial for the congruence of two lattices. On account of the point symmetry of T, these parameters need not be totally independent so that it will be necessary to specify only the minimum number of these parameters for each lattice type.

By definition, each primitive basis vector $\boldsymbol{a}_i$ is a primitive lattice vector (the shortest lattice vector in its direction). Each face $[\boldsymbol{a}_i, \boldsymbol{a}_j]$ formed by two basis vectors $\boldsymbol{a}_i$ and $\boldsymbol{a}_j$ in a primitive basis is primitive in the sense that there is no lattice point on the face except at the vertices. The parallelepiped formed by three basis vectors $[\boldsymbol{a}_1, \boldsymbol{a}_2, \boldsymbol{a}_3]$ of a primitive basis is called a *primitive unit cell* of T in the sense that it contains no lattice point inside or on the faces except at the vertices. From these properties, it is a simple matter to form a primitive basis for a given lattice T. Let $\boldsymbol{a}_1$ be a primitive lattice vector in any given direction, and let $\boldsymbol{a}_2$ be another primitive lattice vector that is not parallel to $\boldsymbol{a}_1$ but has the minimal projection on the line perpendicular to $\boldsymbol{a}_1$. Then, the face $[\boldsymbol{a}_1, \boldsymbol{a}_2]$ is primitive. Finally, let $\boldsymbol{a}_3$ be a primitive lattice vector not on the plane of $\boldsymbol{a}_1$ and $\boldsymbol{a}_2$ but with the minimal height from the plane. Then the set $[\boldsymbol{a}_1, \boldsymbol{a}_2, \boldsymbol{a}_3]$ forms a primitive unit cell.

Let (a_{ix}, a_{iy}, a_{iz}) be the components of a basis vector $\boldsymbol{a}_i$ with respect to an appropriate coordinate system, then one can introduce a matrix expression $\boldsymbol{A}$ for the primitive basis by

$$\boldsymbol{A} = [\boldsymbol{a}_1, \boldsymbol{a}_2, \boldsymbol{a}_3] = \begin{bmatrix} a_{1x} & a_{2x} & a_{3x} \\ a_{1y} & a_{2y} & a_{3y} \\ a_{1z} & a_{2z} & a_{3z} \end{bmatrix} \tag{13.3.1b}$$

where the column vectors of $\boldsymbol{A}$ are defined by the basis vectors following the notation introduced in (5.1.5). In terms of the basis matrix $\boldsymbol{A}$, the lattice translation defined by (13.1.3) is rewritten as

$$\boldsymbol{t} = \boldsymbol{A}\boldsymbol{n}, \qquad \boldsymbol{n} = (n_1, n_2, n_3) \tag{13.3.1c}$$

where $\boldsymbol{n}$ is the column vector with components n_1, n_2 and n_3. The sum of two lattice vectors $\boldsymbol{A}\boldsymbol{n}$ and $\boldsymbol{A}\boldsymbol{m}$ is $\boldsymbol{A}(\boldsymbol{n}+\boldsymbol{m})$. The translation group T is, therefore, completely specified by the matrix $\boldsymbol{A}$.

The choice of a primitive basis of a given lattice T is to some degree quite arbitrary; however, the volume V_0 of the primitive unit cell is independent of the choice and gives the minimum volume of a cell formed by any three lattice vectors that are non-coplanar. To see this, let a set of three lattice vectors be $\boldsymbol{t}_i = \sum_j \boldsymbol{a}_j M_{ji}$ $(i = 1, 2, 3)$ where $M = \|M_{ji}\|$ is a matrix with integral elements. Then we can write

$$\boldsymbol{T} = \boldsymbol{A}M \tag{13.3.2a}$$

where $\boldsymbol{T} = [\boldsymbol{t}_1, \boldsymbol{t}_2, \boldsymbol{t}_3]$ is a matrix defined by the lattice vectors that is analogous to $\boldsymbol{A} = [\boldsymbol{a}_1, \boldsymbol{a}_2, \boldsymbol{a}_3]$. Thus, via the absolute values of the determinants of both sides of (13.3.2a), the cell volume $V = |\det \boldsymbol{T}|$ is related to the primitive cell volume $V_0 = |\det \boldsymbol{A}|$ by

$$V = V_0 |\det M| \tag{13.3.2b}$$

Since $\det M$ is an integer, the minimum value of V must be given by V_0, which corresponds to $\det M = \pm 1$. In the special case in which $[\boldsymbol{t}_1, \boldsymbol{t}_2, \boldsymbol{t}_3]$ forms another primitive basis of T, one should have $\det M = \pm 1$ because both primitive cell volumes must be minimal. A matrix with integral elements and with the determinant equal to ± 1 is called a *unimodular integral matrix*. Then a matrix M that connects two primitive bases of a lattice T is a unimodular integral matrix.

Now let R be an element of the symmetry group K of T, then R must satisfy the compatibility condition (13.2.1). On combining this with (13.3.1a) we obtain

$$R\boldsymbol{a}_i = \sum_j \boldsymbol{a}_j M(R)_{ji}, \qquad R \in K; \qquad i = 1, 2, 3 \tag{13.3.3a}$$

where the transformation matrix $M(R)$ is a unimodular integral matrix, because both the set $[R\boldsymbol{a}_1, R\boldsymbol{a}_2, R\boldsymbol{a}_3]$ and the set $[\boldsymbol{a}_1, \boldsymbol{a}_2, \boldsymbol{a}_3]$ are primitive bases of T. Accordingly, the trace of the matrix $M(R)$ must also be an integer. Moreover, since $M(R)$ is equivalent to the three-dimensional rotation matrix R in the Cartesian coordinate system,[3] we have

$$\operatorname{tr} M(R) = \pm(1 + 2\cos\theta) = \text{an integer} \tag{13.3.3b}$$

where θ is the angle of the rotation R; the $+$ sign is for a proper rotation and the $-$ sign is for a rotation–inversion; cf. (4.3.9a). From (13.3.3b) and the property of the cosine function, the allowed θ is limited to $2\pi/n$ with $n = 1, 2, 3, 4$ or 6. Thus the rotation axes c_n allowed for K are

$$c_1, c_2, c_3, c_4, c_6 \tag{13.3.3c}$$

which was given previously in (13.2.2) without proof. As was mentioned in the introduction, from (13.3.3c) and Lemma 3 given below, we may arrive at the seven crystal systems given by (13.2.3).

The primitive basis $[\boldsymbol{a}_1, \boldsymbol{a}_2, \boldsymbol{a}_3]$ of a Bravais lattice T introduced above does not seem to have any direct correlation to the symmetry group K of the lattice T in any obvious way. To remedy this, we shall look into the relations between the point symmetry K and the lattice vectors of T. The following three simple lemmas hold.

Lemma 1. Let T be a Bravais lattice belonging to a symmetry group K. The K contains an inversion symmetry $\bar{1}$ with respect to a lattice point of T.

Lemma 2.[4] If K contains an n-fold axis of rotation c_n $(n > 1)$, then the lattice T has a lattice vector perpendicular to c_n and also a lattice vector parallel to c_n.

Lemma 3. If K contains a rotation axis c_n $(n > 2)$, it also contains a two-fold axis of rotation c_2' that is perpendicular to c_n (i.e. c_n is two-sided) and parallel to one of the shortest lattice vectors perpendicular to c_n.

[3] In the matrix notation A of the basis, Equation (13.3.3a) is written as $RA = AM(R)$, so that $A^{-1}RA = M(R)$.

[4] A simple proof of Lemma 2 is the following. Let $\boldsymbol{t}$ be any lattice vector of T that is not parallel or perpendicular to c_n. Then, the lattice vector defined by $\boldsymbol{t}_\perp = c_n\boldsymbol{t} - \boldsymbol{t}$ is perpendicular to c_n, while a lattice operation

$$H_n = E + c_n + c_n^2 + \cdots + c_n^{n-1}$$

applied to the vector $\boldsymbol{t}$ produces a lattice vector $\boldsymbol{t}_\parallel = H_n\boldsymbol{t}$ parallel to c_n, for $c_n\boldsymbol{t}_\parallel = \boldsymbol{t}_\parallel$ from $c_nH_n = H_n$; see Figure 13.5. Note that H_n is a projection operator satisfying $H_n^2 = nH_n$.

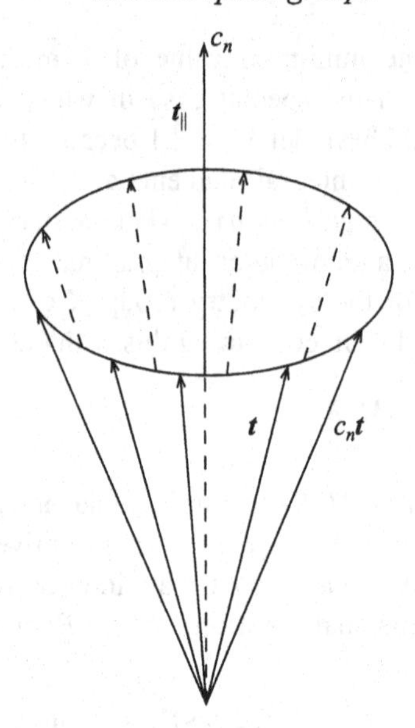

Figure 13.5. A lattice vector $t_{\|}$ parallel to an axis of rotation c_n.

Lemma 1 is obvious since, if $t \in T$, then $-t \in T$, on account of the inversion symmetry of T. Thus, there exists no enantiomorphism (left- and right-handed symmetry) for the Bravais lattice of a crystal, even if it exists for the crystal. Lemma 2 allows us to introduce basis vectors perpendicular to and parallel to a symmetry axis of T. Lemma 3 is interesting: the mere assumption of c_n $(n > 2)$ for T leads to the D_{ni} $(n > 2)$ symmetry for T. These lemmas will be proven through construction of the general algebraic expression for the Bravais lattice types first introduced by Kim (1985) via the projection operator method applied to the lattice vectors of T.

13.3.2 *The projection operators for a Bravais lattice*

According to (4.3.5a), an n-fold axis of rotation c_n is expressed, in terms of the infinitesimal rotation ω about the n-fold axis, by

$$\begin{aligned} c_n &= \exp\left[(2\pi/n)\omega\right] \\ &= E + \omega \sin(2\pi/n) + \omega^2[1 - \cos(2\pi/n)] \end{aligned} \tag{13.3.4a}$$

where E is the unit matrix and ω satisfies $\omega^3 = -\omega$. As was shown in Section 4.3, we have two projection operators, $-\omega^2$ and its dual $(1 + \omega^2)$, both of which satisfy $x^2 = x$, and are mutually orthogonal. Following Kim (1985), we shall derive the general expression for all the possible lattice types of the Bravais lattices by means of these projection operators. Let $\boldsymbol{r}$ be an arbitrary vector in the three-dimensional vector space $V^{(3)}$. Then from (4.3.4) the projections of $\boldsymbol{r}$ perpendicular to and parallel to the axis of rotation c_n are given by

$$\boldsymbol{r}_{\perp} = -\omega^2 \boldsymbol{r}, \qquad \boldsymbol{r}_{\|} = (1 + \omega^2)\boldsymbol{r} \tag{13.3.4b}$$

These projection operators themselves are not symmetry elements of the Bravais lattice T but we can form symmetry operations of T that are proportional to them. Firstly, the one proportional to $-\omega^2$ is defined, from (13.3.4a), by

$$P_n = 2E - c_n - c_n^{-1} = -p_n\omega^2, \qquad p_n = 2[1 - \cos(2\pi/n)] \tag{13.3.5a}$$

where P_n is a symmetry operation of T in the sense that it brings a lattice vector of T to another lattice vector of T. By definition, P_n is involutional, satisfying

$$P_n^2 = p_n P_n \tag{13.3.5b}$$

where the constant coefficient p_n has been defined in (13.3.5a). It depends on n and takes integral values, corresponding to $n = 6$, 4, 3, 2 and 1 allowed for crystal systems K,

$$p_6 = 1, \qquad p_4 = 2, \qquad p_3 = 3, \qquad p_2 = 4, \qquad p_1 = 0 \tag{13.3.6}$$

Secondly, the symmetry operations of T, which is proportional to $(1 + \omega^2)$, are defined by the dual of P_n

$$Q_n = p_n E - P_n = p_n(1 + \omega^2) \tag{13.3.7a}$$

which is also involutional and orthogonal to P_n

$$Q_n^2 = p_n Q_n, \qquad P_n Q_n = 0 \tag{13.3.7b}$$

Now, let $\boldsymbol{t}$ be an arbitrary lattice vector of T that is not perpendicular or parallel to the rotation axis c_n, then the operations of P_n and Q_n on $\boldsymbol{t}$ give us the lattice vectors

$$\boldsymbol{t}_\perp = P_n \boldsymbol{t}, \qquad \boldsymbol{t}_\parallel = Q_n \boldsymbol{t} \tag{13.3.8a}$$

where $\boldsymbol{t}_\perp$ ($\boldsymbol{t}_\parallel$) is perpendicular (parallel) to the axis c_n in view of (13.3.4b). This proves that Lemma 2 holds once again. Moreover, from the involutional properties of P_n and Q_n, we obtain

$$P_n \boldsymbol{t}_\perp = p_n \boldsymbol{t}_\perp, \qquad Q_n \boldsymbol{t}_\parallel = p_n \boldsymbol{t}_\parallel, \qquad P_n \boldsymbol{t}_\parallel = Q_n \boldsymbol{t}_\perp = 0 \tag{13.3.8b}$$

which characterizes $\boldsymbol{t}_\perp$ ($\boldsymbol{t}_\parallel$) as an eigenvector of P_n (Q_n) belonging to the eigenvalue p_n. According to (13.3.6), these eigenvalues p_n are all prime for $n > 2$. When $n = 2$, we have $c_2^{-1} = c_2$ so that from (13.3.5a) and (13.3.7a) the projection operators are reduced to

$$P_2/2 = E - c_2, \qquad Q_2/2 = E + c_2 \tag{13.3.9a}$$

which are still symmetry operations of T. These satisfy $x^2 = 2x$ and also the characteristic relations

$$(P_2/2)\boldsymbol{t}_\perp = 2\boldsymbol{t}_\perp, \qquad (Q_2/2)\boldsymbol{t}_\parallel = 2\boldsymbol{t}_\parallel, \qquad P_2\boldsymbol{t}_\parallel = Q_2\boldsymbol{t}_\perp = 0 \tag{13.3.9b}$$

with the eigenvalues equal to a prime number, namely 2.

The characteristic equations of P_n and Q_n given by (13.3.8b) and (13.3.9b) play the crucial role when we construct the general expression for Bravais lattices. It is also to be noted that the integral value of p_n given by (13.3.6) determines the number of lattice types belonging to the crystal class D_{ni} ($n > 1$), as will be shown in (13.3.16a).

13.3.3 Algebraic expressions for the Bravais lattices

To arrive at the general classification of the lattice types and also prove Lemma 3, we shall introduce a basis that is closely related to the symmetry axes of the Bravais lattice T. Let c_n be the principal axis of rotation for the symmetry group K of the Bravais lattice T and let $[\boldsymbol{e}_1, \boldsymbol{e}_2, \boldsymbol{e}_3]$ be a set of non-coplanar lattice vectors, each of which is a primitive lattice vector. Using Lemma 2, let e_3 be parallel to the principal axis c_n and e_1 and e_2 be on the ω plane of c_n (which is perpendicular to c_n). The latter will be specified further in terms of the symmetry of T. The basis defined by $[e_1, e_2, e_3]$ is called a *conventional lattice basis* of T and the parallelepiped defined by the three basis vectors is called the *Bravais parallelepiped (BP)* of T.

In general, a BP thus defined need not be primitive so that the whole structure of T is described by a set of lattice points inside or on each BP, if there are any, in addition to the sublattice $T^{(c)} = \{t_c\}$ of T defined by

$$t_c = q_1 e_1 + q_2 e_2 + q_3 e_3; \qquad q_1, q_2, q_3 = \text{integers} \tag{13.3.10a}$$

which is obviously an invariant subgroup of the translation group T. In other words, the Bravais lattice T is described by *the conventional sublattice* $T^{(c)}$ decorated by lattice points inside or on each BP, if there are any. Thus a general translation t of T is described by

$$t = xe_1 + ye_2 + ze_3 \qquad (\text{mod } t_c \in T^{(c)}) \tag{13.3.10b}$$

where the *conventional coordinates* x, y and z are rational numbers (mod 1). An allowed set of these points $\{(x, y, z)\}$ for T defines a lattice type belonging to the crystal system K, because two different sets of rational numbers cannot be transformed into each other by a continuous transformation without lowering the point symmetry K. However, two sets connected by permutation or inversion of the coordinates $\{(x, y, z)\}$ belong to the same type. Hereafter, a set of the lattice points defined by $\{(x, y, z)\}$ is referred to as a set of *equivalent points of a BP*: it provides the coset representatives of $T^{(c)}$ in T. Since $T^{(c)}$ is an invariant subgroup of T, the set forms a factor group $T/T^{(c)}$:

$$\{(x, y, z)\} \in T/T^{(c)} \tag{13.3.10c}$$

which may be called the group of the BP under addition as the group multiplication. Thus, if (x_1, y_1, z_1) and $(x_2, y_2, z_2) \in T/T^{(c)}$, then $(x_1 + x_2, y_1 + y_2, z_1 + z_2) \in T/T^{(c)}$. See Section 13.4.

To determine the possible set $\{(x, y, z)\}$ compatible with a given crystal system K, we first consider a BP with the principal axis of rotation c_n with $n > 2$ in view of Lemma 3 (which is to be proved). Let e_1 be one of the shortest lattice vectors on the ω plane of c_n and let $e_2 = c_n e_1$. Then the face $[e_1, e_2]$ of the BP is primitive, because if there were a lattice vector e' inside the face, then at least one of the four vectors e', $e_1 - e'$, $e_2 - e'$, $e_1 + e_2 - e'$ would be shorter than the vector e_1 (see Figure 13.6).

Next, we operate with c_n on t given by (13.3.10b) and subtract it from t to obtain

$$t - c_n t = (x + y)e_1 + [p_n y - (x + y)]e_2 \qquad (\text{mod } t_c \in T^{(c)}) \tag{13.3.11}$$

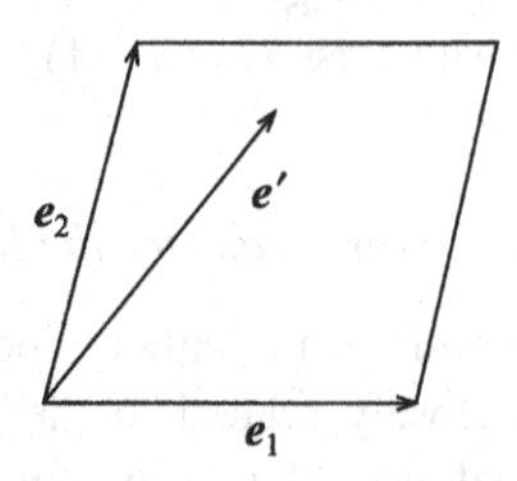

Figure 13.6. A primitive base $[e_1, e_2]$ where $e_2 = c_n e_1$; $n > 2$.

where use has been made of $c_n e_1 = e_2$, $c_n e_3 = e_3$ and $c_n e_2 = 2e_2 - e_1 - p_n e_2$, which follows from $c_n = 2E - c_n^{-1} - P_n$ from (13.3.5a). Since the left-hand side of (13.3.11) is a lattice vector and $[e_1, e_2]$ is a primitive face, (13.3.11) implies that $x + y$ and $p_n y$ are integers; hence we may set $y = m/p_n$ and $x = -m/p_n$, whereupon (13.3.10b) can be rewritten as

$$t = -(m/p_n)e_1 + (m/p_n)e_2 + ze_3 \qquad (\text{mod } t_c \in T^{(c)}) \tag{13.3.12}$$

where the allowed values of m are integers bounded only by $0 \leqslant m < p_n$, for p_n being a prime. Note that this equation is invariant under the exchange of e_1 and $-e_2$, so that there exists a reflection plane $\bar{c}'_2$ in T that is perpendicular to the diagonal lattice vector $e_1 + e_2$ (see Figure 13.7). Combining this with the inversion symmetry $\bar{1}$ of T, we conclude that there exists a two-fold axis of rotation c'_2 along the diagonal $e_1 + e_2$ and hence along e_1 and e_2 given by $c_n^{-1}c'_2$ and $c_n c'_2$, respectively. This proves Lemma 3 since e_1 is one of the shortest lattice vectors of T on the $\omega(c_n)$ plane.

Next, the z-component in (13.3.12) is characterized as follows. By application of the projection operator Q_n of (13.3.7a) to both sides of (13.3.12) and using (13.3.8b), we obtain

$$Q_n t = zp_n e_3 \qquad (\text{mod } t_c \in T^{(c)}) \tag{13.3.13}$$

Since $Q_n t$ is a lattice vector and e_3 is the shortest in its own direction, the coefficient zp_n must be an integer, say m'. Substitution of this result into (13.3.12) yields

$$t = -(m/p_n)e_1 + (m/p_n)e_2 + (m'/p_n)e_3 \qquad (\text{mod } t_c \in T^{(c)}) \tag{13.3.14}$$

where $0 < m,\ m' < p_n$. This equation describes the general structure of a possible lattice type belonging to the crystal system D_{ni} $(n > 2)$.

Before actual construction of the lattice types, we give here some general discussion based on (13.3.14). By definition, the third basis vector which forms a primitive unit cell with e_1 and e_2 is given by a lattice vector t with the smallest z-component in (13.3.14) compatible with a given value of m. Thus, when $m = 0$, we have $m' = 0$ so that a primitive basis $[a_1, a_2, a_3]$ is given by

$$a_1 = e_1, \qquad a_2 = e_2, \qquad a_3 = e_3 \tag{13.3.15}$$

which defines a primitive BP. When $m > 0$, m' cannot be zero because the face $[e_1, e_2]$

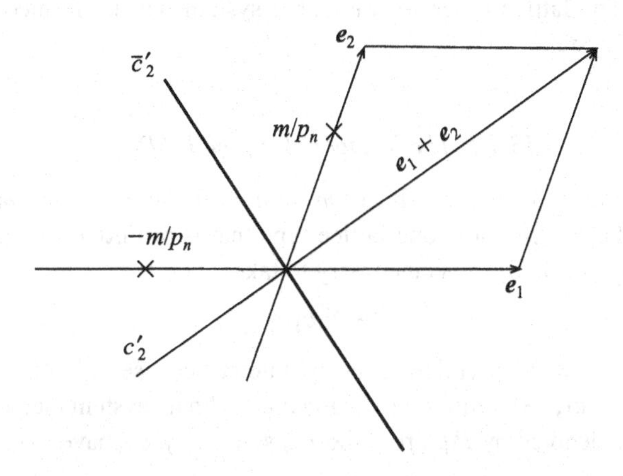

Figure 13.7. A reflection plane perpendicular to a diagonal lattice vector $e_1 + e_2$.

is primitive. Thus the third primitive basic vector $\boldsymbol{a}_3^0$, which forms a primitive basis with $\boldsymbol{e}_1$ and $\boldsymbol{e}_2$, is given by

$$\boldsymbol{a}_3^0 = -(m/p_n)\boldsymbol{e}_1 + (m/p_n)\boldsymbol{e}_2 + (1/p_n)\boldsymbol{e}_3, \qquad 0 < m < p_n \tag{13.3.16a}$$

The notation $\boldsymbol{a}_3$ will be reserved for a more symmetric choice of the primitive basis. Since each allowed integral value of m bounded by $0 < m < p_n$ defines a primitive unit cell $[\boldsymbol{e}_1, \boldsymbol{e}_2, \boldsymbol{a}_3^0]$, the numerical value of p_n determines the number of lattice types belonging to D_{ni} $(n > 2)$, some of which, however, could be equivalent, as will be discussed below. Since $p_n > 1$ in (13.3.16a), the vector $\boldsymbol{a}_3^0$ is inside the BP, so that there may exist one or more extra equivalent points in the BP given by

$$\mu\boldsymbol{a}_3^0 = (-\mu m/p_n, \mu m/p_n, \mu/p_n) \qquad (\text{mod } \boldsymbol{t}_c \in T^{(c)}) \tag{13.3.16b}$$

where μ is an integer bounded by $1 \leqslant \mu < p_n$.

When $n = 2$, the above argument fails since $c_2\boldsymbol{e}_1 = -\boldsymbol{e}_1$. Accordingly, there may but need not exist c_2' perpendicular to c_2; in fact, either case is possible, as will be shown in the next section by construction. Thus, we arrive at the crystal systems D_{2i}, O_i and C_{2v}; here, the system O_i arrives naturally when the three conventional lattice vectors of the D_{2i} system become equivalent. Furthermore, from the reduced projection operators $P_2/2$ and $Q_2/2$ introduced in (13.3.9a), we may proceed as in the case of $n > 2$ with an appropriate choice of the conventional basis $[\boldsymbol{e}_1, \boldsymbol{e}_2, \boldsymbol{e}_3]$ and arrive at the general structure of possible lattice types described by

$$\boldsymbol{t} = (m_1/2)\boldsymbol{e}_1 + (m_2/2)\boldsymbol{e}_2 + (m_3/2)\boldsymbol{e}_3 \qquad (\text{mod } \boldsymbol{t}_c \in T^{(c)}) \tag{13.3.17}$$

where $0 \leqslant m_1, m_2, m_3 < 2$ with a certain constraint characteristic to each system, which will be discussed in the next section for each crystal system; see Sections 13.4.4 and 13.4.6. As a conclusion, we may state that all the Bravais lattice types are described either by the general expression (13.3.14) for $n > 2$ or by (13.3.17) for $n = 2$ (Kim 1985).

13.4 The 14 Bravais lattice types

We shall first construct the lattice types belonging to the crystal systems D_{ni} $(n > 2)$ based on (13.3.14) and then those of the remaining systems through showing that (13.3.17) holds. The lattice types for the cubic system will be formed from those of the orthorhombic system.

13.4.1 The hexagonal system H (D_{6i})

Since $p_6 = 1$ from (13.3.6), we have $m = m' = 0$ from $0 < m$, $m' < p_6 = 1$ in (13.3.14). Thus there exists only one lattice type that is primitive so that the primitive basis is defined by (13.3.15). It is customary to take

$$\boldsymbol{a}_2 = c_3\boldsymbol{a}_1 \tag{13.4.1a}$$

instead of $\boldsymbol{a}_2 = c_6\boldsymbol{a}_1$ without violating the symmetry because $c_3^{-1}\boldsymbol{a}_1$ is parallel to $c_6\boldsymbol{a}_1$. Then, the basis set $[\boldsymbol{a}_1, \boldsymbol{a}_2]$ defines a hexagonal coordinate system (see Figure 13.8(e)). The lattice type is denoted by P_H. From the D_{6i} symmetry, we have

$$|\boldsymbol{a}_1| = |\boldsymbol{a}_2|, \qquad \boldsymbol{a}_1 \wedge \boldsymbol{a}_2 = 120°, \qquad \boldsymbol{a}_3 \perp [\boldsymbol{a}_1, \boldsymbol{a}_2] \tag{13.4.1b}$$

Thus, we need only two lattice parameters, $|\boldsymbol{a}_1|$ and $|\boldsymbol{a}_3|$, to describe any lattice belonging to the lattice type P_{H}.

13.4.2 The tetragonal system Q (D_{4i})

From $0 \leqslant m < p_4 = 2$, we have $m = 0, 1$. When $m = 0$, the BP is primitive and the basis is given by (13.3.15). It is denoted by P_{Q}. From the D_{4i} symmetry, the primitive basis vectors $\boldsymbol{a}_1$, $\boldsymbol{a}_2$ and $\boldsymbol{a}_3$ are perpendicular with respect to each other and $|\boldsymbol{a}_1| = |\boldsymbol{a}_2|$. Thus any lattice belonging to the P_{Q} type is determined by two lattice parameters, $|\boldsymbol{a}_1|$ and $|\boldsymbol{a}_3|$.

When $m = 1$, from (13.3.16a), a primitive basis may be defined by

$$\boldsymbol{a}_1^0 = \boldsymbol{e}_1, \qquad \boldsymbol{a}_2^0 = \boldsymbol{e}_2, \qquad \boldsymbol{a}_3^0 = (\tfrac{1}{2}, \tfrac{1}{2}, \tfrac{1}{2})$$

Thus the BP has only one extra lattice point, given by $\boldsymbol{a}_3^0$, which is at the center of the BP so that it is called the *body-centered BP* and denoted I_{Q}. A more symmetric choice of the primitive basis of I_{Q} is given, by drawing three basis vectors from three vertices of the BP to the center of the BP (see Figure 13.8(d)), by

$$\boldsymbol{a}_i = \boldsymbol{d}_I - \boldsymbol{e}_i; \qquad i = 1, 2, 3 \tag{13.4.2a}$$

where $\boldsymbol{d}_I = (\boldsymbol{e}_1 + \boldsymbol{e}_2 + \boldsymbol{e}_3)/2 = \boldsymbol{a}_1 + \boldsymbol{a}_2 + \boldsymbol{a}_3$ $(= \boldsymbol{a}_3^0)$. Thus the primitive basis $\boldsymbol{A} = [\boldsymbol{a}_1, \boldsymbol{a}_2, \boldsymbol{a}_3]$ is given by

$$\boldsymbol{A} = [(-\boldsymbol{e}_1 + \boldsymbol{e}_2 + \boldsymbol{e}_3)/2, (\boldsymbol{e}_1 - \boldsymbol{e}_2 + \boldsymbol{e}_3)/2, (\boldsymbol{e}_1 + \boldsymbol{e}_2 - \boldsymbol{e}_3)/2] \tag{13.4.2b}$$

Thus, the matrix form of $\boldsymbol{A}$ with respect to the conventional basis is given by

$$\boldsymbol{A} = \begin{bmatrix} -\frac{1}{2} & \frac{1}{2} & \frac{1}{2} \\ \frac{1}{2} & -\frac{1}{2} & \frac{1}{2} \\ \frac{1}{2} & \frac{1}{2} & -\frac{1}{2} \end{bmatrix} \tag{13.4.2c}$$

The three basis vectors are all half of the diagonal lattice vectors of the BP and hence equal in their lengths, so that the primitive unit cell formed by $[\boldsymbol{a}_1, \boldsymbol{a}_2, \boldsymbol{a}_3]$ is rhombic; in fact, from the orthogonality of the conventional basis vectors, we have

$$\boldsymbol{a}_i^2 = (\boldsymbol{e}_1^2 + \boldsymbol{e}_2^2 + \boldsymbol{e}_3^2)/4 = \boldsymbol{d}_I^2; \qquad i = 1, 2, 3 \tag{13.4.2d}$$

Moreover, the edge angles are determined by the scalar products

$$\boldsymbol{a}_i \cdot \boldsymbol{a}_j = \tfrac{1}{2}\boldsymbol{e}_k^2 - \boldsymbol{d}_I^2 \qquad (i, j, k = 1, 2, 3 \text{ cyclic permutations}) \tag{13.4.2e}$$

These satisfy

$$\sum_{i>j} \boldsymbol{a}_i \cdot \boldsymbol{a}_j = -\boldsymbol{d}_I^2; \qquad \text{i.e.} \qquad \sum_{i>j} \cos(\boldsymbol{a}_i {}^\wedge \boldsymbol{a}_j) = -1 \tag{13.4.2f}$$

Finally, from the matrix expression of $\boldsymbol{A}$ and (13.3.1c), the general coordinates of the lattice vector $\boldsymbol{t}$ with respect to the conventional basis are given by

$$\boldsymbol{t} = \boldsymbol{A}\boldsymbol{n} = ((-n_1 + n_2 + n_3)/2, (n_1 - n_2 + n_3)/2, (n_1 + n_2 - n_3)/2) \tag{13.4.2g}$$

Since the sum of any two coordinates is an integer, the three coordinates are either all integers or all half integers; the latter are those of the lattice points at the center of the

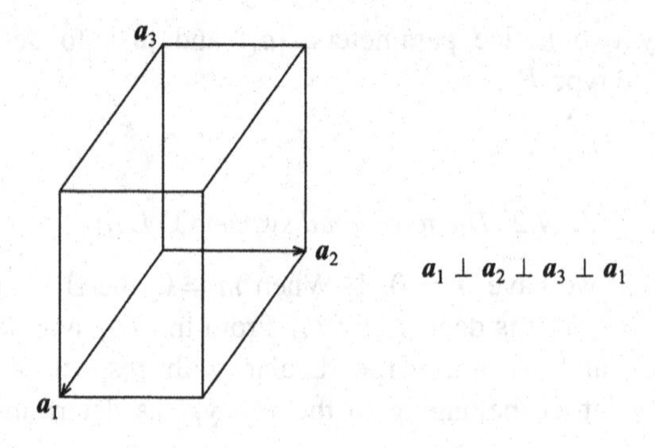

(a) Primitive BP $\in D_{2i}$

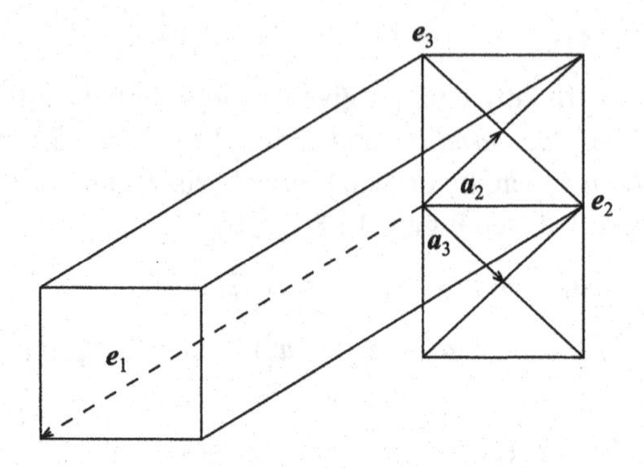

$$a_1 = e_1,\ a_2 = (e_2 + e_3)/2,\ a_3 = (e_2 - e_3)/2$$
$$|a_1| \neq |a_2| = |a_3|$$

(b) Base-centered BP, $A = (0, ½, ½) \in D_{2i}$

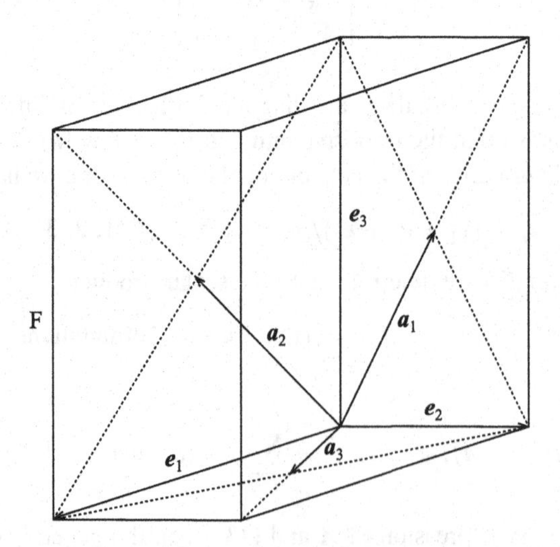

$$a_1 = (e_2 + e_3)/2,\ a_2 = (e_3 + e_1)/2,\ a_3 = (e_1 + e_2)/2$$
$$d_F = (e_1 + e_2 + e_3)/2 = (a_1 + a_2 + e_3)/2$$
$$a_i - a_j = d_F^2 - a_k^2\ (i, j, k \neq;\ 1, 2, 3 \text{ cyclic})$$

(c) Face-centered BP $\in D_{2i}$

Figure 13.8. Graphical representations of BPs.

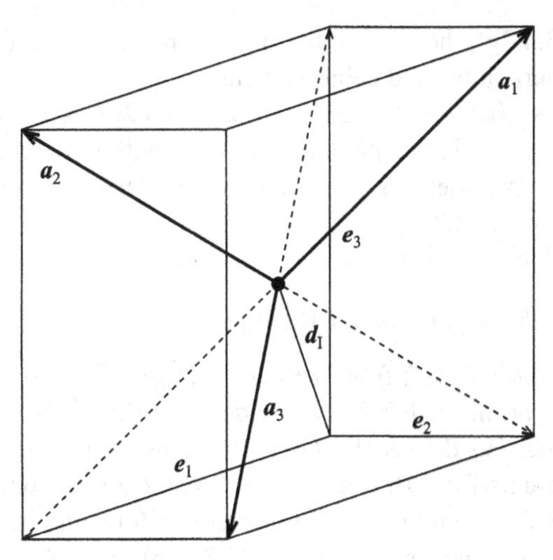

$a_1 = d_I - e_1$, $a_2 = d_I - e_2$, $a_3 = d_I - e_3$
$d_I = (e_1 + e_2 + e_3)/2 = (a_1 + a_2 + a_3)$
$|a_1| = |a_1| = |a_1| = |d_I|$
$\sum_{i>j} \cos (a_i{}^\wedge a_j) = -1$

(d) Body-centered BP $\in D_{2i}$

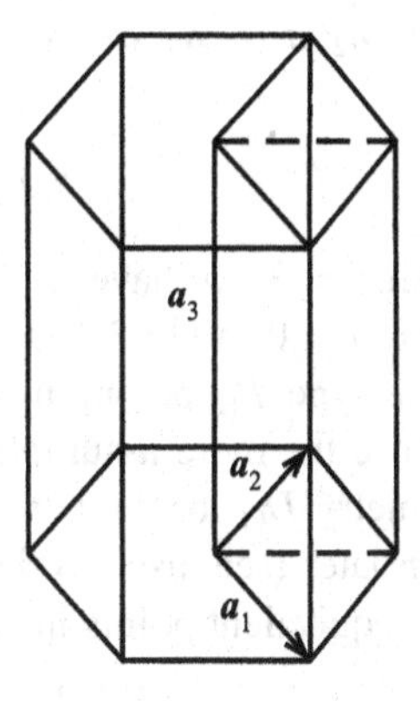

$a_3 \perp [a_1, a_2]$
$a_1{}^\wedge a_2 = 120°$

(e) Hexagonal BP $\in D_{6i}$

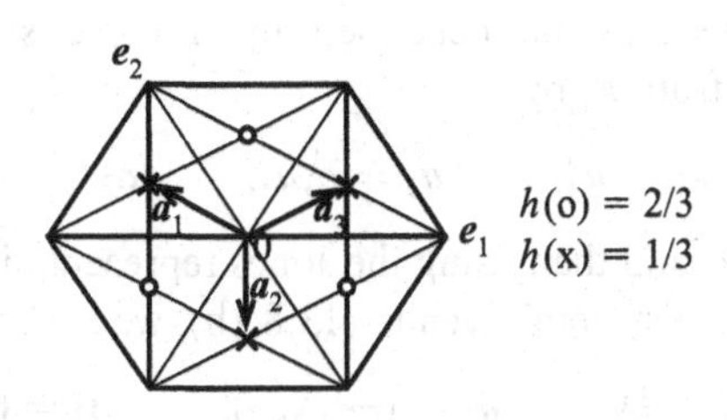

$a_3 = \frac{2}{3}e_1 + \frac{1}{3}e_2 + \frac{1}{3}e_3$; $a_1 = 3_z a_3$, $a_2 = 3_z a_1$

1. The double-centred hexagonal lattice R* projected on the plane $\perp c_3$

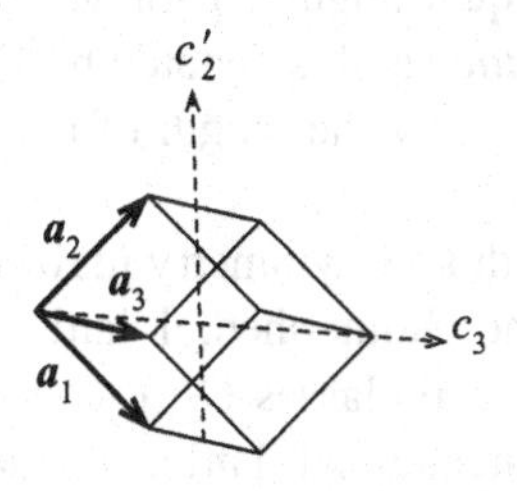

2. The rhombohedral lattice R

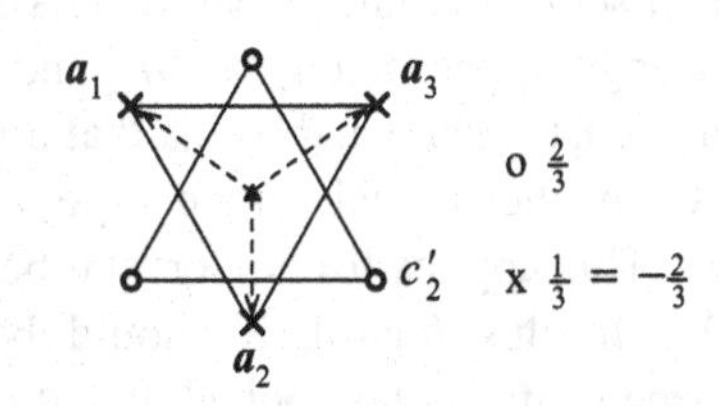

3. The projection of R on the plane $\perp c_3$

$|a_1| = |a_2| = |a_3|$, $a_1{}^\wedge a_2 = a_2{}^\wedge a_3 = a_3{}^\wedge a_1 =$ arbitrary

(f) Rhombohedral BP $\in D_{ei}$

BPs. These relations, (13.4.2b)–(13.4.2g), hold also for all the remaining body-centered lattices belonging to the orthorhombic and cubic systems.

Now, for I_Q, we have $|\boldsymbol{e}_1| = |\boldsymbol{e}_2|$ so that $\boldsymbol{a}_2 \wedge \boldsymbol{a}_3 = \boldsymbol{a}_3 \wedge \boldsymbol{a}_1$ from (13.4.2e). A lattice belonging to I_Q is determined again by two lattice parameters,[5] the length of an edge of the rhombohedron and one angular parameter, say $|\boldsymbol{a}_1|$ and $\boldsymbol{a}_1 \wedge \boldsymbol{a}_2$. The remaining angle ($\boldsymbol{a}_2 \wedge \boldsymbol{a}_3 = \boldsymbol{a}_3 \wedge \boldsymbol{a}_1$) is determined by (13.4.2f).

13.4.3 The rhombohedral system RH (D_{3i})

When $n = 3$, we have $p_3 = 3$ so that $m = 0, 1, 2$ from $0 \leqslant m < p_3$. Since $\boldsymbol{e}_2 = c_3 \boldsymbol{e}_1$, when $m = 0$ in (13.3.16a), we have a primitive BP that coincides with the hexagonal lattice type P_H belonging to the class D_{6i} defined by (13.4.1b). This is interesting because the mere assumption of a c_3-axis for a Bravais lattice T may lead to a point symmetry D_{6i} for the lattice. The physical significance of this result will be discussed again later (see also Section 13.5). Next, when $m = 1$, from (13.3.16b), we have two extra equivalent points inside the BP corresponding to $\mu = 1$ and 2:

$$\boldsymbol{a}_3^0 = \left(\tfrac{2}{3}, \tfrac{1}{3}, \tfrac{1}{3}\right), \qquad 2\boldsymbol{a}_3^0 = \left(\tfrac{1}{3}, \tfrac{2}{3}, \tfrac{2}{3}\right) \qquad (\text{mod } \boldsymbol{t}_c \in T^{(c)}) \tag{13.4.3a}$$

The BP is called the *double-centered hexagonal* BP. The group property (13.3.10c) is satisfied by these equivalent points of the BP, as one can see from $\boldsymbol{a}_3^0 + 2\boldsymbol{a}_3^0 = (0, 0, 0)$ (mod $\boldsymbol{t}_c \in T^{(c)}$). Thirdly, when $m = 2$, we have, from (13.3.16b) with $\mu = 1$ and 2,

$$\boldsymbol{a}_3^{0\prime} = \left(\tfrac{1}{3}, \tfrac{2}{3}, \tfrac{1}{3}\right), \qquad 2\boldsymbol{a}_3^{0\prime} = \left(\tfrac{2}{3}, \tfrac{1}{3}, \tfrac{2}{3}\right)$$

which is, however, equivalent to (13.4.3a) under exchange of $\boldsymbol{e}_1$ and $\boldsymbol{e}_2$. The above double-centered lattice may be described by a more symmetric primitive basis $[\boldsymbol{a}_1, \boldsymbol{a}_2, \boldsymbol{a}_3]$ generated from $\boldsymbol{a}_3^0$ by

$$\boldsymbol{a}_3 = \boldsymbol{a}_3^0, \qquad \boldsymbol{a}_1 = c_3 \boldsymbol{a}_3, \qquad \boldsymbol{a}_2 = c_3 \boldsymbol{a}_1 \tag{13.4.3b}$$

Explicitly, with $c_3 = 3_z$ and then using the Jones representation $3_z = (-y, x - y, z)$ in the hexagonal coordinate system given in (13.8.1b), we have

$$\boldsymbol{a}_3 = \left(\tfrac{2}{3}, \tfrac{1}{3}, \tfrac{1}{3}\right), \qquad \boldsymbol{a}_1 = \left(-\tfrac{1}{3}, \tfrac{1}{3}, \tfrac{1}{3}\right), \qquad \boldsymbol{a}_2 = \left(-\tfrac{1}{3}, -\tfrac{2}{3}, \tfrac{1}{3}\right)$$

Then,

$$|\boldsymbol{a}_1| = |\boldsymbol{a}_2| = |\boldsymbol{a}_3|, \qquad \boldsymbol{a}_1 \wedge \boldsymbol{a}_2 = \boldsymbol{a}_2 \wedge \boldsymbol{a}_3 = \boldsymbol{a}_3 \wedge \boldsymbol{a}_1$$

Thus, the primitive unit cell is a rhombohedron with equal angular parameters (see Figure 13.8(f)). The lattice is called the *rhombohedral lattice* and is denoted by R. The minimum set of the lattice parameters for R may be given by the length of one edge and one angular parameter, say $|\boldsymbol{a}_1|$ and $\boldsymbol{a}_1 \wedge \boldsymbol{a}_2$.

What we have shown above is that a Bravais lattice with a c_3 symmetry may belong either to the hexagonal lattice type $P_H \in D_{6i}$ or to the rhombohedral lattice type $R \in D_{3i}$. This applies for any crystal belonging to the crystal classes C_3, C_{3v}, C_{3i}, D_3 and D_{3i}. In this regard, it should be noted that a hexagonal lattice cannot be transformed into a rhombohedral lattice by an infinitesimal transformation because there is a finite difference between their lattice parameters.

[5] Note that the numbers of lattice parameters for any lattices belonging to a given crystal system are all the same.

Exercise. From (13.4.3b) show that $\boldsymbol{a}_i^2 = (3e_1^2 + e_3^2)/9$ for all i and that $\boldsymbol{a}_i \cdot \boldsymbol{a}_j = (-\frac{3}{2}e_1^2 + e_3^2)/9$ for all pairs (i, j).

Next, we shall construct the Bravais lattice types belonging to the crystal systems D_{2i}, O_i, C_{2i} and C_i using the reduced projection operators introduced by (13.3.9a).

13.4.4 The orthorhombic system O (D_{2i})

We choose all three conventional basis vectors in the directions of the three orthogonal two-fold axes $(2_x, 2_y, 2_z)$ of D_{2i}. Let $\boldsymbol{e}_3 \parallel 2_z$ and then, by applying the reduced projection operator $Q_2/2 = E + 2_z$ to (13.3.10b), we obtain

$$\boldsymbol{t} + 2_z\boldsymbol{t} = 2z\boldsymbol{e}_3 \qquad (\text{mod}\, \boldsymbol{t}_c \in T^{(c)}) \tag{13.4.4a}$$

which yields that $2z$ is an integer. Analogously, we can show that $2x$ and $2y$ are also integers. Thus, the possible lattice vectors are given by

$$\boldsymbol{t} = (m_1/2,\, m_2/2,\, m_3/2) \qquad (\text{mod}\, \boldsymbol{t}_c \in T^{(c)}) \tag{13.4.4b}$$

where $m_i = 0$ or 1 for all i with the condition that $m_1 + m_2 + m_3 \neq 1$, because each basis vector is the shortest in its own direction. From the group requirement (13.3.10c) for a set of equivalent points of a BP, we obtain the four and only four lattice types given below.

(i) The primitive lattice P_O: there are no extra equivalent points of the BP besides (0, 0, 0). The primitive basis is given by $\boldsymbol{a}_i = \boldsymbol{e}_i$ for all i. A lattice of P_O is completely determined by $|\boldsymbol{a}_1|$, $|\boldsymbol{a}_2|$ and $|\boldsymbol{a}_3|$.

(ii) The base-centered lattice A_O: only one base A defined by the face $[\boldsymbol{e}_2, \boldsymbol{e}_3]$ is centered by a lattice point so that the equivalent points of the BP are (0, 0, 0) and $(0, \frac{1}{2}, \frac{1}{2})$. The primitive basis may be defined by (see Figure 13.8(b))

$$\boldsymbol{a}_1 = \boldsymbol{e}_1, \qquad \boldsymbol{a}_2 = (\boldsymbol{e}_2 + \boldsymbol{e}_3)/2, \qquad \boldsymbol{a}_3 = (\boldsymbol{e}_2 - \boldsymbol{e}_3)/2 \tag{13.4.4c}$$

These satisfy, from the orthogonality of the conventional basis vectors,

$$|\boldsymbol{a}_2| = |\boldsymbol{a}_3|, \qquad \boldsymbol{a}_1 \perp [\boldsymbol{a}_2, \boldsymbol{a}_3]$$

The minimum set of lattice parameters may be given by $|\boldsymbol{a}_1|$, $|\boldsymbol{a}_2|$ and $\boldsymbol{a}_2\wedge\boldsymbol{a}_3$. The base-centered lattices B_O with the equivalent base point $(\frac{1}{2}, 0, \frac{1}{2})$ and C_O with $(\frac{1}{2}, \frac{1}{2}, 0)$ may be defined analogously to A_O, but these are equivalent to A_O under permutations of $[\boldsymbol{e}_1, \boldsymbol{e}_2, \boldsymbol{e}_3]$.

(iii) The body-centered lattice I_O: the equivalent points of the BP are (0, 0, 0), $(\frac{1}{2}, \frac{1}{2}, \frac{1}{2})$ (see Figure 13.8(d)). If one defines the primitive basis I_O by (13.4.2b) analogously to I_Q of the tetragonal system, the basis vectors satisfy all the relations (13.4.2c)–(13.4.2f) as in the case of I_Q. The only difference from I_Q is that all three angles, $\boldsymbol{a}_i\wedge\boldsymbol{a}_j$; $i > j$, are different for I_O. The primitive unit cell is again rhombic and the lattice parameters may be determined by one edge length $|\boldsymbol{a}_1|$ and two angular parameters $\boldsymbol{a}_1\wedge\boldsymbol{a}_2$ and $\boldsymbol{a}_2\wedge\boldsymbol{a}_3$ (cf. I_Q).

(iv) The face-centered lattice F_O: the equivalent points of the BP are (0, 0, 0), $(0, \frac{1}{2}, \frac{1}{2})$, $(\frac{1}{2}, 0, \frac{1}{2})$ and $(0, \frac{1}{2}, \frac{1}{2})$. A symmetric set of the primitive basic vectors $A = [\boldsymbol{a}_1, \boldsymbol{a}_2, \boldsymbol{a}_3]$ is defined by the three face-centered lattice points as follows (see Figure 13.8(c)):

$$A = [(e_2 + e_3)/2, (e_3 + e_1)/2, (e_1 + e_2)/2] \tag{13.4.4d}$$

Thus, the matrix form of A with respect to the conventional basis is given by

$$A = \begin{bmatrix} 0 & \frac{1}{2} & \frac{1}{2} \\ \frac{1}{2} & 0 & \frac{1}{2} \\ \frac{1}{2} & \frac{1}{2} & 0 \end{bmatrix} \tag{13.4.4e}$$

If we rewrite (13.4.4d) in the form

$$a_i = d_F - e_i/2; \qquad i = 1, 2, 3; \qquad d_F = (e_1 + e_2 + e_3)/2 = (a_1 + a_2 + a_3)/2$$

where d_F is a half of the diagonal of the BP, then, from the orthogonality of the conventional lattice vectors, it follows that

$$a_i^2 = d_F^2 - e_i^2/4; \qquad i = 1, 2, 3$$
$$a_i \cdot a_j = e_k^2/4 = d_F^2 - a_k^2; \qquad (i, j, k = 1, 2, 3 \text{ cyclic permutations}) \tag{13.4.4f}$$

where $d_F^2 = (e_1^2 + e_2^2 + e_3^2)/4 = (a_1^2 + a_2^2 + a_3^2)/2$. The minimum set of lattice parameters is given by $|a_1|$, $|a_2|$ and $|a_3|$, because all the angular parameters $a_i \wedge a_j$ are completely determined by them through (13.4.4f). Finally, from the matrix expression of A and (13.3.1a), the general coordinates of the lattice vector t with respect to the conventional basis are given by

$$t = An = ((n_2 + n_3)/2, (n_3 + n_1)/2, (n_1 + n_2)/2) \tag{13.4.4g}$$

Since the sum of three coordinates is an integer, either all three coordinates are integers or only one of them is an integer and the remaining two are half integers; the latter are due to the face-centered lattice points of the BP, cf. (13.4.2g). These relations (13.4.4d)–(13.4.4g) hold also for the face-centered lattice of the cubic system.

13.4.5 The cubic system C (O_i)

When the conventional lattice vectors of the orthorhombic system O become equal in length, the Bravais lattice becomes invariant under the point group O_i; in this case, the system is called the cubic system. It has only three types of BP: primitive P_c, face-centered F_c and the body-centered cubic I_c. There is no base-centered BP in the cubic system due to the 3_{xyz} symmetry which permutes e_1, e_2 and e_3. The primitive basis vectors for the types P_c, F_c and I_c are chosen analogously to those as for the orthorhombic system O. Every cubic lattice is, however, determined by only one parameter: the length of a primitive basis vector, because the primitive unit cell is rhombic and the angular parameters of all pairs (i, j) for each lattice type are equal and fixed by

$$\begin{aligned} &P_c\text{:} && a_i \wedge a_j = 90° \\ &F_c\text{:} && a_i \wedge a_j = 60°, \qquad \text{since } a_i \cdot a_j = a_k^2/2 \text{ from (13.4.4f)} \\ &I_c\text{:} && a_i \wedge a_j = 109°28' \text{ (the tetrahedral angle) since } \cos(a_i \wedge a_j) = -\tfrac{1}{3} \text{ from (13.4.2f)} \end{aligned} \tag{13.4.5}$$

13.4.6 The monoclinic system M (C_{2i})

The base $[\boldsymbol{e}_1, \boldsymbol{e}_2]$ perpendicular to the c_2-axis ($\|\boldsymbol{e}_3$) can always be chosen to be primitive. On applying $Q_2/2 = E + 2_z$ and $P_2/2 = E - 2_z$ to (13.3.10b) we obtain, respectively,

$$\boldsymbol{t} + 2_z\boldsymbol{t} = 2z\boldsymbol{e}_3 \qquad (\text{mod}\, \boldsymbol{t}_c \in T^{(c)})$$

$$\boldsymbol{t} - 2_z\boldsymbol{t} = 2x\boldsymbol{e}_1 + 2y\boldsymbol{e}_2 \qquad (\text{mod}\, \boldsymbol{t}_c \in T^{(c)})$$

where $2z$ is an integer, for $\boldsymbol{e}_3$ being primitive. Likewise, $2x$ and $2y$ are integers for the face $[\boldsymbol{e}_1, \boldsymbol{e}_2]$ being primitive. Accordingly, the possible lattice points of the BP are given by

$$\boldsymbol{t} = (m_1/2, m_2/2, m_3/2) \qquad (\text{mod}\, \boldsymbol{t}_c \in T^{(c)})$$

where $m_i = 0$ or 1 for all i excluding $m_1 = m_2 = m_3 = 1$, as will be shown below. From the group requirement (13.3.10c), the possible lattice types are the following.

(i) The primitive lattice P_M: with $\boldsymbol{a}_i = \boldsymbol{e}_i$ for all i, we have $\boldsymbol{a}_3 \perp [\boldsymbol{a}_1, \boldsymbol{a}_2]$. The lattice parameters are determined by $|\boldsymbol{a}_1|$, $|\boldsymbol{a}_2|$, $|\boldsymbol{a}_3|$ and $\boldsymbol{a}_i \wedge \boldsymbol{a}_j$.

(ii) The base-centered lattice A_M or B_M: the C_M lattice type is not possible because the C base $[\boldsymbol{e}_1, \boldsymbol{e}_2]$ is chosen to be primitive. The primitive basis for A_M is defined by

$$\boldsymbol{a}_1 = \boldsymbol{e}_1, \qquad \boldsymbol{a}_2 = (\boldsymbol{e}_2 + \boldsymbol{e}_3)/2, \qquad \boldsymbol{a}_3 = (\boldsymbol{e}_2 - \boldsymbol{e}_3)/2 \tag{13.4.6a}$$

which is analogous to A_0 of the D_{2i} system. Using $\boldsymbol{e}_3 \perp [\boldsymbol{e}_1, \boldsymbol{e}_2]$, we have

$$\boldsymbol{a}_2^2 = \boldsymbol{a}_3^2 = (\boldsymbol{e}_2^2 + \boldsymbol{e}_3^2)/4, \qquad \boldsymbol{a}_2 \cdot \boldsymbol{a}_3 = (\boldsymbol{e}_2^2 - \boldsymbol{e}_3^2)/4$$
$$\boldsymbol{a}_1 \cdot \boldsymbol{a}_2 = \boldsymbol{a}_1 \cdot \boldsymbol{a}_3 = (\boldsymbol{e}_1 \cdot \boldsymbol{e}_2)/2 \tag{13.4.6b}$$

The lattice parameters are determined by $|\boldsymbol{a}_1|$, $|\boldsymbol{a}_2|$ and $\boldsymbol{a}_1 \wedge \boldsymbol{a}_2$ and $\boldsymbol{a}_2 \wedge \boldsymbol{a}_3$.

Note that the body-centered BP with a lattice point $(\frac{1}{2}, \frac{1}{2}, \frac{1}{2})$ in the center of the BP defined by $[\boldsymbol{e}_1, \boldsymbol{e}_2, \boldsymbol{e}_3]$ is reduced to the base-centered lattice A_M with a new basis $[\boldsymbol{e}_1, \boldsymbol{e}_1 + \boldsymbol{e}_2, \boldsymbol{e}_3]$. (See Figure 13.9.)

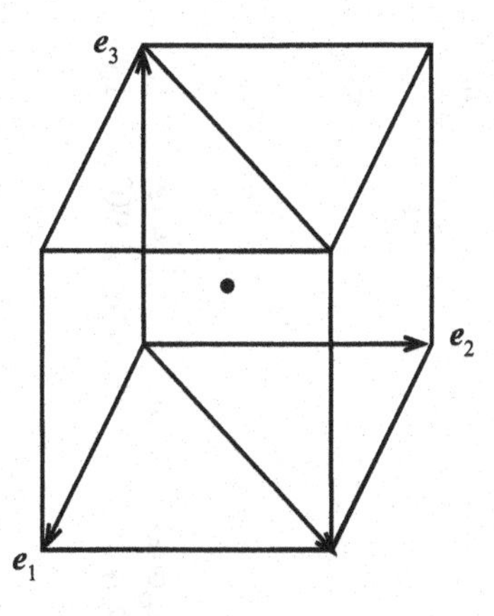

Figure 13.9. The body-centered BP of the monoclinic system is reduced to a base-centered BP.

Table 13.1. *Characteristics of the 14 types of Bravais lattice*

Crystal systems	Conventional lattice basis $[e_1, e_2, e_3]$	Types	Points of BP equivalent to (0, 0, 0)	Primitive basis $[a_1, a_2, a_3]$; $d = (e_1 + e_2 + e_3)/2$	Characteristics of a_1, a_2, a_3; $d_I = a_1 + a_2 + a_3$ $d_F = (a_1 + a_2 + a_3)/2$	Characteristic lattice parameters
Triclinic T (C_i)		P_T		$[e_1, e_2, e_3]$		$\|a_i\|, a_i \hat{} a_j; i \neq j; 1, 2, 3$
Monoclinic M (C_{2i})	All $\|e_i\|$ different; $e_3 \perp [e_1, e_2]$	P_M		$[e_1, e_2, e_3]$	$a_3 \perp [a_1, a_2]$	$\|a_1\|, \|a_2\|, \|a_3\|; a_1 \hat{} a_2$
		A_M	$(0, \frac{1}{2}, \frac{1}{2})$	$[e_1, (e_2 \pm e_3)/2]$	$\|a_2\| = \|a_3\|, a_1 \hat{} a_2 = a_1 \hat{} a_3 \neq 90°$	$\|a_1\|, \|a_2\|; a_1 \hat{} a_2, a_2 \hat{} a_3$
Orthorhombic O (D_{2i})	All $\|e_i\|$ different; all $e_i \hat{} e_j = 90°$	P_O		$[e_1, e_2, e_3]$		$\|a_1\|, \|a_2\|, \|a_3\|$
		A_O	$(0, \frac{1}{2}, \frac{1}{2})$	$[e_1, (e_2 \pm e_3)/2]$	$\|a_2\| = \|a_3\|, a_1 \hat{} a_2 = a_1 \hat{} a_3 \neq 90°$	$\|a_1\|, \|a_2\|; a_2 \hat{} a_3$
		I_O	$(\frac{1}{2}, \frac{1}{2}, \frac{1}{2})$	$a_i = d - e_i$	$\|a_1\| = \|a_2\| = \|a_3\| = \|d_I\|$ $\sum_{i>j} \cos(a_i \hat{} a_j) = -1$	$\|a_1\|; a_1 \hat{} a_2, a_2 \hat{} a_3$
		F_O	$(0, \frac{1}{2}, \frac{1}{2})$ $(\frac{1}{2}, 0, \frac{1}{2})$ $(\frac{1}{2}, \frac{1}{2}, 0)$	$a_i = d - e_i/2$	$a_i \cdot a_j = d_F^2 - a_k^2$ $(i, j, k = \text{cyclic } 1, 2, 3)$ $d_F^2 = (a_1^2 + a_2^2 + a_3^2)/4$	$\|a_1\|, \|a_2\|, \|a_3\|$
Tetragonal Q (D_{4i})	$\|e_1\| = \|e_2\| \neq \|e_3\|$ all $e_i \hat{} e_j = 90°$	P_Q		$[e_1, e_2, e_3]$	$\|a_1\| = \|a_2\| \neq \|a_3\|$ all $a_i \hat{} a_j = 90°$	$\|a_1\|, \|a_3 a\|$
		I_Q	$(\frac{1}{2}, \frac{1}{2}, \frac{1}{2})$	$a_i = d - e_i$	$\|a_1\| = \|a_2\| = \|a_3\| = \|d_I\|$ $a_1 \hat{} a_3 = a_2 \hat{} a_3, \sum_{i>j} \cos(a_i \hat{} a_j) = -1$	$\|a_1\|, a_1 \hat{} a_3$
Hexagonal H (D_{6i})	$\|e_1\| = \|e_2\|$ $e_3 \perp [e_1, e_2]$	P_H		$[e_1, e_2, e_3]$	$\|a_1\| = \|a_2\|, a_3 \perp a_1, a_2$; $a_1 \hat{} a_2 = 120°$	$\|a_1\|, \|a_2\|$
Rhombohedral R (D_{3i})	$e_1 \hat{} e_2 = 120°$	R	$(\frac{2}{3}, \frac{1}{3}, \frac{1}{3})$ $(\frac{1}{3}, \frac{2}{3}, \frac{2}{3})$	a_3 $a_1 = 3_z a_3, a_2 = 3_z^2 a_3$	$a_1 \hat{} a_2 = a_2 \hat{} a_3 = a_3 \hat{} a_1$	$\|a_1\|, a_1 \hat{} a_2$
Cubic C (O_i)	$\|e_1\| = \|e_2\| = \|e_3\|$ all $e_i \hat{} e_j = 90°$	P_C		$[e_1, e_2, e_3]$	All $a_i \hat{} a_j = 90°$	
		I_C	$(\frac{1}{2}, \frac{1}{2}, \frac{1}{2})$	$a_i = d - e_i$	All $a_i \hat{} a_j = 108°27'$	$\|a_1\| = \|a_2\| = \|a_3\|$
		F_C	$(0, \frac{1}{2}, \frac{1}{2})$ $(\frac{1}{2}, 0, \frac{1}{2}), (\frac{1}{2}, \frac{1}{2}, 0)$	$a_i = d - e_i/2$	All $a_i \hat{} a_j = 60°$	

All body-centered lattices I_O, I_Q and I_C, all lattices of the cubic system P_C, F_C and I_C, and the rhombohedral lattice R have the rhombic primitive unit cells; i.e. $|a_1| = |a_2| = |a_3|$.

13.4.7 *The triclinic system* T (C_i)

The BP is always chosen to be primitive since there is no restriction on the lattice parameters. Thus, all six parameters are required in order to define a lattice belonging to this system.

We have constructed a total of 14 Bravais lattice types, which exhausts all the possible lattice types. These are summarized in Table 13.1 and expressed graphically in Figure 13.8.

13.4.8 *Remarks*

Remark 1. In Table 13.1, the lattice types are characterized by the conventional lattice bases $[\boldsymbol{e}_1, \boldsymbol{e}_2, \boldsymbol{e}_3]$ as well as by the primitive bases $[\boldsymbol{a}_1, \boldsymbol{a}_2, \boldsymbol{a}_3]$. Note that the order of the minimum set of the lattice parameters is the same for any lattice type belonging to a particular crystal system: six for triclinic, four for monoclinic, three for orthorhombic, two for the tetragonal, rhombohedral and hexagonal systems, and only one for the cubic system.

Also it should be noted from Table 13.1 that six out of 14 Bravais lattice types are described by rhombic primitive unit cells: all body-centered lattices and all cubic lattices, as well as the rhombohedral lattice R, have the rhombic primitive unit cell, i.e. $|\boldsymbol{a}_1| = |\boldsymbol{a}_2| = |\boldsymbol{a}_3|$. They differ, however, in their three angular parameters; none of them are the same for I_{O}, two of them are the same for I_{Q} and all three of them are the same for R. For the cubic system, all three angular parameters for each lattice type are not only the same but also take a specific value: 90° for P_{c}, 109°28′ for I_{c} and 60° for F_{c}.

Remark 2. The volume V of a Bravais parallelepiped BP is given by $V = NV_0$, where N is the number of equivalent points of the BP and V_0 is the primitive unit cell volume. Thus

$$V_{\mathrm{P}} = V_0, \qquad V_{\mathrm{B}} = V_{\mathrm{I}} = 2V_0, \qquad V_{\mathrm{F}} = 4V_0$$

where V_{P}, V_{B}, V_{I} and V_{F} are the volumes of BPs belonging to the primitive, base-centered, body-centered and face-centered lattices, respectively.

Remark 3. The interrelations between the lattice types based on a rectangular base and a rhombic base. Previously, the cubic system has been introduced as a special case of the orthorhombic system O. Obviously, it can be introduced as a special case of the tetragonal system Q as well; in this system, however, there exist only two lattice types, P_{Q} and I_{Q}, whereas in the cubic system there exist three lattice types, P_{c}, I_{c} and F_{c}. The apparent discrepancy is resolved by introducing a transformation that brings a rectangular base of a BP to a rhombic base.

We shall begin with the orthorhombic system O first. Let $[\boldsymbol{e}_1, \boldsymbol{e}_2]$ be the rectangular base of a Bravais lattice belonging to the system O. If we introduce a rhombic base $[\boldsymbol{e}_1', \boldsymbol{e}_2']$ defined by

$$\boldsymbol{e}_1' = \boldsymbol{e}_1 + \boldsymbol{e}_2, \qquad \boldsymbol{e}_2' = \boldsymbol{e}_1 - \boldsymbol{e}_2 \tag{13.4.7}$$

where $|\boldsymbol{e}_1'| = |\boldsymbol{e}_2'|$, then the lattice types based on the rectangular base transform to those based on the rhombic base as follows:

$$P_0^{\text{Rec}} \to C_0^{\text{Rhomb}}, \qquad C_0^{\text{Rec}} \to P_0^{\text{Rhomb}}, \qquad I_0^{\text{Rec}} \to F_0^{\text{Rhomb}}, \qquad F_0^{\text{Rec}} \to I_0^{\text{Rhomb}} \tag{13.4.8}$$

that is, the primitive lattice P_0^{Rec} based on $[\boldsymbol{e}_1, \boldsymbol{e}_2]$ is transformed to the base-centered lattice C_0^{Rhomb} based on $[\boldsymbol{e}_1', \boldsymbol{e}_2']$ and so on. These transformations may be easily understood from the two-dimensional projections of the lattice types given in Figure 13.10.

Obviously, the above transformation (13.4.7) can be applied to the tetragonal system. In this case, both bases $[\boldsymbol{e}_1, \boldsymbol{e}_2]$ and $[\boldsymbol{e}_1', \boldsymbol{e}_2']$ are square, so that

$$P_{\text{Q}} = C_{\text{Q}}, \qquad I_{\text{Q}} = F_{\text{Q}} \tag{13.4.9}$$

That is, the lattice type P_{Q} can be regarded as C_{Q} whereas I_{Q} can be regarded as F_{Q}. This is not the case for the cubic system, because the transformation (13.4.7) applied for the cubic system lowers the cubic symmetry to the tetragonal symmetry due to the fact that $|\boldsymbol{e}_1'| = |\boldsymbol{e}_2'| = \sqrt{2}|\boldsymbol{e}_1| \neq |\boldsymbol{e}_3|$. Accordingly, I_{c} and F_{c} types are distinct in the cubic system.

Now, application of the deformation $|\boldsymbol{e}_3| \to |\boldsymbol{e}_1| = |\boldsymbol{e}_2|$ to the lattice types P_{Q} and I_{Q} brings out the following lattice types of the cubic system:

$$P_{\text{Q}} \to P_{\text{C}}, \qquad I_{\text{Q}} \to I_{\text{C}}$$

whereas application of the deformation $|\boldsymbol{e}_3| \to |\boldsymbol{e}_1'| = |\boldsymbol{e}_2'|$ to the lattice type F_{Q} brings out F_{C} of the cubic system

$$F_{\text{Q}} \to F_{\text{C}}$$

This dissolves the discrepancy mentioned at the beginning of Remark 3. Hereafter, the lattice types of the orthorhombic system will be referred to the rectangular base unless otherwise specified.

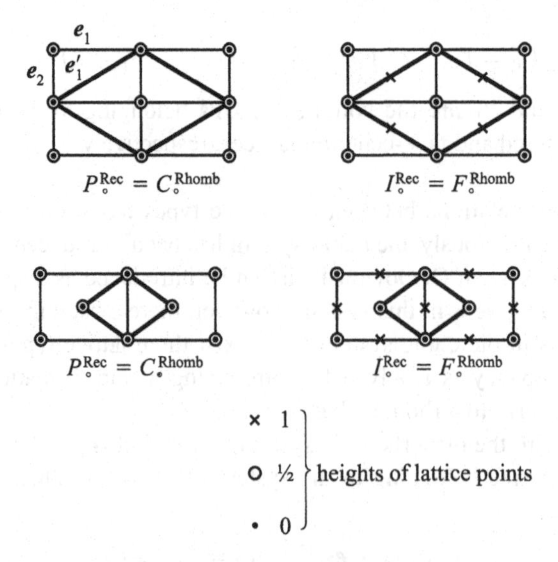

Figure 13.10. Two-dimensional projections of the lattice types of the orthorhombic system: thin lines denote the rectangular bases and thick lines denote the rhombic lattices.

Remark 4. The hierarchy of the crystal systems (Lyubarskii 1960). One says that a crystal system B is subordinate to a crystal system A (this is denoted by $A \Rightarrow B$) if

1. the symmetry point group B is a subgroup of the symmetry group A, and
2. each lattice type belonging to A can be converted into a lattice type of B by an infinitesimal continuous transformation of the basis vectors of A.

From Table 13.1, which contains all the information necessary for the infinitesimal transformations between lattices, we have the following subordination relations between crystal systems:

$$\left.\begin{array}{ll} \text{I.} & O_i \Rightarrow D_{4i} \Rightarrow D_{2i} \\ \text{II.} & O_i \Rightarrow D_{3i} \\ \text{III.} & D_{6i} \Rightarrow D_{2i} \\ \text{IV.} & D_{6i} > D_{3i} \end{array}\right\} \Rightarrow C_{2i} \Rightarrow C_i \tag{13.4.10}$$

It is to be noted here that, if $A > B$, then $A \Rightarrow B$ except for the case in which $A = D_{6i}$ and $B = D_{3i}$, which appears in chain IV. This exception is due to the fact that there are finite differences between their angular parameters for the hexagonal system $P_{\mathrm{H}} \in D_{6i}$: $\boldsymbol{a}_1 \wedge \boldsymbol{a}_2 = 120°$ and $\boldsymbol{a}_2 \wedge \boldsymbol{a}_3 = \boldsymbol{a}_3 \wedge \boldsymbol{a}_1 = 90°$; whereas all angles are the same, $\boldsymbol{a}_1 \wedge \boldsymbol{a}_2 = \boldsymbol{a}_2 \wedge \boldsymbol{a}_3 = \boldsymbol{a}_3 \wedge \boldsymbol{a}_1$, for the rhombohedral system $R \in D_{3i}$, according to Table 13.1.

Exercise. Using Table 13.1, describe the infinitesimal transformations required for the subordinate chain II.

13.5 The 32 crystal classes and the lattice types

So far, we have discussed the Bravais lattices, their point symmetry and the lattice types compatible with the symmetry. Here, we shall discuss the crystal class G which describes the directional symmetry of a crystal and then determine the possible lattice types L allowed for the crystal belonging to a crystal class G. Hereafter, the lattice types L compatible with a crystal class G may be denoted $\{L;\ G\}$. As was mentioned in Section 13.2, a crystal class G of a crystal is a subgroup of the crystal system K which is the symmetry group of the Bravais lattice T of the crystal. Thus from the subgroups of the seven crystal systems K given in (13.2.3), we obtain a total of 32 crystal classes:

$$\begin{gathered} C_1,\ C_i;\ C_2,\ C_s,\ C_{2i};\ D_2,\ C_{2v},\ D_{2i}; \\ C_4,\ C_{2p},\ C_{4i},\ C_{4v},\ D_{2p},\ D_4,\ D_{4i};\ C_3,\ C_{3i},\ D_3,\ C_{3v},\ D_{3i}; \\ C_6,\ C_{3p},\ C_{6i},\ C_{6v},\ D_{3p},\ D_6,\ D_{6i};\ T,\ T_i,\ T_p,\ O,\ O_i \end{gathered} \tag{13.5.1}$$

These are written roughly in increasing order of the orders of principal axes except for the cubic system. Since a particular point group G can be a subgroup of many symmetry groups K, we may assign a crystal class G to the system K of the lowest symmetry for which $G \leqslant K$ holds; thus, we obtain the distribution of the crystal classes among the crystal systems given in Table 13.2.

The assignment of the crystal classes G into the crystal systems K given by Table 13.2 leads us to the possible lattice types L allowed for a crystal belonging to a crystal class G. To see this, we note that the crystal classes G assigned to a crystal system K have a set of common characteristic axes of rotation, given in the third column of

Table 13.2. *The distribution of the crystal classes among the crystal systems*

Crystal system	Crystal classes	Characteristic axes of rotation
Cubic, C	O_i, O, T_p, T_i, T	$2_x, 3_{xyz}$
Hexagonal, H	$D_{6i}, D_6, D_{3p}, C_{6v}, C_{3p}, C_6$	6_z
Rhombohedral, RH	$D_{3i}, D_3, C_{3v}, C_{3i}, C_3$	3_z
Tetragonal, Q	$D_{4i}, D_4, D_{2p}, C_{4v}, C_{4i}, C_{2p}, C_4$	4_z
Orthorhombic, O	D_{2i}, D_2, C_{2v}	$2_z, 2_x$
Monoclinic, M	C_{2i}, C_s, C_2	2_z
Triclinic, T	C_i, C_1	1

1. In the third column we have disregarded the differences between c_n and $\overline{c}_n$ because of the inherent inversion symmetry of any Bravais lattice.
2. A crystal with a direction group assigned to the rhombohedral system RH may have either the hexagonal lattice type or the rhombohedral lattice type.

Table 13.2. Here, we have disregarded the difference between c_n and $\overline{c}_n$ because of the inherent inversion symmetry of any Bravais lattice. Since these characteristic axes determine the allowed lattice types of K as shown in Section 13.4, the required lattice types compatible with a class G are given by the lattice types belonging to the K to which G is assigned. For example, any crystal class belonging to the tetragonal system has a four-fold axis c_4 (or $\overline{c}_4$) as the principal axis of rotation. From Lemmas 1–3, this means that the point symmetry of the Bravais lattice is the D_{4i} symmetry and the allowed lattice types are P_{Q} and I_{Q}, as given in Table 13.1. In particular, a crystal with the principal axis of rotation c_3 (or $\overline{c}_3$) is allowed to have either the hexagonal lattice P_{H} or the rhombohedral lattice R, as has been discussed in Section 13.4.3.

If the Bravais lattice of a crystal has a higher symmetry than that required by the directional symmetry group G of the crystal, it would be unstable because an infinitesimal deformation would lower the symmetry of the Bravais lattice to the lowest level permitted by the crystal class on account of the subordination relations (13.4.10). This also explains the exceptional case of the D_{3i} system, because it is not subordinate to the D_{6i} system.

By combining Tables 13.1 and 13.2, we can directly correlate the lattice types L to a crystal class G and determine the allowed set of lattice types $\{L;\ G\}$ for a crystal belonging to a crystal class G.

In the next section, the 230 space groups are classified by the crystal system K first, then by the crystal class G and then by the lattice type $\{L;\ G\}$ compatible with G. Finally, each space group is defined by a set of the characteristic translations $\boldsymbol{v}_R$ for every rotation $R \in G$ compatible with $\{L;\ G\}$.

Before completing this section, we simply note here that a crystal class G belonging to a crystal system K is an invariant subgroup of K:

$$G \triangleleft K \tag{13.5.2}$$

The physical significance of this relation will be discussed later in connection with the equivalence criteria of space groups.

13.6 The 32 minimal general generator sets for the 230 space groups

13.6.1 Introduction

The present method of constructing the 230 space groups is based on the homomorphism between a space group $\hat{G} = \{\{R|\boldsymbol{t}_R\}\}$ and its crystal class $G = \{R\}$ via the correspondence $\{R|\boldsymbol{t}_R\} \to R$. The kernel of the homomorphism is the invariant subgroup $T = \{\{E|\boldsymbol{t}\}\}$ of $\hat{G}$ so that the factor group $\hat{G}/T$ is isomorphic to G:

$$\hat{G}/T \simeq G \tag{13.6.1}$$

Accordingly, the generator sets of both groups must satisfy the same set of abstract defining relations (or presentation, see Chapters 5 and 11). This is one of the most important basic properties of space groups; since through the presentations of 32 point groups, we can determine all the possible generator sets of the factor groups and hence obtain the 230 space groups.

We shall first describe the one-to-one correspondence for the isomorphism between $\hat{G}/T$ and G. For this purpose, let $(R|\boldsymbol{v}_R)$ be a coset representative of T in $\hat{G}$ defined by

$$(R|\boldsymbol{v}_R) = \{R|\boldsymbol{v}_R\} \qquad (\text{mod}\, \boldsymbol{t} \in T);\ \forall R \in G \tag{13.6.2a}$$

where $\boldsymbol{v}_R$ is either zero or a non-lattice translation that is the minimum translational part associated with the rotation R. Then the set $\{(R|\boldsymbol{v}_R)\}$ provides a faithful representation of $\hat{G}/T$ with the multiplication law

$$(R_1|\boldsymbol{v}_1)(R_2|\boldsymbol{v}_2) = (R_3|\boldsymbol{v}_3) \tag{13.6.2b}$$

where

$$R_1R_2 = R_3, \qquad \boldsymbol{v}_1 + R_1\boldsymbol{v}_2 = \boldsymbol{v}_3 \qquad (\text{mod}\, \boldsymbol{t} \in T)$$

Hereafter, we mean by $\hat{G}/T$ the group $\{(R|\boldsymbol{v}_R)\}$ with this multiplication law. Then the isomorphism (13.6.1) is described by the one-to-one correspondence $(R|\boldsymbol{v}_R) \leftrightarrow R$.

In actual calculation, the translational part $\boldsymbol{v}_R$ is expressed by the conventional coordinates (x_R, y_R, z_R) defined by

$$\boldsymbol{v}_R = x_R\boldsymbol{e}_1 + y_R\boldsymbol{e}_2 + z_R\boldsymbol{e}_3 = (x_R, y_R, z_R) \qquad (\text{mod}\, \boldsymbol{t} \in T) \tag{13.6.3}$$

where $[\boldsymbol{e}_1, \boldsymbol{e}_2, \boldsymbol{e}_3]$ is the conventional lattice basis of a lattice belonging to $\{L; G\}$ (a lattice type L compatible to a crystal class G). In general, the translational parameters x_R, y_R and z_R are rational numbers (mod 1) on account of the discrete nature of the space groups.

To determine all the space groups belonging to a crystal class G from the defining relations, we assume a set of realizations for the generators of $\hat{G}/T$ in terms of $(R|\boldsymbol{v}_R)$ with an undetermined translational part $\boldsymbol{v}_R = (x_R, y_R, z_R)$. Then substitution of these into the defining relations leads to a set of linear equations for x_R, y_R and z_R (mod 1). The solutions of these linear equations for the translational parameters x_R, y_R and z_R with respect to each allowed lattice type L of the class G lead to the generators of the space groups belonging to L. In general, one has more than one solution because of mod $\boldsymbol{t} \in T$ in the definition $(R|\boldsymbol{v}_R)$. Some of the solutions, however, could be equivalent under the lattice transformation $\Lambda = [U|\boldsymbol{s}]$ introduced in (13.1.7) which leaves invariant the lattice type L of the class G. After we have removed these redundant solutions, we obtain the generator sets of the space groups belonging to the set $\{L; G\}$. The generator set for a factor group $\hat{G}/T$ thus obtained may be called the *minimal general generator set (MGGS)* belonging to the crystal class G: it contains a set of parameters that describes the translational parts of the space groups $\hat{G}$ homomorphic to the crystal class G; e.g. see (13.6.10).

To ascertain the inequivalence of the space groups thus formed, we shall introduce a general equivalence criterion in Section 13.7. We shall then construct the MGGSs of the classes belonging to the cubic and rhombohedral systems as the representative examples in the present approach in Sections 13.9 and 13.10, respectively. The 32 minimal general generator sets (one per crystal class) for all of the 230 space groups have been constructed by Kim (1986b) and will be presented in Table 13.3 later.

To show the simplicity of the method, however, it seems worthwhile to discuss the construction of the space groups belonging to the class D_4 through its MGGS as a prototype example, without going through the comprehensive argument concerning their inequivalence. This example will prepare us for introducing the equivalence criteria for the space groups under lattice transformations.

13.6.2 *The space groups of the class* D_4

The defining relations for D_4 are

$$A^4 = B^2 = (AB)^2 = E \tag{13.6.4}$$

The corresponding generators for the factor group $\hat{D}_4/T$ may be expressed by

$$A = (4_z|\xi, \eta, c), \qquad B = (2_x|a, \beta, \gamma), \qquad E = (e|0, 0, 0) \tag{13.6.5}$$

where (ξ, η, c) and (a, β, γ) are the sets of the translational parameters based on the conventional lattice basis of the tetragonal system given in Table 13.1. Before determining these translational parameters, we shall first map off some of the parameters by shifting the origin of the lattice. According to (13.1.8), under a shift $[e|\boldsymbol{s}]$ an element $(R|\boldsymbol{v}_R)$ of $\hat{G}/T$ is transformed to $(R|\boldsymbol{v}'_R)$, where

$$\boldsymbol{v}'_R = \boldsymbol{v}_R - (e - R)\boldsymbol{s} \tag{13.6.6}$$

so that, using the Jones representations $4_z = (\bar{y}, x, z)$, $2_x = (x, \bar{y}, \bar{z})$ given in Table 11.2, the generators A and B are transformed to

$$\begin{aligned} A &\to (4_z|\xi - s_1 - s_2, \eta + s_1 - s_2, c) \\ B &\to (2_x|a, \beta - 2s_2, \gamma - 2s_3) \end{aligned} \tag{13.6.7}$$

where $\boldsymbol{s} = (s_1, s_2, s_3)$. Note that a shift of the origin cannot affect the translational components c and a because they are in the invariant eigenvector spaces of 4_z and 2_x, respectively (cf. Equation (13.1.10)). Now we take $\boldsymbol{s}$ such that

$$\xi - s_1 - s_2 = 0, \qquad \eta + s_1 - s_2 = 0, \qquad \gamma - 2s_3 = 0$$

and set $\beta - 2s_2 = b$; then, the generators are simplified to

$$A = (4_z|0, 0, c), \qquad B = (2_x|a, b, 0) \tag{13.6.8a}$$

The remaining parameters a, b and c will be determined for each lattice type of the class from the defining relations, which state that the following products are all equal to the unit element $E = (e|0, 0, 0)$:

$$\begin{aligned} A^4 = (e|0, 0, 4c), \qquad B^2 = (e|2a, 0, 0) \\ (AB)^2 = (e|a - b, a - b, 0) \end{aligned} \tag{13.6.8b}$$

Now according to Tables 13.1 and 13.2, there exist only two lattice types for the class D_4: the primitive lattice P and the body-centered lattice I.

(i) For the P lattice, from (13.6.4) and (13.6.8b), we have

$$4c \equiv 0, \qquad 2a \equiv 0, \qquad a - b \equiv 0 \quad (\text{mod } 1)$$

which[6] yield

$$c = 0, \tfrac{1}{4}, \tfrac{1}{2}, \tfrac{3}{4}, \qquad a = b = 0, \tfrac{1}{2} \tag{13.6.9a}$$

The space groups with the parameter set (c, a) defined by (13.6.9a) are all inequivalent under any shift $[e|s]$, because the parameters c and a are in the invariant eigenvector spaces of 4_z and 2_x, respectively. This is sufficient to ascertain the inequivalence of these space groups for the class D_4 according to the general criterion for equivalence of space groups which will be introduced in the next section.

(ii) For the I lattice, the parameter sets for all the P lattices given by (13.6.9a) are further reduced to two sets

$$c = 0, \tfrac{1}{4}, \qquad a = b = 0 \tag{13.6.9b}$$

due to the equivalences $c \sim c + \frac{1}{2}$ and $a \sim a + \frac{1}{2}$ for the I lattice, since, under a shift of the origin of the lattice $[e|\frac{1}{2}, 0, \frac{1}{4}]$, we obtain the following transformations of the generators, using (13.6.6),

$$(4_z|0, 0, c) \rightarrow (4_z|\tfrac{1}{2}, \tfrac{1}{2}, c) = (4_z|0, 0, c + \tfrac{1}{2})$$

$$(2_x|a, a, 0) \rightarrow (2_x|a, a, \tfrac{1}{2}) = (2_x|a + \tfrac{1}{2}, a + \tfrac{1}{2}, 0)$$

where the last equalities follow from the equivalence $(\frac{1}{2}, \frac{1}{2}, \frac{1}{2}) \sim (0, 0, 0)$ for the I lattice. The two sets of solutions given by (13.6.9) are inequivalent because c is in the invariant eigenvector space of 4_z.

Summarizing, we obtain a total of ten space groups belonging to the D_4 class expressed by the minimum general generator set (MGGS)

$$(4_z|0, 0, c), (2_x|a, a, 0) \qquad \text{with} \qquad L(c, a) \tag{13.6.10}$$

with the lattice type L and the set of translational parameters (c, a) given by

89. $P(0, 0)$, 90. $P(0, \frac{1}{2})$, 91. $P(\frac{1}{4}, 0)$, 92. $P(\frac{1}{4}, \frac{1}{2})$, 93. $P(\frac{1}{2}, 0)$,
94. $P(\frac{1}{2}, \frac{1}{2})$, 95. $P(\frac{3}{4}, 0)$, 96. $P(\frac{3}{4}, \frac{1}{2})$, 97. $I(0, 0)$, 98. $I(\frac{1}{4}, 0)$

Here, the numbers 89–98 are the space group numbers given in the ITXC.[7] For example, from (13.6.10), the generators of the space group number 98 are given by

$$I(4_z|0, 0, \tfrac{1}{4}), \qquad (2_x|0, 0, 0)$$

where I denotes the body-centered Bravais lattice of the tetragonal system defined in Table 13.1.

[6] For example, $x \equiv y$ (mod 5) reads 'x is congruent to y modulo 5' and means '$x - y$ is divisible by 5'. Thus, $nx \equiv 0$ (mod 1) with a given integer n means that $x \equiv m/n$ (mod 1), where m is an integer bound by $0 \leqslant m < n$.

[7] ITXC denotes *International Tables for X-ray Crystallography* (Kynoch, Birmingham, UK, 1965), Vol. 1 by N. F. M. Henry and K. Lonsdale.

From the MGGS given by (13.6.10), we can easily write down the general elements of the factor group $\hat{D}_4/T$ using the right coset decomposition of $\hat{D}_4$:

$$\hat{D}_4 = \hat{C}_4 + \hat{C}_4\hat{2}_x$$

where $\hat{C}_4$ denotes the space group generated by $\hat{4}_z = (4_z|0, 0, c)$ whereas $\hat{2}_x = (2_x|a, a, 0)$. Thus, the general elements of $\hat{D}_4/T$ are

$$\begin{aligned}
\hat{C}_4:&\quad (e|0, 0, 0), \quad (4_z|0, 0, c), \quad (2_z|0, 0, 2c), \quad (4_z^{-1}|0, 0, -c)\\
\hat{C}_4\hat{2}_x:&\quad (2_x|a, a, 0), \quad (2_{xy}|-a, a, c), \quad (2_y|-a, -a, 2c),\\
&\quad (2_{x\bar{y}}|a, -a, -c)
\end{aligned} \tag{13.6.11}$$

where we have used the Jones representations of the point operations given in Table 11.2. From the parameters c and a given in $L(c, a)$ of (13.6.10), we can reproduce all the elements of each one of the ten space groups numbers 89–98 given in the ITXC.

There are many convenient features for the MGGS of a crystal class:

(1) The number of the generators for every crystal class (or the rank) is less than or equal to three, which is the maximum number for the generators of a point group.
(2) The MGGS for a crystal class provides all the necessary and sufficient information for determining all the (vector or projective) irreducible representations of the space groups belonging to the class (see Chapter 14).
(3) The number of translational parameters required for each class is limited (<5); e.g. only one parameter is required for the octahedral class O (see Table 13.3 later).
(4) The symmetry properties exhibited by a MGGS under lattice tranformations help identify the space groups.
(5) The set of space groups expressed by a MGGS is very effective in describing the group–subgroup relations between the space groups with the same lattice types (see Section 13.11).

Above all, the present method of constructing 32 MGGS provides the simplest algebraic method so far of obtaining the 230 space groups without the help of a computer. This is partly due to the algebraic equivalence criteria for the space groups which will be described in the next section. Furthermore, the MGGSs of the crystal classes lead to similar compact expressions for the extended space groups such as magnetic space groups. This gives us control over the large number of the magnetic space groups (a total of 1421) and helps us to understand the group structure (see Chapter 17).

13.7 Equivalence criteria for space groups

As we have seen in (13.6.9b), it may happen that some of the space groups determined by the present method are equivalent under a lattice transformation $\Lambda = [U|s]$, which leaves the lattice type $\{L;\ G\}$ compatible with the class G invariant. Let the coordinate system be based on the lattice vectors, then the transformation matrix U is a unimodular integral matrix. When the space group has enantiomorphism, U should be a proper unimodular matrix, i.e. $\det U = 1$. Further restrictions on U will be discussed below.

Let $\{R_j|\boldsymbol{v}_j\}$ and $\{R'_j|\boldsymbol{v}'_j\}$ be a pair of solutions obtained from the presentation of $\hat{G}/T$ for a given $\{L;\ G\}$. Then, the pair is equivalent, if and only if there exists a lattice transformation $\Lambda = [U|\boldsymbol{s}]$ that connects the pair by

$$\Lambda^{-1}(R_j|\boldsymbol{v}_j)\Lambda = (R'_j|\boldsymbol{v}'_j) \tag{13.7.1}$$

where, in view of (13.1.7),

$$\begin{aligned} (e - R_j)\boldsymbol{s} &= \boldsymbol{v}_j - U\boldsymbol{v}'_j \qquad (\text{mod}\, \boldsymbol{t} \in T) \\ R'_j &= U^{-1}R_jU; \qquad j = 1, 2, \ldots, r \end{aligned} \tag{13.7.2}$$

Here the set $\{R_1, R_2, \ldots, R_r\}$ is a generator set of G, and $r \leqslant 3$ because the maximum number of the generators of a point group is three. Since $r \leqslant 3$, the set of equations has more unknowns (three for U and three for $\boldsymbol{s}$) than the number of equations; yet, it need not have a solution on account of the unimodular condition imposed on U. Thus, we may restate the problem in a more convenient form for calculation: the pair is equivalent if and only if the set of r equations (13.7.2) has a solution s for any (proper) unimodular matrix U that defines an automorphism of G that leaves the lattice type L invariant.

The solvability of the set of equations (13.7.2) for $\boldsymbol{s}$ for a given U and T is trivial. As was discussed for Equation (13.1.9), if one of the operators R_j is a pure rotation c_n or a reflection m, it has at least one eigenvector ψ_j belonging to the characteristic value 1. Then the jth equation in (13.7.2) has a solution $\boldsymbol{s}_j$ (up to ψ_j) if and only if the right-hand side of the first equation in (13.7.2) is perpendicular to ψ_j. On the other hand, if $R_j \neq c_n$ and m, the corresponding equation has a definite solution $\boldsymbol{s}_j$ for any U. Now the set (13.7.2) has a solution for the given U and T, if all $\boldsymbol{s}_j$ exist and are independent of j.

The crucial part of the problem is to find the allowed set of the unimodular matrices $\{U\}$ which leaves invariant the class G and the lattice type $\{L;\ G\}$ compatible with G, unless we resort to trial and error. The collection of all 3×3 unimodular integral matrices forms a group, which we denote by $GL(3,\ Z)$. Then the allowed set $\{U\}$ is a normalizer $N(G)$ of the class G in $GL(3,\ Z)$, which leaves $\{L;\ G\}$ invariant. Thus, from knowledge of symmetry group K of a crystal system that leaves a crystal class G invariant as given by (13.5.2), we obtain the following normalizer $N(G)$ ($\triangleright G$) for the required sets $\{U\}$ (ignoring enantiomorphism for the moment):

$$\begin{aligned} &O_i \triangleright O,\ T_p,\ T_i,\ T,\ D_{2i},\ D_2 \\ &D_{6i} \triangleright D_6,\ D_{3p},\ C_{6v},\ C_{6i},\ C_6,\ C_{3p} \\ &D_{3i} \triangleright D_3,\ C_{3v},\ C_{3i},\ C_3 \\ &D_{4i} \triangleright D_4,\ D_{2p},\ C_{4v},\ C_{4i},\ C_4,\ C_{2p},\ C_{2v} \\ &M \triangleright C_{2i},\ C_2,\ C_s \\ &TR \triangleright C_i,\ C_1 \end{aligned} \tag{13.7.3}$$

When enantiomorphism exists, the normalizers O_i and D_{ni} given above should be replaced by their proper subgroups O and D_n ($n > 2$), respectively. The sets M and TR will be defined later. It should be noted that the normalizer D_{3i} given above for the rhombohedral system is for the rhombohedral lattice. When the lattice is hexagonal, the normalizer for D_{3i}, D_3, C_{3v} and C_3 is D_{6i}. This does not create any complication in the actual construction of their space groups, since the inequivalence of the allowed solutions for these classes is simply due to the difference in the translational vector component in the invariant eigenvector space of 3_z, the z-component (see the discussion of the rhombohedral system in Section 13.10). The sets M and TR are infinite groups, which will be given later by their generators. Note that some classes of the orthorhombic system are included in the cubic system whereas the others are included in the tetragonal system.

To see the justification of the allowed set given by (13.7.3) for $\{U\}$, we consider, for example, the first set O_i with respect to the space groups belonging to $\{L; D_{2i}\}$, the lattice type L of the crystal class D_{2i}. The conventional basis vectors for the orthorhombic system are mutually orthogonal and differ in their lengths. However, when we discuss the equivalent transformation defined by (13.7.1), the three binary axes of rotation of D_{2i} (or D_2) are regarded as equivalent under their permutation because a permutation of the basis vectors of the orthorhombic system does not change the lattice type $\{L; D_{2i}\}$. (This is consistent with the traditional geometric congruence in classification of their space groups.) Thus, the set of the unimodular transformation matrices $\{U\}$ for this case is described by the set of rotations belonging to the point group O_i, since O_i leaves D_{2i} invariant and describes all permutations of the set of translational parts $\{\boldsymbol{v}_R\} = \{(\pm x_R, \pm y_R, \pm z_R)\}$. In an analogous manner, the case of the class D_2 is justified. It is also noted that O_i is not a normalizer of C_{2v} but D_{4i} is, because the principal axis of rotation c_2 for C_{2v} is pointing in a fixed direction, say the z-axis. The set $\{U\}$ given by (13.7.3) may be referred to as the symmetry groups of the lattice types belonging to the crystal class G.

In the actual calculation, it is sufficient to consider only the coset representatives of G in $N(G)$ for the matrix U in (13.7.3), because the inner automorphism of $\hat{G}$ does not affect $\hat{G}$. Moreover, when there is no enantiomorphism, a proper n-fold rotation c_n can be replaced by the rotation–inversion $\overline{c}_n$ and vice versa. Consequently, *the relevant set of the unimodular transformation* $\{U\}$ for every crystal class may be given by one of the following generator sets, up to a multiplicative factor of an element of the respective class:

$$\{e\}: \quad O_i, O, T_p, D_{6i}, D_6, C_{6v}, D_{3i}, D_3, C_{3v}, D_4, D_{4i}, C_{4v}, D_{2p}$$

$$\{e, c_2'\}: \quad C_{6i}, C_6, C_{3p}, C_{3i}, C_4, C_{2p}$$

$$\{e, 4_z\}: \quad T_i, T, C_{2v}$$

$$\{e, (x, y), (x, z)\}^8: \quad D_{2i}, D_2$$

$$M = \{e, (x, y), (x - y, y, z)\}: \quad C_{2i}, C_2, C_s$$

$$TR = \{e, (x, y), (x, z), (x - y, y, z)\}: \quad C_i, C_1 \tag{13.7.4}$$

Here $c_2' \perp c_n$, (x, y) denotes the interchange of the x- and y-axes, and $(x - y, y, z)$ denotes a linear transformation written in terms of the Jones notation. The coordinate systems are based on the conventional lattice vectors. The last set for TR is academic and has no practical use in the present work.

[8] It is to be noted that, for the space groups of the last seven classes D_{2i} through C_1, there exists no enantiomorphic element because their translational parts $\boldsymbol{v}_R$ will be shown to be binary, i.e. $2\boldsymbol{v}_R \in T$. The same remark applies for T_i and C_{2v}, so that their relevant set $\{4_z\}$ may be replaced by $\{(x, y)\}$.

A striking aspect of the relevant set $\{U\}$ given by (13.7.4) is that it is simpler for a crystal class with higher symmetry. In particular, mere shifts $[e|\boldsymbol{s}]$ of the lattice origin will be sufficient to determine the equivalence or inequivalence for the classes with the relevant set $\{e\}$ in (13.7.4). In general, it is simpler to construct the MGGS of a class with higher symmetry for the present method; this is quite a contrast to the traditional methods based on the solvability of the space groups. From the above criterion, one sees immediately that the space groups of the class D_4 given by (13.6.10) are inequivalent, since the translational parts of those belonging to each lattice type are different in each invariant eigenvector space of 4_z or 2_x. In a special case in which one of the pair of solutions $(R_j|v_j)$ and $(R'_j|v'_j)$ has null translation; say $\boldsymbol{v}'_j = 0$ for all j so that $U\boldsymbol{v}'_j = 0$ in (13.7.2), a mere shift is again sufficient to establish their equivalence or inequivalence.[9]

13.8 Notations and defining relations

13.8.1 Notations

As in the *International Tables for X-ray Crystallography* (ITXC), we shall use the following notations for the lattice type L: primitive P, base-centered A, B or C, face-centered F, body-centered I, and rhombohedral R. When the lattice R is regarded as the double-centered hexagonal lattice, it is denoted by R^*. The coordinate systems will be based on the conventional lattice vectors. Note, however, that the rhombohedral coordinates are used for the R lattice whereas hexagonal coordinates are used for the R^* lattice.

Next, we shall introduce the notations for the pure point operations in terms of the Jones faithful representations, which conveniently replace their three-dimensional matrix representations (see Section 5.1.3). These are given here to the extent that is needed for the multiplications of the group generators and for the lattice transformations.

(i) For Cartesian coordinates:

$$2_x = (x, \bar{y}, \bar{z}), \qquad 2_y = (\bar{x}, y, \bar{z}), \qquad 2_z = (\bar{x}, \bar{y}, z)$$

$$2_{xy} = (y, x, \bar{z}), \qquad 2_{x\bar{y}} = (\bar{y}, \bar{x}, \bar{z})$$

$$3_{xyz} = (z, x, y), \qquad 3_{\bar{x}\bar{y}\bar{z}} = (y, z, x)$$

$$4_z = (\bar{y}, x, z), \qquad \bar{1} = (\bar{x}, \bar{y}, \bar{z}), \qquad m_z = \bar{2}_x$$

$$m_{xy} = \bar{2}_{xy}, \qquad \bar{n} = \bar{1}n \tag{13.8.1a}$$

See also Table 5.2.

[9] We have stated that, when R_j is a proper rotation P or a reflection m for an element $(R_j|\boldsymbol{v}_j)$ of $\hat{G}$, the translational component of $\boldsymbol{v}_j$ in the invariant vector space ψ_j of R_j cannot be changed by a mere shift $[e|\boldsymbol{s}]$ of the lattice's origin. This statement can be slightly extended, i.e. it holds even under a lattice transformation $[U|\boldsymbol{s}]$ provided that U commutes with R_j and $U\psi_j = \psi_j$. This follows again from the fact that the left-hand side of the first equation in (13.7.2) is perpendicular to ψ_j so that

$$0 = (\psi_j, (e - R_j)\boldsymbol{s}) = (\psi_j, (\boldsymbol{v}_j - U\boldsymbol{v}'_j)) = (\psi_j, (\boldsymbol{v}_j - \boldsymbol{v}'_j))$$

where U is an orthogonal matrix. For example, the translational parameter c in $(2_z|a, b, c)$ is invariant under $[4_z|\boldsymbol{s}]$ with any shift $\boldsymbol{s}$.

(ii) For hexagonal coordinates:

$$2_z = (\bar{x}, \bar{y}, z), \qquad 3_z = (\bar{y}, x - y, z), \qquad 6_z = (x - y, x, z)$$
$$u_0 = (x - y, \bar{y}, \bar{z}), \qquad u_1 = (x, x - y, \bar{z})$$
$$\bar{1} = (\bar{x}, \bar{y}, \bar{z}), \qquad m_0 = \bar{u}_0, \qquad m_1 = \bar{u}_1 \tag{13.8.1b}$$

Here u_ν is a binary rotation in the x, y plane about an axis that makes an angle $\nu\pi/6$ with the x-axis.

(iii) For rhombohedral coordinates:

$$3_{xyz} = (z, x, y), \qquad u_{z\bar{x}} = (\bar{z}, \bar{y}, \bar{x}), \qquad \bar{1} = (\bar{x}, \bar{y}, \bar{z}) \tag{13.8.1c}$$

Here the coordinate system is taken such that the first two operators 3_{xyz} and $u_{z,\bar{x}}$ correspond to the operators 3_z and u_0 in the hexagonal coordinates for the R^* lattice, respectively (see Figure 13.8(f)).

13.8.2 Defining relations of the crystal classes

For the sake of generality and also to prepare for the double-valued representations of the space groups, we shall base our argument on the double space groups from the outset (cf. Chapter 11). This means that the rotational part R of an element $\{R|\boldsymbol{t}_R\}$ of a space group $\hat{G}$ should be regarded as a direct product operation

$$R \to R(\boldsymbol{\theta}) \times S(\boldsymbol{\theta}) \in G \times G' \tag{13.8.2}$$

where $R(\boldsymbol{\theta})$ is an ordinary rotation in the point group G and $S(\boldsymbol{\theta}) = \exp(-\mathrm{i}\boldsymbol{\theta} \cdot \boldsymbol{\sigma}/2)$ is a spinor transformation belonging to the double group G', both of which are defined by the same set of rotation vectors $\boldsymbol{\theta}$ chosen properly according to (11.1.3) and (11.1.4). Since only the rotational part $R(\boldsymbol{\theta})$ acts on the translational part $\boldsymbol{t}_R$, the generalization to the double space groups does not require any more effort than that required for the ordinary space groups. It is necessary only to interpret the rotational part of a space group $\hat{G} = \{R|\boldsymbol{t}_R\}$ via (13.8.2), then the space group becomes its projective set belonging to the factor system defined by (11.1.2). Hereafter, we mean by a point group or space group the respective double group, unless specified otherwise. Since there exists a one-to-one correspondence between a space group and its projective set representing the double group via the interpretation (13.8.2), we may use the same notation $\hat{G} = \{\{R|\boldsymbol{t}_R\}\}$ for a space group and its projective set. Thus, for example, the factor groups for $\hat{D}_4/T$ defined by (13.6.11) may well be regarded as the projective sets representing the corresponding double groups.

A point group is isomorphic either to a proper point group P or to a rotation inversion group $P_i = P \times C_i$, where C_i is the group of inversion. The presentation of P has been given in (11.2.2) by

$$A^n = B^m = (AB)^l = E', \qquad E'^2 = E \tag{13.8.3a}$$

where A and B are abstract generators, E' corresponds to the 2π rotation that commutes with all the elements, and E is the identity. The set of integers $\{n, m, l\}$ is given by $\{n, 0, 0\}$ for a cyclic group C_n, by $\{n, 2, 2\}$ for D_n, by $\{3, 3, 2\}$ for T, and by $\{4, 3, 2\}$ for O. The presentation of P_i is given by

$$X \in P; \qquad I^2 = [X, I] = E \tag{13.8.3b}$$

where $[X, I] = XIX^{-1}I^{-1}$ is the commutator.

Now, through the isomorphism $\hat{G}/T \simeq G$ the presentation of a space group $\hat{G}/T$ is also given by (13.8.3); here, the realization of an operator is given by $(R|\boldsymbol{v}_R)$ for $\hat{G}/T$ instead of R of G. In particular, E' and E for $\hat{G}/T$ are

$$E' = (e'|0), \qquad E = (e|0)$$

where e' and e stand for the 2π rotation and the identity for G, respectively. The realization of inversion translation for $\hat{G}/T$ is expressed by $I = (\bar{1}|\boldsymbol{v}_i)$, where $\bar{1}$ is the pure inversion.

Let us discuss the commutator in (13.8.3b) in some detail. Let $X_1 = (R_1|\boldsymbol{v}_1) \in \hat{P}/T$, then

$$[X_1, I] = (e|2\boldsymbol{v}_1 + R_1\boldsymbol{v}_i - \boldsymbol{v}_i) \tag{13.8.4}$$

In a special case in which $\boldsymbol{v}_1$ is binary (i.e. $2\boldsymbol{v}_1$ is a primitive translation of T) or, more generally, if $\boldsymbol{v}_1$ and $R_1\boldsymbol{v}_i - \boldsymbol{v}_i$ are linearly independent, we have

$$R_1\boldsymbol{v}_i = \boldsymbol{v}_i \qquad (\text{mod}\, \boldsymbol{t} \in T) \tag{13.8.5}$$

which characterizes $\boldsymbol{v}_i$ (mod $\boldsymbol{t} \in T$) as an invariant eigenvector of R_1. Such a characterization of $\boldsymbol{v}_i$ for a generator R_1 of P frequently reduces the labor involved in determining $\boldsymbol{v}_i$ from the commutators with the remaining generators of P.

For further illustration of the present approach, we shall explicitly construct the MGGSs for the classes belonging to the cubic system and the rhombohedral system in the next sections. The cubic system is chosen because it is of the highest symmetry, whereas the rhombohedral system is chosen because its space groups are described by the primitive hexagonal lattice or by the rhombohedral lattice (see Table 13.1). A complete listing of the 32 MGGSs of the 230 space groups is given in Table 13.3. For the complete derivation see Kim (1986b).

13.9 The space groups of the cubic system

The characteristic generator of the cubic system is $B = (3_{xyz}|a, b, c)$. However, this generator can always be transformed into the pure rotation $(3_{xyz}|0)$ by a shift of the origin of the lattice. Firstly, using (13.8.1a) we have

$$B^3 = (e'|a+b+c,\ a+b+c,\ a+b+c) = (e'|0)$$

so that $a+b+c = 0$ (mod 1) for any lattice of type P, F or I; here, the condition for the I lattice is obtained by beginning with $B = (3_{xyz}|a+\frac{1}{2}, b+\frac{1}{2}, c+\frac{1}{2})$. Secondly, under a shift $[e|\boldsymbol{s}]$ we have, using (13.6.6),

$$B \to (3_{xyz}|a - s_1 + s_3,\ b - s_2 + s_1,\ c - s_3 + s_2)$$

which is reduced to $(3_{xyz}|0)$ with $s_1 = 0$, $s_2 = b$ and $s_3 = -a$. It is also to be noted that $(3_{xyz}|0)$ is invariant under a shift $[e|\alpha, \alpha, \alpha]$ in the diagonal (1.1.1) direction of the coordinate system, which is in the invariant eigenvector space of 3_{xyz}. Now, from (13.7.4), the relevant set of the unimodular transformations $\{U\}$ is given by $\{e, 4_z\}$ for the classes T and T_i and by $\{e\}$ for the classes O, T_p and O_i. It turns out also for the class T that one needs only a shift in order to ascertain the inequivalence of the solutions, since there are only one or two solutions for each lattice type of the class.

Table 13.3. *The 32 minimal general generator sets of the 230 space groups*

A. The cubic system

§T. $(2_z|c, 0, c)$, $(3_{xyz}|0)$ with $L(c)$: 195. $P(0)$, 196. $F(0)$, 197. $I(0)$, 198. $P(\frac{1}{2})$, 199. $I(\frac{1}{2})$.

§T_i (= $T_{\rm h}$). $(2_z|c+a, a, c)$, $(3_{xyz}|0)$, $(\mathrm{i}|0)$ with $L(c, a)$: 200. $P(0, 0)$, 201. $P(0, \frac{1}{2})$, 202. $F(0, 0)$, 203. $F(0, \frac{1}{4})$, 204. $I(0, 0)$, 205. $P(\frac{1}{2}, 0)$, 206. $I(\frac{1}{2}, 0)$.

§O. $(4_z|0, -c, c)$, $(3_{xyz}|0)$ with $L(c)$: 207. $P(0)$, 208. $P(\frac{1}{2})$, 209. $F(0)$, 210. $F(\frac{1}{4})$, 211. $I(0)$, 212. $P(\frac{3}{4})$, 213. $P(\frac{1}{4})$, 214. $I(\frac{1}{4})$.

§T_p (= $T_{\rm d}$). $(\bar{4}_z|c, -c, c)$, $(3_{xyz}|0)$ with $L(c)$: 215. $P(0)$, 216. $F(0)$, 217. $I(0)$, 218. $P(\frac{1}{2})$, 219. $F(\frac{1}{2})$, 220. $I(\frac{1}{4})$.

§O_i (= $O_{\rm h}$). $(4_z|-a, -c, c)$, $(3_{xyz}|0)$, $(\mathrm{i}|0)$ with $L(a, c)$: 221. $P(0, 0)$, 222. $P(\frac{1}{2}, 0)$, 223. $P(\frac{1}{2}, \frac{1}{2})$, 224. $P(0, \frac{1}{2})$, 225. $F(0, 0)$, 226. $F(\frac{1}{2}, 0)$, 227. $F(\frac{1}{2}, \frac{1}{4})$, 228. $F(0, \frac{1}{4})$, 229. $I(0, 0)$, 230. $I(\frac{1}{4}, \frac{1}{4})$.

B. The hexagonal system

§C_6. $P(6_z|0, 0, c)$ with c: 168. 0, 169. $\frac{1}{6}$, 170. $\frac{5}{6}$, 171. $\frac{1}{3}$, 172. $\frac{2}{3}$, 173. $\frac{1}{2}$.

§C_{3p} (= $C_{3\rm h}$). 174. $P(\bar{6}_z|0)$.

§C_{6i} (= $C_{6\rm h}$). $P(6_z|0, 0, c)$, $(\mathrm{i}|0)$ with c: 175. 0, 176. $\frac{1}{2}$.

§D_6. $P(6_z|0, 0, c)$, $(u_0|0)$ with c: 177. 0, 178. $\frac{1}{6}$, 179. $\frac{5}{6}$, 180. $\frac{1}{3}$, 181. $\frac{2}{3}$, 182. $\frac{1}{2}$.

§C_{6v}. $P(6_z|0, 0, c)$, $(m_0|0, 0, c')$ with (c, c'): 183. $(0, 0)$, 184. $(0, \frac{1}{2})$, 185. $(\frac{1}{2}, \frac{1}{2})$, 186. $(\frac{1}{2}, 0)$.

§D_{3p} (= $D_{3\rm h}$). $P(\bar{6}_z|0, 0, 0)$, $(m_0|0, 0, c)$ or $(u_0|0, 0, c)$ with $(m; c)$ or $(u; c)$: 187. $(m; 0)$, 188. $(m; \frac{1}{2})$, 189. $(u; 0)$, 190. $(u; \frac{1}{2})$.

§D_{6i} (= $D_{6\rm h}$). $P(6_z|0, 0, c)$, $(u_0|0, 0, c')$, $(\mathrm{i}|0)$ with (c, c'): 191. $(0, 0)$, 192. $(0, \frac{1}{2})$, 193. $(\frac{1}{2}, \frac{1}{2})$, 194. $(\frac{1}{2}, 0)$.

C. The rhombohedral system

§C_3. $(3_z|0, 0, c)$ with $L(c)$: 143. $P(0)$, 144. $P(\frac{1}{3})$, 145. $P(\frac{2}{3})$, 146. $R^*(0)$.

§C_{3i}. $(3_z|0)$, $(\mathrm{i}|0)$ with L: 147. P, 148. R^*.

§D_3. $(3_z|0, 0, c)$, $(u_v|0)$ with $L(u_v; c)$: 149. $P(u_1; 0)$, 150. $P(u_0; 0)$, 151. $P(u_1; \frac{1}{3})$, 152. $P(u_0; \frac{1}{3})$, 153. $P(u_1; \frac{2}{3})$, 154. $P(u_0; \frac{2}{3})$, 155. $R^*(u_0; 0)$.

§C_{3v}. $(3_z|0)$, $(m_v|0, 0, c)$ with $L(m_v; c)$: 156. $P(m_0; 0)$, 157. $P(m_1; 0)$, 158. $P(m_0; \frac{1}{2})$, 159. $P(m_1; \frac{1}{2})$, 160. $R^*(m_0; 0)$, 161. $R^*(m_0; \frac{1}{2})$.

§D_{3i} (= $D_{3\rm d}$). $(3_z|0)$, $(u_v|0)$, $(\mathrm{i}|0, 0, c)$ with $L(u_v; c)$: 162. $P(u_1; 0)$, 163. $P(u_1; \frac{1}{2})$, 164. $P(u_0; 0)$, 165. $P(u_0; \frac{1}{2})$, 166. $R_0^*(u_0; 0)$, 167. $R_0^*(u_0; \frac{1}{2})$.

D. The tetragonal system

§C_4. $(4_z|0, 0, c)$ with $L(c)$: 75. $P(0)$, 76. $P(\frac{1}{4})$, 77. $P(\frac{1}{2})$, 78. $P(\frac{3}{4})$, 79. $I(0)$, 80. $I(\frac{1}{4})$.

§C_{2p} (= S_4).$(\bar{4}_z|0)$ with L: 81. P, 82. I.

§C_{4i} (= $C_{4\rm h}$). $(4_z, a, b, c)$ $(\mathrm{i}|0)$ with $L(a, b, c)$: 83. $P(0, 0, 0)$, 84. $P(0, 0, \frac{1}{2})$, 85. $P(\frac{1}{2}, 0, 0)$, 86. $P(\frac{1}{2}, 0, \frac{1}{2})$, 87. $I(0, 0, 0)$, 88. $I(\frac{1}{4}, \frac{1}{4}, \frac{1}{4})$.

§D_4. $(4_z|0, 0, c)$, $(2_x|a, a, 0)$ with $L(c, a)$: 89. $P(0, 0)$, 90. $P(0, \frac{1}{2})$, 91. $P(\frac{1}{4}, 0)$, 92. $P(\frac{1}{4}, \frac{1}{2})$, 93. $P(\frac{1}{2}, 0)$, 94. $P(\frac{1}{2}, \frac{1}{2})$, 95. $P(\frac{3}{4}, 0)$, 96. $P(\frac{3}{4}, \frac{1}{2})$, 97. $I(0, 0)$, 98. $I(\frac{1}{4}, 0)$.

§C_{4v}. $(4_z|0, 0, c)$, $(m_x|a+2c, a, c')$ with $L(c, a, c')$: 99. $P(0, 0, 0)$, 100. $P(0, \frac{1}{2}, 0)$, 101. $P(\frac{1}{2}, 0, \frac{1}{2})$, 102. $P(\frac{1}{2}, \frac{1}{2}, \frac{1}{2})$, 103. $P(0, 0, \frac{1}{2})$, 104. $P(0, \frac{1}{2}, \frac{1}{2})$, 105. $P(\frac{1}{2}, 0, 0)$, 106. $P(\frac{1}{2}, \frac{1}{2}, 0)$, 107. $I(0, 0, 0)$, 108. $I(0, 0, \frac{1}{2})$, 109. $I(\frac{1}{4}, 0, 0)$, 110. $I(\frac{1}{4}, \frac{1}{2}, 0)$.

§D_{2p} (= D_{2d}). $(\overline{4}_x|0)$, $(2_z$ or $m_x|a+2c, a, c)$ with $L(2$ or m; $a, c)$: 111. $P(2; 0, 0)$, 112. $P(2; 0, \frac{1}{2})$, 113. $P(2; \frac{1}{2}, 0)$, 114. $P(2; \frac{1}{2}, \frac{1}{2})$, 115. $P(m; 0, 0)$, 116. $P(m; 0, \frac{1}{2})$, 117. $P(m; \frac{1}{2}, 0)$, 118. $P(m; \frac{1}{2}, \frac{1}{2})$, 119. $I(m; 0, 0)$, 120. $I(m; 0, \frac{1}{2})$, 121. $I(2; 0, 0)$, 122. $I(2; \frac{1}{2}, \frac{1}{4})$.

§D_{4i} (= D_{4h}). $(4_z|0, 0, c)$, $(2_x|a, a, 0)$, $(\mathrm{i}|\alpha, \alpha+2c, \gamma)$ or $(4_z|\alpha+c, c, c)$, $(2_x|a, a+\alpha+2c, \gamma)$, $(\mathrm{i}|0)$ with $L(c, a; \alpha, \gamma)$:

		(α, γ)			
L	(c, a)	(0, 0)	$(0, \frac{1}{2})$	$(\frac{1}{2}, 0)$	$(\frac{1}{2}, \frac{1}{2})$
P	(0, 0)	123	124	125	126
	$(0, \frac{1}{2})$	127	128	129	130
	$(\frac{1}{2}, 0)$	131	132	133	134
	$(\frac{1}{2}, \frac{1}{2})$	135	136	137	138
I	(0, 0)	139	140		
	$(\frac{1}{4}, 0)$	142	141		

E. The orthorhombic system

§D_2. $(2_z|c, 0, c)$, $(2_x|a, a, 0)$ with $L(c, a)$: 16. $P(0, 0)$, 17. $P(\frac{1}{2}, 0)$, 18. $P(0, \frac{1}{2})$, 19. $P(\frac{1}{2}, \frac{1}{2})$, 20. $C(\frac{1}{2}, 0)$, 21. $C(0, 0)$, 22. $F(0, 0)$, 23. $I(0, 0)$, 24. $I(\frac{1}{2}, \frac{1}{2})$.

§C_{2v}. $(2_z|a', b, c+c')$, $(m_x|0, b, c)$, $(m_y|a', 0, c')$ with $L(b, c; a', c')$:

		(a', c')			
L	(b, c)	(0, 0)	$(0, \frac{1}{2})$	$(\frac{1}{2}, 0)$	$(\frac{1}{2}, \frac{1}{2})$
P	(0, 0)	25	26	28	31
	$(0, \frac{1}{2})$		27	29	30
	$(\frac{1}{2}, 0)$			32	33
	$(\frac{1}{2}, \frac{1}{2})$				34
C	(0, 0)	35	36		
	$(0, \frac{1}{2})$		37		
A	(0, 0)	38	40		
	$(0, \frac{1}{2})$	39	41		
I	(0, 0)	44	46		
	$(0, \frac{1}{2})$		45		
F	$b=c=a'=c'=d$; 42. $d=0$, 43. $d=\frac{1}{4}$				

Table 13.3 (*cont.*)

§D_{2i} (= D_{2h}). $(2_z|c, 0, c)$, $(2_x|a, a, 0)$, $(\mathrm{i}|\alpha, \beta, \gamma)$ or $(2_z|c+\alpha, \beta, c)$, $(2_x|a, a+\beta, \gamma)$, $(\mathrm{i}|0)$ with $L(c, a; \alpha, \beta, \gamma)$:

L	(c, a)	(α, β, γ)			
P	(0, 0)	47. (0, 0, 0)	48. ($\frac{1}{2}$, $\frac{1}{2}$, $\frac{1}{2}$)	49. (0, 0, $\frac{1}{2}$)	50. ($\frac{1}{2}$, $\frac{1}{2}$, 0)
	($\frac{1}{2}$, 0)	51. ($\frac{1}{2}$, 0, 0)	52. (0, $\frac{1}{2}$, 0)	53. (0, 0, 0)	54. ($\frac{1}{2}$, $\frac{1}{2}$, 0)
	(0, $\frac{1}{2}$)	55. (0, 0, 0)	56. ($\frac{1}{2}$, $\frac{1}{2}$, $\frac{1}{2}$)	57. (0, $\frac{1}{2}$, 0)	58. (0, 0, $\frac{1}{2}$)
		59. ($\frac{1}{2}$, $\frac{1}{2}$, 0)	60. (0, $\frac{1}{2}$, $\frac{1}{2}$)		
	($\frac{1}{2}$, $\frac{1}{2}$)	61. (0, 0, 0)	62. (0, 0, $\frac{1}{2}$)		
C	(0, 0)	65. (0, 0, 0)	66. (0, 0, $\frac{1}{2}$)	67. (0, $\frac{1}{2}$, 0)	68. (0, $\frac{1}{2}$, $\frac{1}{2}$)
	($\frac{1}{2}$, 0)	63. (0, $\frac{1}{2}$, 0)	64. (0, 0, 0)		
F	(0, 0)	69. (0, 0, 0)	70. ($\frac{1}{4}$, $\frac{1}{4}$, $\frac{1}{4}$)		
I	(0, 0)	71. (0, 0, 0)	72. (0, 0, $\frac{1}{2}$)		
	($\frac{1}{2}$, $\frac{1}{2}$)	73. (0, 0, 0) .	74. (0, 0, $\frac{1}{2}$)		

F. The monoclinic system

§C_2. $(2_z|0, 0, c)$ with $L(c)$: 3. $P(0)$, 4. $P(\frac{1}{2})$, 5. $B(0)$.

§(C_{1p}) (= C_s). $(m_z|0, b, 0)$ with $L(b)$: 6. $P(0)$, 7. $P(\frac{1}{2})$, 8. $B(0)$, 9. $B(\frac{1}{2})$.

§(C_{2i}) (= C_{2h}). $(2_z|0, b, c)$, $(\mathrm{i}|0)$ with $L(b, c)$: 10. $P(0, 0)$, 11. $P(0, \frac{1}{2})$, 12. $B(0, 0)$, 13. $P(\frac{1}{2}, 0)$, 14. $P(\frac{1}{2}, \frac{1}{2})$, 15. $B(\frac{1}{2}, 0)$.

G. The triclinic system

§C_1. 1. $P(e'|0)$. §C_i. 2. $P(e'|0)$, $(i|0)$

1. L($a, \ldots$) means the lattice type L with the parameters $a, \ldots$.
2. The number assigned to each space group is in accordance with *The International* Tables for X-ray Crystallography (1965) Volume 1.
3. The crystal classes are expressed in the present notation as well as in the Schönflies notation.
4. The inversion is expressed by i instead of $\bar{1}$ for simplicity.

13.9.1 The class T

$A^2 = B^3 = E'$, $(AB)^3 = E$. Adjusting the origin of the lattice, one may set

$$A = (2_z|a, 0, c), \qquad B = (3_{xyz}|0) \tag{13.9.1a}$$

where a and c are parameters to be determined for each lattice type from the defining relations and the products

$$A^2 = (e'|0, 0, 2c), \qquad (AB)^3 = (e|a-c, -a+c, -a+c) \tag{13.9.1b}$$

(i) For the P lattice, the allowed values of the parameters are

$$c = a = 0, \tfrac{1}{2} \tag{13.9.1c}$$

which define two space groups with the generators

$$(2_z|a, 0, a), (3_{xyz}|0); \qquad a = 0, \tfrac{1}{2} \tag{13.9.1d}$$

These are inequivalent because a shift cannot affect the translational component a in the invariant eigenvector space of 2_z.

(ii) For the F lattice, the above solutions (13.9.1d) are reduced to one solution with the parameter

$$c = a = 0 \tag{13.9.1e}$$

(iii) For the I lattice, directly from (13.9.1b) we obtain

$$2c \equiv 0, \qquad a - c \equiv 0, \tfrac{1}{2} \pmod 1$$

which yield $a, c = 0, \frac{1}{2}$. Since, however, $c \sim c + \frac{1}{2}$ under $[e|\frac{1}{4}, \frac{1}{4}, \frac{1}{4}]$, we may set $a = c$ to obtain (13.9.1d) again for the I lattice.

The five space groups obtained above for the class T are summarized by the following MGGS,

$$(2_z|a, 0, a), (3_{xyz}|0) \qquad \text{with} \qquad L(a): \tag{13.9.1f}$$

195. $P(0)$, 196. $F(0)$, 197. $I(0)$, 198. $P(\frac{1}{2})$, 199. $I(\frac{1}{2})$

These are also listed in Table 13.3, which contains all 32 MGGSs of the 230 space groups.

From (13.9.1f), one can easily write down the general elements of the space groups belonging to the class T using the right coset decomposition

$$\hat{T} = \hat{D}_2 + \hat{D}_2\hat{3}_{xyz} + \hat{D}_2\hat{3}_{\bar{x}\bar{y}\bar{z}}$$

where $\hat{3}_{xyz} = (3_{xyz}|0)$. From $\hat{2}_z = (2_z|a, 0, a)$, the elements of $\hat{D}_2$ are obtained by $\hat{2}_x = \hat{3}_{xyz}\hat{2}_z\hat{3}^{-1}_{xyz}$, $\hat{2}_y = \hat{3}_{xyz}\hat{2}_x\hat{3}^{-1}_{xyz}$, then the rest follows from the multiplication tables Tables 5.6 or 11.2:

$$\begin{array}{lllll}
\hat{D}_2: & (e|0) & (2_z|a, 0, a) & (2_x|a, a, 0) & (2_y|0, a, a) \\
\hat{D}_2\hat{3}_{xyz}: & (3_{xyz}|0) & (3_{x\bar{y}\bar{z}}|a, 0, a) & (3_{\bar{x}y\bar{z}}|a, a, 0) & (3_{\bar{x}\bar{y}z}|0, a, a) \\
\hat{D}_2\hat{3}_{\bar{x}\bar{y}\bar{z}}: & (3_{\bar{x}\bar{y}\bar{z}}|0) & (3_{x\bar{y}z}|a, 0, a) & (3_{xy\bar{z}}|a, a, 0) & (3_{\bar{x}yz}|0, a, a)
\end{array} \tag{13.9.1g}$$

The set reproduces all the elements of the seven space groups belonging to the class T given in the ITXC with use of MGGS (13.9.1f). The above set may be regarded either as an ordinary space group as it stands or as the projective set of the space group with the interpretation (13.8.2), in view of (11.1.3).

13.9.2 The class T_i (=T_h)

$A, B \in T$; $I^2 = [A, I] = [B, I] = E$. The space groups of this class may be constructed by augmenting those of the class T with the inversion translation I. Using (13.8.5) with $R_1 = 3_{xyz}$, we may set $\boldsymbol{v}_i = (\alpha, \alpha, \alpha)$. Then

$$A = (2_z|a, 0, a), \qquad B = (3_{xyz}|0), \qquad I = (\bar{1}|\alpha, \alpha, \alpha) \tag{13.9.2a}$$

where $a = 0, \frac{1}{2}$ for the class T. From the commutators

$$[A, I] = (e|-2\alpha, -2\alpha, 0), \qquad [B, I] = (e|0) \tag{13.9.2b}$$

obtained by (13.8.4), we determine the parameter α for each lattice type. Here the augmentation does not affect the original parameter a. By the shift $[e|\alpha/2, \alpha/2, \alpha/2]$ of the coordinate origin to the center of the inversion, we transform the generator set (13.9.2a) into the form

$$(2_z|a + \alpha, \alpha, a), (3_{xyz}|0), (\bar{1}|0) \tag{13.9.2c}$$

The shift does not affect the commutators because they are equal to the identity element.

(i) For the P lattice, from (13.9.1d) we have $a = 0, \frac{1}{2}$ and from (13.9.2b) we have $\alpha = 0, \frac{1}{2}$. These provide the ordered pair (a, α) with the four sets of parameters

$$(0, 0), (0, \tfrac{1}{2}), (\tfrac{1}{2}, 0) \sim (\tfrac{1}{2}, \tfrac{1}{2}) \tag{13.9.2d}$$

of which the last two sets are equivalent. To see this we write down explicitly the generators of the corresponding four space groups

$$(2_z|0, 0, 0), (2_z|\tfrac{1}{2}, \tfrac{1}{2}, 0), (2_z|\tfrac{1}{2}, 0, \tfrac{1}{2}), (2_z|0, \tfrac{1}{2}, \tfrac{1}{2}) \tag{13.9.2e}$$

in addition to the common generators $(3_{xyz}|0)$ and $(\bar{1}|0)$. Note that the first two sets are inequivalent to the last two sets because the translational parts are different in the invariant eigenvector space of 2_z, which commutes with 4_z in the relevant set $\{4_z\}$ of T_i. Secondly, the first two sets are mutually inequivalent because $(2_z|\frac{1}{2}, \frac{1}{2}, 0)$ cannot be brought into coincidence with $(2_z|0, 0, 0)$ by $[2_z|\frac{1}{2}, \frac{1}{2}, \frac{1}{2}]$, the only shift which leaves the common generators $(3_{xyz}|0)$ and $(\bar{1}|0)$ invariant. Finally, the last two sets are equivalent because, under the lattice transformation[10] $[2_{x\bar{y}}|0]$, we have

$$(2_z|\tfrac{1}{2}, 0, \tfrac{1}{2}) \to (2_z|0, \tfrac{1}{2}, \tfrac{1}{2}), (3_{xyz}|0) \to (3_{\bar{x}\bar{y}\bar{z}}|0)$$

while $(\bar{1}|0)$ remains the same. Thus, there exist three independent solutions given by the parameter sets $(a, \alpha) = (0, 0), (0, \frac{1}{2}), (\frac{1}{2}, 0)$.

(ii) For the F lattice, from (13.9.1e), (13.9.2b) and (13.9.2c) we obtain

$$a = 0, \qquad \alpha = 0, \tfrac{1}{4} \tag{13.9.2f}$$

which yield two independent solutions.

(iii) For the I lattice, from the result for the I lattice of the class T and (13.9.2b), we have $a, \alpha = 0, \frac{1}{2}$; however, since $\alpha \sim \alpha + \frac{1}{2}$ for the I lattice from (13.9.2a), we obtain two independent solutions characterized by

$$a = 0, \tfrac{1}{2}, \qquad \alpha = 0 \tag{13.9.2g}$$

[10] Note that $2_{x\bar{y}} = e'2_y4_{\bar{z}}$ is used in the place of the relevant transformation $U = 4_z$ given in (13.7.4). Also note that inversion of the basis vectors leaves the space group invariant.

The seven space groups obtained above for the class T_i are summarized by the MGGS

$$(2_z|a+\alpha,\ \alpha,\ a),\ (3_{xyz}|0),\ (\bar{1}|0) \quad \text{with} \quad L(a,\ \alpha): \tag{13.9.2h}$$

200. $P(0,\ 0)$, 201. $P(0,\ \frac{1}{2})$, 202. $F(0,\ 0)$, 203. $F(0,\ \frac{1}{4})$,

204. $I(0,\ 0)$, 205. $P(\frac{1}{2},\ 0)$, 206. $I(\frac{1}{2},\ 0)$

Note that the MGGS of the class T_i given above reduces to the MGGS of the class T given by (13.9.1f) with $\alpha = 0$ and deleting $(\bar{1}|0)$. Thus, every space group of the class T is a subgroup of a space group of the class T_i.

Exercise. Write down the general space group elements of $\hat{T}_i$ from the MGGS (13.9.2h).

13.9.3 The class O

$A^4 = B^3 = (AB)^2 = E'$. Let

$$A = (4_z|0,\ b,\ c), \qquad B = (3_{xyz}|0) \tag{13.9.3a}$$

Then

$$A^4 = (e'|0,\ 0,\ 4c), \qquad (AB)^2 = (e'|0,\ b+c,\ b+c) \tag{13.9.3b}$$

(i) For the P lattice, the allowed values of the parameters are

$$-b = c = 0,\ \tfrac{1}{4},\ \tfrac{1}{2},\ \tfrac{3}{4} \tag{13.9.3c}$$

which yield four independent solutions.

(ii) For F and I lattices, under the shift of the lattice origin by $[e|\frac{1}{4},\ \frac{1}{4},\ \frac{1}{4}]$, we have $(4_z|0,\ b,\ c) \rightarrow (4_z|\frac{1}{2},\ b,\ c)$, which yields the equivalences $b \sim b+\frac{1}{2}$ and $c \sim c+\frac{1}{2}$ for the F lattice and $(b,\ c) \sim (b+\frac{1}{2},\ c+\frac{1}{2})$ for the I lattice. Thus, from (13.9.3b) we obtain

$$-b = c = 0,\ \tfrac{1}{4} \tag{13.9.3d}$$

which provides two independent solutions each for both lattices.

The MGGS for the eight space groups belonging to the class O is given by

$$(4_z|0,\ -c,\ c),\ (3_{xyz}|0) \quad \text{with} \quad L(c): \tag{13.9.3e}$$

207. $P(0)$, 208. $P(\frac{1}{2})$, 209. $F(0)$, 210. $F(\frac{1}{4})$,

211. $I(0)$, 212. $P(\frac{3}{4})$, 213. $P(\frac{1}{4})$, 214. $I(\frac{1}{4})$

From the MGGS (13.9.3e), one can write down the space group elements of $\hat{O}/T$ using

$$\hat{O} = \hat{T} + \hat{4}_z\hat{T}; \qquad \hat{4}_z = (4_z|0,\ -c,\ c)$$

Starting from $(4_z|0,\ -c,\ c)^2 = (2_z|c,\ -c,\ 2c)$, we first construct the space group $\hat{T}$ as in (13.9.1g) and then, by augmenting it with $\hat{4}_z$, we obtain the space group $\hat{O}$, using the multiplication tables Table 5.6 or Table 11.2:

$$\hat{D}_2: \quad (e|\boldsymbol{v}_1),\ (2_z|\boldsymbol{v}_2),\ (2_x|\boldsymbol{v}_3),\ (2_y|\boldsymbol{v}_4)$$

$$\hat{D}_2\hat{3}_{xyz}: \quad (3_{xyz}|\boldsymbol{v}_1),\ (3_{\bar{x}y\bar{z}}|\boldsymbol{v}_2),\ (3_{\bar{x}y\bar{z}}|\boldsymbol{v}_3),\ (3_{\bar{x}\,\bar{y}z}|\boldsymbol{v}_4)$$

$$\hat{D}_2\hat{3}_{xyz}: \quad (3_{\bar{x}\,\bar{y}\,\bar{z}}|\boldsymbol{v}_1),\ (3_{x\bar{y}z}|\boldsymbol{v}_2),\ (3_{xy\bar{z}}|\boldsymbol{v}_3),\ (3_{\bar{x}yz}|\boldsymbol{v}_4)$$

$$\hat{4}_z\hat{T}: \quad (4_z|\boldsymbol{v}_1'),\ (4_{\bar{z}}|\boldsymbol{v}_2'),\ (2_{xy}|\boldsymbol{v}_3'),\ (2_{x\bar{y}}|\boldsymbol{v}_4')$$

$$(2_{yz}|\boldsymbol{v}_1'),\ (4_x|\boldsymbol{v}_2'),\ (4_{\bar{x}}|\boldsymbol{v}_3'),\ (2_{z\bar{y}}|\boldsymbol{v}_4')$$

$$(4_{\bar{y}}|\boldsymbol{v}_1'),\ (2_{zx}|\boldsymbol{v}_2'),\ (4_y|\boldsymbol{v}_3'),\ (2_{z\bar{x}}|\boldsymbol{v}_4')$$

where

$$\boldsymbol{v}_1 = (0,\ 0,\ 0), \qquad \boldsymbol{v}_2 = (c,\ -c,\ 2c), \qquad \boldsymbol{v}_3 = (2c,\ c,\ -c), \qquad \boldsymbol{v}_4 = (-c,\ 2c,\ c)$$

$$\boldsymbol{v}_1' = (0,\ -c,\ c), \qquad \boldsymbol{v}_2' = (c,\ 0,\ 3c), \qquad \boldsymbol{v}_3' = (-c,\ c,\ 0), \qquad \boldsymbol{v}_4' = (-2c,\ -2c,\ 2c)$$

Note that $\boldsymbol{v}_i' = \hat{4}_z\boldsymbol{v}_i$ for $i = 1, \ldots, 4$.

13.9.4 The class T_p ($=T_d$)

$A^4 = B^3 = (AB)^2 = E'$. Let

$$A = (\bar{4}_z|a,\ 0,\ c), \qquad B = (3_{xyz}|0) \tag{13.9.4a}$$

Then

$$A^4 = (e'|0), \qquad (AB)^2 = (e'|2a,\ -c,\ c) \tag{13.9.4b}$$

(i) For the P lattice, the allowed values of the parameters are

$$a = 0,\ \tfrac{1}{2}, \qquad c = 0 \tag{13.9.4c}$$

which yield two independent solutions of (13.9.4a).

(ii) For the F lattice, we may set one of the parameters a or c of A equal to zero since both of them are binary[11] from (13.9.4b) and since $(\bar{4}_z|a,\ 0,\ c) \sim (\bar{4}_z|a+\tfrac{1}{2},\ 0,\ c+\tfrac{1}{2})$. Thus we obtain again two independent solutions given by (13.9.4c) for the F lattice.

(iii) For the I lattice, from (13.9.4b) we obtain

$$a = c = 0 \qquad \text{or} \qquad 2a = c = \tfrac{1}{2} \tag{13.9.4d}$$

which give two independent solutions of (13.9.4a). See below.

It is interesting to note that all these six space groups obtained above for T_p can be expressed in the following form with one parameter a:

$$(\bar{4}_z|a,\ -a,\ a), \qquad (3_{xyz}|0) \tag{13.9.4e}$$

This follows since, under a shift $[e|a/2,\ a/2,\ a/2]$, we have

$$(\bar{4}_z|a,\ 0,\ c) \rightarrow (\bar{4}_z|a,\ -a,\ c-a) = (\bar{4}_z|a,\ -a,\ a)$$

using $c = 2a$, which holds for all three lattice types.

[11] A lattice parameter a is binary if $a = 0,\ \frac{1}{2}$.

The MGGS for the T_p class is given by

$$(\bar{4}_z|a, -a, a), \qquad (3_{xyz}|0) \qquad \text{with} \qquad L(a): \tag{13.9.4f}$$

215. $P(0)$, 216. $F(0)$, 217. $I(0)$, 218. $P(\frac{1}{2})$,
219. $F(\frac{1}{2})$, 220. $I(\frac{1}{4})$

13.9.5 The class O_i (=O_h)

$A, B \in O$, $I^2 = [A, I] = [B, I] = E$. By augmenting the class O with I we set

$$A = (4_z|0, -c, c), \qquad B = (3_{xyz}|0), \qquad (\bar{1}|\alpha, \alpha, \alpha) \tag{13.9.5a}$$

Then, the required commutators are, using (13.8.4),

$$[A, I] = (e| - 2\alpha, -2c, 2c), \qquad [B, I] = (e|0, 0, 0) \tag{13.9.5b}$$

Since $[A, I]$ depends on c, some of the previous values $\{0, \frac{1}{2}, \frac{1}{4}, \frac{3}{4}\}$ of c given by (13.9.3c) for the class O may not be allowed for the class O_i, being incompatible with the inversion symmetry I.

Before calculating the parameters, we first shift the origin of the lattice to the center of symmetry by $[e|\alpha/2, \alpha/2, \alpha/2]$ and obtain

$$(4_z| - \alpha, -c, c), (3_{xyz}|0), (\bar{1}|0) \tag{13.9.5c}$$

Obviously, the shift leaves the commutators given by (13.9.5b) invariant, because they are equal to the identity element E.

(i) For the P lattice, from (13.9.5b), the allowed values of the parameters are

$$c, \alpha = 0, \tfrac{1}{2} \tag{13.9.5d}$$

which yield four independent solutions of (13.9.5a) from the inequivalence criteria with (13.7.4), because they are invariant under the shift $[e|\frac{1}{2}, \frac{1}{2}, \frac{1}{2}]$ which is the only one that leaves $(3_{xyz}|0)$ and $(\bar{1}|0)$ invariant.

(ii) For the F lattice, from (13.9.5b) we obtain

$$c = 0, \tfrac{1}{4}, \qquad \alpha = 0, \tfrac{1}{2} \tag{13.9.5e}$$

which provide four independent solutions of (13.9.5a).

(iii) For the I lattice, from (13.9.5a) we have $\alpha \sim \alpha + \frac{1}{2}$ and hence $c \sim c + \frac{1}{2}$ from (13.9.5c). Thus, from (13.9.5b), we obtain the two sets of parameters

$$c = \alpha = 0, \tfrac{1}{4} \tag{13.9.5f}$$

which yield two independent solutions of (13.9.5a).

The ten space groups determined above for the class O_i are defined by the MGGS

$$(4_z| - \alpha, -c, c), (3_{xyz}|0), (\bar{1}|0) \qquad \text{with} \qquad L(\alpha, c): \tag{13.9.5g}$$

221. $P(0, 0)$, 222. $P(\frac{1}{2}, 0)$, 223. $P(\frac{1}{2}, \frac{1}{2})$, 224. $P(0, \frac{1}{2})$,
225. $F(0, 0)$, 226. $F(\frac{1}{2}, 0)$, 227. $F(\frac{1}{2}, \frac{1}{4})$, 228. $F(0, \frac{1}{4})$,
229. $I(0, 0)$, 230. $I(\frac{1}{4}, \frac{1}{4})$

These are tabulated in Table 13.3.

In terms of the above MGGS one can write down the general space group elements of $\hat{O}_i/T$ as follows, analogously to the case of $\hat{O}/T$:

$\hat{D}_2$:	$(e\|0)$, $(2_z\|\boldsymbol{v}_2)$, $(2_x\|\boldsymbol{v}_3)$, $(2_y\|\boldsymbol{v}_4)$
$\hat{D}_2\hat{3}_{xyz}$:	$(3_{xyz}\|0)$, $(3_{x\bar{y}\bar{z}}\|\boldsymbol{v}_2)$, $(3_{\bar{x}y\bar{z}}\|\boldsymbol{v}_3)$, $(3_{\bar{x}\bar{y}z}\|\boldsymbol{v}_4)$,
$\hat{D}_2\hat{3}_{xyz}$:	$(3_{\bar{x}\bar{y}\bar{z}}\|0)$, $(3_{x\bar{y}z}\|\boldsymbol{v}_2)$, $(3_{xy\bar{z}}\|\boldsymbol{v}_3)$, $(3_{\bar{x}yz}\|\boldsymbol{v}_4)$
$\hat{4}_z\hat{T}$:	$(4_z\|\boldsymbol{v}_1')$, $(4_{\bar{z}}\|\boldsymbol{v}_2')$, $(2_{xy}\|\boldsymbol{v}_3')$, $(2_{x\bar{y}}\|\boldsymbol{v}_4')$
	$(2_{yz}\|\boldsymbol{v}_1')$, $(4_x\|\boldsymbol{v}_2')$, $(4_{\bar{x}}\|\boldsymbol{v}_3')$, $(2_{z\bar{y}}\|\boldsymbol{v}_4')$
	$(4_{\bar{y}}\|\boldsymbol{v}_1')$, $(2_{zx}\|\boldsymbol{v}_2')$, $(4_y\|\boldsymbol{v}_3')$, $(2_{z\bar{x}}\|\boldsymbol{v}_4')$
$\hat{O}\hat{I}$:	$\{\{\overline{R}\|\boldsymbol{v}_R\}\}$

where $\overline{R} = R\bar{1}$ for all $R \in 0$ and $\boldsymbol{v}_R$ is the characteristic translation of each R given by (0, 0, 0) or

$$\boldsymbol{v}_2 = (c - \alpha, -c - \alpha, 2c), \qquad \boldsymbol{v}_3 = (2c, c - \alpha, -c - \alpha),$$
$$\boldsymbol{v}_4 = (-c - \alpha, 2c, c - \alpha), \qquad \boldsymbol{v}_1' = (-\alpha, -c, c), \qquad \boldsymbol{v}_2' = (c, -\alpha, -c),$$
$$\boldsymbol{v}_3' = (-c, c, -\alpha), \qquad \boldsymbol{v}_4' = (-2c - \alpha, -2c - \alpha, 2c - \alpha)$$

13.10 The space groups of the rhombohedral system

The hexagonal coordinate system is used for the primitive hexagonal lattice P and for the double-centered hexagonal lattice R^*, whereas the rhombohedral coordinate system is used when the latter is regarded as the rhombohedral lattice R defined by (13.4.3b). On account of the difference between the coordinate systems of P and R lattices we shall discuss their space groups separately. Then we simply write down the results for the R^* lattice from those for the R lattice. It is to be noted that there are two types of realizations for the binary generators of the classes D_3, C_{3v} and D_{3i} with the hexagonal P lattice (but not with the rhombohedral R (or R^*) lattice), because each one of these classes has only one equivalence set of binary axes, which must coincide with one of two equivalence sets of binary axes of the Bravais lattice $\in D_{6i}$ (cf. Section 5.2). See Figure 13.11. The MGGSs of this system have been given in Table 13.3.

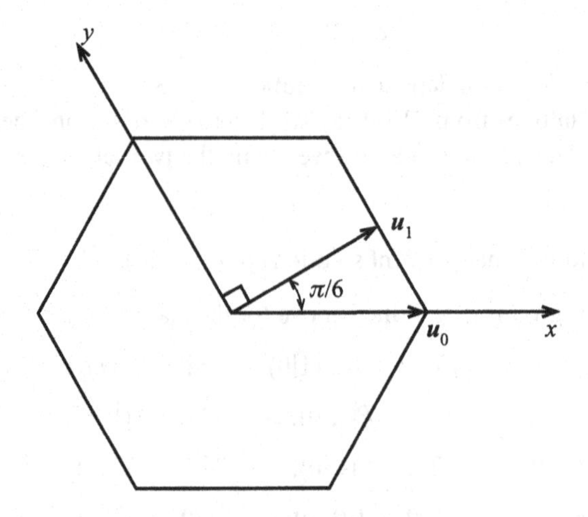

Figure 13.11. Two types of realizations of the binary generators u_0 and u_1 for the classes D_3, C_{3v} and D_{3i}.

13.10.1 The class C_3

$A^3 = E'$.

(i) For the P lattice

$$A = (3_z|0, 0, c); \qquad c = 0, \tfrac{1}{3}, \tfrac{2}{3} \tag{13.10.1a}$$

(ii) For the R lattice

$$A = (3_{xyz}|0) \tag{13.10.1b}$$

and for the R^* lattice

$$A = (3_z|0) \tag{13.10.1c}$$

Note that 3_z in the hexagonal coordinate system corresponds to 3_{xyz} in the rhombohedral coordinate system, as was discussed for (13.8.1c).

13.10.2 The class C_{3i}

$A^3 = E'$, $I^2 = [A, I] = E$.

(i) For the P lattice, the class C_{3i} can be generated by a single generator $\overline{3}_z$. We may set

$$B = (\overline{3}_z|0) \tag{13.10.2a}$$

because there exists no invariant eigenvector for $\overline{3}_z$. In terms of B, the generators A and I are given by

$$A = E'B^4 = (3_z|0), \qquad I = E'B^3 = (\overline{1}|0) \tag{13.10.2b}$$

(ii) For the R lattice, in terms of the rhombohedral coordinates and using (13.8.5) with $R_1 = 3_{xyz}$, we obtain only one space group characterized by

$$A = (3_{xyz}|0), \qquad I = (\overline{1}|0) \tag{13.10.2c}$$

which may be rewritten for the R^* lattice as follows:

$$(3_z|0), \qquad (\overline{1}|0) \tag{13.10.2d}$$

13.10.3 The class D_3

$A^3 = B^2 = (AB)^2 = E'$.

(i) For the P lattice, there exist two types of realizations for the binary generator B, corresponding to two equivalence sets of binary axes for the hexagonal P lattice $\in D_{6i}$. We set

$$A = (3_z|0, 0, c), \qquad B_\nu = (u_\nu|a, b, 0), \qquad \nu = 0, 1 \tag{13.10.3a}$$

where $u_0 \parallel x$ and $u_1 \perp y$ with the x- and y-axes forming a hexagonal coordinate system (see Figure 13.11). Here, $u_1 = c_6 u_0$ so that either u_0 or u_1 belongs to D_3 but not both, because c_6 of the hexagonal P lattice is not contained in D_3.

For the AB_0 type we have, using (13.8.1b),

$$A^3 = (e'|0, 0, 3c), \qquad B_0^2 = (e'|2a - b, 0, 0)$$

$$(AB_0)^3 = (e'|a - 2b,\ a - 2b,\ 0) \tag{13.10.3b}$$

which yield

$$b = 2a, \qquad c,\ a = 0, \tfrac{1}{3}, \tfrac{2}{3} \tag{13.10.3c}$$

For each given value of c, however, we have $a = 0$ since $(u_0|a, 2a, 0) \sim (u_0|0, 0, 0)$ via $[e|-a, a, 0]$, which leaves $(3_z|0, 0, c)$ invariant. Thus we are left with only three inequivalent sets of the parameters for AB_0 type:

$$a = b = 0, \qquad c = 0, \tfrac{1}{3}, \tfrac{2}{3} \tag{13.10.3d}$$

For the AB_1 type, we have

$$B_1^2 = (e'|2a, a, 0), \qquad (AB_1)^2 = (e'|a - b,\ 2a - 2b,\ 0) \tag{13.10.3e}$$

which immediately yield the same parameters sets as those given by (13.10.3d).

Thus, there exist altogether six independent space groups belonging to the P lattice given by

$$(3_z|0, 0, c),\ (u_\nu|0); \qquad \nu = 0, 1, \qquad c = 0, \tfrac{1}{3}, \tfrac{2}{3} \tag{13.10.3f}$$

(ii) For the R lattice, since the Bravais lattice does not have c_6 symmetry, in contrast to the P hexagonal lattice, we need to consider only one type of binary axis of rotation as a generator besides the three-fold axis. In terms of the rhombohedral coordinates, using the correspondence $u_{z\bar{x}} \leftrightarrow u_0$ as was stated for (13.8.1c), we set

$$A = (3_{xyz}|0), \qquad B = (u_{z\bar{x}}|a, b, c) \tag{13.10.3g}$$

Then, using (13.8.1c), we obtain

$$B^2 = (e'|a - c,\ 0,\ -a + c), \qquad (AB)^2 = (e'|0,\ a - b,\ -a + b) \tag{13.10.3h}$$

which yield $a = b = c$. Consequently, we can reduce (13.10.3g), via $[e|a/2, a/2, a/2]$, to the form

$$(3_{xyz}|0),\ (u_{z\bar{x}}|0) \tag{13.10.3i}$$

which may be rewritten for the R^* lattice as follows:

$$(3_z|0),\ (u_0|0) \tag{13.10.3j}$$

13.10.4 The class C_{3v}

$A^3 = B^2 = (AB)^2 = E'$.

(i) For the P lattice, on replacing the binary rotations u_ν of D_3 by the reflection $m_\nu = \bar{u}_\nu$ we arrive at the class C_{3v}. We set

$$A = (3_z|0, 0, c), \qquad B_\nu = (m_\nu|a', b', c'), \qquad \nu = 0, 1 \tag{13.10.4a}$$

Then, for the AB_0 type we have

$$A^3 = (e'|0, 0, 3c)$$
$$B_0^2 = (e'|b', 2b', 2c')$$
$$(AB_0)^2 = (e'|-a', a', 2c + 2c') \tag{13.10.4b}$$

which yield the two inequivalent sets of parameters

$$c = a' = b' = 0, \qquad c' = 0, \tfrac{1}{2} \tag{13.10.4c}$$

For the AB_1 type

$$B_1^2 = (e'|0, 2b' - a', 2c'), \qquad (AB_1)^2 = (e'|-a' - b', 0, 2c + 2c') \tag{13.10.4d}$$

which yield

$$-a' = b' = 0, \tfrac{1}{3}, \tfrac{2}{3}, \qquad c = 0, \qquad c' = 0, \tfrac{1}{2}$$

However, $(m_1|-b', b', c') \sim (m_1|0, 0, c')$ under a shift $[e|b', -b', 0]$ that leaves $(3_z|0)$ invariant so that we have again the same parameter sets for AB_1 as given by (13.10.4c). Thus, we obtain altogether four independent solutions for the P lattice

$$(3_z|0), (m_\nu|0, 0, c'); \qquad \nu = 0, 1, \qquad c' = 0, \tfrac{1}{2} \tag{13.10.4e}$$

(ii) For the R lattice, let

$$A = (3_{xyz}|0), \qquad B = (m_{z\bar{x}}|a, b, c) \tag{13.10.4f}$$

Then, from

$$B^2 = (e'|a + c, 2b, a + c), \qquad (AB)^2 = (e'|2c, a + b, a + b) \tag{13.10.4g}$$

we obtain two inequivalent sets of the parameters

$$a = b = c = 0, \tfrac{1}{2} \tag{13.10.4h}$$

which define two independent solutions of (13.10.4f). These may be rewritten for the R^* lattice as follows

$$(3_z|0), (m_0|0, 0, c); \qquad c = 0, \tfrac{1}{2} \tag{13.10.4i}$$

13.10.5 The class D_{3i} ($=D_{3d}$)

$A, B \in D_3, I^2 = [A, I] = [B, I] = E.$

(i) For the P lattice, on augmenting the class D_3 we may set, using (13.10.3f) and (13.8.5) with $R_1 = 3_z$,

$$A = (3_z|0, 0, c); \qquad B_\nu = (u_\nu|0), \qquad \nu = 0, 1; \qquad I = (\bar{1}|0, 0, \gamma) \tag{13.10.5a}$$

where the previous values $\{0, \tfrac{1}{3}, \tfrac{2}{3}\}$ of c for the class D_3 and the new parameter γ are to be specified further by the commutators

$$[A, I] = (e|0, 0, 2c), \qquad [B_\nu, I] = (e|0, 0, -2\gamma) \tag{13.10.5b}$$

obtained from (13.8.4). Since the commutator $[A, I]$ depends on the parameter c, some of the previous values $\{0, \tfrac{1}{3}, \tfrac{2}{3}\}$ of c for the class D_3 may not be allowed. In fact, the allowed values of parameters are

$$c = 0, \qquad \gamma = 0, \tfrac{1}{2} \tag{13.10.5c}$$

Thus we obtain two independent solutions each for both types,

$$(3_z|0), (u_\nu|0), (\bar{1}|0, 0, \gamma); \qquad \nu = 0, 1, \qquad \gamma = 0, \tfrac{1}{2} \tag{13.10.5d}$$

(ii) For the R lattice, from (13.10.3i) we may set

$$A = (3_{xyz}|0), \qquad B = (u_{z\bar{x}}|0), \qquad I = (\bar{1}|\alpha, \alpha, \alpha) \tag{13.10.5e}$$

Then, from (13.8.4),

$$[A, I] = (e|0), \qquad [B, I] = (e|-2\alpha, -2\alpha, -2\alpha) \tag{13.10.5f}$$

which yield two independent solutions for the R lattice characterized by

$$\alpha = 0, \tfrac{1}{2} \tag{13.10.5g}$$

The solutions may be rewritten for the R^* lattice as follows:

$$(3_z|0), (u_0|0), (\bar{1}|0, 0, \alpha); \qquad \alpha = 0, \tfrac{1}{2} \tag{13.10.5h}$$

The MGGSs thus obtained for the rhombohedral system are contained in Table 13.3.

13.11 The hierarchy of space groups in a crystal system

Previously, we have discussed the symmetry hierarchy of crystal systems. In this section, we shall discuss the group–subgroup relations between the space groups *with the same lattice type* but belonging to different crystal classes within a crystal system. This will be discussed in terms of the 32 MGGSs for the 230 space groups given in Table 13.3.

Let G_r and G_s be point groups representing two crystal classes belonging to a crystal system and let G_r be a supergroup of G_s, i.e. $G_r > G_s$. Then a space group belonging to G_r is always a supergroup of a certain space group belonging to G_s, because by eliminating some of the generators of G_r we obtain G_s on account of the solvability of the 32 point groups. The converse is true for most cases but not always. The reason is that some of the space groups belonging to G_s may not have any supergroup belonging to G_r, because the translational part of the former in a certain case may not be allowed to the space groups belonging to G_r due to the additional symmetry requirements for G_r. Let $\hat{G}_r = \{\hat{G}_r^{(i)}\}$ and $\hat{G}_s = \{\hat{G}_r^{(j)}\}$ be the sets of space groups belonging to the classes G_r and G_s, respectively. Then a space group $\hat{G}_s^{(j)}$ without any superspace group in $\hat{G}_r$ is said to be non-subordinate to $\hat{G}_r$. If every member of the set $\{\hat{G}_s^{(j)}\}$ is a subgroup of some member of the set $\{\hat{G}_r^{(i)}\}$, then we say that $\hat{G}_s$ is *subordinate* to $\hat{G}_r$ and denoted by $\hat{G}_s < \hat{G}_r$. For example, the group–subgroup relations between two classes O_i and O are given as follows. From Table 13.3 there exist ten space groups (numbers 221–30) for the class O_i and eight space groups (numbers 207–14) for the class O. From the MGGS of the class O_i

$$O_i: \qquad (4_z|-a, -c, c), (3_{xyz}|0), (\bar{1}|0) \qquad \text{with} \qquad L(c, a)$$

we eliminate $(\bar{1}|0)$ and set $a = 0$ (or eliminate a by a shift $[e|-a/2, -a/2, -a/2]$) and arrive at the MGGS of the class O given by

$$O: \qquad (4_z|0, -c, c), (3_{xyz}|0) \qquad \textit{with} \qquad L(c)$$

with some exceptions, because the parameters $c = \frac{1}{4}, -\frac{1}{4}$ allowed for the P lattice of the class O are not allowed for the P lattice of the class O_i. Thus, the following two space groups of the class O

$$212.\quad P(4_z|0, \tfrac{1}{4}, -\tfrac{1}{4}), (3_{xyz}|0)$$

$$213.\quad P(4_z|0, -\tfrac{1}{4}, \tfrac{1}{4}), (3_{xyz}|0) \qquad (13.11.1)$$

are not subgroups of any space group of the class O_i. This implies that these two groups are not compatible with the inversion symmetry of O_i. Note that the above two space groups numbers 212 and 213 are mutually antipodal whereas such enantiomorphism is obviously absent for the space groups of the class O_i. From Table 13.3, it follows that the remaining space groups of the class O are subgroups of one or more space groups of the class O_i:

$$221, 222 > 207; \qquad 223, 224 > 208;$$

$$225, 226 > 202; \qquad 227, 228 > 210;$$

$$229 > 211; \qquad 230 > 214 \qquad (13.11.2)$$

In an analogous manner we can show that the space group

$$205.\quad P(2_z|\tfrac{1}{2}, 0, \tfrac{1}{2}), \qquad (3_{xyz}|0) \in \hat{T}_i$$

is not subordinate to $\hat{O}_i$. These space groups numbers 212, 213 and 205 are the only space groups without supergroups in the cubic system (except for $\hat{O}_i$ itself, obviously). Since these are subordinate to no space groups, they may be called *ancestorial space groups*. In the following, we shall give all the non-subordinate space groups, based on Table 13.3.

13.11.1 The cubic system

$\hat{O}_i > \hat{O}$ (Except numbers 212, 213 $\in \hat{O}$)
$\hat{O}_i > \hat{T}_i$ (Except number 205 $\in \hat{T}_i$)
$\hat{O}_i > \hat{T}_p$.
$\hat{O}, \hat{T}_p, \hat{T}_i > \hat{T}$

Here, note that all non-subordinate space groups are with P lattices. This statement holds for the remaining systems, too. It can be shown that the subordinate relations of space groups are transitive, i.e. if $A > B$ and $B > C$ then $A > C$.

13.11.2 The hexagonal system

$\hat{D}_{6i} > \hat{D}_6$ (Except numbers 178, 179, 180, 181 $\in \hat{D}_6$)
$\hat{D}_{6i}, \hat{C}_{6i}, \hat{C}_{6v} > \hat{C}_6$ (Except numbers 169, 170, 171, 172 $\in \hat{C}_6$)
$\hat{D}_{6i} > \hat{D}_{3p}, \hat{C}_{6i}, \hat{C}_{6v}$
$\hat{D}_6 > \hat{C}_6$
$\hat{D}_{3p}, \hat{C}_{6i} > \hat{C}_{3p}$

13.11.3 The rhombohedral system

$\hat{D}_{3i} > \hat{D}_3$ (Except numbers 151, 152, 153, 154 $\in \hat{D}_3$)
$\hat{D}_{3i}, \hat{C}_{3i}, \hat{C}_{3v} > \hat{C}_3$ (Except numbers 144, 145 $\in \hat{C}_3$)
$\hat{D}_{3i} > \hat{C}_{3i}, \hat{C}_{3v}$
$\hat{D}_3 > \hat{C}_3$

Note that the space groups with the rhombohedral lattice are all subordinate to some supergroups. For a hexagonal lattice, we can also consider the subordinate relations with the space groups belonging to the hexagonal system; the non-subordinate space groups are again limited to those given above. Note also that $\hat{D}_6 > \hat{D}_3$ and $\hat{C}_6 > \hat{C}_3$ for the hexagonal lattice.

13.11.4 *The tetragonal system*

$\hat{D}_{4i} > \hat{D}_4$ (Except numbers 91, 92, 95, 96 $\in \hat{D}_4$)
$\hat{D}_{4i} > \hat{C}_{4i}$ (Except numbers 84, 86 $\in \hat{C}_{4i}$)
$\hat{D}_{4i}, \hat{C}_{4i}, \hat{C}_{4v} > \hat{C}_4$ (Except numbers 76, 78 $\in \hat{C}_4$)
$\hat{D}_{4i} > \hat{C}_{4v}, \hat{D}_{2p}$; $\hat{D}_{2p} > \hat{C}_{2p}$; $\hat{D}_4 > \hat{C}_4$

Altogether there exists a total of 25 space groups that are non-subordinate (excluding the holohedral space groups). Some of their general features are

1) all non-subordinate space groups have P lattices and
2) all space groups belonging to the crystal systems with symmetry less than or equal to that of the orthorhombic system are subordinate.

Also, note that $\hat{C}_n$ is always subordinate to $\hat{D}_n$, but not completely subordinate to $\hat{C}_{ni}$ or $\hat{C}_{nv}$ for the P lattice. The subordination relations between a proper rotation P and the corresponding rotation–inversion group P_i can easily be understood from the fact that non-subordination occurs when the translational parts are not equivalent under inversion. The subordination relations among space groups within a crystal system or between different crystal systems provide valuable information on structural phase transitions in solid state physics.

13.12 Concluding remarks

Using the defining relations of point groups, we have discussed a method of constructing the 230 space groups through the 32 minimal general generator sets (MGGSs) introduced by Kim (1986b). The algebraic equivalence criteria of the space groups with respect to the lattice transformation $[U|s]$ are completed by introducing the relevant set (13.7.4) of the unimodular matrices $\{U\}$ for each crystal class. Since the order of each relevant set is one or two, it greatly reduces the labor involved in removing redundant solutions in order to arrive at the independent space groups. According to relevant sets (13.7.4), mere shifts $[e|s]$ of the lattice origin are necessary and sufficient to determine the equivalence or inequivalence for almost all classes of high symmetry (with minor exceptions). In fact, it is simpler to construct the space groups belonging to a class with higher symmetry with the present method, quite in contrast to the traditional methods based on the solvability of the space groups. The symmetry properties of MGGSs with respect to lattice transformations play the essential role in identifying a space group and also in constructing extended space groups like the magnetic space groups.

As one can see from the total results given in Table 13.3, the number of the translational parameters of a MGGS is very limited: the maximum numbers for each crystal system are two for cubic, one for hexagonal, rhombohedral and monoclinic, four for tetragonal, five for orthorhombic, and zero for the triclinic system. Obviously these numbers roughly measure the number of independent space groups belonging to the crystal system.

The present method is easily extended to construct the magnetic space groups. In terms of MGGSs of the space groups belonging to each crystal class, one can construct similar general expressions for the magnetic space groups, as will be shown in Chapter 17.

14

Representations of the space groups

We shall begin with the unirreps of the translation group T which is an Abelian invariant subgroup of the space group $\hat{G}$ of a crystal under consideration. Since T is Abelian, its unirreps are one-dimensional. On the basis of this simple fact, we shall analyze the reciprocal lattice structure of the crystal and then construct the unirreps of the space group $\hat{G}$, following the general theory of induced representations introduced in Chapter 8 and the theory of projective representations introduced in Chapter 12. We shall present a more or less self-contained treatment of the representations of the space groups. The representation theory developed here will be applied to the theory of energy bands of the electron in a solid and the symmetry coordinates of vibration of a crystal in the next chapter.

14.1 The unirreps of translation groups

We shall begin with a general discussion of the translation group $T = \{\{e|\boldsymbol{t}\}\}$ in three dimensions, where $\{e|\boldsymbol{t}\}$ is the Seitz notation for a pure translation introduced in Chapter 13. Since T is Abelian, there exists a complete set of simultaneous eigenfunctions of all elements of T. Let $\varphi(\boldsymbol{r})$ be a function of the space variable $\boldsymbol{r}$ which is an eigenfunction of T such that

$$\{e|\boldsymbol{t}\}^{\circ}\varphi(\boldsymbol{r}) = \varphi(\boldsymbol{r} - \boldsymbol{t}) = \tau(\boldsymbol{t})\varphi(\boldsymbol{r}) \tag{14.1.1}$$

for any element $\{e|\boldsymbol{t}\}$ of T, then $\tau(\boldsymbol{t})$ is an eigenvalue of the element $\{e|\boldsymbol{t}\}$ which provides the representative of the element in T. To solve the eigenvalue problem, let us describe the translation of $\varphi(\boldsymbol{r})$ in terms of the differential operator $\nabla = (\partial/\partial x, \partial/\partial y, \partial/\partial z)$ using Taylor's theorem as follows:

$$\varphi(\boldsymbol{r} - \boldsymbol{t}) = \mathrm{e}^{-\boldsymbol{t}\cdot\nabla}\varphi(\boldsymbol{r}) = \tau(\boldsymbol{t})\varphi(\boldsymbol{r}) \tag{14.1.2}$$

Since the translation operator $\exp(-\boldsymbol{t}\cdot\nabla)$ is an analytic function of the operator ∇, the problem is reduced to the eigenvalue problem of the infinitesimal operator ∇:

$$\nabla\varphi(\boldsymbol{r}) = \mathrm{i}\boldsymbol{k}\varphi(\boldsymbol{r}) \tag{14.1.3}$$

where we have introduced an imaginary eigenvalue $\mathrm{i}\boldsymbol{k}$ with a real vector $\boldsymbol{k}$ in order to make $\varphi(\boldsymbol{r})$ be a basis of a unitary representation of T. The solution of (14.1.3) may be expressed by

$$\varphi^{(\boldsymbol{k})}(\boldsymbol{r}) = \mathrm{e}^{\mathrm{i}\boldsymbol{k}\cdot\boldsymbol{r}} \tag{14.1.4a}$$

so that, from (14.1.1), the representation of $T = \{\{e|\boldsymbol{t}\}\}$ is given by

$$\tau_{\boldsymbol{k}}(\boldsymbol{t}) = \mathrm{e}^{-\mathrm{i}\boldsymbol{k}\cdot\boldsymbol{t}}; \qquad \forall\, \boldsymbol{t} \in T \tag{14.1.4b}$$

which is indeed unitary since $\tau_{\boldsymbol{k}}^{*}(\boldsymbol{t})\tau_{\boldsymbol{k}}(\boldsymbol{t}) = 1$ for any $\boldsymbol{k}$. Here, the vector $\boldsymbol{k}$ identifies the unirrep $\tau_{\boldsymbol{k}}(\boldsymbol{t})$ of T and is called *a wave vector* on account of the oscillatory nature of the basis function $\exp(\mathrm{i}\boldsymbol{k}\cdot\boldsymbol{r})$. In quantum mechanics, $\varphi^{(\boldsymbol{k})}(\boldsymbol{r})$ is the momentum eigenfunction belonging to an eigenvalue $\hbar\boldsymbol{k}$ of a particle. The set $\{\varphi^{(\boldsymbol{k})}(\boldsymbol{r}); -\infty < k_x, k_y, k_z < \infty\}$ is an orthonormal complete set satisfying

$$\int_{-\infty}^{+\infty} \varphi^{(\boldsymbol{k})}(\boldsymbol{r})^* \varphi^{(\boldsymbol{k}')}(\boldsymbol{r})\,\mathrm{d}^3\boldsymbol{r} = (2\pi)^3\delta(\boldsymbol{k}-\boldsymbol{k}') \tag{14.1.5}$$

where $\delta(\boldsymbol{k}-\boldsymbol{k}')$ is the Dirac delta function.

For a discrete translation group T of the Bravais lattice of a crystal, the general element $t_n = \{e|\boldsymbol{t}_n\}$ is described by a primitive basis $[\boldsymbol{a}_1, \boldsymbol{a}_2, \boldsymbol{a}_3]$ of T as follows:

$$\boldsymbol{t}_n = n_1\boldsymbol{a}_1 + n_2\boldsymbol{a}_2 + n_3\boldsymbol{a}_3, \qquad n_j = \pm \text{ integers or zero} \tag{14.1.6a}$$

Thus, from (14.1.4b), the unirrep of a discrete translation group $T = \{\{e|\boldsymbol{t}_n\}\}$ is given by

$$\tau_{\boldsymbol{k}}(t_n) = \exp(-\mathrm{i}\boldsymbol{k}\cdot\boldsymbol{t}_n), \qquad \forall\, t_n = \{e|\boldsymbol{t}_n\} \in T \tag{14.1.6b}$$

On account of the discrete nature of T, the wave vector $\boldsymbol{k}$ is no longer determined uniquely by the representation $\tau_{\boldsymbol{k}}(t_n)$. To see this, let us introduce the so-called reciprocal lattice basis $[\boldsymbol{b}_1, \boldsymbol{b}_2, \boldsymbol{b}_3]$ which is orthogonal to the primitive basis $[\boldsymbol{a}_1, \boldsymbol{a}_2, \boldsymbol{a}_3]$ such that

$$\boldsymbol{b}_i\cdot\boldsymbol{a}_j = 2\pi\delta_{ij} \qquad (i, j = 1, 2, 3) \tag{14.1.7}$$

This set of equations for $\{\boldsymbol{b}_i\}$ has a unique solution given by the vector products as follows:

$$\boldsymbol{b}_1 = 2\pi[\boldsymbol{a}_2\times\boldsymbol{a}_3]/V_0, \qquad \boldsymbol{b}_2 = 2\pi[\boldsymbol{a}_3\times\boldsymbol{a}_1]/V_0, \qquad \boldsymbol{b}_3 = 2\pi[\boldsymbol{a}_1\times\boldsymbol{a}_2]/V_0 \tag{14.1.8}$$

where $V_0 = \boldsymbol{a}_1\cdot[\boldsymbol{a}_2\times\boldsymbol{a}_3]$ is the primitive cell volume of the lattice T. These basis vectors $\boldsymbol{b}_1, \boldsymbol{b}_2, \boldsymbol{b}_3$ are ordinary polar vectors obeying the ordinary vector transformation law, if one takes into consideration the change in sign of the cell volume under inversion. The two bases $[\boldsymbol{a}_1, \boldsymbol{a}_2, \boldsymbol{a}_3]$ and $[\boldsymbol{b}_1, \boldsymbol{b}_2, \boldsymbol{b}_3]$ are said to be mutually *dual* in the sense that they are mutually reciprocal.

The vector space spanned by the basis $[\boldsymbol{b}_1, \boldsymbol{b}_2, \boldsymbol{b}_3]$ is called *the reciprocal space* of the lattice T. In particular, a set of vectors spanned by

$$\boldsymbol{g}_l = l_1\boldsymbol{b}_1 + l_2\boldsymbol{b}_2 + l_3\boldsymbol{b}_3, \qquad l_i = \pm \text{ integers or zero} \tag{14.1.9}$$

is called *the reciprocal lattice* T' of T. These two lattices are mutually dual, satisfying

$$\boldsymbol{g}_l\cdot\boldsymbol{t}_n = 2\pi(l_1n_1 + l_2n_2 + l_3n_3) = 2\pi\times\text{integer} \tag{14.1.10}$$

so that $\exp(\mathrm{i}\boldsymbol{g}_l\cdot\boldsymbol{t}_n) = 1$. Accordingly, if two wave vectors $\boldsymbol{k}$ and $\boldsymbol{k}'$ differ by a reciprocal lattice vector $\boldsymbol{g}_l$,

$$\boldsymbol{k}' = \boldsymbol{k} + \boldsymbol{g}_l, \qquad \boldsymbol{g}_l \in T' \tag{14.1.11}$$

then two representations $\tau_{\boldsymbol{k}'}(T)$ and $\tau_{\boldsymbol{k}}(T)$ are equal:

$$\tau_{\boldsymbol{k}'}(t_n) = \tau_{\boldsymbol{k}}(t_n), \qquad \forall\, t_n = \{e|\boldsymbol{t}_n\} \in T \tag{14.1.12}$$

Conversely, from (14.1.12), (14.1.11) follows via (14.1.10). Two wave vectors $\boldsymbol{k}'$ and $\boldsymbol{k}$ that differ by a reciprocal lattice vector $\boldsymbol{g}_l$ are said to be equivalent with respect to the reciprocal lattice T' and denoted by $\boldsymbol{k}' \sim \boldsymbol{k}$.

It is always profitable to make the translation group T finite without modifying the physics. For this purpose, we introduce the *Born–von Karman* cyclic boundary condition for the Bravais lattice with a period $N_j\boldsymbol{a}_j$ along the direction of each primitive basis vector $\boldsymbol{a}_j$ ($j = 1, 2, 3$), where N_j is a large number of the order of Avogadro's number. Then the order of T becomes finite and equal to $|T| = N_1N_2N_3$. The cyclic boundary condition imposes the following conditions for the representation:

$$\tau_{\boldsymbol{k}}(\{e|N_j\boldsymbol{a}_j\}) = \exp(-\mathrm{i}N_j\boldsymbol{k}\cdot\boldsymbol{a}_j) = 1; \qquad j = 1, 2, 3 \tag{14.1.13}$$

so that the allowed wave vectors must satisfy

$$N_j\boldsymbol{k}\cdot\boldsymbol{a}_j = 2\pi h_j, \qquad h_j = \pm \text{ an integer or zero}$$

To solve this set of equations for the wave vector $\boldsymbol{k}$, we express $\boldsymbol{k}$ as a linear combination of $\boldsymbol{b}_1$, $\boldsymbol{b}_2$ and $\boldsymbol{b}_3$, and then using (14.1.7) we obtain a total of $N_1N_2N_3$ wave vectors $\boldsymbol{k}$ defined by

$$\boldsymbol{k} = \frac{h_1}{N_1}\boldsymbol{b}_1 + \frac{h_2}{N_2}\boldsymbol{b}_2 + \frac{h_3}{N_3}\boldsymbol{b}_3; \qquad \pm h_j = 1, 2, \ldots, N_j \; (j = 1, 2, 3) \tag{14.1.14}$$

These wave vectors may be regarded as 'continuously' distributed over a primitive unit cell in the reciprocal lattice, because the N_j are assumed to be very large numbers. Moreover, these are all inequivalent with respect to T' for, since they are all in a primitive unit cell of the reciprocal lattice T', no two of them can differ by a reciprocal lattice vector $\boldsymbol{g}_l \in T'$. Now, the unirreps $\tau_{\boldsymbol{k}}(T)$ of the translation group T satisfy the orthogonality relations

$$\sum_{\tau_n}\tau^*_{\boldsymbol{k}}(t_n)\tau_{\boldsymbol{k}'}(t_n) = \prod_{j=1}^{3}\sum_{n_j=1}^{N_j}\exp[2\pi\mathrm{i}(h_j - h'_j)n_j/N_j] = N_1N_2N_3\delta_{\boldsymbol{k}\boldsymbol{k}'} \tag{14.1.15a}$$

via (14.1.14), on account of Fourier's theorem. The set of all unirreps $\{\tau_{\boldsymbol{k}}(T)\}$ with $\boldsymbol{k}$ over the primitive unit cell of the reciprocal lattice is complete because the set satisfies the completeness condition

$$\sum_{\boldsymbol{k}}|\tau_{\boldsymbol{k}}(T)|^2 = N_1N_2N_3 = |T| \tag{14.1.15b}$$

where $|\tau_{\boldsymbol{k}}(T)| = 1$, the dimensionality of the unirrep $\tau_{\boldsymbol{k}}(T)$.

14.2 The reciprocal lattices

14.2.1 General discussion

The vector space spanned by the reciprocal basis $[\boldsymbol{b}_1, \boldsymbol{b}_2, \boldsymbol{b}_3]$ is called *the reciprocal space* of the direct lattice T. Then a wave vector $\boldsymbol{k}$ defined by (14.1.14) is a vector in the reciprocal space. To discuss the symmetry properties of the reciprocal space, let us introduce the matrices of the basis vectors for the direct lattice T as well as for the reciprocal lattice T', as in (13.3.1b):

$$\boldsymbol{A} = [\boldsymbol{a}_1, \boldsymbol{a}_2, \boldsymbol{a}_3], \qquad \boldsymbol{B} = [\boldsymbol{b}_1, \boldsymbol{b}_2, \boldsymbol{b}_3] \tag{14.2.1a}$$

where the column vectors of the matrices are given by the respective basis vectors following the notation introduced by (5.1.5). Then, the defining relation (14.1.7) of the reciprocal basis is expressed in the matrix form

$$B^{\sim}A = 2\pi\mathbf{1} \tag{14.2.1b}$$

where $B^{\sim}$ is the transpose of the matrix B. The matrix solution of this equation for the matrix B is given by

$$B = 2\pi A^{\#}; \qquad A^{\#} = A^{\sim-1} \tag{14.2.1c}$$

where $A^{\#}$ is called the contragredient matrix of A: it obeys the product law $(A_1A_2)^{\#} = A_1{}^{\#}A_2{}^{\#}$. In the special case of an orthogonal matrix R we have $R^{\#} = R$. If the matrix A is symmetric, we have $B = 2\pi A^{-1}$, which is also symmetric. This is the case for the cubic system.

From (14.2.1c) or (14.2.1b), we shall first show that the symmetry point group of the reciprocal lattice T' coincides with the symmetry point group K of the direct lattice T. If a rotation R belongs to the symmetry point group K of $T = \{t_n\}$, the transformed basis $\{Ra_j\}$ is also a primitive basis of T so that $\{Rb_i\}$ must also be a primitive basis of the reciprocal lattice T' of T, because

$$RB = 2\pi(RA)^{\#}$$

that is RB represents the reciprocal lattice vector of the transformed primitive basis $RA \in T$. Secondly, the primitive cell volume V_0' of T' is given by the primitive cell volume V_0 of T as follows:

$$V_0' = (2\pi)^3/V_0 \tag{14.2.2}$$

This is shown by taking the determinants of both sides of (14.2.1b) and using $V_0 = \det A$ and $V_0' = \det B$. Thirdly, the direct lattice T and reciprocal lattice T' defined by (14.1.6a) and (14.1.9), respectively, are expressed by

$$t_n = An, \qquad g_l = Bl \tag{14.2.3a}$$

where $n = (n_1, n_2, n_3)$ and $l = (l_1, l_2, l_3)$ are column vectors. Thus, their scalar product yields

$$g_l \cdot t_n = (Bl) \cdot A_n = l \cdot B^{\sim}An = 2\pi l \cdot n \tag{14.2.3b}$$

which is (14.1.10).

Since a lattice T and its reciprocal lattice T' belong to the same crystal system K, the allowed lattice types for T are the same as those of T. This does not necessarily mean that the lattice type of a specific lattice T' is the same as that of the corresponding T, unless only one lattice type is allowed for the crystal system K. For example, there exist three lattice types for the cubic system: simple cubic P_c, face-centered cubic F_c and body-centered cubic I_c lattices. As will be shown later, the reciprocal lattice P_c' of P_c is also P_c whereas the reciprocal lattice F_c' of F_c is I_c, and conversely that of I_c' is F_c, because T and T' are mutually dual.

In an actual determination of the reciprocal lattice T' of a given lattice T, it is always convenient to introduce a Cartesian basis $[e_1^0, e_2^0, e_3^0]$ for the given lattice T. It may be taken parallel to the conventional lattice basis $[e_1, e_2, e_3]$ given in Table 13.1, if the latter is an orthogonal set. Then, through the Cartesian components of the primitive basis $[a_1, a_2, a_3]$ of T, those of the reciprocal lattice basis $[b_1, b_2, b_3]$ of T' are determined by (14.1.8) or (14.2.1c). Thus, a reciprocal lattice vector $g_l \in T'$ may be expressed in terms of the Cartesian basis as follows:

$$g_l = Bl = 2\pi a^{-1}[l_1^0 e_1^0 + l_2^0 e_2^0 + l_3^0 e_3^0] \equiv 2\pi a^{-1} l^0 \tag{14.2.4}$$

where a is an appropriate lattice parameter of T. The relation between the reciprocal lattice indices $\boldsymbol{l} = (l_1, l_2, l_3)$ and Cartesian indices $\boldsymbol{l}^0 = [l_1^0, l_2^0, l_3^0]$ depends on the lattice type T. Here and hereafter the Cartesian indices are expressed by square brackets [...] whereas the reciprocal lattice indices are expressed by round brackets (...). In the following, the method of constructing the reciprocal lattices will be exemplified for the cubic system.

14.2.2 Reciprocal lattices of the cubic system

From Table 13.1, we may set the Cartesian basis $[\boldsymbol{e}_1^0, \boldsymbol{e}_2^0, \boldsymbol{e}_3^0]$ proportional to the conventional basis $[\boldsymbol{e}_1, \boldsymbol{e}_2, \boldsymbol{e}_3]$ of the cubic system such that

$$\boldsymbol{e}_i = a\boldsymbol{e}_i^0 \qquad (i = 1, 2, 3) \tag{14.2.5}$$

where a is the lattice spacing of a cubic crystal. Then the primitive lattice basis $\boldsymbol{A}$ of each lattice type, the reciprocal lattice basis $\boldsymbol{B}$ and the Cartesian indices $\boldsymbol{l}^0$ of a reciprocal lattice vector $\boldsymbol{g}_l$ are given in terms of the Cartesian basis as follows.

(i) For the simple cubic lattice P_c, the primitive basis is expressed by $\boldsymbol{A} = [\boldsymbol{a}_1, \boldsymbol{a}_2, \boldsymbol{a}_3] = a\mathbf{1}$, where $\mathbf{1}$ is the unit matrix, so that the matrix $\boldsymbol{B}$ is given by $\boldsymbol{B} = 2\pi\boldsymbol{A}^{-1} = 2\pi a^{-1}\mathbf{1}$, which defines also a simple cubic lattice P_c' with the lattice spacing $2\pi a^{-1}$. Thus, the reciprocal lattice vector $\boldsymbol{g}_l \in P_c'$ is given by

$$\boldsymbol{g}_l = \boldsymbol{B}\boldsymbol{l} = 2\pi a^{-1}\boldsymbol{l} \tag{14.2.6}$$

(ii) For the face-centered cubic lattice F_c, from (13.4.4e) and (14.2.1c), the primitive bases are expressed by

$$\boldsymbol{A} = (a/2)\begin{bmatrix} 0 & 1 & 1 \\ 1 & 0 & 1 \\ 1 & 1 & 0 \end{bmatrix}, \qquad \boldsymbol{B} = 2\pi a^{-1}\begin{bmatrix} -1 & 1 & 1 \\ 1 & -1 & 1 \\ 1 & 1 & -1 \end{bmatrix} \tag{14.2.7a}$$

where $\boldsymbol{B}$ defines a *body-centered cubic lattice*, i.e. $F_c' = I_c$ with the lattice spacing $4\pi a^{-1}$, cf. Equation (13.4.2c). From (14.2.4), the reciprocal lattice vectors are given by $\boldsymbol{g}_l = \boldsymbol{B}\boldsymbol{l} = 2\pi a^{-1}\boldsymbol{l}^0$ with the Cartesian indices

$$\boldsymbol{l}^0 = [-l_1 + l_2 + l_3,\ l_1 - l_2 + l_3,\ l_1 + l_2 - l_3] \tag{14.2.7b}$$

where the sum of any two components of $\boldsymbol{l}^0$ is even so that the three components are all odd or even, including zero, cf. Equation (13.4.2g).

(iii) For a body-centered lattice I_c, the primitive bases are expressed, from (13.4.2c) and (14.2.1c), by

$$\boldsymbol{A} = (a/2)\begin{bmatrix} -1 & 1 & 1 \\ 1 & -1 & 1 \\ 1 & 1 & -1 \end{bmatrix}, \qquad \boldsymbol{B} = 2\pi a^{-1}\begin{bmatrix} 0 & 1 & 1 \\ 1 & 0 & 1 \\ 1 & 1 & 0 \end{bmatrix} \tag{14.2.8a}$$

where $\boldsymbol{B}$ describes the *face-centered cubic lattice* (i.e. $I_c' = F_c$) with the lattice spacing $4\pi a^{-1}$. The reciprocal lattice vector is given by $\boldsymbol{g}_l = 2\pi a^{-1}\boldsymbol{l}^0$ with the Cartesian indices

$$\boldsymbol{l}^0 = [l_2 + l_3,\ l_3 + l_1,\ l_1 + l_2] \tag{14.2.8b}$$

where the sum of the three components l_1^0, l_2^0 and l_3^0 is even so that they are all even or only two of them are odd (cf. 13.4.4g).

For a complete list of the correspondences between the types of the crystal lattices and their reciprocal lattices, see e.g. Bir and Pikus (1974).

14.2.3 The Miller indices

Let us consider the geometric structure of a direct lattice T relative to its reciprocal T' based on the dual relation (14.1.10). For this purpose, we note that the equation of a plane perpendicular to a unit vector $\boldsymbol{u}$ is defined by

$$\boldsymbol{u} \cdot \boldsymbol{x} = d \tag{14.2.9}$$

where $\boldsymbol{x}$ is a point on the plane and d is the normal distance to the plane from the coordinate origin O (see Figure 14.1). If we compare (14.1.10) with (14.2.9), we may arrive at the conclusion that the whole direct lattice $T = \{\boldsymbol{t}_n\}$ can be described as a series of parallel lattice planes perpendicular to any given reciprocal lattice vector $\boldsymbol{g}_l$ in T'. To see this more clearly we introduce a primitive lattice vector $\boldsymbol{g}_m$ in the reciprocal lattice T' defined by

$$\boldsymbol{g}_m = m_1\boldsymbol{b}_1 + m_2\boldsymbol{b}_2 + m_3\boldsymbol{b}_3 \equiv (m_1, m_2, m_3) \tag{14.2.10}$$

where (m_1, m_2, m_3) is a set of integers without a common integral devisor (i.e. they are mutually prime) and is called a set of *the Miller indices*. The primitive lattice vector $\boldsymbol{g}_m$ itself may also be called *the Miller vector*. In terms of $\boldsymbol{g}_m$, (14.1.10) is rewritten as

$$\boldsymbol{g}_m \cdot \boldsymbol{t}_n = 2\pi p, \qquad p = m_1 n_1 + m_2 n_2 + m_3 n_3 \tag{14.2.11a}$$

where p may take the value 0 or any positive or negative integer, $0, \pm 1, \pm 2, \ldots$, because the Miller indices are mutually prime. Now, let $\boldsymbol{u}_m = (\boldsymbol{g}_m/|\boldsymbol{g}_m|)$ be the unit vector parallel to $\boldsymbol{g}_m$, then (14.2.11a) is rewritten in the standard form of the equation of a lattice plane perpendicular to $\boldsymbol{u}_m$

$$\boldsymbol{u}_m \cdot \boldsymbol{t}_n = p d_m, \qquad d_m = 2\pi/|\boldsymbol{g}_m| \tag{14.2.11b}$$

with the normal distance pd_m from the lattice origin O. Since p changes by unity, d_m is the interplanar distance. Therefore, the set of equations with $p = 0, \pm 1, \pm 2, \ldots$ describes the whole set of lattice points $\{\boldsymbol{t}_n\}$ of T as a series of parallel lattice planes perpendicular to a given primitive lattice vector $\boldsymbol{g}_m$ in T'. It is customary to refer to a lattice plane perpendicular to the Miller vector $\boldsymbol{g}_m$ as the (m_1, m_2, m_3) plane.

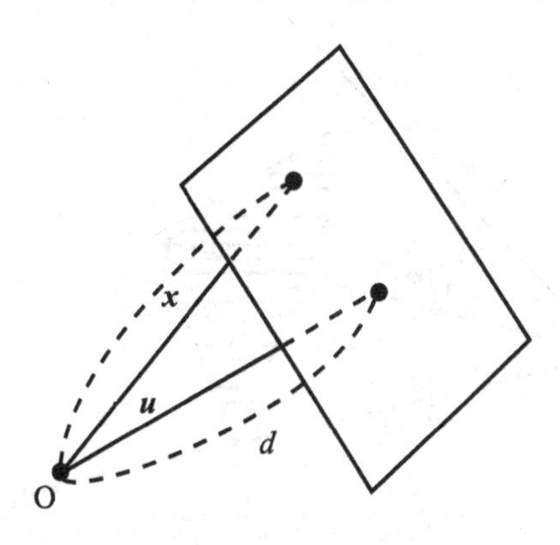

Figure 14.1. A plane perpendicular to a unit vector $\boldsymbol{u}$ with the normal distance d described by $\boldsymbol{u} \cdot \boldsymbol{x} = d$.

The geometric construction of the (m_1, m_2, m_3) plane nearest to the lattice origin O is interesting. The plane is characterized by (14.2.11a) with $p = 1$ so that a general point $\boldsymbol{x} = \sum_i x_i \boldsymbol{a}_i$ on a continuous plane defined by $\boldsymbol{u}_m \cdot \boldsymbol{x} = d_m$ is characterized by

$$m_1 x_1 + m_2 x_2 + m_3 x_3 = 1 \tag{14.2.12}$$

Here the x_i are not necessarily integers, being on the continuous plane. At the point where the plane intersects with the primitive basis vector $\boldsymbol{a}_1$, we have $x_2 = x_3 = 0$ so that $x_1 = 1/m_1$ from (14.2.12). Likewise, the points of the intersections of the plane with $\boldsymbol{a}_2$ and $\boldsymbol{a}_3$ are $\boldsymbol{a}_2/m_2$ and $\boldsymbol{a}_3/m_3$, respectively. Accordingly, the (m_1, m_2, m_3) plane nearest to the origin O can be drawn in terms of the intersections $\{\boldsymbol{a}_1/m_1, \boldsymbol{a}_2/m_2, \boldsymbol{a}_3/m_3\}$ as shown in Figure 14.2.

14.2.4 The density of lattice points on a plane

Let ρ be the lattice density defined by the average number of lattice points per unit volume of a given lattice T and let ρ_m be the plane density defined by the average number of lattice points per unit area of the lattice plane perpendicular to Miller's vector $\boldsymbol{g}_m$. Then we have, from (14.2.11b),

$$\rho_m = \rho d_m = 2\pi\rho/|\boldsymbol{g}_m| \tag{14.2.13a}$$

where d_m is the interplanar distance in the direction of $\boldsymbol{g}_m$ and

$$|\boldsymbol{g}_m| = 2\pi a^{-1}(m_1^{02} + m_2^{02} + m_3^{02})^{1/2} \tag{14.2.13b}$$

in terms of the Cartesian indices of $\boldsymbol{g}_m$ introduced by (14.2.4). Thus, the density ρ_m on a plane (m_1, m_2, m_3) is higher if the Miller vector $|\boldsymbol{g}_m|$ is smaller. For example, for a simple cubic crystal we have $m_j = m_j^0$ so that the (100) plane has a higher plane density of lattice points than does the (111) plane by a factor of $\sqrt{3}$. For a face-centered cubic crystal, the (100) and (111) planes have the same density of lattice points because $|g_{100}^{\mathrm{fcc}}| = |g_{111}^{\mathrm{fcc}}| = 2\pi a^{-1}\sqrt{3}$, from (14.2.7b). The lattice planes with the smaller $\boldsymbol{g}_m$ play the dominant role, for example, for the diffraction of X-rays by a crystal.

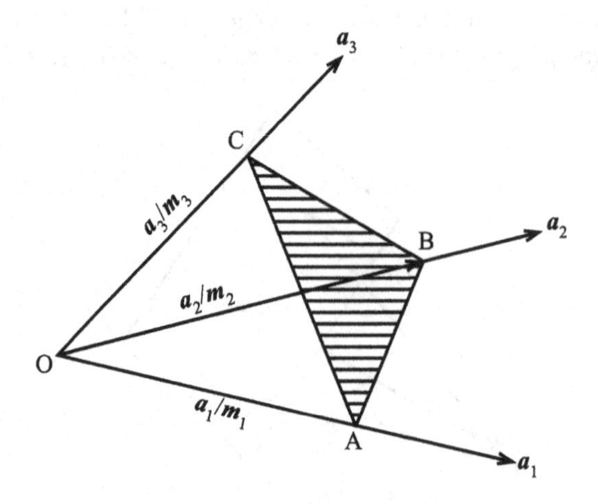

Figure 14.2. Construction of the (m_1, m_2, m_3) Miller plane nearest to O: $\mathrm{OA} = \boldsymbol{a}_1/m_1$, $\mathrm{OB} = \boldsymbol{a}_2/m_2$ and $\mathrm{OC} = \boldsymbol{a}_3/m_3$.

14.3 Brillouin zones

14.3.1 General construction of Brillouin zones

Previously, in (14.1.14), the domain of wave vectors $\boldsymbol{k}$ (or $\boldsymbol{k}$-space) has been defined by a primitive unit cell of the reciprocal lattice T' under consideration. This is, however, not always convenient because the choice of a primitive unit cell is quite arbitrary and thus it need not be invariant under the symmetry group of the lattice. We shall introduce a symmetrized unit cell of T' that is centered by a lattice point of T' and is invariant under the symmetry point group K of the lattice and has the volume equal to the volume of the primitive unit cell of T'. The domain of $\boldsymbol{k}$-space thus defined is known as the *first Brillouin zone* or simply *the Brillouin zone*[1] of the Bravais lattice T: it need not be a parallelepiped in general but a certain polyhedron centered by a lattice point of T'.

More explicitly, the Brillouin zone (abbreviated as B-zone) of a Bravais lattice T is defined as follows. Choose a lattice point of T' as the origin and then, for a given primitive lattice vector $\boldsymbol{g}_m$ (or Miller vector) from the origin O, draw a plane that is perpendicular to $\boldsymbol{g}_m$ and bisects $\boldsymbol{g}_m$ (see Figure 14.3). By definition a point $\boldsymbol{X}$ on the plane thus formed satisfies an equation of the plane defined by

$$\boldsymbol{X} \cdot \boldsymbol{u}_m = |\boldsymbol{g}_m|/2, \qquad \boldsymbol{u}_m = \boldsymbol{g}_m/|\boldsymbol{g}_m| \tag{14.3.1}$$

where $|\boldsymbol{g}_m|/2$ is the perpendicular distance to the plane from the coordinate origin O. By varying $\boldsymbol{g}_m$ over the Miller indices (m_1, m_2, m_3), we obtain an analytic expression for the B-zone. The polyhedron bounded by the planes nearest to O defines the first Brillouin zone. Analogously second, third and higher Brillouin zones are defined by perpendicular bisecting planes of reciprocal lattice vectors $\boldsymbol{g}_l$. The wave vectors in the higher zones are, however, equivalent to the wave vectors in the first Brillouin zone with respect to T'. The first zone plays a central role in subsequent discussions, because all basis functions of the translation group T can be classified in terms of the wave vectors in the first Brillouin zone. Hereafter, the Brillouin zone means the first B-zone unless specified otherwise.

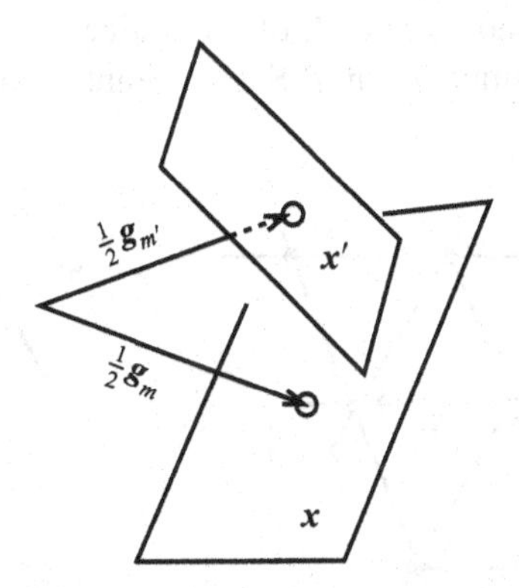

Figure 14.3. Construction of the Brillouin zone.

[1] It is customary that the Brillouin zone is referred to the direct lattice T rather than to the reciprocal lattice T', even though it is defined in the reciprocal lattice space.

The B-zone thus defined is invariant under the point group K because the reciprocal lattice vectors $\boldsymbol{g}_m$, which define the B-zone, permute among themselves under the symmetry operations of the lattice T' at the lattice point O in the center of the B-zone. The volume of the B-zone thus defined equals the primitive unit cell volume of T' because the whole reciprocal lattice T' is completely filled by the B-zones drawn one for each lattice point in T'.

Moreover, a face of the B-zone perpendicular to $\boldsymbol{g}_m$ is parallel to the whole set of parallel lattice planes of T perpendicular to $\boldsymbol{g}_m$. As an example, the Bravais parallelepiped and the B-zone of the hexagonal lattice $P_{\rm H}$ are shown by their projections on the plane perpendicular to the principal axis of rotation 6_z of the lattice in Figure 14.4. Note that each face of the B-zone is parallel to a set of the lattice planes of the hexagonal lattice. Furthermore, from the defining equation (14.3.1), the smaller $\boldsymbol{g}_m$ are required for the first Brillouin zone. Accordingly, the faces of B-zone are parallel to those lattice planes of T with higher lattice point densities. For example, for the cubic system, the Miller indices for the faces of the B-zone are given by $\pm(m_1, m_2, m_3)$ with $m_i = 1$ or 0, excluding (0, 0, 0), as will be shown in Section 14.3.3. Thus the number of faces of a B-zone for the cubic system is less than or equal to $14 = 2(2^3 - 1)$ (cf. Equation (14.3.7a)).

It is of interest to note that the equation of the plane (14.3.1) corresponds to the condition for the Bragg reflection of X-rays, in the first order, at a set of planes perpendicular to $\boldsymbol{g}_m$. To see this let us identify the vector $\boldsymbol{X}$ as the wave vector $\boldsymbol{k}$ of the X-ray beam with $|\boldsymbol{k}| = 2\pi/\lambda$, where λ is the wave length and let the angle between $\boldsymbol{k}$ and the unit vector $\boldsymbol{u}_m$ be $\frac{1}{2}\pi - \theta$. Then (14.3.1) is reduced to

$$(2\pi/\lambda)\sin\theta = |\boldsymbol{g}_m|/2 \tag{14.3.2}$$

This may be rewritten by (14.2.11b) in the familiar form

$$2d_m \sin\theta = \lambda$$

14.3.2 The wave vector point groups

The symmetry group of a wave vector $\boldsymbol{k}$ in the Brillouin zone (B-zone) is defined by the subgroup of the symmetry group K of the lattice T that leaves $\boldsymbol{k}$ equivalent with respect to the reciprocal lattice T' of T. Such a point group is called the *wave vector*

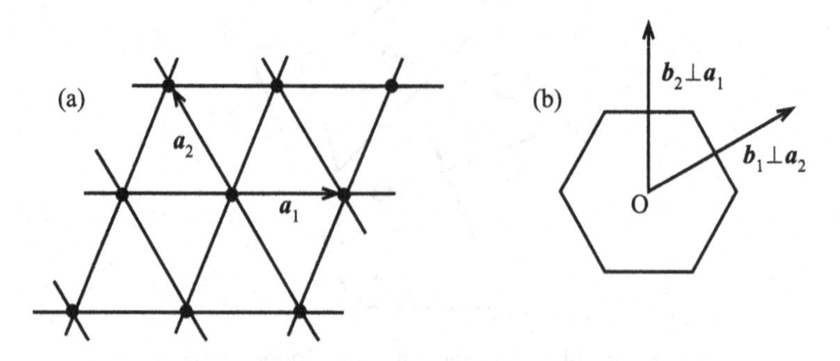

Figure 14.4. For the hexagonal lattice $P_{\rm H}$ projected onto the x, y plane: (a) the direct lattice T and (b) the Brillouin zone. The vectors $\boldsymbol{a}_1$ and $\boldsymbol{a}_2$ are the primitive basic vectors in the direct lattice T, whereas $\boldsymbol{b}_1$ and $\boldsymbol{b}_2$ are those of the reciprocal lattice T'. The unit of the length is not drawn to scale.

point group of a given $\boldsymbol{k}$ and denoted $K_{\boldsymbol{k}}$. Let R be a rotation belonging to $K_{\boldsymbol{k}}$, then by definition

$$R\boldsymbol{k} = \boldsymbol{k} + \boldsymbol{g}_l, \qquad \boldsymbol{g}_l \in T' \tag{14.3.3}$$

When $\boldsymbol{k}$ is an interior point of the B-zone, we have $R\boldsymbol{k} = \boldsymbol{k}$ for all $R \in K_{\boldsymbol{k}}$, because two wave vectors inside the B-zone cannot differ by a reciprocal lattice vector. By definition $K_{\boldsymbol{k}} \leqslant K$ and $K_{\boldsymbol{k}}$ coincides with K at the lattice origin where $\boldsymbol{k} = 0$.

In the following, we shall construct the Brillouin zones for the cubic system and discuss the wave vector point group $K_{\boldsymbol{k}}$ at the points of high symmetry in the B-zone. For this purpose, it is always profitable to introduce the Cartesian coordinates of the wave vector $\boldsymbol{k}$ by

$$\boldsymbol{k} = 2\pi a^{-1}[\xi e_1^0 + \eta e_2^0 + \zeta e_3^0] \equiv 2\pi a^{-1}[\xi, \eta, \zeta] \tag{14.3.4}$$

analogous to (14.2.4). For a wave vector $\boldsymbol{k}$ inside the B-zone and perpendicular to a zone boundary that bisects a Miller vector $\boldsymbol{g}_m$, we have from (14.2.13b)

$$|\boldsymbol{k}| \leqslant |\boldsymbol{g}_m|/2 = 2\pi a^{-1}[m_1^{02} + m_2^{02} + m_3^{02}]^{1/2}/2 \tag{14.3.5}$$

14.3.3 *The Brillouin zones of the cubic system*

(i) For the simple cubic lattice P_c, the Cartesian indices $\boldsymbol{m}^0 = [m_1^0, m_2^0, m_3^0]$ of a Miller vector $\boldsymbol{g}_m$ are the same as the Miller indices $\boldsymbol{m} = (m_1, m_2, m_3)$, from (14.2.6). There exist six smallest non-zero $\boldsymbol{g}_m = 2\pi a^{-1}\boldsymbol{m}$ with $\boldsymbol{m}$ given by

$$\pm[100], \pm[010], \pm[001] \tag{14.3.6a}$$

each of which is parallel to the coordinate axis x, y or z. Each face of the B-zone is perpendicular to one of these so that the B-zone is simple cubic, as shown in Figure 14.5. From (14.3.5) the perpendicular distance of each zone boundary from the zone center O equals $\frac{1}{2}$ in the unit of $2\pi a^{-1}$. The Cartesian coordinates $[\xi, \eta, \zeta]$ of $\boldsymbol{k}$ at some high symmetry points and their point symmetries $K_{\boldsymbol{k}}$ are

$$\begin{aligned}
&\varGamma = [000], \quad R = [\tfrac{1}{2}\tfrac{1}{2}\tfrac{1}{2}] \in O_i^z; \quad X = [00\tfrac{1}{2}], \quad m = [\tfrac{1}{2}\tfrac{1}{2}0] \in D_{4i}^z \\
&\varDelta = [00\zeta], \quad T = [\tfrac{1}{2}\tfrac{1}{2}\zeta] \in C_{4v}^z; \quad \varLambda = [\xi\xi\xi] \in C_{3v}^{xyz} \\
&\Sigma = [\xi\xi 0] \in C_{2v}^{xy}; \quad S = [\xi\tfrac{1}{2}\xi] \in C_{2v}^{xz}; \quad Z = [\xi\tfrac{1}{2}0] \in C_{2v}^{x}
\end{aligned} \tag{14.3.6b}$$

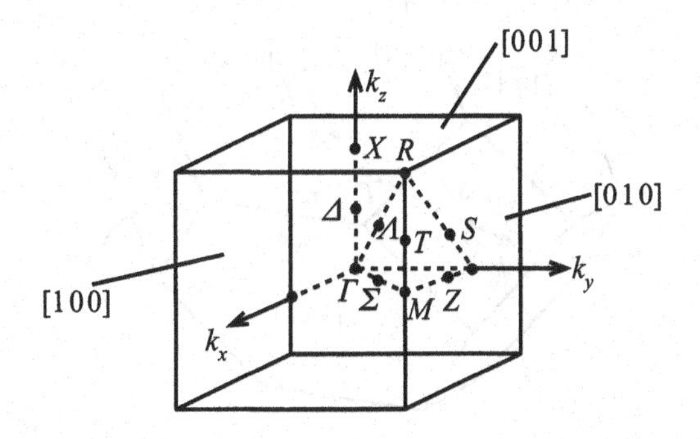

Figure 14.5. The Brillouin zone of the simple cubic lattice with the symmetry points. The Miller vector $\boldsymbol{g}_m$ perpendicular to each zone boundary is denoted by the Cartesian indices $[m_1^0, m_2^0, m_3^0]$.

These notations for the symmetry points are due to Bouchaert, Smoluchowski and Wigner (Bouchaert *et al.* 1936). Their symmetry point groups $K_{\boldsymbol{k}}$ given above may be defined by their generators as follows:

$$D_{4i}^z = \{4_z, 2_x, \bar{1}\}, \qquad C_{4v}^z = \{4_z, \bar{2}_x\}, \qquad C_{3v}^{xyz} = \{3_{xyz}, \bar{2}_{x\bar{y}}\}$$
$$C_{2v}^{xy} = \{2_{xy}, \bar{2}_z\}, \qquad C_{2v}^{xz} = \{2_{xz}, \bar{2}_y\}, \qquad C_{2v}^{x} = \{2_x, \bar{2}_z\} \qquad (14.3.6c)$$

The superscript for each $K_{\boldsymbol{k}}$ denotes the direction of the principal axis of the point group.

(ii) For the face-centered cubic lattice F_c, construction of the B-zone requires a set of 14 Miller vectors $\boldsymbol{g}_m = \boldsymbol{Bm}$ with the indices $\boldsymbol{m} = (m_1 m_2 m_3)$:

$$\pm(100), \pm(010), \pm(001), \pm(111), \pm(011), \pm(101), \pm(110) \qquad (14.3.7a)$$

Note that each Miller index is equal either to zero or to unity, as was mentioned before. The corresponding Cartesian indices $[m_1^0, m_2^0, m_3^0]$ of $\boldsymbol{g}_m = 2\pi \boldsymbol{m}^0$ are given, via (14.2.7b), by

$$\pm[\bar{1}11], \pm[1\bar{1}1], \pm[11\bar{1}], \pm[111]; \pm[200], \pm[020], \pm[002] \qquad (14.3.7b)$$

The 14 faces of the B-zone formed by these indices via (14.3.1) are presented in Figure 14.6. According to the matrix $\boldsymbol{B}$ given in (14.2.7a), the reciprocal lattice type is the body-centered cubic I_c with the primitive basis defined by the Cartesian indices $[\bar{1}11]$, $[1\bar{1}1]$ and $[11\bar{1}]$. Note that there exist two types of faces: six regular hexagonal faces and four square faces. The former are perpendicular to three-fold axes c_3 whereas the latter are perpendicular to four-fold axes c_4. From (14.3.5) their perpendicular distances from the origin O are $3^{1/2}/2$ and 1, respectively, in the unit of $2\pi a^{-1}$. The Cartesian coordinates $[\xi, \eta, \zeta]$ of the wave vectors $\boldsymbol{k}$ at the high symmetry points and their point symmetries $K_{\boldsymbol{k}}$ are

$$X = [001] \in D_{4i}^z; \qquad \Delta = [00\zeta] \in C_{4v}^z; \qquad L = [\tfrac{1}{2}\tfrac{1}{2}\tfrac{1}{2}] \in D_{3i}^{xyz}$$
$$W = [\tfrac{1}{2}01] \in D_{2p}^x; \qquad \Sigma = [\xi\xi 0] \in C_{2v}^{xy}; \qquad K = [\tfrac{3}{4}\tfrac{3}{4}0] \in C_{2v}^{xy} \qquad (14.3.7c)$$

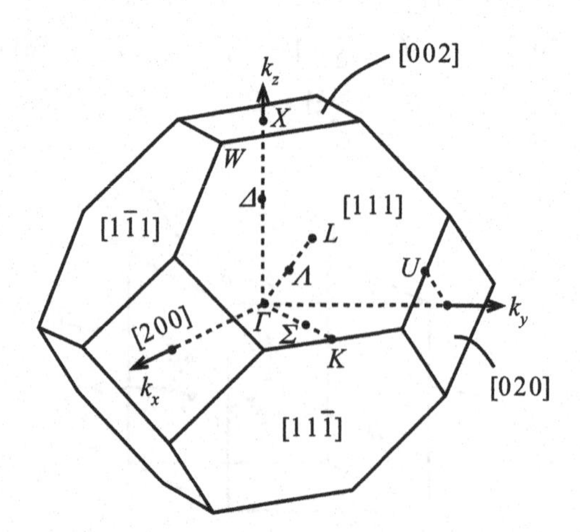

Figure 14.6. The Brillouin zone of the face-centered cubic lattice with the symmetry points. Each face is characterized by the Cartesian indices $[m_1^0, m_2^0, m_3^0]$ of $\boldsymbol{g}_m$.

where $|\zeta| < 1$. The generators of the point groups are

$$D_{3i}^{xyz} = \{3_{xyz}, 2_{x\bar{y}}, \bar{1}\}, \qquad D_{2p}^{x} = \{\bar{4}_x, 2_{yz}\} \tag{14.3.7d}$$

The remaining point groups in (14.3.7c) have been defined already in (14.3.6c). It should be noted that the symmetry point groups of the interior points of the B-zones are the same for all Bravais lattice types of a given crystal system K, because an interior point $\boldsymbol{k}$ is invariant with respect to the wave vector point group $K_{\boldsymbol{k}}$.

Remark. The wave vector point groups $K_{\boldsymbol{k}}$ given in (14.3.7c) are easily shown by applying the generators of $K_{\boldsymbol{k}}$ to the wave vector $\boldsymbol{k}$ using the Jones representations given in (13.8.1). Here are two examples.

1. $L \in D_{3i}^{xyz}$ is shown, using $3_{xyz} = (z, x, y)$ and $2_{x\bar{y}} = (\bar{y}, \bar{x}, \bar{z})$, by

$$L - 3_{xyz}L = [000], \qquad L - 2_{x\bar{y}}L = [111] \sim \boldsymbol{b}_1 + \boldsymbol{b}_2 + \boldsymbol{b}_3$$
$$L - \bar{1}L = [111] \sim \boldsymbol{b}_1 + \boldsymbol{b}_2 + \boldsymbol{b}_2 \tag{14.3.7e}$$

   where $\boldsymbol{B} = [\boldsymbol{b}_1, \boldsymbol{b}_2, \boldsymbol{b}_3]$ is the primitive basis for the reciprocal lattice F'_{c} given in (14.2.7a).

2. Analogously, $W \in D_{2p}^{x}$ is shown, using $\bar{4}_x = (\bar{x}, z, \bar{y})$, by

$$W - \bar{4}_x W = [1\bar{1}1] \sim \boldsymbol{b}_2, \qquad W - 2_{yz}W = [1\bar{1}1] \sim \boldsymbol{b}_2 \tag{14.3.7f}$$

(iii) For the body-centered cubic lattice I_{c}, construction of the B-zone requires the following 12 Miller vectors $\boldsymbol{g}_m = \boldsymbol{Bm} \in I'_{\mathrm{c}}$ with the indices $\boldsymbol{m} = (m_1, m_2, m_3)$:

$$\pm(100), \pm(010), \pm(001), \pm(0\bar{1}1), \pm(\bar{1}01), \pm(\bar{1}10) \tag{14.3.8a}$$

The corresponding Cartesian indices $[m_1^0, m_2^0, m_3^0]$ of $\boldsymbol{g}_m = 2\pi\boldsymbol{m}^0$ are given, from (14.2.8b), by

$$\pm[011], \pm[101], \pm[110], \pm[01\bar{1}], \pm[10\bar{1}], \pm[1\bar{1}0] \tag{14.3.8b}$$

The Brillouin zone formed via (14.3.1) is presented in Figure 14.7. According to

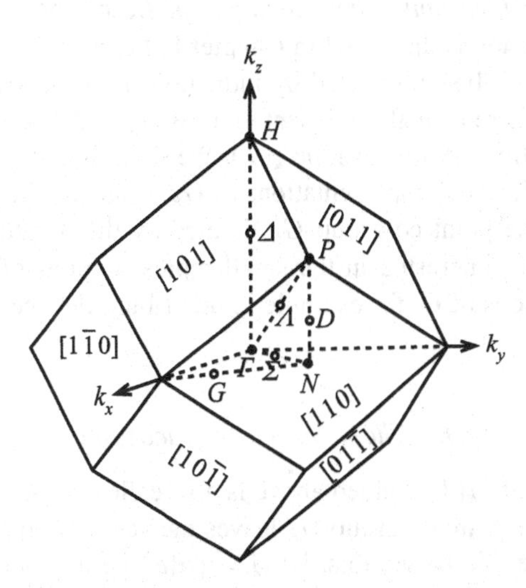

Figure 14.7. The Brillouin zone of the body-centered cubic lattice with the symmetry points. Each face is denoted by the Cartesian indices $[m_1^0, m_2^0, m_3^0]$ of $\boldsymbol{g}_m$.

(14.2.8a) the reciprocal lattice is F_c with the primitive basis defined by the Cartesian indices [011], [101] and [110]. There exists only one type of face that is rhombic. Each face is perpendicular to a two-fold axis of rotation and the perpendicular distance from the origin O equals $2\pi a^{-1}/\sqrt{2}$ via (14.3.5). The $[\xi, \eta, \zeta]$ coordinates of the wave vectors $\boldsymbol{k}$ at the symmetry points denoted in Figure 14.7 are given, together with their symmetry groups, by

$$H = [001] \in O_i; \qquad N = [\tfrac{1}{2}\tfrac{1}{2}0] \in D_{2i}^{xy}; \qquad P = [\tfrac{1}{2}\tfrac{1}{2}\tfrac{1}{2}] \in T_p^z;$$

$$D = [\tfrac{1}{2}\tfrac{1}{2}\zeta] \in C_{2v}^z \tag{14.3.8c}$$

where $D_{2i}^{xy} = \{2_{xy}, 2_z, \bar{1}\}$, $C_{2v}^z = \{2_z, \bar{2}_{x\bar{y}}\}$. The symmetry of $P \in T_p^z$ can be shown just like for $W \in D_{2p}^x$ in (14.3.7f).

The Brillouin zones for the cubic system described above are relatively simple due to the fact that only one lattice parameter is required in order to describe any lattice type belonging to the cubic system. As was shown in Table 13.1, one requires more than one lattice parameter in order to describe the lattices of the remaining crystal systems. As a result there may exist more than one type of Brillouin zone for a given Bravais lattice, depending on the relative magnitude of the lattice parameters. For example, for the tetragonal system, the Cartesian components of the Miller vector $\boldsymbol{g}_m$ for the body-centered lattice are given by

$$\boldsymbol{g}_m = 2\pi a^{-1}[m_2 + m_3,\ m_3 + m_1,\ (m_1 + m_2)a/c] \tag{14.3.9}$$

where a and c are the lattice parameters. Thus the perpendicular distance $|\boldsymbol{g}_m|/2$ depend on the ratio a/c. For the complete list of the Brillouin zones for all crystal systems see, e.g., Zak *et al.* (1969).

14.4 The small representations of wave vector space groups

Let $\hat{G}$ be a space group and T be its translation group and let $\tau_{\boldsymbol{k}}(T)$ be a unirrep of T that is one-dimensional. Let $\hat{G}_{\boldsymbol{k}}$ be *the little group of* $\tau_{\boldsymbol{k}}(T)$ *which is the subgroup of* $\hat{G}$ *that leaves* $\tau_{\boldsymbol{k}}(T)$ *invariant under conjugation.* According to the general theorems on induced representations discussed in Chapter 8, the complete set of the unirreps of the space group $\hat{G}$ can be constructed by induction from *the small representations of the little group* $\hat{G}_{\boldsymbol{k}}$: here a small representation of $\hat{G}_{\boldsymbol{k}}$ is defined by any unirrep of $\hat{G}_{\boldsymbol{k}}$ that subduces a multiple of the unirrep $\tau_{\boldsymbol{k}}(T)$. We shall first discuss the little group $G_{\boldsymbol{k}}$ and then show that the small representations of $\hat{G}_{\boldsymbol{k}}$ can be constructed by the projective representations of the point co-group $G_{\boldsymbol{k}}$ defined by the rotational part of $\hat{G}_{\boldsymbol{k}}$. More often than not we are interested in the small representations of $\hat{G}_{\boldsymbol{k}}$ in physics rather than the representations of $\hat{G}$; for example in describing the energy band structure of a solid.

14.4.1 *The wave vector space groups* $\hat{G}_{\boldsymbol{k}}$

The little group $\hat{G}_{\boldsymbol{k}}$ of $\tau_{\boldsymbol{k}}(T)$ defined above is also called *the group of the wave vector* $\boldsymbol{k}$ in the sense that its point co-group $G_{\boldsymbol{k}}$ leaves the vector $\boldsymbol{k}$ equivalent with respect to the reciprocal lattice T'. To see this, let $A = \{\alpha|\boldsymbol{a}\}$ be a general element of a space group $\hat{G}$, where $\boldsymbol{a}$ is a translation associated with the rotation α. Then the conjugate transform of $t_n = \{e|\boldsymbol{t}_n\} \in T$ by A is given by

$$A^{-1}t_nA = \{\alpha|\boldsymbol{a}\}^{-1}\{e|\boldsymbol{t}_n\}\{\alpha|\boldsymbol{a}\} = \{e|\alpha^{-1}\boldsymbol{t}_n\} \tag{14.4.1}$$

Note that only the rotational part α of A affects the translation. Thus, the conjugate representation of $\tau_{\boldsymbol{k}}(T)$ by A is given by

$$\tau_{\boldsymbol{k}}(A^{-1}t_nA) = \exp(-\mathrm{i}\alpha\boldsymbol{k}\cdot\boldsymbol{t}_n) = \tau_{\alpha\boldsymbol{k}}(t_n) \tag{14.4.2a}$$

Since by definition $\hat{G}_{\boldsymbol{k}} = \{\{\beta|\boldsymbol{b}\}\}$ is the subgroup of the space group $\hat{G}$ that leaves $\tau_{\boldsymbol{k}}(T)$ invariant under conjugation, we have

$$\tau_{\beta\boldsymbol{k}}(t_n) = \tau_{\boldsymbol{k}}(t_n), \qquad \forall\, t_n \in T \tag{14.4.2b}$$

This means that the point co-group $G_{\boldsymbol{k}} = \{\beta\}$ of the wave vector group $\hat{G}_{\boldsymbol{k}}$ leaves $\boldsymbol{k}$ equivalent with respect to the reciprocal lattice T' of T.

Obviously $G_{\boldsymbol{k}}$ is a subgroup of the point co-group (or the crystal class) G of the space group $\hat{G}$, and it is also a subgroup of the wave vector point group $K_{\boldsymbol{k}}$ of the crystal lattice discussed in the previous section. Moreover, if $G_{\boldsymbol{k}}$ is known then the corresponding wave vector space group $\hat{G}_{\boldsymbol{k}}$ is determined by simply adding the translation parts of the corresponding elements in the space group $\hat{G}$ because $\hat{G}_{\boldsymbol{k}}$ is a subgroup of the space group $\hat{G}$. For the same reason, the translation group T of $\hat{G}_{\boldsymbol{k}}$ is the same as that of the space group $\hat{G}$ and it is an Abelian invariant subgroup of $\hat{G}_{\boldsymbol{k}}$. There exists an isomorphism

$$\hat{G}_{\boldsymbol{k}}/T \simeq G_{\boldsymbol{k}} \tag{14.4.3}$$

corresponding to $\hat{G}/T \simeq G$. From this and also with the fact that the unirreps $\tau_{\boldsymbol{k}}(T)$ of T are one-dimensional we shall show that the small representations of $\hat{G}_{\boldsymbol{k}}$ can be constructed by the projective representations of the point co-group $G_{\boldsymbol{k}}$.

As a preparation, we first discuss the induced representations of $\hat{G}_{\boldsymbol{k}}$ from the unirreps $\{\tau_{\boldsymbol{k}}(T)\}$ of T. Let the left coset decomposition of $\hat{G}_{\boldsymbol{k}}$ by T be

$$\hat{G}_{\boldsymbol{k}} = \sum_{\beta_\lambda \in G_{\boldsymbol{k}}} \{\beta_\lambda|\boldsymbol{b}_\lambda\}T \equiv \sum B_\lambda T \tag{14.4.4}$$

where $B_\lambda = \{\beta_\lambda|\boldsymbol{b}_\lambda\}$; $\lambda = 1, 2, \ldots, |G_{\boldsymbol{k}}|$. Then the induced representation of $\hat{G}_{\boldsymbol{k}} = \{B = \{\beta|\boldsymbol{b}\}\}$ from $\tau_{\boldsymbol{k}}(T)$ is given, from the general definition (8.2.5), by

$$\Lambda_{\boldsymbol{k}}(B)_{\lambda\lambda'} = \tau_{\boldsymbol{k}}^{\uparrow}(B)_{\lambda\lambda'} = \sum_{t_{n'}\in T} \tau_{\boldsymbol{k}}(t_{n'})\delta(t_{n'}, B_\lambda^{-1}BB_{\lambda'})$$

$$\forall\, B \in \hat{G}_{\boldsymbol{k}}; \qquad \lambda, \lambda' = 1, 2, \ldots, |G_{\boldsymbol{k}}| \tag{14.4.5}$$

It subduces a multiple of $\tau_{\boldsymbol{k}}(T)$ onto T

$$\Lambda_{\boldsymbol{k}}(t_n)_{\lambda\lambda'} = \tau_{\boldsymbol{k}}(t_n)\delta_{\lambda\lambda'} \tag{14.4.6a}$$

Formally, it is expressed by

$$\Lambda_{\boldsymbol{k}}^{\downarrow}(T) = \Lambda_{\boldsymbol{k}}(T) = \mathbf{1}_{\boldsymbol{k}} \times \tau_{\boldsymbol{k}}(T) \equiv f_{\boldsymbol{k}}\tau_{\boldsymbol{k}}(T) \tag{14.4.6b}$$

which is a constant matrix for a given $\boldsymbol{k}$ because $\tau_{\boldsymbol{k}}(T)$ is one-dimensional. The dimensionality equals $f_{\boldsymbol{k}} = |\hat{G}_{\boldsymbol{k}}/T| = |G_{\boldsymbol{k}}|$. This is consistent with Corollary 8.5.2.

According to Theorem 8.5.3, any unirrep of $\hat{G}_{\boldsymbol{k}}$ contained in the induced representation $\Lambda_{\boldsymbol{k}}(\hat{G}_{\boldsymbol{k}}) = \tau_{\boldsymbol{k}}^{\uparrow}(\hat{G}_{\boldsymbol{k}})$ is a *small representation* and conversely any small representation of $\hat{G}_{\boldsymbol{k}}$ is always contained in the induced representation $\tau_{\boldsymbol{k}}^{\uparrow}(\hat{G}_{\boldsymbol{k}})$. Here, we give a direct verification of the theorem once more for the present simple special case as an illustration. Let $\Lambda_k^{(i)}(\hat{G}_k)$ be a unirrep of $\hat{G}_k$ contained in $\Lambda_k(\hat{G}_k)$ with a frequency F_k^i, i.e.

$$\tau_k^{\uparrow}(B) = \Lambda_k(B) \sim \sum_i \oplus F_k^i \Lambda_k^{(i)}(B); \qquad \forall B \in \hat{G}_k \tag{14.4.7a}$$

Since from (14.4.6b) the left-hand side of this equation subduces a multiple of $\tau_k(T)$, each submatrix $\Lambda_k^{(i)}$ on the right-hand side must also subduce some multiple of $\tau_k(T)$:

$$\Lambda_k^{(i)\downarrow}(T) \sim \mathbf{1}_k^i \times \tau_k(T) \equiv f_k^i \tau_k(T), \qquad f_k^i = |\Lambda_k^{(i)}| \tag{14.4.7b}$$

This means that the unirrep $\Lambda_k^{(i)}(\hat{G}_k)$ is indeed a small representation of $\hat{G}_k$. Now, from the Frobenius reciprocity theorem (Theorem 8.5.1), the two frequencies F_k^i and f_k^i are equal:

$$F_k^i = f_k^i \tag{14.4.8}$$

The theorem is proven since if $F_k^i \neq 0$ then $f_k^i \neq 0$ and vice versa, i.e. if there is a small representation that satisfies (14.4.7b) then it is contained in $\tau_k^{\uparrow}(\hat{G}_k)$. Q.E.D.

By equating the dimensions of the two sides of (14.4.7a) and using $F_k^i = |\Lambda_k^{(i)}|$ we obtain the completeness relations for the small representations of $\hat{G}_k$:

$$\sum_i |\Lambda_k^{(i)}(\hat{G}_k)|^2 = |\Lambda_k| = |\hat{G}_k|/|T| = |G_k| \tag{14.4.9}$$

which is a special case of (8.5.10a) because $|\gamma(H)| = 1$ for the present case.

14.4.2 Small representations of $\hat{G}_k$ via the projective representations of G_k

In general, it is not a simple matter to reduce the induced representation $\Lambda_k(\hat{G}_k)$ to obtain the small representations $\Lambda_k^{(i)}(\hat{G}_k)$ contained in it. For a wave vector group $\hat{G}_k$, however, there is a very effective way of constructing a small representation $\Lambda_k^{(i)}(\hat{G}_k)$ by regarding it as a projective unirrep of the point co-group G_k, using the fact that $\tau_k(T)$ is one-dimensional. To see this, let us write down the multiplication law for the coset representatives $\{\beta_\lambda|\boldsymbol{b}_\lambda\}$ of T in $\hat{G}_k$:

$$\{\beta_1|\boldsymbol{b}_1\}\{\beta_2|\boldsymbol{b}_2\} = \{e|\boldsymbol{t}(1,2)\}\{\beta_3|\boldsymbol{b}_3\} \tag{14.4.10}$$[2]

where $\beta_1\beta_2 = \beta_3$ and $\boldsymbol{t}(1, 2)$ is a lattice vector defined by

$$\boldsymbol{t}(1,2) = \beta_1\boldsymbol{b}_2 + \boldsymbol{b}_1 - \boldsymbol{b}_3 \in T$$

It is a lattice vector because a pure translation belonging to $\hat{G}_k$ must be a lattice translation. The representations of both sides of (14.4.10) via the small representation $\Lambda_k^{(i)}(\hat{G}_k)$ lead to

$$\Lambda_k^{(i)}(\{\beta_1|\boldsymbol{b}_1\})\Lambda_k^{(i)}(\{\beta_2|\boldsymbol{b}_2\}) = \mathrm{e}^{-\mathrm{i}\boldsymbol{k}\cdot\boldsymbol{t}(1,2)}\Lambda_k^{(i)}(\{\beta_3|\boldsymbol{b}_3\}) \tag{14.4.11}$$

where a numerical factor $\exp[-\mathrm{i}\boldsymbol{k}\cdot\boldsymbol{t}(1,2)]$ comes in because the small representation $\Lambda_k^{(i)}$ subduces a constant matrix $\tau_k(T)\mathbf{1}_k^i$, as shown by (14.4.7b). Now in view of the isomorphism $\hat{G}_k/T \simeq G_k$ we may set

[2] When we were constructing the space group in Chapter 13, we introduced a coset representative $(\beta_\lambda|\boldsymbol{b}_\lambda) = (\beta_\lambda|\boldsymbol{b}_\lambda)$ (mod $\boldsymbol{t}_n \in T$). This quantity is, however, not useful for the present purpose of constructing the irrep of $\hat{G}_k$.

$$\Delta_{\boldsymbol{k}}^{(i)}(\{\beta_\lambda|\boldsymbol{b}_\lambda\}) \equiv \breve{\Delta}_{\boldsymbol{k}}^{(i)}(\beta_\lambda), \qquad \forall \beta_\lambda \in G_{\boldsymbol{k}} \tag{14.4.12}$$

then Equation (14.4.11) takes the form

$$\breve{\Delta}_{\boldsymbol{k}}^{(i)}(\beta_1)\breve{\Delta}_{\boldsymbol{k}}^{(i)}(\beta_2) = \lambda_{\boldsymbol{k}}(\beta_1, \beta_2)\breve{\Delta}_{\boldsymbol{k}}^{(i)}(\beta_3) \tag{14.4.13a}$$

where $\lambda_{\boldsymbol{k}}(\beta_1, \beta_2)$ is a numerical factor defined by

$$\lambda_{\boldsymbol{k}}(\beta_1, \beta_2) = \exp[-\mathrm{i}\boldsymbol{k} \cdot \boldsymbol{t}(1, 2)], \qquad \boldsymbol{t}(1, 2) \in T \tag{14.4.13b}$$

This implies that a small representation $\{\Delta_{\boldsymbol{k}}^{(i)}(\{\beta_\lambda|b_\lambda\})\}$ of the coset representative of T in $\hat{G}_{\boldsymbol{k}}$ can be regarded as a projective representation $\{\breve{\Delta}_{\boldsymbol{k}}^{(i)}(\beta_\lambda)\}$ of the point co-group $G_{\boldsymbol{k}}$ belonging to the factor system defined by (14.4.13b). The projective representation $\breve{\Delta}_{\boldsymbol{k}}^{(i)}(G_{\boldsymbol{k}})$ defined by (14.4.12) is irreducible since the small representation $\Delta_{\boldsymbol{k}}^{(i)}(\hat{G}_{\boldsymbol{k}})$ is irreducible. It should be noted that the notation $\breve{\Delta}_{\boldsymbol{k}}^{(i)}(\beta_\lambda)$ can be deceiving: it still depends on $\boldsymbol{b}_\lambda$ (which is fixed for a given β_λ), as one can see from the left-hand side of (14.4.12); see also (14.4.17b) given below.

Remark. The above equations (14.4.11) and (14.4.13a) are analogous to (12.4.3) and (12.4.5b), respectively. However, the wave vector group $\hat{G}_{\boldsymbol{k}}$ is not a covering group of $G_{\boldsymbol{k}}$ simply because the translation group T is not in the center of $\hat{G}_{\boldsymbol{k}}$, as one can see from (14.4.1). The reason that the vector unirrep $\Delta_{\boldsymbol{k}}^{(i)}(\hat{G}_{\boldsymbol{k}})$ provides a projective representation $\breve{\Delta}_{\boldsymbol{k}}(G_{\boldsymbol{k}})$ are that T is an Abelian invariant subgroup of $\hat{G}_{\boldsymbol{k}}$ and that a small representation of $\hat{G}_{\boldsymbol{k}}$ subduces an integral multiple of a one-dimensional representation $\tau_{\boldsymbol{k}}(T)$ that provides the factor system (cf. Equation (12.4.3)).

Now substitution of (14.4.12) into the completeness relation (14.4.9) for the small representations of $\hat{G}_{\boldsymbol{k}}$ yields

$$\sum_i |\breve{\Delta}_{\boldsymbol{k}}^{(i)}(G_{\boldsymbol{k}})|^2 = |G_{\boldsymbol{k}}| \tag{14.4.14}$$

which is nothing but the completeness condition (12.3.7) for the projective unirreps of $G_{\boldsymbol{k}}$ belonging to a factor system. This means that construction of the small representations of $\hat{G}_{\boldsymbol{k}}$ requires the complete set of the projective unirreps of the point group $G_{\boldsymbol{k}}$ with the factor system defined by (14.4.13). The latter has already been constructed in Chapter 12 for all crystallographic point groups and presented in Table 12.4.

Once a projective unirrep $\breve{\Delta}_{\boldsymbol{k}}^{(i)}(G_{\boldsymbol{k}})$ has been determined, the corresponding small representations $\Delta_{\boldsymbol{k}}^{(i)}(\hat{G}_{\boldsymbol{k}})$ are given, via (14.4.12), by

$$\Delta_{\boldsymbol{k}}^{(i)}(\{\beta_\lambda|\boldsymbol{b}_\lambda + \boldsymbol{t}_n\}) = \mathrm{e}^{-\mathrm{i}\boldsymbol{k}\cdot\boldsymbol{t}_n}\breve{\Delta}_{\boldsymbol{k}}^{(i)}(\beta_\lambda); \qquad \forall \beta_\lambda \in G_{\boldsymbol{k}} \tag{14.4.15}$$

Moreover, if we assume that the representation is unitary, as usual, we have the following orthogonality relations:

$$\sum_{\beta_\lambda \in G_{\boldsymbol{k}}} \Delta_{\boldsymbol{k}}^{(i)}(\{\beta_\lambda|\boldsymbol{b}_\lambda\})_{rs}^* \Delta_{\boldsymbol{k}}^{(i')}(\{\beta_\lambda|\boldsymbol{b}_\lambda\})_{r's'} = \sum_{\beta_\lambda \in G_{\boldsymbol{k}}} \breve{\Delta}_{\boldsymbol{k}}^{(i)}(\beta_\lambda)_{rs}^* \breve{\Delta}_{\boldsymbol{k}}^{(i')}(\beta_\lambda)_{r's'}$$

$$= (|G_{\boldsymbol{k}}|/d_i)\delta_{ii'}\delta_{rr'}\delta_{ss'} \tag{14.4.16}$$

Here the summation over the lattice translations $\boldsymbol{t}_n \in T$ has been eliminated by the complex conjugation. Note also that the above orthogonality relations hold for the projective unirreps belonging to the same factor system (cf. Theorem 12.3.2).

In the important special case in which the wave vector $\boldsymbol{k}$ is invariant with respect to $G_{\boldsymbol{k}}$, i.e.

$$\beta_\lambda \boldsymbol{k} = \boldsymbol{k} \qquad \forall \beta_\lambda \in G_{\boldsymbol{k}} \tag{14.4.17a}$$

the factor system (14.4.13b) becomes a trivial factor system because

$$\lambda_{\boldsymbol{k}}(\beta_1, \beta_2) = \mathrm{e}^{-\mathrm{i}\boldsymbol{k}\cdot(\boldsymbol{b}_1+\boldsymbol{b}_2-\boldsymbol{b}_3)} = \mathrm{e}^{\mathrm{i}\boldsymbol{k}\cdot\boldsymbol{b}_3}/\mathrm{e}^{\mathrm{i}\boldsymbol{k}\cdot\boldsymbol{b}_1}\mathrm{e}^{\mathrm{i}\boldsymbol{k}\cdot\boldsymbol{b}_2}$$

so that, from (14.4.13a), *the projective unirrep $\breve{\Lambda}_{\boldsymbol{k}}^{(i)}(G_{\boldsymbol{k}})$ becomes p-equivalent to a vector unirrep $\dot{\Lambda}_{\boldsymbol{k}}^{(i)}(G_{\boldsymbol{k}})$ of $G_{\boldsymbol{k}}$* given by

$$\Lambda_{\boldsymbol{k}}^{(i)}(\{\beta_\lambda|\boldsymbol{b}_\lambda\}) = \breve{\Lambda}_{\boldsymbol{k}}^{(i)}(\beta_\lambda) = \mathrm{e}^{-\mathrm{i}\boldsymbol{k}\cdot\boldsymbol{b}_\lambda}\dot{\Lambda}_{\boldsymbol{k}}^{(i)}(\beta_\lambda); \qquad \forall \beta_\lambda \in G_{\boldsymbol{k}} \tag{14.4.17b}$$

The condition (14.4.17a) holds for every interior point of the Brillouin zone, but does not hold for the high-symmetry points on the surface in general, with some exceptions (see Section 15.4). Note also that (14.4.17b) holds for every symmorphic space group simply because $\boldsymbol{b}_\lambda = 0$ for all $\beta_\lambda \in G_{\boldsymbol{k}}$, even if $\boldsymbol{k}$ is not invariant with respect to $G_{\boldsymbol{k}}$.

Finally, we shall determine the number of the small representations of $\hat{G}_{\boldsymbol{k}}$ contained in the induced representation $\Lambda_{\boldsymbol{k}}(\hat{G}_{\boldsymbol{k}}) = \tau_{\boldsymbol{k}}^{\uparrow}(\hat{G}_{\boldsymbol{k}})$ from $\tau_{\boldsymbol{k}}(T)$. Let the corresponding projective representation $\breve{\Lambda}_{\boldsymbol{k}}(G_{\boldsymbol{k}})$ of $G_{\boldsymbol{k}}$ be, analogously to (14.4.12),

$$\Lambda_{\boldsymbol{k}}(\{\beta_\lambda|\boldsymbol{b}_\lambda\}) \equiv \breve{\Lambda}_{\boldsymbol{k}}(\beta_\lambda), \qquad \forall \beta_\lambda \in G_{\boldsymbol{k}} \tag{14.4.17c}$$

Since both $\breve{\Lambda}_{\boldsymbol{k}}^{i}(G_{\boldsymbol{k}})$ and $\breve{\Lambda}_{\boldsymbol{k}}(G_{\boldsymbol{k}})$ belong to the same factor system $\{\lambda_{\boldsymbol{k}}(\beta_1, \beta_2)\}$ defined by (14.4.13b), the number of times $n_{\boldsymbol{k}}^{i}$ that a unirrep $\breve{\Lambda}_{\boldsymbol{k}}^{i}(G_{\boldsymbol{k}})$ is contained in $\breve{\Lambda}_{\boldsymbol{k}}(G_{\boldsymbol{k}})$ is given, in terms of their characters, by

$$n_{\boldsymbol{k}}^{i} = \frac{1}{|G_{\boldsymbol{k}}|}\sum_{\beta_\lambda \in G_{\boldsymbol{k}}} \breve{\chi}_{\boldsymbol{k}}^{(i)}(\beta_\lambda)^{*}\breve{\chi}_{\boldsymbol{k}}(\beta_\lambda) \tag{14.4.17d}$$

In the special case in which the wave vector $\boldsymbol{k}$ is invariant with respect to $G_{\boldsymbol{k}}$ we have, analogously to (14.4.17b),

$$\Lambda_{\boldsymbol{k}}(\{\beta_\lambda|\boldsymbol{b}_\lambda\}) = \breve{\Lambda}_{\boldsymbol{k}}(\beta_\lambda) = \mathrm{e}^{-\mathrm{i}\boldsymbol{k}\cdot\boldsymbol{b}_\lambda}\dot{\Lambda}_{\boldsymbol{k}}(\beta_\lambda) \tag{14.4.17e}$$

where $\{\dot{\Lambda}_{\boldsymbol{k}}(\beta_\lambda)\}$ is a vector representation of the point group $G_{\boldsymbol{k}} = \{\beta_\lambda\}$. As a result the frequency $n_{\boldsymbol{k}}^{i}$ takes the form

$$n_{\boldsymbol{k}}^{i} = \frac{1}{|G_{\boldsymbol{k}}|}\sum_{\beta_\lambda} \dot{\chi}_{\boldsymbol{k}}^{(i)}(\beta_\lambda)^{*}\dot{\chi}_{\boldsymbol{k}}(\beta_\lambda) \tag{14.4.17f}$$

where $\dot{\chi}_{\boldsymbol{k}}^{(i)}(\beta_\lambda)$ and $\dot{\chi}_{\boldsymbol{k}}(\beta_\lambda)$ are the characters of the vector representations $\dot{\Lambda}_{\boldsymbol{k}}^{(i)}(\beta_\lambda)$ and $\dot{\Lambda}_{\boldsymbol{k}}(\beta_\lambda)$ of the point group $G_{\boldsymbol{k}}$. The former is well known for every point group but the latter is to be calculated from $\Lambda_{\boldsymbol{k}}(\{\beta_\lambda|\boldsymbol{b}_\lambda\})$ by (14.4.17e) as follows:

$$\dot{\chi}_{\boldsymbol{k}}(\beta_\lambda) = \mathrm{e}^{\mathrm{i}\boldsymbol{k}\cdot\boldsymbol{b}_\lambda}\,\mathrm{tr}\,\Lambda_{\boldsymbol{k}}(\{\beta_\lambda|\boldsymbol{b}_\lambda\}) \tag{14.4.17g}$$

which may still depend on $\boldsymbol{b}_\lambda$; e.g. see (15.2.2e).

14.4.3 Examples of the small representations of $\hat{G}_{\boldsymbol{k}}$

We shall discuss how to determine the small representations of a wave vector space group $\hat{G}_{\boldsymbol{k}}$ from the projective unirreps of the corresponding point group $G_{\boldsymbol{k}}$. It is necessary only to find the appropriate gauge transformation which makes the generators of $\hat{G}_{\boldsymbol{k}}$ satisfy the defining relations of the corresponding representation group given in (12.5.9). Let us illustrate the procedure through two typical examples of $\hat{G}_{\boldsymbol{k}}$ in the following; cf. Kim (1983a).

Example 1. O_i^2 (number 222) at $R(\boldsymbol{k} = (\pi, \pi, \pi))$ of the Brillouin zone defined in (14.3.6b); see Miller and Love (1967) and Zak *et al.* (1969). The wave vector space group $\hat{O}_i(\boldsymbol{k})$ has the following generator set, from Table 13.3:

$$P. \quad \{4_z|\tfrac{1}{2}00\}, \{3_{xyz}|000\}, \{\bar{1}|000\} \tag{14.4.18a}$$

From (12.5.9c), the representation group of the point group O_i is defined by

$$A^4 = B^3 = (AB)^2 = E', \qquad IAI = \beta A, \qquad IBI = B, \qquad I^2 = E'^2 = E \tag{14.4.18b}$$

where β $(= \pm 1)$ is the parameter which determines the p-equivalence class $K(\beta)$ of factor systems for the projective unirreps of the point group O_i. We find that the one-to-one correspondence

$$A \leftrightarrow \{4_z|\tfrac{1}{2}00\}, \qquad B \leftrightarrow \{3_{xyz}|000\}, \qquad I \leftrightarrow \{\bar{1}|000\}, \qquad E' \leftrightarrow \{e'|000\} \tag{14.4.18c}$$

satisfies the defining relations (14.4.18b) with $\beta = -1$. To see this, let us calculate IAI using the correspondence (14.4.18c) and obtain

$$IAI = \{4_z| -\tfrac{1}{2}00\} = \{e| -100\}A$$

then, from (14.4.13b), the representative of the lattice translation $\{e| - 100\} \in T$ determines β with $\boldsymbol{k} = (\pi, \pi, \pi)$ as follows:

$$\beta = \tau_{\boldsymbol{k}}(\{e| -100\}) = \exp(\pi \mathrm{i}) = -1$$

The projective unirreps $\check{A}_{\boldsymbol{k}}^{(i)}$ of O_i belonging to the class of the factor system $K(\beta = -1)$ are given, according to Table 12.4, by

$$K(-1); \qquad \check{D}(A_1, A_2; 1), \check{D}(E; \pm\sigma_y), \check{D}(T_1, T_2; 1), \check{D}(E_{\frac{1}{2}}, E'_{\frac{1}{2}}; 1), \check{D}(Q; \pm\sigma_0 \times \sigma_y)$$

The corresponding bases can also be written down in terms of the bases of the unirreps of the halving subgroup using (12.6.10b).

Example 2. $\hat{D}_{6i}^4$ (number 194) at $\boldsymbol{k} = (0, 0, \pi)$ of the Brillouin zone; see Miller and Love (1967). The generators of the wave vector group $\hat{D}_{6i}(\boldsymbol{k})$ are given, according to Table 13.3, by

$$P. \qquad \{6_z|00\tfrac{1}{2}\}, \{u_0|000\}, \{\bar{1}|000\} \tag{14.4.19a}$$

and the representation group of the point group D_{6i} is defined, from (12.5.9b), by

$$A^6 = B^2 = (AB)^2 = E', \qquad IA = \beta AI, \qquad IB = \gamma BI, \qquad I^2 = E'^2 = E \tag{14.4.19b}$$

where β $(= \pm 1)$ and γ $(= \pm 1)$ determine the p-equivalence class $K(\beta, \gamma)$ of factor systems for the projective unirreps of D_{6i}. We find that the one-to-one correspondences

$$A \leftrightarrow \{6_z|00\tfrac{1}{2}\}, \qquad B \leftrightarrow \mathrm{i}\{u_0|000\}, \qquad I \leftrightarrow \{\bar{1}|000\}, \qquad E' \leftrightarrow -\{e'|000\} \tag{14.4.19c}$$

satisfy the defining relations (14.3.19b) with $\beta = -1$ and $\gamma = 1$. The negative sign for the correspondence $E' \leftrightarrow -\{e'|000\}$ in (14.4.19c) means that the integral (or half-integral) representations of the space group $\hat{D}_{6i}$ are given by the half-integral (or integral) projective representations of the point group D_{6i}.

The projective unirreps of D_{6i} belonging to the class of factor systems $K(-1, 1)$ are given, according to Table 12.4, by

$$\check{D}(A, B_1; 1), \check{D}(A_2, B_2; 1), \check{D}(E_1, E_2; \sigma_z), \check{D}(E_{1/2}, E_{5/2}; \sigma_z), \check{D}(E_{3/2}; \pm\sigma_z) \tag{14.4.19d}$$

of which the first three describe the half-integral representations of the wave vector group $\hat{D}_{6i}(\boldsymbol{k})$ corresponding to $D(\{e'|000\}) = -1$ and $\check{D}(E') = 1$ whereas the remaining three representations are the integral representations of $\hat{D}_{6i}(\boldsymbol{k})$ corresponding to $D(\{e'|000\}) = 1$ and $\check{D}(E') = -1$. This kind of exchange between integral and half-integral representations occurs quite frequently (Miller and Love 1967, Zak *et al.* 1969).

14.5 The unirreps of the space groups

Following Theorem 8.5.4 on the induced representations, we shall construct the complete set of the unirreps of a space group $\hat{G}$ by induction from the small representations of the wave vector space groups $\hat{G}_{\boldsymbol{k}}$. Let the induced representation of $\hat{G}$ from a small representation $\Lambda_{\boldsymbol{k}}^{(i)}(G_{\boldsymbol{k}})$ be

$$\Gamma_{\boldsymbol{k}}^{(i)}(\hat{G}) \equiv \Lambda_{\boldsymbol{k}}^{(i)\uparrow}(\hat{G}) = [\Lambda_{\boldsymbol{k}}^{(i)}(\hat{G}_{\boldsymbol{k}})\uparrow\hat{G}]$$

and let the left coset decomposition of $\hat{G}$ by $\hat{G}_{\boldsymbol{k}}$ be

$$\hat{G} = \sum_{\nu=1}^{s} \{\alpha_\nu|\boldsymbol{a}_\nu\}\hat{G}_{\boldsymbol{k}} = \sum_{\nu=1}^{s} A_\nu\hat{G}_{\boldsymbol{k}} \tag{14.5.1}$$

where $A_\nu = \{\alpha_\nu|\boldsymbol{a}_\nu\}$ and $s = |\hat{G}|/|\hat{G}_{\boldsymbol{k}}| = |G|/|G_{\boldsymbol{k}}|$ is the index of $G_{\boldsymbol{k}}$ in G. The induced representation $\Gamma_{\boldsymbol{k}}^{(i)}(\hat{G})$ of $\hat{G} = \{A\}$ is given, from the general definition (8.2.5), by

$$\Gamma_{\boldsymbol{k}}^{(i)}(A)_{\nu\mu} = \Lambda_{\boldsymbol{k}}^{(i)\uparrow}(A)_{\nu\mu} = \sum_{B\in\hat{G}_{\boldsymbol{k}}} \Lambda_{\boldsymbol{k}}^{(i)}(B)\delta(B, A_\nu^{-1}AA_\mu)$$

$$\forall A \in \hat{G}; \qquad \nu, \mu = 1, 2, \ldots, s \tag{14.5.2}$$

where the sum is over all elements $B = \{\beta_\lambda|\boldsymbol{b}_\lambda + \boldsymbol{t}_n\}$ of $\hat{G}_{\boldsymbol{k}}$.

The representations of $\hat{G}$ induced from the small representations of $\hat{G}_{\boldsymbol{k}}$ given above are irreducible according to Theorem 8.5.4. Here we shall give a direct simple proof available for this case.

From Theorem 8.2.3 on the general irreducibility criterion for induced representations and Theorem 8.1.2 on subduced representations, the induced representation $\Lambda_{\boldsymbol{k}}^{(i)\uparrow}(\hat{G})$ is irreducible if the conjugate representations of the subduced representation $\Lambda_{\boldsymbol{k}}^{(i)\downarrow}(T)$ are orthogonal with respect to the coset representatives A_ν of $\hat{G}_{\boldsymbol{k}}$ in $\hat{G}$. From (14.4.7b) and (14.4.2a), the conjugate representations are given by

$$\Lambda_{\boldsymbol{k}}^{(i)}(A_\nu^{-1}t_nA_\nu) = \boldsymbol{1}_{\boldsymbol{k}}^i\tau_{\boldsymbol{k}_\nu}(t_n); \qquad \nu = 1, 2, \ldots, s = |G|/|G_{\boldsymbol{k}}| \tag{14.5.3}$$

where $\boldsymbol{k}_\nu = \alpha_\nu\boldsymbol{k}$ and $\boldsymbol{1}_{\boldsymbol{k}}^i$ is the unit matrix of the order $|\Lambda_{\boldsymbol{k}}^{(i)}|$. These are obviously orthogonal over $T = \{t_n\}$ for different values of ν from the orthogonality relations (14.1.15a) for $\{\tau_{\boldsymbol{k}}(T)\}$. Q.E.D.

To understand the structure of the unirrep $\Gamma_{\boldsymbol{k}}^{(i)}(\hat{G}) = \Lambda_{\boldsymbol{k}}^{(i)\uparrow}(\hat{G})$ given by (14.5.2) we shall calculate its subduced representation onto T in the following exercise.

Exercise. Show that the subduced representation of $\Gamma_{\boldsymbol{k}}^{(i)}(\hat{G})$ onto T is given by

$$\Gamma_{\boldsymbol{k}}^{(i)\downarrow}(T) = \Gamma_{\boldsymbol{k}}^{(i)}(T) = \mathbf{1}_{\boldsymbol{k}}^{i} \times \boldsymbol{O}_{\boldsymbol{k}}$$

where $|\mathbf{1}_{\boldsymbol{k}}^{i}| = |\Lambda_{\boldsymbol{k}}^{(i)}|$ and $\boldsymbol{O}_{\boldsymbol{k}}$ is the direct sum of the unirreps in the orbit of $\tau_{\boldsymbol{k}}(T)$ relative to $\hat{G}$. (This is a special case of the Clifford theorem, Theorem 8.5.5.)

Solution. From (14.5.2) and (14.5.3) we have

$$\begin{aligned}\Gamma_{\boldsymbol{k}}^{(i)}(t_n)_{\nu\mu} &= \Lambda_{\boldsymbol{k}}^{(i)}(A_\nu^{-1} t_n A_\nu)\delta_{\nu\mu} \\ &= \mathbf{1}_{\boldsymbol{k}}^{i} \times \exp(-\mathrm{i}\boldsymbol{k}_\nu \cdot \boldsymbol{t}_n)\,\delta_{\nu\mu}; \qquad \nu, \mu = 1, 2, \ldots, s = |G|/|G_{\boldsymbol{k}}|\end{aligned}$$

If we rewrite these in a matrix form, we obtain the required result:

$$\begin{aligned}\Gamma_{\boldsymbol{k}}^{(i)}(t_n) &= \mathbf{1}_{\boldsymbol{k}}^{i} \times \mathrm{diag}\,(\mathrm{e}^{-\mathrm{i}\boldsymbol{k}_1\cdot\boldsymbol{t}_n}, \mathrm{e}^{-\mathrm{i}\boldsymbol{k}_2\cdot\boldsymbol{t}_n}, \ldots, \mathrm{e}^{-\mathrm{i}\boldsymbol{k}_s\cdot\boldsymbol{t}_n}) \\ &= \mathbf{1}_{\boldsymbol{k}}^{i} \times \boldsymbol{O}_{\boldsymbol{k}}\end{aligned} \tag{14.5.4}$$

14.5.1 The irreducible star

To describe the completeness of the induced unirreps $\{\Gamma_{\boldsymbol{k}}^{(i)}(\hat{G})\}$ of a space group $\hat{G}$, one additional concept is useful. In view of (14.5.1), the left coset decomposition of the point co-group $G = \{\alpha\}$ by $G_{\boldsymbol{k}} = \{\beta\}$ is given by

$$G = \sum_{\nu=1}^{s} \alpha_\nu G_{\boldsymbol{k}}; \qquad s = |G|/|G_{\boldsymbol{k}}| \tag{14.5.5a}$$

Since $G_{\boldsymbol{k}}$ is the subgroup of G which leaves $\boldsymbol{k}$ equivalent with respect to the reciprocal lattice T', the set of wave vectors that are inequivalent with respect to T' is given by

$$S(\boldsymbol{k}) = \{\boldsymbol{k}_1, \boldsymbol{k}_2, \ldots, \boldsymbol{k}_s\}; \qquad |S(\boldsymbol{k})| = s = |G|/|G_{\boldsymbol{k}}| \tag{14.5.5b}$$

where $\boldsymbol{k}_\nu = \alpha_\nu \boldsymbol{k}$ with $\boldsymbol{k}_1 = \boldsymbol{k}$. The set is called *the irreducible star* of the wave vector $\boldsymbol{k}$. There exists a one-to-one correspondence between the star $S(\boldsymbol{k})$ and the orbit $\{\tau_{\boldsymbol{k}_\nu}(T)\}$ of an unirrep $\tau_{\boldsymbol{k}}(T)$ of T relative to $\hat{G}$. Moreover, the star $S(\boldsymbol{k})$ characterizes every unirrep $\Gamma_{\boldsymbol{k}}^{(i)}(\hat{G})$ of $\hat{G}$ induced by a unirrep $\Lambda_{\boldsymbol{k}}^{(i)}(\hat{G}_{\boldsymbol{k}})$ of $\hat{G}_{\boldsymbol{k}}$, as one can see from its subduced form $\Gamma_{\boldsymbol{k}}^{(i)}(T)$ given by (14.5.4).

An irreducible star $S(\boldsymbol{k})$ is invariant under the point co-group $G = \{\alpha\}$ because $\alpha\boldsymbol{k}_\nu = \alpha\alpha_\nu\boldsymbol{k} = \alpha_\mu\beta\boldsymbol{k} \sim \boldsymbol{k}_\mu \in S(\boldsymbol{k})$ for some $\alpha_\mu \in G$ and some $\beta \in G_{\boldsymbol{k}}$. The irreducible star $S(\boldsymbol{k})$ may also be regarded as the star of any member $\boldsymbol{k}_\nu$ of the star because from $\boldsymbol{k} = \alpha_\nu^{-1}\boldsymbol{k}_\nu$ follows $\boldsymbol{k}_\mu = \alpha_\mu\alpha_\nu^{-1}\boldsymbol{k}_\nu$ for every $\boldsymbol{k}_\mu$. Moreover, the point group $G_{\boldsymbol{k}_\nu}$ of $\boldsymbol{k}_\nu \in S(\boldsymbol{k})$ is isomorphic to $G_{\boldsymbol{k}}$ via the conjugation

$$G_{\boldsymbol{k}_\nu} = \alpha_\nu G_{\boldsymbol{k}} \alpha_\nu^{-1} \tag{14.5.6}$$

because if $\beta\boldsymbol{k} \sim \boldsymbol{k}$ then $\alpha_\nu\beta\alpha_\nu^{-1}\boldsymbol{k}_\nu = \alpha_\nu\beta\boldsymbol{k} \sim \boldsymbol{k}_\nu$ for all $\beta \in G_{\boldsymbol{k}}$. Hence one can speak of the little group of an irreducible star up to an isomorphism. Thus it is only necessary to construct one complete set of the small representations of the little group corresponding to one wave vector chosen arbitrarily from a given star. This conclusion is easily understood from the fact that all the conjugate representations of $\tau_k(T)$ by the coset representatives A_ν have already been incorporated in defining the induced representation $\Gamma_k^{(i)}(\hat{G}) = [\Lambda_k^{(i)}(\hat{G}_k)\uparrow\hat{G}]$, as one can see from (14.5.4).

Suppose that there exists a wave vector $\boldsymbol{k}'$ that is not contained in the star of $\boldsymbol{k}$, then one can construct another star $S(\boldsymbol{k}') = \{\boldsymbol{k}'_\mu\}$ of $\boldsymbol{k}'$ by the coset representatives $\{\alpha'_\mu\}$ of $G_{k'}$ in $G\,(= \sum_\mu \alpha'_\mu G_{k'})$.

In this way we can group all of the $\boldsymbol{k}$ in the Brillouin zone into irreducible stars that are disjoint with respect to G in the same way as we classify all the unirreps of T into disjoint orbits of T relative to $\hat{G}$. Let $|S(\boldsymbol{k})|$ be the order of a star $S(\boldsymbol{k})$ then

$$\sum_{(\boldsymbol{k})} |S(\boldsymbol{k})| = |T| \tag{14.5.7}$$

where the summation over $(\boldsymbol{k})$ means summation over one wave vector per irreducible star. This is so because each member of a star defines an inequivalent unirrep $\tau_{k_\nu}(T)$ and also the total number of the unirreps of T equals the order $|T|$ of T, as shown in (14.1.15b).

Finally, we shall give here a direct proof that the induced unirreps $\Gamma_k^{(i)}(\hat{G}) = [\Lambda_k^{(i)}(\hat{G}_k)\uparrow\hat{G}]$ of the small group $\hat{G}$ by the small representations of $\hat{G}_k$ are complete (cf. Theorem 8.5.4). From the general theorem on the completeness condition it is necessary only to show that the dimensionalities $|\Gamma_k^{(i)}(\hat{G})|$ satisfy

$$\sum_{(\boldsymbol{k})}\sum_{i} |\Gamma_k^{(i)}(\hat{G})|^2 = |\hat{G}| \tag{14.5.8}$$

where i is over all small representations of the wave vector group $\hat{G}_k$ and $(\boldsymbol{k})$ is over the wave vector $\boldsymbol{k}$, one per irreducible star $S(\boldsymbol{k})$. From (14.5.2) the dimensionality $|\Gamma_k^{(i)}(\hat{G})|$ is given by

$$|\Gamma_k^{(i)}(\hat{G})| = |\Lambda_k^{(i)}|\,|S(\boldsymbol{k})|, \qquad |S(\boldsymbol{k})| = |\hat{G}|/|\hat{G}_k| \tag{14.5.9}$$

On substituting this into the left-hand side of (14.5.8) and using the completeness condition (14.4.9) for the small representations $\Lambda_k^{(i)}(G_k)$, we have

$$\sum_{(\boldsymbol{k})i} |\Gamma_k^{(i)}(\hat{G})|^2 = \sum_{(\boldsymbol{k})} |S(\boldsymbol{k})|^2 |\hat{G}_k|/|T| = |\hat{G}| \sum_{(\boldsymbol{k})} |S(\boldsymbol{k})|/|T| = |\hat{G}|$$

where we have used $|\hat{G}_k|\,|S(\boldsymbol{k})| = |\hat{G}|$ and (14.5.7). This proves the completeness condition (14.5.8).

14.5.2 A summary of the induced representations of the space groups

It is worthwhile to summarize the method of constructing the unirreps $\Gamma_k^{(i)}(\hat{G}) = \Lambda_k^{(i)\uparrow}(\hat{G})$ of a space group $\hat{G}$ by induction from the small representations $\Lambda_k^{(i)}(\hat{G}_k)$ of the wave vector group $\hat{G}_k$ of a unirrep $\tau_k(T)$ of the translation group T.

1. Let $\tau_k(T)$ be a unirrep of T characterized by a wave vector $\boldsymbol{k}$.
2. Group the wave vectors in the Brillouin zone into different irreducible stars, and arbitrarily select one wave vector $\boldsymbol{k}$ from each star.

3. Determine the wave vector space group $\hat{G}_{\boldsymbol{k}}$ for each selected wave vector $\boldsymbol{k}$ from the point co-group $G_{\boldsymbol{k}}$ and the space group $\hat{G}$.
4. Find all the small representations $\Lambda_{\boldsymbol{k}}^{(i)}$ of the wave vector space group $\hat{G}_{\boldsymbol{k}}$ via the projective unirreps of the point group $G_{\boldsymbol{k}} \simeq \hat{G}_{\boldsymbol{k}}/T$.
5. Construct the unirreps of $\hat{G}$ by the induced representations $\Gamma_{\boldsymbol{k}}^{(i)}(\hat{G}) = [\Lambda_{\boldsymbol{k}}^{(i)}(\hat{G}_{\boldsymbol{k}}) \uparrow \hat{G}]$ from $\Lambda_{\boldsymbol{k}}^{(i)}(\hat{G}_{\boldsymbol{k}})$.
6. The set of all $\Gamma_{\boldsymbol{k}}^{(i)}(\hat{G})$ thus constructed provides the complete set of the unirreps of $\hat{G}$.

15

Applications of unirreps of space groups to energy bands and vibrational modes of crystals

In the previous chapter (Chapter 14), we have shown that a unirrep of a wave vector space group $\hat{G}_{\boldsymbol{k}}$ can be constructed by a projective unirrep of the point co-group $G_{\boldsymbol{k}}$ through Table 12.4. With this in mind, we shall describe the band energy eigenfunction of the electron in a solid and the symmetry coordinates of vibration of a crystal.

15.1 Energy bands and the eigenfunctions of an electron in a crystal

The energy spectrum of an electron in a crystal is described by the Schrödinger equation

$$H\psi = E\psi \tag{15.1.1}$$

with the Hamiltonian H defined by

$$H = -\frac{\hbar^2}{2m}\nabla^2 + V(\boldsymbol{r})$$

where $V(\boldsymbol{r})$ is a periodic potential that is invariant with respect to the crystal symmetry described by the space group $\hat{G}$ of the crystal. Accordingly, the energy eigenstates can be classified by the unirreps of the space group $\hat{G}$. Let $T = \{\{e|\boldsymbol{t}_n\}\}$ be the translational symmetry of the crystal. Since T is Abelian and its element commutes with the Hamiltonian H, there exist the simultaneous eigenfunctions for every element of T and H. These will be classified first by the small representations of the wave vector space group $\hat{G}_{\boldsymbol{k}}$. Then the overall symmetry of the energy eigenstates is described by the unirreps of $\hat{G}$ induced by the small representations of the wave vector group $\hat{G}_{\boldsymbol{k}}$ as described in Section 14.5.

Let $\psi^{\boldsymbol{k}}(\boldsymbol{r})$ be a simultaneous eigenfunction of $\{e|\boldsymbol{t}_n\} \in T$ and the Hamiltonian H such that

$$\{e|\boldsymbol{t}_n\}\psi^{\boldsymbol{k}}(\boldsymbol{r}) = \mathrm{e}^{-\mathrm{i}\boldsymbol{k}\cdot\boldsymbol{t}_n}\psi^{\boldsymbol{k}}(\boldsymbol{r})$$

$$H\psi^{\boldsymbol{k}}(\boldsymbol{r}) = E(\boldsymbol{k})\psi^{\boldsymbol{k}}(\boldsymbol{r}) \tag{15.1.2a}$$

where $\boldsymbol{k}$ is a wave vector in the first Brillouin zone. The eigenfunction $\psi^{\boldsymbol{k}}(\boldsymbol{r})$ is known as the Bloch function belonging to the wave vector $\boldsymbol{k}$. Since the set of eigenvalues $\{\exp(-\mathrm{i}\boldsymbol{k}\cdot\boldsymbol{t}_n)\}$ defines a unirrep $\tau_{\boldsymbol{k}}(T)$, the Bloch function is also the eigenfunction simultaneously belonging to the unirrep $\tau_{\boldsymbol{k}}(T)$ and the energy $E(\boldsymbol{k})$. The energy $E(\boldsymbol{k})$ as a function of a 'continuous' variable $\boldsymbol{k}$ describes a *band structure of the energy* and there exist many such bands for a Hamiltonian with a periodic potential $V(\boldsymbol{r})$. The wave function $\psi^{\boldsymbol{k}}(\boldsymbol{r})$ may also depend on many discrete indices of the quantum numbers due to further symmetry of the Hamiltonian. We, however, assume that we are on one particular energy band so that the energy eigenvalues depend only on the wave vector $\boldsymbol{k}$.

Remark. A wave function belonging to the unirrep $\tau_{\boldsymbol{k}}(T)$ must have the following form, in view of (14.1.4a):

$$\psi^{\boldsymbol{k}}(\boldsymbol{r}) = \mathrm{e}^{\mathrm{i}\boldsymbol{k}\cdot\boldsymbol{r}} u_{\boldsymbol{k}}(\boldsymbol{r}) \tag{15.1.2b}$$

where $u_{\boldsymbol{k}}(\boldsymbol{r})$ is periodic with respect to the lattice translations $\{e|\boldsymbol{t}_n\} \in T$, i.e. $u_{\boldsymbol{k}}(\boldsymbol{r}) = u_{\boldsymbol{k}}(\boldsymbol{r}+\boldsymbol{t}_n)$. On substituting the above expression into the Schrödinger equation we obtain the differential equation for $u_{\boldsymbol{k}}(\boldsymbol{r})$

$$\left(-\frac{\hbar^2}{2m}(\boldsymbol{\nabla}+\mathrm{i}\boldsymbol{k})^2 + V(\boldsymbol{r})\right) u_{\boldsymbol{k}}(\boldsymbol{r}) = E(\boldsymbol{k}) u_{\boldsymbol{k}}(\boldsymbol{r}) \tag{15.1.2c}$$

which needs to be solved for one unit cell of the crystal with the periodic boundary condition. Since $\boldsymbol{k}$ is a coefficient in the differential equation, the eigenvalue $E(\boldsymbol{k})$ must be a continuous function of $\boldsymbol{k}$. Note that the label $\boldsymbol{k}$ of $E(\boldsymbol{k})$ is determined by the label $\boldsymbol{k}$ of the unirrep $\tau_{\boldsymbol{k}}(T)$; cf. Peierls (1955). Hereafter, however, we shall find it more convenient to discuss the symmetry properties of the energy $E(\boldsymbol{k})$ as well as its wave function $\psi^{\boldsymbol{k}}$ in terms of (15.1.2a) instead of (15.1.2c).

Let us first discuss the transformation of the basic equations (15.1.2a) under a general element $A \equiv \{\alpha|\boldsymbol{a}\}$ of the space group $\hat{G}$ of the crystal. We have

$$\begin{aligned} \{e|\boldsymbol{t}_n\}^{\circ}[\mathring{A}\psi^{\boldsymbol{k}}] &= \mathrm{e}^{\mathrm{i}\alpha\boldsymbol{k}\cdot\boldsymbol{t}_n}[\mathring{A}\psi^{\boldsymbol{k}}] \\ H[\mathring{A}\psi^{\boldsymbol{k}}] &= E(\boldsymbol{k})[\mathring{A}\psi^{\boldsymbol{k}}] \end{aligned} \tag{15.1.3}$$

which are obtained by applying the following commutation relations (cf. 14.4.1) to $\psi^{\boldsymbol{k}}$:

$$\begin{aligned} \{e|\boldsymbol{t}_n\}\{\alpha|\boldsymbol{a}\} &= \{\alpha|\boldsymbol{a}\}\{e|\alpha^{-1}\boldsymbol{t}_n\} \\ H\{\alpha|\boldsymbol{a}\} &= \{\alpha|\boldsymbol{a}\}H \end{aligned} \tag{15.1.4}$$

Since the eigenvector $\mathring{A}\psi^{\boldsymbol{k}}$ belongs to the wave vector $\alpha\boldsymbol{k}$, the energy eigenvalue as a function of the wave vector $\boldsymbol{k}$ must satisfy the following symmetry relation, in comparison of (15.1.3) with (15.1.2a):

$$E(\alpha\boldsymbol{k}) = E(\boldsymbol{k}), \qquad \forall \alpha \in G \tag{15.1.5a}$$

where G is the point co-group of $\hat{G}$. This equation means that the band energy $E(\boldsymbol{k})$ as a function of the wave vector $\boldsymbol{k}$ possesses the full point symmetry G of the crystal itself. If there is no other symmetry for the Hamiltonian H besides the space group symmetry $\hat{G}$, the whole set of the degenerate eigenfunctions belonging to $E(\boldsymbol{k})$ is given, from an eigenfunction $\psi^{\boldsymbol{k}}$, by,

$$\{\{\alpha|\boldsymbol{a}\}^{\circ}\psi^{\boldsymbol{k}};\ \forall\{\alpha|\boldsymbol{a}\} \in \hat{G}/T\} \tag{15.1.5b}$$

Here we have excluded the lattice translations $\{e|\boldsymbol{t}_n\} \in T$ from the operations since they simply change the phase factors of the eigenfunctions, as one can see from (15.1.3).

We shall classify these eigenfunctions (15.1.5b) belonging to $E(\boldsymbol{k})$ first by the small representations of the wave vector space group $\hat{G}_{\boldsymbol{k}}$ that leaves the unirrep $\tau_{\boldsymbol{k}}(T)$ invariant and then by those unirreps of the space group $\hat{G}$ which are induced by the small representations of $\hat{G}_{\boldsymbol{k}}$. For this purpose, we introduce the left coset decomposition of $\hat{G}$ relative to T in two steps, in view of $T \triangleleft \hat{G}_{\boldsymbol{k}} < \hat{G}$,

$$\hat{G}_{\boldsymbol{k}} = \sum_{\lambda=1}^{t} \{\beta_\lambda | \boldsymbol{b}_\lambda\} T, \qquad t = |\hat{G}_{\boldsymbol{k}}|/|T| \tag{15.1.6a}$$

$$\hat{G} = \sum_{\nu=1}^{s} \{\alpha_\nu | \boldsymbol{a}_\nu\} \hat{G}_{\boldsymbol{k}}, \qquad s = |S(\boldsymbol{k})| = |\hat{G}|/|\hat{G}_{\boldsymbol{k}}| \tag{15.1.6b}$$

as in (14.4.4) and (14.5.1), respectively. Then the set of the Bloch functions belonging to the wave vector $\boldsymbol{k}$ is given by a set of the equivalent Bloch functions with respect to $\hat{G}_{\boldsymbol{k}}$

$$\{\{\beta_\lambda | \boldsymbol{b}_\lambda\}^\circ \psi^{\boldsymbol{k}};\ \forall\, \{\beta_\lambda | \boldsymbol{b}_\lambda\} \in \hat{G}_{\boldsymbol{k}}/T\} \tag{15.1.7a}$$

which exhausts all the solutions of the Schrödinger equation (15.1.2a) for a given wave vector $\boldsymbol{k}$. Let d be the number of linearly independent functions contained in the set, then

$$d \leqslant |\hat{G}_{\boldsymbol{k}}|/|T| = |G_{\boldsymbol{k}}| \tag{15.1.7b}$$

where $G_{\boldsymbol{k}}$ is the point co-group of $\hat{G}_{\boldsymbol{k}}$. The set provides a basis for a d-dimensional representation $\Delta_{\boldsymbol{k}}^{(d)}(\hat{G}_{\boldsymbol{k}})$ of the wave vector space group $\hat{G}_{\boldsymbol{k}}$. In the special case in which $d = |G_{\boldsymbol{k}}|$, the representation $\Delta_{\boldsymbol{k}}^{(d)}$ becomes equivalent to the induced representation $\Lambda_{\boldsymbol{k}}(\hat{G}_{\boldsymbol{k}})$ from $\tau_{\boldsymbol{k}}(T)$ introduced in (14.4.5).

Let $\Lambda_{\boldsymbol{k}}^{(i)}(\hat{G}_{\boldsymbol{k}})$ be a unirrep contained in $\Delta_{\boldsymbol{k}}^{(d)}(\hat{G}_{\boldsymbol{k}})$, then it is a small representation of $\hat{G}_{\boldsymbol{k}}$ because the representation $\Delta_{\boldsymbol{k}}^{(d)}(\hat{G}_{\boldsymbol{k}})$ subduces an integral multiple of $\tau_{\boldsymbol{k}}(T)$ onto T in view of (15.1.3). Then, using the isomorphism $\hat{G}_{\boldsymbol{k}}/T \simeq G_{\boldsymbol{k}}$, we introduce their projective representations $\check{\Lambda}_{\boldsymbol{k}}^{(i)}(G_{\boldsymbol{k}})$ and $\check{\Delta}_{\boldsymbol{k}}^{(d)}(G_{\boldsymbol{k}})$ of the point group $G_{\boldsymbol{k}}$ by

$$\check{\Lambda}_{\boldsymbol{k}}^{(i)}(\beta_\lambda) \equiv \Lambda_{\boldsymbol{k}}^{(i)}(\{\beta_\lambda | \boldsymbol{b}_\lambda\}), \qquad \check{\Delta}_{\boldsymbol{k}}^{(d)}(\beta_\lambda) \equiv \Delta_{\boldsymbol{k}}^{(d)}(\{\beta_\lambda | \boldsymbol{b}_\lambda\}) \tag{15.1.8}$$

both of which belong to the same factor system (14.4.13b), analogously to (14.4.12) and (14.4.17c), respectively. In Section 14.4.3, we have shown how to determine the projective unirrep $\check{\Lambda}_{\boldsymbol{k}}^{(i)}(G_{\boldsymbol{k}})$ by the unirreps of the representation group $G'_{\boldsymbol{k}}$. The number of times that a small representation $\Lambda_{\boldsymbol{k}}^{(i)}$ of $\hat{G}_{\boldsymbol{k}}$ is contained in $\Delta_{\boldsymbol{k}}^{(d)}$ is determined by the frequency rule (14.4.17d) or (14.4.17f).

Next, we shall discuss how to form a basis of the projective unirrep $\check{\Lambda}_{\boldsymbol{k}}^{(i)}(G_{\boldsymbol{k}})$ (or the *band eigenfunction* in short) by the symmetry-adapted linear combinations (SALCs) of the equivalent Bloch functions given in (15.1.7a). Since $\check{\Lambda}_{\boldsymbol{k}}^{(i)}(G_{\boldsymbol{k}})$ is defined by the unirrep $\Lambda_{\boldsymbol{k}}^{(i)}(\hat{G}_{\boldsymbol{k}})$, the required SALCs are formed by (6.9.3), using the generating operator method, as follows:

$$\psi_j^{\boldsymbol{k},i} = (|\check{\Lambda}_{\boldsymbol{k}}^{(i)}|/|G_{\boldsymbol{k}}|) \sum_{\beta_\lambda \in G_{\boldsymbol{k}}} \check{\Lambda}_{\boldsymbol{k}}^{(i)}(\beta_\lambda)_{jp}^{*} \{\beta_\lambda | \boldsymbol{b}_\lambda\}^\circ \psi^{\boldsymbol{k}}(\boldsymbol{r}); \qquad j = 1, 2, \ldots, |\check{\Lambda}_{\boldsymbol{k}}^{(i)}| \tag{15.1.9a}$$

where $|\check{\Lambda}_{\boldsymbol{k}}^{(i)}|$ is the dimension of the representation $\check{\Lambda}_{\boldsymbol{k}}^{(i)}$, $\psi^{\boldsymbol{k}}$ is an appropriate Bloch function in (15.1.7a) and the column number p of the unirrep $\check{\Lambda}_{\boldsymbol{k}}^{(i)}(\beta_\lambda)_{jp}$ should be chosen appropriately to obtain non-null bases for the unirreps contained in $\check{\Delta}_{\boldsymbol{k}}^{(d)}(G_{\boldsymbol{k}})$. Note here that the phase factors due to the lattice translations $\{e|\boldsymbol{t}_n\} \in T$ are cancelled out in the complex conjugation in (15.1.9a). Obviously, if $\beta_\lambda \boldsymbol{k} = \boldsymbol{k}$ for all $\beta_\lambda \in G_{\boldsymbol{k}}$,

Equation (15.1.9a) is simplified with use of (14.4.17b), i.e. $\breve{\Lambda}_k^{(i)}(\beta_\lambda) = \dot{\Lambda}^{(i)}(\beta_\lambda)\mathrm{e}^{-\mathrm{i}\boldsymbol{k}\cdot\boldsymbol{b}_\lambda}$ where $\{\dot{\Lambda}^{(i)}(\beta_\lambda)\}$ is a vector unirrep of G_k. Moreover, for the special case of the identity representation, for which $\dot{\Lambda}^{(1)}(\beta_\lambda) = 1$ for all $\beta_\lambda \in G_k$, the basis SALC is given by

$$\psi^{k,1} = (1/|G_k|) \sum_{\beta_\lambda \in G_k} \mathrm{e}^{\mathrm{i}\boldsymbol{k}\cdot\boldsymbol{b}_\lambda} \{\beta_\lambda|\boldsymbol{b}_\lambda\}^0 \psi^k(\boldsymbol{r}) \qquad (15.1.9b)$$

which is nothing other than the weighted sum of the equivalent functions given in (15.1.7a). In the actual calculation of the basis SALCs in the free-electron approximation introduced in the next section, we shall use the correspondence theorem introduced in Chapter 7 together with the invariant basis given above.

Finally, let $\{\Gamma_k^{(i)}(\hat{G})\}$ be the induced unirreps of the space group $\hat{G}$ from the small representation $\{\Lambda_k^{(i)}(\hat{G}_k)\}$ defined by

$$\Gamma_k^{(i)}(\hat{G}) = [\Lambda_k^{(i)}(\hat{G}_k) \uparrow \hat{G}] \qquad (15.1.10a)$$

Then its basis is given, in view of (15.1.6b), by

$$\psi_j^{k_\nu,i} \equiv \{\alpha_\nu|\boldsymbol{a}_\nu\}\psi_j^{k,i}; \qquad \nu = 1, 2, \ldots, |S(\boldsymbol{k})| \qquad (15.1.10b)$$

which still belongs to $E(\boldsymbol{k})$ from (15.1.3). The degenerate eigenfunctions given by (15.1.5b) are then described by these induced bases. If there exists no accidental degeneracy, i.e. $\Delta_k^{(d)}$ contains only one small representation of $\hat{G}_k$ (this is true in general for a realistic periodic potential $V(\boldsymbol{r})$), then the total degeneracy of $E(\boldsymbol{k})$ is given by the dimensionality of the unirrep $\Gamma_k^{(i)}(\hat{G})$, which is determined by two factors, in view of (15.1.10b):

$$|\Gamma_k^{(i)}(\hat{G})| = |\Lambda_k^{(i)}| \cdot |S(\boldsymbol{k})| \qquad (15.1.11)$$

The first factor arises from the band degeneracy at the point $\boldsymbol{k}$ due to the symmetry with respect to the wave vector group $\hat{G}_k$ given by (15.1.9a), whereas the second factor is due to the equivalence under $\{\alpha_\nu|\boldsymbol{a}_\nu\} \in \hat{G}$ of the wave vectors $\boldsymbol{k}$ corresponding to different points in the star of the wave vector $\boldsymbol{k}$. Equation (15.1.11) is identical to (14.5.9) given in the previous section, as it should be.

15.2 Energy bands and the eigenfunctions for the free-electron model in a crystal

In the previous section, we have classified the energy band eigenfunctions of an electron in a crystal in terms of the small representations of the wave vector space groups. To exemplify the classification, we shall determine the possible energy bands and their eigenfunctions given by the symmetry-adapted wave functions in the free-electron approximation. 'This is far from a trivial calculation,' as stated by Jones (1975): it gives deep insight into the possible band structure of an actual crystal belonging to the same space group. There may occur accidental degeneracies in the free-electron approximation, which, however, may be removed by introducing a realistic periodic potential of the crystal as a perturbation. Here the symmetry-adapted free-electron wave functions provide the correct zeroth-order linear combinations for the perturbation. We shall see that the correspondence theorem on the basis functions introduced in Chapter 7 greatly simplifies the construction of the symmetry-adapted Bloch functions.

In the free-electron approximation, the Hamiltonian of the system is given by

$$H^{(0)} = -[\hbar^2/(2m)]\nabla^2$$

so that a simultaneous eigenfunction of the translation group T and the Hamiltonian $H^{(0)}$ is described by an elementary Bloch function belonging to a wave vector, say $\boldsymbol{k}_1$, in the extended zone scheme and the energy E such that

$$\psi(\boldsymbol{r}) = \mathrm{e}^{\mathrm{i}\boldsymbol{k}_1\cdot\boldsymbol{r}} \in E = [\hbar^2/(2m)]|\boldsymbol{k}_1|^2$$

In general, the wave vector $\boldsymbol{k}_1$ need not be in the first Brillouin zone, but is related to a wave vector $\boldsymbol{k}$ in the first Brillouin zone by

$$\boldsymbol{k}_1 = \boldsymbol{k} - \boldsymbol{g}_1$$

where $\boldsymbol{g}_1$ is a reciprocal lattice vector. Thus the energy and its eigenfunction are expressed as

$$\psi_1^{\boldsymbol{k}}(\boldsymbol{r}) = \mathrm{e}^{\mathrm{i}\boldsymbol{k}_1\cdot\boldsymbol{r}} = \mathrm{e}^{\mathrm{i}(\boldsymbol{k}-\boldsymbol{g}_1)\cdot\boldsymbol{r}} \in E(\boldsymbol{k}) = [\hbar^2/(2m)](\boldsymbol{k} - \boldsymbol{g}_1)^2 \tag{15.2.1a}$$

Here the factor $\exp(-\mathrm{i}\boldsymbol{g}_1 \cdot \boldsymbol{r})$ corresponds to $u_{\boldsymbol{k}}(\boldsymbol{r})$ in (15.1.2b), being periodic with respect to the lattice translation $\{e|\boldsymbol{t}_n\} \in T$.

Under a symmetry operation $\{\beta_\lambda|\boldsymbol{b}_\lambda\} \in \hat{G}_{\boldsymbol{k}}$, the free-electron wave function transforms according to

$$\{\beta_\lambda|\boldsymbol{b}_\lambda\}^0\psi_1^{\boldsymbol{k}}(\boldsymbol{r}) = \exp[\mathrm{i}\beta_\lambda\boldsymbol{k}_1 \cdot (\boldsymbol{r} - \boldsymbol{b}_\lambda)] \tag{15.2.1b}$$

Since the translational part $\boldsymbol{b}_\lambda$ introduces only a phase factor, the linearly independent set of Bloch functions equivalent with respect to $\hat{G}_{\boldsymbol{k}}$ is expressed by

$$\check{\Delta}_{\boldsymbol{k}}^{(d)}(G_{\boldsymbol{k}}): \qquad \{\psi_l^{\boldsymbol{k}}(\boldsymbol{r}) = \exp(\mathrm{i}\boldsymbol{k}_l \cdot \boldsymbol{r});\ l = 1, 2, \ldots, d\} \in E(\boldsymbol{k}) \tag{15.2.1c}$$

where $\boldsymbol{k}_l = \beta_\lambda\boldsymbol{k}_1 = \boldsymbol{k} - \boldsymbol{g}_l$ with some reciprocal lattice vector $\boldsymbol{g}_l$. Here $\{\boldsymbol{k}_l\}$ is a linearly independent equivalent set of wave vectors with respect to the point group $G_{\boldsymbol{k}}$ whereas the set of reciprocal lattice vectors $\{\boldsymbol{g}_l\}$ is linearly independent but need not be given by $\{\beta_\lambda\boldsymbol{g}_1\}$ unless $\boldsymbol{k}$ is invariant under $G_{\boldsymbol{k}}$. Obviously, the set of Bloch functions given by (15.2.1c) belongs to the same common energy $E(\boldsymbol{k})$, which may be rewritten as

$$E(\boldsymbol{k}) = [\hbar^2/(2m)](\boldsymbol{k} - \boldsymbol{g}_l)^2; \qquad l = 1, 2, \ldots, d \tag{15.2.1d}$$

This expression defines the band energy of an electron as a quadratic function of $\boldsymbol{k}$; e.g. see Figure 15.1. Conversely, from this expression, we can determine the allowed energies $E(\boldsymbol{k})$ for a given $\boldsymbol{k}$ with a compatible set of linearly independent reciprocal lattice vectors $\{\boldsymbol{g}_l\}$ for each band, which then lead to the linearly independent set of the equivalent wave functions (15.2.1c), with due caution for the rare case of band energy crossing, see Figure 15.1. This gives a convenient way of finding the linear independent set (15.2.1c) belonging to an allowed energy $E(\boldsymbol{k})$.

The equivalent basis set (15.2.1c) provides a $d \times d$ matrix representation $\Delta_{\boldsymbol{k}}^{(d)}(\hat{G}_{\boldsymbol{k}})$ that also defines the projective representation $\check{\Delta}_{\boldsymbol{k}}^{(d)}(G_{\boldsymbol{k}})$ in accordance with (15.1.8); in fact, from

$$\{\beta_\lambda|\boldsymbol{b}_\lambda\}^\circ\psi_l^{\boldsymbol{k}}(\boldsymbol{r}) = \exp[\mathrm{i}\beta_\lambda\boldsymbol{k}_l \cdot (\boldsymbol{r} - \boldsymbol{b}_\lambda)] = \sum_{l'} \psi_{l'}^{\boldsymbol{k}}(\boldsymbol{r})\Delta_{\boldsymbol{k}}^{(d)}(\beta_\lambda|\boldsymbol{b}_\lambda)_{l'l}$$

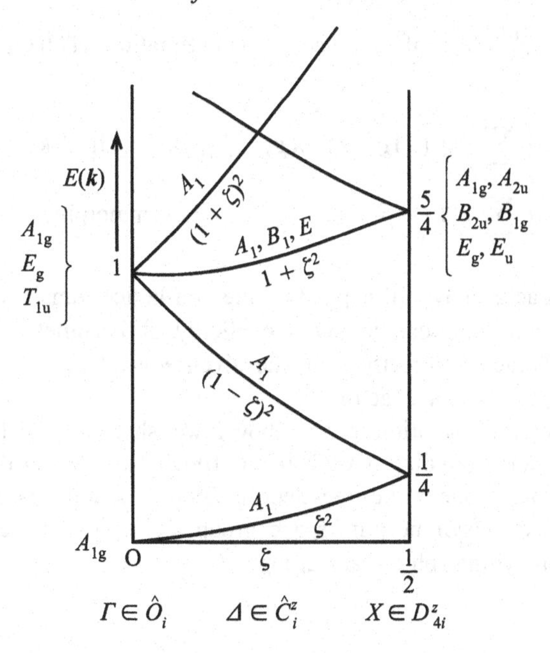

Figure 15.1. Free-electron energy bands along the z-axis of the Brillouin zone of the simple cubic lattice.

we have

$$\check{\Delta}_{\boldsymbol{k}}^{(d)}(\beta_\lambda)_{l'l} = \Delta_{\boldsymbol{k}}^{(d)}(\{\beta_\lambda|\boldsymbol{b}_\lambda\})_{l'l}$$
$$= \exp(-\mathrm{i}\boldsymbol{k}_{l'} \cdot \boldsymbol{b}_\lambda)\delta(\boldsymbol{k}_{l'} - \beta_\lambda \boldsymbol{k}_l) \qquad (15.2.2a)$$

which is a weighted permutation representation of $G_{\boldsymbol{k}}$ based on the equivalent set of wave vectors $\{\boldsymbol{k}_l\}$.

In the special case in which $\beta_\lambda \boldsymbol{k} = \boldsymbol{k}$ for all $\beta_\lambda \in G_{\boldsymbol{k}}$, the projective representation is p-equivalent to a vector representation $\dot{\Delta}_{\boldsymbol{k}}^{(d)}(G_{\boldsymbol{k}})$. In fact, from (14.4.17e) the representation $\Delta_{\boldsymbol{k}}^{(d)}$ can be expressed in the following three ways:

$$\Delta_{\boldsymbol{k}}^{(d)}(\{\beta_\lambda|\boldsymbol{b}_\lambda\}) = \check{\Delta}_{\boldsymbol{k}}^{(d)}(\beta_\lambda) = \mathrm{e}^{-\mathrm{i}\boldsymbol{k}\cdot\boldsymbol{b}_\lambda}\dot{\Delta}_{\boldsymbol{k}}^{(d)}(\beta_\lambda); \qquad \forall \beta_\lambda \in G_{\boldsymbol{k}} \qquad (15.2.2b)$$

Accordingly, from (15.2.2a) with $\beta_\lambda \boldsymbol{k} = \boldsymbol{k}$,

$$\dot{\Delta}_{\boldsymbol{k}}^{(d)}(\beta_\lambda)_{l'l} = \exp(+\mathrm{i}\boldsymbol{g}_{l'} \cdot \boldsymbol{b}_\lambda)\,\delta(\boldsymbol{g}_{l'} - \beta_\lambda \boldsymbol{g}_l) \qquad (15.2.2b')$$

which is a weighted permutation representation of $G_{\boldsymbol{k}}$ based on the equivalent set of the reciprocal lattice vectors $\{\boldsymbol{g}_l\}$.

In the case of a symmorphic space group we have, from (15.2.2a) with $\boldsymbol{b}_\lambda = 0$,

$$\dot{\Delta}_{\boldsymbol{k}}^{(d)}(\beta_\lambda)_{l'l} = \check{\Delta}_{\boldsymbol{k}}^{(d)}(\beta_\lambda)_{l'l} = \delta(\boldsymbol{k}_{l'l} - \beta_\lambda \boldsymbol{k}_l) \qquad (15.2.2c)$$

which is an ordinary permutation representation of $G_{\boldsymbol{k}}$ based on $\{\boldsymbol{k}_l\}$, as expected.

Finally, the number of times a small representation $\Delta_{\boldsymbol{k}}^{(i)}$ is contained in $\Delta_{\boldsymbol{k}}^{(d)}$ is determined by the frequency rule (14.4.17d) or (14.4.17f). The required character $\check{\chi}_{\boldsymbol{k}}^{(d)}(G_{\boldsymbol{k}})$ of the projective representation $\check{\Delta}_{\boldsymbol{k}}^{(d)}(G_{\boldsymbol{k}})$ is given, from (15.2.2a), by

$$\check{\chi}_{\boldsymbol{k}}^{(d)}(\beta_\lambda) = \chi_{\boldsymbol{k}}^{(d)}(\{\beta_\lambda|\boldsymbol{b}_\lambda\}) = \sum_l \exp(-\mathrm{i}\boldsymbol{k}_l \cdot \boldsymbol{b}_\lambda)\,\delta(\boldsymbol{k}_l - \beta_\lambda \boldsymbol{k}_l) \qquad (15.2.2d)$$

or the character $\dot{\chi}_{\boldsymbol{k}}^{(d)}(G_{\boldsymbol{k}})$ of the vector representation $\dot{\Delta}_{\boldsymbol{k}}^{(d)}(G_{\boldsymbol{k}})$ is given, from (15.2.2b′), by

$$\dot{\chi}_{\boldsymbol{k}}^{(d)}(\beta_\lambda) = \begin{cases} \sum_l \exp(-\mathrm{i}\boldsymbol{g}_l \cdot \boldsymbol{b}_\lambda)\,\delta(\boldsymbol{g}_l - \beta_\lambda \boldsymbol{g}_l), & \text{if } \beta_\lambda \boldsymbol{k} = \boldsymbol{k} \\ \sum_l \delta(\boldsymbol{k}_l - \beta_\lambda \boldsymbol{k}_l), & \text{if } G_{\boldsymbol{k}} \text{ is symmorphic} \end{cases} \tag{15.2.2e}$$

Note that the character may still depend on the non-lattice translation $\boldsymbol{b}_\lambda$. When a set of equivalent Bloch functions is given explicitly, it becomes a trivial matter to calculate these characters directly from the given equivalent set of vectors $\{\boldsymbol{k}_l\}$ or $\{\boldsymbol{g}_l\}$ with the respective phase factor.

With the general preparation given above, we shall now determine the band energies, and the corresponding SALCs of the Bloch functions at the high-symmetry points in the Brillouin zone in the free-electron approximation; first for a simple cubic crystal belonging to a symmorphic space group and then for the diamond crystal belonging to a non-symmorphic space group.

15.2.1 *The notations for a small representation of $\hat{G}_{\boldsymbol{k}}$*

When $\boldsymbol{k}$ is invariant with respect to the point co-group $G_{\boldsymbol{k}}$, it has been shown by (14.4.17b) that a small representation $\Lambda_{\boldsymbol{k}}^{(i)}$ of $\hat{G}_{\boldsymbol{k}}$ can be expressed in the following three ways:

$$\Lambda_{\boldsymbol{k}}^{(i)}(\{\beta_\lambda|\boldsymbol{b}_\lambda\}) = \check{\Lambda}_{\boldsymbol{k}}^{(i)}(\beta_\lambda) = \mathrm{e}^{-\mathrm{i}\boldsymbol{k}\cdot\boldsymbol{b}_\lambda}\dot{\Lambda}_{\boldsymbol{k}}^{(i)}(\beta_\lambda); \qquad \forall \beta_\lambda \in G_{\boldsymbol{k}}$$

where $\{\dot{\Lambda}_{\boldsymbol{k}}^{(i)}(\beta_\lambda)\}$ is a unirrep of the point group $G_{\boldsymbol{k}}$ that is independent of $\{\boldsymbol{b}_\lambda\}$. Accordingly, a small representation of $\hat{G}_{\boldsymbol{k}}$ may be expressed by the familiar notation of a unirrep of the point group $G_{\boldsymbol{k}}$ with some indication of the $\boldsymbol{k}$-dependence. For example, a small representation of a cubic space group $\hat{O}_i$ corresponding to the identity representation $A_{1\mathrm{g}}$ of O_i may be denoted by

$$A_{1\mathrm{g}}^{\boldsymbol{k}}(\{\beta_\lambda|\boldsymbol{b}_\lambda + \boldsymbol{t}_n\}) \equiv \mathrm{e}^{-\mathrm{i}\boldsymbol{k}\cdot(\boldsymbol{b}_\lambda+\boldsymbol{t}_n)}A_{1\mathrm{g}}(\beta_\lambda) \tag{15.2.2f}$$

if $\beta_\lambda \boldsymbol{k} = \boldsymbol{k}$ for all $\beta_\lambda \in O_i$, which is also applicable for a unirrep of a symmorphic space group. For a general case, an analogous notation will be used for the respective halving subgroup which is symmorphic, then the projective unirrep of $G_{\boldsymbol{k}}$ corresponding to a small representation of $\hat{G}_{\boldsymbol{k}}$ is described through the unirreps of the halving subgroup given in Table 12.4.

15.2.2 *Example 1. A simple cubic lattice*

Consider a crystal belonging to the space group O_i^1 (number 221, $Pm3m$), which is a symmorphic space group with a simple cubic lattice. The Brillouin zone of the simple cubic lattice has been given in Figure 14.5. The wave vector groups at the high-symmetry points along the z-axis are, from (14.3.6b),

$$\Gamma = [000] \in \hat{O}_i, \quad \Delta = [00\zeta] \in \hat{C}_{4v}^z, \quad X = [00\tfrac{1}{2}] \in D_{4i}^z \tag{15.2.3a}$$

where $0<\zeta<\frac{1}{2}$. In terms of the Cartesian coordinates of the wave vector $\boldsymbol{k}$ in the Brillouin zone defined by (14.3.4), i.e.

$$\boldsymbol{k} = 2\pi a^{-1}[\xi \boldsymbol{e}_1^0 + \eta \boldsymbol{e}_2^0 + \zeta \boldsymbol{e}_3^0] \equiv 2\pi a^{-1}[\xi, \eta, \zeta]$$

the energy $E(\boldsymbol{k})$ along the z-axis is given, in the unit of $\hbar^2/(2m)$ and with the lattice spacing $a = 1$, by

$$E(\boldsymbol{k}) = (\boldsymbol{k} - \boldsymbol{g}_l)^2 = l_1^2 + l_2^2 + (\zeta - l_3)^2; \qquad 0 \leqslant \zeta \leqslant \tfrac{1}{2} \tag{15.2.3b}$$

where l_1, l_2 and l_3 are integers compatible with a given $E(\boldsymbol{k})$ in accordance with (15.2.1d). The corresponding set of the Bloch functions is given, from (15.2.1c) with $\boldsymbol{k}_1 = \boldsymbol{k} - \boldsymbol{g}_l = \boldsymbol{k} - 2\pi \boldsymbol{l}$, by

$$\dot{\Delta}_{\boldsymbol{k}}^{(d)}(G_{\boldsymbol{k}}): \qquad \{\exp[2\pi \mathrm{i}(-l_1 x - l_2 y + (\zeta - l_3)z)];\ \forall\, \boldsymbol{l} \in E(\boldsymbol{k})\} \tag{15.2.3c}$$

which belongs to the permutation representation $\dot{\Delta}_{\boldsymbol{k}}^{(d)}(G_{\boldsymbol{k}})$ of the point co-group $G_{\boldsymbol{k}}$ as given by (15.2.2c). The energy levels $E(\boldsymbol{k})$ given by (15.2.3b) are presented graphically in Figure 15.1.

At the lowest energy $E_\Gamma = 0$, which occurs at $\zeta = 0$ with $\boldsymbol{l} = 0$, the Bloch function is given by

$$\psi_\Gamma = 1 \in A_{1\mathrm{g}}^\Gamma \text{ of } \hat{O}_i \tag{15.2.4a}$$

where the notation $A_{1\mathrm{g}}^\Gamma$ denotes the small representation of $\hat{O}_i$ corresponding to the identity representation $A_{1\mathrm{g}}$ of the point group O_i in accordance with (15.2.2f).

Starting from this point moving along $\Delta(0<\zeta<\frac{1}{2})$ in the B-zone we have, still in the first band with $\boldsymbol{l} = 0$,

$$E_\Delta = \zeta^2, \qquad \psi_\Delta = \exp(+2\pi \mathrm{i}\zeta z) \in A_1^\Delta \text{ of } \hat{C}_{4\mathrm{v}}^z$$

At the end point $X(\zeta = \frac{1}{2}) \in \hat{D}_{4i}$ of the first zone, we have, from (15.2.3b) and (15.2.3c) with $\boldsymbol{l} = [0, 0, 0], [0, 0, 1]$,

$$E_x = \tfrac{1}{4}, \qquad \psi_x = \{\mathrm{e}^{\mathrm{i}\pi z}, \mathrm{e}^{-\mathrm{i}\pi z}\} \in \hat{D}_{4i} \tag{15.2.4b}$$

This set of two equivalent Bloch functions closes under the point group D_{4i} as expected and thus provides a permutation representation $\dot{\Delta}^{(2)}$ of D_{4i} in accordance with (15.2.2c). Using (15.2.2e) for the symmorphic group and the frequency rule (14.4.17d), we conclude that it contains two unirreps:

$$\dot{\Delta}^{(2)} = \underset{1}{A_{1\mathrm{g}}} + \underset{z}{A_{2\mathrm{u}}}$$

Here, 1 and z are the elementary bases of $A_{1\mathrm{g}}$ and $A_{2\mathrm{u}}$ of D_{4i}, respectively (see the character table in the Appendix). The symmetry-adapted bases for the wave vector group $\hat{D}_{4i}$ are given by

$$A_{1\mathrm{g}}^{\mathrm{x}}: \qquad \psi^1 = \cos(\pi z)$$

$$A_{2\mathrm{u}}^{\mathrm{x}}: \qquad \psi_z = \sin(\pi z) \tag{15.2.4c}$$

where the superscript x ($\boldsymbol{k} = 2\pi(00\frac{1}{2})$) denotes the $\boldsymbol{k}$-dependence of the unirreps for the translational degree of freedom.

Note that ψ^1 in (15.2.4c) is a totally symmetric function with respect to the point group D_{4i}, whereas ψ_z transforms like z under D_{4i}. The former is obtained by summing up the members of the equivalent set of functions following (15.1.9b) whereas the latter is obtained by operating on ψ^1 with $\partial/\partial z$. This is based on the simple *correspondence theorem* which states that the differential operators $\partial/\partial x$, $\partial/\partial y$ and $\partial/\partial z$ obey the same transformation law as do the Cartesian coordinates x, y and z under a rotation but are invariant under translation. This simple correspondence theorem will be used to simplify the construction of SALCs in the present chapter.

For the second band given in Figure 15.1 along $\Delta(\frac{1}{2} > \zeta > 0) \in \hat{C}_{4v}$, the energy and the corresponding Bloch function with $l = [0, 0, 1]$ are described by

$$E_\Delta = (1 - \zeta)^2, \qquad \psi^\Delta = \exp[-2\pi i(1 - \zeta)z] \in A_1^\Delta \text{ of } \hat{C}_{4v}$$

At the end point $\Gamma(\zeta = 0) \in \hat{O}_i$ of this band, with $l = [0, 0, \pm 1], [0, \pm 1, 0]\ [\pm 1, 0, 0]$ we have

$$E_\Gamma = 1, \qquad \{e^{\pm 2\pi i z}, e^{\pm 2\pi i y}, e^{\pm 2\pi i x}\} \in \hat{O}_i \tag{15.2.5a}$$

The permutation representation $\dot{\Delta}^{(6)}$ of the point group O_i based on these functions or the equivalent set of six vectors $\{l\}$ contains the following unirreps of O_i:

$$\begin{aligned} \dot{\Delta}_\Gamma^{(6)} = {} & A_{1g} + \quad E_g \quad + \quad T_{1u} \\ & \ \ 1 \qquad [u, v] \quad [x, y, z] \end{aligned} \tag{15.2.5b}$$

The corresponding symmetry-adapted functions for the space group $\hat{O}_i$ are given by

$$\begin{aligned}
A_{1g}^\Gamma:&\quad \psi^1 = \cos(2\pi z) + \cos(2\pi y) + \cos(2\pi x)\\
E_g^\Gamma:&\quad \psi_u = 2\cos(2\pi z) - \cos(2\pi y) - \cos(2\pi x)\\
&\quad \psi_v = \sqrt{3}[\cos(2\pi x) - \cos(2\pi y)]\\
T_{1u}^\Gamma:&\quad \psi_x = \sin(2\pi x), \qquad \psi_y = \sin(2\pi y), \qquad \psi_z = \sin(2\pi z)
\end{aligned} \tag{15.2.5c}$$

where $\psi^1(\boldsymbol{r})$ is a totally symmetric function with respect to O_i obtained by simply summing up all the equivalent Bloch functions in (15.2.5a). The remaining bases are obtained by use of the correspondence theorem as follows:

$$\psi_u = u(\nabla)\psi^1, \qquad \psi_v = v(\nabla)\psi^1; \qquad \psi_x = \partial_x\psi^1$$

etc., where

$$u(\nabla) = 2\partial_z^2 - \partial_x^2 - \partial_y^2, \qquad v(\nabla) = \sqrt{3}(\partial_x^2 - \partial_y^2)$$

On the third band along $\Delta(0 < \zeta < \frac{1}{2}) \in \hat{C}_{4v}$, the energy and the corresponding Bloch functions are described by $l = [\pm 1, 0, 0], [0, \pm 1, 0]$, i.e.

$$E_\Delta = 1 + \zeta^2, \qquad \{e^{i2\pi(\zeta z \pm x)}, e^{i2\pi(\zeta z \pm y)}\} \in \hat{C}_{4v} \tag{15.2.6a}$$

The representation $\dot{\Delta}_\Delta^{(4)}$ of C_{4v} based on the equivalent set of the wave vectors $\{[\pm 1, 0, \zeta], [0, \pm 1, \zeta]\}$ contains the following unirreps:

$$\begin{aligned} \dot{\Delta}_\Delta^{(4)} = {} & A_1 + \quad B_1 \quad + \ E \\ & \ 1 \quad\ \ x^2 - y^2 \quad [x, y] \end{aligned}$$

Starting from the totally symmetric function and using the correspondence theorem, the basis functions for the unirreps are given by

$$A_1^\Delta: \qquad \psi^1 = \exp(i2\pi\zeta z)[\cos(2\pi x) + \cos(2\pi y)]$$

$$B_1^\Delta: \qquad \psi_{x^2-y^2} = \exp(\mathrm{i}2\pi\zeta z)[\cos(2\pi x) - \cos(2\pi y)]$$

$$E^\Delta: \qquad \psi_x = \exp(\mathrm{i}2\pi\zeta z)\sin(2\pi x), \qquad \psi_y = \exp(\mathrm{i}2\pi\zeta z)\sin(2\pi y) \tag{15.2.6b}$$

As one can see from Figure 15.1, there exists one more band along Δ, which stems from $E_\Gamma = 1$:

$$E_\Delta = (1+\zeta)^2, \qquad \psi^\Delta = \exp[2\pi\mathrm{i}(\zeta-1)z] \in A_1^\Delta \text{ of } \hat{C}_{4v}$$

The relations of compatibility between the states along the symmetry axis $\Delta \in C_{4v}$ and the states at both ends $\Gamma \in O_i$ and $X \in D_{4i}$ are exhibited in Figure 15.1. These are easily understood from the compatibility tables for the corresponding point groups given by Koster *et al.* (1963):

$$\{A_{1g},\ E_g,\ T_{1u} \text{ of } O_i\} \leftrightarrow \{A_1,\ A_1+B_1,\ A_1+E \text{ of } C_{4v}\}$$
$$\{A_{1g},\ A_{2u} \text{ of } D_{4i}\} \leftrightarrow \{A_1,\ A_1 \text{ of } C_{4v}\} \tag{15.2.7}$$

The energy bands and their eigenfunctions in the free-electron approximation provide the starting point for the further refinements of the theory with a realistic potential, as has been discussed by Jones (1975). As the potential energy $V(\boldsymbol{r})$ changes slowly from a constant value to the actual potential for a real crystal, the energy bands shown in Figure 15.1 change accordingly. In particular, the bands with the different symmetry types will separate so that there remains only the essential degeneracy due to the unirreps of the wave vector groups.

15.2.3 Example 2. The diamond crystal

It is a face-centered cubic crystal belonging to the space group O_i^7 (number 227, $fd3m$). From Table 13.3 of the space groups, the space group $\hat{O}_i$ $(= O_i^7)$ may be expressed by the left coset decomposition relative to the halving subgroup $\hat{T}_p$ which is symmorphic:

$$\hat{O}_i = \hat{T}_p + \{\bar{1}|\tfrac{1}{4}\tfrac{1}{4}\tfrac{1}{4}\}\hat{T}_p \tag{15.2.8}$$

with the lattice spacing $a = 1$. To compare this with the case of the symmorphic space group O_i^1 discussed in Example 1, we again consider the energy bands along the z-axis of the Brillouin zone given in Figure 14.6. The wave vector groups at the high-symmetry points are, from (14.3.7c),

$$\Gamma = [000] \in \hat{O}_i, \qquad \Delta = [00\zeta] \in \hat{C}_{4v}^z, \qquad X = [001] \in \hat{D}_{4i}^z \tag{15.2.9}$$

where $0 < \zeta < 1$ and the subgroups of $\hat{O}_i$ are defined by

$$\hat{C}_{4v}^z = \hat{C}_{2v}^z + \{4_z|\tfrac{1}{4}\tfrac{1}{4}\tfrac{1}{4}\}\hat{C}_{2v}^z, \qquad C_{2v}^z = \{2_z,\ \bar{2}_{x\bar{y}}\}$$
$$\hat{D}_{4i}^z = \hat{D}_{2p}^z + [\bar{1}|\tfrac{1}{4}\tfrac{1}{4}\tfrac{1}{4}\}\hat{D}_{2p}^z, \qquad D_{2p}^z = \{\bar{4}_z,\ 2_x\} \tag{15.2.10}$$

Here the halving subgroups are all symmorphic.

In terms of the Cartesian coordinates of the wave vector $\boldsymbol{k} = 2\pi[0,\ 0,\ \zeta]$ and the Cartesian indices of the reciprocal lattice vector $\boldsymbol{g}_l$, the free-electron energy (in the unit $\hbar^2/(2m)$) along the z-axis and the corresponding set of the equivalent Bloch functions are given, from (15.2.1d) and (15.2.1c), by

$$E(\boldsymbol{k}) = l_1^{02} + l_2^{02} + (l_3^0 - \zeta)^2$$

$$\check{\Delta}_{\boldsymbol{k}}^{(d)}(G_{\boldsymbol{k}}): \qquad \{\exp[-2\pi\mathrm{i}(l_1^0 x + l_2^0 y + (l_3^0 - \zeta)z)];\ \forall\, l^0 \in E(\boldsymbol{k})\} \tag{15.2.11}$$

Here l_1^0, l_2^0 and l_3^0 are either all odd or all even integers, including zero, according to (14.2.7b). The energy bands described by (15.2.11) are presented graphically in Figure 15.2. The SALCs of the Bloch functions belonging to the unirreps contained in $\Delta_k^{(d)}(\hat{G}_k)$ will be constructed by applying the correspondence theorem to the invariant function of G_k. Since the symmetry points Γ and Δ are inside the B-zone, the projective representation $\hat{\Delta}_k^{(d)}(G_k)$ is p-equivalent to the vector representation of G_k given by (15.2.2b). Thus, the number n_k^i of times a small representation of $\hat{G}_k$ is contained in the equivalent representation $\Delta_k^{(d)}$ is determined by (14.4.17f), using the character given by (15.2.2e) and the character table of the point co-group G_k. The symmetry point X, however, is located on a face of the B-zone and is not invariant with respect to D_{4i} so that the unirreps of the space groups $\hat{D}_{4i}$ will be determined by the projective unirreps of the point group D_{4i} following the method developed in Section 14.4. The present approach should be compared with the classical work of Jones (1975) on the same subject.

15.2.3.1 The unirreps and the bases at $\Gamma(\boldsymbol{k}=0) \in \hat{O}_i$

Since $\boldsymbol{k}=0$, we have $\Delta_k^{(d)}(\{\beta_\lambda|\boldsymbol{b}_\lambda\}) = \check{\Delta}_k^{(d)} = \dot{\Delta}_k^{(d)}(\beta_\lambda)$ from (15.2.2b).

(i) At the lowest energy state $E_\Gamma = 0$, the Bloch function is given by

$$\psi = 1 \in A_{1g}^{\Gamma} \text{ of } \hat{O}_i \qquad (15.2.12a)$$

(ii) The next higher energy at Γ is given by $E_\Gamma = 3$ from (15.2.11) with $l^0 = \{[\pm 1, \pm 1, \pm 1]\}$, so that there exist eight Bloch functions:

$$\dot{\Delta}_\Gamma^{(8)}: \qquad \{\exp[2\pi i(\pm x \ \pm y \ \pm z)]\} \in \hat{O}_i(\Gamma) \qquad (15.2.12b)$$

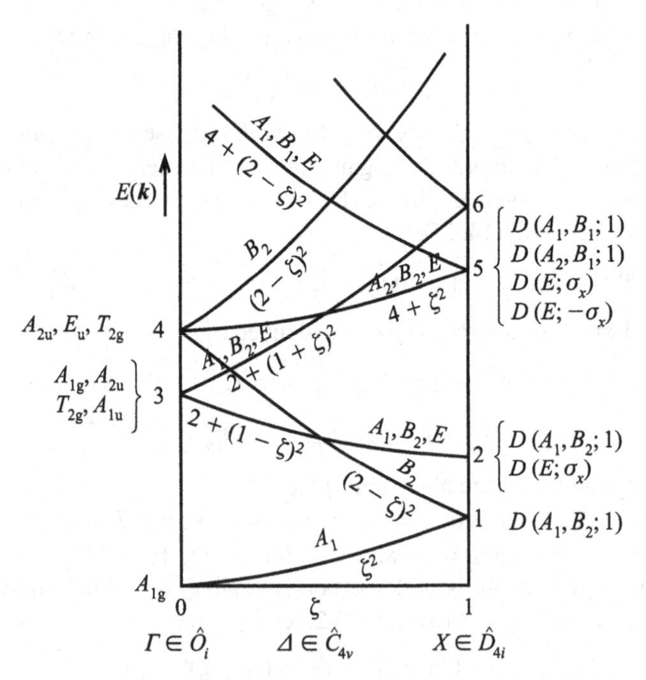

Figure 15.2. Free-electron energy band along the z-axis of the Brillouin zone of the face-centered cubic lattice.

The representation[1] of $\hat{O}_i(\Gamma)$ based on these contains the following small representations of the wave vector group, from the frequency rule (14.4.17f),

$$\begin{array}{llllll} \dot{\Delta}_\Gamma^{(8)} = & A_{1g} + & A_{2u} + & T_{2g} & + & T_{1u} \\ & 1 & xyz & [yz, zx, xy] & & [x, y, z] \end{array} \tag{15.2.12c}$$

where the elementary bases are those of the point co-group O_i. The symmetry-adapted basis functions for $\hat{O}_i$ are

$$A_{1g}^\Gamma: \quad \psi^1 = \cos(2\pi x)\cos(2\pi y)\cos(2\pi z) + \sin(2\pi x)\sin(2\pi y)\sin(2\pi z)$$

$$A_{2u}^\Gamma: \quad \psi_{xyz} = \cos(2\pi x)\cos(2\pi y)\cos(2\pi z) - \sin(2\pi x)\sin(2\pi y)\sin(2\pi z)$$

$$T_{1u}^\Gamma: \quad \psi_x = -\sin(2\pi x)\cos(2\pi y)\cos(2\pi z) + \cos(2\pi x)\sin(2\pi y)\sin(2\pi z)$$

$$\psi_y = -\cos(2\pi x)\sin(2\pi y)\cos(2\pi z) + \sin(2\pi x)\cos(2\pi y)\sin(2\pi z)$$

$$\psi_z = -\cos(2\pi x)\cos(2\pi y)\sin(2\pi z) + \sin(2\pi x)\sin(2\pi y)\cos(2\pi z)$$

$$T_{2g}^\Gamma: \quad \psi_{yz} = \cos(2\pi x)\sin(2\pi y)\sin(2\pi z) + \sin(2\pi x)\cos(2\pi y)\cos(2\pi z)$$

$$\psi_{zx} = \sin(2\pi x)\cos(2\pi y)\sin(2\pi z) + \cos(2\pi x)\sin(2\pi y)\cos(2\pi z)$$

$$\psi_{xy} = \sin(2\pi x)\sin(2\pi y)\cos(2\pi z) + \cos(2\pi x)\cos(2\pi y)\sin(2\pi z)$$

Here the totally symmetric basis $\psi^1 \in A_{1g}^\Gamma$ of $\hat{O}_i$ is obtained by (15.1.9b), which takes the form, for the present case,

$$\psi^1 = [(e|0)^\circ + (\bar{1}|\tfrac{1}{4}\tfrac{1}{4}\tfrac{1}{4})^\circ] \sum_{\beta_\lambda \in T_p} \mathring{\beta}_\lambda \exp[2\pi \mathrm{i}(x + y + z)]$$

The remaining bases are constructed via the correspondence theorem $(\partial x, \partial y, \partial z) \leftrightarrow (x, y, z)$; e.g. $\psi_{xyz} \propto \partial x\, \partial y\, \partial z\, \psi^1$.

(iii) For $E_\Gamma = 4$ (the next higher energy at Γ), the equivalent set of Bloch functions is given by a total of six Bloch functions, with $l^0 = \{[\pm 2, 0, 0], [0, \pm 2, 0], [0, 0, \pm 2]\}$,

$$\dot{\Delta}_\Gamma^{(6)}: \quad \{e^{\pm \mathrm{i}4\pi z}, \mathrm{e}^{\pm \mathrm{i}4\pi x}, \mathrm{e}^{\pm \mathrm{i}4\pi y}\} \in \hat{O}_i(\Gamma) \tag{15.2.13a}$$

The representation of $\hat{O}_i(\Gamma)$ based on these contains the following unirreps, from (15.2.2e) and the frequency rule (14.4.17f):

$$\begin{array}{lllll} \dot{\Delta}_\Gamma^{(6)} = & A_{2u} + & E_u & + & T_{2g} \\ & xyz & [u, v]xyz & & [x, y, z]xyz \end{array} \tag{15.2.13b}$$[2]

[1] For convenience, we may give here the character of $\dot{\Delta}_\Gamma^{(8)}(O_i)$:

e	$8 \times 3_{xyz}$	$3 \times 2_z$	$6 \times \bar{4}_z$	$6 \times \bar{2}_{x\bar{y}}$	$(\bar{1}\vert\frac{1}{4}\frac{1}{4}\frac{1}{4})T_p$
8	2	0	0	4	0

There is no contribution from the elements which involve non-lattice translations for this example.

[2] From (15.2.2e), the character of $\dot{\Delta}_\Gamma^{(6)}(O_i)$ is given by

e	$8 \times 3_{xyz}$	$3 \times 2_z$	$6 \times \bar{4}_z$	$6 \times \bar{2}_{x\bar{y}}$	$\hat{I}$	$8 \times \hat{I}3_{xyz}$	$3 \times \hat{I}2_z$	$6 \times \hat{I}\bar{4}_z$	$6 \times \hat{I}\bar{2}_{x\bar{y}}$
6	0	2	0	2	0	0	−4	−2	0

with $\hat{I} = (\bar{1}|\frac{1}{4}\frac{1}{4}\frac{1}{4})$.

Their bases are

$$A^{\Gamma}_{2u}: \quad \psi_{xyz} = \cos(4\pi x) + \cos(4\pi y) + \cos(4\pi z)$$

$$E^{\Gamma}_{u}: \quad u(\nabla)\psi_{xyz} \propto 2\cos(4\pi z) - \cos(4\pi y) - \cos(4\pi x)$$

$$v(\nabla)\psi_{xyz} \propto \sqrt{3}[\cos(4\pi x) - \cos(4\pi y)]$$

$$T^{\Gamma}_{2g}: \quad [\partial_x, \partial_y, \partial_z]\psi_{xyz} \propto [\sin(\pi x), \sin(\pi y), \sin(\pi z)]$$

Here, the $\psi_{xyz} \in A^{\Gamma}_{2u}$ of $\hat{O}_i(\Gamma)$ is obtained from (15.1.9a), which takes the form

$$\psi_{xyz} \sim (\{e|0\}^{\circ} - \{\bar{1}\tfrac{1}{4}\tfrac{1}{4}\tfrac{1}{4}\}^{\circ}) \sum_{\beta_\lambda \in T_p} \mathring{\beta}_\lambda e^{i4\pi z}$$

The remaining bases are obtained by application of the correspondence theorem to ψ_{xyz} following the elementary bases given in (15.2.13b).

15.2.3.2 The unirreps and the bases along $\Delta(\boldsymbol{k} = 2\pi[00\zeta]) \in \hat{C}^z_{4v}$

The generators of the factor group $\hat{C}_{4v}/T$ have been given by (15.2.10). Since the symmetry point Δ is inside the Brillouin zone, a projective representation of $\hat{C}^z_{4v}(\Delta)$ is expressed by the corresponding vector representation given by (15.2.2b). The first few bands given in Figure 15.2 will be described below.

(i) For $E_\Delta = \zeta^2$ with $l^0 = [0, 0, 0]$, the Bloch function is given, from (15.2.11), by

$$\psi = e^{2\pi i \zeta z} \in A^{\Delta}_1 \text{ of } \hat{C}_{4v}(\Delta) \tag{15.2.14a}$$

Note that $\breve{A}^{\Delta}_1(\beta_\lambda) = A^{\Delta}_1(\{\beta_\lambda|\boldsymbol{b}_\lambda\}) = \exp(-i\boldsymbol{k}\cdot\boldsymbol{b}_\lambda]$ in accordance with (15.2.2b).

(ii) For $E_\Delta = (2-\zeta)^2$ with $l^0 = [0, 0, 2]$,

$$\psi = e^{2\pi i(\zeta-2)z} \in B^{\Delta}_2 \text{ of } \hat{C}_{4v}(\Delta)$$

(iii) For $E_\Delta = 2 + (1-\zeta)^2$ with $l^0 = \{[\pm 1, \pm 1, 1]\}$, there exist four equivalent Bloch functions given, from (15.2.11), by

$$\dot{\Delta}^{(4)}_{\Delta}: \quad \{\exp[2\pi i(\pm x \pm y + (\zeta-1)z)]\} \in C_{4v}(\Delta) \tag{15.2.14b}$$

The unirreps of C_{4v} contained in the representation $\dot{\Delta}^{(4)}_{\Delta}$ based on these are given, using (15.2.2e) or directly from the basis, by

$$\dot{\Delta}^{(4)}_{\Delta} = \underset{1}{A_1} + \underset{xy}{B_2} + \underset{[x,\,y]}{E} \tag{15.2.14c}$$

As usual, the totally symmetric basis is obtained by (15.1.9b), which takes the form

$$\psi^1 = [e + (4_z|\tfrac{1}{4}\tfrac{1}{4}\tfrac{1}{4})^{\circ} e^{i\pi\zeta/2}] \sum_{\beta_\lambda \in C_{2v}} \mathring{\beta}_\lambda \exp\{2\pi i[x + y + (\zeta-1)z]\}$$

whereas the remaining bases are determined by the correspondence theorem. The results are

$$A^{\Delta}_1: \quad \psi^1 = \{\cos[2\pi(x+y)] + i\cos[2\pi(x-y)]\}\exp[2\pi i(\zeta-1)z]$$

$$B^{\Delta}_2: \quad \psi_{xy} = \{\cos[2\pi(x+y)] - i\cos[2\pi(x-y)]\}\exp[2\pi i(\zeta-1)z]$$

$$E^{\Delta}: \quad \psi_x = \{\sin[2\pi(x+y)] + \mathrm{i}\sin[2\pi(x-y)]\}\exp[2\pi\mathrm{i}(\zeta-1)z]$$
$$\psi_y = \{\sin[2\pi(x+y)] - \mathrm{i}\sin[2\pi(x-y)]\}\exp[2\pi\mathrm{i}(\zeta-1)z] \quad (15.2.14\text{d})$$

Since the basis functions are complex, if the time-reversal symmetry of the crystal is taken into consideration the dimensions of these representations will be doubled in their co-representations.

(iv) For $E_{\Delta} = 2 + (1+\zeta)^2$ with $l^0 = \{[\pm 1, \pm 1, -1]\}$, it can be shown that the band eigenfunctions are classified by A_1^{Δ}, B_2^{Δ} and E^{Δ} of C_{4v}^{Δ} analogous to (15.2.14c). From (15.1.9b) the totally symmetric basis $\psi^1 \in A_1^{\Delta}$ is given by

$$A_1^{\Delta}: \quad \psi^1 = [\cos(x+y) - \mathrm{i}\cos(x-y)]\exp[2\pi\mathrm{i}(\zeta+1)z]$$

while $\partial_x\partial_y\psi^1$ and $(\partial_x\psi^1, \partial_y\psi^1)$ provide the bases of B_2^{Δ} and E^{Δ}, respectively.

15.2.3.3 The unirreps and the bases at $X(\boldsymbol{k} = 2\pi[0, 0, 1]) \in \hat{D}_{4i}^{z}$

The point X is on a face of the Brillouin zone and is not invariant with respect to the point co-group D_{4i}^{z}. Thus the small representations of $\hat{D}_{4i}^{z}$ will be determined by the projective unirreps of the point group D_{4i} via the vector unirreps of the halving subgroup D_{2p} as discussed in Section 14.4. The generators of $\hat{D}_{4i}/T$ are given, from (15.2.10), by

$$a = \{\bar{4}_z|0\}, \qquad b = \{2_x|0\}, \qquad \hat{I} = \{\bar{1}|\tfrac{1}{4}, \tfrac{1}{4}, \tfrac{1}{4}\}$$

These satisfy the defining relations

$$a^4 = b^2 = (ab)^2 = e', \qquad a\hat{I} = -\hat{I}a, \qquad b\hat{I} = -\hat{I}b \quad (15.2.15\text{a})$$

Comparing these with (12.5.9b), the class of factor systems for D_{4i} is $K(\beta = -1, \gamma = -1)$ and the required projective unirreps are given, from Table 12.4, by

$$\check{D}(A_1, B_2; 1),\ \check{D}(A_2, B_1; 1),\ \check{D}(E; -\sigma_x),\ \check{D}(E; \sigma_z) \quad (15.2.15\text{b})$$

where A_1, A_2, B_1, B_2 and E are the unirreps of the symmorphic halving subgroup $\hat{D}_{2p}$. The unirreps of the wave vector space group $\hat{D}_{4i}^{z}$ and their bases are expressed, using (12.6.10), by

$$X_1 \equiv D(A_1, B_2; 1)^{\mathrm{x}}: \quad [\psi^{A_1}, \hat{I}^{\circ}\psi^{A_1}]$$
$$X_2 \equiv D(A_2, B_1; 1)^{\mathrm{x}}: \quad [\psi^{A_2}, \hat{I}^{\circ}\psi^{A_2}]$$
$$X_3 \equiv D(E; -\sigma_x)^{\mathrm{x}}: \quad \psi^E - \hat{I}^{\circ}\psi^E\sigma_x \equiv \psi^{E-}$$
$$X_4 \equiv D(E; \sigma_x)^{\mathrm{x}}: \quad \psi^E + \hat{I}^{\circ}\psi^E\sigma_x \equiv \psi^{E+} \quad (15.2.16)$$

where the superscript x ($\boldsymbol{k} = 2\pi[0, 0, 1]$) denotes the $\boldsymbol{k}$-dependence of the representation while ψ^{A_1}, ψ^{A_2} and ψ^E are the basis functions belonging to the unirreps A_1^{x}, A_2^{x} and E^{x} of $\hat{D}_{2p}$, respectively. As usual, from (15.2.11) we find the set of equivalent Bloch functions belonging to $\Delta_{\boldsymbol{k}}^{(d)}(\hat{D}_{4i})$ at X. Then, the number of times any small representation of $\hat{D}_{4i}$ is contained in $\Delta_{\boldsymbol{k}}^{(d)}$ may be found by application of the frequency rule (14.4.17d). However, it is more convenient to find the unirreps of $\hat{D}_{2p}$ contained in the subduced representation $\Delta_{\boldsymbol{k}}^{(d)\downarrow}(\hat{D}_{2p})$ and their bases because $\hat{D}_{2p}$ is symmorphic. Then from these follow the unirreps of $\hat{D}_{4i}$ contained in $\Delta_{\boldsymbol{k}}^{(d)}(\hat{D}_{4i})$ as well as their bases via (15.2.16). We may need the following elementary bases of the point group D_{2p}:

$$\begin{matrix} A_1, & A_2, & B_1, & B_2, & E \\ 1 & z(x^2-y^2) & x^2-y^2 & z \text{ or } xy & [x, y] \end{matrix} \tag{15.2.16$'$}$$

(i) For energy $E_x = 1$, we have $l^0 = [0, 0, 0]$, $[0, 0, 2]$ and the corresponding equivalent Bloch functions for $\hat{D}^z_{4i}$ are given, from (15.2.11), by

$$\Delta^{(2)}_x\text{: } \{\exp(\mathrm{i}2\pi z), \exp(-\mathrm{i}2\pi z)\} \in \hat{D}^z_{4i} \tag{15.2.17a}$$

The unirreps of D_{2p} contained in the vector representation $\dot{\Delta}^{(2)\downarrow}_x(D_{2p})$ based on these are

$$\dot{\Delta}^{(2)\downarrow}_x(D_{2p}) = \underset{1}{A_1} + \underset{z}{B_2}$$

This means that only one unirrep X_1 of $\hat{D}^z_{4i}$ defined in (15.2.16) is contained in $\Delta^{(2)}_x$ of $\hat{D}^z_{4i}$. Thus we obtain

$$X_1 = D(A_1, B_2; 1)^{\mathrm{x}}\text{:} \qquad [\cos(2\pi z), \sin(2\pi z)] \tag{15.2.17b}$$

where $\cos(\pi z)$ is the totally symmetric function with respect to D_{2p} and its partner is obtained by $\hat{I}^{\circ}\cos(2\pi z)$, from (15.2.16). Note that the simple application of the correspondence theorem by $\partial_z \cos(2\pi z) = -2\pi \sin(2\pi z)$ does not give the correct linear coefficient for the partner.

(ii) For $E_x = 2$, we have $l^0 = \{[\pm 1, \pm 1, 1]\}$. The corresponding Bloch functions are given by

$$\Delta^{(4)}_x\text{:} \qquad \{\mathrm{e}^{\pm \mathrm{i}2\pi(x+y)}, \mathrm{e}^{\pm \mathrm{i}2\pi(x-y)}\} \in \hat{D}^z_{4i} \tag{15.2.18a}$$

The vector representation $\dot{\Delta}^{(4)\downarrow}_x(D_{2p})$ based on these contains the following unirreps:

$$\dot{\Delta}^{(4)\downarrow}(D_{2p}) = \underset{1}{A_1} + \underset{xy}{B_2} + \underset{[x,\, y]}{E} \tag{15.2.18b}$$

Thus, from (15.2.16), the required small representations of $\hat{D}_{4i}$ are given by X_1 and either X_3 or X_4. Now, the totally symmetric wave function belonging to A_1 of D_{2p} formed by (15.1.9b) is given by

$$\psi^1 = \cos(2\pi x)\cos(2\pi y) \in A^{\mathrm{x}}_1 \text{ of } D_{2p}$$

from which the bases of the small representations of $\hat{D}^z_{4i}$ contained in $\check{\Delta}^{(4)}_x$ are given, using (15.2.16), by

$$X_1 = D(A_1, B_2; 1)^{\mathrm{x}}\text{:} \qquad [\cos(2\pi x)\cos(2\pi y), \sin(2\pi x)\sin(2\pi y)]$$

$$X_4 = D(E; \sigma_x)^{\mathrm{x}}\text{:} \qquad [\sin(2\pi x)\cos(2\pi y), \cos(2\pi x)\sin(2\pi y)] \equiv \psi^{E+} \tag{15.2.18c}$$

where the basis of X_1 is obtained as in (15.2.17b), whereas the basis of X_4 is obtained as follows. Firstly, the basis ψ^E of $E \in \hat{D}_{2p}$ is given by

$$\psi^E = [\partial_x \psi^1, \partial_y \psi^1] \sim [\sin(2\pi x)\cos(2\pi y), \cos(2\pi x)\sin(2\pi y)]$$

Then operating $\hat{I} = \{\bar{1}|\tfrac{1}{4}, \tfrac{1}{4}, \tfrac{1}{4}\}$ on ψ^E and using the Pauli spin matrix σ_x we have

$$\hat{I}^{\circ}\psi^E\sigma_x = [\sin(2\pi x)\cos(2\pi y), \cos(2\pi x)\sin(2\pi y)]$$

which is equal to the original ψ^E. Accordingly, we have

$$\psi^E + \hat{I}^\circ \psi^E \sigma_x = \psi^{E+} \in D(E; \sigma_x)^\times$$

$$\psi^E - \hat{I}^\circ \psi^E \sigma_x = 0$$

The last equation means that $D(E; -\sigma_x)$ is not contained in $\Delta_x^{(4)}$.

(iii) For $E_x = 5$, we have $l^\circ = \{[\pm 2, 0, 0], [\pm 2, 0, 2], [0, \pm 2, 0], [0, \pm 2, 2]\}$ from (15.2.11). The corresponding eight Bloch functions for $\hat{D}_{4i}^z$ are

$$\Delta_x^{(8)}: \qquad \{\exp[2\pi i(\pm 2x \pm z)], \exp[2\pi i(\pm 2y \pm z)]\} \in \hat{D}_{4i}^z$$

The unirreps contained in the representation $\dot{\Delta}^{(8)}$ of the halving subgroup D_{2p} based on these Bloch functions are

$$\dot{\Delta}_x^{(8)\downarrow}(D_{2p}) = \underset{1}{A_1} + \underset{z(x^2-y^2)}{A_2} + \underset{x^2-y^2}{B_1} + \underset{z}{B_2} + \underset{[x,\, y]}{2E} \tag{15.2.19}$$

which contain all the vector unirreps of D_{2p} according to (15.2.16′). From the totally symmetric Bloch function with respect to D_{2p}

$$\psi^1 = [\cos(4\pi x) + \cos(4\pi y)]\cos(2\pi z) \in A_1 \text{ of } D_{2p}$$

the symmetry adapted functions belonging to the unirreps of $\hat{D}_{4i}^z$ are given, using (15.2.16) together with the correspondence theorem, by

$$X_1 = D(A_1, B_2; 1)^\times: \qquad [\cos(4\pi x) + \cos(4\pi y)][\cos(2\pi z), -\sin(2\pi z)]$$

$$X_2 = D(A_2, B_1; 1)^\times: \qquad [\cos(4\pi x) - \cos(4\pi y)][\sin(2\pi z), -\cos(2\pi z)]$$

$$X_4 = D(E; \sigma_x)^\times: \qquad [\sin(4\pi x)\cos(2\pi z) + \sin(4\pi y)\sin(2\pi z),$$
$$\sin(4\pi x)\sin(2\pi z) + \sin(4\pi y)\cos(2\pi z)]$$

$$X_3 = D(E; -\sigma_x)^\times: \qquad [\sin(4\pi x)\cos(2\pi z) - \sin(4\pi y)\sin(2\pi z),$$
$$-\sin(4\pi x)\sin(2\pi z) + \sin(4\pi y)\cos(2\pi z)]$$

The relations of compatibility between the states at $\Gamma \in \hat{O}_i$ and at $\Delta \in \hat{C}_{4v}$ exhibited in Figure 15.2 are easily understood in terms of the relations of compatibility between the corresponding point groups O_i and C_{4v} given by Koster *et al.* (1963):

O_i:	A_{1g},	A_{2u},	T_{2g},	T_{1u},	A_{2u},	E_u,	T_{2g}
C_{4v}:	A_1,	B_2,	$B_2 + E$,	$A_1 + E$,	B_2,	$A_2 + B_2$,	$B_2 + E$

The compatibilities between the states at $\Delta \in \hat{C}_{4v}$ and $X \in \hat{D}_{4i}$ are obvious from Figure 15.2.

15.3 Symmetry coordinates of vibration of a crystal

15.3.1 General discussion

The normal coordinates of vibration of a molecule or crystal are formed by the symmetry adapted linear combinations (SALCs) of the atomic displacements from their equilibrium positions in the molecule or crystal (see Chapter 7). Here the linear coefficients define the corresponding normal mode of vibrations. In the case of a crystal, the symmetry coordinates are characterized by the SALCs of Fourier transforms of the atomic displacements on account of the translational periodicity of the crystal.

Let $r_s^n = r_s^0 + t_n$ be the equilibrium position of the sth atom in the nth primitive unit cell of the Bravais lattice T of the crystal and let $u(r_s^n) = (u_x(r_s^n), u_y(r_s^n), u_z(r_s^n))$ be the atomic displacement from the equilibrium position r_s^n. In the classical theory, the circular frequencies of vibration ω of the crystal and the normal coordinates of vibration belonging to each ω are determined by the following eigenvalue problem:

$$\omega^2 M_s u(r_s^n) = \sum_{s',n'} \Phi(r_s^n - r_{s'}^{n'}) \cdot (r_{s'}^{n'}); \qquad s = 1, 2, \ldots, d; \qquad n = 1, 2, \ldots, N \tag{15.3.1}$$

where d is the number of atoms in a primitive unit cell, N is the number of primitive unit cells in the crystal, M_s is the mass of the sth atom in a unit cell, and the force constants $\Phi(r_s^n - r_{s'}^{n'})$ are 3×3 tensors given by the second-order derivative of the potential energy with respect to the displacements at the equilibrium positions. Note that $\Phi(r_s^n - r_{s'}^{n'}) = \Phi(r_{s'}^{n'} - r_s^n)$. Since Φ are translational invariants, the set of equations (15.3.1) will be simplified by introducing the Fourier transforms

$$u_s^k = (M_s/N)^{1/2} \sum_n^N e^{ik \cdot r_s^n} u(r_s^n)$$

$$F_{ss'}^k = (M_s M_{s'}/N)^{1/2} \sum_{n,n'}^N e^{ik \cdot (r_s^n - r_{s'}^{n'})} \Phi(r_s^n - r_{s'}^{n'}) \tag{15.3.2}$$

where k is a wave vector in the Brillouin zone of the Bravais lattice T of the crystal. In fact, in terms of these, the eigenvalue problem (15.3.1) is reduced to a d-dimensional problem:

$$\omega^2 u_s^k = \sum_{s'} F_{ss'}^k \cdot u_{s'}^k; \qquad s = 1, 2, \ldots, d \tag{15.3.3}$$

The force constants $\Phi(r_s^n - r_{s'}^{n'})$ defined by the equilibrium positions of the atoms are invariant under all symmetry operations of the space groups $\hat{G}$ of the crystal such that

$$\Phi(gr_s^n - gr_{s'}^{n'}) = \Phi(r_s^n - r_{s'}^{n'}), \qquad \forall g \in \hat{G}$$

However, the set of Fourier transforms $\{F^k\}$ is invariant under the wave vector space group $\hat{G}_k$ ($\leqslant \hat{G}$) which leaves the wave vector k equivalent (with respect to the reciprocal lattice T'). Accordingly, we can classify the eigenvectors of F^k by the bases of the unirreps of $\hat{G}_k$ corresponding to the respective eigenvalue ω^2 as in the cases of the Schrödinger equation. The irreducible bases formed by the SALCs of $\{u_s^k; s = 1, 2, \ldots, d\}$ are called *the symmetry coordinates of vibration*. For later use one may introduce the following notation for the spatial components of each Fourier transform:

$$u_s^k = (x_s^k, y_s^k, z_s^k); \qquad s = 1, 2, \ldots, d \tag{15.3.4}$$

As usual we shall first determine the representation of $\hat{G}_k$ based on the set $\{u_s^k\}$. Then the symmetry coordinates of vibration will be formed by the SALCs of the set belonging to the unirreps contained in the representation. Let $A = \{\alpha|a\}$ be a general element of the space group $\hat{G}$. Under the operation A, the displacement vector $\boldsymbol{u}(\boldsymbol{r}_s^n) = \{u_j(\boldsymbol{r}_s^n); j = 1, 2, 3\}$ transforms according to

$$\mathring{A}u_j(\boldsymbol{r}_s^n) = \sum_{i=1}^{3} u_i(A\boldsymbol{r}_s^n)D_{ij}^{(1)}(\alpha); \qquad j = 1, 2, 3 \tag{15.3.5a}$$

where $D^{(1)}(\alpha)$ is the 3×3 rotation matrix which describes the vector transformation in the three-dimensional space. This means that, under the space group $\hat{G}$, the set $\{\boldsymbol{u}(\boldsymbol{r}_s^n)\}$ transforms like a set of the spatial derivatives of scalar functions $\{\boldsymbol{\nabla}\phi(\boldsymbol{r} - \boldsymbol{r}_s^n)\}$ where $\{\phi(\boldsymbol{r} - \boldsymbol{r}_s^n)\}$ is a set of equivalent scalars defined on the set of equivalent points $\{\boldsymbol{r}_s^n\}$ with respect to $\hat{G}$; in fact, under $A = \{\alpha|\boldsymbol{a}\} \in \hat{G}$ we have

$$\mathring{A}\partial_j\phi(\boldsymbol{r} - \boldsymbol{r}_s^n) = \sum_{i=1}^{3} \partial_i\phi(\boldsymbol{r} - A\boldsymbol{r}_s^n)D_{ij}^{(1)}(\boldsymbol{\alpha}) \tag{15.3.5b}$$

Moreover, if we introduce the Fourier transform of the set $\{\phi(\boldsymbol{r} - \boldsymbol{r}_s^n)\}$ by

$$\phi_s^{\boldsymbol{k}}(\boldsymbol{r}) = \frac{1}{\sqrt{N}}\sum_n \mathrm{e}^{\mathrm{i}\boldsymbol{k}\cdot\boldsymbol{r}_s^n}\phi(\boldsymbol{r} - \boldsymbol{r}_s^n); \qquad s = 1, 2, \ldots, d \tag{15.3.6}$$

Then both Fourier transforms $\{\boldsymbol{u}_s^{\boldsymbol{k}}\}$ and $\{\boldsymbol{\nabla}\phi_s^{\boldsymbol{k}}(\boldsymbol{r})\}$ obey the same transformation law under the wave vector space group $\hat{G}_{\boldsymbol{k}}$ with the correspondence

$$(x_s^{\boldsymbol{k}}, y_s^{\boldsymbol{k}}, z_s^{\boldsymbol{k}}) \leftrightarrow (\partial_x\phi_s^{\boldsymbol{k}}, \partial_y\phi_s^{\boldsymbol{k}}, \partial_z\phi_s^{\boldsymbol{k}}) \tag{15.3.7}$$

The characterization of the transformation of the set $\{\boldsymbol{u}_s^{\boldsymbol{k}}\}$ by the set $\{\boldsymbol{\nabla}\phi_s^{\boldsymbol{k}}(\boldsymbol{r})\}$ is convenient and effective, simply because $\{\phi_s^{\boldsymbol{k}}(\boldsymbol{r})\}$ is a set of functions of the continuous space variable $\boldsymbol{r}$, whereas the set $\{\boldsymbol{u}_s^{\boldsymbol{k}}\}$ lacks such a variable. We may determine the transformation properties of the set $\{\boldsymbol{u}_s^{\boldsymbol{k}}\}$ under the group $\hat{G}_{\boldsymbol{k}}$ through those of $\{\boldsymbol{\nabla}\phi_s^{\boldsymbol{k}}(\boldsymbol{r})\}$ and thus apply the correspondence theorem on the basis functions developed in Chapter 7 to the present problem of constructing the symmetry coordinates of vibration of a solid.

The set of functions $\{\phi_s^{\boldsymbol{k}}(\boldsymbol{r}); s = 1, 2, \ldots, d\}$ defined by (15.3.6) may be called *a set of equivalent Bloch functions* since it belongs to the unirrep $\tau_{\boldsymbol{k}}(T)$ of the translation group $T = \{\{e|\boldsymbol{t}_n\}\}$ of the crystal:

$$\{e|\boldsymbol{t}_n\}^0\phi_s^{\boldsymbol{k}}(\boldsymbol{r}) = \mathrm{e}^{-\mathrm{i}\boldsymbol{k}\cdot\boldsymbol{t}_n}\phi_s^{\boldsymbol{k}}(\boldsymbol{r}); \qquad s = 1, 2, \ldots, d \tag{15.3.8}$$

as was shown in (7.2.17a) by the correspondence theorem. Since the differential operator is invariant under translation, via the correspondence (15.3.7), the set $\{\boldsymbol{u}_s^{\boldsymbol{k}}\}$ is also a set of the Bloch functions belonging to $\tau_{\boldsymbol{k}}(T)$ and also to the eigenvalue ω^2 of $\boldsymbol{F}^{\boldsymbol{k}}$ given by (15.3.3). This is analogous to the case of an energy eigenfunction $\psi^{\boldsymbol{k}}$ that satisfies (15.1.2a). The present approach should be compared with the classical work of Bir and Pikus (1974) on the same subject.

15.3.2 The small representations of the wave vector groups $\hat{G}_{\boldsymbol{k}}$ based on the equivalent Bloch functions

First, we shall discuss the transformation of the equivalent Bloch functions $\{\phi_s^{\boldsymbol{k}}(\boldsymbol{r}); s = 1, 2, \ldots, d\}$ under the wave vector group $\hat{G}_{\boldsymbol{k}}$. Let us assume that the set is linearly independent and let $B = \{\beta|\boldsymbol{b}\}$ be a general element of $\hat{G}_{\boldsymbol{k}}$, then observe that the set of equivalent scalar functions $\{\phi(\boldsymbol{r} - \boldsymbol{r}_s^n)\}$ transforms according to

$$\mathring{B}\phi(\boldsymbol{r}-\boldsymbol{r}_s^n)=\phi(\boldsymbol{r}-B\boldsymbol{r}_s^n)$$
$$=\phi(\boldsymbol{r}-\beta\boldsymbol{t}_n-B\boldsymbol{r}_s^0) \qquad (15.3.9)$$

for $B\boldsymbol{r}_s^n=\beta(\boldsymbol{t}_n+\boldsymbol{r}_s^0)+\boldsymbol{b}=\beta\boldsymbol{t}_n+B\boldsymbol{r}_s^0$. Thus, the set of the Bloch functions (15.3.6) transforms as follows:

$$\mathring{B}\phi_s^{\boldsymbol{k}}(\boldsymbol{r})=\frac{1}{\sqrt{N}}\sum_n \mathrm{e}^{\mathrm{i}\boldsymbol{k}\cdot(\boldsymbol{t}_n+\boldsymbol{r}_s^0)}\phi(\boldsymbol{r}-\beta\boldsymbol{t}_n-B\boldsymbol{r}_s^0)$$
$$=\frac{1}{\sqrt{N}}\mathrm{e}^{\mathrm{i}\boldsymbol{k}\cdot(\boldsymbol{r}_s^0-B\boldsymbol{r}_s^0)}\sum_n \mathrm{e}^{\mathrm{i}\boldsymbol{k}\cdot(\boldsymbol{t}_n+B\boldsymbol{r}_s^0)}\phi(\boldsymbol{r}-\beta\boldsymbol{t}_n-B\boldsymbol{r}_s^0)$$
$$=\frac{1}{\sqrt{N}}\mathrm{e}^{\mathrm{i}\boldsymbol{k}\cdot(\boldsymbol{1}-B)\boldsymbol{r}_s^0}\sum_{n'} \mathrm{e}^{\mathrm{i}\boldsymbol{k}\cdot(\boldsymbol{t}_{n'}+B\boldsymbol{r}_s^0)}\phi(\boldsymbol{r}-\boldsymbol{t}_{n'}-B\boldsymbol{r}_s^0)$$

where we have set $\boldsymbol{t}_{n'}=\beta\boldsymbol{t}_n$ and used the equivalence $\beta\boldsymbol{k}\sim\boldsymbol{k}$ with respect to the reciprocal lattice T'. Now, we may set $B\boldsymbol{r}_s^0=\boldsymbol{r}_{s'}^0$ (mod $\boldsymbol{t}_n\in T$) for some s' because $B=\{\beta|\boldsymbol{b}\}$ is a symmetry element of the crystal. Then we arrive at the following representation of $\hat{G}_{\boldsymbol{k}}$ based on the equivalent Bloch functions:

$$\mathring{B}\phi_s^{\boldsymbol{k}}(\boldsymbol{r})=\sum_{s'=1}^{d}\phi_{s'}^{\boldsymbol{k}}(\boldsymbol{r})\Delta_{\boldsymbol{k}}^{(d)}(B)_{s's}; \qquad \forall B\in\hat{G}_{\boldsymbol{k}} \qquad (15.3.10a)$$

where $s=1,2,\ldots,d$ and

$$\Delta_{\boldsymbol{k}}^{(d)}(B)_{s's}=\exp\left[\mathrm{i}\boldsymbol{k}\cdot(\boldsymbol{1}-B)\boldsymbol{r}_s^0\right]\delta(\boldsymbol{r}_{s'}^0\sim B\boldsymbol{r}_s^0)$$
$$\delta(\boldsymbol{r}_{s'}^0\sim B\boldsymbol{r}_s^0)=\begin{cases}1, & \text{if } \boldsymbol{r}_{s'}^0=B\boldsymbol{r}_s^0 \quad (\text{mod } \boldsymbol{t}_n\in T)\\ 0, & \text{otherwise}\end{cases} \qquad (15.3.10b)$$

which is *a weighted permutation representation* of $\hat{G}_{\boldsymbol{k}}$. Its character $\chi_{\boldsymbol{k}}^{(d)}(\hat{G}_{\boldsymbol{k}})$ is given by

$$\chi_{\boldsymbol{k}}^{(d)}(B)=\sum_{s=1}^{m}\mathrm{e}^{\mathrm{i}\boldsymbol{k}\cdot(\boldsymbol{1}-B)\boldsymbol{r}_s^0}\delta(\boldsymbol{r}_s^0\sim B\boldsymbol{r}_s^0) \qquad (15.3.11)$$

where every atom in a primitive unit cell contributes to the character quite independently from each other.

Next, we note that a unirrep contained in $\Delta_{\boldsymbol{k}}^{(d)}(\hat{G}_{\boldsymbol{k}})$ is a small representation of $\hat{G}_{\boldsymbol{k}}$ because the representation subduces a multiple of the unirrep $\tau_{\boldsymbol{k}}(T)$ of T, as one can see from (15.3.8) or directly from the representation $\Delta_{\boldsymbol{k}}^{(d)}(\hat{G}_{\boldsymbol{k}})$ given by (15.3.10b):

$$\Delta_{\boldsymbol{k}}^{(d)}(\{e|\boldsymbol{t}_n\})_{s's}=\exp(-\mathrm{i}\boldsymbol{k}\cdot\boldsymbol{t}_n)\delta_{s's}$$

Accordingly, a unirrep $\Delta_{\boldsymbol{k}}^{(i)}(\hat{G}_{\boldsymbol{k}})$ contained in $\Delta_{\boldsymbol{k}}^{(d)}(\hat{G}_{\boldsymbol{k}})$ is also a small representation of $\hat{G}_{\boldsymbol{k}}$. An analogous statement holds also for the direct product representation $\Lambda_{\boldsymbol{k}}=D^{(1)}\times\Delta_{\boldsymbol{k}}^{(d)}$ because $D^{(1)}(e)=\boldsymbol{1}$, the 3×3 unit matrix.

Now let $B_\lambda=\{\beta_\lambda|\boldsymbol{b}_\lambda\}$ be a coset representative of T in $\hat{G}_{\boldsymbol{k}}$, then from the isomorphism $\hat{G}_{\boldsymbol{k}}/T\simeq G_{\boldsymbol{k}}$, we arrive at the projective representation of $G_{\boldsymbol{k}}$ given, following (14.4.17c), by

$$\Delta_{\boldsymbol{k}}^{(d)}(\{\beta_\lambda|\boldsymbol{b}_\lambda\})=\breve{\Delta}_{\boldsymbol{k}}^{(d)}(\beta_\lambda), \qquad \forall\beta_\lambda\in G_{\boldsymbol{k}} \qquad (15.3.12)$$

which belongs to the factor system defined by (14.4.13b). Therefore, we can use all the results obtained in Section 14.4 on the projective representations for the present case.

In the special case in which the wave vector $\boldsymbol{k}$ is invariant under the wave vector point group $G_{\boldsymbol{k}} = \{\beta_\lambda\}$, we have

$$\exp[\mathrm{i}\boldsymbol{k}\cdot(\mathbf{1} - B_\lambda)\boldsymbol{r}_s^0] = \exp[\mathrm{i}\boldsymbol{k}\cdot(\boldsymbol{r}_s^0 - \beta_\lambda \boldsymbol{r}_s^0 - \boldsymbol{b}_\lambda)] = \exp(\mathrm{i}\boldsymbol{k}\cdot\boldsymbol{b}_\lambda)$$

Accordingly, from (15.3.10b) and (15.2.2b), the vector representation $\dot{\Delta}_{\boldsymbol{k}}^{(d)}(G_{\boldsymbol{k}})$ corresponding to $\Delta_{\boldsymbol{k}}^{(d)}(\hat{G}_{\boldsymbol{k}})$ is given by

$$\dot{\Delta}_{\boldsymbol{k}}^{(d)}(\beta_\lambda)_{s's} = \delta(\boldsymbol{r}_{s'}^0 \sim B_\lambda \boldsymbol{r}_s^0), \qquad \forall \beta_\lambda \in G_{\boldsymbol{k}} \tag{15.3.13a}$$

which may still depend on $\boldsymbol{b}_\lambda$ through B_λ. The character of this vector representation is given by

$$\dot{\chi}_{\boldsymbol{k}}^{(d)}(\beta_\lambda) = \sum_{s=1}^{d} \delta(\boldsymbol{r}_s^0 \sim B_\lambda \boldsymbol{r}_s^0) \tag{15.3.13b}$$

i.e. only those atoms which remain equivalent with respect to T under $B_\lambda = \{\beta_\lambda|\boldsymbol{b}_\lambda\}$ contribute to the character. The number of times a vector unirrep $\dot{\Delta}_{\boldsymbol{k}}^{(i)}(G_{\boldsymbol{k}})$ is contained in $\dot{\Delta}^{(d)}(G_{\boldsymbol{k}})$ is given by their characters obtained by the frequency rule (14.4.17f), as usual.

If the wave vector group $\hat{G}_{\boldsymbol{k}}$ is symmorphic, the projective representation $\check{\Delta}_{\boldsymbol{k}}^{(d)}(G_{\boldsymbol{k}})$ defined by (15.3.12) becomes a vector representation and is given, from (15.3.10b), by

$$\check{\Delta}_{\boldsymbol{k}}^{(d)}(\beta_\lambda)_{s's} = \dot{\Delta}_{\boldsymbol{k}}^{(d)}(\beta_\lambda)_{s's} = \exp[\mathrm{i}\boldsymbol{k}\cdot(\mathbf{1} - \beta_\lambda)\boldsymbol{r}_s^0]\delta(\boldsymbol{r}_{s'}^0 \sim \beta_\lambda \boldsymbol{r}_s^0); \qquad \forall \beta_\lambda \in G_{\boldsymbol{k}} \tag{15.3.13c}$$

which is not a special case of (15.3.13a) unless $\boldsymbol{k}$ is invariant with respect to $G_{\boldsymbol{k}}$. By direct matrix multiplication one can easily verify that this representation is indeed a vector representation of $G_{\boldsymbol{k}}$.

Finally, we shall discuss how to determine the unirreps (small representations) of $\hat{G}_{\boldsymbol{k}}$ contained in the direct product $\Delta_{\boldsymbol{k}} = D^{(1)} \times \Delta_{\boldsymbol{k}}^{(d)}$. One way is to reduce each factor of the direct product separately into its irreducible components:

$$D^{(1)} = \sum_i \oplus D_i^{(1)}, \qquad \Delta_{\boldsymbol{k}}^{(d)} = \sum_j \oplus \Delta_{\boldsymbol{k}}^{(j)}$$

Then, we have the following decomposition:

$$D^{(1)} \times \Delta_{\boldsymbol{k}}^{(d)} \sim \sum_{i,j} \oplus D_i^{(1)} \times \Delta_{\boldsymbol{k}}^{(j)} \tag{15.3.14}$$

Now, the basis of the irreducible component $D_i^{(1)}$ is described by the SALCs of the operators ∂_x, ∂_y and ∂_z whereas the basis of $\Delta_{\boldsymbol{k}}^{(j)}$ is given by the SALCs of the equivalent Bloch functions $\{\phi_s^{\boldsymbol{k}};\ s = 1, 2, \ldots, d\}$. Thus the basis of the direct product $D_i^{(1)} \times \Delta_{\boldsymbol{k}}^{(j)}$ is given by the direct product of the two SALCs. The direct product representation $D_i^{(1)} \times \Delta_{\boldsymbol{k}}^{(j)}$ is irreducible if one of the factors is one-dimensional; otherwise, the basis of the unirreps contained in $D_i^{(1)} \times \Delta_{\boldsymbol{k}}^{(j)}$ is constructed with the help of the coupling efficients. Thus, via the correspondence (15.3.7), the symmetry coordinates of vibration are given by the SALCs of the displacement vectors (x_s^k, y_s^k, z_s^k) belonging to the unirreps contained in $D^{(1)} \times \Delta_k^{(d)}$. These will be worked out explicitly for the diamond crystal in the next section; cf. Bir and Pikus (1974).

15.4 The symmetry coordinates of vibration for the diamond crystal

15.4.1 General discussion

The diamond structure belongs to the space group O_i^7 (number 227) defined by

$$\hat{O}_i = \hat{T}_p + \{\bar{1}|\tfrac{1}{4}\tfrac{1}{4}\tfrac{1}{4}\}\hat{T}_p \tag{15.4.1a}$$

with the lattice spacing $a = 1$, where $\hat{T}_p$ is the symmorphic halving subgroup (cf. Equation (15.2.8)). The distribution of the carbon atoms in the crystal is presented by the two-dimensional projection in Figure 15.3. The primitive unit cell of this crystal contains two carbon atoms at the sites given by the Cartesian coordinates

$$r_1^0 = (0, 0, 0), \qquad r_2^0 = (\tfrac{1}{4}, \tfrac{1}{4}, \tfrac{1}{4})$$

We shall determine the symmetry coordinates of vibration for the crystal at the following symmetry points in the Brillouin zone of the f.c.c. lattice given in Figure 14.6:

$$\Gamma = [000], \qquad \Delta = [00\zeta], \qquad X = [001], \qquad \Lambda = [\zeta\zeta\zeta], \qquad L = [\tfrac{1}{2}\tfrac{1}{2}\tfrac{1}{2}]$$
$$\Sigma = [\xi\xi 0], \qquad K = [\tfrac{3}{4}\tfrac{3}{4}0] \tag{15.4.1b}$$

where $0 < \xi < 1$ and $0 < \zeta < 1$; X, L and K are surface points and the remaining points are interior points of the B-zone.

First, we shall determine the weighted permutation representation $\Delta_k^{(2)}(\hat{G}_k)$ introduced in (15.3.10b) from the equivalent Bloch functions $[\phi_1^k(r), \phi_2^k(r)]$ of two atoms

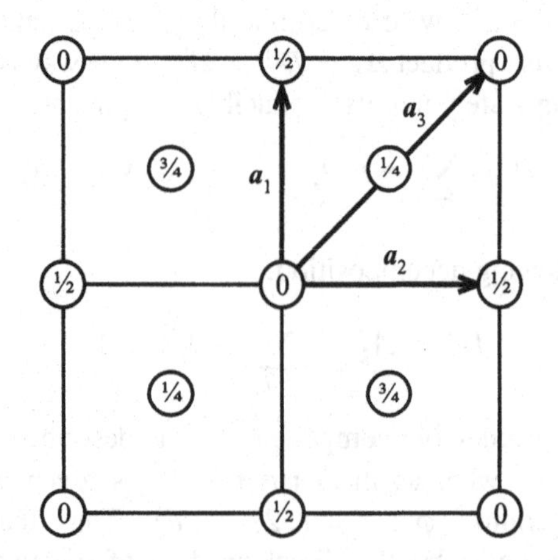

Figure 15.3. The atomic structure of the diamond crystal with the lattice spacing $a = 1$. The Cartesian coordinates of the particles 1 and 2 are (000) and $(\frac{1}{4}\frac{1}{4}\frac{1}{4})$, respectively. A fractional number in a circle denotes the z-coordinate of the particle in the circle. The primitive basis vectors are given by $a_1 = (0, \frac{1}{2}, \frac{1}{2})$, $a_2 = (\frac{1}{2}, 0, \frac{1}{2})$ and $a_3 = (\frac{1}{2}, \frac{1}{2}, 0)$.

located at $\boldsymbol{r}_1^0 = (0, 0, 0)$ and $\boldsymbol{r}_2^0 = (\frac{1}{4}, \frac{1}{4}, \frac{1}{4})$, respectively, in a primitive unit cell following (15.3.10a). For this purpose, we observe that the inversion $\{\bar{1}|\tau\}$ with the non-lattice translation $\boldsymbol{\tau} = (\frac{1}{4}, \frac{1}{4}, \frac{1}{4})$ exchanges two positions:

$$\{\bar{1}|\boldsymbol{\tau}\}\boldsymbol{r}_1^0 = \boldsymbol{r}_2^0, \qquad \{\bar{1}|\boldsymbol{\tau}\}\boldsymbol{r}_2^0 = \boldsymbol{r}_1^0 \tag{15.4.2a}$$

whereas a point operation $\alpha \in T_p$ leaves $\boldsymbol{r}_1^0$ invariant but $\boldsymbol{r}_2^0$ equivalent with respect to the face-centered cubic lattice F_c,

$$\alpha \boldsymbol{r}_1^0 = \boldsymbol{r}_1^0, \qquad \alpha \boldsymbol{r}_2^0 \sim \boldsymbol{r}_2^0; \qquad \alpha \in T_p \tag{15.4.2b}$$

which may be checked by the generators $\bar{4}_z$ and 3_{xyz} of T_p. On combining these we obtain

$$\{\overline{\alpha}|\boldsymbol{\tau}\}\boldsymbol{r}_1^0 = \boldsymbol{r}_2^0, \qquad \{\overline{\alpha}|\boldsymbol{\tau}\}\boldsymbol{r}_2^0 \sim \boldsymbol{r}_1^0 \tag{15.4.2c}$$

where $\overline{\alpha} = \bar{1} \times \alpha;\ \alpha \in T_p$.

Now, a wave vector group $\hat{G}_{\boldsymbol{k}}$ is either symmorphic or non-symmorphic.

(i) If it is *symmorphic*, $G_{\boldsymbol{k}}$ must be a subgroup of T_p; hence, from (15.3.13c) and (15.4.2b), the representation $\dot{\Delta}_{\boldsymbol{k}}^{(2)}(G_{\boldsymbol{k}})$ is diagonal. Thus, for $\beta \in G_{\boldsymbol{k}} \leqslant T_p$, we have two one-dimensional representations $\dot{\Delta}'_{\boldsymbol{k}}(\beta)$ and $\dot{\Delta}''_{\boldsymbol{k}}(\beta)$ with the respective bases, from (15.3.10a) and (15.3.13c),

$$\phi_1^{\boldsymbol{k}}(\boldsymbol{r}) \in \dot{\Delta}'_{\boldsymbol{k}}(\beta) = 1$$
$$\phi_2^{\boldsymbol{k}}(\boldsymbol{r}) \in \dot{\Delta}''_{\boldsymbol{k}}(\beta) = \exp[\mathrm{i}\boldsymbol{k} \cdot (\mathbf{1} - \beta)\boldsymbol{r}_2^0] \tag{15.4.3}$$

The latter, $\dot{\Delta}''_{\boldsymbol{k}}(\beta)$, is tabulated for all $\beta \in T_p$ in Table 15.1 at the high-symmetry points given in (15.4.1b): it takes either 1 or -1; cf. Bir and Pikus (1974).

(ii) If $\hat{G}_{\boldsymbol{k}}$ is *non-symmorphic*, it must have the following group structure, in view of (15.4.1a):

$$\hat{G}_{\boldsymbol{k}} = \hat{H}_{\boldsymbol{k}} + q\hat{H}_{\boldsymbol{k}}, \qquad q = \{\overline{\boldsymbol{\alpha}}|\boldsymbol{\tau}\} \tag{15.4.4}$$

where $\hat{H}_{\boldsymbol{k}} = \{\{h|\boldsymbol{t}_n\}\}$ is a subgroup of $\hat{T}_p$ and $\overline{\alpha} = \bar{1} \times \alpha(\alpha \in T_p)$. Since $\hat{H}_{\boldsymbol{k}}$ is symmorphic, the subduced representation $\Delta_{\boldsymbol{k}}^{(2)\downarrow}(\hat{H}_{\boldsymbol{k}})$ onto $\hat{H}_{\boldsymbol{k}}$ is again diagonal, as given by (15.4.3), whereas the representative of the augmentor q is off-diagonal in view of (15.4.2c). Thus, from (15.3.10), we have the following basic representation of a non-symmorphic space group $\hat{G}_{\boldsymbol{k}}$ $(< \hat{O}_i)$:

$$\mathring{B}[\phi_1^{\boldsymbol{k}}, \phi_2^{\boldsymbol{k}}] = [\phi_1^{\boldsymbol{k}}, \phi_2^{\boldsymbol{k}}]\Delta_{\boldsymbol{k}}^{(2)}(B), \qquad \forall\, B \in \hat{G}_{\boldsymbol{k}} \tag{15.4.5a}$$

where

$$\Delta_{\boldsymbol{k}}^{(2)}(h) = \begin{bmatrix} 1 & 0 \\ 0 & \exp[\mathrm{i}\boldsymbol{k} \cdot (\mathbf{1} - h)\boldsymbol{r}_2^0)] \end{bmatrix}, \qquad \forall\, h \in H_{\boldsymbol{k}}$$

$$\Delta_{\boldsymbol{k}}^{(2)}(q) = \begin{bmatrix} 0 & \mathrm{e}^{\mathrm{i}\boldsymbol{k}\cdot(\mathbf{1}-q)\boldsymbol{r}_2^0} \\ \mathrm{e}^{-\mathrm{i}\boldsymbol{k}\cdot\boldsymbol{r}_2^0} & 0 \end{bmatrix} \tag{15.4.5b}$$

Table 15.1. *The phase factor* $\exp[\mathrm{i}\boldsymbol{k}\cdot(\mathbf{1}-\beta)\boldsymbol{r}_2^0]$ *for* $\beta\in G_{\boldsymbol{k}}\leqslant T_p$; $\boldsymbol{r}_2^0=(\frac{1}{4}\frac{1}{4}\frac{1}{4})$ ($\{\boldsymbol{a}_1=(0\frac{1}{2}\frac{1}{2}),\ \boldsymbol{a}_2=(\frac{1}{2}0\frac{1}{2}),\ \boldsymbol{a}_3=(\frac{1}{2}\frac{1}{2}0)\}$ *is the primitive basis of the face-centred cubic lattice with lattice spacing* $\boldsymbol{a}=1$)

		$\exp[\mathrm{i}\boldsymbol{k}\cdot(\mathbf{1}-\beta)\boldsymbol{r}_2^0]$; $\boldsymbol{k}=2\pi[\xi,\mu,\zeta]$					
$\alpha\in\mathrm{T_p}$	$\exp[\mathrm{i}\boldsymbol{k}\cdot(\mathbf{1}-\alpha)\boldsymbol{r}_2^0]$	Γ [000]	X [001]	Δ $[00\zeta]$	Λ $(\xi\xi\xi)$	L $(\frac{1}{2}\frac{1}{2}\frac{1}{2})$	$\Sigma=[\xi\xi0)$ $K=[\frac{3}{4}\frac{3}{4}0)$
$e=(xyz)$	1	1	1	1	1	1	1
$2_x=(x\bar y\bar z)$	$\exp(\mathrm{i}\boldsymbol{k}\cdot\boldsymbol{a}_1)$	1	-1				
$2_y=(\bar x y\bar z)$	$\exp(\mathrm{i}\boldsymbol{k}\cdot\boldsymbol{a}_2)$	1	-1				
$2_z=(\bar x\bar y z)$	$\exp(\mathrm{i}\boldsymbol{k}\cdot\boldsymbol{a}_3)$	1	1				
$\bar 4_x=(\bar x z\bar y)$	$\exp(\mathrm{i}\boldsymbol{k}\cdot\boldsymbol{a}_2)$	1					
$\bar 4_y=(\bar z\bar y x)$	$\exp(\mathrm{i}\boldsymbol{k}\cdot\boldsymbol{a}_3)$	1					
$\bar 4_z=(y\bar x\bar z)$	$\exp(\mathrm{i}\boldsymbol{k}\cdot\boldsymbol{a}_1)$	1	-1				
$\bar 4_{\bar x}=(\bar x\bar z y)$	$\exp(\mathrm{i}\boldsymbol{k}\cdot\boldsymbol{a}_3)$	1					
$\bar 4_{\bar y}=(z\bar y\bar x)$	$\exp(\mathrm{i}\boldsymbol{k}\cdot\boldsymbol{a}_1)$	1					
$\bar 4_{\bar z}=(\bar y x\bar z)$	$\exp(\mathrm{i}\boldsymbol{k}\cdot\boldsymbol{a}_2)$	1	-1				
$3_{xyz}=(zxy)$	1	1			1	1	
$3_{x\bar y\bar z}=(\bar z\bar x y)$	$\exp(\mathrm{i}\boldsymbol{k}\cdot\boldsymbol{a}_3)$	1					
$3_{\bar x y\bar z}=(z\bar x\bar y)$	$\exp(\mathrm{i}\boldsymbol{k}\cdot\boldsymbol{a}_1)$	1					
$3_{\bar x\bar y z}=(\bar z x\bar y)$	$\exp(\mathrm{i}\boldsymbol{k}\cdot\boldsymbol{a}_2)$	1					
$3_{\bar x\bar y\bar z}=(yzx)$	1	1			1	1	
$3_{\bar x yz}=(\bar y z\bar x)$	$\exp(\mathrm{i}\boldsymbol{k}\cdot\boldsymbol{a}_2)$	1					
$3_{x\bar y z}=(\bar y\bar z x)$	$\exp(\mathrm{i}\boldsymbol{k}\cdot\boldsymbol{a}_3)$	1					
$3_{xy\bar z}=(y\bar z\bar x)$	$\exp(\mathrm{i}\boldsymbol{k}\cdot\boldsymbol{a}_1)$	1					
$\bar 2_{\bar y z}=(xzy)$	1	1			1	1	
$\bar 2_{yz}=(x\bar z\bar y)$	$\exp(\mathrm{i}\boldsymbol{k}\cdot\boldsymbol{a}_1)$	1					
$\bar 2_{\bar x z}=(zyx)$	1	1			1	1	
$\bar 2_{xz}=(\bar z y\bar x)$	$\exp(\mathrm{i}\boldsymbol{k}\cdot\boldsymbol{a}_2)$	1					
$\bar 2_{x\bar y}=(yxz)$	1	1	1		1	1	1
$\bar 2_{xy}=(\bar y\bar x z)$	$\exp(\mathrm{i}\boldsymbol{k}\cdot\boldsymbol{a}_3)$	1	1	1			

(iii) In the important special case in which the wave vector $\boldsymbol{k}$ is *invariant* under $G_{\boldsymbol{k}}$, from (15.2.2b) the vector representation $\dot\Delta_{\boldsymbol{k}}^{(2)}(G_{\boldsymbol{k}})$ corresponding to (15.4.5b) takes the form

$$\dot\Delta_{\boldsymbol{k}}^{(2)}(h)=\begin{bmatrix}1&0\\0&1\end{bmatrix},\qquad \dot\Delta_{\boldsymbol{k}}^{(2)}(\overline{\alpha})=\begin{bmatrix}0&1\\1&0\end{bmatrix} \tag{15.4.6a}$$

which is reduced to two one-dimensional unirreps of $G_{\boldsymbol{k}}$

$$\dot\Delta_{\boldsymbol{k}}^{(\pm)}(h)=1,\qquad \dot\Delta_{\boldsymbol{k}}^{(\pm)}(\overline{\alpha})=\pm1 \tag{15.4.6b}$$

with the respective bases

$$\phi_+^{\boldsymbol{k}}=(\phi_1^{\boldsymbol{k}}+\phi_2^{\boldsymbol{k}})\in\dot\Delta_{\boldsymbol{k}}^{(+)},\qquad \phi_-^{\boldsymbol{k}}=(\phi_1^{\boldsymbol{k}}-\phi_2^{\boldsymbol{k}})\in\dot\Delta_{\boldsymbol{k}}^{(-)} \tag{15.4.6c}$$

This special case applies for all interior points of the Brillouin zone and for some surface points, as in the case of the symmetry point K (see below).

In the above, for when $\hat{G}_{\boldsymbol{k}}$ is symmorphic or $\boldsymbol{k}$ is invariant under $G_{\boldsymbol{k}}$, we have shown that the unirreps $\Delta_{\boldsymbol{k}}^{(j)}$ contained in the weighted permutation representation $\Delta_{\boldsymbol{k}}^{(2)}(\hat{G}_{\boldsymbol{k}})$ given in (15.3.10b) become *one-dimensional*; hence, the direct product representations $D_i^{(1)} \times \Delta_{\boldsymbol{k}}^{(j)}$ contained in the direct product representation $D^{(1)} \times \Delta_{\boldsymbol{k}}^{(2)}$ of $\hat{G}_{\boldsymbol{k}}$ are all irreducible. A similar simplification for $\hat{G}_{\boldsymbol{k}}$ on a surface point of the Brillouin zone will be achieved by subducing $D^{(1)} \times \Delta_{\boldsymbol{k}}^{(2)}$ onto the halving subgroup of $\hat{G}_{\boldsymbol{k}}$ which is symmorphic.

15.4.2 Construction of the symmetry coordinates of vibration

The general discussion given above will now be applied to construct the symmetry coordinates of vibration of the diamond crystal at each symmetry point in the Brillouin zone given in (15.4.1b).

1. At $\Gamma(\boldsymbol{k}=0) \in \hat{O}_i$, since Γ is an interior point with $\boldsymbol{k}=0$, we have $\check{\Delta}_{\Gamma}^{(2)}(O_i) = \dot{\Delta}_{\Gamma}^{(2)}(O_i)$, which is reduced to two one-dimensional vector unirreps, according to (15.4.6b):

$$\dot{\Delta}_{\Gamma}^{(\pm)}(h) = 1, \qquad \dot{\Delta}_{\Gamma}^{(\pm)}(\bar{1}) = \pm 1 \in O_i \tag{15.4.7a}$$

where $h \in T_p$. From the character table of the point group O_i, these unirreps are identified as A_{1g} and A_{2u}, respectively. Their bases are, from (15.4.6c),

$$\begin{aligned} \phi_+^{\Gamma} &= (\phi_1^{\Gamma} + \phi_2^{\Gamma}) \in A_{1g}^{\Gamma}; \qquad 1 \\ \phi_-^{\Gamma} &= (\phi_1^{\Gamma} - \phi_2^{\Gamma}) \in A_{2u}^{\Gamma}; \qquad xyz \end{aligned} \tag{15.4.7b}$$

Here 1 and xyz are the elementary bases of the point co-group O_i. See (15.2.2f) for the notation of the unirreps.

From the character table of the point group O_i, the elementary basis $[x, y, z]$ belongs to $D^{(1)} = T_{1u} \in O_i$. Hence, the direct products $T_{1u} \times \dot{\Delta}_{\Gamma}^{(\pm)}$ are irreducible and given by

$$T_{1u} \times A_{1g}^{\Gamma} = T_{1u}^{\Gamma}, \quad T_{1u} \times A_{2u}^{\Gamma} = T_{2g}^{\Gamma} \in \hat{O}_i$$

Correspondingly, the normal coordinates at $\Gamma \in \hat{O}_i$ which transform according to these direct products are given by the direct product bases

$$\begin{aligned} T_{1u}^{\Gamma}&: \quad [\partial_x, \partial_y, \partial_z]\phi_+^{\Gamma} \sim [x_1^{\Gamma} + x_2^{\Gamma}, y_1^{\Gamma} + y_2^{\Gamma}, z_1^{\Gamma} + z_2^{\Gamma}] \\ T_{2g}^{\Gamma}&: \quad [\partial_x, \partial_y, \partial_z]\phi_-^{\Gamma} \sim [x_1^{\Gamma} - x_2^{\Gamma}, y_1^{\Gamma} - y_2^{\Gamma}, z_1^{\Gamma} - z_2^{\Gamma}] \end{aligned} \tag{15.4.7c}$$

using the correspondence given by (15.3.7). Note that the first symmetry coordinates in (15.4.7c) transform like the components of a polar vector and thus belong to the acoustical mode whereas the second coordinates belong to the optical mode. Since $\boldsymbol{k}=0$ at Γ, from the definition (15.3.2), the Fourier transforms $(x_s^{\Gamma}, y_s^{\Gamma}, z_s^{\Gamma})$ are proportional to the displacement of the center of mass of the sth atoms in the unit cells of the crystal:

$$u_s^\Gamma = (M_s/N)^{1/2} \sum_n u(r_s^{(n)})$$

2. At Δ ($\boldsymbol{k} = 2\pi[0, 0, \zeta]$) $\in \hat{C}_{4v}^z$, the wave vector group has been defined by (15.2.10). Since the point Δ is again inside the Brillouin zone, the representation $\Delta_\Delta^{(2)}(\hat{C}_{4v})$ is reduced to one-dimensional vector unirreps, according to (15.4.6b),

$$\dot{\Delta}_\Delta^{(\pm)}(h) = 1, \qquad \dot{\Delta}_\Delta^{(\pm)}(4_z) = \pm 1 \in C_{2v};\ h \in C_{2v} = \{2_z, \overline{2}_{x\overline{y}}\}$$

From the character table of C_{4v}, these unirreps $\Delta_\Delta^{(+)}$ and $\Delta_\Delta^{(-)}$ are identified as the unirreps A_1^Δ and B_2^Δ, respectively, and their bases are classified, from (15.4.6c), by

$$\begin{aligned} \phi_+^\Delta &= \phi_1^\Delta + \phi_2^\Delta \in A_1^\Delta \text{ of } \hat{C}_{4v}; \qquad 1 \\ \phi_-^\Delta &= \phi_1^\Delta - \phi_2^\Delta \in B_2^\Delta \text{ of } \hat{C}_{4v}; \qquad xy \end{aligned} \tag{15.4.8}$$

following the notation introduced by (15.2.2f). Now, the reduction of the representation $D^{(1)}$ with respect to C_{4v}^z is given by

$$D^{(1)} = \underset{z}{A_1} + \underset{[x,\, y]}{E} \in C_{4v}$$

Thus, the direct products $D^{(1)} \times \{A_1, B_2\}$ are reduced to

$$A_1 \times A_1^\Delta = A_1^\Delta, \qquad A_1 \times B_2^\Delta = B_2^\Delta, \qquad E \times A_1^\Delta = E^\Delta, \qquad E \times B_2^\Delta = E^\Delta \in C_{4v}$$

so that the symmetry coordinates of vibration at Δ belonging to the unirreps of $\hat{C}_{4v}$ are given by

$$\begin{aligned} A_1^\Delta:&\quad \partial_z\phi_+^\Delta \sim (z_1^\Delta + z_2^\Delta) \\ B_2^\Delta:&\quad \partial_z\phi_-^\Delta \sim (z_1^\Delta - z_2^\Delta) \\ 2E^\Delta:&\quad [\partial_x, \partial_y]\phi_+^\Delta \sim (x_1^\Delta + x_2^\Delta, y_1^\Delta + y_2^\Delta) \\ &\quad [\partial_y, \partial_x]\phi_-^\Delta \sim (y_1^\Delta - y_2^\Delta, x_1^\Delta - x_2^\Delta) \end{aligned}$$

These two bases of E^Δ are mutually orthogonal but belong to exactly the same matrix representation[3] (not up to a similarity transformation).

3. At X ($\boldsymbol{k} = 2\pi[0, 0, 1]$) $\in \hat{D}_{4i}^z = \hat{D}_{2p}^z + \hat{I}\hat{D}_{2p}^z$. Previously, in (15.2.15a), we have shown that the representation of $\hat{D}_{4i}^z$ at X is described by the projective unirreps of D_{4i}^z belonging to the class of the factor system $K(-1, +1)$. Since the projective unirreps of D_{4i}^z are described by (15.2.16) in terms of the unirreps of the symmorphic halving subgroup $\hat{D}_{2p}^z = \{\overline{4}_z, 2_x\}$, we first classify the bases ϕ_1^x and ϕ_2^x by the unirreps of the subgroup D_{2p}^z and obtain, from (15.4.3),

$$\phi_1^x \in \dot{\Delta}' = \underset{1}{A_1}, \qquad \phi_2^x \in \dot{\Delta}'' = B_2 \in \underset{xy}{D_{2p}} \tag{15.4.9a}$$

using Table 15.1 and the character table of D_{2p}^z. Next, the reduction of $D^{(1)}$ with respect to D_{2p} yields

$$D^{(1)} = \underset{z}{B_2} + \underset{[x,\, -y]}{E} \in D_{2p}$$

[3] From the correspondence theorem developed in Chapter 7 we have $[x, y] \sim [\partial_y, \partial_x]xy \sim [\partial_y, \partial_x]\phi_-^\Delta$, where $xy \in B_2$ of C_{4v}. Here, $A \sim B$ means that 'A and B belong to the same matrix representation.'

Therefore, the direct product $D^{(1)} \times \{A_1, B_2\}$ is reduced to

$$B_2 \times A_1 = B_2, \qquad B_2 \times B_2 = A_1, \qquad E \times A_1 = E, \qquad E \times B_2 = E \in D_{2p}$$

The corresponding direct product bases for $\hat{D}_{2p}$ are

$$\begin{aligned} A_1: &\quad \partial_z \phi_2^{\mathrm{x}} \sim z_2^{\mathrm{x}}, \qquad B_2: \quad \partial_z \phi_1^{\mathrm{x}} \sim z_1^{\mathrm{x}} \\ 2E: &\quad [\partial_x, -\partial_y]\phi_1^{\mathrm{x}} \sim [x_1^{\mathrm{x}}, -y_1^{\mathrm{x}}] \equiv \psi_1^E \\ &\quad [\partial_y, -\partial_x]\phi_2^{\mathrm{x}} \sim [y_2^{\mathrm{x}}, -x_2^{\mathrm{x}}] \equiv \psi_2^E \end{aligned} \qquad (15.4.9\text{b})^4$$

where both bases of E belong to the same unirrep based on the elementary bases $[x, -y]$. Accordingly, from (15.2.16), the bases of the unirreps of the wave vector group $\hat{D}_{4i}^z$ compatible with the above unirreps of $\hat{D}_{2p}$ are given, with the use of $(\bar{1}|\tau)\phi_2^{\mathrm{x}} = \mathrm{i}\phi_1^{\mathrm{x}}$ obtained from (15.4.5), by

$$\begin{aligned} X_1 = D(A_1, B_2; 1)^{\mathrm{x}}: &\quad [z_2^{\mathrm{x}}, -\mathrm{i}z_1^{\mathrm{x}}] \\ X_3 = D(E; -\sigma_x)^{\mathrm{x}}: &\quad [x_1^{\mathrm{x}} + \mathrm{i}y_2^{\mathrm{x}}, -y_1^{\mathrm{x}} - \mathrm{i}x_2^{\mathrm{x}}] \\ X_4 = D(E; \sigma_x)^{\mathrm{x}}: &\quad [x_1^{\mathrm{x}} - \mathrm{i}y_2^{\mathrm{x}}, -y_1^{\mathrm{x}} + \mathrm{i}x_2^{\mathrm{x}}] \end{aligned} \qquad (15.4.9\text{c})$$

4. At Λ ($\boldsymbol{k} = 2\pi[\xi, \xi, \xi]$) $\in \hat{C}_{3v}^{xyz}$, the wave vector group is symmorphic and defined in (14.3.6c) by the generators $\{3_{xyz}, \bar{2}_{x\bar{y}}\}$ and also Λ is an interior point of the B-zone. Moreover, all elements of C_{3v} leave both points $\boldsymbol{r}_1^0$ and $\boldsymbol{r}_2^0$ in their places so that, from (15.4.3), we have $\dot{\Delta}_\lambda' = \dot{\Delta}_\lambda'' = 1$. Thus, from the character table of C_{3v}, we have

$$\phi_1^\Lambda, \phi_2^\Lambda \in A_1^\Lambda \text{ of } \hat{C}_{3v}^{xyz}$$

The decomposition of the 3×3 representation $D^{(1)}$ with respect to C_{3v}^{xyz} is given by

$$D^{(1)} = \underset{z'}{A_1} + \underset{[x',\, y']}{E}$$

where x', y' and z' are the new coordinates defined below in (15.4.11), of which z' is pointing diagonally with respect to the original coordinates. Thus the direct product $D^{(1)} \times A_1$ is reduced to

$$A_1 \times A_1 = A_1, \qquad E \times A_1 = E \quad \in C_{3v}$$

so that the SALCs belonging to the unirreps of $\hat{C}_{3v}^{xyz}$ are given by

$$\begin{aligned} 2A_1^\Lambda: &\quad \partial_{z'}\phi_1^\Lambda \sim z_1' = (x_1 + y_1 + z_1)^\Lambda/\sqrt{3} \\ &\quad \partial_{z'}\phi_2^\Lambda \sim z_2' = (x_2 + y_2 + z_2)^\Lambda/\sqrt{3} \\ 2E^\Lambda: &\quad [\partial_{x'}, \partial_{y'}]\phi_1^\Lambda \sim [x_1', y_1']^\Lambda = [(x_1 - y_1)/\sqrt{2}, (x_1 + y_1 - 2z_1)/\sqrt{6}]^\Lambda \\ &\quad [\partial_{x'}, \partial_{y'}]\phi_2^\Lambda \sim [z_2', y_2']^\Lambda = [(x_2 - y_2)/\sqrt{2}, (x_2 + y_2 - 2z_2)/\sqrt{6}]^\Lambda \end{aligned} \qquad (15.4.10)$$

[4] $[\partial_y, -\partial_x]xy \simeq [x, -y]$, where $xy \in B_2$ of D_{2p}.

where x', y' and z' are defined by

$$z' = (x+y+z)/\sqrt{3}, \qquad y' = (x+y-2z)/\sqrt{6}, \qquad x' = (x-y)/\sqrt{2} \quad (15.4.11)$$

due to the fact that the principal axis of C_{3v}^{xyz} is pointing diagonally with respect to the x, y and z coordinate axes.[5] These degenerate bases are orthogonal if their overlap integrals are neglected.

5. At $L\,(\boldsymbol{k} = 2\pi[\frac{1}{2}, \frac{1}{2}, \frac{1}{2}]) \in \hat{D}_{3i}^{xyz}$, the wave vector group is defined by

$$\hat{D}_{3i}^{xyz} = \hat{C}_{3v}^{xyz} + \{\bar{1}|\boldsymbol{\tau}\}\hat{C}_{3v}^{xyz}, \qquad C_{3v}^{xyz} = \{3_{xyz}, \bar{2}_{x\bar{y}}\}$$

The point L is a surface point of the Brillouin zone, as shown by Figure 14.6. It is not invariant under D_{3i} because $\bar{1}\boldsymbol{k} = -\boldsymbol{k}$. However, the generators of the halving subgroup $\hat{C}_{3v}^{xyz}$ commute with the augmentor $\hat{I} = \{\bar{1}|\tau\}$ so that the projective representations of D_{3i} become the vector representations belonging to the class $K(1, 1)$, as one can see from (12.5.9b). Thus, for $\{\beta_\lambda|\boldsymbol{b}_\lambda\} \in \hat{D}_{3i}$

$$\Delta_L^{(2)}(\{\beta_\lambda|\boldsymbol{b}_\lambda\}) = \check{\Delta}_L^{(2)}(\beta_\lambda) = \dot{\Delta}_L^{(2)}(\beta_\lambda), \qquad \forall \beta_\lambda \in D_{3i} \qquad (15.4.12)$$

which may be compared with (14.4.17b). From (15.4.5) and $h\boldsymbol{r}_2^0 = \boldsymbol{r}_2^0$ for all $h \in C_{3v}$, the representation is given, with $\bar{h} = \bar{1}h\ (h \in C_{3v})$, by

$$\dot{\Delta}_L^{(2)}(h \in C_{3v}) = \begin{bmatrix} 1 & 0 \\ 0 & 1 \end{bmatrix}, \qquad \dot{\Delta}_L^{(2)}(\bar{h}) = \begin{bmatrix} 0 & \exp(\frac{3}{4}\pi \mathrm{i}) \\ \exp(-\frac{3}{4}\pi \mathrm{i}) & 0 \end{bmatrix}$$

which indeed forms a vector representation of D_{3i}. Since $\dot{\Delta}_L^{(2)}(\bar{h})$ is involutional, unitary and Hermitian (IUH) and also anti-commutes with the Pauli spin σ_z, it is diagonalized by an involutional transformation, from Lemma 2.1.1,

$$Y\dot{\Delta}_L^{(2)}(h)Y = \begin{bmatrix} 1 & 0 \\ 0 & 1 \end{bmatrix}, \qquad Y\dot{\Delta}_L^{(2)}(\bar{h})Y = \begin{bmatrix} 1 & 0 \\ 0 & -1 \end{bmatrix}$$

with $Y = 2^{-1/2}[\dot{\Delta}^{(2)}(\bar{h}) + \sigma_z]$. Thus, the representation is reduced to two one-dimensional representations defined by

$$\dot{\Delta}_L^{(\pm)}(h) = 1, \qquad \dot{\Delta}_L^{(\pm)}(\bar{h}) = \pm 1; \qquad \forall\, h \in C_{2v}$$

These are characterized by the unirreps of D_{3i} as follows:

$$\dot{\Delta}_l^{(+)} \in A_{1\mathrm{g}}^L, \qquad \dot{\Delta}_L^{(-)} \in A_{2\mathrm{u}}^L$$

via the character table. The corresponding bases are determined by

$$[\phi_1^k, \phi_2^k]Y = [\psi_+^L, \psi_-^L]$$

where

$$\psi_+^L = \phi_1^L + \rho\phi_2^L \in A_{1\mathrm{g}}^L \text{ of } \hat{D}_{3i}^{z'}; \qquad 1, z'^2$$

$$\psi_-^L = \rho^{-1}\phi_1^L - \phi_2^L \sim \phi_1^L - \rho\phi_2^L \in A_{2\mathrm{u}}^L \text{ of } \hat{D}_{3i}^{z'}; \qquad z'$$

Here $\rho = \exp(-3\pi \mathrm{i}/4)$ and z' has been defined in (15.4.11). The same result could have been obtained by the projection operator method using (15.1.9a).

[5] Note that $\boldsymbol{k}' = (\boldsymbol{i}+\boldsymbol{j}+\boldsymbol{k})/\sqrt{3}$, $\boldsymbol{j}' = (\boldsymbol{i}+\boldsymbol{j}-2\boldsymbol{k})/\sqrt{6}$ and $\boldsymbol{i}' = (\boldsymbol{i}-\boldsymbol{j})/\sqrt{2}$. Thus e.g. $z' = \boldsymbol{r}\cdot\boldsymbol{k}' = \boldsymbol{r}\cdot(\boldsymbol{i}+\boldsymbol{j}+\boldsymbol{k})/\sqrt{3} = (x+y+z)/\sqrt{3}$.

Now the reduction of the 3×3 representation $D^{(1)}$ is given by

$$D^{(1)} = \underset{z'}{A_{2u}} + \underset{[x',\, y']}{E_u} \in D_{3i}^{z'}$$

Then, the direct products $D^{(1)} \times \dot{\Delta}_L^{(\pm)}$ are reduced to

$$A_{2u} \times A_{1g}^L = A_{2u}^L, \qquad A_{2u} \times A_{2u}^L = A_{1g}^L, \qquad E_u \times A_{1g}^L = E_u^L, \qquad E_u \times A_{2u}^L = E_g^L$$

so that their bases give the following symmetry coordinates of vibration belonging to $\hat{D}_{3i}^{z'}$ at L:

$$\begin{aligned}
A_{1g}^L&: \quad \partial_{z'}\psi_-^L \sim (z_1' - \rho z_2')^L \qquad (\rho \equiv \exp(-3\pi i/4))\\
A_{2u}^L&: \quad \partial_{z'}\psi_+^L \sim (z_1' + \rho z_1')^L\\
E_g^L&: \quad [\partial_{y'}, -\partial_{x'}]\psi_-^L \sim [y_1' - \rho y_2', -x_1' + \rho x_2']^L\\
E_u^L&: \quad [\partial_{x'}, \partial_{y'}]\psi_+^L \sim [x_1' + \rho x_2', y_1' + \rho y_2']^L
\end{aligned}$$

where x', y' and z' are defined by (15.4.11), and $[y', -x']\, z' \in E_g$ of D_{3i}.

6. Finally, we consider Σ ($\boldsymbol{k} = 2\pi[\xi, \xi, 0]$) and K ($\boldsymbol{k} = 2\pi[\frac{3}{4}, \frac{3}{4}, 0]$) $\in \hat{C}_{2v}^{xy}$, $C_{2v}^{xy} = \{2_{xy}, \overline{2}_z\}$ from (14.3.6c). In view of (15.4.1a) and Table 15.1, the elements of $\hat{C}_{2v}^{xy}/T$ are

$$\{e|0\}, \{\overline{2}_{x\overline{y}}|0\}, \{2_{xy}|\boldsymbol{\tau}\}, \{\overline{2}_z|\boldsymbol{\tau}\}; \qquad \boldsymbol{\tau} = (\tfrac{1}{4}, \tfrac{1}{4}, \tfrac{1}{4})$$

Since the point group C_{2v} is isomorphic to the proper point group D_2, the projective unirreps of C_{2v} are p-equivalent to the vector unirreps of C_{2v} (Theorem 12.5.1). Moreover, both points Λ and K are invariant with respect to the point group C_{2v}^{xy} so that the unirreps are described by the vector unirreps $\dot{\Delta}_{\boldsymbol{k}}^{(\pm)}$ given by (15.4.6b), i.e. through the correspondence $2_{z'} \leftrightarrow \{2_{xy}|\tau\}, \overline{2}_{y'} \leftrightarrow \{\overline{2}_z|\tau\}$, we have

$$\begin{aligned}
\phi_+^{\boldsymbol{k}} \in \dot{\Delta}_{\boldsymbol{k}}^{(+)} &= A_1^{\boldsymbol{k}} \text{ of } \hat{C}_{2v}; \qquad 1\\
\phi_-^{\boldsymbol{k}} \in \dot{\Delta}^{(-)} &= B_2^{\boldsymbol{k}} \text{ of } \hat{C}_{2v}; \qquad y'
\end{aligned}$$

using the character table of C_{2v}. Here $\boldsymbol{k}$ stands for Σ or K. The reduction of $D^{(1)}$ for $C_{2v}^{z'}$ is given by

$$D^{(1)} = \underset{z'}{A_1} + \underset{x'}{B_1} + \underset{y'}{B_2}$$

where

$$x' = (x - y)/\sqrt{2}, \qquad y' = z, \qquad z' = (x + y)/\sqrt{2} \tag{15.4.13a}$$

Thus, from the unirreps contained in the direct products $D^{(1)} \times \dot{\Delta}_{\boldsymbol{k}}^{(\pm)}$ for C_{2v}

$$\begin{aligned}
A_1 \times A_1^{\boldsymbol{k}} &= A_1^{\boldsymbol{k}}, \qquad & A_1 \times B_2^{\boldsymbol{k}} &= B_2^{\boldsymbol{k}}\\
B_1 \times A_1^{\boldsymbol{k}} &= B_1^{\boldsymbol{k}}, \qquad & B_1 \times B_2^{\boldsymbol{k}} &= A_2^{\boldsymbol{k}}\\
B_2 \times A_1^{\boldsymbol{k}} &= B_2^{\boldsymbol{k}}, \qquad & B_2 \times B_2^{\boldsymbol{k}} &= A_1^{\boldsymbol{k}}
\end{aligned} \tag{15.4.13b}$$

we obtain the following symmetry coordinates of vibration at Σ or K:

$$2A_1^{\boldsymbol{k}}: \quad \partial_{z'}\phi_+ \sim (z_1' + z_2')^{\boldsymbol{k}} = (x_1 + y_1 + x_2 + y_2)^{\boldsymbol{k}}/\sqrt{2}$$

$$\partial_{y'}\phi_- \sim (y_1' - y_2')^{\boldsymbol{k}} = (z_1 - z_2)^{\boldsymbol{k}}$$

$$A_2^{\boldsymbol{k}}: \quad \partial_{x'}\phi_- \sim (x_1' - x_2')^{\boldsymbol{k}} = (x_1 - y_1 - x_2 + y_2)^{\boldsymbol{k}}/\sqrt{2}$$

$$B_1^{\boldsymbol{k}}: \quad \partial_{x'}\phi_+ \sim (x_1' + x_2')^{\boldsymbol{k}} = (x_1 - y_1 + x_2 - y_2)^{\boldsymbol{k}}/\sqrt{2}$$

$$2B_2^{\boldsymbol{k}}: \quad \partial_{y'}\phi_+ \sim (y_1' + y_2')^{\boldsymbol{k}} = (z_1 + z_2)^{\boldsymbol{k}}$$

$$\partial_{z'}\phi_- \sim (z_1' - z_2')^{\boldsymbol{k}} = (x_1 + y_1 - x_2 - y_2)^{\boldsymbol{k}}/\sqrt{2}$$

where $\boldsymbol{k}$ stands for Σ and K.

Concluding remarks. In this chapter, it has been demonstrated through examples that the energy band eigenfunctions for the electron in a crystal and also the vibrational coordinates of a crystal can be calculated with ease, by using the correspondence theorem on the SALCs developed in Chapter 7 and the projective representations of the point groups given in Table 12.4.

16

Time reversal, anti-unitary point groups and their co-representations

16.1 Time-reversal symmetry, classical

16.1.1 General introduction

We shall begin with the time-reversal symmetry of Newton's equation of motion for a point mass in a conservative field of force. For simplicity, we consider one-dimensional motion because the dimensionality of the space is irrelevant to the discussion. For a particle with unit mass in a force field $f(x)$, the equation of motion is given by

$$\frac{\mathrm{d}^2 x}{\mathrm{d}t^2} = f(x) \tag{16.1.1a}$$

where x is the position coordinate and t is the time. Since the equation is described by the second-order derivative of x with respect to time, it is invariant under time reversal $t \to \tau = -t$

$$\mathrm{d}^2 x/\mathrm{d}\tau^2 = f(x) \tag{16.1.1b}$$

Therefore, if $x(t) = x(x_0, v_0;\ t)$ is a possible solution of the equation of motion, where x_0 and v_0 are the initial position and velocity, respectively, then the time-reversed motion defined by

$$x^\tau(t) = x(x_0, v_0;\ -t) \equiv x(-t) \tag{16.1.1c}$$

is also a solution of the equation of motion. Thus, if $x(t)$ describes the forward motion, then $x^\tau(t) = x(-t)$ describes the corresponding reverse motion, as time t proceeds in the positive direction (see Example 1).

Let the velocity of the forward motion be

$$v(t) = \mathrm{d}x(t)/\mathrm{d}t \equiv v(x_0, v_0;\ t)$$

then the velocity of the reverse motion is given, via direct differentiation of (16.1.1c) with respect to t, by

$$v^\tau(t) = \mathrm{d}x^\tau(t)/\mathrm{d}t = -v(x_0, v_0;\ -t) \equiv -v(-t)$$

In particular, at $t = 0$ we have

$$x^\tau(0) = x(0), \qquad v^\tau(0) = -v(0) \tag{16.1.2}$$

which define the initial condition for the reverse motion. Since the initial condition determines the solution of the equation of motion completely, the reverse motion is expressed in the following two ways:

$$x^\tau(t) = x(x_0, v_0;\ -t) = x(x_0, -v_0;\ t) \tag{16.1.3a}$$

Accordingly, the velocity of the reverse motion is also expressed in the following two ways:

$$v^\tau(t) = -v(x_0, v_0; -t) = v(x_0, -v_0; t) \tag{16.1.3b}$$

See (16.1.5b) for the formal proof.

Example 1. The motion of a falling body with unit mass in the gravitational field $f(x) = -g$ is described by Newton's equation of motion $\ddot{x} = -g$, which has the solution

$$x(t) = x_0 + v_0 t - \tfrac{1}{2}gt^2$$

$$v(t) = v_0 - gt$$

where x_0 and v_0 are the initial position and velocity of the body. The reverse motion is described by

$$x^\tau(t) = x(-t) = x_0 - v_0 t - \tfrac{1}{2}gt^2$$

$$v^\tau(t) = \mathrm{d}x^\tau/\mathrm{d}t = -v_0 - gt$$

These are indeed, in accordance with (16.1.3a) and (16.1.3b), respectively.

In general, there are two kinds of physical quantities under time reversal or time inversion: the position-like quantities and the velocity-like quantities. The position coordinates and the kinetic energy are of the first kind: the reversal of the direction of motion has no effect on these quantities. The velocity, linear momentum and angular momentum (and the spin in quantum mechanics) are of the second kind: these change their signs upon reversal of the direction of motion. Those of the first kind are said to be *even with respect to time reversal*, whereas those of the second kind are *odd with respect to time reversal*. Physical quantities that are even (odd) functions of momenta are obviously even (odd) with respect to time reversal. Naturally, there are physical quantities that are not of either kind. In such a case the quantity can be separated into the even and odd parts (see (16.2.13)). For the time being we shall not be concerned with quantities of this nature. Thus, for example, for a physical quantity $Q = Q(x, p)$ expressed by a function of the position x and the momentum p of a particle we assume

$$Q^\tau = Q(x, -p) = \varepsilon_Q Q(x, p) \tag{16.1.3c}$$

where $\varepsilon_Q = 1\ (-1)$ if Q is even (or odd) with respect to p.

Let us discuss the temporal development of a physical quantity $Q = Q(x, p)$, where x and p are the conjugate pair of the position and momentum of a particle. Firstly, the pair obeys Hamilton's equation of motion

$$\dot{x} = \partial H/\partial p, \qquad \dot{p} = -\partial H/\partial x$$

where $H = H(x, p)$ is the Hamiltonian of the system defined by $H = p^2/2 + V(x)$ with the potential of the force $V(x)$. Thus, the equation of motion of the physical quantity $Q = Q(x, p)$ is described by

$$\begin{aligned}\frac{\mathrm{d}Q}{\mathrm{d}t} &= \dot{x}\frac{\partial Q}{\partial x} + \dot{p}\frac{\partial Q}{\partial p}\\ &= \frac{\partial H}{\partial p}\frac{\partial Q}{\partial x} - \frac{\partial H}{\partial x}\frac{\partial Q}{\partial p} \equiv \mathrm{i}LQ\end{aligned} \tag{16.1.4}$$

where $L(x, p)$ is the *Liouville operator* defined by

$$iL(x, p) = \frac{\partial H}{\partial p}\frac{\partial}{\partial x} - \frac{\partial H}{\partial x}\frac{\partial}{\partial p}$$

Since $H(x, p)$ is an even function of p, the Liouville operator $L(x, p)$ is an odd function of p. Thus the equation of motion is invariant under $t \to -t$, $p \to -p$, i.e. *the classical equation of motion is invariant under time reversal $t \to -t$ provided that velocity-like quantities are reversed simultaneously.* The content of this statement is called *the principle of dynamical reversibility.*

Let $x(t)$ and $p(t)$ be the temporal developments of the position and momentum of the system, respectively, then the motion of a physical quantity $Q = Q(x, p)$ is described by $Q(t) = Q(x(t), p(t))$ whereas the time-reversed motion of $Q^\tau(t)$ is described by

$$Q^\tau(t) = Q(x^\tau(t), p^\tau(t)) = Q(x(-t), -p(-t)) = \varepsilon_Q Q(x(-t), p(-t)) \equiv \varepsilon_Q Q(-t) \tag{16.1.5a}$$

where we have used (16.1.3c) for the third equality. Let $\{x_0, p_0\}$ be the initial condition for the forward motion, then the initial condition $\{x_0^\tau, p_0^\tau\}$ for the reverse motion is given by

$$x_0^\tau = x_0, \qquad p_0^\tau = -p_0$$

Since the initial condition completely determines the temporal development of the system via the equation of motion, if $Q(t) = Q(x_0, p_0; t)$ then $Q^\tau(t) = Q(x_0, -p_0; t)$. Combining this with (16.1.5a) we see that the time-reversed quantity $Q^\tau(t)$ is described in the following two ways:

$$Q^\tau(t) = Q(x_0, -p_0; t) = \varepsilon_Q Q(x_0, p_0; -t) \qquad (16.1.5b)^1$$

It is stressed here again that the only difference between $Q(t)$ and $Q^\tau(t)$ occurs in their initial conditions.

16.1.2 The time correlation function

Let $\overline{Q(t)}$ be the ensemble average of $Q(t)$ defined by

$$\overline{Q(t)} = \int \rho(x_0, p_0) Q(x_0, p_0; t)\, dx_0\, dp_0$$

where $\rho(x_0, p_0)$ is the ensemble density over the phase space. If p_0 and $-p_0$ are equally probable in the ensemble, i.e. $\rho(x_0, -p_0) = \rho(x_0, p_0)$, then we have

$$\overline{Q(t)} = \int \rho(x_0, p_0) Q(t_0, -p_0, t)\, dx_0\, dp_0 = \varepsilon_Q \overline{Q(-t)}$$

[1] The formal proof of this basic relation based on the Liouville equation of motion is as follows: From the formal solution of (16.1.4)

$$Q(t) = \exp[iL(x_0, p_0)t]\, Q(x_0, p_0)$$

we have

$$\begin{aligned} Q^\tau(t) &= Q(x_0, -p_0; t) = \exp[iL(x_0, -p_0)t] Q(x_0, -p_0) \\ &= \exp[-iL(x_0, p_0)t]\, \varepsilon_Q Q(x_0, p_0) = \varepsilon_Q Q(x_0, p_0; -t) \end{aligned}$$

where in the third equality we have used the fact that $L(x, p)$ is an odd function of p and also that $Q(x_0, -p_0) = \varepsilon_Q Q(x_0, p_0)$.

with use of (16.1.5b) in the second equality. Accordingly, we have *the time-reversal symmetry*

$$\overline{Q(t)} = \varepsilon_Q \overline{Q(-t)} \tag{16.1.6a}$$

i.e. the ensemble average is either even or odd with respect to time. Let us apply this relation to *the time correlation function* $\overline{Q(t)} = \overline{A(t)B(0)}$ for the correlation between two physical quantities A and B. Then, from (16.1.6a) with $\varepsilon_Q = \varepsilon_A \varepsilon_B$, we have

$$\overline{A(t)B(0)} = \varepsilon_A \varepsilon_B \overline{A(-t)B(0)} \tag{16.1.6b}$$

where $\varepsilon_Q = \varepsilon_A \varepsilon_B$ comes in because the even–oddness of Q depends on those of A and B. This relation (16.1.6b) plays the crucial role in deriving *Onsager's reciprocal relations* (Onsager 1931) for transport coefficients.

In particular, when $t \to 0$, from (16.1.6a)

$$\overline{Q(+0)} = \varepsilon_Q \overline{Q(-0)}$$

Thus, if Q is position-like ($\varepsilon_Q = 1$), the two averages converge to the same limit, whereas when Q is velocity-like ($\varepsilon_Q = -1$) the two averages differ in their signs (see Figure 16.1 for $v(t)$ and $v(t)^\tau$).

16.1.3 Onsager's reciprocity relation for transport coefficients

For example, the flow $\boldsymbol{J} = (J_1, J_2, J_3)$ of electricity in a metal under an electric field $\boldsymbol{E} = (E_1, E_2, E_3)$ is described by

$$J_i = \sum_{j=1}^{3} \sigma_{ij} E_j; \qquad i = 1, 2, 3$$

where σ_{ij} is called the electric conductivity tensor. It has been firmly established by experiments that the conductivity tensor is symmetric:

$$\sigma_{ij} = \sigma_{ji}; \qquad i, j = 1, 2, 3$$

even if the metal is completely asymmetric. This is a typical example of *Onsager's reciprocity relation* (Onsager 1931) for transport coefficients originating from the time-reversal symmetry. In the following we shall prove the reciprocity relation for transport coefficients in general.

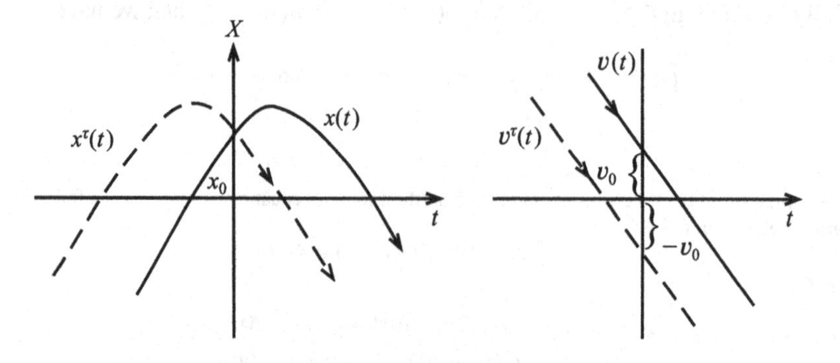

Figure 16.1. The forward and reverse motion of a falling body with $x_0 > 0$, $v_0 > 0$.

Consider a small subsystem surrounded by a large bath in equilibrium. Let $(a_1, a_2, \ldots, a_n)$ be the deviations from the equilibrium state variables of the subsystem (like the energy, the volume and the number of molecules of each species), then the deviation of the entropy S of the subsystem from the maximum equilibrium value is given by

$$\Delta S = \Delta S(a_1, a_2, \ldots) = \frac{1}{2}\sum_{i,j} \frac{\partial^2 S}{\partial a_i\, \partial a_j} a_i a_j$$

$$\equiv -\frac{1}{2}\sum_{ij} g_{ij} a_i a_j \leqslant 0, \qquad g_{ij} = g_{ji} \tag{16.1.7}$$

According to the statistical mechanical principle, the probability distribution of the fluctuations $(a_1, a_2, \ldots, a_n)$ is described by

$$P \propto \exp \Delta S = \exp\left(-\frac{1}{2}\sum_{ij} g_{ij} a_i a_j\right)$$

The driving force for the ith fluctuation is defined by

$$X_i = \partial \Delta S / \partial a_i = -\sum_j g_{ij} a_j \tag{16.1.8}$$

We shall first establish that the force X_i and the fluctuation a_j are not correlated unless $i = j$, i.e.

$$\overline{a_i x_j} = -\delta_{ij} = \begin{cases} -1, & \text{if } i = j \\ 0, & \text{only if } i \neq j \end{cases} \tag{16.1.9}$$

where the correlation $\overline{a_i X_j}$ is defined by

$$\overline{a_i X_j} = \int a_i \frac{\partial \Delta S}{\partial a_j} P\, \mathrm{d}\tau$$

Here $\mathrm{d}\tau = \mathrm{d}a_1\, \mathrm{d}a_2 \ldots \mathrm{d}a_n$. The proof is trivial, from $\Delta S = \ln P +$ constant, we have

$$\overline{a_j X_j} = \int a_i \frac{\partial \ln P}{\partial a_j} P\, \mathrm{d}\tau$$

$$= \int a_i\, \partial P / \partial a_j\, \mathrm{d}\tau$$

$$= a_i P \Big|_{-\infty}^{+\infty} - \delta_{ij} \int P\, \mathrm{d}\tau = -\delta_{ij}$$

which is the required proof.

Next consider the flow[2] of the ith variable a_i defined by the time derivative $\dot{a}_i$. Since it must be a function of the fluctuations $(a_1, a_2, \ldots, a_n)$, it must also be a function of the forces $(X_1, X_2, \ldots, X_n)$, through (16.1.8),

$$\dot{a}_i = \dot{a}_i(a_1, a_2, \ldots, a_n) = \dot{a}_i(X_1, X_2, \ldots, X_n)$$

[2] For example, let $\boldsymbol{Q} = \sum_i e_i \boldsymbol{r}_i$ be the electric moment of the subsystem, where e_i and $\boldsymbol{r}_i$ are the electric charge and position of the ith particle. Then the electric flow is defined by $\dot{\boldsymbol{Q}} = \sum_i e_i \dot{\boldsymbol{r}}_i$.

In the region of the linear approximation we have

$$\dot{a}_i = \sum_j L_{ij} X_j \tag{16.1.10}$$

Following Onsager we shall show that the following *reciprocity relation* holds:

$$L_{ij} = L_{ji} \tag{16.1.11}$$

Proof. Firstly, L_{ij} is expressed by the correlation

$$\overline{\dot{a}_i a_k} = \sum_j L_{ij} \overline{X_j a_k} = -L_{ik} \tag{16.1.12}$$

where use of (16.1.9) for the second equality has been made. Now, from the time-reversal symmetry (16.1.6b) we have

$$\overline{a_i(t)a_k(0)} = \overline{a_i(-t)a_k(0)} = \overline{a_i(0)a_k(t)} = \overline{a_k(t)a_i(0)}$$

where we have assumed that the state variables a_i are all even with respect to time reversal and commute with each other, being classical quantities. By differentiation of the above relation with respect to time t we have

$$\overline{\dot{a}_i(t)a_k} = \overline{\dot{a}_k(t)a_i(0)}$$

Now, on letting $t \to 0$, we obtain

$$\overline{\dot{a}_i a_k} = \overline{\dot{a}_k a_i}$$

i.e.

$$L_{ik} = L_{ki} \tag{16.1.13}$$

which are Onsager's reciprocity relations. Q.E.D.

16.2 Time-reversal symmetry, quantum mechanical

16.2.1 General introduction

Analogously to classical mechanics, the principle of the dynamical reversibility holds also for quantum mechanics. Consider an isolated conservative system, the Hamiltonian of which is independent of time. We shall base the argument on the coordinate representation of quantum mechanical quantities because the time reversal has no effect on the position variables. Let $\Psi(x, t)$ be the wave function which represents a state of the system at time t. Then, it satisfies the Schrödinger equation

$$\mathrm{i}\frac{\partial}{\partial t}\Psi(x, t) = \mathfrak{H}\Psi(x, t) \tag{16.2.1a}$$

This equation is not invariant under time reversal $t \to -t$ alone, being a first-order differential equation in time. First, we consider the simple Schrödinger theory which does not involve the spin. Then the Hamiltonian $\mathfrak{H}$ is real[3] so

[3] In general, the Hamiltonian which involves the spin is not real. For example, spin–orbit coupling in an atom is given by

$$\mathfrak{H}'_{\mathrm{so}} = C\boldsymbol{S}\cdot\boldsymbol{L}, \qquad C = \text{a real constant} \tag{16.2.1c}$$

where $\boldsymbol{S}$ and $\boldsymbol{L}$ are the spin and orbital angular momentum operators of the atom, respectively.

that the equation becomes invariant under $t \to -t$ together with complex conjugation:

$$i\frac{\partial}{\partial t}\Psi^*(x, -t) = \mathfrak{H}\Psi^*(x, -t) \tag{16.2.1b}$$

Thus if $\Psi(x, t)$ is a possible solution of the Schrödinger equation, then the 'time-reversed state' defined by $\Psi^\tau(x, t) = \Psi^*(x, -t)$ also provides a possible solution of the Schrödinger equation; in fact, both solutions exhibit the same temporal development, given by

$$\Psi(x, t) = e^{-i\mathfrak{H}t}\psi(x), \qquad \Psi^\tau(x, t) = \Psi^*(x, -t) = e^{-i\mathfrak{H}t}\psi^*(x) \tag{16.2.2a}$$

the only difference being in their initial states $\Psi(x, 0) = \psi(x)$ and $\Psi^\tau(x, 0) = \psi^*(x)$, analogously to classical mechanics.

To see the meaning of complex conjugations, consider a special case of the momentum eigenfunction $\psi(x) = e^{ikx}$ belonging to the momentum eigenvalue $\hbar k$. Then $\psi^*(x) = e^{-ikx}$ describes the state with the reverse momentum $-\hbar k$. In general, a state function $\psi(x)$ is formed by a superposition of momentum eigenfunctions, such that

$$\psi(x) = \int a(k)e^{ikx}dk, \qquad \psi^\tau(x) \equiv \psi(x)^* = \int a(k)^* e^{-ikx}dk$$

where complex conjugation of $\psi(x)$ reverses all momentum states in $\psi(x)$. Moreover, the probability density $W^\tau(x, t)$ described by $\Psi^\tau(x, t)$ at a time t equals the probability density $W(x, -t)$ described by $\Psi^*(x, -t)$ at the earlier time $-t$:

$$W^\tau(x, t) = \Psi^{\tau*}(x, t)\Psi^\tau(x, t) = \Psi(x, -t)\Psi^*(x, -t) = W(x, -t)$$

Therefore, we may conclude that the time-reversed state of $\Psi(x, t)$ is described by $\Psi^\tau(x, t) = \Psi^*(x, -t)$ at least in the simple Schrödinger theory which does not involve the spin. In particular, at the instant of time reversal, $t = 0$, we have

$$\psi^\tau(x) = \psi^*(x) \tag{16.2.2b}$$

To arrive at the dynamical reversibility in quantum mechanics which involves the spin we again classify the physical quantities into the position-like and velocity-like quantities analogous to the classical counterparts, if they exist. The position coordinates x and the Hamiltonian $\mathfrak{H}$ are of the first kind whereas linear and angular momenta including the spins are of the second kind. Following Wigner (1962), let us introduce the time-reversal operator $\boldsymbol{\theta}$ which brings a physical quantity A to its time-reversed operator A^τ such that

$$A^\tau = \boldsymbol{\theta} A \boldsymbol{\theta}^{-1} = \varepsilon_A A \tag{16.2.3}$$

where $\varepsilon_A = 1$ if A is position-like and $\varepsilon_A = -1$ if A is velocity-like. Since any quantum mechanical observable is a function of positional coordinates, linear momenta and the spins, it is necessary and sufficient to determine $\boldsymbol{\theta}$ by the following three requirements:

$$\boldsymbol{\theta r \theta}^{-1} = \boldsymbol{r}, \qquad \boldsymbol{\theta \hat{p} \theta}^{-1} = -\hat{\boldsymbol{p}}, \qquad \boldsymbol{\theta s \theta}^{-1} = -\boldsymbol{s} \tag{16.2.4}$$

where $\boldsymbol{r} = (x, y, z)$ and $\hat{\boldsymbol{p}} = (-i\partial/\partial x, -i\partial/\partial y, -i\partial/\partial z)$ are the position and the momentum of a particle, respectively, and $s = (\frac{1}{2}\sigma_x, \frac{1}{2}\sigma_y, \frac{1}{2}\sigma_z)$ is the spin vector defined by the Pauli matrices

$$\sigma_x = \begin{bmatrix} 0 & 1 \\ 1 & 0 \end{bmatrix}, \qquad \sigma_y = \begin{bmatrix} 0 & -\mathrm{i} \\ \mathrm{i} & 0 \end{bmatrix}, \qquad \sigma_z = \begin{bmatrix} 1 & 0 \\ 0 & -1 \end{bmatrix}$$

Firstly, the first two requirements in (16.2.4) are satisfied by

$$\boldsymbol{\theta} = K_0$$

where K_0 is the complex conjugation operator. This operator is satisfactory for the simple Schrödinger theory which does not involve spin. However, $\boldsymbol{\theta} = K_0$ does not satisfy the third condition, because σ_x and σ_z are real but σ_y is imaginary so that

$$s_x^* = s_x, \qquad s_y^* = -s_y, \qquad s_z^* = s_z$$

Hence the Hamiltonian which involves the spin need not be invariant under complex conjugation (see for example $\mathfrak{H}'_{\mathrm{so}}$ defined by (16.2.1c)). To satisfy all three conditions we may, following Wigner (1962), set

$$\boldsymbol{\theta} = K_0 U \tag{16.2.5a}$$

where U is a non-singular unitary matrix, which is assumed to be independent of the position coordinates and thus commutes with the position coordinates and momenta; hence, it is determined by the third requirement in (16.2.4):

$$U\sigma_x U^{-1} = -\sigma_x, \qquad U\sigma_y U^{-1} = \sigma_y, \qquad U\sigma_z U^{-1} = -\sigma_z$$

The simplest solution for U which anticommutes with σ_x and σ_z and commutes with σ_y is given by

$$U = -\mathrm{i}\sigma_y = \begin{bmatrix} 0 & -1 \\ 1 & 0 \end{bmatrix}$$

where the phase factor $-\mathrm{i}$ is introduced for convenience to make U real so that it commutes with K_0. The matrix U is real and unitary. The final expression for the time-reversal operator $\boldsymbol{\theta}$ which satisfies all three requirements in (16.2.4) is given by

$$\boldsymbol{\theta} = -\mathrm{i}\sigma_y K_0 \tag{16.2.5b}$$

So far, we have discussed a one-electron system. For a system of n electrons, $\boldsymbol{\theta}$ may be defined by the direct product

$$\boldsymbol{\theta} = UK_0; \qquad U = (-\mathrm{i}\sigma_y)_1(-\mathrm{i}\sigma_y)_2 \ldots (-\mathrm{i}\sigma_y)_n \tag{16.2.5c}$$

For simplicity we shall confine the discussion to the one-electron problem, the extension to many-electron problems being straightforward.

In general, the Hamiltonian $\mathfrak{H}$ of an isolated conservative system is expressed by the position coordinates, momenta and spins, and it is even with respect to time reversal such that

$$\mathfrak{H}(x, -\hat{p}_x, -s_x, \ldots) = \mathfrak{H}(x, \hat{p}_x, s_x, \ldots)$$

Accordingly,

$$\boldsymbol{\theta}\mathfrak{H}\boldsymbol{\theta}^{-1} = \mathfrak{H} \qquad \text{or} \qquad \boldsymbol{\theta}\mathfrak{H} = \mathfrak{H}\boldsymbol{\theta} \tag{16.2.6a}$$

If this is the case, the Schrödinger equation (16.2.1a) is invariant under time reversal $t \to -t$ followed by the application of the operator $\boldsymbol{\theta} = -\mathrm{i}\sigma_y K_0$:

$$i\frac{\partial}{\partial t}\boldsymbol{\theta}\Psi(x, -t) = \mathfrak{H}\boldsymbol{\theta}\Psi(x, -t) \tag{16.2.6b}$$

Thus we may define the time-reversed state of $\Psi(x, t)$ by

$$\Psi^{\tau}(x, t) = \boldsymbol{\theta}\Psi(x, -t) = e^{-i\mathfrak{H}t}\boldsymbol{\theta}\psi(x)$$

where $\psi(x) = \Psi(x, 0)$. At $t = 0$, the time-reversed state is given by

$$\Psi^{\tau}(x, 0) = \boldsymbol{\theta}\psi(x) \equiv \psi^{\tau}(x) \tag{16.2.7}$$

If $\{\psi_n\}$ is a complete set of the energy eigenfunctions, then so is the set $\{\boldsymbol{\theta}\psi_n\}$, i.e. if $\mathfrak{H}\psi_n = E_n\psi_n$ then $\mathfrak{H}(\boldsymbol{\theta}\psi_n) = E_n(\boldsymbol{\theta}\psi_n)$ for any energy eigenstate ψ_n. Thus, the time-reversal symmetry of the Hamiltonian doubles the degeneracy of the energy level E_n, if ψ_n and $\boldsymbol{\theta}\psi_n$ are linearly independent. The doubling of the degeneracy will be discussed extensively later on in Section 16.4.

In the simplest special case of elementary spinors ξ_1 and ξ_2 defined by

$$\xi_1 = \begin{bmatrix} 1 \\ 0 \end{bmatrix}, \qquad \xi_2 = \begin{bmatrix} 0 \\ 1 \end{bmatrix} \tag{16.2.8a}$$

where ξ_1 is the spin-up state and ξ_2 is the spin-down state, we have

$$\boldsymbol{\theta}\xi_1 = \xi_2, \qquad \boldsymbol{\theta}\xi_2 = -\xi_1 \tag{16.2.8b}$$

This result coincides with the phase convention first introduced by Kramers and later by Wigner and then by Condon and Shortley. The set $\{\xi_1, \xi_2 = \boldsymbol{\theta}\xi_1\}$ is often called the set of Kramers' conjugate spinors.

So far, we have shown the dynamical reversibility (16.2.6b) for a conservative system. However, we can extend the principle further to a system that is not in a conservative force field. Suppose that the system is in a constant external magnetic field $\boldsymbol{H}$. Then (16.2.6a) should be replaced by

$$\boldsymbol{\theta}\mathfrak{H}(-\boldsymbol{H})\boldsymbol{\theta}^{-1} = \mathfrak{H}(\boldsymbol{H}) \tag{16.2.9}$$

because a magnetic field $\boldsymbol{H}$ is a velocity-like quantity, its direction is reversed when the direction of the electric current which produces the magnetic field is reversed. Note that $\boldsymbol{\theta}$ does not act on the field $\boldsymbol{H}$. For example, when a spin is placed in the magnetic field $\boldsymbol{H}$, the Hamiltonian $\mathfrak{H}(\boldsymbol{H})$ contains a term

$$\mathfrak{H}'(\boldsymbol{H}) = -\mu_0\boldsymbol{s}\cdot\boldsymbol{H}$$

where $\mu_0\boldsymbol{s}$ is the magnetic moment of the spin $\boldsymbol{s}$. Obviously, the term $\mathfrak{H}'(\boldsymbol{H})$ satisfies (16.2.9). To see the effect of the dynamical reversibility for such a case, consider the eigenvalue problem of $\mathfrak{H}(\boldsymbol{H})$:

$$\mathfrak{H}(\boldsymbol{H})\psi_n(\boldsymbol{H}) = E_n(\boldsymbol{H})\psi_n(\boldsymbol{H})$$

Application of $\boldsymbol{\theta}$ to both sides followed by $\boldsymbol{H} \to -\boldsymbol{H}$ yields

$$\mathfrak{H}(\boldsymbol{H})[\boldsymbol{\theta}\psi_n(-\boldsymbol{H})] = E_n(-\boldsymbol{H})[\boldsymbol{\theta}\psi_n(-\boldsymbol{H})]$$

Thus, if $\psi_n(\boldsymbol{H})$ is an eigenfunction of $\mathfrak{H}(\boldsymbol{H})$ belonging to an eigenvalue $E_n(\boldsymbol{H})$, then $\boldsymbol{\theta}\psi_n(-\boldsymbol{H})$ is also an eigenfunction of $\mathfrak{H}(\boldsymbol{H})$ belonging to an eigenvalue $E_{n'}(\boldsymbol{H})$ of some quantum number n' such that

$$E_{n'}(\boldsymbol{H}) = E_n(-\boldsymbol{H}), \qquad \psi_{n'}(H) = \boldsymbol{\theta}\psi_n(-\boldsymbol{H})$$

Accordingly, the time-reversal operator $\boldsymbol{\theta}$ couples two different wave functions $\psi_{n'}(\boldsymbol{H})$

and $\psi_n(-\boldsymbol{H})$ belonging to the same energy $E_{n'}(\boldsymbol{H}) = E_n(-\boldsymbol{H})$. Thus the time-reversal symmetry doubles the degeneracy of the energy, if $\psi_n(\boldsymbol{H})$ and $\boldsymbol{\theta}\psi_n(\boldsymbol{H})$ are linearly independent.

Example. For $\mathfrak{H}'(\boldsymbol{H}) = -\mu_0 s_z H$, we have

$$\mathfrak{H}'(\boldsymbol{H})\xi_1 = -\tfrac{1}{2}\mu_0 H\xi_1 = E_1(\boldsymbol{H})\xi_1, \qquad \mathfrak{H}'(\boldsymbol{H})\xi_2 = +\tfrac{1}{2}\mu_0 H\xi_2 = E_2(\boldsymbol{H})\xi_2$$

where

$$E_2(\boldsymbol{H}) = \tfrac{1}{2}\mu_0 H = E_1(-\boldsymbol{H}), \qquad \xi_2 = \boldsymbol{\theta}\xi_1$$

16.2.2 *The properties of the time-reversal operator* $\boldsymbol{\theta}$

(i) If we apply the time-reversal operator $\boldsymbol{\theta}$ twice on any operator A, then the operator must return to the original operator, i.e. $\boldsymbol{\theta}^2 A\boldsymbol{\theta}^{-2} = A$. Thus $\boldsymbol{\theta}^2$ must commute with any operator and hence must be a constant; in fact, it equals '2π rotation e' belonging to $SU(2)$':

$$\boldsymbol{\theta}^2 = e' \in SU(2) \tag{16.2.10}$$

This is so because $K_0^2 = 1$ and $-\mathrm{i}\sigma_y$ is an element of $SU(2)$ that is interpreted as the spinor representation of the two-fold rotation $2y$ about the y-axis. Note that the 2π rotation e' has the eigenvalue 1 for an integral representation and has the eigenvalue -1 for a half-integral representation of the group $SU(2)$.

(ii) When $\boldsymbol{\theta}$ acts on a linear combination of two wave functions (or spinors) ψ and φ, we have

$$\boldsymbol{\theta}(a\psi + b\varphi) = a^*\boldsymbol{\theta}\psi + b^*\boldsymbol{\theta}\varphi \tag{16.2.11}$$

where a^* and b^* are the complex conjugates of the constant coefficients. An operator of this kind is said to be *antilinear.* It is also *anti-unitary* in the sense that

$$\langle\psi, \varphi\rangle = \langle\boldsymbol{\theta}\psi, \boldsymbol{\theta}\varphi\rangle^* \tag{16.2.12}$$

where $\langle\psi, \varphi\rangle$ is a scalar product of two wave vectors. This follows from

$$\langle\psi, \varphi\rangle = \langle K_0\psi, K_0\varphi\rangle^* = \langle UK_0\psi, UK_0\varphi\rangle^*$$

U being a unitary operator. From (16.2.12) we may state that the time-reversal operator $\boldsymbol{\theta}$ leaves invariant the probability of transition between any two states ψ and φ

$$|\langle\psi, \varphi\rangle| = |\langle\boldsymbol{\theta}\psi, \boldsymbol{\theta}\varphi\rangle|$$

Hereafter, (16.2.12) may be referred to as *the time-reversal symmetry of a scalar product.*

(iii) When an operator V is neither even nor odd we can separate V into even and odd parts. To see this, let us introduce an operator $\boldsymbol{O}$ defined by

$$V^\tau = \boldsymbol{\theta}V\boldsymbol{\theta}^{-1} \equiv \boldsymbol{O}V$$

then $\boldsymbol{O}$ is an antilinear operator but is involutional, satisfying $\boldsymbol{O}^2 V = V$. As a result, its eigenvalues are $+1$ and -1,

$$\mathbf{O}V_+ = V_+, \qquad \mathbf{O}V_- = -V_-$$

with the eigenvectors $V_\pm = \frac{1}{2}(\mathbf{1} \pm \mathbf{O})V$ projected out from V. Since $V = V_+ + V_-$, any operator can be separated into even and odd parts under time reversal. Hereafter, every operator is assumed to be either even or odd under $\boldsymbol{\theta}$ such that

$$V^\tau = \boldsymbol{\theta} V \boldsymbol{\theta}^{-1} = \varepsilon_v V, \qquad \varepsilon_v = \pm 1 \tag{16.2.13}$$

in accordance with (16.2.3). Thus the time-reversal symmetry of the Heisenberg operator $V(t) = \exp(\mathrm{i}\mathfrak{H}t)V\exp(-\mathrm{i}\mathfrak{H}t)$ is described by

$$\begin{aligned} V^\tau(t) = \boldsymbol{\theta} V(t) \boldsymbol{\theta}^{-1} &= \exp(-\mathrm{i}\mathfrak{H}t)\boldsymbol{\theta} V \boldsymbol{\theta}^{-1} \exp(\mathrm{i}\mathfrak{H}t) \\ &= \varepsilon_V V(-t) \end{aligned} \tag{16.2.14}$$

which corresponds to the classical result (16.1.5a).

16.2.3 The time-reversal symmetry of matrix elements of a physical quantity

Let us define a matrix element of an operator V relative to two wave functions (or spinors) by

$$V_{12} = \langle \psi_1, V\psi_2 \rangle = \int \psi_1^* V \psi_2 \, \mathrm{d}\tau$$

Then, from (16.2.12) and (16.2.13), there follows the time-reversal symmetry

$$\begin{aligned} V_{12} = \langle \boldsymbol{\theta}\psi_1, \boldsymbol{\theta} V \boldsymbol{\theta}^{-1} \boldsymbol{\theta}\psi_2 \rangle^* &= \varepsilon_V \langle \psi_1^\tau, V\psi_2^\tau \rangle^* \\ &= \varepsilon_V \langle \psi_2^\tau, V^\dagger \psi_1^\tau \rangle \end{aligned} \tag{16.2.15}$$

where $V^\dagger$ is the Hermitian conjugate of V. This relation imposes a condition on the selection rule of the matrix element of the operator V (cf. Section 17.8). For example, when $\psi_2 = \boldsymbol{\theta}\psi_1 = \psi_1^\tau$ we have $\psi_2^\tau = e'\psi_1$ so that, for a Hermitian operator V,

$$V_{12} = \varepsilon_V e' \langle \psi_1, V\psi_2 \rangle = \varepsilon_V e' V_{12}$$

Accordingly,

$$V_{12} = 0 \qquad \text{if} \qquad \varepsilon_V e' = -1$$

Thus, for an even operator V, the matrix element between Kramers conjugate spinors $(\xi_1, \boldsymbol{\theta}\xi_1)$ vanishes, whereas for an odd operator V, the matrix element between a complex conjugate pair (ψ, ψ^*) of wave functions vanishes in the simple Schrödinger theory which does not involve the spin. The selection rule under time-reversal symmetry will be discussed using (16.2.15) in Section 17.8.

In the special case in which $\psi_1 = \psi_2 = \psi$, from Equation (16.2.14) the time-reversal symmetry of the expectation value of the Heisenberg operator $V(t)$ is described by

$$\langle \psi, V(t)\psi \rangle = \langle \psi^\tau, V^\tau(t)\psi^\tau \rangle^* = \varepsilon_V \langle \psi^\tau, V(-t)\psi^\tau \rangle \tag{16.2.16}$$

where V is assumed Hermitian. Let us apply this result for the canonical ensemble average of the operator $V(t)$

$$\langle\langle V(t)\rangle\rangle = \operatorname{tr}\rho(\mathfrak{H})\, V(t)$$
$$= \sum_n \rho(E_n)\langle\psi_n,\ V(t)\psi_n\rangle$$

where $\{\psi_n\}$ is the complete set of the energy eigenfunctions and the canonical distribution is described by

$$\rho(E_n) = \exp(-\beta E_n) \Big/ \sum_{n'} \exp(-\beta E_{n'})$$

From (16.2.16) we have

$$\langle\psi_n,\ V(t)\psi_n\rangle = \varepsilon_V\langle\psi_n^\tau,\ V(-t)\psi_n^\tau\rangle$$

Since $\{\psi_n^\tau\}$ forms a complete set of energy eigenfunctions, we have the time-reversal symmetry

$$\langle\langle V(t)\rangle\rangle = \varepsilon_V\langle\langle V(-t)\rangle\rangle \tag{16.2.17}$$

which corresponds to the classical result (16.1.6a).

Example. We shall show using the time-reversal symmetry that the electric conductance tensor $\sigma_{\mu\nu}$ (μ, $\nu = 1, 2, 3$) is a symmetric tensor. Let $\boldsymbol{Q}$ be the electric moment of a conducting medium defined by

$$\boldsymbol{Q} = \sum_i e_i \boldsymbol{r}_i$$

where e_i and $\boldsymbol{r}_i$ are the charge and the position of the ith particle in the medium. Then the perturbation of the Hamiltonian due to the electric field $\boldsymbol{E}(t)$ introduced adiabatically at $t = -\infty$ on the medium is given by

$$\mathfrak{H}_1(t) = -\boldsymbol{Q}\cdot\boldsymbol{E}(t); \quad \boldsymbol{E}(t) = \boldsymbol{E}^0\cos(\omega t)\exp(\epsilon t)$$

where ϵ is an infinitesimal positive quantity such that $\boldsymbol{E}(t) \to 0$ as $t \to -\infty$. According to the theory of linear response in statistical mechanics (Kubo 1957), the average electric flow

$$\dot{\boldsymbol{Q}} = \sum_i e_i \dot{\boldsymbol{r}}_i$$

in the μth direction caused by the νth component of the electric field is given by

$$J_\mu(t) = \langle\langle \dot{Q}_\mu(t)\rangle\rangle = \mathrm{i}\int_{-\infty}^{t} \operatorname{tr}\rho[\dot{Q}_\mu(t-t'),\ Q_\nu]\, E_\nu(t')\,\mathrm{d}t'$$

where $[\dot{Q}_\mu,\ Q_\nu]$ is the commutator. This is so, since the non-commuting part of the commutator up to time t contributes to the flow at time t. The flow may be written in the form

$$J_\mu(t) = \operatorname{Re}\sigma_{\mu\nu}(\omega)\, E_\nu^0 \mathrm{e}^{\mathrm{i}\omega t}; \qquad \mu,\ \nu = 1, 2, 3$$

where $\sigma_{\mu\nu}(\omega)$ is the frequency-dependent conductance tensor defined by

$$\sigma_{\mu\nu}(\omega) = \mathrm{i}\int_0^\infty \operatorname{tr}\rho[\dot{Q}_\mu(t),\ Q_\nu]\exp(-\mathrm{i}\omega t)\,\mathrm{d}t \tag{16.2.18}$$

Now, we shall show that $\sigma_{\mu\nu}(\omega)$ is a symmetric tensor

$$\sigma_{\mu\nu}(\omega) = \sigma_{\nu\mu}(\omega); \qquad \nu, \mu = 1, 2, 3 \tag{16.2.19}$$

which is a special case of Onsager's reciprocity relation of transport coefficients applied to electrical conduction. For this purpose, let us introduce the so-called *response function* defined by

$$\phi_{\mu\nu}(t) = \mathrm{i}\,\mathrm{tr}\,\rho[\dot{Q}_\mu(t), Q_\nu]; \qquad \mu, \nu = 1, 2, 3 \tag{16.2.20}$$

Then the conductance tensor defined by (16.2.18) is expressed by the half-interval Fourier transform of the response function

$$\sigma_{\mu\nu}(\omega) = \int_0^\infty \phi_{\mu\nu}(t) \exp(-\mathrm{i}\omega t)\,\mathrm{d}t \tag{16.2.21}$$

The properties of the response function are as follows.

(i) The response function is real, because the operator $V(t)_{\mu\nu} = \mathrm{i}[\dot{Q}_\mu(t), Q_\nu]$ is Hermitian.

(ii) It is invariant under translation in time, i.e.

$$\phi_{\mu\nu}(t) = \mathrm{i}\,\mathrm{tr}\,\rho[\dot{Q}_\mu(t+t'), Q_\nu(t')]$$

which follows from the invariance of the trace under a similarity transformation by $\exp(\mathrm{i}\mathfrak{H}t')$.

(iii) It is an even function of time and symmetric with respect to ν and μ: because Q_ν as well as $\mathrm{i}\dot{Q}_\mu = -[\mathfrak{H}, Q_\mu]$ are even with respect to time reversal so that, from (16.2.17),

$$\phi_{\mu\nu}(t) = \phi_{\mu\nu}(-t) = \phi_{\nu\mu}(t) \tag{16.2.22}$$

where the last equality follows from the translational invariance of the correlation $\mathrm{tr}\,\rho[Q_\mu, Q_\nu]$ in time, i.e. $\mathrm{tr}\,\rho[\dot{Q}_\mu, Q_\nu] + \mathrm{tr}\,\rho[Q_\mu, \dot{Q}_\nu] = 0$.

From (16.2.21) and (16.2.22) we obtain the required reciprocity relation (16.2.19) of the conductance tensor $\sigma_{\mu\nu}(\omega)$.

16.3 Anti-unitary point groups

16.3.1 General discussion

For convenience, we mean by the rotation group its double group $SU(2)$ or more generally the direct product group $SO(3, r) \times SU(2)$. Let us denote it by G_s and its improper group by G_{si}, which is the direct product of G_s and the group of inversion C_i. We may extend the improper group G_{si} by augmenting it with the time-reversal operator $\boldsymbol{\theta} = -\mathrm{i}\sigma_y K_0$ such that

$$G^e_{si} = G_{si} + \boldsymbol{\theta} G_{si}$$

This is possible because $\boldsymbol{\theta}^2 = e' \in G_s$ and $\boldsymbol{\theta}$ commutes with any element of G_{si}. Since $\boldsymbol{\theta}$ is anti-unitary, whereas an element of G_{si} is unitary, any element contained in the coset $\boldsymbol{\theta} G_{si}$ is anti-unitary. Hereafter the ordinary point group which is a subgroup of G_{si} may be called *a unitary point group* whereas a subgroup of G^e_{si} that contains both unitary and anti-unitary elements is called *an anti-unitary point group*.

Since a product of two anti-unitary operators is unitary whereas the product of an anti-unitary operator and a unitary operator is anti-unitary, an anti-unitary point group always has a halving subgroup that is unitary. Accordingly, any anti-unitary point group is formed by augmenting a unitary point group H with an anti-unitary operator a as follows:

$$H^z \equiv H + aH, \qquad a = \boldsymbol{\theta} z \tag{16.3.1a}$$

where z is an element of G_{si} that is not contained in H. This notation H^z for an anti-unitary point group was first introduced by Kim (1983c). According to this notation, two groups H^z and $H^{z'}$ are identical if $z' \in zH$ because z is a coset representative of H that is defined up to a multiplicative factor $h \in H$. Moreover, the halving subgroup H is an invariant subgroup of H^z so that the augmentor has to satisfy the compatibility condition

$$z^2, \qquad z^{-1}hz \in H; \qquad \forall\, h \in H \tag{16.3.1b}$$

On account of this condition, only a very limited number ($\leqslant 6$) of z is allowed for a given H, obviously up to a multiplicative factor $h \in H$. As in the case of improper point groups, *the maximal set* of possible z for a given H occurs when $H = C_n$ (a uniaxial group). We choose the following set of augmentors z for C_n together with their representative symbols:

$$\begin{aligned} z = e,\ \bar{1},\ c_{2n},\ \bar{c}_{2n},\ c_2',\ \bar{c}_2' \\ e,\ i,\ q,\ p,\ u,\ v \end{aligned} \tag{16.3.2a}$$

where e is the identity operator, $i = \bar{1}$ is the inversion; $q = c_{2n}$ is the $2n$-fold rotation axis parallel to the principal axis c_n and $p = \bar{q}$ is the rotation–inversion; $u = c_2'$ is a binary rotation perpendicular to c_n and $v = \bar{u}$. Thus we arrive at six anti-unitary point groups denoted by

$$C_n^z; \qquad C_n^e,\ C_n^i,\ C_n^q,\ C_n^p,\ C_n^u,\ C_n^v \tag{16.3.2b}$$

For a point group H with higher symmetry than C_n, only a subset of the maximal set $\{z\}$ given by (16.3.2a) suffices. Thus, for a dihedral group D_n, there exist only four types of anti-unitary groups denoted by

$$D_n^e,\ D_n^i,\ D_n^q,\ D_n^p \tag{16.3.3a}$$

Here we have excluded D_n^u and D_n^v because they are expressed by

$$D_n^u = D_n^e \text{ or } D_n^q, \qquad D_n^v = D_n^i \text{ or } D_n^p \tag{16.3.3b}$$

The reason is as follows. Since D_n already contains a binary rotation $u_0 \perp c_n \in D_n$, the allowed u in D_n^u is given by u_0 or $c_{2n}u_0$ on account of the compatibility condition (16.3.1b). This explains the first relation in (16.3.3b). The second relation follows analogously, since the allowed v in D_n^v is given by $\bar{u}_0$ or $\bar{c}_{2n}u_0$. For the tetrahedral group T we have

$$T^e,\ T^i,\ T^q,\ T^p \tag{16.3.3c}$$

where $q = 4_z \parallel 2_z \in T$ and $p = \bar{4}_z$. For $H = O$ or Y we have only two anti-unitary groups for each:

$$O^e,\ O^i; \qquad Y^e,\ Y^i \tag{16.3.3d}$$

For an improper halving subgroup $H = P_{\bar{x}} = P + \bar{x}P$, where P is a proper point group, the augmentor $\bar{x}$ is given by $\bar{1}$, $p = \bar{q}$ or $v = \bar{u}$. For this case, the allowed set of

z for H^z is even more limited because the augmentor z has an arbitrary multiplicative element of H so that

$$P^z_{\overline{x}} = P^{z\overline{x}}_{\overline{x}} \tag{16.3.4a}$$

Accordingly, when $H = C_{ni}$, for example, we have only three anti-unitary groups:

$$C^e_{ni},\ C^q_{ni},\ C^u_{ni} \tag{16.3.4b}$$

Analogously, for $H = C_{np}$ or C_{nv} we have

$$C^e_{np},\ C^i_{np},\ C^u_{np};\qquad C^e_{nv},\ C^i_{nv},\ C^q_{nv},\ C^p_{nv} \tag{16.3.4c}$$

For $H = D_{ni}$ or D_{np},

$$D^e_{ni},\ D^q_{ni};\qquad D^e_{np},\ D^i_{np} \tag{16.3.4d}$$

Here D^u_{np} is excluded, for example, because

$$D^u_{np} = D^e_{np} \text{ or } D^i_{np} \tag{16.3.4e}$$

analogously to (16.3.3b). Finally, for $H = T_i$, T_p, O_i and Y_i we have

$$T^e_i,\ T^q_i;\qquad T^e_p,\ T^i_p;\qquad O^e_i;\qquad Y^e_i \tag{16.3.4f}$$

All anti-unitary point groups are formed in this way and presented in Table 16.1, excluding 13 types of groups H^e (called the *gray groups*) because H^e occurs for every point group H. The table is described by the present system of symbols, and also by the symbols of Shubnikov *et al.* (1964) and the international symbols. Note that Table 16.1 classifies all of the anti-unitary groups into $31 + 13 = 44$ general types.

Now we shall describe the isomorphisms between the improper and proper anti-unitary point groups. The present system of symbols is very effective for this. Making the one-to-one correspondences

$$\overline{\boldsymbol{\theta}} \leftrightarrow \boldsymbol{\theta},\qquad \overline{c}_{2n} \leftrightarrow c_{2n},\qquad \overline{c}'_2 \leftrightarrow c'_2 \tag{16.3.5}$$

where $\overline{\boldsymbol{\theta}} = \overline{1}\boldsymbol{\theta}$, we obtain the following isomorphisms:

$$H^i \simeq H^e,\qquad H^p \simeq H^q,\qquad H^v \simeq H^u,\qquad P^z_i \simeq P^z \times C_i \tag{16.3.6a}$$

$$C^z_{np} \simeq C^z_{2n},\qquad C^z_{nv} \simeq D^z_n,\qquad D^z_{np} \simeq D^z_{2n},\qquad T^z_p \simeq O^z \tag{16.3.6b}$$

where z stands for an operator compatible with the corresponding halving subgroup H. On account of these isomorphisms, the group structures of all of the anti-unitary point groups are described by the following three types:

$$P^e,\ P^u,\ P^q \tag{16.3.7}$$

and their direct products with C_i. Here P is a proper point group. These isomorphisms are described explicitly for every type of the anti-unitary groups in the fifth column of Table 16.1 except for the 13 types of the gray groups. Obviously, these isomorphisms greatly reduce the labor of constructing the representations of the anti-unitary point groups.[4]

[4] We have excluded in (16.3.6) the possible isomorphism between an anti-unitary group H^z and a unitary group $H_z = H + zH$, because their 'matrix representations' are different due to the antilinear nature of $\boldsymbol{\theta}z$; see Section 16.4.

Table 16.1. *The anti-unitary point groups*

No.	H^z	Shubnikov	International	Isomorphism	$G = H_z$
1	G_s^i	$\infty/\infty\cdot\underline{m}$	$\infty\infty m'$	G_s^e	G_{si}
2	C_∞^i	$\infty:\underline{m}$	∞/m'	C_∞^e	$C_{\infty i}$
3	C_∞^u	$\infty:2$	$\infty 2'$	C_∞^u	D_∞
4	C_∞^v	$\infty\cdot\underline{m}$	$\infty m'$	C_∞^u	$C_{\infty v}$
5	$C_{\infty i}^u$	$\underline{m}\infty:m$	∞/mm'	$C_\infty^u\times C_i$	$D_{\infty i}$
6	$C_{\infty v}^i$	$m\cdot\infty:m$	$\infty/m'm$	D_∞^e	$D_{\infty i}$
7	D_∞^i	$\underline{m}\cdot\infty:\underline{m}$	$\infty/m'm'$	D_∞^e	$D_{\infty i}$
8	C_n^i	$\underline{\overline{2n}}_o$, n_e: $\underline{m}$	$\overline{n}'_o$, n_e/m'	C_n^e	C_{ni}
9	C_n^q	$\underline{2n}$	$(2n)'$	C_n^q	C_{2n}
10	C_n^p	$\underline{\overline{2n}}_e$, $\overline{n}_o$	$(\overline{2n})'$	C_n^q	C_{np}
11	C_n^u	$n:2$	$n_e 2'2'$, $n_o 2'$	C_n^u	D_n
12	C_n^v	$n\cdot\underline{m}$	$n_e m'm'$, $n_o m'$	C_n^u	C_{nv}
13	C_{ni}^q	$\underline{2n}_e:m$, $\underline{2n}_o:m$	$(2n_e)'/m$, $(2n_o)'/m'$	$C_n^q\times C_i$	C_{2ni}
14	C_{ni}^u	$\underline{m}\cdot n_e:m$, $\overline{2n}_o\cdot\underline{m}$	$n_e/mm'm'$, $\overline{n}_o m'$	$C_n^u\times C_i$	D_{ni}
15	C_{np}^i	$\underline{2n}_e:\underline{m}$, $\underline{2n}_o:m$	$(2n_e)'/m'$, $(2n_o)'/m$	C_{2n}^e	C_{2ni}
16	C_{np}^u	$\overline{2n}_e\cdot\underline{m}$, $\underline{m}\cdot n_o:m$	$\overline{2n}_e 2'm'$, $\overline{2n}_o m'2'$	C_{2n}^u	D_{np}
17	C_{nv}^i	$m\cdot n_e:\underline{m}$, $\underline{\overline{2n}}_o\cdot m$	$n_e/m'mm$, $\overline{n}_o'm$	D_n^e	D_{ni}
18	C_{nv}^q	$\underline{2n}\cdot m$	$(2n)'m'm$	D_n^q	C_{2nv}
19	C_{nv}^p	$\underline{\overline{2n}}_e\cdot m$, $m\cdot n_o:\underline{m}$	$\overline{2n}'_e 2'm$, $\overline{2n}'_o m2'$	D_n^q	D_{np}
20	D_n^i	$\underline{m}\cdot n_e:\underline{m}$, $\underline{\overline{2n}}_o\cdot\underline{m}$	$n_e/m'm'm'$, $\overline{n}_o'm'$	D_n^e	D_{ni}
21	D_n^q	$\underline{2n}:2$	$(2n)'2'2$	D_n^q	D_{2n}
22	D_n^p	$\underline{\overline{2n}}_e\cdot m$, $\underline{m}\cdot n_o:\underline{m}$	$\overline{2n}_e'2m'$, $\overline{2n}_o'm'2$	D_n^q	D_{np}
23	D_{ni}^q	$m\cdot\underline{2n}_e:m$, $m\cdot 2n_o:\underline{m}$	$(2n_e)'/mm'm$, $(2n_o)'/m'm'm$	$D_n^q\times C_i$	D_{2ni}
24	D_{np}^i	$m\cdot\underline{2n}_e:\underline{m}$, $m\cdot\underline{2n}_o:m$	$(2n_e)'/m'm'm$, $(2n_o)'/mm'm$	D_{2n}^e	D_{2ni}
25	T^i	$\overline{6}/2$	$m'3$	T^e	T_i
26	T^q	$3/4$	$4'32'$	T^q	O
27	T^p	$3/\underline{\overline{4}}$	$\overline{4}'3m'$	T^q	T_p
28	T_i^q	$\underline{\overline{6}}/4$	$m3m'$	$T^q\times C_i$	O_i
29	T_p^i	$\underline{\overline{6}}/4$	$m'3m$	O^e	O_i
30	O^i	$\underline{\overline{6}}/4$	$m'3m'$	O^e	O_i
31	Y^i	$3/\underline{\overline{10}}$	$5'3m'$	Y^e	Y_i

1. The gray groups H^e are not listed.
2. $n_e(n_o)$ denotes even (odd) n; thus, equivalence of the notations means that, e.g, $C^i_{n_o} \leftrightarrow \underline{\overline{2n}}_o$, $C^i_{n_e} \leftrightarrow n_e:\underline{m}$ for No. 8.
3. $n > 1$ for Nos. 17–24.
4. $G = H + zH = H_z$ is defined in (16.3.8b).

For a graphical representation of H^z we may follow Shubnikov *et al.* (1964) and regard the time-reversal operator $\boldsymbol{\theta}$ as an operator that changes color from black to white and white to black, provided that $\boldsymbol{\theta}^2 = e$. Then an anti-unitary point group may be described by the symmetry of a system consisting of white and black balls. For

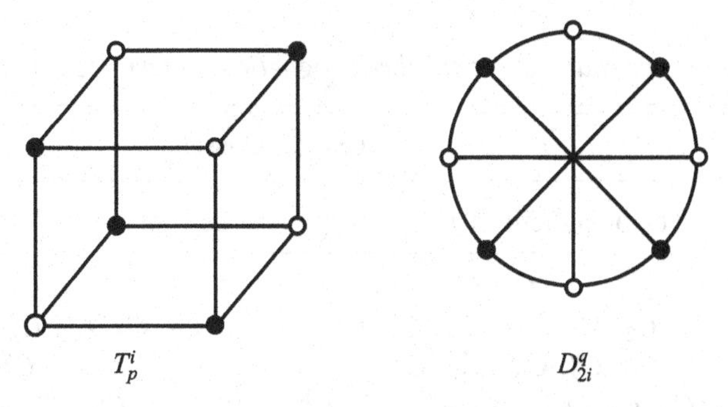

Figure 16.2. The graphical presentations of the anti-unitary groups T_p^i and D_{2i}^q.

example, such systems belonging to T_p^i and D_{2i}^q are presented in Figure 16.2. In this description, H^z with $z \neq e$ is called a *black and white group* whereas H^e is called a *gray group*. Table 16.1 contains a total of 31 types of black and white groups. Note that the time-reversal operator $\boldsymbol{\theta}$ is an operation that reverses the direction of a magnetic moment, because $\boldsymbol{\theta}$ anticommutes with the spin, as given in (16.2.4). Thus, for example, the point symmetry of a paramagnetic crystal in the absence of an external magnetic field is described by a gray point group H^e, because all spins are randomly oriented. In general, the symmetry point group of a spin system is called a *magnetic point group*. It is either a unitary point group or an anti-unitary point group that leaves the spin system invariant. See the next section for the symmetry point groups of ferromagnetic crystals.

Remark. Let G be a unitary point group and let H be a halving subgroup of G. Then an anti-unitary point group formed by G is expressed in the so-called Schönflies notation as follows:

$$G(H) = H + \boldsymbol{\theta}(G - H) \tag{16.3.8a}$$

This system of notation cannot express the gray groups H^e. Otherwise, this notation is equivalent to the present one. Let the left coset decomposition of G with respect to H be

$$G = H + zH \equiv H_z, \qquad z \notin H \tag{16.3.8b}$$

then we arrive at $G(H) = H^z$ since $G - H = zH$. For the sake of comparison, we have listed $G = H_z$ in the last column of Table 16.1. As one can see from Table 16.1, the classification of the anti-unitary point groups by the Schönflies notation $G(H)$ or by the notation of Shubnikov *et al.* or the international notations is not quite effective for describing their isomorphisms.

16.3.2 The classification of ferromagnetics and ferroelectrics

As an application of magnetic point groups we shall discuss the point symmetries of ferromagnetic and ferroelectric solids.

16.3.2.1 Ferromagnetic crystals

A ferromagnetic crystal is characterized by an orderly distribution of magnetic moments (spins) of atoms or ions in the crystal in the absence of any external magnetic

Table 16.2. *The crystal classes for ferromagnetic and ferroelectric crystals*

	Crystal classes	
Crystal systems	Ferromagnetic $< C^u_{\infty i} = C^v_{\infty i}$	Ferroelectric $< C^e_{\infty v} = C^v_{\infty v}$
Triclinic	C_1, C_i	C_1, C^e_1
Monoclinic	$C_2, C_{1p}, C_{2i}, C^u_1, C^v_1, C^u_i$	$C_2, C_v, C^e_2, C^e_v, C^q_1, C^v_1$
Orthorhombic	$C^u_2, C^v_2, C^u_{1p}, C^u_{2i}$	$C_{2v}, C^e_{2v}, C^q_{1v}, C^v_2$
Tetragonal	$C_4, C_{2p}, C_{4i}, C^u_4, C^v_4, C^u_{2p}, C^u_{4i}$	$C_4, C_{4v}, C^e_4, C^e_{4v}, C^q_2, C^q_{2v}, C^v_4$
Rhombohedral	$C_3, C_{3i}, C^u_3, C^v_3, C^u_{3i}$	$C_3, C_{3v}, C^e_3, C^e_{3v}, C^v_3$
Hexagonal	$C_6, C_{3p}, C_{6i}, C^u_6, C^v_6, C^u_{3p}, C^u_{6i}$	$C_6, C_{6v}, C^e_6, C^e_{6v}, C^q_3, C^q_{3v}, C^v_6$

$C^u_{\infty i} > C^u_{ni}, C^u_{up}, C^u_n, C_{ni}, C_{np}, C_n$
$C^e_{\infty v} > C^e_{nv}, C^e_n, C^q_{nv}, C^q_n, C^v_n, C_{nv}, C_n$

field. It possesses a resultant magnetic moment that is a pseudovector (or axial vector) belonging to the symmetry group $C^u_{\infty i}$: it has a cylindrical symmetry c_∞ about the magnetic moment and, being a spin vector is invariant under the inversion; moreover, it is invariant under an anti-unitary operation $\boldsymbol{\theta} u$ (where u is a binary rotation perpendicular to c_∞) because each one of $\boldsymbol{\theta}$ and u inverts the direction of the spin. Thus ferromagnetism is possible for those crystals in which the class of magnetic symmetry is one of the subgroups of the symmetry group $C^u_{\infty i}$. Since $C^u_{\infty i} = C^v_{\infty i}$, the subgroups of finite orders are

$$C^u_{\infty i} > C^u_{ni}, C^u_{np}, C^u_n, C^v_n, C_{ni}, C_{np}, C_n$$

where C_{np} comes in because it is a subgroup of C_{2ni} $(= C_{np} + \bar{1}C_{np})$. Note that the first four types of the subgroups are anti-unitary whereas the last three types are unitary. Since the allowed n-fold axes for a crystal class are with $n = 1, 2, 3, 4$ and 6, we obtain Table 16.2 that contains 31 *magnetic crystal classes*. Note that we have only $n = 1, 2, 3$ for C_{np} and C^u_{np} for the obvious reason that $p = \overline{2n}$. These classes are classified by the six crystal systems in Table 16.2. Corresponding to these point groups, there exists a total of 275 magnetic space groups that will be discussed in the next chapter.

16.3.2.2 Ferroelectric crystals

Analogously to a ferromagnetic crystal, a ferroelectric crystal is characterized by a spontaneous resultant electric moment that is a polar vector (i.e. an ordinary vector). An electric moment vector belongs to an anti-unitary group $C^e_{\infty v}$: it is invariant under time reversal, being a position-like quantity, and also under the vertical reflection plane v which contains the symmetry axis c_∞ of the moment. Thus the symmetry group of a ferroelectric crystal is one of the subgroups of $C^e_{\infty v}$. Since $C^e_{\infty v} = C^v_{\infty v}$, the required subgroups of finite orders are

$$C^e_{\infty v} > C^e_{nv}, C^e_n, C^q_{nv}, C^q_n, C^v_n, C_{nv}, C_n$$

In Table 16.2, we have classified a total of 31 *ferroelectric crystal classes* by the six crystal systems. Correspondingly, there exists a total of 275 anti-unitary space groups analogous to those for the ferromagnetic crystals.

16.4 The co-representations of anti-unitary point groups

16.4.1 General discussion

Following Wigner (1962), we shall discuss the so-called co-representations of an anti-unitary point group H^z. Let $\psi = [\psi_1, \psi_2, \ldots, \psi_d]$ be a row vector basis that closes under an anti-unitary group H^z such that, under an element g of H^z, it transforms according to

$$\mathring{g}\psi = \psi S(g), \qquad \forall\, g \in H^z \tag{16.4.1}$$

where $S(g)$ is a $d \times d$ matrix representative of the element $g \in H^z$. Let h be a unitary element belonging to H and a be an anti-unitary element belonging to $\boldsymbol{\theta} zH$. Then their matrix representatives satisfy

$$S(ah) = S(a)S(h)^* \tag{16.4.2}$$

where $s(h)^*$ is the complex conjugate of $S(h)$. This follows from (16.4.1) and the fact that a is an anti-unitary operator:

$$\mathring{a}\mathring{h}\psi = \mathring{a}(\psi S(h)) = (\mathring{a}\psi)S^*(h) = \psi S(a)S^*(h)$$

In an analogous manner we can show that

$$S(h_1 h_2) = S(h_1)S(h_2), \qquad S(ha) = S(h)S(a)$$

$$S(ah) = S(a)S(h)^*, \qquad S(a_1 a_2) = S(a_1)S(a_2)^* \tag{16.4.3}$$

where h_1 and h_2 are unitary operators belonging to H whereas a_1 and a_2 are anti-unitary operators belonging to $\boldsymbol{\theta} zH$. Accordingly, a matrix system $S(H^z) = \{S(g);\ \forall\, g \in H^z\}$ introduced by (16.4.1) is not a representation of H^z in the usual sense on account of the complex conjugate sign in (16.4.3). The matrix system $S(H^z)$ may be called a *co-representation* of H^z to remind one of the complex conjugate sign in (16.4.3).

A matrix $S(g)$ of the co-representation will be *unitary* if the basis vectors ψ_i are mutually orthonormal. This follows from the anti-unitarity of a:

$$\langle a\psi_i, a\psi_j\rangle = \langle \psi_i, \psi_j\rangle^* = \delta_{ij}$$

Accordingly, we have $S(g)^{-1} = S(g)^\dagger$ for all $g \in H^z$; hence, from (16.4.3)

$$S(h^{-1}) = S(h)^\dagger, \qquad S(a^{-1}) = [S(a)^*]^{-1} = S(a)^\sim \tag{16.4.4}$$

for all $h \in H$ and all $a \in \boldsymbol{\theta} zH$. In the special case in which $a = \boldsymbol{\theta}$, we have $\boldsymbol{\theta}^{-1} = e'\boldsymbol{\theta}$ so that

$$S(\boldsymbol{\theta})^\sim = S(e')S(\boldsymbol{\theta}) = \pm S(\boldsymbol{\theta}) \tag{16.4.5}$$

where $S(e') = +\ (-)$ for an integral (half-integral) representation. Thus $S(\boldsymbol{\theta})$ is symmetric (antisymmetric) for an integral (half-integral) representation; for example, see (16.5.5).

Next, we shall discuss the unitary transformation of the co-representation. Let ψ' be a new basis that is related to the original basis ψ by a unitary matrix T such that

$$\psi' = \psi T \tag{16.4.6}$$

then the new co-representation S' based on ψ' satisfies

$$S(h)' = T^{-1}S(h)T$$
$$S(a)' = T^{-1}S(a)T^* \tag{16.4.7}$$

of which the first one is obvious whereas the second one follows from

$$\mathring{a}\psi' = (\mathring{a}\psi)T^* = \psi S(a)T^* = \psi' T^{-1}S(a)T^* = \psi' S(a)'$$

In general, two co-representations S and S' are called equivalent if they can be transformed into each other by a unitary transformation as defined by (16.4.7). A co-representation $S(H^z)$ is called irreducible if it cannot be brought into a reduced form by a unitary transformation (16.4.7).

In the special case in which T $(= \omega\mathbf{1})$ is a constant unitary matrix, we have from (16.4.7)

$$S(h)' = S(h), \qquad S(a)' = \omega^{*2}S(a) \tag{16.4.8}$$

where $T^{-1}T^* = \omega^{*2}$. Accordingly, two co-representations of H^z are equivalent if their $S(h)$ are the same while their $S(a)$ are different by a common phase factor.

16.4.2 Three types of co-unirreps

The basic theory of *unitary irreducible co-representations (co-unirreps)* of an anti-unitary point group has been worked out by Wigner (1962). We shall reconstruct the co-unirreps in a form that is more transparent and better suited for constructing the co-unirreps of the anti-unitary space groups following the approach developed by Kim (1983c). Since the defining relation of H^z is given by the left coset decomposition with respect to the halving subgroup H, the co-unirreps may be constructed by induced representation from the unirreps of H. However, since the coset representative $\boldsymbol{\theta}z$ is anti-unitary, a certain modification of the ordinary theory of induced representation in order to obtain the co-unirreps of H^z is inevitable.

Let us assume that there is given a complete set of unirreps $\{\Delta^{(\nu)}(H)\}$ for the halving subgroup H. Let $\psi^\nu = [\psi_1^\nu, \psi_2^\nu, \ldots, \psi_{d_\nu}^\nu]$ be a basis of a d_ν-dimensional unirrep $\Delta^{(\nu)}(H)$. Then the basis $\mathring{a}\psi^\nu$ induced by the anti-unitary augmentor $a = \boldsymbol{\theta}z$ also provides a basis of a representation of the halving subgroup H:

$$\mathring{h}(\mathring{a}\psi^\nu) = \mathring{a}(a^{-1}ha)^0\psi^\nu = (\mathring{a}\psi^\nu)\Delta^{(\nu)}(a^{-1}ha)^*, \qquad \forall\, h \in H \tag{16.4.9}$$

because $a^{-1}ha \in H$ from the compatibility condition. The matrix representation of H thus induced must be equivalent to a member of the assumed set $\{\Delta^{(\nu)}(H)\}$ of the unirreps of H, say, the $\bar{\nu}$th unirrep $\Delta^{(\bar{\nu})}(H)$. This means that there exists a unitary matrix $N(a)$ such that

$$\Delta^{(\nu)}(a^{-1}ha)^* = N(a)^{-1}\Delta^{(\bar{\nu})}(h)N(a), \qquad \forall\, h \in H \tag{16.4.10a}$$

Substitution of this into (16.4.9) yields that $(\mathring{a}\psi^\nu)N(a)^{-1} \equiv \phi^{\bar{\nu}}$ is a basis of the unirrep $\Delta^{(\bar{\nu})}(H)$, so that

$$\mathring{a}\psi^\nu = \phi^{\bar{\nu}}N(a) \tag{16.4.10b}$$

Here $\phi^{\bar{\nu}}$ need not be linearly dependent on ψ^ν even if $\nu = \bar{\nu}$. The relation between two unirreps $\Delta^{(\nu)}(H)$ and $\Delta^{(\bar{\nu})}(H)$ is reciprocal; in fact, if we repeat the above argument again starting from $\phi^{\bar{\nu}} \in \Delta^{(\bar{\nu})}(H)$, we obtain

$$\Delta^{(\bar{\nu})}(a^{-1}ha)^* = \overline{N}(a)^{-1}\Delta^{(\nu)}(h)\overline{N}(a) \tag{16.4.11a}$$

$$\mathring{a}\phi^{\bar{\nu}} = \psi^\nu\overline{N}(a) \tag{16.4.11b}$$

Now, by applying the augmentor a once more to (16.4.10b) and (16.4.11b), we obtain

$$\overline{N}(a)N(a)^* = \Delta^{(\nu)}(a^2), \qquad N(a)\overline{N}(a)^* = \Delta^{(\bar{\nu})}(a^2) \tag{16.4.11c}$$

so that $\overline{N}(a) = \Delta^{(\nu)}(a^2)N(a)^\sim$, where $N(a)^\sim$ is the transpose of $N(a)$.

Since the augmentor a connects two bases ψ^ν and $\phi^{\bar{\nu}}$ of H, if they are linearly independent, the combined basis row vector defined by $\Psi^{(\nu,\bar{\nu})} = [\psi^\nu, \phi^{\bar{\nu}}]$ provides a basis of a $2d_\nu$-dimensional co-representation $S^{(\nu,\bar{\nu})}$ of H^z given by

$$S^{(\nu,\bar{\nu})}(h) = \begin{bmatrix} \Delta^{(\nu)}(h) & 0 \\ 0 & \Delta^{(\bar{\nu})}(h) \end{bmatrix}$$

$$S^{(\nu,\bar{\nu})}(a) = \begin{bmatrix} 0 & \overline{N}(a) \\ N(a) & 0 \end{bmatrix} \tag{16.4.12}$$

which is unitary because $N(a)$ and $\overline{N}(a)$ as well as $\Delta^{(\nu)}(H)$ and $\Delta^{(\bar{\nu})}(H)$ are all unitary. The co-representation should be compared with the induced representation $D^{(\nu,\bar{\nu})}$ of the unitary group H_z given by (12.6.5). In the special case of a gray group H^e, we have $a = \boldsymbol{\theta}$ and $\Delta^{(\nu)}(\boldsymbol{\theta}^2 = e') = \pm 1$ so that $\overline{N}(\boldsymbol{\theta}) = \pm N(\boldsymbol{\theta})^\sim$; hence, $S^{(\nu,\bar{\nu})}(\boldsymbol{\theta})$ given by (16.4.12) is indeed symmetric or antisymmetric in accordance with (16.4.5).

There exist three types of co-unirreps for H^z that can be obtained from the general co-representation $S^{(\nu,\bar{\nu})}(H^z)$ given above. It is convenient to discuss the third type first.

16.4.2.1 Type c

$\nu \neq \bar{\nu}$, i.e. $\Delta^{(\nu)}(H)$ and $\Delta^{(\bar{\nu})}(H)$ are inequivalent so that $S^{(\nu,\bar{\nu})}(H^z)$ given by (16.4.12) is irredicible[5] because the augmentor a connects the bases ψ^ν and $\phi^{\bar{\nu}}$ belonging to inequivalent unirreps $\Delta^{(\nu)}(H)$ and $\Delta^{(\bar{\nu})}(H)$ of the halving subgroup H, respectively. The dimensionality of the co-unirrep $S^{(\nu,\bar{\nu})}(H^z)$ is $2d_\nu$.

[5] The formal proof is that, since $S^{(\nu,\bar{\nu})}(H)$ is already in a reduced form, if $S^{(\nu,\bar{\nu})}(H^z)$ is reducible, $S^{(\nu,\bar{\nu})}(a)$ must be reduced through $T^{-1}S^{(\nu,\bar{\nu})}(a)T^*$ with a matrix T that commutes with $S^{(\nu,\bar{\nu})}(h)$ for all $h \in H$. From Schur's lemma such a matrix T must be of the form

$$\begin{bmatrix} T_1 & 0 \\ 0 & T_2 \end{bmatrix}$$

However, a diagonal block matrix T cannot reduce the off-diagonal block matrix $S^{(\nu,\bar{\nu})}(a)$ because $T^{-1}S^{(\nu,\bar{\nu})}(a)T^*$ remains of off-diagonal block form, as one can see by direct matrix multiplication. See Section 16.8 for the irreducibility criteria for a co-representation.

Next, when $\nu = \overline{\nu}$ we have, from (16.4.10a) and (16.4.11a),

$$N(a)^{-1}\Delta^{(\nu)}(h)N(a) = \overline{N}(a)^{-1}\Delta^{(\nu)}(h)\overline{N}(a), \qquad \forall\, h \in H \tag{16.4.13a}$$

and from (16.4.11c)

$$\overline{N}(a)N(a)^* = N(a)\overline{N}(a)^* = \Delta^{(\nu)}(a^2) \tag{16.4.13b}$$

Thus, from (16.4.13a), $\overline{N}(a)N(a)^{-1}$ commutes with the unirrep $\Delta^{(\nu)}(h)$ for all $h \in H$ so that $\overline{N}(a)N(a)^{-1} = \omega\mathbf{1}$ is a constant unitary matrix. On substituting $\overline{N}(a) = \omega N(a)$ into (16.4.13b), we obtain $\omega^2 = 1$ and $N(a)N(a)^* = \omega\Delta^{(\nu)}(a^2)$. Thus there exist two cases for $\nu = \overline{\nu}$ corresponding to $\omega = \pm 1$, i.e.

$$\overline{N}(a) = \pm N(a), \qquad N(a)N(a)^* = \pm\Delta^{(\nu)}(a^2) \tag{16.4.14}$$

where + (−) sign is for type a (b), as will be defined below.

16.4.2.2 Type b
$\nu = \overline{\nu}$,

$$\overline{N}(a) = -N(a), \qquad N(a)N(a)^* = -\Delta^{(\nu)}(a^2) \tag{16.4.15a}$$

The general expression $S^{(\nu,\overline{\nu})}(H^z)$ given by (16.4.12) is simplified to the following direct product expressions, with the basis $\Phi^{(\nu,\nu)} = [\psi^\nu, \phi^\nu]$:

$$S^{(\nu,\nu)}(h) = \begin{bmatrix} \Delta^{(\nu)}(h) & 0 \\ 0 & \Delta^{(\nu)}(h) \end{bmatrix} = \mathbf{1}_2 \times \Delta^{(\nu)}(h)$$

$$S^{(\nu,\nu)}(a) = \begin{bmatrix} 0 & -N(a) \\ N(a) & 0 \end{bmatrix} = \boldsymbol{\theta} \times N(a)^* \tag{16.4.15b}$$

where we have used $-i\sigma_y = \boldsymbol{\theta}K_0$. This $2d_\nu$-dimensional co-representation is also irreducible under any similarity transformation defined by (16.4.7). Since $S^{(\nu,\nu)}(H)$ is already reduced, it suffices to show that $S^{(\nu,\nu)}(a)$ is irreducible under the similarity transformation by a matrix of the form $T = U \times \mathbf{1}$ where U is a 2×2 unitary matrix and $\mathbf{1}$ is the d_ν-dimensional unit matrix. We set $U = \omega U_0$, where ω is the phase factor and U_0 is a special unitary matrix that always commutes with the time-reversal operator $\boldsymbol{\theta}$, then

$$T^{-1}S^{(\nu,\nu)}(a)T^* = [U^{-1}\boldsymbol{\theta}U^*] \times N(a)^* = \omega^{*2}S^{(\nu,\nu)}(a) \tag{16.4.16}$$

which proves the irreducibility, in view of (16.4.8). See Section 16.8 for a simpler proof based on the general criteria for irreducibility of a co-representation.

16.4.2.3 Type a
$\nu = \overline{\nu}$,

$$\overline{N}(a) = N(a), \qquad N(a)N(a)^* = \Delta^{(\nu)}(a^2) \tag{16.4.17a}$$

The general expression (16.4.12) is simplified to

$$S^{(\nu,\nu)}(h) = \mathbf{1}_2 \times \Delta^{(\nu)}(h), \qquad S^{(\nu,\nu)}(a) = \sigma_x \times N(a)$$

where σ_x is the Pauli spin matrix in the x-direction. Since σ_x is involutional, it is reduced to σ_z by an involutional matrix $Y = (\sigma_x + \sigma_z)/\sqrt{2}$ that is real too (Lemma 2.1.1). Thus, by use of a transformation matrix $T = Y \times \mathbf{1}$, the co-representation $S^{(\nu,\nu)}(H^z)$ is reduced to $S^{\nu+}$ and $S^{\nu-}$ defined by

$$S^{(\nu\pm)}(h) = \Delta^{(\nu)}(h), \qquad S^{(\nu\pm)}(a) = \pm N(a) \tag{16.4.17b}$$

These two co-unirreps are, however, equivalent since the only difference is in the phase factors of $S^{(\nu,\pm)}(a)$. Thus either one of them provides the required co-unirrep. The corresponding bases are given by

$$\Psi^{(\nu\pm)} = \psi^{\nu} \pm \phi^{\nu}, \qquad \phi^{\nu} = (a\psi^{\nu})N(a)^{-1}$$

Let $\Psi^{\nu} = \Psi^{\nu+}$, then we have, from (16.4.17b),

$$h\Psi^{\nu} = \Psi^{\nu}\Delta^{(\nu)}(h), \qquad a\Psi^{\nu} = \Psi^{\nu}N(a) \tag{16.4.17c}$$

That is, *there exists a basis Ψ^{ν} of a unirrep $\Delta^{(\nu)}(H)$ that also provides a basis of the anti-unitary group H^z. Conversely, if there exists such a basis, we have $\nu = \overline{\nu}$ and $N(a)N(a)^* = \Delta^{(\nu)}(a^2)$ so that the co-unirrep is of type a.* Thus, the existence of such a basis provides the criterion for type a.

Remark 1. The physical significance of the anti-unitary operator a is as follows. Let $H = \{h\}$ be the unitary symmetry group of the Hamiltonian $\mathfrak{H}$ of a physical system. Then the degeneracy of an energy level E_ν is given by the dimensionality d_ν of the unirrep $\Delta^{(\nu)}(H)$. If $\mathfrak{H}$ is also invariant under an anti-unitary operator a, then its degeneracy doubles for types b and c. These would have been regarded as accidental degeneracies, had the anti-unitary symmetry $a = \boldsymbol{\theta} z$ not been recognized.

Remark 2. The three types of the co-unirreps given above are explicit in terms of the unirreps $\Delta^{(\nu)}$ and $\Delta^{(\overline{\nu})}$ of the unitary halving subgroup H and the transformation matrix $N(a)$. In general, $N(a)$ is determined from (16.4.10a) or (16.4.10b). In particular, the latter is convenient when the bases of the unitary halving subgroup H are all known, as in the case of the point groups discussed in Chapters 10 and 11. In the important special case in which $\Delta^{(\nu)}(H)$ is one-dimensional, we have $\Delta^{\overline{\nu}}(h) = \Delta^{(\nu)}(a^{-1}ha)^*$ from (16.4.10a) because one-dimensional matrices always commute. In this case we may take $N(a) = 1$. This choice simplifies $S^{(\nu,\overline{\nu})}(a)$ given by (16.4.12) with $\overline{N}(a) = \Delta^{(\nu)}(a^2)$, and also $\phi^{\overline{\nu}} = \mathring{a}\psi^{\nu}$ from (16.4.10b).

Remark 3. When $\nu = \overline{\nu}$ and $N(a) = \mathbf{1}$, the criterion (16.4.14) is simplified to

$$\Delta^{(\nu)}(a^2) = \begin{cases} +\mathbf{1} & \text{for type a} \\ -\mathbf{1} & \text{for type b} \end{cases} \tag{16.4.18}$$

In particular, when $a = \boldsymbol{\theta}$, we have $\Delta^{(\nu)}(a^2 = e') = \mathbf{1}\ (-\mathbf{1})$ so that the co-unirrep induced by an integral (half-integral) representation belongs to type a (b).

16.5 Construction of the co-unirreps of anti-unitary point groups

We shall now explicitly construct all the co-unirreps of each anti-unitary point group H^z by induction from the unirreps of the respective halving subgroup H using the general expressions given in the previous section. We introduce the following notations for the three types of co-unirreps:

$$S(\Delta^{(\nu)}, N(a)) \qquad \text{for } S^{(\nu+)} \text{ of (16.4.17b)} \in \text{type a}$$

$$S(\Delta^{(\nu)}, \Delta^{(\nu)}; N(a)) \qquad \text{for } S^{(\nu,\nu)} \text{ of (16.4.15b)} \in \text{type b}$$

$$S(\Delta^{(\nu)}, \Delta^{(\bar{\nu})}; N(a)) \qquad \text{for } S^{(\nu,\bar{\nu})} \text{ of (16.4.12)} \in \text{type c} \qquad (16.5.1)^6$$

On account of the isomorphism (16.3.6), it is necessary to construct only the co-unirreps of the following 12 types of proper anti-unitary point groups, excluding Y^e for convenience:

$$G_s^e; \quad C_\infty^e, C_\infty^u; \quad C_n^e, C_n^u, C_n^q; \quad D_\infty^e; \quad D_n^e, D_n^q; \quad T^e, T^q; \quad O^e \qquad (16.5.2)$$

Now, following Kim (1983c), we shall construct their co-unirreps explicitly, beginning with the gray group G_s^e.

16.5.1 G_s^e

Following the notation given in Chapter 10, the basis of the $(2j+1)$-dimensional unirrep $D^{(j)}$ of $G_s = SU(2)$ is given by a row-vector basis $\Phi^{(j)} = [\phi(j, j), \phi(j, j-1), \ldots, \phi(j, -j)]$, where

$$\phi(j, m) \equiv \phi(\xi_1, \xi_2; j, m) = \xi_1^{j+m}\xi_2^{j-m}/[(j+m)!(j-m)!]^{1/2};$$
$$m = j, j-1, \ldots, -j \qquad (16.5.3)$$

for a given j $(= 0, \frac{1}{2}, 1, \ldots)$. Here, ξ_1 and ξ_2 are the elementary spinors. From the transformations $\boldsymbol{\theta}\xi_1 = \xi_2$ and $\boldsymbol{\theta}\xi_2 = -\xi_1$ we have

$$\mathring{\boldsymbol{\theta}}\phi(j, m) \equiv \phi(\boldsymbol{\theta}\xi_1, \boldsymbol{\theta}\xi_2; j, m) = (-1)^{j-m}\phi(j, -m); \qquad m = j, j-1, \ldots, -j \qquad (16.5.4a)^7$$

This means that the basis $\Phi^{(j)}$ of the unirrep $D^{(j)}(G_s)$ of the halving subgroup G_s satisfies (16.4.17c) with the transformation matrix $N(\boldsymbol{\theta})$ given by

$$N^{(j)}(\boldsymbol{\theta})_{nm} = (-1)^{j-m}\delta(n, -m); \qquad n, m = j, j-1, \ldots, -j \qquad (16.5.4b)$$

for all j $(= 0, \frac{1}{2}, 1, \ldots)$. Thus all the co-unirreps of G_s^e are of type a and may be expressed, following the notation (16.5.1), by

$$S(D^{(j)}; N^{(j)}(\boldsymbol{\theta})) \qquad (16.5.5a)$$

[6] The use of the notation S for the induced co-unirreps for H^z here is to avoid the confusion with the notation D for the induced unirreps of the unitary group H_z introduced in (12.6.5) and (12.6.10a),

$$D^{(\nu,\bar{\nu})} = D(\Gamma^{(\nu)}, \Gamma^{(\bar{\nu})}; N), \qquad D^{(\nu\pm)} = D(\Gamma^{(\upsilon)}; \pm N)$$

Here $D^{(\nu,\bar{\nu})}$ has a similar structure to that of $S^{(\nu,\bar{\nu})}$ except for $N(z)$ in the place of $N(a)^*$ in $S^{(\nu,\bar{\nu})}$, and $D^{(\nu,\nu)}$ is always reduced to two inequivalent unirreps $D^{(\nu+)}$ and $D^{(\nu-)}$ (see Equation 12.6.6)), whereas $S^{(\nu,\nu)}$ cannot be reduced.

[7] In the Schödinger theory which does not involve spins, (16.5.4a) is equivalent to the well-known property of spherical harmonics $Y(l, m)$,

$$Y(l, m)^* = (-1)^{l-m}Y(l, -m)$$

The explicit form of $N^{(j)}(\boldsymbol{\theta})$ is given, from (16.5.4b), by

$$N^{(j)}(\boldsymbol{\theta}) = \begin{bmatrix} 0 & \dots & 0 & 1 \\ 0 & \dots & -1 & 0 \\ \vdots & & \vdots & \vdots \\ \pm 1 & \dots & 0 & 0 \end{bmatrix} \tag{16.5.5b}$$

where the anti-diagonal elements alternate between $+1$ and -1. This was first obtained by Wigner. The matrix is symmetric if j is an integer and antisymmetric if j is half integer. This is consistent with (16.4.5).

Since $\boldsymbol{\theta}$ commutes with any element of G_s, we have, from (16.4.10a) with $\nu = \bar{\nu}$ and identifying $\Delta^{(\nu)}(H)$ with $D^{(j)}(G_s)$,

$$D^{(j)}(h)^* = N(\boldsymbol{\theta})^{-1} D^{(j)}(h) N(\boldsymbol{\theta}), \qquad \forall\, h \in G_s$$

i.e. the complex conjugate representation $D^{(j)}(G_s)^*$ is equivalent to the representation $D^{(j)}(G_s)$. Their matrix elements satisfy, via (16.5.4b),

$$D^{(j)}(h)^*_{m'm} = (-1)^{m-m'} D^{(j)}(h)_{-m',-m} \tag{16.5.5c}$$

which is consistent with the previous result (10.4.16c).

16.5.2 C_n^e, C_n^q and C_n^u

As was given by (11.3.3b), there exist $2n$ one-dimensional representations $\{M_m\}$ for C_n:

$$M_m(c_n^k) = \exp[-2\pi \mathrm{i} mk/n]; \qquad k = 0, 1, \dots, n-1 \tag{16.5.6a}$$

where m is an integer or half integer limited by $-n/2 < m \leqslant n/2$ because $M_m = M_{m+n}$. The basis of M_m is $\phi(j, m)$ given by (16.5.3). Since M_m is one-dimensional, one may take $N(\boldsymbol{\theta}) = 1$.

For C_n^e, we observe that $a = \boldsymbol{\theta}$ transforms the basis $\phi(j, m)$ of M_m into the basis $(-1)^{j-m}\phi(j, -m) \in M_{-m}$, as shown in (16.5.4a). When $m \neq 0$ or $n/2$, we have $M_m \neq M_{-m}$ so that the co-unirreps of C_n^e are of type c, given by

$$S(M_m, M_{-m}; 1);\ m = \tfrac{1}{2}, 1, \dots, \tfrac{1}{2}(n-1) \tag{16.5.6b}$$

When $m = 0$ or $n/2$, on the other hand, the two bases $\phi(j, m)$ and $\phi(j, -m)$ belong to the same unirrep M_m of C_n so that the corresponding co-unirreps must belong to type a or b. We set $N(\boldsymbol{\theta}) = 1$, the representations $M_m(C_n)$ being one-dimensional, then the types are determined by the criteria (16.4.18). Now, since $M_m(e') = (-1)^{2m}$ from (16.5.6a) with $k = n$, we have

$$S(M_0; 1),\ S(M_{n/2}; 1)_{n=\text{even}},\ S(M_{n/2}, M_{n/2}; 1)_{n=\text{odd}} \tag{16.5.6c}$$

following the notation introduced in (16.5.1). These results, (16.5.6b) and (16.5.6c), are presented in Table 16.3.

For C_n^q, we set $q = c_{2n} = \exp\{-\mathrm{i}[\pi/(2n)]\sigma_z\} \in SU(2)$, then, using $\sigma_z\xi_1 = \xi_1$ and $\sigma_z\xi_2 = -\xi_2$, we obtain

$$[\boldsymbol{\theta} c_{2n}]^\circ \phi(j, m) = (-1)^{j-m} \exp(-\mathrm{i}\pi m/n)\, \phi(j, -m) \tag{16.5.7a}$$

Thus we obtain $S(M_m, M_{-m}, 1)$ since $M_m \neq M_{-m}$ except for $m = 0$, $n/2$. From

Table 16.3. *The co-unirreps of anti-unitary point groups*

1. C_∞: $M_0, M_{\pm m}$; $m = \frac{1}{2}, 1, \ldots, \infty$
   C^e_∞: $S(M_0, 1), S(M_m, M_{-m}; 1)$
   C^u_∞: $S(M_0; 1), S(M_{\pm m}; 1)$

2. C_n: $M_0, M_{n/2}, M_{\pm m}$; $m = \frac{1}{2}, 1, \ldots, \frac{1}{2}(n-1)$; $M_m = M_{m+n}$
   C^e_n: $S(M_0; 1), S(M_{n/2}; 1)_{n=\text{even}}, S(M_{n/2}, M_{n/2}; 1)_{n=\text{odd}}, S(M_m, M_{-m}; 1)$
   C^u_n: $S(M_0; 1), S(M_{n/2}; 1), S(M_{\pm m}; 1)$
   C^q_n: $S(M_0; 1), S(M_{n/2}, M_{n/2}; 1)_{n=\text{even}}, S(M_{n/2}; 1)_{n=\text{odd}}, S(M_m, M_{-m}; 1)$

3. D_∞: A_1, A_2, E_m; $m = \frac{1}{2}, 1, \ldots, \infty$
   D^e_∞: $S(A_1; 1), D(A_2; 1), S(E_m; Y_m)$

4. D_n: A_1, A_2, B_1, B_2, E_m; $m = \frac{1}{2}, 1, \ldots, \frac{1}{2}(n-1)$
   D^e_n: $S(A_1; 1), S(A_2; 1), S(B_1; 1)_{n=\text{even}}, S(B_2; 1)_{n=\text{even}}, S(B_1, B_2; 1)_{n=\text{odd}}$, $S(E_m; Y_m)$
   D^q_n: $S(A_1; 1), S(A_2; 1), S(B_1, B_2; 1)_{n=\text{even}}, S(B_1; 1)_{n=\text{odd}}, S(B_2; 1)_{n=\text{odd}}$, $S(E_m; Z^{(n)}_m)$

5. T: $A, A', A'', T, E_{1/2}, E'_{1/2} = A' \times E_{1/2}, E''_{1/2} = A'' \times E_{1/2}$
   T^e: $S(A; 1), S(A', A''; 1), S(T; \mathbf{1}_3), S(E_{1/2}; \sigma_y), S(E'_{1/2}, E''_{1/2}; \sigma_y)$
   T^q: $S(A, 1), S(A', 1), S(A'', 1), S(T; \|4_z\|), S(E_{1/2}; Z), S(E'_{1/2}; Z), S(E''_{1/2}; Z)$

6. O: $A_1, A_2, E, T_1, T_2, E_{1/2}, E'_{1/2} = A_2 \times E_{1/2}, Q = E \times E_{1/2}$
   O^e: $S(A_1; 1), S(A_2; 1), S(E; \mathbf{1}_2), S(T_1; \mathbf{1}_3), S(T_2; \mathbf{1}_3), S(E_{1/2}; \sigma_y), S(E'_{1/2}; \sigma_y)$, $S(Q; \mathbf{1}_2 \times \sigma_y)$

1. All unirreps of the proper point groups given in 1–6 are defined in Chapter 11.
2. The transformation matrices Y_m and $Z^{(n)}_m$ are defined by (16.5.11a) and (16.5.11b), respectively:

$$Y_m = \begin{cases} \mathbf{1}_2, & \text{for an integral } m \\ \sigma_y, & \text{for a half-integral } m \end{cases}$$

$$Z^{(n)}_m = Y_m \begin{bmatrix} \cos(\pi m/n) & -\sin(\pi m/n) \\ \sin(\pi m/n) & \cos(\pi m/n) \end{bmatrix}$$

3. The transformation matrix $\|4_z\|$ is the representation of 4_z based on $[x, y, z]$ or $[\boldsymbol{i}, \boldsymbol{j}, \boldsymbol{k}]$ and Z is defined by (16.5.11b):

$$\|4_z\| = [\boldsymbol{j}, -\boldsymbol{i}, \boldsymbol{k}] = \begin{bmatrix} 0 & -1 & 0 \\ 1 & 0 & 0 \\ 0 & 0 & 1 \end{bmatrix}, \qquad Z = Z^{(2)}_{1/2} = \mathrm{i}2^{-1/2}\begin{bmatrix} -1 & -1 \\ 1 & -1 \end{bmatrix}$$

4. All co-unirreps of G^e_s, C^u_∞, C^u_n, D^e_∞, T^q and O^e belong to type a.
5. There exist only two co-unirreps belonging to type b: $S(M_{n/2}, M_{n/2}; 1)_{n=\text{odd}}$ of C^e_n and $S(M_{n/2}, M_{n/2}; 1)_{n=\text{even}}$ of C^q_n.

$M_m[(\boldsymbol{\theta} c_{2n})^2] = (-1)^{2m} \exp(-\mathrm{i}2\pi m/n)$, which equals 1 for $m = 0$ and $(-1)^{n+1}$ for $m = n/2$, we obtain via the criteria (16.4.18)

$$S(M_0;\ 1),\ S(M_{n/2},\ M_{n/2};\ 1)_{n=\text{even}},\ S(M_{n/2};\ 1)_{n=\text{odd}} \tag{16.5.7b}$$

as given in Table 16.3.

For C_n^u, we may take u parallel to the x-axis, then $u = c_2' = -\mathrm{i}\sigma_x$ and $\boldsymbol{\theta} c_2' = \sigma_y \sigma_x K_0 = -\mathrm{i}\sigma_z K_0$ so that

$$[\boldsymbol{\theta} c_2']^\circ \phi(j,\ m) = \mathrm{i}^{-2m} \phi(j,\ m) \tag{16.5.8}$$

This means that every basis $\phi(j,\ m)$ of C_n provides a basis of C_n^u. Thus, from (16.4.17c), all co-unirreps of C_n^u belong to type a as given in Table 16.3.

By extending the above results for C_n^e and C_n^u to C_∞^e and C_∞^u by putting $n \to \infty$, we obtain

$$\begin{aligned} C_\infty^e: &\quad S(M_0,\ 1),\ S(M_m,\ M_{-m};\ 1) \\ C_\infty^u: &\quad S(M_0;\ 1),\ S(M_{\pm m};\ 1) \end{aligned}$$

as given in Table 16.3.

16.5.3 D_n^e and D_n^q

As was given in Table 11.5, there exist four one-dimensional unirreps A_1, A_2, B_1 and B_2 and $n-1$ two-dimensional unirreps E_m, $m = \frac{1}{2}, 1, \ldots, \frac{1}{2}(n-1)$, for D_n. For convenience we shall redefine the spinor bases $\phi_\pm(j,\ m)$ of D_n introduced in Table 11.5 with an extra common phase factor i^{j+m} (which does not affect the unirreps of D_n) as follows:

$$\begin{aligned} \phi_+(j,\ m) &= 2^{-1/2}[\phi(j,\ m) + \phi(j,\ -m)]\mathrm{i}^{j+m} \\ \phi_-(j,\ m) &= 2^{-1/2}[\phi(j,\ m) - \phi(j,\ -m)]\mathrm{i}^{j+m-1} \end{aligned} \tag{16.5.9a}$$

Under the time reversal $\boldsymbol{\theta}$, these transform according to

$$\mathring{\boldsymbol{\theta}} \phi_\pm(j,\ m) = \begin{cases} \phi_\pm(j,\ m) & \text{for an integral } m \\ \pm \mathrm{i}\phi_\mp(j,\ m) & \text{for a half-integral } m \end{cases} \tag{16.5.9b}$$

The extra factor i^{j+m} was introduced in (16.5.9a) to eliminate an extra phase factor $(-1)^{j+m}$ that would appear in (16.5.9b) otherwise.

Now, we shall first show that the bases of two-dimensional unirreps E_m of D_n defined by

$$\Phi(j,\ m) = [\phi_+(j,\ m),\ \phi_-(j,\ m)]; \qquad m = \tfrac{1}{2}, 1, \ldots, \tfrac{1}{2}(n-1) \tag{16.5.10}$$

also provide the bases of D_n^e and D_n^q; hence they induce the type a co-unirreps according to (16.4.17c). Firstly, under $\boldsymbol{\theta} \in D_n^e$, the basis transforms according to, from (16.5.9b),

$$\mathring{\boldsymbol{\theta}} \Phi(j,\ m) = \Phi(j,\ m) Y_m; \qquad Y_m = \begin{cases} \mathbf{1}_2 & \text{for an integral } m \\ \sigma_y & \text{for a half-integral } m \end{cases} \tag{16.5.11a}$$

where the transformation matrix $N(\boldsymbol{\theta})$ is given by Y_m, which is symmetric (antisymmetric) when m is integral (half integral). Secondly, under $\boldsymbol{\theta} c_{2n} \in D_n^q$ we have

$$(\boldsymbol{\theta} c_{2n})^\circ \Phi(j,\ m) = \Phi(j,\ m) Z_m^{(n)}$$

where the transformation matrix $N(\boldsymbol{\theta}c_{2n})$ is given by

$$Z_m^{(n)} = Y_m E_m(c_{2n}) = Y_m \begin{bmatrix} \cos(\pi m/n) & -\sin(\pi m/n) \\ \sin(\pi m/n) & \cos(\pi m/n) \end{bmatrix} \tag{16.5.11b}$$

Here $E_m(c_{2n})$ is simply the matrix representative of $c_{2n} \| c_n \in D_n$. Thus we arrive at the type a co-unirreps given by

$$S(E_m;\, Y_m) \in D_n^e, \qquad S(E_m;\, Z_m^{(n)}) \in D_n^q$$

where the allowed values of m are defined in (16.5.10).

Next, we see that (16.5.11a) and (16.5.11b) hold also for $m = n/2$ and n. Firstly, for the unirreps B_1 and B_2 of D_n with the respective bases $\phi_+(j_e,\, n/2)$ and $\phi_-(j_e,\, n/2)$ given by Table 11.5, the transformation matrices for the combined basis $\Phi(j_e,\, n/2)$ are given, consistently with (16.5.11a) and (16.5.11b) with $m = n/2$, by

$$Y_{n/2} = \begin{cases} \mathbf{1}_2 & \text{for an even } n \\ \sigma_y & \text{for an odd } n \end{cases}, \qquad Z_{n/2}^{(n)} = \begin{cases} -\mathrm{i}\sigma_y & \text{for an even } n \\ -\mathrm{i}\mathbf{1}_2 & \text{for an odd } n \end{cases} \tag{16.5.12a}$$

where $\mathbf{1}_2$ is diagonal but σ_y is off-diagonal and antisymmetric. The diagonal transformation matrices lead to type a co-unirreps of D_n^e and D_n^q, whereas the off-diagonal matrices lead to type c co-unirreps of D_n^e and D_n^q induced from B_1 and B_2 of D_n as given in Table 16.3. Secondly, for the unirreps A_1 and A_2 of D_n with the bases $\phi_+(j_e,\, n)$ and $\phi_-(j_e,\, n)$ given by Table 11.5, the transformation matrices for $\Phi(j_e,\, n)$ are given, consistently with (16.5.11a) and (16.5.11b) with $m = n$, by

$$Y_m = \mathbf{1}_2, \qquad Z_n^{(n)} = -\mathbf{1}_2 \tag{16.5.12b}$$

both of which are diagonal. Thus we have $S(A_1;\, 1)$ and $S(A_2;\, 1)$ for both D_n^e and D_n^q as given in Table 16.3. It should be noted that in general the identity representation of any point group H induces the type a co-unirreps $S(A;\, 1)$ for any anti-unitary group H^z.

16.5.4 The cubic groups

According to Table 11.7 the unirreps A' and A'' of the tetragonal group T are mutually complex conjugate with the respective bases $\phi(2,\, 0) + \mathrm{i}\phi_+(2,\, 2)$ and $\phi(2,\, 0) - \mathrm{i}\phi_+(2,\, 2)$, which are connected by the time-reversal operator $\boldsymbol{\theta}$ as follows:

$$\mathring{\boldsymbol{\theta}}[\phi(2,\, 0) + \mathrm{i}\phi_+(2,\, 2)] = [\phi(2,\, 0) - \mathrm{i}\phi_+(2,\, 2)]$$

Thus, we obtain $S(A',\, A'';\, 1)$ for T^e. Here, $\phi(2,\, 0)$ and $\phi_+(2,\, 2)$ are the original spinor bases defined in Table 11.7. These bases are invariant under $\boldsymbol{\theta}4_z$ so that we obtain two co-unirreps $S(A';\, 1)$ and $S(A'';\, 1)$ for T^q.

Next, the basis $\Phi^T = [\phi_-(1,\, 1),\, -\phi_+(1,\, 1),\, \mathrm{i}\phi(1,\, 0)]$ or $[x,\, y,\, z]$ of the unirrep T of the group T is invariant under $\boldsymbol{\theta}$. Accordingly, under $\boldsymbol{\theta}4_z \in T^q$ we obtain

$$(\boldsymbol{\theta}4_z)^\circ \Phi^T = \Phi^T \|4_z\|$$

where $\|4_z\|$ is the 3×3 matrix of the rotation 4_z given in the footnote of Table 16.3. Thus we obtain the co-unirreps $S(T;\, \mathbf{1}_3)$ for T^e while $S(T;\, \|4_z\|)$ for T^q. Next, for the unirrep $E_{1/2}$ of T we have used the basis $\Phi(\frac{1}{2},\, \frac{1}{2})$ defined by (16.5.10) so that the transformation matrices are given by $N(\boldsymbol{\theta}) = Y_{1/2} = \sigma_y$ from (16.5.11a) and

$N(\boldsymbol{\theta}4_z) = Z^{(2)}_{1/2} \equiv Z$ from (16.5.11b), where Z is given in the footnote of Table 16.3. Accordingly we arrive at

$$S(E_{1/2}; \sigma_y) \in T^e, \qquad S(E_{1/2}; Z) \in T^q$$

Analogously, we have determined the remaining co-unirreps of T^e and T^q and also those of O^e given in Table 16.3. All the co-unirreps of G^e_s, C^u_∞, D^e_∞, T^q and O^e are of type a according to Table 16.3. This table has been extended further to the projective co-unirreps of the magnetic point groups of infinite order by Kim (1984c).

16.6 Complex conjugate representations

Let $\Delta^{(\nu)}(H) = \{\Delta^{(\nu)}(h); h \in H\}$ be a unirrep of a unitary group H, then the complex conjugate set $\Delta^{(\nu)}(H)^* = \{\Delta^{(\nu)}(h)^*\}$ also provides a unirrep of H, because, if $\Delta^{(\nu)}(h_1)\Delta^{(\nu)}(h_2) = \Delta^{(\nu)}(h_1 h_2)$ then $\Delta^{(\nu)}(h_1)^*\Delta^{(\nu)}(h_2)^* = \Delta^{(\nu)}(h_1 h_2)^*$. There exist three cases for $\Delta^{(\nu)}(H)$.

1. $\Delta^{(\nu)}(H)$ is real or can be brought to real form by a unitary transformation.
2. $\Delta^{(\nu)}(H)$ is equivalent to $\Delta^{(\nu)}(H)^*$ but cannot be brought to real form.
3. $\Delta^{(\nu)}(H)$ is not equivalent to $\Delta^{(\nu)}(H)^*$.

For the first two cases their characters $\chi^{(\nu)}(H)$ are real but for the third case the character is complex (not real).

For example, a one-dimensional complex representation of any group belongs to case 3 for the obvious reason that a one-dimensional representation is invariant under any similarity transformation. According to (10.3.9), the character of every unirrep of $SU(2)$ is real and hence a unirrep of $SU(2)$ belongs either to case 1 or to case 2. The same is true for a unirrep of any group G of which every class is ambivalent, i.e. every element g and its inverse g^{-1} are in the same class, since then their characters are all real from $\chi(g) = \chi(g^{-1}) = \chi(g)^*$.

The above classification of the unirreps $\{\Delta^{(\nu)}(H)\}$ of a point group H by their complex conjugation is closely correlated to the types of the co-unirreps of the gray group H^e. Firstly, from (16.4.9) with $a = \boldsymbol{\theta}$, we have

$$\mathring{h}(\mathring{\boldsymbol{\theta}}\psi^\nu) = (\mathring{\boldsymbol{\theta}}\psi^\nu)\Delta^{(\nu)}(h)^*; \qquad \forall\, h \in H \tag{16.6.1}$$

which is also understood from the fact that $\boldsymbol{\theta}$ commutes with any unitary operator $h \in G_{si}$. Thus, if ψ^ν is a basis of $\Delta^{(\nu)}(H)$, then the basis of $\Delta^{(\nu)}(H)^*$ is given by the time-reversed basis $\mathring{\boldsymbol{\theta}}\psi^\nu$. Secondly, from (16.4.10a) with $a = \boldsymbol{\theta}$, we have

$$\Delta^{(\nu)}(h)^* = N(\boldsymbol{\theta})^{-1}\Delta^{(\overline{\nu})}(h)N(\boldsymbol{\theta}); \qquad \forall\, h \in H \tag{16.6.2}$$

so that $\chi^{(\nu)}(H)^* = \chi^{(\overline{\nu})}(H)$, i.e. the complex conjugate unirrep $\Delta^{(\nu)}(H)^*$ is equivalent to a unirrep $\Delta^{(\overline{\nu})}(H)$ contained in the assumed complete set of the unirreps of H. From this, we can determine the three cases of the unirreps $\Delta^{(\nu)}(H)$ by the three types of co-unirreps of H^e induced from $\Delta^{(\nu)}(H)$, since the latter have been completely worked out for the point groups, as given in Table 16.3.

It is obvious that either $\nu \neq \overline{\nu}$ or $\nu = \overline{\nu}$ in (16.6.2).

(i) When $\nu \neq \overline{\nu}$, we have $\chi^{(\nu)*} = \chi^{(\overline{\nu})} \neq \chi^{(\nu)}$ so that the character of $\Delta^{(\nu)}(H)$ cannot be real. Accordingly, $\Delta^{(\nu)}(H)$ belongs to case 3, whereas the co-unirrep of H^e

induced by $\Delta^{(\nu)}(H)$ is of type c. This means that there exists the correspondence

$$\text{case 3} \leftrightarrow \text{type c} \tag{16.6.3}$$

Thus, the co-unirrep of H^e can be given simply by $S(\Delta^{(\nu)}, \Delta^{(\nu)*}; \mathbf{1})$ with the basis $[\psi^\nu, \boldsymbol{\theta}\psi^\nu]$ from (16.6.1) instead of $S(\Delta^{(\nu)}, \Delta^{(\overline{\nu})}; N(\boldsymbol{\theta}))$. Obviously, these two co-unirreps are equivalent. For example, the unirreps $M_{\pm m}$ $(m = \frac{1}{2}, 1, \ldots, \frac{1}{2}(n-1))$ of the uniaxial group C_n are complex and thus belong to case 3 (the character is also complex, being one-dimensional) whereas the corresponding co-unirreps are of type c and are given by $S(M_m, M_{-m}; 1)^c$ as presented in Table 16.3. Here $M_{-m} = M_m^*$.

(ii) When $\nu = \overline{\nu}$, we have, from (16.6.2),

$$\Delta^{(\nu)}(h)^* = N(\boldsymbol{\theta})^{-1}\Delta^{(\nu)}(h)N(\boldsymbol{\theta}); \qquad \forall\, h \in H \tag{16.6.4a}$$

Hence $\Delta^{(\nu)}(H)$ belongs to case 1 or to case 2 and the character is real. Now, from the type criteria (16.4.14) with $a = \boldsymbol{\theta}$ and $\boldsymbol{\theta}^2 = e'$, the transformation matrix $N(\boldsymbol{\theta})$ satisfies

$$N(\boldsymbol{\theta})N(\boldsymbol{\theta})^* = \pm\Delta^{(\nu)}(e') \tag{16.6.4b}$$

where the $+$ sign stands for type a and the $-$ sign stands for type b. Moreover, from $\Delta^{(\nu)}(e') = \pm\mathbf{1}$, we have $N(\boldsymbol{\theta})N(\boldsymbol{\theta})^* = \pm\mathbf{1}$ for both types. Since $N(\boldsymbol{\theta})$ is unitary, if $N(\boldsymbol{\theta})N(\boldsymbol{\theta})^* = \mathbf{1}$ we have $N(\boldsymbol{\theta}) = N(\boldsymbol{\theta})^\sim$, i.e. $N(\boldsymbol{\theta})$ is symmetric, whereas, if $N(\boldsymbol{\theta})N(\boldsymbol{\theta})^* = -\mathbf{1}$, we have $N(\boldsymbol{\theta}) = -N(\boldsymbol{\theta})^\sim$, i.e. $N(\boldsymbol{\theta})$ is antisymmetric. Now, the case criterion is closely correlated to the type criterion because, if $N(\boldsymbol{\theta})$ is symmetric (antisymmetric), then $\Delta^{(\nu)}(H)$ belongs to case 1 (case 2) according to the following basic theorem (Wigner 1962).

Theorem 16.6.1. When a unirrep $\Delta(H)$ is equivalent to $\Delta^*(H)$, through a unitary matrix N such that $\Delta^*(h) = N^{-1}\Delta(h)N$ for all $h \in H$, then $\Delta(H)$ can be brought to real form by a unitary matrix U if and only if the matrix N is symmetric.

Proof. Suppose that $\Delta(H)$ is brought to a real representation $\overline{\Delta}(H)$ by a unitary matrix U such that $\Delta(h) = U\overline{\Delta}(h)U^\dagger$ for all $h \in H$, then substitution of this relation into the assumed relation $\Delta^*(h) = N^{-1}\Delta(h)N$ yields

$$U^*\overline{\Delta}(h)U^\sim = N^{-1}U\overline{\Delta}(h)U^\dagger N; \qquad \forall\, h \in H$$

from which it follows that $U^\dagger NU^*$ commutes with the unirrep $\overline{\Delta}(H)$ so that $U^\dagger NU^* = c\mathbf{1}$, a constant matrix from Schur's lemma. Thus, we can write

$$N = cUU^\sim$$

which means that N is indeed a symmetric matrix. Conversely, if the unitary matrix N is symmetric then $\Delta(h)$ can be made real by a unitary transformation. To see this we use the lemma that a unitary symmetric matrix N can be diagonalized by a real orthogonal matrix R (see Section 1.5, Remark 2) such that

$$RNR^\sim = \Lambda, \qquad RR^\sim = \mathbf{1}$$

where Λ is a diagonal matrix, whose diagonal elements are of modulus unity, because N is unitary. Now, by the similarity transformation of the assumed relation $\Delta^*(h) = N^{-1}\Delta(h)N$ with R we obtain

$$R\Delta^*(h)R^\sim = \Lambda^* R\Delta(h)R^\sim \Lambda, \qquad \Lambda^*\Lambda = 1$$

Next, we write the diagonal matrix Λ as the square of another diagonal matrix Λ_1 such that $\Lambda = \Lambda_1^2$. Then the modulus of the diagonal elements of Λ_1 is also unity so that $\Lambda_1^{-1} = \Lambda_1^*$. Then the above equation is rewritten as

$$\Lambda_1 R\Delta^*(h)R^\sim \Lambda_1^* = \Lambda_1^* R\Delta(h)R^\sim \Lambda_1 \equiv \Gamma(h); \quad \forall\, h \in H$$

where the first and second expressions of $\Gamma(h)$ are mutually complex conjugate so that $\Gamma(h)$ is real. This means that the representation $\{\Delta(h)\}$ is transformed into a real representation $\{\Gamma(h)\}$ by a unitary matrix $\Lambda_1^* R$. Q.E.D.

On combining this theorem with the type criteria given by (16.6.4b), we have the following correlations between the *case* criteria for a unirrep $\Delta^{(\nu)}(H)$ with a real character and the *type* criteria of the co-unirrep of H^e induced from $\Delta^{(\nu)}(H)$.

(iia) If $\Delta^{(\nu)}(H)$ is an integral unirrep,

$$\text{case } 1 \leftrightarrow \text{type a}; \qquad \text{case } 2 \leftrightarrow \text{type b} \tag{16.6.5}$$

(iib) If $\Delta^{(\nu)}(H)$ is a half-integral unirrep,

$$\text{case } 2 \leftrightarrow \text{type a}; \qquad \text{case } 1 \leftrightarrow \text{type b} \tag{16.6.6}$$

Remark 1. The second correspondence in (16.6.5) is academic for a point group because all integral co-unirreps of a gray point group H^e are of type a according to Table 16.3. Thus, we may conclude that an integral unirrep of a point group with a real character belongs to case 1, i.e. it is always transformed to real form by a symmetric unitary matrix.

Remark 2. Analogously to the above, the second correspondence in (16.6.6) is almost academic because a type b half-integral co-unirrep of H^e occurs once and only once according to Table 16.3: it is given by $S(M_{n/2}, M_{n/2}; 1)_{n=\text{odd}}$ of C_n^e, where $M_{n/2}(c_n^k) = (-1)^k$ is real and hence belongs to case 1. Except for this case, a half-integral unirrep $\Delta^{(\nu)}(H)$ with a real character belongs to case 2, corresponding to the type a co-unirrep of H^e. For such a case the dimensionality d_ν of $\Delta^{(\nu)}(H)$ must be even, since, from $N(\boldsymbol{\theta})^\sim = -N(\boldsymbol{\theta})$, the determinant satisfies $\det N(\boldsymbol{\theta}) = (-1)^{d_\nu} \det N(\boldsymbol{\theta})$ so that d_ν is even since it is null otherwise.

Example 1. According to (10.3.9), the characters of the unirreps of $G_s = SU(2)$ are all real and the co-unirreps of the gray group G_s^e belong to type a according to (16.5.5a). Therefore, from (16.6.5) and (16.6.6), all integral unirreps of G_s belong to case 1 whereas all half-integral unirrreps of G_s belong to case 2. This means that the ordinary single-valued unirreps of the three-dimensional rotation group can be made real whereas its double-valued unirreps are complex and cannot be made real.

Example 2. According to Table 16.3, all two-dimensional unirreps E_m of D_n form type a co-unirreps $S(E_m; Y_m)$ of D_n^e so that the characters of E_m are all real from (16.6.4a). When m is an integer, E_m belongs to case 1 from (16.6.5), whereas, when m is a half integer, E_m belongs to case 2 from (16.6.6). For the latter case, E_m should be complex and cannot be transformed to real form. These are explicitly verified by the table of the representations of D_n given by Table 11.5.

Example 3. All co-unirreps of O^e are of type a so that their characters are all real. The integral unirreps belong to case 1 whereas the half-integral unirreps belong to case 2. The dimensionalities of their representations are all even for the latter.

Example 4. There exist three half-integral unirreps $E_{1/2}$, $E'_{1/2}$ and $E''_{1/2}$ for the group T, all in two dimensions: $E_{1/2}$ forms a type a co-unirrep $S(E_{1/2}; \sigma_y)$ of T^e whereas $E'_{1/2}$ and $E''_{1/2}$ form a type c co-unirrep $S(E'_{1/2}, E''_{1/2}; \sigma_y)$ of T^e. Thus $E_{1/2}$ belongs to case 2 whereas $E'_{1/2}$ and $E''_{1/2}$ as a pair belong to case 3. Since $E''_{1/2}$ is equivalent to $E'^*_{1/2}$, the type c co-unirrep can be expressed also by $S(E_{1/2}, E'^*_{1/2}; 1)$.

16.7 The orthogonality theorem on the co-unirreps

To extend the orthogonality theorem on the unirreps of a unitary group to the co-unirreps of an anti-unitary group, we shall first extend Schur's lemma to an anti-unitary group. It will be shown that only the second half, B, of Theorem 6.5.2 on Schur's lemma has to be modified (Dimmock 1963, Kim 1984b). Let an anti-unitary group H^z be written as

$$H^z = H + a_0 H, \qquad a_0 = \boldsymbol{\theta} z$$

where $\boldsymbol{\theta}$ is the time-reversal operator. Then an anti-unitary operator a belongs to $a_0 H$.

Theorem 16.7.1. (Schur's lemma on anti-unitary groups.) Let $S^{(\alpha)}(H^z)$ and $S^{(\beta)}(H^z)$ be two co-unirreps of H^z with the dimensionalities d_α and d_β, respectively. If there exists a $d_\alpha \times d_\beta$ intertwining matrix M such that

$$\begin{aligned} S^{(\alpha)}(h)M &= MS^{(\beta)}(h); \qquad \forall\, h \in H \\ S^{(\alpha)}(a)M^* &= MS^{(\beta)}(a); \qquad \forall\, a \in a_0 H \end{aligned} \tag{16.7.1}$$

then we have

A$'$. M is either a null matrix or a square non-singular matrix.
B$'$. M is a real constant matrix when $\alpha = \beta$, provided that M is Hermitian.

This theorem differs from Schur's lemma given by Theorem 6.5.2 on the appearance of M^* in (16.7.1) and the Hermiticity condition of M in B$'$. We shall give here the proof only for the second half, B$'$, because the first half, A$'$, proceeds exactly as A in the case of a unitary group. Let us assume that A$'$ has been proven. Then to prove B$'$ we may set $\alpha = \beta$ in (16.7.1) and obtain

$$\begin{aligned} S^{(\alpha)}(h)M &= MS^{(\alpha)}(h); \qquad \forall\, h \in H \\ S^{(\alpha)}(a)M^* &= MS^{(\alpha)}(a); \qquad \forall\, a \in a_0 H \end{aligned} \tag{16.7.2}$$

Now, let $c\mathbf{1}$ be a real constant matrix of the dimensionality d_α, then it commutes with any $d_\alpha \times d_\alpha$ matrix so that

$$\begin{aligned} S^{(\alpha)}(h)[M - c\mathbf{1}] &= [M - c\mathbf{1}]S^{(\alpha)}(h) \\ S^{(\alpha)}(a)[M - c\mathbf{1}]^* &= [M - c\mathbf{1}]S^{(\alpha)}(a) \end{aligned} \tag{16.7.3}$$

which have the form of (16.7.2) with M replaced by $[M - c\mathbf{1}]$. Let M be Hermitian and c be one of its eigenvalues, then $[M - c\mathbf{1}]$ is a singular matrix that must be the null matrix from the first half A′ of the theorem. Thus $M = c\mathbf{1}$. Q.E.D.

Theorem 16.7.2 (The orthogonality theorem). Let $S^{(\alpha)}$ and $S^{(\beta)}$ be two co-unirreps of H^z that are assumed to be inequivalent when $\alpha \neq \beta$. Then, for $\alpha \neq \beta$, we have

$$\sum_{h\in H} S^{(\alpha)}(h)_{\mu\nu} S^{(\beta)}(h)^*_{\mu'\nu'} = 0$$

$$\sum_{a\in a_0 H} S^{(\alpha)}(a)_{\mu\nu} S^{(\beta)}(a)^*_{\mu'\nu'} = 0 \tag{16.7.4}$$

where $\mu, \nu = 1, 2, \ldots, d_\alpha$; $\mu', \nu' = 1, 2, \ldots, d_\beta$. For $\alpha = \beta$ we have

$$\sum_{h\in H} S^{(\alpha)}(h)_{\mu\nu} S^{(\alpha)}(h)^*_{\mu'\nu'} + \sum_{a\in a_0 H} S^{(\alpha)}(a)_{\mu\nu'} S^{(\alpha)}(a)^*_{\mu'\nu} = (2|H|/d_\alpha)\delta_{\mu\mu'}\delta_{\nu\nu'} \tag{16.7.5}$$

where $|H|$ is the order of H. Note that the indices ν and ν' in the second sum in (16.7.5) are interchanged relative to those of the first sum.

Proof. Analogously to (6.5.9), an intertwining matrix for $S^{(\alpha)}$ and $S^{(\beta)}$ that satisfies (16.7.1) is given by

$$M^{(\alpha,\beta)} \equiv \sum_h S^{(\alpha)}(h) X S^{(\beta)}(h)^\dagger + \sum_a S^{(\alpha)}(a) X^* S^{(\beta)}(a)^\dagger \tag{16.7.6}$$

where X is an arbitrary $d_\alpha \times d_\beta$ matrix: this is verified by making a simple substitution of $M = M^{(\alpha,\beta)}$ into (16.7.1). When $\alpha \neq \beta$, $S^{(\alpha)}$ and $S^{(\beta)}$ are inequivalent by assumption so that $M^{(\alpha,\beta)}$ must be a null matrix from Theorem 16.7.1. Now we set $X_{\nu\nu'} = 1$ for a particular element of X in (16.7.6) and the remaining elements equal to zero to obtain

$$\sum_h S^{(\alpha)}(h)_{\mu\nu} S^{(\beta)}(h)^*_{\mu'\nu'} + \sum_a S^{(\alpha)}(a)_{\mu\nu} S^{(\beta)}(a)^*_{\mu'\nu'} = 0 \tag{16.7.7a}$$

Next, we set $X_{\nu\nu'} = \mathrm{i}$ and all other elements zero to obtain

$$\sum_h S^{(\alpha)}(h)_{\mu\nu} S^{(\beta)}(h)^*_{\mu'\nu'} - \sum_a S^{(\alpha)}(a)_{\mu\nu} S^{(\beta)}(a)^*_{\mu'\nu'} = 0 \tag{16.7.7b}$$

By adding and subtracting these two equations we obtain (16.7.4), the first half of Theorem 16.7.2.

When $\alpha = \beta$, the matrix $M^{(\alpha,\alpha)}$ defined by (16.7.6) is a constant matrix according to Theorem 16.7.1, since $M^{(\alpha,\alpha)}$ is made Hermitian by taking X to be Hermitian. Thus, we set $X_{\nu\nu'} = X_{\nu'\nu} = 1$ and the remaining elements zero in (16.7.6) with $\alpha = \beta$ and obtain

$$\sum_h \{S^{(\alpha)}(h)_{\mu\nu} S^{(\alpha)}(h)^*_{\mu'\nu'} + S^{(\alpha)}(h)_{\mu\nu'} S^{(\alpha)}(h)^*_{\mu'\nu}\}$$

$$+ \sum_h \{S^{(\alpha)}(a)_{\mu\nu} S^{(\alpha)}(a)^*_{\mu'\nu'} + S^{(\alpha)}(a)_{\mu\nu'} S^{(\alpha)}(a)^*_{\mu'\nu}\} = c(\nu, \nu')\delta_{\mu\mu'} \tag{16.7.8}$$

where $c(\nu, \nu')$ is a constant independent of μ, μ' but may depend on ν and ν' on account of the particular choice $X_{\nu\nu'} = 1$. To determine $c(\nu, \nu')$, we set $\mu = \mu'$ in (16.7.8) and sum up both sides of the equations over μ and obtain, via the unitarity of the matrix $S^{(\alpha)}$,

$$c(\nu, \nu') = \delta_{\nu\nu'}4|H|/d_\alpha$$

Analogous to (16.7.8), we may set $X_{\nu\nu'} = -X_{\nu'\nu} = \mathrm{i}$ and the remaining elements of X zero in (16.7.6) with $\alpha = \beta$ and obtain

$$\sum_h \{S^{(\alpha)}(h)_{\mu\nu} S^{\alpha}(h)^*_{\mu'\nu'} - S^{(\alpha)}(h)_{\mu\nu'} S^{(\alpha)}(h)^*_{\mu'\nu}\}$$

$$+\sum_a \{-S^{(\alpha)}(a)_{\mu\nu} S^{(\alpha)}(a)^*_{\mu'\nu'} + S^{(\alpha)}(a)_{\mu\nu'} S^{(\alpha)}(a)^*_{\mu'\nu}\} = c'(\nu, \nu') d_{\mu\mu'} \qquad (16.7.9)$$

In this case $c'(\nu, \nu') = 0$: this follows if we set $\mu = \mu'$ in (16.7.9) and sum up the equations over μ. Now, by combining (16.7.8) and (16.7.9), we obtain the second half of the orthogonality theorem given by (16.7.5).

16.8 Orthogonality relations for the characters, the irreducibility condition and the type criteria for co-unirreps

16.8.1 Orthogonality relations for the characters of co-unirreps

The character of a co-unirrep is defined by its trace as usual. We set $\mu' = \mu$ and $\nu' = \nu$ in the orthogonality relations of co-unirreps given by (16.7.4) and (16.7.5), and then sum them over μ and ν to obtain

$$\sum_h \kappa^{(\alpha)}(h)\kappa^{(\beta)}(h)^* = 0; \qquad \alpha \neq \beta \qquad (16.8.1a)$$

$$\sum_a \kappa^{(\alpha)}(a)\kappa^{(\beta)}(a)^* = 0; \qquad \alpha \neq \beta \qquad (16.8.1b)$$

$$\sum_h |\kappa^{(\alpha)}(h)|^2 + \sum_a \kappa^{(\alpha)}(a^2) = 2|H| \qquad (16.8.1c)$$

where $\kappa^{(\alpha)}$ and $\kappa^{(\beta)}$ are the characters of co-unirreps $S^{(\alpha)}$ and $S^{(\beta)}$ of H^z, respectively. Since a^2 belong to H, the first and third equations depend only on the elements belonging to the unitary halving subgroup H of H^z. It will be shown below that the last relation (16.8.1c) provides the necessary and sufficient condition for the co-representation $S^{(\alpha)}(H^z)$ to be irreducible.

16.8.2 Irreducibility criteria for co-unirreps

Let S be a unitary co-representation of H^z and κ be its character. Let f_α be the number of times a co-unirrep $S^{(\alpha)}$ appears in the reduced form of S. Then

$$\kappa(h) = \sum_\alpha f_\alpha \kappa^{(\alpha)}(h); \qquad \forall\, h \in H$$

$$\kappa(a) = \sum_\alpha f_\alpha \kappa^{(\alpha)}(a); \qquad \forall\, a \in a_0 H \qquad (16.8.2)$$

where f_α is a positive integer or zero and may be expressed, using the orthogonality relations, in the following two ways:

$$f_\alpha = \sum_h \kappa(h)\kappa^{(\alpha)}(h)^* \Big/ \sum_h |\kappa^{(\alpha)}(h)|^2$$
$$= \sum_a \kappa(a)\kappa^{(\alpha)}(a)^* \Big/ \sum_a |\kappa^{(\alpha)}(a)|^2 \qquad (16.8.3)$$

In view of (16.8.1c), let us define a sum $Z(S)$ by

$$Z(S) = \frac{1}{2|H|}\left(\sum_h |\kappa(h)|^2 + \sum_a \kappa(a^2)\right) \qquad (16.8.4)$$

which depends on the co-representation $S(H^z)$ through its character κ. Then substituting the first relation of (16.8.2) into $\kappa(h)$ and $\kappa(a^2)$ in $Z(S)$, and then using (16.8.1a) and (16.8.1c), we obtain

$$Z(S) = \frac{1}{2|H|}\sum_\alpha\left[(f_\alpha^2 - f_\alpha)\left(\sum_h |\kappa^{(\alpha)}(h)|^2\right) + 2|H|f_\alpha\right]$$
$$\geqslant \sum_\alpha f_\alpha \geqslant 1$$

where the inequalities follow from $f_\alpha^2 \geqslant f_\alpha$ because f_α is a positive integer or zero and also from the fact that at least one of f_α must be unity. Now the co-unirrep $S(H^z)$ is irreducible if and only if only one of f_α equals unity and the remaining f_α are equal to zero. Thus *the irreducibility condition* of $S(H^z)$ is given by

$$Z(S) = 1 \qquad \text{(the minimum value of } Z(S)) \qquad (16.8.5)$$

or explicitly

$$\sum_h |\kappa(h)|^2 + \sum_a \kappa(a^2) = 2|H|$$

which coincides with (16.8.1c).

This irreducibility condition (16.8.5) for a co-unirrep $S(H^z)$ is a simple extension of the ordinary irreducibility condition for a unitary representation. The above criterion was first obtained by Kim (1984b) and has been generalized further to a projective co-representation. The condition is convenient since it requires only knowledge of the characters of the halving subgroup H.

16.8.3 The type criterion for a co-unirrep

By the irreducibility condition (16.8.5), we shall reconfirm the irreducibility of the three types of co-unirrep of H^z defined by (16.5.1) and then deduce the well-known type criterion first introduced by Dimmock and Wheeler (1962). Let us first calculate the sum $Z(S)$ defined by (16.8.4) for each of the three types of co-unirreps $S^{(\nu+)}$, $S^{(\nu,\nu)}$ and $S^{(\nu,\bar{\nu})}$ given in (16.5.1). Using the orthogonality relations of characters $\chi^{(\nu)}(H)$ ($= \kappa^{(\nu)}(H)$) of the unirreps of the halving subgroup H, we obtain

$$\text{type a:} \qquad Z(S^{(\nu+)}) = \frac{1}{2} + \frac{1}{2|H|}\sum_{a} \chi^{(\nu)}(a^2) \tag{16.8.6a}$$

$$\text{type b:} \qquad Z(S^{(\nu,\nu)}) = 2 + \frac{1}{|H|}\sum_{a} \chi^{(\nu)}(a^2) \tag{16.8.6b}$$

$$\text{type c:} \qquad Z(S^{(\nu,\overline{\nu})}) = 1 + \frac{1}{2|H|}\sum_{a} [\chi^{(\nu)}(a^2) + \chi^{(\nu)}(a^2)^*] \tag{16.8.6c}$$

where we have used $\chi^{(\overline{\nu})}(a^2) = \chi^{(\nu)}(a^2)^*$ obtained from (16.4.10a). Thus, from the irreducibility condition $Z(S) = 1$, we obtain the following irreducibility conditions for the three types:

$$\text{type a:} \qquad \sum_{a} \chi^{(\nu)}(a^2) = |H| \tag{16.8.7a}$$

$$\text{type b:} \qquad \sum_{a} \chi^{(\nu)}(a^2) = -|H| \tag{16.8.7b}$$

$$\text{type c:} \qquad \sum_{a} [\chi^{(\nu)}(a^2) + \chi^{(\nu)}(a^2)^*] = 0 \tag{16.8.7c}$$

To prove these we shall first rewrite $\Delta^{(\nu)}(a^2)$, for $a = a_0 h$, in the form

$$\begin{aligned}
\Delta^{(\nu)}((a_0 h)^2) &= \Delta^{(\nu)}(a_0^2 a_0^{-1} h a_0 h) \\
&= \Delta^{(\nu)}(a_0^2)\Delta^{(\nu)}(a_0^{-1} h a_0)\Delta^{(\nu)}(h) \\
&= \Delta^{(\nu)}(a_0^2) N(a_0)^{*-1} \Delta^{(\overline{\nu})}(h)^* N(a_0)^* \Delta^{(\nu)}(h)
\end{aligned} \tag{16.8.8}$$

where (16.4.10a) has been used for the third equality. Then, for type c for which $\nu \neq \overline{\nu}$, the orthogonality relations between the unirreps $\Delta^{(\nu)}(H)$ and $\Delta^{(\overline{\nu})}(H)$ lead to

$$\sum_{h} \Delta^{(\nu)}((a_0 h)^2) = 0$$

so that its trace yields

$$\sum_{h} \chi^{(\nu)}((a_0 h)^2) = 0 \tag{16.8.9a}$$

from which (16.8.7c) follows. Next, for types a and b, we have $\nu = \overline{\nu}$ and $\Delta^{(\nu)}(a_0^2) = \pm N(a_0) N(a_0)^*$ so that (16.8.8) is simplified to

$$\Delta^{(\nu)}((a_0 h)^2) = \pm N(a_0) \Delta^{(\nu)}(h)^* N(a_0)^* \Delta^{(\nu)}(h) \tag{16.8.9b}$$

where the + (−) sign is for type a (b). Now, using the orthogonality relations for the unirrep $\Delta^{(\nu)}(H)$, and the fact that $N(a_0)$ is a $d_\nu \times d_\nu$ unitary matrix, we obtain

$$\sum_{h} \Delta^{(\nu)}((a_0 h)^2) = \pm(|H|/d_\nu)\mathbf{1}$$

where **1** is the $d_\nu \times d_\nu$ unit matrix. Then, the traces of both sides of this equation lead to (16.8.7a) or (16.8.7b).

We may rewrite (16.8.7a)–(16.8.7c), in view of (16.8.9a), in the form

$$\frac{1}{|H|}\sum_{h\in H}\chi^{(\nu)}((a_0h)^2)=\begin{cases}1; & \text{type a}\\ -1; & \text{type b}\\ 0; & \text{type c}\end{cases} \tag{16.8.10}$$

These provide the irreducibility conditions for the three types of co-unirreps as well as for their type criteria. This criterion is convenient because it depends only on the characters of the unirreps of the halving subgroup H of H^z. The above criterion has been extended also to any projective co-unirreps by Kim (1984b).

17

Anti-unitary space groups and their co-representations

17.1 Introduction

An anti-unitary space group is formed by augmenting a unitary space group with an anti-unitary element, analogously to the anti-unitary point group. Belov, Neronova and Smirnova (Belov *et al.* 1955) provided a listing of all 1651 magnetic space groups (or Shubnikov groups), which include the 230 unitary space groups and 1421 anti-unitary space groups. They also provided the method of deriving them through geometric consideration. Extensive reviews on the subject have been published by Opechowski and Guccione (1965) and by Bradley and Cracknell (1972).

Here we shall construct the anti-unitary space groups via the algebraic method introduced by Kim (1986c). Since we have already constructed a total of 230 unitary space groups through 32 minimal general generator sets (MGGSs) in Chapter 13, we can construct 1421 anti-unitary space groups through 38 MGGSs by augmenting the 32 MGGSs of the unitary space groups with anti-unitary operators. It requires a minor amount of additional work on the space groups, because the time-reversal operator $\boldsymbol{\theta}$ commutes with any operator of the space groups.

For a given space group $\hat{G}$ there exist two kinds of anti-unitary space groups denoted by $\hat{G}^e$ and $\hat{H}^z$, respectively. The first kind $\hat{G}^e$ is formed by augmenting $\hat{G}$ by a time-reversal translation operator $\hat{\boldsymbol{\theta}} = (\boldsymbol{\theta}|\boldsymbol{v}_{\boldsymbol{\theta}})$ (where $\boldsymbol{v}_{\boldsymbol{\theta}}$ is the minimum translation characteristic to $\boldsymbol{\theta}$), i.e.

$$\hat{G}^e = \hat{G} + \hat{\boldsymbol{\theta}}\hat{G} \tag{17.1.1}$$

The factor group of the first kind $\hat{G}^e/T$ with respect to the translation group T of $\hat{G}$ is isomorphic to the gray point group G^e discussed in the previous chapter. Here the so-called *magnetic Bravais lattices* are introduced by decorating the Bravais lattice of a space group $\hat{G}$ with the time-reversal translation operator $\hat{\boldsymbol{\theta}}$. Next, when $\hat{G}$ has a halving subgroup $\hat{H}$ such that $\hat{G} = \hat{H} + \hat{z}\hat{H}$, where $\hat{z} = (z|\boldsymbol{v}_z)$ is a coset representative of $\hat{H}$ in $\hat{G}$, by augmenting $\hat{H}$ with the anti-unitary operator $\boldsymbol{\theta}\hat{z}$ we form the second kind $\hat{H}^z$, as follows:

$$\hat{H}^z = \hat{H} + \boldsymbol{\theta}\hat{z}\hat{H}; \qquad z \notin H \tag{17.1.2}$$

Obviously there is an isomorphism between the factor group $\hat{H}^z/T$ and the anti-unitary point group H^z introduced in the previous chapter.

All anti-unitary space groups of the second kind $\hat{H}^z$ from a given space group $\hat{G}$ are formed via the coset decompositions of $\hat{G}$ with respect to all the possible halving subgroups $\hat{H}$ of $\hat{G}$. These will be formed via simple algebraic lemmas on how to form the halving subgroups H of the point co-group G of $\hat{G}$. In cases in which G does not have a halving subgroup there is no anti-unitary space group of the second kind which is formed from the space group $\hat{G}$. This is the case, for example, for the space groups $\hat{C}_n$ with an odd n, and for the space groups $\hat{T}$ of the tetrahedral group T. By definition, however, there exists at least one anti-unitary space group $\hat{G}^e$ of the first kind for a given space group $\hat{G}$. It should be noted that $\hat{G}$ is a halving subgroup of $\hat{G}^e$ as well as $\hat{H}$ is a halving subgroup of $\hat{H}^z$. Consequently, every anti-unitary space group can be expressed by $\hat{H}^z$, where $\hat{H}$ is a halving subgroup of $\hat{H}^z$: it includes $\hat{G}^e$ as a special case in which $\hat{H} = \hat{G}$ and $z \in H$. This extended view of the notation $\hat{H}^z$ is convenient and becomes important when we construct the co-unirreps of the anti-unitary space groups in Section 17.7. Unless stated otherwise we shall use the same notations as in Chapter 13 and we mean by a point group or space group the respective double group defined through (13.8.2).

Analogously to anti-unitary point groups, if we regard the time-reversal operator $\boldsymbol{\theta}$ as an operator that changes color from white to black and from black to white, then an anti-unitary space group may describe the symmetry of a system of black and white balls. This is possible except for *the gray space groups*, for which the augmentor is the pure time-reversal operator $\hat{\boldsymbol{\theta}} = (\boldsymbol{\theta}|000)$. Excluding this case, all the anti-unitary space groups are also called *black–white space groups* and may be classified further into the first and second kinds.

Now, consider a *magnetic crystal*, which is a crystal formed by atoms or ions with non-vanishing magnetic moments. For simplicity, the system may be called a spin system even though there may be an orbital contribution to the magnetic moment. Now a spin is an axial vector under a point operation and reverses its direction under time reversal. The *magnetic space group* of a magnetic crystal is the symmetry group of the crystal which leaves the spin arrangement invariant: it could be anti-unitary as well as unitary. For example, a *paramagnetic* crystal in the absence of an external magnetic field will be invariant under a gray space group because all spins are randomly oriented. If the temperature of a paramagnetic crystal is lowered, it is possible that some long-range magnetic order will set in below a certain temperature such that the crystal may become *antiferromagnetic, ferromagnetic* or *ferrimagnetic*. These are described by the anti-unitary groups of the first or the second kind. The first kind $\hat{G}^e$ with an anti-unitary translation $\hat{\boldsymbol{\theta}} = (\boldsymbol{\theta}|\boldsymbol{v}_{\boldsymbol{\theta}} \neq 0)$ must necessarily be antiferromagnetic because there exist equal numbers of spins $\boldsymbol{s}$ and reversed spins $-\boldsymbol{s}$ separated by the pure translation $\boldsymbol{v}_{\boldsymbol{\theta}} \neq 0$ (see Figure 17.1). An anti-unitary space group of the second kind $\hat{H}^z$ is antiferromagnetic if the anti-unitary operator $\boldsymbol{\theta}z$ of the augmenting operator $(\boldsymbol{\theta}z|\boldsymbol{v}_z)$ reverses the direction of spin (i.e. the point operation z leaves the spin invariant) since then a spin $\boldsymbol{s}$ is cancelled by $-\boldsymbol{s}$ separated by $\boldsymbol{v}_z$ ($\neq 0$). Thus, for example, magnetic space groups belonging to the magnetic classes H^i, C^q_n, C^p_n and C^q_{ni} (or their subgroups) may be antiferromagnetic. In particular, for the last three types, all spins are aligned along the axis of rotation c_n, half of which are parallel and the remaining half are antiparallel. On the other hand, if the point operation z reverses the spin, then $\boldsymbol{\theta}z$ will leave the spin invariant so that the corresponding space group $\hat{H}^z$ may describe a *ferromagnetic crystal*. In fact, as discussed in Section 16.3.2, the point group of a ferromagnetic crystal is one of 31 subgroups of $C^u_{\infty i} = C^v_{\infty i}$ that leaves the spin invariant. To these there correspond 275 magnetic space groups, some of which are unitary and some of which are anti-unitary. Note, however, that these 275 magnetic space groups are not necessarily all ferromagnetic. Their separate moments may cancel out to be antiferromagnetic by interlocking of ferromagnetic spin arrangements, for example, in the directions perpendicular to the principal axis of rotation. If the separate moments only partially cancel, the resulting structure will be *ferrimagnetic*. See Opechowski and Guccione (1965) for further detail.

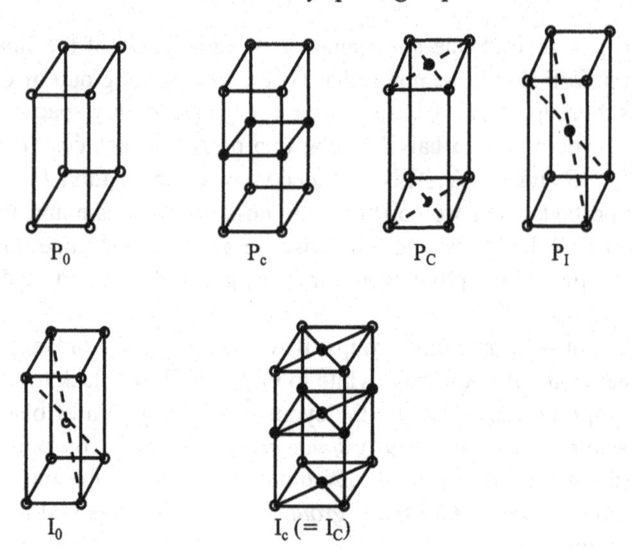

Figure 17.1. The black and white Bravais parallelepipeds of the tetragonal system. Here the time-reversal operator $\boldsymbol{\theta}$ is regarded as an operator which changes the color black to white and white to black, following Belov *et al.* (1955).

17.2 Anti-unitary space groups of the first kind

Let $\hat{G}$ be a space group and T be its translation subgroup. Then $\hat{G}$ is completely defined by T and a set of the coset representatives $\{(R|\boldsymbol{v}_R)\}$ of T in $\hat{G}$, where $\boldsymbol{v}_R$ is the minimum translation characteristic to the rotation R defined in (13.6.2a). Then an anti-unitary space group of the first kind is defined by augmenting $\hat{G}$ with a time-reversal translation operator $\hat{\boldsymbol{\theta}}$ as follows:

$$\hat{G}^e = \hat{G} + \hat{\boldsymbol{\theta}}\hat{G}; \qquad \hat{\boldsymbol{\theta}} = (\boldsymbol{\theta}|\boldsymbol{v}_{\boldsymbol{\theta}}) \tag{17.2.1}$$

where $\boldsymbol{\theta}$ is the time-reversal operator and $\boldsymbol{v}_{\boldsymbol{\theta}}$ is the minimum translation characteristic to $\boldsymbol{\theta}$. The time-reversal operator $\boldsymbol{\theta}$ commutes with any rotational part of $\hat{G}$ and leaves any translational part invariant, because a translational part behaves like a position-like quantity that is even under time reversal. As a result, the factor group $\hat{G}^e/T$ is isomorphic to the gray point group G^e.

Since $\hat{G}$ is an invariant subgroup of $\hat{G}^e$, being a halving subgroup, the augmenting operator $\boldsymbol{\theta}$ must satisfy the compatibility condition

$$\hat{\boldsymbol{\theta}}^2, \boldsymbol{\theta}^{-1}\hat{R}\hat{\boldsymbol{\theta}} \in \hat{G}; \qquad \forall \hat{R} \in \hat{G} \tag{17.2.2}$$

Since

$$\hat{\boldsymbol{\theta}}^2 = (e'|2\boldsymbol{v}_{\boldsymbol{\theta}}) = (e'|000) \tag{17.2.3a}$$

$$\hat{\boldsymbol{\theta}}^{-1}\hat{R}\hat{\boldsymbol{\theta}} = (R|\boldsymbol{v}_R + R\boldsymbol{v}_{\boldsymbol{\theta}} - \boldsymbol{v}_{\boldsymbol{\theta}}) = (R|\boldsymbol{v}_R) \tag{17.2.3b}$$

we have

$$2\boldsymbol{v}_{\boldsymbol{\theta}} \equiv 0 \qquad (\text{mod}\, \boldsymbol{t} \in T) \tag{17.2.4a}$$

$$R\boldsymbol{v}_{\boldsymbol{\theta}} \equiv \boldsymbol{v}_{\boldsymbol{\theta}} \qquad (\text{mod}\, \boldsymbol{t} \in T); \qquad \forall R \in G \tag{17.2.4b}$$

That is, the characteristic translation $\boldsymbol{v}_{\boldsymbol{\theta}}$ of $\boldsymbol{\theta}$ is a binary translation belonging to the invariant eigenvector space of all $R \in G$ (mod $\boldsymbol{t} \in T$).

All the possible augmenting operators $(\boldsymbol{\theta}|\boldsymbol{v_\theta})$ are formed from (17.2.4a) and (17.2.4b) by determining the translational parameters of $\boldsymbol{v_\theta} = (\xi, \eta, \zeta)$ for each lattice type L of the crystal class G. An immediate conclusion from the first condition (17.2.4a) is that these parameters are all binary, i.e. they are limited to

$$\xi, \eta, \zeta = 0, \tfrac{1}{2} \qquad (\text{mod}\, \boldsymbol{t} \in T) \tag{17.2.5}$$

for any lattice type L. For further specification of these parameters by the second condition (17.2.4b), it suffices to limit the rotations R to being the generators of G and also the orders of the generators to being higher than two, because a binary operation simply reproduces (17.2.5). Thus we are left with one characteristic generator R_c for each crystal class G introduced in Table 13.2. We may assume further that R_c is proper; the improper generator $\overline{R}_c$ leads to the same restriction on the parameters, because (17.2.4b) means that $R_c\boldsymbol{v_\theta} = \pm\boldsymbol{v_\theta}$ (mod $\boldsymbol{t} \in T$) from $2\boldsymbol{v_\theta} \in T$. According to Table 13.2, such a generator R_c of a crystal class G is characteristic of the crystal system K to which the class G belongs. Accordingly, R_c is a property of the crystal system K common to all classes belonging to the system. As a result, the condition (17.2.4b) plays no role for the orthorhombic and monoclinic systems, because their generators are all binary.

Obviously, it is possible that some of the $\hat{G}^e$ thus obtained may be equivalent under a lattice transformation $\Lambda = [U|\boldsymbol{s}]$ that leaves invariant the halving subgroup $\hat{G}$ of $\hat{G}^e$. Since, from (13.1.7),

$$\Lambda(\boldsymbol{\theta}|\boldsymbol{v_\theta})\Lambda^{-1} = (\boldsymbol{\theta}|U\boldsymbol{v_\theta}) \tag{17.2.6a}$$

the condition for equivalence of two operations $(\boldsymbol{\theta}|\boldsymbol{v_\theta})$ and $(\boldsymbol{\theta}|\boldsymbol{v'_\theta})$ under Λ is given by

$$\boldsymbol{v'_\theta} = U\boldsymbol{v_\theta} \qquad (\text{mod}\, \boldsymbol{t} \in T) \tag{17.2.6b}$$

which is independent of the shift $\boldsymbol{s}$. The set of unimodular transformations $\{U\}'$ which establishes the equivalence of $\hat{G}^e$ must leave $\hat{G}$ invariant under $[U|\boldsymbol{s}]$, so that it is a subset of the relevant set $\{U\}$ given by (13.7.4) that establishes the equivalence of $\hat{G}$. Since the equivalence condition (17.2.6b) for $\hat{G}^e$ is independent of the shift, the condition imposes no constraint on the crystal classes of high symmetry, for which the relevant lattice transformation Λ is a pure shift $[e|\boldsymbol{s}]$ according to (13.7.4). In fact, as will be shown below, the equivalence condition (17.2.6b) becomes a constraint only for those crystal classes belonging to the orthorhombic and monoclinic systems.

For actual construction of $\hat{G}^e$ we may regard it as a semi-direct product of two groups. To see this we rewrite (17.2.1) in the form $\hat{G}^e = (T + \hat{\boldsymbol{\theta}}T)\hat{G}/T$, using $T\hat{G}T^{-1} = \hat{G}$; then we obtain the following semi-direct product:

$$\hat{G}^e = T^e \wedge \hat{G}/T; \qquad T^e = T + \hat{\boldsymbol{\theta}}T \tag{17.2.7a}$$

where T^e is an invariant subgroup of $\hat{G}^e$ and may be called *the magnetic Bravais lattice* (denoted M_L). It is a Bravais lattice decorated with $\hat{\boldsymbol{\theta}} = (\boldsymbol{\theta}|\xi, \eta, \zeta)$; see e.g. Figure 17.1. It plays the role of the Bravais lattice for $\hat{G}^e$ that the Bravais lattice T plays for $\hat{G}$; in fact, there exists the isomorphism

$$\hat{G}^e/T^e \simeq \hat{G}/T \simeq G \tag{17.2.7b}$$

where G is the point co-group of $\hat{G}$. Since the unitary space group $\hat{G}/T$ is known, the construction of $\hat{G}^e$ requires only the possible magnetic Bravais lattices M_L compatible with each given space group $\hat{G}$.

Let L be a lattice type that is allowed for the crystal class G of a space group $\hat{G}$. Then a magnetic Bravais lattice M_L may be expressed by L and $\hat{\boldsymbol{\theta}}$ jointly as $L\hat{\boldsymbol{\theta}} = L(\boldsymbol{\theta}|\xi, \eta, \zeta)$. Then all the possible M_L are expressed, in view of (17.2.5), by

$$\begin{aligned} &L_0 = L(\boldsymbol{\theta}|0, 0, 0), \quad &&L_a = L(\boldsymbol{\theta}|\tfrac{1}{2}, 0, 0), \quad &&L_b = L(\boldsymbol{\theta}|0, \tfrac{1}{2}, 0)\\ &L_c = L(\boldsymbol{\theta}|0, 0, \tfrac{1}{2}), \quad &&L_A = L(\boldsymbol{\theta}|0, \tfrac{1}{2}, \tfrac{1}{2}), \quad &&L_B = L(\boldsymbol{\theta}|\tfrac{1}{2}, 0, \tfrac{1}{2})\\ &L_C = L(\boldsymbol{\theta}|\tfrac{1}{2}, \tfrac{1}{2}, 0), \quad &&L_I = L(\boldsymbol{\theta}|\tfrac{1}{2}, \tfrac{1}{2}, \tfrac{1}{2}) \end{aligned} \tag{17.2.8a}$$

There exists a total of 48 M_L since L may take any one of the six conventional lattice types (P, A, B, C, F and I) introduced in Section 13.4. Obviously, some of them may be equal or equivalent or may not be allowed for a certain crystal system. According to (17.2.8a), each allowed L_x introduces an additional set of equivalent points with the time inversion $\boldsymbol{\theta}$ into the conventional unit cell of the Bravais lattice L of G. The above notations are in accordance with those introduced by Belov *et al.* (1955) for the black and white lattices defined by figures. They introduced 36 topologically independent M_L through geometric consideration. Following Kim (1986c), we shall construct them algebraically through further restriction of the above set (17.2.8a) by (17.2.4b) with the characteristic generator R_c for each crystal system and removing possible redundancy through equivalence (17.2.6b).

Before explicit construction of the M_L for each crystal system, we shall first remove the M_L that are redundant due to the inherent symmetry of the lattice type L described by

$$L(\boldsymbol{\theta}|\boldsymbol{v_\theta}) = L(\boldsymbol{\theta}|\boldsymbol{v_\theta} + \boldsymbol{t}_L) \qquad (\text{mod}\, \boldsymbol{t} \in T) \tag{17.2.8b}$$

where $\boldsymbol{t}_L$ is an equivalent point of the Bravais parallelepiped of the lattice type L. Thus, for example, there exists a maximum of four types of M_L for the A base-centered lattice:

$$A_0 = A_A, \qquad A_a = A_I, \qquad A_b = A_c, \qquad A_B = A_C \tag{17.2.9a}$$

Simultaneous cyclic permutations of $\{A, B, C\}$ and (a, b, c) on (17.2.9a) yield

$$B_0 = B_B, \qquad B_b = B_I, \qquad B_c = B_a, \qquad B_C = B_A \tag{17.2.9b}$$

$$C_0 = C_C, \qquad C_c = C_I, \qquad C_a = C_b, \qquad C_A = C_B \tag{17.2.9c}$$

Analogously, there exist two types of the face-centered lattice F:

$$F_0 = F_A = F_B = F_C, \qquad F_I = F_a = F_b = F_c \tag{17.2.9d}$$

and four types of the body-centered lattice I:

$$I_0 = I_I, \qquad I_a = I_A, \qquad I_b = I_B, \qquad I_c = I_C \tag{17.2.9e}$$

Further restrictions imposed by (17.2.4b) and (17.2.6b) on the allowed M_L will be discussed for each lattice type L beginning with the simplest case of the cubic system.

17.2.1 The cubic system

The characteristic proper point operation R_c for this system is 3_{xyz} according to Table 13.2. Using the Jones representation $3_{xyz} = (z, x, y)$ given in Section 13.8.1, we have, from (17.2.4b) with $R = R_c$ and $\boldsymbol{v_\theta} = (\xi, \eta, \zeta)$,

$$3_{xyz}\boldsymbol{v_\theta} - \boldsymbol{v_\theta} = (\zeta - \xi, \xi - \eta, \eta - \zeta) = (0, 0, 0)$$

which yields

$$\xi = \eta = \zeta = 0, \tfrac{1}{2}$$

Thus for the lattice types of this system, $L = P$, F and I, we obtain five independent M_L:

$$\{P_0, P_I\}, \{F_0, F_I\}, I_0 \qquad (17.2.10)^1$$

17.2.2 The hexagonal system

Using $R_c = 6_z = (x - y, x, z)$ in the hexagonal coordinate system for R in (17.2.4b), we obtain

$$\xi = \eta = 0, \qquad \zeta = 0, \tfrac{1}{2}$$

Thus, for $L = P$, which is the only lattice type of this system, we obtain two M_L:

$$P_0, P_c \qquad (17.2.11)$$

17.2.3 The rhombohedral system

With $R_c = 3_{xyz} = (z, x, y)$ in the rhombohedral coordinate system for R of (17.2.4b), we obtain

$$R_0, R_I \qquad (17.2.12a)$$

as in the case of the cubic system. These may be reexpressed by

$$R_0^*, R_c^* \qquad (17.2.12b)$$

in the hexagonal coordinates for the double-centered hexagonal lattice R^*. See (13.8.1c). For the primitive hexagonal lattice P of the system, we obtain again two M_L just like in (17.2.11):

$$P_0, P_c \qquad (17.2.12c)$$

17.2.4 The tetragonal system

With $R_c = 4_z = (\bar{y}, x, z)$ for R of (17.2.4b), we obtain

$$\xi = \eta = 0, \tfrac{1}{2}, \qquad \zeta = 0, \tfrac{1}{2}$$

which yield

$$\{P_0, P_c, P_C, P_I\}, \{I_0, I_c\} \qquad (17.2.13)$$

These magnetic Bravais lattices are presented in Figure 17.1 by the black and white Bravais parallelepipeds.

17.2.5 The orthorhombic system

Since there is no characteristic R_c of order higher than two for this system, all the parameter sets given by (17.2.5) are allowed with possible redundancies depending on

[1] Belov *et al.* (1955) used F_s for F_I.

the symmetry of the space group $\hat{G}$, due to the equivalence condition (17.2.6b). Since the relevant set $\{U\}'$ is a subset of the set $\{U\}$ given by (13.7.4), which establishes the equivalence of $\hat{G}$, we need to consider only the permutations of the x-, y- and z-axes followed by appropriate shifts for this system.

As a preparation, let us look into the symmetry property of the space groups belonging to the class D_2 as a typical example for removing the $\hat{G}^e$ which are redundant. According to Table 13.3 of the space groups, the MGGS of the class D_2 is given by

$$(2_z|c,\, 0,\, c)(2_x|a,\, a,\, 0)(2_y|c-a,\, -a,\, c) \text{ with } L(c,\, a):$$

$$\begin{array}{llll} 16.\ P(0,0), & 17.\ P(\tfrac{1}{2},0), & 18.\ P(0,\tfrac{1}{2}), & 19.\ P(\tfrac{1}{2},\tfrac{1}{2}) \end{array}$$

$$\begin{array}{lllll} 20.\ C(\tfrac{1}{2},0), & 21.\ C(0,0), & 22.\ F(0,0), & 23.\ I(0,0), & 24.\ I(\tfrac{1}{2},\tfrac{1}{2}) \end{array} \tag{17.2.14}$$

The MGGS is invariant under a lattice transformation $[(x,\, y)|c/2,\, c/2,\, c/2]$, where $(x,\, y)$ denotes the interchanges of the x- and y-axes. Moreover, the MGGS is invariant under the cyclic permutation $(x,\, y,\, z)$ of the three axes, provided that $a = c$ except for the C lattice. Thus the space groups numbers 16, 19 and 22–4 are invariant under $(x,\, y,\, z)$ whereas the remaining space groups, numbers 17, 18, 20 and 21, are symmetric under $[(x,\, y)|c/2,\, c/2,\, c/2]$.

Analogously, the symmetry of MGGS $\in D_{2i}$ can be treated, because it is defined by augmenting $\hat{D}_2(a,\, c)$ with the inversion $(\mathrm{i}|\alpha,\, \beta,\, \gamma)$ according to Table 13.3: the space groups with $a = c$ and $\alpha = \beta = \mathrm{c}$ are invariant under the cyclic permutation $(x,\, y,\, z)$ whereas the remaining space groups are invariant under $[(x,\, y|c/2,\, c/2,\, c/2)]$ or inequivalent with respect to the x-, y- and z-axes.

Finally, the MGSS $\in C_{2v}$ is defined by

$$(m_x|0,\, b,\, c),\ (m_y|a',\, 0,\, c') \in C_{2v}$$

It is invariant under the interchange $(x,\, y)$ if $(b,\, c) = (a',\, c')$, but is inequivalent with respect to the x-, y- and z-axes, otherwise.

We are now ready to construct $\hat{G}^e$ for this system.

(i) If the three axes of the coordinate system are inequivalent for a given $\hat{G}$, we have the following independent sets of M_L corresponding to $L = P,\, C,\, A,\, F$ and I, from (17.2.8a)–(17.2.9e):

$$\begin{aligned} M(8P) &= \{P_0,\, P_a,\, P_b,\, P_c,\, P_A,\, P_B,\, P_C,\, P_I\} \\ M(C) &= \{C_0,\, C_b,\, C_I,\, C_B\} \\ M(A) &= \{A_0,\, A_b,\, A_I,\, A_B\} \\ M(F) &= \{F_0,\, F_I\} \\ M(4I) &= \{I_0,\, I_a,\, I_b,\, I_c\} \end{aligned} \tag{17.2.15a}$$

Note that the lattice type B is not included here, simply because it does not appear for the orthorhombic system in Table 13.3 of the 32 MGGSs.

(ii) If $\hat{G}$ is invariant under $\Lambda = [(x,\, y)|s]$, where $(x,\, y)$ denotes the interchange of the x- and y-axes, we may characterize the independent parameters of $\boldsymbol{v}_\theta$ by

$$\xi \leqslant \eta = 0,\, \tfrac{1}{2} \qquad \zeta = 0,\, \tfrac{1}{2} \tag{17.2.15b}$$

Then we obtain, for $L = P, I, C$ and F,

$$M(6P) = \{P_0, P_b, P_c, P_A, P_C, P_I\}$$
$$M(3I) = \{I_0, I_b, I_c\},\ M(C),\ M(F) \qquad (17.2.15c)$$

where the interchange (x, y) does not affect $M(C)$ and $M(F)$ given in (17.2.15a), whereas $L = A$ does not come in because it is not invariant under (x, y).

(iii) If $\hat{G}$ is invariant under $\Lambda = [(x, y, z)|s]$, where (x, y, z) denotes the cyclic permutations of the x-, y- and z-axes, we may set

$$\xi \leqslant \eta \leqslant \zeta = 0, \tfrac{1}{2} \qquad (17.2.15d)$$

because two of the three parameters are always equal for each M_L from (17.2.8a). Thus, the independent M_L for $L = P, I$ and F are given by

$$M(4P) = \{P_0, P_c, P_A, P_I\}$$
$$M(2I) = \{I_0, I_c\},\ M(F) \qquad (17.2.15e)$$

The symmetry does not affect $M(F)$ of (17.2.15a), and is absent for $L = C$ and A.

17.2.6 The monoclinic system

Let the binary axis of rotation be parallel to the z-axis. Then the angle between the x- and y-axes is arbitrary. As in the previous Section 17.2.5, the possible redundancy of $\hat{G}^e$ may be removed by the lattice transformation (17.2.6b) with U: $(x, y, z) \rightarrow (y, x, z)$ or $(x \pm y, y, z)$. If the x- and y-axes are inequivalent for $\hat{G}$, we have, for $L = P$ and B,

$$M(6P) = \{P_0, P_a, P_b, P_c, P_A, P_B\}$$
$$M(B) = \{B_0, B_b, B_c\} \qquad (17.2.16a)$$

Note here the equivalence relations $P_c \simeq P_b$, $P_I \simeq P_A$ and $B_c \simeq B_b$ under the transformation $(x + y, y, z)$. If $\hat{G}$ is invariant under $[(x, y)|s]$ we have only four M_L, given by

$$M(4P) = \{P_0, P_a, P_c, P_A\} \qquad (17.2.16b)$$

Note that $P_b \simeq P_a$ and $P_B \simeq P_A$ under the interchange (x, y) whereas the B lattice is not invariant under (x, y).

17.2.7 The triclinic system

Since all three angles of the coordinate system are arbitrary, we have only the two M_L given by

$$M(P) = \{P_0, P_s\} \qquad (17.2.17)$$

where $P_s = P(\boldsymbol{\theta}|0, 0, \tfrac{1}{2})$, taking the z-axis in the direction of (ξ, η, ζ).

Thus we have constructed altogether 48 M_L, of which only 36 are for the holohedral space groups, in agreement with the result of Belov *et al.* (1955). In Table 17.1 we have presented the anti-unitary space groups of the first kind $\hat{G}^e$ in terms of the M_L and MGGS of each crystal class. It contains also the magnetic space groups of the second kind $\hat{H}^z$, which will be discussed in the next section.

Table 17.1. *The 38 assemblies of the general generator sets for the 1421 anti-unitary space groups*

A. The cubic system

L = P, F, I; $M_L = \{P_O, P_I\}, \{F_O, F_I\}, I_O$

T (195–9): $M_L(2_z|c, 0, c)(3_{xyz}|0)$

T_i (200–6): $M_L\,(2_z|c+a, a, c)(3_{xyz}|0)(\mathrm{i}|0)$; L{23i′}

O (207–14): $M_L(4_z|0, -c, c)(3_{xyz}|0)$; L{4′3}

T_p (215–20): $M_L(\bar{4}_z|c, -c, c)(3_{xyz}|0)$; L{$\bar{4}$′3}

O_i (221–30): $M_L(4_z|a, -c, c)(3_{xyz}|0)(\mathrm{i}|0)$; L. {43i′}, {$\bar{4}$3i}, {4′3i}

B. The hexagonal system

L = P; $M_L = \{P_O, P_c\}$

C_6 (168–73): $M_L(6_z|0, 0, c)$; P{6′}

C_{3p} (174): $M_L(\bar{6}_z|0)$; P{$\bar{6}$′}

C_{6i} (175–6): $M_L(6_z|0, 0, c)(\mathrm{i}|0)$; P. {6′i}, {6i′}, {$\bar{6}$i′}

D_6 (177–82): $M_L(6_z|0, 0, c)(u_0|0)(u_1|0, 0, c)^\dagger$; P. {6′$u_0$}, {6′$u_1$}, {6$u_0'$}

C_{6v} (183–6): $M_L(6_z|0, 0, c)(m_0|0, 0, c')(m_1|0, 0, c+c')^\dagger$; P. {6′$m_0$}, {6′$m_1$}, {6$m_0'$}

D_{3p} (187–8): $M_L(\bar{6}_z|0)(m_0|0, 0, c)(u_1|0, 0, c)^\dagger$; P. {$\bar{6}$′$m_0$}, {$\bar{6}$′$u_1$}, {$\bar{6}m_0'$}

(189–90): $M_L(\bar{6}_z|0)(u_0|0, 0, c)(m_1|0, 0, c)^\dagger$; P. {$\bar{6}$′$u_0$}, {$\bar{6}$′$m_1$}, {$\bar{6}u_0'$}

D_{6i} (191–4): $M_L(6_z|0, 0, c)(u_0|0, 0, c')(u_1|0, 0, c+c')^\dagger(\mathrm{i}|0)$; P. {6′$u_0$i}, {6′$u_1$i}, {6$u_0'$i}, {6$u_0$i′}, {6$\bar{u}_0$i′}, {$\bar{6}u_0$i′}, {$\bar{6}u_1$i′}

C. The rhombohedral system

L = P, R*; $M_L = \{P_0, P_c\}, \{R_0^*, R_c^*\}$

C_3 (143–6): $M_L(3_z|0, 0, c)$

C_{3i} (147–8): $M_L(3_z|0)(\mathrm{i}|0)$; L{3i′}

D_3 (149–55): $M_L(3_z|0, 0, c)(u_v|0)$; L{3u_v'}

C_{3v} (156–61): $M_L(3_z|0)(m_v|0, 0, c)$; L{3m_v'}

D_{3i} (162–7): $M_L(3_z|0)(u_v|0, 0, c)(m_v|0, 0, c)^\dagger(\mathrm{i}|0)$; L. {3$u_v'$}, {3$u_v$i′}, {3$m_v$i′}

D. The tetragonal system

L = P, I; $M_L = \{P_0, P_c, P_C, P_I\}, \{I_0, I_c\}$

C_4 (75–80): $M_L(4_z|0, 0, c)$; L{4′}

C_{2p} (81–2): $M_L(4_z|0)$; L{$\bar{4}$′}

C_{4i} (83–8): $M_L(4_z|a, b, c)(\mathrm{i}|0)$; L. {4′i}, {4i′}, {$\bar{4}$i′}

D_4 (89–98): $M_L(4_z|0, 0, c)(2_x|a, a, 0)(2_{xy}|a, a, c)^\dagger$; L. {4′$2_x$}, {4′$2_{xy}$}, {4$2_x'$}

C_{4v} (99–110): $M_L(4_z|0, 0, c)(m_x|a+2c, a, c')(m_{xy}|a, a+2c, c+c')^\dagger$; L. {4′$m_x$}, {4′$m_{xy}$}, {4$m_x'$}

D_{2p} (111–14, 121–2): $M_L(4_z|0)(2_x|a+2c, a, c)(m_{xy}|a, a+2c, -c)^\dagger$; L. {$\bar{4}$′2}, {$\bar{4}$′$m$}, {$\bar{4}$2′}

(115–20): $M_L(4_z|0)(m_x|a+2c, a, c)(2_{xy}|a, a+2c, -c)^\dagger$; L. {$\bar{4}$′$m$}, {$\bar{4}$′2}, {$\bar{4}m'$}

D_{4i} (123–42): $M_L(4_z|\alpha+c, c, c)(2_x|a, a+2c+\alpha, \gamma)(2_{xy}|a-c, a+c, c+\gamma)^\dagger(\mathrm{i}|0)$; L. {4′$2_x$i}, {4′$2_{xy}$i}, {4$2_x'$i}, {4$2_x$i′}, {4$\bar{2}_x$i′}, {$\bar{4}2_x$i′}, {$\bar{4}2_{xy}$i′}

E. The orthorhombic system
L = P, C, A, F, I;
$M(4\mathrm{P}) = \{P_0, P_\mathrm{c}, P_\mathrm{A}, P_\mathrm{I}\}$, $M(6\mathrm{P}) = \{P_0, P_\mathrm{b}, P_\mathrm{c}, P_\mathrm{A}, P_\mathrm{c}, P_\mathrm{I}\}$,
$M(8\mathrm{P}) = \{P_0, P_\mathrm{a}, P_\mathrm{b}, P_\mathrm{c}, P_\mathrm{A}, P_\mathrm{B}, P_\mathrm{C}, P_\mathrm{I}\}$
$M(\mathrm{C}) = \{C_0, C_\mathrm{b}, C_\mathrm{I}, C_\mathrm{B}\}$
$M(\mathrm{A}) = \{A_0, A_\mathrm{b}, A_\mathrm{I}, A_\mathrm{B}\}$
$M(\mathrm{F}) = \{F_0, F_\mathrm{I}\}$
$M(2\mathrm{I}) = \{I_0, I_\mathrm{c}\}$, $M(3\mathrm{I}) = \{I_0, I_\mathrm{b}, I_\mathrm{c}\}$, $M(4\mathrm{I}) = \{I_0, I_\mathrm{a}, I_\mathrm{b}, I_\mathrm{c}\}$

D_2 (16, 19; 22; 23–4): $M(4\mathrm{P}; \mathrm{F}; 2\mathrm{I})(2_z|c, 0, c)(2_x|c, c, 0)$; L(P; F; I)$\{22'\}$
(17–18; 20–1): $M(6\mathrm{P}; \mathrm{C})(2_z|c, 0, c)(2_x|a, a, 0)$; L(P, C). $\{2'2\}, \{22'\}$
C_{2v} (25, 27, 32, 34; 35, 37; 42–3; 44–5): $M(6\mathrm{P}; \mathrm{C}; \mathrm{F}; 3\mathrm{I})(2_z|b, b, 0)(m_x|0, b, c)$; L(P; C; F; I). $\{2'm\}, \{2m'\}$
(26, 28–31, 33; 36; 38–41; 46): $M(8\mathrm{P}; \mathrm{C}; \mathrm{A}; 4\mathrm{I})(2_z|a', b, c + c')(m_x|0, b, c)(m_y|a', 0, c')^\dagger$; L(P; C; A; I), $\{2'm_x\}, \{2'm_y\}, \{2m_x'\}$
D_{2i} (47–8, 61; 69–70; 71, 73): $M(4\mathrm{P}; \mathrm{F}; 2\mathrm{I})(2_z|c + \alpha, \alpha, c)(2_x|c, c + \alpha, \alpha)(\mathrm{i}|0)$; L(P; F; I). $\{2'2\mathrm{i}\}, \{22\mathrm{i}'\}, \{2\bar{2}\mathrm{i}'\}$
(49–50, 55–6, 58–9; 65–8; 72, 74): $M(6\mathrm{P}; \mathrm{C}; 3\mathrm{I})(2_z|c + \alpha, \beta, c)(2_x|a, a + \beta, \gamma)(\mathrm{i}|0)$; L(P; C; I). $\{2'2\mathrm{i}\}, \{22'\mathrm{i}\}, \{22\mathrm{i}'\}, \{2\bar{2}\mathrm{i}'\}, \{\bar{2}2\mathrm{i}'\}$
(51–4, 57, 60, 62; 63–4): $M(8\mathrm{P}; \mathrm{C})(2_z|c + \alpha, \beta, c)(2_x|a, a + \beta, \gamma)(2_y|a + c + \alpha, a, c + \gamma)^\dagger(\mathrm{i}|0)$; L(P; C). $\{2_z'2_x\mathrm{i}\}, \{2_z2_x'\mathrm{i}\}, \{2_z'2_y\mathrm{i}\}, \{2_z2_x\mathrm{i}'\}, \{2_z\bar{2}_x\mathrm{i}'\}, \{\bar{2}_z2_y\mathrm{i}'\}$

F. The monoclinic system
L = P, B; $M(4\mathrm{P}) = \{P_0, P_\mathrm{a}, P_\mathrm{c}, P_\mathrm{A}\}$, $M(6\mathrm{P}) = \{P_0, P_\mathrm{a}, P_\mathrm{b}, P_\mathrm{c}, P_\mathrm{A}, P_\mathrm{B}\}$,
$M(\mathrm{B}) = \{B_0, B_\mathrm{b}, B_\mathrm{c}\}$
C_2 (3–4; 5): $M(4\mathrm{P}; \mathrm{B})(2_z|0, 0, c)$; L(P; B)$\{2'\}$
C_s (6; 7; 8–9): $M(4\mathrm{P}; 6\mathrm{P}; \mathrm{B})(m_z|0, b, 0)$; L(P; P; B)$\{m'\}$
C_{2i} (10–11; 12, 15; 13–14): M(4P; B; 6P)$(2_z|0, b, c)(\mathrm{i}|0)$; L(P; B; P). $\{2'\mathrm{i}\}, \{2\mathrm{i}'\}, \{\bar{2}\mathrm{i}'\}$

G. The triclinic system
L = P; $M(\mathrm{P}) = \{P_0, P_\mathrm{s}\}$
$C_1(1)$: $M(\mathrm{P})(\bar{e}|0)$
$C_i(2)$: $M(\mathrm{P})(\bar{e}|0)(\mathrm{i}|0)$; P$\{\bar{e}\mathrm{i}'\}$

---

1. The translational parameters $a, b, \ldots, \gamma$ are given in Table 13.3.
2. The numbers in the parentheses after the class symbols are the space group numbers.
3. The superscript † denotes auxiliary generators needed to describe $\hat{H}^z$.
4. Abbreviations of the symbols are used whenever no confusion exists.
5. In G, $\bar{e} = 2\pi$ rotation.

17.3 Anti-unitary space groups of the second kind

Let $\hat{H}$ be a halving subgroup of a space group $\hat{G}$, then the coset decomposition of $\hat{G}$ with respect to $\hat{H}$ yields

$$\hat{G} = \hat{H} + \hat{z}\hat{H} \equiv \hat{H}_z; \qquad \hat{z} = (z|\boldsymbol{v}_z) \notin \hat{H} \tag{17.3.1}$$

where the coset representative $\hat{z}$ is a generator of $\hat{G}$. To each coset decomposition of $\hat{G}$ defined by a pair $\{\hat{z},\ \hat{H}\}$, there corresponds an anti-unitary space group of the second kind defined by

$$\hat{H}^z = \hat{H} + \hat{z}'\hat{H}, \qquad \hat{z}' = \boldsymbol{\theta}\hat{z} \tag{17.3.2}$$

Since the halving subgroup $\hat{H}$ is an invariant subgroup of $\hat{G}$, the pair $\{\hat{z},\ \hat{H}\}$ must satisfy the following compatibility conditions expressed by their angular parts:

$$z^2,\ zHz^{-1} \in H \tag{17.3.3}$$

Since H is the halving subgroup of the point co-group G of $\hat{G}$ corresponding to $\hat{H}$ of $\hat{G}$, from the isomorphism between the factor group $\hat{G}/T$ and the point co-group G, all the possible compatible pairs $\{\hat{z},\ \hat{H}\}$ of $\hat{G}$ are determined by the compatible pairs $\{z,\ H\}$ of G, which satisfy (17.3.3) and thus define the coset decompositions of G, $G = H + zH$.

As was shown by (13.8.3a), the abstract generators of $\hat{G}/T$ and G are characterized by the same defining relations

$$A^n = B^m = (AB)^l = E', \qquad I^2 = E, \qquad E'^2 = E \tag{17.3.4}$$

where the set of orders $\{n,\ m,\ l\}$ is characteristic to G: $\{n,\ 0,\ 0\}$ for C_n, $\{n,\ 2,\ 2\}$ for D_n, $\{3,\ 3,\ 2\}$ for T, $\{4,\ 3,\ 2\}$ for O and $\{5,\ 3,\ 2\}$ for Y. Here an element with an odd (even) order with respect to the 2π rotation E' is simply called an odd (even) element. When there exist two generators A, B ($\neq I$) for G, the above defining relations can be written in the following standard form:

$$A^n = B^m = (AB)^2 = E' \qquad \text{(17.3.5a)}^2$$

A generator set $\{A,\ B\}$ that satisfies this standard form is called *a canonical set* of generators of G.

Now, from the compatibility condition (17.3.3) and the defining relations of G, we obtain the following lemmas, which play the essential role in determining all the possible compatible pairs $\{z,\ H\}$ of G which identify the compatible pairs $\{\hat{z},\ \hat{H}\}$ of $\hat{G}$.

Lemma 1. An odd generator of G is not acceptable for z. The generator of a cyclic group of an even order and the inversion operator I of any G are always acceptable for z.

Lemma 2. If there exist two generators A, B ($\neq I$) for G, choose them such that $(AB)^2 = E'$. Then, an even one, say A, is acceptable for z. The corresponding halving subgroup H is generated by A^2, B and I (if I is contained in G).

[2] We may define two generators of G by $\alpha = A^{-1}$, $\beta = AB$; then from (17.3.5a) we have

$$\alpha^n = \beta^2 = (\alpha\beta)^m$$

Accordingly, the set $(\alpha,\ \beta)$ cannot be a canonical generator set of G unless $m = 2$.

Proof for Lemma 1. Let A be an odd generator of G such that $A^{2r+1} = E'$, where r is an integer, then $A = E'(A^2)^{-r}$. Now, if A were acceptable for z, then $A^2 \in H$ from the compatibility condition (17.3.3) so that $A \in H$. This is contradictory because z should not be contained in H. The remaining statement in Lemma 1 is trivially true.

Proof for Lemma 2. From $(AB)^2 = E'$ we have $ABA^{-1} = E'B^{-1}A^{-2} \in H$ because H is defined by A^2, B and I (if I is contained in G). This means that A is acceptable for z satisfying the compatibility condition (17.3.3). To show that the generating set of H forms a group, let $n = 2r$ be the order of A. Then, from (17.3.5a), the generator set $\{\alpha = A^2, B, I\}$ of H satisfies, from $AB^{-1}A^{-1} = E'\alpha B$,

$$\alpha^r = B^m = (E'\alpha B)^m = E' \tag{17.3.5b}$$

which can be shown to define one of the point groups defined by (17.3.5a). For example, for a dihedral group D_{2r}, we have $m = 2$ so that the halving subgroup H is the dihedral group D_r. If G is the octahedral group O, we have $r = 2$ and $m = 3$ so that the halving subgroup H is the tetrahedral group T.

From the lemmas, it follows that, if a canonical set of generators of a point group G does not have an even generator, then there is no halving subgroup for G. Thus the tetrahedral group T and the icosahedral group Y have no halving subgroup. Obviously, the uniaxial group C_n with an odd n has no halving subgroup. Consequently, there is no anti-unitary space group of the second kind formed from $\hat{T}$ and $\hat{C}_n$ with an odd n. Except for these cases, every space group has one or more halving subgroups. Hereafter we may take a generator set of G to be canonical whenever there exist two generators A, B ($\neq I$) for G, unless stated otherwise.

Now, according to (17.3.4) the generator set of a space group $\hat{G}$ with a given lattice type L can be expressed as one of the following five kinds:

$$L\{A\},\ L\{I\},\ L\{A, I\},\ L\{A, B\},\ L\{A, B, I\} \tag{17.3.6}$$

Here A and B are not the inversion I. The corresponding anti-unitary groups of the second kind $\hat{H}^z$ are immediately constructed by using the lemmas. In the special case of $L\{A, B\}$ belonging to the class T, there exists no $\hat{H}^z$ since both generators are odd. Now, for the even dihedral crystal systems (orthorhombic, tetragonal and hexagonal), both generators A and B are even. In such a case, we obtain the following seven $\hat{H}^z$ for the most general space group $L\{A, B, I\}$:

$$L.\quad \{A', B, I\}, \{A', (AB), I\}, \{A, B', I\}, \{A, B, I'\}, \{A, \overline{B}, I'\}, \{\overline{A}, B, I'\},$$
$$\{\overline{A}, (AB), I'\} \qquad (17.3.7)^3$$

provided that A, B and AB are all inequivalent under the lattice transformation which leaves the space group $L\{A, B, I\}$ invariant. Here $X' = \boldsymbol{\theta} X$ and $\overline{X} = XI$, and, for simplicity, the lattice type L is not repeated. If one only of the generators A or B is even, say A, then one obtains the following $\hat{H}^z$ from $L\{A, B, I\}$:

$$L.\quad \{A', B, I\}, \{A, B, I'\}, \{\overline{A}, B, I'\}$$

This case is applicable to the classes D_{3i} and O_i. Thus, we have exhausted all the possible anti-unitary space groups of the second kind $\hat{H}^z$ arising from a given space

[3] Note that $L\{A, (AB)'\} = L\{A, B'\}$ because $A^{-1}(AB)' = B'$.

group $\hat{G}$ since the remaining cases of (17.3.6) may be regarded as special cases of $L\{A, B, I\}$.

The completeness of the set of anti-unitary groups given by (17.3.7) for a space group $L\{A, B, I\}$ can be seen also from the fact that these sets are equal to the following sets, in the order written:

$$L.\quad \{A', B, I\}, \{A', B', I\}, \{A, B', I\}, \{A, B, I'\}, \{A, B', I'\}, \{A', B, I'\}, \{A', B', I'\} \tag{17.3.8}$$

which are obtained by distributing the $\boldsymbol{\theta}$ in every possible way to the given set $\{A, B, I\}$. In fact, these expressions given by (17.3.8) for $\hat{H}^z$ provide the basis for constructing the magnetic space groups expressed in terms of the international notation of the space groups, as has been carried out by Belov *et al.* (1955). We prefer the expressions given by (17.3.7), since these are directly related to their irreducible co-representations (co-irreps) which will be determined through the projective co-irreps of their point co-groups $\hat{H}^z$.

Some of the magnetic space groups $\hat{H}^z$ thus formed may be redundant, if some of the generators are equivalent under the lattice transformations $\Lambda = [U|s]$ which leave the space group $\hat{G}$ $(= H_z)$ invariant. For example, for a space group $L\{A, B\}$, we have three $\hat{H}^z$ given by

$$L\{A', B\}, L\{A', AB\}, L\{A, B'\} \tag{17.3.9}$$

provided that A, B and AB are all inequivalent under the lattice transformations Λ. If, for example, A and B are equivalent, then $L\{A', B\}$ and $L\{A, B'\}$ are equivalent. Likewise if A and AB are equivalent, then $L\{A', AB\}$ and $L\{A, B'\}$ are also equivalent for $L\{A, B'\} = L\{A, (AB)'\}$. This kind of situation occurs, however, only for the orthorhombic system D_{2i}, for which the orders of the generators are all equal and even. For example, consider the space groups numbers 20 and 21 belonging to the class D_2 defined in (17.2.14) with $a = 0$:

$$C.\quad (2_z|c0c), (2_x|000), (2_y|c0c); \qquad c = 0, \tfrac{1}{2}$$

Here, $(2_x|000)$ and $(2_y|c0c)$ are equivalent under $[(x, y)|c/2, c/2, c/2]$ as was stated in regard to (17.2.14). Following (17.3.9) we may form three $\hat{H}^z$ expressed, using obvious abbreviation of the generators, by

$$C\{2_z', 2_x\} \sim C\{2_z', 2_y\}, C\{2_z, 2_x'\}$$

of which the first two groups are equivalent, whereas the third group is inequivalent, because, for a C lattice, the x- and y-axes are equivalent but the z- and x-axes are inequivalent.

It should be noted that one assembly of $\hat{H}^z$ given by (17.3.7) corresponds to one minimal general generator set (MGGS) of a crystal class except for the classes D_{3p} and D_{2p} and all three classes D_2, C_{2v} and D_{2i} of the orthorhombic system. For the former two classes, this is because there are two inequivalent realizations for the generator set of the space group $\hat{G}$ itself. For the latter three classes, this is due to the possible equivalence of some or all of the three binary axes of $\hat{G}$, as discussed in Section 17.2.5. Thus, two or three assemblies of $\hat{H}^z$ correspond to a crystal class for these cases.

All the magnetic space groups of the second kind $\hat{H}^z$ thus determined are also given in Table 17.1 together with those of the first kind $\hat{G}^e$ constructed in Section 17.2 (Kim 1983c). Here, for convenience, we have expressed a space group $L\{A, B\}$ or $L\{A, B, I\}$ by

$$L\{A, B, (AB)^{\dagger}\} \qquad \text{or} \qquad L\{A, B, (AB)^{\dagger}, I\} \tag{17.3.10}$$

with an auxiliary generator (AB), in cases in which the product (AB) is needed to describe $\hat{H}^z$ (the superscript dagger on $(AB)^{\dagger}$ simply means that (AB) is an auxiliary generator).

17.3.1 Illustrative examples

On account of the compact nature of Table 17.1, it seems worthwhile to give some illustrative examples for its use. For each example, we first write down a space group $\hat{G}$ in terms of its generators from Table 13.3. Then we give the corresponding anti-unitary space groups $\hat{G}^e$ and $\hat{H}^z$ from Table 17.1. These are reexpressed in terms of the notations introduced by Belov, Neronova and Smirnova (BNS) for comparison.

Example 1. Number 230. $I(4_z|\frac{1}{4}, -\frac{1}{4}, \frac{1}{4}), (3_{xyz}|0), (\mathrm{i}|0) \in O_i$. Since $L = I$, we have $M_L = I_0$ from Table 17.1. Then, from the line O_i (221–30) we obtain

$$I_0\{43\mathrm{i}\}; \qquad I\{43\mathrm{i}'\}, I\{\bar{4}3\mathrm{i}'\}, I\{4'3\mathrm{i}\} \tag{17.3.11a}$$

where 4, 3 and i are obvious abbreviations for the generators given for space group number 230. The first group $I_0\{43\mathrm{i}\}$ given above is of the first kind and the remaining three groups are of the second kind. These correspond, in the order given, to

$$Ia3'd; \qquad Ia'3d', Ia'3d, Ia3d' \tag{17.3.11b}$$

in the BNS notation, respectively.

Example 2. Number 70. $F(2_z|\frac{1}{4}, \frac{1}{4}, 0), (2_x|0, \frac{1}{4}, \frac{1}{4}), (\mathrm{i}|0) \in D_{2i}$. Since $L = F$, from Table 17.1, we have $M(F) = \{F_0, F_I\}$ and then, from number 70 of D_{2i},

$$F_0\{22\mathrm{i}\}, F_I\{22\mathrm{i}\}; \qquad F\{2'2\mathrm{i}\}, F\{22\mathrm{i}'\}, F\{2\bar{2}\mathrm{i}'\} \tag{17.3.12a}$$

where the first two groups are of the first kind and the remaining groups are of the second kind. These correspond to

$$Fddd1', F_sddd; \qquad Fd'd'd, Fd'd'd', Fd'dd \tag{17.3.12b}$$

in the notation of BNS.

These illustrative examples show that the present table is much more explicit than are those given by BNS.

17.3.2 Concluding remarks

Using the general generator sets of the space group $\hat{G}$ given in Table 13.3 and the defining relations of the point groups, we have rigorously constructed all the generator sets of the anti-unitary space groups with ease. These are presented in Table 17.1. It describes all the generators of the 1421 anti-unitary space groups by a mere 38 assemblies of general expressions with the translational parameters that are predetermined for the space groups $\hat{G}$. In almost all cases, one assembly corresponds to one crystal class. The exceptions occur only for the five classes D_{3p}, D_{2p}, D_2, C_{2v} and D_{2i}, for which two or three assemblies of the generators sets are required to describe the $\hat{H}^z$ for each class.

As will be shown in the next section, the generator sets given in Table 17.1 are essential and sufficient for constructing all the irreducible co-representations (co-irreps) through the isomorphisms $\hat{G}^e/T \simeq G^e$ and $\hat{H}^z/T \simeq H^z$ from the projective co-irreps of the gray point groups G^e and the magnetic point groups H^z, respectively. These general generator sets given in Table 17.1 are also very convenient for discussing their symmetry properties, as we have discussed for those of the space groups $\hat{G}$. Undoubtedly, the present compact presentation of the magnetic space groups gives us control over their large number and thus helps us to study the symmetry properties of the magnetically ordered crystalline solids even more systematically.

17.4 The type criteria for the co-unirreps of anti-unitary space groups and anti-unitary wave vector groups

By using the type criteria for the co-unirreps of an anti-unitary point group introduced in Section 16.8, we shall discuss the type criteria for those of anti-unitary space groups. For convenience in the discussion, we may use the notation $\hat{H}^z$ for any anti-unitary space group for which $\hat{H}$ is the halving subgroup of $\hat{H}^z$ such that

$$\hat{H}^z = \hat{H} + \hat{a}\hat{H}; \qquad \hat{a} = (\boldsymbol{\theta} z|\boldsymbol{v}_a) \tag{17.4.1}$$

with the compatibility condition

$$\hat{a}^2,\ \hat{a}\hat{h}\hat{a}^{-1} \in \hat{H}; \qquad \forall\, \hat{h} \in \hat{H}$$

Here, in (17.4.1), $z = e$ and $\boldsymbol{v}_a = \boldsymbol{v}_{\boldsymbol{\theta}}$ for the first kind, whereas $z \notin H$ and $\boldsymbol{v}_a = \boldsymbol{v}_z$ for the second kind. For either kind, the translation group of $\hat{H}$ is the same as the translation group T of $\hat{H}^z$.

As a preparation, we shall first construct the unirreps of a space group $\hat{H}$ by induction from the unirreps of the wave vector space groups $\hat{H}_{\boldsymbol{k}}$ contained in $\hat{H}$. Let the left coset decomposition of $\hat{H}$ by $\hat{H}_{\boldsymbol{k}}$ be expressed by

$$\hat{H} = \sum_j \hat{h}_j \hat{H}_{\boldsymbol{k}}; \qquad \hat{H}_{\boldsymbol{k}} = \{\hat{q}\} \tag{17.4.2}$$

where $\hat{h}_j$ is a coset representative of $\hat{H}_{\boldsymbol{k}}$ in $\hat{H}$. Let $\Delta_{\boldsymbol{k}}^{(\nu)}(\hat{H}_{\boldsymbol{k}})$ be a small representation of $\hat{H}_{\boldsymbol{k}}$ that subduces $\tau_{\boldsymbol{k}}(T)$ onto the translation group T. Then, as was given by (14.5.2), we can induce a unirrep of $\hat{H}$ from the small representation $\Delta_{\boldsymbol{k}}^{(\nu)}$ as follows:

$$D_{\boldsymbol{k}}^{(\nu)}(\hat{h})_{ji} = \sum_{\hat{q}\in\hat{H}_{\boldsymbol{k}}} \Delta_{\boldsymbol{k}}^{(\nu)}(\hat{q})\delta(\hat{q},\ \hat{h}_j^{-1}\hat{h}\hat{h}_i); \qquad j, i = 1, 2, \ldots, |H|/|H_{\boldsymbol{k}}| \tag{17.4.3a}$$

for all $\hat{h} \in \hat{H}$. A co-unirrep of $\hat{H}^z$ is obtained by a further induction of $D_{\boldsymbol{k}}^{(\nu)}(\hat{H})$ with the anti-unitary operator $\hat{a}$ as in Section 16.4.2. As a result, a co-unirrep of $\hat{H}^z$ is obtained by successive inductions from $\Delta_{\boldsymbol{k}}^{(\nu)}$ as follows:

$$S(\hat{H}^z) = [D_{\boldsymbol{k}}^{(\nu)}(\hat{H}) \uparrow^* \hat{H}^z] = [[\Delta_{\boldsymbol{k}}^{(\nu)}(\hat{H}_{\boldsymbol{k}}) \uparrow \hat{H}] \uparrow^* \hat{H}^z] \tag{17.4.3b}$$

where $\uparrow^*$ denotes the induction by an anti-unitary operator $\hat{a} = (\boldsymbol{\theta} z|\boldsymbol{v}_a)$. The type of the co-unirrep thus formed is determined by the characters of the unirrep $D_{\boldsymbol{k}}^{(\nu)}(\hat{H})$ of the halving subgroup $\hat{H}$ via the type criteria (16.8.10). This is so because the criteria are applicable to any anti-unitary group of finite order.

Let the characters of the two representations in (17.4.3a) be denoted by $\chi_D(\hat{h}) = \operatorname{tr} D_{\boldsymbol{k}}^{(\nu)}(\hat{h})$ and $\chi_\Delta(\hat{q}) = \operatorname{tr} \Delta_{\boldsymbol{k}}^{(\nu)}(\hat{q})$, then we have

$$\chi_D(\hat{h}) = \sum_j \sum_{\hat{q}\in \hat{H}_{\boldsymbol{k}}} \chi_\Delta(\hat{q})\delta(\hat{q},\, \hat{h}_j^{-1}\hat{h}\hat{h}_j) = \sum_j \mathring{\chi}_\Delta(\hat{h}_j^{-1}\hat{h}\hat{h}_j) \tag{17.4.4}$$

where

$$\mathring{\chi}_\Delta(\hat{h}_j^{-1}\hat{h}\hat{h}_j) \equiv \begin{cases} \chi_\Delta(\hat{h}_j^{-1}\hat{h}\hat{h}_j); & \text{if } \hat{h}_j^{-1}\hat{h}\hat{h}_j \in \hat{H}_{\boldsymbol{k}} \\ 0; & \text{otherwise} \end{cases}$$

Now, from (16.8.10), the type criteria for the co-unirreps of $\hat{H}^z$ induced from the unirrep $D_{\boldsymbol{k}}^{(\nu)}(\hat{H})$ take the form

$$\begin{aligned} \Xi_{\boldsymbol{k}}^{(\nu)} &= \frac{1}{|\hat{H}|}\sum_{\hat{h}\in\hat{H}} \chi_D[(\hat{a}\hat{h})^2] = \frac{1}{|\hat{H}|}\sum_{\hat{h}\in\hat{H}}\sum_j \mathring{\chi}_\Delta[(\hat{h}_j^{-1}\hat{a}\hat{h}\hat{h}_j)^2] \\ &= \begin{cases} 1 & \text{for type a} \\ -1 & \text{for type b} \\ 0 & \text{for type c} \end{cases} \end{aligned} \tag{17.4.5}$$

To simplify these relations we shall carry out the double summations in (17.4.5). We note first that, as $\hat{h}$ sweeps through all elements of $\hat{H}$, the element $\hat{h}_j^{-1}\hat{a}\hat{h}\hat{h}_j$ for any given h_j sweeps through all anti-unitary elements belonging to $\hat{a}\hat{H}$, because $\hat{h}_j^{-1}\hat{a}\hat{h}\hat{h}_j = \hat{a}\hat{a}^{-1}\hat{h}_j^{-1}\hat{a}\hat{h}\hat{h}_j \in \hat{a}\hat{H}$ via the condition for compatibility between $\hat{a}$ and $\hat{H}$. Thus,

$$\Xi_{\boldsymbol{k}}^{(\nu)} = \frac{1}{|\hat{H}_{\boldsymbol{k}}|}\sum_{\hat{h}\in\hat{H}} \mathring{\chi}_\Delta[(\hat{a}\hat{h})^2] \tag{17.4.6}$$

This expression will be simplified further by summing over the lattice translation $\boldsymbol{t} \in T < \hat{H}$. We write

$$\hat{h} = \{e|\boldsymbol{t}\}\{h|\boldsymbol{v}_h\} \equiv \{e|\boldsymbol{t}\}\overline{h}; \qquad \overline{h} = \{h|\boldsymbol{v}_h\} \in \hat{H}/T$$

where $\boldsymbol{v}_h$ is the minimum translation characteristic to $h \in H$. If we use $\hat{a}\{e|\boldsymbol{t}\}\hat{\boldsymbol{a}}^{-1} = \{e|z\boldsymbol{t}\} = \{e|\boldsymbol{t}'\}$, where $\boldsymbol{t}' = z\boldsymbol{t} \in T$, we have

$$(\hat{a}\hat{h})^2 = (\{e|\boldsymbol{t}'\}\hat{a}\overline{h})^2 = \{e|\boldsymbol{t}' + zh\boldsymbol{t}'\}(\hat{a}\overline{h})^2$$

On substituting this into (17.4.6) and replacing $\boldsymbol{t}'$ by $\boldsymbol{t}$, we obtain

$$\begin{aligned} \Xi_{\boldsymbol{k}}^{(\nu)} &= \frac{1}{|\hat{H}_{\boldsymbol{k}}|}\sum_{h\in\hat{H}/T}\sum_{\boldsymbol{t}\in T} \mathring{\chi}_\Delta[(\hat{a}\overline{h})^2]\mathrm{e}^{-\mathrm{i}(\boldsymbol{k}+h^{-1}z^{-1}\boldsymbol{k})\cdot\boldsymbol{t}} \\ &= \frac{1}{|H_{\boldsymbol{k}}|}\sum_{h\in\hat{H}/T} \mathring{\chi}_\Delta[(\hat{a}\overline{h})^2]\delta(zh\boldsymbol{k} \sim -\boldsymbol{k}) \end{aligned} \tag{17.4.7}$$

where $|H_{\boldsymbol{k}}|$ is the order of the point co-group $H_{\boldsymbol{k}}$ of $\hat{H}_{\boldsymbol{k}}$ and the Kronecker delta is defined by

$$\delta(zh\boldsymbol{k} \sim -\boldsymbol{k}) = \begin{cases} 1, & \text{if } zh\boldsymbol{k} \sim -\boldsymbol{k} \\ 0, & \text{otherwise} \end{cases} \tag{17.4.8}$$

Here $\sim$ means the equivalence with respect to the reciprocal lattice T' of T. There are two cases.

(i) There exists no element h in the point co-group H of $\hat{H}$ that satisfies $zh\boldsymbol{k} \sim -\boldsymbol{k}$. In this case, we have $\Xi_{\boldsymbol{k}}^{(\nu)} = 0$ so that the co-unirreps of $\hat{H}^z$ induced by $D_{\boldsymbol{k}}^{(\nu)}(\hat{H})$ are of type c.

(ii) There exists at least one element h in H that satisfies $zh\boldsymbol{k} \sim -\boldsymbol{k}$, then $(zh)^2\boldsymbol{k} \sim \boldsymbol{k}$, which means that $(zh)^2$ belongs to $H_{\boldsymbol{k}}$. Accordingly, $(\hat{a}\bar{h})^2 \in \hat{H}_{\boldsymbol{k}}$ so that $\mathring{\chi}_\Delta[(\hat{a}\hat{h})^2] = \chi_\Delta[(\hat{a}\bar{h})^2]$ in (17.4.7). Thus we have

$$\Xi_{\boldsymbol{k}}^{(\nu)} = \frac{1}{|H_{\boldsymbol{k}}|} \sum_{h \in \hat{H}/T} \operatorname{tr} \Delta_{\boldsymbol{k}}^{(\nu)}[(\hat{a}\bar{h})^2]\, \delta(zh\boldsymbol{k} \sim -\boldsymbol{k})$$

$$= \begin{cases} 1 & \text{for type a} \\ -1 & \text{for type b} \\ 0 & \text{for type c} \end{cases} \tag{17.4.9}$$

which gives the type criteria for a co-unirrep of $\hat{H}^z$ induced by the unirrep $D_{\boldsymbol{k}}^{(\nu)}(\hat{H})$. Note that case (i) is obviously included in (17.4.9).

Since z contains an arbitrary multiplicative factor that is an element of $\hat{H}$, we may redefine z in (17.4.1) for case (ii) such that $z\boldsymbol{k} \sim -\boldsymbol{k}$; e.g., $\hat{H}^e = \hat{H}^i$, if $\hat{I} \in \hat{H}$. Then $a = \boldsymbol{\theta} z$ leaves the unirrep $\tau_{\boldsymbol{k}}(T) = \{\exp(-\mathrm{i}\boldsymbol{k} \cdot \boldsymbol{t})\}$ invariant because $\boldsymbol{\theta} z(\mathrm{i}\boldsymbol{k}) \sim \mathrm{i}\boldsymbol{k}$. Accordingly, for case (ii), we can define an *anti-unitary wave vector space group* $\hat{H}_{\boldsymbol{k}}^z$ by

$$\hat{H}_{\boldsymbol{k}}^z = \hat{H}_{\boldsymbol{k}} + \hat{a}\hat{H}_{\boldsymbol{k}}; \qquad \hat{a} = (\boldsymbol{\theta} z|\boldsymbol{v}_a) \tag{17.4.10}$$

Obviously, the augmentor $\hat{a}$ is compatible with $\hat{H}_{\boldsymbol{k}}$ because we have, from $z\boldsymbol{k} \sim -\boldsymbol{k}$,

$$z^2, zqz^{-1} \in H_{\boldsymbol{k}}; \qquad \forall q \in H_{\boldsymbol{k}}$$

Consequently, the type criteria (17.4.9) for case (ii) are rewritten

$$\Xi_{\boldsymbol{k}}^{(\nu)} = \frac{1}{|H_{\boldsymbol{k}}|} \sum_{\bar{h} \in \hat{H}_{\boldsymbol{k}}/T} \operatorname{tr} \Delta_{\boldsymbol{k}}^{(\nu)}((\hat{a}\bar{h})^2)$$

$$= \begin{cases} 1 & \text{for type a} \\ -1 & \text{for type b} \\ 0 & \text{for type c} \end{cases} \tag{17.4.11}$$

This means that the type criteria (17.4.9) for the co-unirrep of $\hat{H}^z$ induced by the unirrep $D_{\boldsymbol{k}}^{(\nu)}(\hat{H})$ can also be regarded as the type criteria for the co-unirrep of the anti-unitary wave vector group $\hat{H}_{\boldsymbol{k}}^z$ induced by the small representation $\Delta_{\boldsymbol{k}}^{(\nu)}(\hat{H}_{\boldsymbol{k}})$. Consequently, both are of the same type. Here, $D_{\boldsymbol{k}}^{(\nu)}(\hat{H})$ is the induced unirrep of $\hat{H}$ from the small representation $\Delta_{\boldsymbol{k}}^{(\nu)}(\hat{H}_{\boldsymbol{k}})$ defined by (17.4.3a).

Now, the anti-unitary wave vector group $\hat{H}_{\boldsymbol{k}}^z$ is a subgroup of $\hat{H}^z$ such that the left coset decomposition of $\hat{H}^z$ by $\hat{H}_{\boldsymbol{k}}^z$ is given by

$$\hat{H}^z = \sum_j \hat{h}_j \hat{H}^z_{\boldsymbol{k}} \tag{17.4.12}$$

with the same set of coset representatives $\{\hat{h}_j\}$ as in (17.4.2). To see this we rewrite (17.4.1) in the form

$$\hat{H}^z = \hat{H} + \hat{H}\hat{a}$$

which is allowed because a halving subgroup $\hat{H}$ is always an invariant subgroup. Then substitution of (17.4.2) into this leads to (17.4.12) on account of (17.4.10). The decomposition (17.4.12) is not possible for case (i), simply because there exists no *anti-unitary wave vector group* for case (i).

Corresponding to two alternative schemes of the coset decompositions of $\hat{H}^z$ given by (17.4.1) and (17.4.2) or by (17.4.10) and (17.4.12), there exist two alternative ways of inducing the co-unirreps of a given anti-unitary space group $\hat{H}^z$ from a given small representation $\Delta^{(\nu)}_{\boldsymbol{k}}(\hat{H}_{\boldsymbol{k}})$. The first way is through (17.4.3b), which has been discussed already. When there exists an anti-unitary wave vector group $\hat{H}^z_{\boldsymbol{k}}$ for a given wave vector $\boldsymbol{k}$ (case (ii)), we may induce the co-unirreps of $\hat{H}^z$ via the co-unirreps of $\hat{H}^z_{\boldsymbol{k}}$ as follows:

$$S'(\hat{H}^z) = [[\Delta^{(\nu)}_{\boldsymbol{k}}(\hat{H}_{\boldsymbol{k}}) \uparrow^* \hat{H}^z_{\boldsymbol{k}}] \uparrow \hat{H}^z] \tag{17.4.13}$$

These two induced co-unirreps $S(\hat{H}^z)$ and $S'(\hat{H}^z)$ are not only of the same type but also equivalent to each other, as can be shown by their characters. In the analysis of the energy band structure of a solid, the second way is more useful even though it is limited to case (ii). In the later sections we shall construct the co-unirreps of the anti-unitary wave vector group $\hat{H}^z_{\boldsymbol{k}}$ via the projective unirreps of the magnetic point co-group $\hat{H}^z_{\boldsymbol{k}}$, quite analogously to the small representations of $\hat{G}_{\boldsymbol{k}}$ obtained via the projective unirreps of the wave vector point group $G_{\boldsymbol{k}}$.

Example. Let C^e_i be an anti-unitary point group of the inversion group $C_i = \{E, I\}$. An anti-unitary wave vector group that leaves a unirrep $\tau_{\boldsymbol{k}}(T)$ with an arbitrary $\boldsymbol{k}$ is given by $E^i = E + \boldsymbol{\theta} IE$, where E is the identity group, which obviously leaves any wave vector $\boldsymbol{k}$ invariant. The coset decomposition of C^e_i by E^i is given by

$$C^e_i = C^i_i = E^i + \mathrm{i}E^i \tag{17.4.14}$$

Compare this with the coset decomposition $C_i = E + \mathrm{i}E$. This gives a simple example of (17.4.12) based on an anti-unitary point group.

17.5 The representation groups of anti-unitary point groups

In Section 14.4.2, we have shown that the representations of a wave vector space group $\hat{G}_{\boldsymbol{k}}$ can be formed via the projective unirreps of the point co-group $\hat{G}_{\boldsymbol{k}}$. Analogously, the co-representations of an anti-unitary wave vector space group $\hat{H}^z_{\boldsymbol{k}}$ can be formed via the projective co-representations of the corresponding anti-unitary point group $\hat{H}^z_{\boldsymbol{k}}$ (Kim 1984a). According to Section 12.4, a projective representation of a group is p-equivalent to a vector representation of its representation group. Since an analogous theorem also holds for co-representations of an anti-unitary group, we shall first construct the representation groups of anti-unitary point groups H^z extending those of

Table 17.2. *The representation groups of the unitary and anti-unitary point groups (of finite order)*

1.	C'_n:	$x^n = e', e'^2 = e$
2.	$C^{e\prime}_n$:	$C'_n; xa = ax, a^2 = \tau e'\ \tau^2 = e$
		$(\tau = 1$, if n is odd)
3.	$C^{q\prime}_n$:	$C'_n; xa = ax, a^2 = \tau e'x; \tau^2 = e$
		$(\tau = 1$, if n is even)
4.	$C^{u\prime}_n$:	$C'_n; xax = \xi a, a^2 = \tau e; \xi^2 = \tau^2 = e$
		$(\xi = 1$, if n is odd)
5.	C'_{ni}:	$x^n = e', e'^2 = e, xI = \beta Ix, I^2 = e; \beta^2 = e$
		$(\beta = 1$, if n is odd)
6.	$C^{e\prime}_{ni}$:	$C'_{ni}(\beta); xa = ax, Ia = \zeta aI, a^2 = \tau e'; \zeta^2 = \tau^2 = e$
		$(\beta = \tau = 1$, if n is odd)
7.	$C^{q\prime}_{ni}$:	$C'_{ni}(\beta = 1); xa = ax, Ia = \zeta aI, a^2 = \tau e'x; \zeta^2 = \tau^2 = e$
		$(\tau = 1$, if n is even)
8.	$C^{u\prime}_{ni}$:	$C'_{ni}(\beta); xax = \xi a, Ia = \zeta aI, a^2 = \tau e; \xi^2 = \zeta^2 = \tau^2 = e$
		$(\beta = \xi = 1$, if n is odd)
9.	D'_n:	$x^n = y^2 = (xy)^2 = e', e'^2 = e$
10.	$D^{e\prime}_{2r}$:	$D'_{2r}; xa = \xi ax, ya = \eta ay, a^2 = \tau e'; \xi^2 = \eta^2 = \tau^2 = e$
11.	$D^{e\prime}_{2r+1}$:	$D'_{2r+1}; xa = ax, ya = ay, a^2 = \tau e'; \tau^2 = e$
12.	$D^{q\prime}_{2r}$:	$D'_{2r}; xa = ax, ya = ayx, a^2 = \tau e'x; \tau^2 = e$
13.	$D^{q\prime}_{2r+1}$:	$D'_{2r+1}; xa = ax, ya = \eta ayx, a^2 = \tau e'x; \eta^2 = \tau^2 = e$
14.	D'_{ni}:	$x^n = y^2 = (xy)^2 = e', e'^2 = e, xI = \beta Ix, yI = \gamma Iy, I^2 = e;$
		$\beta^2 = \gamma^2 = e\ (\beta = 1$, if n is odd)
15.	$D^{e\prime}_{2r,i}$:	$D'_{2r,i}(\beta, \gamma); xa = \xi ax, ya = \eta ay, Ia = \zeta aI, a^2 = \tau e';$
		$\xi^2 = \eta^2 = \zeta^2 = \tau^2 = e$
16.	$D^{e\prime}_{2r+1,i}$:	$D'_{2r+1,i}(\gamma); xa = ax, ya = ay, Ia = \zeta aI, a^2 = \tau e'; \zeta^2 = \tau^2 = e$
17.	$D^{q\prime}_{2r,i}$:	$D'_{2r,i}(\beta = 1, \gamma); xa = ax, ya = ayx, Ia = \zeta aI, a^2 = \tau e'x; \zeta^2 = \tau^2 = e$
18.	$D^{q\prime}_{2r+i,i}$:	$D'_{2r+i,i}(\gamma); xa = ax, ya = \eta ayx, Ia = \zeta aI, a^2 = \tau e'x; \eta^2 = \zeta^2 = \tau^2 = e$
19.	T':	$x^2 = y^3 = e', (xy)^3 = e, e'^2 = e$
20.	$T^{e\prime}$:	$T'; xa = ax, ya = ay, a^2 = \tau e'; \tau^2 = e$
21.	$T^{q\prime}$:	$T'; xa = ax, yay = ax, a^2 = \tau e'x; \tau^2 = e$
22.	T'_i:	$T'; Ix = xI, Iy = yI, I^2 = e$
23.	$T^{e\prime}_i$:	$T'_i; xa = ax, ya = ay, Ia = \zeta aI, a^2 = \tau e'; \zeta^2 = \tau^2 = e$
24.	$T^{q\prime}_i$:	$T'_i; xa = ax, ya = axy^{-1}, Ia = \zeta aI, a^2 = \tau e'x; \zeta^2 = \tau^2 = e$
25.	O':	$x^4 = y^3 = (xy)^2 = e', e'^2 = e$
26.	$O^{e\prime}$:	$O'; xa = \xi ax, ya = ay, a^2 = \tau e'; \xi^2 = \tau^2 = e$
27.	O'_i:	$x^4 = y^3 = (xy)^2 = e', e'^2 = e, xI = \beta Ix, yI = Iy, I^2 = e; \beta^2 = e$
28.	$O^{e\prime}_i$:	$O'_i(\beta); xa = \xi ax, ya = ay, Ia = \zeta aI, a^2 = \tau e'; \xi^2 = \zeta^2 = \tau^2 = e$
29.	Y':	$x^5 = y^3 = (xy)^2 = e', e'^2 = e$
30.	$Y^{e\prime}$:	$Y'; xa = ax, ya = ay, a^2 = \tau e'; \tau^2 = e$
31.	Y'_i:	$Y'_i; Ix = xI, Iy = yI, I^2 = e$
32.	$Y^{e\prime}_i$:	$Y'_i; xa = ax, ya = ay, Ia = \zeta aI, a^2 = \tau e'; \zeta^2 = \tau^2 = e$

1. e is the identity element, e' is 2π rotation, I is the inversion.
2. The second-order elements $\beta, \gamma, \xi, \eta, \zeta, \tau$ and e' are all in the center of the respective $H^{z\prime}$.

the unitary point groups introduced in Section 12.5. Then, through their co-unirreps, we shall show how to construct the co-unirreps of the anti-unitary wave vector space groups $\hat{H}^z_{\boldsymbol{k}}$ in the next section.

On account of the isomorphism for anti-unitary point groups discussed in Section 16.3, it is sufficient to construct the representation groups of the following three kinds of anti-unitary point groups:

$$H^e,\ H^q,\ H^u \tag{17.5.1}$$

where H is a proper rotation group P or a rotation–inversion group $P_i = P \times C_i$. Hereafter the representation group of H^z shall be denoted by $H^{z\prime}$. In general $P^z_i \simeq P^z \times C_i$. However, $P^{z\prime}_i$ need not be equivalent to $P^{z\prime} \times C_i$ because the inversion I need not commute with $a = \boldsymbol{\theta}z$ in the representation group. For convenience, we have presented a total of 22 kinds of H^z, in Table 17.2, postponing their constructions. These are described by the defining relations in terms of the abstract group generators, which are common to all groups isomorphic to each other. Also included in Table 17.2 are the representation groups of the unitary groups determined in Section 12.5. Thus, Table 17.2 provides all the representation groups of all finite point groups (unitary or anti-unitary) through isomorphism.

Construction of the representation groups $H^{z\prime}$ of H^z is as straightforward as it was in the case of those of the unitary point group G given in Section 12.5. The only difference is that there exists an anti-unitary operator $a = \boldsymbol{\theta}z$ for H^z. We shall begin with the most general projective representation of the most general anti-unitary point group P^z_i, which may be expressed by the defining relations with three unitary generators x, y and I and an anti-unitary generator $a = \boldsymbol{\theta}z$ as follows:

$$x^n = y^m = (xy)^l = e', \qquad I^2 = e'^2 = e \tag{17.5.2a}$$

$$IxI = \beta x, \qquad IyI = \gamma y \tag{17.5.2b}$$

$$a^{-1}xa = \xi x^z, \qquad a^{-1}ya = \eta y^z, \qquad a^{-1}Ia = \zeta I, \qquad a^2 = \tau e' z^2 \tag{17.5.2c}$$

where the 2π rotation e' is in the center of the whole group and $X^z = z^{-1}Xz$ is the conjugate operator of X by z; β, γ, ξ, η, ζ and τ are the projective factors with modulus unity whose values define the class of factor systems for a projective co-representation of the point group. These values are determined self-consistently by the gauge transformations of the generators or by the mutual conjugations of the generators, in particular, with respect to the anti-unitary generator a. These will be shown to satisfy the same quadratic equation

$$x^2 = 1$$

analogously to the case of the unitary groups given in (12.5.8b).

The representation group $P^{z\prime}_i$ is then defined by the defining relations (17.5.2a)–(17.5.2c) with the original generators of P^z_i and the projective factors β, γ, ξ, η, ζ and τ regarded as abstract generators of order two in the center of the group.

We shall first discuss some of the general properties of the projective factors. Previously, in (12.5.8b), we have shown that the projective factors β and γ for the rotation inversion groups P_i satisfy

$$\beta^2 = \gamma^2 = 1 \tag{17.5.2d}$$

where $\beta = 1$ if n is odd and $\gamma = 1$ if m is odd. Some of the negative roots may not be allowed on account of the compatibility condition with respect to the anti-unitary augmentor $a = \boldsymbol{\theta} z$ for P_i. Now, by the similarity transformation of (17.5.2a) with the anti-unitary operator a, we obtain, using (17.5.2c),

$$\xi^n = \eta^m = (\xi\eta)^l = 1, \qquad \zeta^2 = 1 \tag{17.5.3a}$$

Next, by the conjugation of $a^2 = \tau e' z^2$ with a, we obtain $a^{-1}z^2 a = \tau^2 z^2$, so that

$$\tau^2 = 1 \text{ for } H^{e'} \text{ or } H^{u'}; \qquad \tau^2 = \xi \text{ for } H^{q'} \tag{17.5.3b}$$

because $z^2 = e$ for $H^{e'}$, $z^2 = e'$ for $H^{u'}$ and $z^2 = x$ for $H^{q'}$. Eventually we will establish that $\tau^2 = 1$ for every $H^{z'}$. Furthermore, by the conjugation of $a^{-1}Ia = \zeta I$ with a, we obtain $a^{-2}Ia^2 = I$, which leads to $x^{-1}Ix = I$ with $z^2 = x$ for $P_i^{q\prime}$, so that, from (17.5.2b),

$$\beta = 1 \text{ for } P_i^{q\prime} \tag{17.5.3c}$$

which holds even if n is even; cf. (17.5.2d). With the general preparation given above we shall give a further specification of the projective factors for each type of the representation groups:

(i) $C_n^{e\prime}$: $x^n = e'$, $e'^2 = e$, $a^{-1}xa = \xi x$ and $a^2 = \tau e'$, where $\xi = 1$ and $\tau^2 = 1$ ($\tau = 1$ when n is odd).

Proof. From (17.5.3a) and (17.5.3b) we have $\xi^n = \tau^2 = 1$. To show that $\xi = 1$, we rewrite $a^{-1}xa = \xi x$ in the form $a^{-1}\xi^{1/2}xa = \xi^{1/2}x$ and then introduce a gauge transformation, using $\xi^{n/2} = \pm 1$,

$$\xi^{1/2}x \to x, \qquad \xi^{n/2}e' \to e', \qquad \xi^{-n/2}\tau \to \tau$$

which maps off ξ from $C_n^{e\prime}$ as follows:

$$x^n = e', \qquad e'^2 = e, \qquad a^{-1}xa = x, \qquad a^2 = \tau e'$$

Next, for an odd n, we can also map off τ by a further gauge transformation, $x \to \tau x$, $\tau e' \to e'$, because $\tau^n = \tau$ for an odd n.

(ii) $C_n^{q\prime}$: $x^n = e'$, $e'^2 = e$, $a^{-1}xa = x$ and $a^2 = \tau e' x$, where $\tau^2 = 1$ ($\tau = 1$ when n is even).

Proof. Here $\xi = 1$ is shown as in (i) so that $\tau^2 = 1$ from (17.5.3b). When n is even, τ is mapped off by a gauge transformation $\tau x \to x$.

(iii) $C_n^{u\prime}$: $x^n = e'$, $e'^2 = e$, $a^{-1}xa = \xi x^{-1}$ and $a^2 = \tau e$, where $\xi^2 = \tau^2 = 1$ ($\xi = 1$ when n is odd).

Proof. From (17.5.3a) and (17.5.3b) we have $\xi^n = \tau^2 = 1$. Moreover, the similarity transformation of $a^{-1}xa = \xi x^{-1}$ by a yields $\xi^2 = 1$, so that $\xi = 1$ for an odd n.

(iv) $D_n^{e\prime}$: $x^n = y^2 = (xy)^2 = e'$, $e'^2 = e$, $a^{-1}xa = \xi x$, $a^{-1}ya = \eta y$ and $a^2 = \tau e'$, where $\xi^2 = \eta^2 = \tau^2 = 1$ ($\xi = \eta = 1$ when n is odd).

Proof. From (17.5.3a) and (17.5.3b) we have

$$\xi^n = \xi^2 = \eta^2 = \tau^2 = 1$$

which immediately yields $\xi = 1$ for an odd n. To show that $\eta = 1$ for an odd n, we map off $\eta = -1 = \mathrm{i}^2$ via $a^{-1}\mathrm{i}ya = \mathrm{i}y$ and the gauge transformation $\mathrm{i}y \to y$, $-x \to x$, $-e' \to e'$, $-\tau \to \tau$.

(v) $D_n^{q\prime}$: $x^n = y^2 = (xy)^2 = e'$, $e'^2 = e$, $a^{-1}xa = x$, $a^{-1}ya = \eta yx$ and $a^2 = \tau e'x$, where $\eta^2 = \tau^2 = 1$ ($\eta = 1$ when n is even).

Proof. The proof follows from (17.5.3a) and (17.5.3b) except that $\eta = 1$ for an even n; this is shown by mapping off $\eta = -1$ by a gauge transformation, $-y \to y$, $-x \to x$ and $-\tau \to \tau$.

(vi) $P^{e'}$ ($P = T, O, Y$): $x^n = y^3 = (xy)^2 = e'$, $e'^2 = e$, $a^{-1}xa = \xi x$, $a^{-1}ya = \eta y$ and $a^2 = \tau e'$, where $\xi^2 = \eta = 1$ ($\xi = 1$ when n is odd) and $\tau^2 = 1$.

Proof. From (17.5.3a) and (17.5.3b) we have

$$\eta = \xi^2, \qquad \xi^{6-n} = 1; \qquad \tau^2 = 1$$

For O, for which $n = 4$, we have $\xi^2 = 1$ so that $\eta = 1$. For Y, for which $n = 5$, we have $\xi = 1$ and hence $\eta = 1$. For T, for which $n = 3$, we have $\xi^3 = \eta^3 = (\xi\eta)^2 = 1$ directly from (17.5.3a), so that some of the defining relations are rewritten as

$$a^{-1}\xi^2 xa = \xi^2 x, \qquad a^{-1}\eta^2 ya = \eta^2 y$$

Accordingly, the gauge transformation $\xi^2 x \to x$, $\eta^2 y \to y$ maps off the parameters ξ and η in the defining relations completely.

(vii) $T^{q\prime}$: $x^2 = y^3 = e'$, $(xy)^3 = e$, $e'^2 = e$, $a^{-1}xa = \xi x$, $a^{-1}ya = \eta xy^{-1}$ and $a^2 = \tau e'x$, where $\xi = \eta = \tau^2 = 1$. Here, we have taken $x = 2_z$ and $y = 3_{xyz}$ for convenience of the group representation so that $q = 4_z$.

Proof. From (17.5.3a) we obtain $\xi = \eta^3 = 1$ so that $\tau^2 = 1$ from (17.5.3b). Then the similarity transformation[4] of $a^{-1}ya = \eta xy^{-1}$ by a yields

$$x^{-1}yx \Leftarrow a^{-2}ya^2 = \eta^* x\eta^* yx^{-1} = \eta^{*2}xyx^{-1}$$

Since $x^{-1}yx = xyx^{-1}$ from $x^2 = e'$, we obtain $\eta^2 = 1$, which yields $\eta = 1$ together with $\eta^3 = 1$.

(viii) $P_i^{z\prime}(\beta, \gamma; \xi, \eta, \zeta; \tau)$: here the parameters are determined from those of $P_i'(\beta, \gamma)$ and of $P^{z\prime}(\xi, \eta; \tau)$; also $\zeta^2 = 1$, from (17.5.3a), and $\beta = 1$ for $P_i^{q\prime}$, from (17.5.3c). Note that $P_i^{z\prime} \not\simeq P^{z\prime} \times C_i$ unless $\beta = \gamma = \zeta = 1$.

The representation groups of the anti-unitary point groups thus determined have been presented in Table 17.2, which was written down first by Kim (1984a) without proof. In the next section, we shall determine all the co-unirreps of these representation groups which provide all the projective co-unirreps of the anti-unitary point groups. From these we will determine the co-unirreps of the anti-unitary wave vector space groups.

[4] $y^z = xy^{-1}$ is obtained from Table 11.2.

17.6 The projective co-unirreps of anti-unitary point groups

By construction, every projective unirrep of a group G is given by a vector unirrep of the representation group G' up to p-equivalence. An analogous theorem also holds for the projective co-unirreps of an anti-unitary group H^z. Each coefficient set $\{\alpha_i\}$ in a representation group $H^{z\prime}$ given by Table 17.2 defines a class of factor systems for the anti-unitary point groups H^z of finite orders. Thus, the number of p-equivalence classes (or the order of the multiplicator) for a given H^z equals 2^ν, where ν is the number of the binary coefficients in $H^{z\prime}$. According to Table 17.2, the maximum number of classes for a given H^z equals 64 ($= 2^6$), which occurs for $D^{e\prime}_{2r,i}$, whereas the minimum number equals unity, which occurs for $C^{e\prime}_{2r+1}$ and $C^{q\prime}_{2r}$. According to the present classification, there exist in total 180 p-equivalence classes for all H^z and 13 p-inequivalence classes for the unitary point groups.

We shall construct the co-unirreps of the representation groups $H^{z\prime}$ listed in Table 17.2 (the icosahedral groups are excluded) using the modified theory of co-representations introduced in Section 16.4. For convenience, we reproduce here some of the basic results of the theory with some modifications for the notations of the co-unirreps.

Let H' be the unitary halving subgroup of $H^{z\prime}$ such that

$$H^{z\prime} = H' + aH'; \qquad a = \boldsymbol{\theta} z$$

and let $\{\Delta^{(\nu)}(H')\}$ be a complete set of the unirreps of H'. Then, from (16.4.10a), there exists a unitary transformation matrix $N(a)$ such that

$$\Delta^{(\nu)}(a^{-1}h'a)^* = N(a)^{-1}\Delta^{(\bar{\nu})}(h')N(a); \qquad \forall\, h' \in H' \tag{17.6.1}$$

From this, it has been shown in Section 16.4.2 that there exist three types of co-unirreps for $H^{z\prime}$.

For type a, $\nu = \bar{\nu}$ and $N(a)N(a)^* = \Delta^{(\nu)}(a^2)$. There exist two equivalent co-unirreps:

$$S^{(\nu\pm)}(h') = \Delta^{(\nu)}(h'), \qquad S^{(\nu\pm)}(a) = \pm N(a) \tag{17.6.2a}$$

Either one of them provides the required co-unirrep.

For type b, $\nu = \bar{\nu}$ and $N(a)N(a)^* = -\Delta^{(\nu)}(a^2)$. The co-unirrep is

$$S^{(\nu,\nu)}(h') = \begin{bmatrix} \Delta^{(\nu)}(h') & 0 \\ 0 & \Delta^{(\nu)}(h') \end{bmatrix}$$

$$S^{(\nu,\nu)}(a) = \begin{bmatrix} 0 & -N(a) \\ N(a) & 0 \end{bmatrix} \tag{17.6.2b}$$

For type c, $\nu \neq \bar{\nu}$. The co-unirrep is

$$S^{(\nu,\bar{\nu})}(h') = \begin{bmatrix} \Delta^{(\nu)}(h') & 0 \\ 0 & \Delta^{(\bar{\nu})}(h') \end{bmatrix}$$

$$S^{(\nu,\bar{\nu})}(a) = \begin{bmatrix} 0 & \Delta^{(\nu)}(a^2)N(a)^{\sim} \\ N(a) & 0 \end{bmatrix} \tag{17.6.2c}$$

where $N(a)^{\sim}$ is the transpose of $N(a)$. These three types of co-unirreps are denoted by the following notations.

$$S(\Delta^{(\nu)};\ N(a)) \qquad \text{for } S^{(\nu+)} \text{ of (17.6.2a)} \in \text{type a}$$
$$S(\Delta^{(\nu)},\ \Delta^{(\nu)};\ N(a)) \qquad \text{for } S^{(\nu,\nu)} \text{ of (17.6.2b)} \in \text{type b}$$
$$S(\Delta^{(\nu)},\ \Delta^{(\overline{\nu})};\ N(a)) \qquad \text{for } S^{(\nu,\overline{\nu})} \text{ of (17.6.2c)} \in \text{type c} \tag{17.6.3}$$

The co-unirreps of $H^{z\prime}$ given by (17.6.2) are completely explicit in terms of the unirreps $\Delta^{(\nu)}(H')$ of H' except for the transformation matrix $N(a)$. When $\Delta^{(\nu)}$ is one-dimensional, one can take $N(a) = 1$. In such a trivial case we shall delete N from (17.6.3). For a higher-dimensional case, one determines $N(a)$ from (17.6.1) for each class of factor systems by using the defining relations of the group generators given in Table 17.2. Let us illustrate how to determine $N(a)$ for the representation group $H^{z\prime}$ of the most general anti-unitary point group defined by (17.5.2). It has ten generators, $\{x,\ y,\ I;\ \beta,\ \gamma;\ \xi,\ \eta,\ \zeta;\ a,\ \tau\}$, and its halving subgroup H' of $H^{z\prime}$ is defined by a direct product group,

$$H' = P_i'(x,\ y,\ I;\ \beta,\ \gamma) \times C_\xi \times C_\eta \times C_\zeta \tag{17.6.4}$$

where P_i' is the representation group of a rotation–inversion group P_i and $C_\alpha = \{e,\ \alpha\}$ is a group of order 2 with $\alpha = \xi,\ \eta$ or ζ. Thus all the unirreps of H' follow from the unirreps of the group P_i' given in Table 12.4 for each allowed set $(\beta,\ \gamma)$ and the one-dimensional representations of C_ξ, C_η and C_ζ. Now, from (17.6.1) and (17.5.2c), the transformation matrix $N(a)$ is determined by the set of simultaneous equations

$$\xi\Delta^{(\nu)}(x^z)^* = N(a)^{-1}\Delta^{(\overline{\nu})}(x)N(a)$$
$$\eta\Delta^{(\nu)}(y^z)^* = N(a)^{-1}\Delta^{(\overline{\nu})}(y)N(a)$$
$$\zeta\Delta^{(\nu)}(I)^* = N(a)^{-1}\Delta^{(\overline{\nu})}(I)N(a) \tag{17.6.5}$$

for a class of factor systems defined by an allowed set of values of ξ, η and ζ given in Table 17.2. Obviously, this is the most complicated case; usually there are fewer generators in a point group. Calculation of $N(a)$ from (17.6.5) is straightforward, as will be exemplified later. For almost all $H^{z\prime}$, the transformation matrices are expressed by the unit matrix $\mathbf{1}_d$ with appropriate dimensionality d, by the Pauli spin matrices (σ_x, σ_y and σ_z) or by their direct products except for a few cases in which the anti-unitary operator a involves the operator q ($= c_{2n}$) (see, for example, Table 17.3 (12)). Since the coefficients ξ, η and ζ in (17.6.5) are ± 1, frequently the anticommutation relations $\sigma_j\sigma_i\sigma_j = -\sigma_i$ ($i \neq j$) of the spin matrices play the crucial role for determining $N(a)$, see Examples 2 and 3 below.

Finally, from $a^2 = \tau e' z^2$, we have

$$\Delta^{(\nu)}(a^2) = \tau\Delta^{(\nu)}(e')\Delta^{(\nu)}(z^2) \tag{17.6.6a}$$

where $\tau = \pm 1$ and $\Delta^{(\nu)}(e') = 1\ (-1)$ for an integral (half-integral) unirrep of H'. Note that the coefficient τ is the only coefficient which is not involved in the determination of the transformation matrix $N(a)$. As a result, when $\nu = \overline{\nu}$, the sign change of τ ($= \pm 1$) simply interchanges the type a and type b co-unirreps, because, according to the type criteria given in (17.6.2a) and (17.6.2b), we have

$$\Delta^{(\nu)}(a^2) = \tau\Delta^{(\nu)}(e')\Delta^{(\nu)}(z^2) = \pm N(a)N(a)^* \tag{17.6.6b}$$

where the $+$ ($-$) sign is for type a (b). The change of sign does not affect type c. This fact will be utilized later to simplify the tabulation of the projective co-unirreps.

Table 17.3. *The projective unirreps and co-unirreps of the unitary and anti-unitary point groups (of finite order)*

---

1. $C_n(K^0)$
   K^0: M_m; $m = m^0$, $-\frac{1}{2}n < m^0 \leqslant \frac{1}{2}n$

2. $C_n^e(K; \tau = 1$, if n is odd)
   K: $S(M_0)$, $S(M_{n_e/2})$, $S(M_{n_o/2}, M_{n_o/2})$, $S(M_m, M_{-m})$;
   $m = m^* = \frac{1}{2}, 1, \ldots, \frac{1}{2}(n-1)$

3. $C_n^q(K; \tau = 1$, if n is even)
   K: $S(M_0)$, $S(M_{n_o/2})$, $S(M_{n_e/2}, M_{n_e/2})$, $S(M_m, M_{-m})$; $m = m^*$

4. $C_n^u(K_t; t = \{\xi\}; \xi = 1$ if n is odd)
   K_1: $S(M_m)$; $m = m^0$
   $K_2(n = 2r)$: $S(M_m, M_{m-r})$; $m = m' = \frac{1}{2}, 1, \ldots, r$

5. $C_{ni}(K_s^0; s = \{\beta\}; \beta = 1$, if n is odd)
   K_1^0: $C_n(K^0) \times C_i$; $M_m^{\pm}$; $m = m^0$
   $K_2^0(n = 2r)$: $D_m = D(M_m, M_{m-r})$; $m = m'$

6. $C_{ni}^e(K_{st}; s = \{\beta\}, t = \{\zeta\}; \beta = \tau = 1$, if n is odd)
   K_{11}: $C_n^e(K) \times C_i$
   K_{12}: $S(M_m^+, M_{-m}^-)$; $m = m^0$
   $K_{21}(n = 2r)$: $S(D_r; 1_2)$, $S(D_{r_e/2}; \sigma_x)$, $S(D_{r_o/2}, D_{r_o/2}; \sigma_x)$,
   $S(D_m, D_{r-m}; \sigma_x)$; $m = m^\dagger = \frac{1}{2}, 1, \ldots, \frac{1}{2}(r-1)$
   $K_{22}(n = 2r)$: $S(D_r; \sigma_z)$, $S(D_{r_e/2}, D_{r_e/2}; \sigma_y)$, $S(D_{r_o/2}; \sigma_y)$
   $S(D_m, D_{r-m}; \sigma_y)$; $m = m^\dagger$

7. $C_{ni}^q(K_t; t = \{\zeta\}; \tau = 1$, if n is even)
   K_1: $C_n^q(K) \times C_i$
   K_2: $S(M_m^+, M_{-m}^-)$; $m = m^0$

8. $C_{ni}^u(K_{st}; s = \{\beta\}, t = \{\xi, \zeta\}; \xi = 1$, if n is odd)
   K_{11}: $C_n^u(K_1) \times C_i$
   K_{12}: $S(M_m^+, M_m^-)$; $m = m^0$
   $K_{13}(n = 2r)$: $C_{2r}^u(K_2) \times C_i$
   $K_{14}(n = 2r)$: $S(M_m^+, M_{m-r}^-)$; $m = m^0$, $-r < m^0 \leqslant r$
   $K_{21}(n = 2r)$: $S(D_m; 1_2)$; $m = m' = \frac{1}{2}, 1, \ldots, r$
   $K_{22}(n = 2r)$: $S(D_m; \sigma_z)$; $m = m'$
   $K_{23}(n = 2r)$: $S(D_m; \sigma_x)$; $m = m'$
   $K_{24}(n = 2r)$: $S(D_m, D_m; \sigma_y)$; $m = m'$

9. $D_n(K^0)$
   K^0: A_1, A_2, B_1, B_2, E_m; $m = m^* = \frac{1}{2}, 1, \ldots, \frac{1}{2}(n-1)$

10. $D^e_{2r}(K_t;\ t = \{\xi, \eta\})$
K_1: $S(A_1)$, $S(A_2)$, $S(B_1)$, $S(B_2)$, $S(E_m;\ 1_2;\ \sigma_y)$; $m = m^* = \frac{1}{2}, 1, \ldots, r - \frac{1}{2}$
K_2: $S(A_1, A_2)$, $S(B_1, B_2)$, $S(E_m, E_m;\ \sigma_y, 1_2)$; $m = m^*$
K_3: $S(A_1, B_1)$, $S(A_2, B_2)$, $S(E_{r_e/2};\ \sigma_z)$, $S(E_{r_o/2}, E_{r_o/2};\ \sigma_x)$,
$S(E_m, E_{r-m};\ \sigma_z, \sigma_x)$; $m = m^\dagger = \frac{1}{2}, 1, \ldots, \frac{1}{2}(r-1)$
K_4: $S(A_1, B_2)$, $S(A_2, B_1)$, $S(E_{r_e/2};\ \sigma_x)$, $S(E_{r_o/2}, E_{r_o/2};\ \sigma_z)$,
$S(E_m, E_{r-m};\ \sigma_x, \sigma_z)$; $m = m^\dagger$

11. $D^e_{2r+1}(K)$
K_1: $S(A_1)$, $S(A_2)$, $S(B_1, B_2)$, $S(E_m;\ 1_2, \sigma_y)$; $m = m' = \frac{1}{2}, 1, \ldots, r$

12. $D^q_{2r}(K)$
K: $S(A_1)$, $S(A_2)$, $S(B_1, B_2)$, $S(E_m;\ R_m, \sigma_y R_m)$; $m = m^* = \frac{1}{2}, 1, \ldots, r - \frac{1}{2}$

13. $D^q_{2r+1}(K_t;\ t = \{\eta\})$
K_1: $S(A_1)$, $S(A_2)$, $S(B_1)$, $S(B_2)$, $S(E_m;\ R_m, \sigma_y R_m)$; $m = m'$
K_2: $S(A_1, A_2)$, $S(B_1, B_2)$, $S(E_m, E_m;\ \sigma_y R_m, R_m)$; $m = m'$

14. $D_{ni}(K^0_s;\ s = \{\beta, \gamma\};\ \beta = 1$, if n is odd$)$
K^0_1: $D_n(K^0) \times C_i$; $A^\pm_1$, $A^\pm_2$, $B^\pm_1$, $B^\pm_2$, $B^\pm_m$; $m = m^* = \frac{1}{2}, 1, \ldots, \frac{1}{2}(n-1)$
K^0_2: $D_A = D(A_1, A_2)$, $D_B = D(B_1, B_2)$, $D^{\pm y}_m = D(E_m;\ \pm\sigma_y)$; $m = m^*$
$K^0_3(n = 2r)$: $D_{11} = D(A_1, B_1)$, $D_{22} = D(A_2, B_2)$, $D^{\pm z}_{r/2} = D(E_{r/2};\ \pm\sigma_z)$,
$D^z_{m,r-m} = D(E_m, E_{r-m};\ \sigma_z)$; $m = m^\dagger = \frac{1}{2}, 1, \ldots, \frac{1}{2}(r-1)$
$K^0_4(n = 2r)$: $D_{12} = D(A_1, B_2)$, $D_{21} = D(A_2, B_1)$, $D^{\pm x}_{r/2} = D(E_{r/2};\ \pm\sigma_x)$;
$D^x_{m,r-m} = D(E_m, E_{r-m};\ \sigma_x)$; $m = m^\dagger$

15. $D^e_{2r,i}(K_{st};\ s = \{\beta, \gamma\},\ t = \{\xi, \eta, \zeta\})$
K_{11}: $D^e_{2r}(K_1) \times C_i$
K_{12}: $S(A^+_1, A^-_1)$, $S(A^+_2, A^-_2)$, $S(B^+_1, B^-_1)$, $S(B^+_2, B^-_2)$,
$S(E^+_m, E^-_m;\ 1_2, \sigma_y)$; $m = m^* = \frac{1}{2}, 1, \ldots, r - \frac{1}{2}$
K_{13}: $D^e_{2r}(K_2) \times C_i$
K_{14}: $S(A^\pm_1, A^\pm_2)$, $S(B^\pm_1, B^\pm_2)$, $S(E^+_m, E^-_m;\ \sigma_y, 1_2)$; $m = m^*$
K_{15}: $D^e_{2r}(K_3) \times C_i$
K_{16}: $S(A^\pm_1, B^\mp_1)$, $S(A^\pm_2, B^\mp_2)$, $S(E^+_m, E^-_{r-m};\ \sigma_z, \sigma_x)$; $m = m^*$
K_{17}: $D^e_{2r}(K_4) \times C_i$
K_{18}: $S(A^\pm_1, B^\mp_2)$, $S(A^\pm_2, B^\mp_1)$, $S(E^+_m, E^-_{r-m};\ \sigma_x, \sigma_z)$; $m = m^*$
K_{21}: $S(D_A;\ 1_2)$, $S(D_B;\ 1_2)$, $S(D^{+y}_m, D^{-y}_m;\ 1_2, \sigma_y)$; $m = m^*$
K_{22}: $S(D_A;\ \sigma_z)$, $S(D_B;\ \sigma_z)$, $S(D^{\pm y}_m;\ 1_2, \sigma_y)$; $m = m^*$
K_{23}: $S(D_A;\ \sigma_x)$, $S(D_B;\ \sigma_x)$, $S(D^{+y}_m, D^{-y}_m;\ \sigma_y, 1_2)$; $m = m^*$
K_{24}: $S(D_A, D_A;\ \sigma_y)$, $S(D_B, D_B;\ \sigma_y)$, $S(D^{\pm y}_m, D^{\pm y}_m;\ \sigma_y, 1_2)$; $m = m^*$
K_{25}: $S(D_A, D_B;\ \sigma_x)$, $S(D^{\pm y}_{r_e/2};\ \sigma_z)$, $S(D^{\pm y}_{r_o/2}, D^{\pm y}_{r_o/2};\ \sigma_x)$,
$S(D^{\pm y}_m, D^{\pm y}_{r-m};\ \sigma_z, \sigma_x)$; $m = m^\dagger = \frac{1}{2}, 1, \ldots, \frac{1}{2}(r-1)$
K_{26}: $S(D_A, D_B;\ \sigma_y)$, $S(D^{+y}_m, D^{-y}_{r-m};\ \sigma_z, \sigma_x)$; $m = m^*$
K_{27}: $S(D_A, D_B;\ 1_2)$, $S(D^{\pm y}_{r_e/2};\ \sigma_x)$, $S(D^{\pm y}_{r_o/2}, D^{\pm y}_{r_o/2};\ \sigma_z)$,
$S(D^{\pm y}_m, D^{\pm y}_{r-m};\ \sigma_x, \sigma_z)$; $m = m^\dagger$
K_{28}: $S(D_A, D_B;\ \sigma_z)$, $S(D^{+y}_m, D^{-y}_{r-m};\ \sigma_x, \sigma_z)$; $m = m^*$

Table 17.3. (*cont.*)

K_{31}: $S(D_{11};\ 1_2)$, $S(D_{22};\ 1_2)$, $S(D^{\pm z}_{r_e/2};\ 1_2)$, $S(D^{+z}_{r_o/2},\ D^{-z}_{r_o/2};\ \sigma_y)$,
$S(D^{z}_{m,r-m};\ 1_4,\ \sigma_z\times\sigma_y)$; $m=m^{\dagger}=\frac{1}{2},1,\ldots,\frac{1}{2}(r-1)$
K_{32}: $S(D_{11};\ 1_2)$, $S(D_{22};\ \sigma_z)$, $S(D^{+z}_{r_e/2},\ D^{-z}_{r_e/2};\ 1_2)$, $S(D^{\pm z}_{r_o/2};\ \sigma_y)$,
$S(D^{z}_{m,r-m};\ \sigma_z\times\mathbf{1}_2,\ \mathbf{1}_2\times\sigma_y)$; $m=m^{\dagger}$,
K_{33}: $S(D_{11},\ D_{22};\ 1_2)$, $S(D^{+z}_{r_e/2},\ D^{-z}_{r_e/2};\ \sigma_y)$, $S(D^{\pm z}_{r_o/2},\ D^{\pm z}_{r_o/2};\ 1_2)$,
$S(D^{z}_{m,r-m},\ D^{z}_{m,r-m};\ \sigma_z\times\sigma_y,\ 1_4)$; $m=m^{\dagger}$
K_{34}: $S(D_{11},\ D_{22};\ \sigma_z)$, $S(D^{\pm z}_{r_e/2},\ D^{\pm z}_{r_e/2};\ \sigma_y)$, $S(D^{+z}_{r_o/2},\ D^{-z}_{r_o/2};\ 1_2)$,
$S(D^{z}_{m,r-m},\ D^{z}_{m,r-m};\ 1_2\times\sigma_y,\ \sigma_z\times 1_2)$; $m=m^{\dagger}$
K_{35}: $S(D_{11};\ \sigma_x)$, $S(D_{22};\ \sigma_x)$, $S(D^{\pm z}_{r_e/2};\ \sigma_z)$, $S(D^{+z}_{r_o/2},\ D^{-z}_{r_o/2};\ \sigma_x)$,
$S(D_{m,r-m};\ \sigma_x\times\sigma_z,\ \sigma_y\times\sigma_x)$; $m=m^{\dagger}$
K_{36}: $S(D_{11},\ D_{11};\ \sigma_y)$, $S(D_{22},\ D_{22};\ \sigma_y)$, $S(D^{+z}_{r_e/2},\ D^{-z}_{r_e/2};\ \sigma_z)$,
$S(D^{\pm z}_{r_o/2},\ D^{\pm z}_{r_0/2};\ \sigma_x)$, $S(D^{z}_{m,r-m},\ D^{z}_{m,r-m};\ \sigma_y\times\sigma_z,\ \sigma_x\times\sigma_z)$; $m=m^{\dagger}$
K_{37}: $S(D_{11},\ D_{22};\ \sigma_x)$, $S(D^{+z}_{r_e/2},\ D^{-z}_{r_e/2};\ \sigma_x)$, $S(D^{\pm z}_{r_o/2},\ D^{\pm z}_{r_0/2};\ \sigma_z)$,
$S(D^{z}_{m,r-m},\ D^{z}_{m,r-m};\ \sigma_y\times\sigma_x,\ \sigma_x\times\sigma_z)$; $m=m^{\dagger}$
K_{38}: $S(D_{11},\ D_{22};\ \sigma_y)$, $S(D^{\pm z}_{r_e/2};\ \sigma_x)$, $S(D^{+z}_{r_o/2},\ D^{-z}_{r_o/2};\ \sigma_z)$,
$S(D^{z}_{m,r-m};\ \sigma_x\times\sigma_x,\ \sigma_y\times\sigma_z)$; $m=m^{\dagger}$
K_{41}: $S(D_{12};\ 1_2)$, $S(D_{21};\ 1_2)$, $S(D^{\pm x}_{r_e/2};\ 1_2)$, $S(D^{+x}_{r_o/2},\ D^{-x}_{r_o/2};\ \sigma_y)$,
$S(D^{x}_{m,r-m};\ 1_4,\ \sigma_z\times\sigma_y)$; $m=m^{\dagger}$
K_{42}: $S(D_{12};\ \sigma_z)$, $S(D_{21};\ \sigma_z)$, $S(D^{+x}_{r_e/2},\ S(D^{-x}_{r_e/2};\ 1_2)$, $S(D^{\pm x}_{r_o/2};\ \sigma_y)$,
$S(D^{x}_{m,r-m};\ \sigma_z\times 1_2,\ 1_2\times\sigma_y)$; $m=m^{\dagger}$
K_{43}: $S(D_{12},\ D_{21};\ 1_2)$, $S(D^{+x}_{r_e/2},\ D^{-x}_{r_e/2};\ \sigma_y)$, $S(D^{\pm x}_{r_o/2};\ 1_2)$,
$S(D^{x}_{m,r-m},\ D^{x}_{m,r-m};\ \sigma_z\times\sigma_y,\ 1_4)$; $m=m^{\dagger}$
K_{44}: $S(D_{12},\ D_{21};\ \sigma_z)$, $S(D^{\pm x}_{r_e/2},\ D^{\pm x}_{r_e/2};\ \sigma_y)$, $S(D^{+x}_{r_o/2},\ D^{-x}_{r_o/2};\ 1_2)$,
$S(D^{x}_{m,r-m},\ D^{x}_{m,r-m};\ 1_2\times\sigma_y,\ \sigma_z\times 1_2)$; $m=m^{\dagger}$
K_{45}: $S(D_{12},\ D_{21};\ \sigma_x)$, $S(D^{+x}_{r_e/2},\ D^{-x}_{r_e/2};\ \sigma_z)$, $S(D^{\pm x}_{r_o/2},\ D^{\pm x}_{r_o/2};\ \sigma_x)$,
$S(D^{x}_{m,r-m},\ D^{x}_{m,r-m};\ \sigma_y\times\sigma_z,\ \sigma_x\times\sigma_x)$; $m=m^{\dagger}$
K_{46}: $S(D_{12},\ D_{21};\ \sigma_y)$, $S(D^{\pm x}_{r_e/2};\ \sigma_z)$, $S(D^{+x}_{r_o/2},\ D^{-x}_{r_o/2};\ \sigma_x)$,
$S(D^{x}_{m,r-m};\ \sigma_x\times\sigma_z,\ \sigma_y\times\sigma_x)$; $m=m^{\dagger}$
K_{47}: $S(D_{12};\ 1_2)$, $S(D_{21};\ \sigma_x)$, $S(D^{\pm x}_{r_e/2};\ \sigma_x)$, $S(D^{+x}_{r_o/2},\ D^{-x}_{r_o/2};\ \sigma_z)$,
$S(D^{x}_{m,r-m};\ \sigma_x\times\sigma_x,\ \sigma_y\times\sigma_x)$; $m=m^{\dagger}$
K_{48}: $S(D_{12},\ D_{12};\ \sigma_y)$, $S(D_{21},\ D_{21};\ \sigma_y)$, $S(D^{+x}_{r_e/2},\ D^{-x}_{r_e/2};\ \sigma_x)$,
$S(D^{\pm x}_{r_o/2},\ D^{\pm x}_{r_o/2};\ \sigma_z)$, $S(D^{x}_{m,r-m},\ D^{x}_{m,r-m};\ \sigma_y\times\sigma_x,\ \sigma_x\times\sigma_z)$; $m=m^{\dagger}$

16. $D^{e}_{2r+1,i}(K_{st};\ s=\{\gamma\},\ t=\{\zeta\})$
K_{11}: $D^{e}_{2r+1}(K_1)\times C_i$
K_{12}: $S(A_1^+,\ A_1^-)$, $S(A_2^+,\ A_2^-)$, $S(B_1^{\pm},\ B_2^{\mp})$, $S(E_m^+,\ E_m^-;\ 1_2,\ \sigma_y)$; $m=m'=\frac{1}{2},1,\ldots,r$
K_{21}: $S(D_A,\ 1_2)$, $S(D_B,\ D_B;\ \sigma_x)$, $S(D_m^{+y},\ D_m^{-y};\ 1_2,\ \sigma_y)$; $m=m'$
K_{22}: $S(D_A;\ \sigma_z)$, $S(D_B;\ \sigma_y)$, $S(D_m^{\pm y};\ 1_2,\ \sigma_y)$; $m=m'$

17. $D^{q}_{2r,i}(K_{st};\ s=\{\gamma\},\ t=\{\zeta\})$
K_{11}: $D^{q}_{2r}(K)\times C_i$
K_{12}: $S(A_1^+,\ A_1^-)$, $S(A_2^+,\ A_2^-)$, $S(B_1^{\pm},\ B_2^{\mp})$, $S(E_m^+,\ E_m^-;\ R_m,\ \sigma_yR_m)$; $m=m^*=\frac{1}{2},1,$
$\ldots,r-\frac{1}{2}$
K_{21}: $S(D_A;\ 1_2)$, $S(D_B,\ D_B;\ \sigma_x)$, $S(D_m^{+y},\ D_m^{-y};\ R_m,\ \sigma_yR_m)$; $m=m^*$
K_{22}: $S(D_A;\ \sigma_z)$, $S(D_B;\ \sigma_y)$, $S(D_m^{\pm y};\ R_m,\ \sigma_yR_m)$; $m=m^*$

18. $D^q_{2r+1,i}(K_{st};\ s=\{\gamma\},\ t=\{\eta,\zeta\})$
K_{11}: $D^q_{2r+1}(K_1)\times C_i$
K_{12}: $S(A_1^+, A_1^-)$, (A_2^+, A_2^-), $S(B_1^+, B_1^-)$, (B_2^+, B_2^-),
$S(E_m^+, E_{m'}^-;\ \sigma_y R_m)$; $m = m' = \frac{1}{2}, 1, \ldots, r$
K_{13}: $D^q_{2r+1}(K_2)\times C_i$
K_{14}: $S(A_1^\pm, A_2^\mp)$, $S(B_1^\pm, B_2^\mp)$, $S(E_m^+, E_{m'}^-;\ \sigma_y R_m, R_m)$; $m = m'$
K_{21}: $S(D_A;\ 1_2)$, $S(D_B;\ 1_2)$, $S(D_m^{+y}, D_{m'}^{-y};\ R_m, \sigma_y R_m)$; $m = m'$
K_{22}: $S(D_A;\ \sigma_z)$, $S(D_B;\ \sigma_z)$, $S(D_m^{\pm y};\ R_m, \sigma_y R_m)$; $m = m'$
K_{23}: $S(D_A;\ \sigma_x)$, $S(D_B;\ \sigma_x)$, $S(D_m^{+y}, D_{m'}^{-y};\ \sigma_y R_m, R_m)$; $m = m'$
K_{24}: $S(D_A, D_A;\ \sigma_y)$, $S(D_B, D_B;\ \sigma_y)$, $S(D_m^{\pm y}, D_{m'}^{\pm y};\ \sigma_y R_m, R_m)$; $m = m'$

19. $T(K^0)$
K^0: $A, A', A'', T, E_{1/2}, E'_{1/2} = E_{1/2}\times A', E''_{1/2} = E_{1/2}\times A''$

20. $T^e(K)$
K: $S(A)$, $S(A', A'')$, $S(T;\ 1_3)$, $S(E_{1/2};\ \sigma_y)$, $S(E'_{1/2}, E''_{1/2};\ \sigma_y)$

21. $T^q(K)$
K: $S(A)$, $S(A')$, $S(A'')$, $S(T;\ 4_z)$, $S(E_{1/2};\ Z)$, $S(E'_{1/2};\ Z)$, $S(E''_{1/2};\ Z)$

22. $T_i(K^0)$
K^0: $T(K^0)\times C^i$; $A^\pm, A'^\pm, A''^\pm, T^\pm, E^\pm_{1/2}, E'^\pm_{1/2}, E''^\pm_{1/2}$

23. $T_i^e(K_t;\ t=\{\eta\})$
K_1: $T^e(K)\times C^i$
K_2: $S(A^+, A^-)$, $S(A'^\pm, A''^\mp)$, $S(T^+, T^-;\ 1_3)$, $S(E^+_{1/2}, E^-_{1/2};\ \sigma_y)$,
$S(E'^\pm_{1/2}, E''^\mp_{1/2};\ \sigma_y)$

24. $T_i^q(K_t;\ t=\{\eta\})$
K_1: $T^q\times C_i$
K_2: $S(A^+, A^-)$, $S(A'^+, A'^-)$, $S(A''^+, A''^-)$, $S(T^+, T^-;\ 4_z)$,
$S(E^+_{1/2}, E^-_{1/2};\ Z)$, $S(E'^+_{1/2}, E'^-_{1/2};\ Z)$, $S(E''^+_{1/2}, E''^-_{1/2};\ Z)$

25. $O(K^0)$
K^0: $A_1, A_2, E, T_1, T_2, E_{1/2}, E'_{1/2} = E_{1/2}\times A_2, Q = E_{1/2}\times E$

26. $O^e(K_t;\ t=\{\xi\})$
K_1: $S(A_1)$, $S(A_2)$, $S(E;\ 1_2)$, $S(T_1;\ 1_3)$, $S(T_2;\ 1_3)$, $S(E_{1/2};\ \sigma_y)$,
$S(E'_{1/2};\ \sigma_y)$, $S(Q;\ \sigma_y\times 1_2)$
K_2: $S(A_1, A_2)$, $S(E, E;\ \sigma_y)$, $S(T_1, T_2;\ 1_3)$, $S(E_{1/2}, E'_{1/2};\ \sigma_y)$,
$S(Q, Q;\ \sigma_y\times\sigma_y)$

27. $O_i(K_s^0;\ s=\{\beta\})$
K_1^0: $O(K^0)\times C_i$; $A_1^\pm, A_2^\pm, E^\pm, T_1^\pm, T_2^\pm, E^\pm_{1/2}, E'^\pm_{1/2}, Q^\pm$
K_2^0: $D_A = D(A_1, A_2)$, $D_E^{\pm y} = D(E;\ \pm\sigma_y)$, $D_T = D(T_1, T_2;\ 1_3)$,
$D_{1/2,1/2} = D(E_{1/2}, E'_{1/2};\ 1_2)$, $D_Q^{\pm y} = D(Q;\ \pm 1_2\times\sigma_y)$

Table 17.3. (*cont.*)

28. $O_i^e(K_{st};\ s=\{\beta\},\ t=\{\xi,\zeta\})$
  K_{11}: $O^e(K_1)\times C_i$
  K_{12}: $S(A_1^+, A_1^-)$, $S(A_2^+, A_2^-)$, $S(E^+, E^-; 1_2)$, $S(T_1^+, T_1^-; 1_3)$, $S(T_2^+, T_2^-; 1_3)$, $S(E_{1/2}^+, E_{1/2}^-; \sigma_y)$, $S(E'^+_{1/2}, E'^-_{1/2}; \sigma_y)$, $S(Q^+, Q^-; \sigma_y\times 1_2)$
  K_{13}: $O^e(K_2)\times C_i$
  K_{14}: $S(A_1^\pm, A_2^\mp)$, $S(E^+, E^-; \sigma_y)$, $S(T_1^\pm, T_2^\mp; 1_3)$, $S(E_{1/2}^\pm, E'^\mp_{1/2}; \sigma_y)$, $S(Q^+, Q^-; \sigma_y\times\sigma_y)$
  K_{21}: $S(D_A; 1_2)$, $S(D_E^{+y}, D_E^{-y}; 1_2)$, $S(D_T; 1_6)$, $S(D_{1/2,1/2}; 1_2\times\sigma_y)$, $S(D_Q^{+y}, D_Q^{-y}; \sigma_y\times 1_2)$
  K_{22}: $S(D_A; \sigma_z)$, $S(D_E^{\pm y}; 1_2)$, $S(D_T; \sigma_z\times 1_3)$, $S(D_{1/2,1/2}; \sigma_z\times\sigma_y)$, $S(D_Q^{\pm y}; \sigma_y\times 1_2)$
  K_{23}: $S(D_A; \sigma_x)$, $S(D_E^{+y}, D_E^{-y}; \sigma_y)$, $S(D_T; \sigma_x\times 1_3)$, $S(D_{1/2,1/2}; \sigma_x\times\sigma_y)$, $S(D_Q^{+y}, D_Q^{-y}; \sigma_y\times\sigma_y)$
  K_{24}: $S(D_A, D_A; \sigma_y)$, $S(D_E^{\pm y}, D_E^{\pm y}; \sigma_y)$, $S(D_T, D_T; \sigma_y\times 1_3)$, $S(D_{1/2,1/2}, D_{1/2,1/2}; \sigma_y\times\sigma_y)$, $S(D_Q^{\pm y}, D_Q^{\pm y}; \sigma_y\times\sigma_y)$

1. All unirreps given for the ordinary unitary point groups are defined in Chapter 11. All co-unirreps given in this table are for $\tau = 1$.
2. m^0, m^*, m' and $m^\dagger$ are integers or half integers defined by $-\frac{1}{2}n < m^0 \leqslant \frac{1}{2}n$; $m^* = \frac{1}{2}, 1, \ldots, \frac{1}{2}(n-1)$; $m' = \frac{1}{2}, 1, \ldots, r$; $m^\dagger = \frac{1}{2}, 1, \ldots, \frac{1}{2}(r-1)$ for a given integer n or r.
3. n_0 (n_e) and r_o (r_e) are odd (even) integers.
4. When two transformation matrices are given for a set of co-unirreps such as in $S(E_m; 1_2, \sigma_y)$ with $m = m^*$ for K_1 of D_{2r}^e in (10), the first one is for every integral m and the second one is for every half-integral m.
5. In (14) $D_m^{\pm y} = D(E_m; \pm\sigma_y)$ means $D_m^{+y} = D(E_m; \sigma_y)$ and $D_m^{-y} = D(E_m; -\sigma_y)$. $S(A_1^\pm, A_2^\mp)$ means $S(A_1^+, A_2^-)$ and $S(A_1^-, A_2^+)$; the ($\pm$) in the remaining notations should be understood similarly.
6. The transformation matrix R_m in (12), (13), (17) and (18) is defined by

$$R_m = \begin{bmatrix} \cos(\pi m/n) & -\sin(\pi m/n) \\ \sin(\pi m/n) & \cos(\pi m/n) \end{bmatrix}$$

7. The transformation matrix Z in (21) and (24) is defined by $Z = 2^{-1/2}(\sigma_y - i1_2)$ if the basis of $E_{1/2}$ is $[\phi_+(\frac{1}{2},\frac{1}{2}), \phi_-(\frac{1}{2},\frac{1}{2})]$, and $Z = \mathrm{i}2^{-1/2}(\sigma_x - \sigma_y)$ if the basis is $[\phi(\frac{1}{2},\frac{1}{2}), \phi(\frac{1}{2},-\frac{1}{2})]$. The rotation matrix 4_z in (21) and (24) is defined, in terms of the basis-vector notation given in Table 5.1, by

$$4_z = [\boldsymbol{j}, -\boldsymbol{i}, \boldsymbol{k}] = \begin{bmatrix} 0 & -1 & 0 \\ 1 & 0 & 0 \\ 0 & 0 & 1 \end{bmatrix}$$

8. The above table differs from Table II given by the author in *J. Math. Phys.* **25**, 189 (1984) for those related to the unirreps B_1 and B_2 of the group D_n due to the exchange of their definitions in Table 11.5 from Table II.

The projective co-unirreps of H^z given by the vector co-unirreps of $H^{z\prime}$ thus determined are given in Table 17.3 together with the projective unirreps of the unitary point groups determined previously in Table 12.4. These are classified by the classes of factor systems specified by the representation of the coefficient set $\{\alpha_i\}$ in the representation groups. It is to be noted, however, that only those co-unirreps belonging to the classes (90 of them) with $\tau = 1$ are given in Table 17.3. Let us call two classes mutually dual if they differ only in the coefficient τ. Then the co-unirreps belonging to the class K' with $\tau = -1$ are obtained from those of its dual K with $\tau = 1$ by interchanging type a and type b, and leaving type c unchanged. In general, a class K and its dual K' are p-inequivalent. In cases for which $K \sim K'$ (p-equivalent), we have set $\tau = 1$ by mapping off $\tau = -1$, e.g. see $C^{e\prime}_{2r+1}$ in Table 17.2 (2).

Since several coefficients α_i are involved in specifying a class of factor systems, it is necessary to introduce a convenient system of notation for expressing a class. For example, according to Table 17.2 (15), the representation group $D^{e\prime}_{2r,i}$ has a set of six coefficients $\{\beta, \gamma; \xi, \eta, \zeta; \tau\}$, where the subset $s = \{\beta, \gamma\}$ characterizes the unitary representation group $D'_{2r,i}$, whereas the subset $t = \{\xi, \eta, \zeta\}$ characterizes the conjugations of the generators x, y and I by a, and finally τ characterizes a^2. As one can see from Tables 17.2 and 17.3, this is the most complicated case. Usually fewer coefficients are contained in each subset s or t and often there exists only one subset s or t or none besides τ. In the extreme cases even τ is fixed to unity. In any case, each subset s or t may be identified by a number such that, for a one-member subset $\{\alpha_1\}$

$$1 = \{1\}, \qquad 2 = \{-1\} \tag{17.6.7a}$$

for a two-member subset $\{\alpha_1, \alpha_2\}$

$$1 = \{1, 1\}, \qquad 2 = \{1, -1\}, \qquad 3 = \{-1, 1\}, \qquad 4 = \{-1, -1\} \tag{17.6.7b}$$

and for a three-member subset $\{\alpha_1, \alpha_2, \alpha_3\}$

$$\begin{aligned} &1 = \{1, 1, 1\}, \quad 2 = \{1, 1, -1\}, \quad 3 = \{1, -1, 1\}, \quad 4 = \{1, -1, -1\} \\ &5 = \{-1, 1, 1\}, \quad 6 = \{-1, 1, -1\}, \quad 7 = \{-1, -1, 1\}, \quad 8 = \{-1, -1, -1\} \end{aligned} \tag{17.6.7c}$$

There exists no subset with more than three members. Now, a class specified by the subsets s and t and $\tau = 1$ is denoted by K_{st} and its dual with $\tau = -1$ is denoted by K'_{st}. Analogously, a class involved with one subset t is denoted by K_t, a class with no subset besides τ is denoted by K and their duals are denoted by K'_t and K', respectively. Finally, the classes of the unitary point groups are denoted by K^0_s or K^0. Obviously, no dual can exist for the unitary classes. Table 17.3 contains a total of 90 classes of factor systems for anti-unitary point groups H^z and a total of 13 classes of factor systems for the unitary point groups.

It is worthwhile to illustrate Table 17.3 through an example, postponing its construction. According to Table 17.3(10), the group D^e_{2r} has a total of eight classes of factor systems given by K_t and K'_t, where $t = \{\xi, \eta\}$. The co-unirreps belonging to one of them, say K_2 ($\xi = 1$, $\eta = -1$; $\tau = 1$), are given by

$$S(A_1, A_2),\ S(B_1, B_2),\ S(E_m, E_m; \sigma_y, \mathbf{1}_2); \qquad m = m^* = \tfrac{1}{2}, \ldots, r - \tfrac{1}{2} \tag{17.6.8}$$

in terms of the notation introduced by (17.6.3). The class structure (or the type distribution of the class) may conveniently be denoted by

$$K_2 = (c_2^2 b_4^{r-1} | b_4^r) \in D_{2r}^e \tag{17.6.9}$$

where c and b denote the types of the co-unirreps, their subscripts denote the dimensions of representation, their superscripts denote the numbers of the respective types, and, finally, the left-hand half of the bracket contains the integral co-unirreps for which $\Delta^{(\nu)}(e') = 1$, and the right-hand half contains the half-integral co-unirreps for which $\Delta^{(\nu)}(e') = -1$. It should also be noted that the last $2r - 1$ co-unirreps $S(E_m, E_m; \sigma_y, \mathbf{1}_2)$ given in (17.6.8) contain two transformation matrices σ_y and $\mathbf{1}_2$; in such a case, the first one σ_y is for every integral m and the second one $\mathbf{1}_2$ is for every half-integral m. One can immediately write down the co-unirreps belonging to the dual class K_2' from (17.6.8) as follows by changing type b to type a:

$$S(A_1, A_2),\ S(B_1, B_2),\ S(E_m; \sigma_y, \mathbf{1}_2); \qquad m = m^* \tag{17.6.10}$$

keeping in mind that $\Delta^{(\nu)}(a^2) = -\Delta^{(\nu)}(e')$. The type distribution is given by $(c_2^2 a_2^{r-1} | a_2^r)$.

Let us comment on the dimensions of the projective co-representations of H^z. According to Table 17.3, one-dimensional projective co-unirreps of H^z occur only for the vector co-representations. Excluding these trivial cases, the dimensions of the projective co-unirreps are all even and limited to 2, 4, 6, 8 and 12. The highest dimension 12 occurs for O_i^e. For the projective unirreps of the ordinary unitary groups, the dimensions are limited to 1, 4 and 6 except for the trivial cases of the vector unirreps, for which the dimensions are limited to 1, 2 and 3 (the icosahedral group is excluded). In the following, we shall illustrate the method of constructing Table 17.3 through a few typical examples.

17.6.1 Examples for the construction of the projective co-unirreps of H^z

Example 1. For the projective co-unirreps of $C_n^e \in K(\tau = 1)$, the defining relations taken from Table 17.2(2) are

$$x^n = e', \qquad x^a = x, \qquad a^2 = e'$$

where $x^a = a^{-1}xa$. The unirreps of C_n are

$$M_m(x) = \exp(-2\pi i m/n), \qquad -n/2 < m \leqslant n/2 \tag{17.6.11a}$$

where m is an integer or a half integer. Since the representation is one-dimensional, we may take $N(a) = 1$ in (17.6.1) and obtain

$$M_m(x)^* = \exp(2\pi i m/n) = M_{-m}(x) \tag{17.6.11b}$$

Here $M_m \neq M_{-m}$, unless $m = 0$ or $n/2$, so that, from (17.6.3), we arrive at type c co-unirreps given, with $N(a) = 1$, by

$$S(M_m, M_{-m}); \qquad m = \tfrac{1}{2}, 1, \ldots, \tfrac{1}{2}(n-1) \tag{17.6.11c}$$

When $m = 0$ or $n/2$, we have $M_m = M_{-m}$ so that we obtain either type a or type b co-unirreps, from (17.6.6b) with $\tau = 1$:

$$S(M_0),\ S(M_{n_e/2}),\ S(M_{n_o/2}, M_{n_o/2}) \tag{17.6.11d}$$

where n_e (n_o) is an even (odd) n. These provide all the projective co-unirreps of $C_n^e \in K(\tau = 1)$ as presented in Table 17.3(2). Their type distributions for $n \geqslant 2$ are

$$\{a_1^2, c_2^{n/2-1} | c_2^{n/2}\}_{n=\text{even}},\ \{a_1, c_2^{(n-1)/2} | b_2, c_2^{(n-1)/2}\}_{n=\text{odd}}$$

For the dual $K'(\tau = -1)$ we have

$$\{b_2^2 c_2^{n/2-1} | c_2^{n/2}\}_{n=\text{even}}, \{b_2, c_2^{(n-1)/2} | a_1, c_2^{(n-1)/2}\}_{n=\text{odd}}$$

Here, $K' \sim K$ (p-equivalent) for an odd n consistent with Table 17.2(2), where we have set $\tau = 1$. Note, however, that one has p-equivalence between the integral and half-integral representations due to the change in sign of τ.

Example 2. For the projective co-unirreps of $C^e_{2r,i} \in K_{22}(\beta = -1; \zeta = -1; \tau = 1)$ from Table 17.2(5, 6) we have

$$x^{2r} = e', \qquad x^I = -x, \qquad x^a = x, \qquad I^a = -I, \qquad a^2 = e'$$

which implies that $\xi = 1$ and $\eta = -1$. Firstly, the projective unirreps $\check{D}(M_m, M_{m-r}; 1)$ of the double point group $C'_{2r,i}$ for $K^0_2(\beta = -1)$ given in Table 12.4 are reexpressed by

$$D_m(x) = \exp(-im\pi/r)\sigma_z, \qquad D_m(I) = \sigma_x; \qquad m = \tfrac{1}{2}, 1, \ldots, r \qquad (17.6.12a)$$

where σ_z and σ_x are the Pauli spin matrices. Since

$$D_m(x)^* = -D_{r-m}(x), \qquad D_m(I)^* = D_{r-m}(I) \qquad (17.6.12b)$$

from (17.6.5) with $\xi = 1$ and $\zeta = -1$, we obtain $N(a) = \sigma_y$. Thus, from (17.6.3) we arrive at the type c co-unirreps of $C^{e\prime}_{2r,i}$:

$$S(D_m, D_{r-m}; \sigma_y); \qquad m = \tfrac{1}{2}, 1, \ldots, \tfrac{1}{2}(r-1)$$

excluding $m = r$ or $\frac{1}{2}r$. When $m = r$ we have, from (17.6.12a),

$$D_r(x) = -\sigma_z, \qquad D_r(I) = \sigma_x$$

Since these are real, we obtain $N(a) = \sigma_z$ via (17.6.5). Thus, from the type criteria (17.6.6b) with $\tau = 1$, we arrive at a type a co-unirrep given by

$$S(D_r; \sigma_z)$$

Analogously, for $m = r/2$ we have $D_{r/2}(x) = -i\sigma_z$, $D_{r/2}(I) = \sigma_x$ so that via (17.6.5) we obtain $N(a) = \sigma_y$. Thus, from the type criteria (17.6.6b), we arrive at

$$S(D_{r_e/2}, D_{r_e/2}; \sigma_y), S(D_{r_o/2}; \sigma_y)$$

where r_e (r_o) is an even (odd) r. These provide the projective co-unirreps of $C^e_{2r,i} \in K_{22}$ given in Table 17.3(6). Their type distributions are

$$\{a_2, b_4, c_4^{\frac{1}{2}r-1} | c_4^{\frac{1}{2}r}\}_{r=\text{even}}, \{a_2, c_4^{(r-1)/2} | a_2 c_4^{(r-1)/2}\}_{r=\text{odd}}$$

Example 3. For D^q_{2r}, the defining relations are, from Table 17.2(12),

$$D'_{2r}; \qquad x^a = x, \qquad y^a = yx, \qquad a^2 = \tau e' x; \qquad \tau^2 = 1 \qquad (17.6.13)$$

which imply that $\xi = \eta = 1$. There exist only one class $K(\tau = 1)$ and its dual $K'(\tau = -1)$. Firstly, from Table 11.5, the unirreps of D_{2r} are given by

	$x = n_z$	$y = 2_x$	
A_1	1	1	
A_2	1	−1	
B_1	−1	1	
B_2	−1	−1	
E_m	$c_m - \mathrm{i}\sigma_y s_m$	$(-\mathrm{i})^{2m}\sigma_z$;	$m = \frac{1}{2}, 1, \ldots, r - \frac{1}{2}$

where $c_m = \cos(2\pi m/n)$ and $s_m = \sin(2\pi m/n)$. Note that $E_m(x)$ is real whereas $E_m(y)$ is real for an integral m, but imaginary for a half-integral m.

From the unirreps of D_{2r} given above we shall form the projective co-unirreps of D_{2r}^q through the table of construction given below for the class K ($\tau = 1$), from (17.6.5) and (17.6.13):

$h \in D_{2r}$	x	y	
h^a	x	yx	
$A_1(h^a)^*$	1	$1 \to A_1(h)$;	$S(A_1)$
$A_2(h^a)^*$	1	$-1 \to A_2(h)$;	$S(A_2)$
$B_1(h^a)^*$	−1	$-1 \to B_2(h)$;	$S(B_1, B_2)$
$E_m(h^a)^*$	$c_m - \mathrm{i}\sigma_y s_m$	$\pm(-\mathrm{i})^{2m}\sigma_z E_m(x)$	
$E_m(h)$	$c_m - \mathrm{i}\sigma_y s_m$	$(-\mathrm{i})^{2m}\sigma_z$;	$S(E_m;\, R_m,\, \sigma_y R_m)$

Here the + (−) sign is for an integral (half-integral) m. Only the last co-unirrep $S(E_m;\, R_m,\, \sigma_y R_m)$ may require some explanation: $E_m(h^a)^*$ is equivalent to $E_m(h)$ via (17.6.5) with $N(a) = R_m$ for an integral m and $\sigma_y R_m$ for a half-integral m, where R_m is defined by

$$R_m = c_{m/2} - \mathrm{i}\sigma_y s_{m/2}, \qquad R_m^{-1} = c_{m/2} + \mathrm{i}\sigma_y s_{m/2}$$

which is real and commutes with σ_y and hence with $E_m(x)$, and satisfies

$$R_m^{-1}\sigma_z R_m = \sigma_z E_m(x), \qquad (\sigma_y R_m)^{-1}\sigma_z(\sigma_y R_m) = -\sigma_z E_m(x); \qquad R_m^2 = E_m(x)$$

Since $E_m(a^2) = E_m(e'x) = N(a)N(a)^*$ for an integral as well as for a half-integral m, we obtain a type a co-unirrep for either case from (17.6.6b):

$$S(E_m;\, R_m) \quad \text{for an integral } m$$

$$S(E_m;\, \sigma_y R_m) \quad \text{for a half-integral } m$$

These are combined into one formula $S(E_m;\, R_m,\, \sigma_y R_m)$ as given in Table 17.3(12). The type distribution for the class $K(\tau = 1)$ is given by

$$\{a_1^2,\, c_2,\, a_2^{r-1} | a_2^r\}$$

17.7 The co-unirreps of anti-unitary wave vector space groups

As in the case of a unitary space group, the co-unirreps of an anti-unitary space group of wave vector $\hat{H}_k^z$ can be regarded as the projective co-unirreps of the corresponding anti-unitary point group belonging to a certain factor system. The present results given in Tables 17.2 and 17.3 specialized for the crystallographic point groups provide all the co-unirreps of any $\hat{H}_k^z$. It is necessary only to determine the appropriate gauge transformations which connect the generators of $\hat{H}_k^z$ with those of the corresponding

Table 17.4. *The crystallographic anti-unitary point groups (the gray groups are not listed)*

No.	International	Present	No.	International	Present	No.	International	Present
1	$\bar{1}'$	C_1^i	21	$4m'm'$	C_4^v	41	$6'2'2'$	D_3^q
2	$2'$	$C_1^q\ (C_1^u)$	22	$\bar{4}'2'm$	C_{2v}^p	42	$62'2'$	C_6^u
3	m'	$C_1^p\ (C_1^v)$	23	$\bar{4}'2m'$	D_2^p	43	$6'm'm$	C_{3v}^q
4	$2/m'$	C_2^i	24	$\bar{4}2'm'$	C_{2p}^u	44	$6m'm'$	C_6^v
5	$2'/m'$	$C_i^u\ (C_i^q)$	25	$4/m'mm$	C_{4v}^i	45	$\bar{6}'m'2$	D_3^p
6	$2'/m$	C_{1p}^i	26	$4'/mm'm$	D_{2i}^q	46	$\bar{6}'m2'$	C_{3v}^p
7	$22'2'$	C_2^u	27	$4'/m'm'm$	D_{2p}^i	47	$\bar{6}m'2'$	C_{3p}^u
8	$m'm'2$	C_2^v	28	$4/mm'm'$	C_{4i}^u	48	$6/m'mm$	C_{6v}^i
9	$m'm2'$	C_{1p}^u	29	$4/m'm'm'$	D_4^i	49	$6'/mm'm$	D_{3p}^i
10	$m'mm$	C_{2v}^i	30	$\bar{3}'$	C_3^i	50	$6'/m'm'm$	D_{3i}^q
11	$m'm'm$	C_{2i}^u	31	$32'$	C_3^u	51	$6/mm'm'$	C_{6i}^u
12	$m'm'm'$	D_2^i	32	$3m'$	C_3^v	52	$6/m'm'm'$	D_6^i
13	$4'$	C_2^q	33	$\bar{3}'m$	C_{3v}^i	53	$m'3$	T^i
14	$\bar{4}'$	C_2^p	34	$\bar{3}'m'$	D_3^i	54	$4'32'$	T^q
15	$4'/m$	C_{2i}^q	35	$\bar{3}m'$	C_{3i}^u	55	$\bar{4}'3m'$	T^p
16	$4/m'$	C_4^i	36	$6'$	C_3^q	56	$m'3m$	T_p^i
17	$4'/m'$	C_{2p}^i	37	$\bar{6}'$	C_3^p	57	$m3m'$	T_i^q
18	$4'2'2$	D_2^q	38	$6'/m$	C_{3p}^i	58	$m'3m'$	O^i
19	$42'2'$	C_4^u	39	$6/m'$	C_6^i			
20	$4'm'm$	C_{2v}^q	40	$6'/m'$	C_{3i}^q			

representation group $H^{z\prime}$ given in Table 17.2. In order to utilize the isomorphisms between the anti-unitary point groups and limit the discussion to the three types H^e, H^q and H^u given in (17.5.1), it is necessary to classify the crystallographic anti-unitary groups in terms of the present system of notations H^z. For convenience, we have expressed their international symbols in terms of the present system in Table 17.4. This table is a special case of the more comprehensive one given by Table 16.1.

To illustrate the procedure of obtaining the co-unirreps of $\hat{H}_k^z$ from Tables 17.2, 17.3 and 17.4, we shall use some typical examples taken from the tables of irreducible co-representations of magnetic space groups given by Miller and Love 1967 (ML). We shall follow the notations used by them for the magnetic space groups as well as for the special points of the Brillouin zone. In using their tables, caution should be exercised, since the generator sets of $\hat{H}_k^z$ given by ML are in general not in agreement with those given in Table 17.2. It is also to be noted that ML have given the matrix co-representations explicitly for the generators only whereas the present general expressions given in Table 17.3 provide the co-representations for all the unitary elements and the anti-unitary augmenting operator a in terms of the unirreps of the point groups given in section 10 of Table 17.3.

Example 1. Group (76) 9; $P4_1'$. From Table 17.4, the point co-group $4'$ of this space group is identified to be C_2^q. It is an anti-unitary space group of the second kind $\hat{C}_2^q$

formed from the unitary space group $\hat{C}_4$ (number 76); $P4_1$ and defined by the generator $P(4'_z|00\frac{1}{4})$ according to Tables 17.1 and 13.3. For convenience, we may express the generator set of $\hat{C}_2^q$ as follows:

$$P\{2_z|00\tfrac{1}{2}\}, \{\boldsymbol{\theta}4_z|00\tfrac{1}{4}\} \qquad (17.7.1a)$$

which includes the generator $\{2_z|00\frac{1}{2}\}$ of the halving subgroup $\hat{C}_2$. The defining relations of the representation group $C_2^{q\prime}$ are given, from Table 17.2(3), by

$$x^2 = e', \qquad xa = ax, \qquad a^2 = e'x, \qquad \tau = 1 \qquad (17.7.1b)$$

where $a = \boldsymbol{\theta}4_z$ and $x = 2_z$ belong to C_2^q. This is one of the most simple cases in which there exists only one class of factor system K ($\sim K'$). The projective co-unirreps of $C_2^q \in K$ are given, from Table 17.3(3), by

$$S(M_0), S(M_1, M_1), S(M_{1/2}, M_{-1/2}) \qquad (17.7.1c)$$

with the type distribution $(a_1 b_2|c_2)$.

Now, let us consider the wave vector group $\hat{C}_2^q(\boldsymbol{k})$ at each of the high-symmetry points $GM = (000)$, $M = (\frac{1}{2}\frac{1}{2}0)$, $A = (\frac{1}{2}\frac{1}{2}\frac{1}{2})$ and $Z = (00\frac{1}{2})$ of the Brillouin zone of the primitive tetragonal lattice P_Q. Then the generators of $\hat{C}_2^q(\boldsymbol{k})$ given by (17.7.1a) satisfy the defining relations (17.7.1b) via the following one-to-one correspondences:

$$\{\boldsymbol{\theta}4_z|00\tfrac{1}{4}\} \leftrightarrow a, \qquad \{2_z|00\tfrac{1}{2}\} \leftrightarrow x, \qquad \{E'|000\} \leftrightarrow e' \qquad \text{at } GM \text{ or } M$$

$$\{\boldsymbol{\theta}4_z|00\tfrac{1}{4}\} \leftrightarrow a, \qquad \{2_z|00\tfrac{1}{2}\} \leftrightarrow -x, \qquad \{E'|000\} \leftrightarrow -e' \qquad \text{at } A \text{ or } Z$$

Through these correspondences, the co-unirreps of $\hat{C}_2^q$ are determined by the projective co-unirreps of C_2^q given by (17.7.1c). Here E' is the 2π rotation for the space group $\hat{C}_2^q$ whereas e' is the 2π rotation for the point group C_2^q. The correspondence $\{E'|000\} = -e'$ means that the integral and half-integral representations are interchanged between $\hat{C}_2^q$ and C_2^q. The type distribution of the co-unirreps at GM and at M is given by $(a_1 b_2|c_2)$ whereas that at A and at Z is given by $(c_2|a_1 b_2)$ due to the correspondence $\{E'|000\} \leftrightarrow -e'$. These results are equivalent to those given by ML. On account of the isomorphism $\hat{C}_2^q \simeq \hat{C}_2^p$, a similar treatment may be given for $\hat{C}_2^p(\boldsymbol{k})$.

Example 2. Group (194) 166; $P6'_3/mm'c$: From Table 17.4 the magnetic point co-group is D_{3p}^i, which is isomorphic to the gray group D_6^e according to (16.3.6a). It is an anti-unitary space group of the second kind $\hat{D}_{3p}^i$ isomorphic to the unitary space group $\hat{D}_{6i}$. Let us consider the wave vector space group $\hat{D}_{3p}^i(\boldsymbol{k})$ at $A = (00\frac{1}{2})$ in the Brillouin zone. From the space group tables (Tables 13.3 and 17.1), the generators of $\hat{D}_{3p}^i$ are

$$P. \quad \{\bar{6}_z|00\tfrac{1}{2}\}, \{2_x|000\}, \{\boldsymbol{\theta}\bar{1}|000\} \qquad (17.7.2a)$$

These satisfy the defining relations of $D_{3p}^{i\,\prime}$ ($\simeq D_6^{e\prime}$} given in Table 17.2(10):

$$x^6 = y^2 = (xy)^2 = e', \qquad e'^2 = e; \qquad xa = \xi ax, \qquad ya = \eta ay, \qquad a^2 = \tau e'$$

via the correspondences

$$\{\bar{6}_z|00\tfrac{1}{2}\} \leftrightarrow \mathrm{i}x, \qquad \{2_x|000\}, \leftrightarrow \mathrm{i}y, \qquad \{\boldsymbol{\theta}\bar{1}|000\} \leftrightarrow a, \qquad \{E|000\} \leftrightarrow -e' \qquad (17.7.2b)$$

with the projective factors

$$(\xi, \eta) = (1, -1), \qquad \tau = -1 \qquad (17.7.2c)$$

which identifies K_2' (the dual of K_2) as the class of the factor systems for the projective representations of D_{3p}^i. Thus, from K_2 given in Table 17.3(10) the projective co-unirreps of D_{3p}^i belonging to K_2' are given by

$$S(A_1, A_2),\ S(B_1, B_2),\ S(E_1; 1_2),\ S(E_2; 1_2),\ S(E_{1/2}; \sigma_y),\ S(E_{3/2}; \sigma_y),\ S(E_{5/2}; \sigma_y) \tag{17.7.2d}$$

with the type distribution $(c_2^2 a_2^2 | a_2^3)$. From (17.7.2b) and (17.7.2d) follow all the co-unirreps of $\hat{D}_{3p}^i(\boldsymbol{k})$ at A with the type distribution $(a_2^3 | c_2^2 a_2^2)$ since $\{E'|000\} = -e'$. Analogous treatments may be given for the magnetic space groups belonging to D_{3p}^e, D_6^i, C_{6v}^i and C_{6v}^e, all of which are isomorphic to each other.

Example 3. Group (222) 99; $Pm3'm$: The corresponding anti-unitary point co-group is a gray group O_i^e so that it is an anti-unitary space group of the first kind. The generators of $\hat{O}_i^e$ are, from Tables 17.1 and 13.3,

$$P_0\{4_z|\tfrac{1}{2}00\},\ \{3_{xyz}|000\},\ \{\bar{1}|000\} \tag{17.7.3a}$$

where P_0 denotes $P(\boldsymbol{\theta}|000)$. Let us consider the wave vector space group $\hat{O}_i^e(\boldsymbol{k})$ at $R = (\frac{1}{2}\frac{1}{2}\frac{1}{2})$. The unirreps of the space group (222) at R had previously been discussed in (14.4.18a). The generators of $\hat{O}_i^e$ given above satisfy the defining relations of $O_i^{e\prime}$ given in Table 17.2(28) via the one-to-one correspondences

$$\{4_z|\tfrac{1}{2}00\} \leftrightarrow x, \qquad \{3_{xyz}|000\} \leftrightarrow y, \qquad \{\bar{1}|000\} \leftrightarrow I, \qquad \{\boldsymbol{\theta}|000\} \leftrightarrow a,$$
$$\{E'|000\} \leftrightarrow e' \tag{17.7.3b}$$

with the projective factors

$$\beta = -1, \qquad (\xi, \zeta) = (1, 1), \qquad \tau = 1 \tag{17.7.3c}$$

Thus, the co-unirreps belonging to the class K_{21} of O_i^e given in Table 17.3(28) provide all the co-unirreps of $\hat{O}_i^e(\boldsymbol{k})$ with the correspondence (17.7.3b) without any gauge factors. The type distribution is given by $(a_2 c_4 a_6 | a_4 c_8)$, which is in agreement with that given by ML.

17.7.1 Concluding remarks

We have described the representation groups of all finite point groups by those of ten unitary and 22 anti-unitary point groups in Table 17.2. These are given in terms of the abstract group generators which are common to all point groups that are isomorphic to each other. Then, by means of the modified theory of induced representations introduced in Section 16.4, we have presented in Table 17.3 the general expressions of all p-inequivalent projective co-unirreps of all the anti-unitary point groups in terms of the unirreps of the proper point groups previously determined in Chapter 11.

The present results are more than sufficient to find all the unirreps or co-unirreps of any space group (unitary or anti-unitary) of a wave vector through simple gauge transformations. Here it is essential to identify the point groups corresponding to respective space groups in terms of the present system of notations H^z. Table 17.4 identifies the international notations of the crystallographic magnetic point groups in terms of the present system of notations H^z. As one can see from Table 17.4, it is hardly possible to recognize their isomorphism from the international symbols. The present work has been extended further to the projective co-unirreps of the magnetic point groups of infinite order by Kim (1984c).

17.8 Selection rules under an anti-unitary group

17.8.1 General discussion

In Section 6.10 we discussed the selection rules for a quantum mechanical system that is invariant under a unitary group. If the Hamiltonian $\boldsymbol{H}_0$ of the system is invariant under an anti-unitary symmetry group H^z, the invariance of the transition moment under the unitary as well as under the anti-unitary operations belonging to H^z should be discussed, provided that the group is a volume-preserving one. Accordingly, the eigenfunctions of $\boldsymbol{H}_0$ should be classified by the co-unirreps of the anti-unitary group H^z. Let $\boldsymbol{S}^{(\nu)}$ be a co-unirrep of H^z induced from a unirrep $\Delta^{(\nu)}$ of the halving subgroup H: here, $\boldsymbol{S}^{(\nu)}$ stands for $S^{(\nu+)}$, $S^{(\nu,\nu)}$ or $S^{(\nu,\bar{\nu})}$ defined in (17.6.3). Let $\Phi^\nu = \{\Phi_i^\nu\}$ and $\Phi^\mu = \{\Phi_j^\mu\}$ be two sets of the eigenfunctions of $\boldsymbol{H}_0$ belonging to the co-unirrep $\boldsymbol{S}^{(\nu)}$ and $\boldsymbol{S}^{(\mu)}$, respectively. Then the set of transition moments for transitions between two sets of states due to a set of perturbations $V = \{V_l\}$ imposed on the system is defined by

$$\boldsymbol{M}^{\nu\mu} = \langle \Phi^\nu, V\Phi^\mu \rangle = \{M_{ijl}^{\nu\mu}\}; \qquad M_{ijl}^{\nu\mu} = \langle \Phi_i^\nu, V_l \Phi_j^\mu \rangle \tag{17.8.1}$$

The absolute square $|M_{ijl}^{\nu\mu}|^2$ determines the probability of transition between two eigenstates Φ_i^ν and Φ_j^μ.

Each transition moment is a constant, being a definite integral, and is invariant under a unitary operation $h \in H$ and an anti-unitary operation $a = a_0 h = \boldsymbol{\theta} z h$ expressed as follows:

$$\boldsymbol{M}^{\nu\mu} = \langle h\Phi^\nu, (hVh^{-1})h\Phi^\mu \rangle \equiv \bar{h}\boldsymbol{M}^{\nu\mu} \tag{17.8.2a}$$

$$\boldsymbol{M}^{\nu\mu} = \langle a\Phi^\mu, (aVa^{-1})a\Phi^\nu \rangle \equiv \bar{a}\boldsymbol{M}^{\nu\mu} \tag{17.8.2b}$$

in view of (16.2.15), assuming V to be Hermitian. Here the operators $\bar{h}$ and $\bar{a}$ are introduced to denote the invariance of $\boldsymbol{M}^{\nu\mu}$ under h and a, both of which act on the integrands. The crucial aspect of the operator $\bar{a}$ is that it is linear even though a is anti-unitary, i.e.

$$\bar{a}(c\boldsymbol{M}^{\nu\mu}) = c\bar{a}\boldsymbol{M}^{\nu\mu} \tag{17.8.3}$$

where c is an arbitrary complex constant. The operators $\bar{h}$ and $\bar{a}$ satisfy the ordinary group property

$$\bar{a}\bar{h} = \overline{ah}, \qquad \bar{h}\bar{a} = \overline{ha}, \qquad \bar{a}_1\bar{a}_2 = \overline{a_1 a_2} \tag{17.8.4}$$

where $a_1 = a_0 h_1$ and $a_2 = a_0 h_2$ for $h_1, h_2 \in H$. Thus the set $\{\bar{h}, \bar{a}_0\bar{h}; \forall\, h \in H\}$ forms a unitary group isomorphic to the anti-unitary group H^z via the one-to-one correspondences $\bar{h} \leftrightarrow h$ and $\bar{a} \leftrightarrow a$. Since this group is formed by augmenting the unitary group $\bar{H} = \{\bar{h}\}$ with $\bar{a}$, it may be denoted by $\bar{H}_{\bar{a}}$ and may be referred to as *the unitary group associated with the anti-unitary group* H^z. It is the group which describes the invariance of $\boldsymbol{M}^{\nu\mu}$ under H^z. When $z \notin H$, there exists another unitary group H_z defined by

$$H_z = H + zH \tag{17.8.5}$$

which is also isomorphic to $\overline{H}_{\overline{a}}$. This group is needed to describe the transformation of the set of the perturbation operators $V = \{V_l\}$.

Under a symmetry operation of H^z, the basis functions in the integrand transform according to the co-unirreps of H^z, whereas the perturbation is assumed to transform according to a unirrep $D^{(\lambda)}$ of H_z in addition to its time-reversal symmetry. However, the set of transition moments $\{M^{\nu\mu}_{ijl}\}$ as a whole provides a basis of a representation of a unitary group $\overline{H}_{\overline{a}}$ that must be a unitary representation. Since each transition moment is a constant, it is non-zero if and only if the invariant part of the integrand with respect to H^z is non-zero. Under the selection rules, we discuss what the linear relations between transition moments are, which moments differ from zero and how many linearly independent non-zero elements there are. We have established three selection rules in Section 6.10 under a unitary symmetry group. These rules hold with little modification under the anti-unitary group H^z, the reason being that the matrix system $\boldsymbol{M}^{\nu\mu}$ transforms according to a representation of the unitary group $\overline{H}_{\overline{a}} = \{\overline{h}, \overline{a}\}$.

In the following, we shall determine the representation of $\overline{H}_{\overline{a}}$ based on the set $\boldsymbol{M}^{\nu\mu}$ in terms of the co-unirreps $\boldsymbol{S}^{(\nu)}$ and $\boldsymbol{S}^{(\mu)}$ of H^z and a unirrep $D^{(\lambda)}$ of H^z that describes the symmetry of the perturbation V. Then we calculate the number of linearly independent non-zero moments in the set.

17.8.2 Transitions between states belonging to different co-unirreps

When $\nu \neq \mu$, the two states Φ^ν and Φ^μ introduced in the set of transition moments (17.8.1) belong to different energy levels except for the case of accidental degeneracy. We discuss this case first and then specialize it to the case in which $\nu = \mu$.

The invariance of the set $\boldsymbol{M}^{\nu\mu} = \{M^{\nu\mu}_{ijl}\}$ under $\overline{H}_{\overline{a}}$ has been described by (17.8.2a) and (17.8.2b): the basis functions Φ^ν and Φ^μ in the integrands transform according to the co-unirreps $\boldsymbol{S}^{(\nu)}$ and $\boldsymbol{S}^{(\mu)}$ of H^z, whereas the perturbation $V = \{V_l\}$ transforms according to a unirrep $D^{(\lambda)}$ of H_z besides the time-reversal symmetry such that

$$aV_l a^{-1} = \varepsilon_v zhV_l(zh)^{-1} = \varepsilon_v \sum_{l'} V_{l'} D^{(\lambda)}_{l'l}(zh) \tag{17.8.6}$$

where we have used $\boldsymbol{\theta} V_l \boldsymbol{\theta}^{-1} = \varepsilon_v V_l$ with $\varepsilon_v = 1\ (-1)$ from (16.2.13). Thus we have, from (17.8.2a) and (17.8.2b),

$$\begin{aligned}
\overline{h}M^{\nu\mu}_{ijl} &= \sum_{i',j',l'} M^{\nu\mu}_{i'j'l'} \boldsymbol{S}^{(\nu)}_{i'i}(h)^* \boldsymbol{S}^{(\mu)}_{j'j}(h) D^{(\lambda)}_{l'l}(h) \\
&\equiv \sum_{i',j',l'} M^{\nu\mu}_{i'j'l'} \boldsymbol{P}^{(\nu\mu)}_{i'j'l',ijl}(\overline{h}) \\
\overline{a}M^{\nu\mu}_{ijl} &= \sum_{i',j',l'} M^{\mu\nu}_{j',i',l'} \varepsilon_v \boldsymbol{S}^{(\mu)}_{j'j}(a)^* \boldsymbol{S}^{(\nu)}_{i'i}(a) D^{(\lambda)}_{l'l}(zh) \\
&\equiv \sum_{i',j',l'} M^{\mu\nu}_{j'i'l'} \boldsymbol{P}^{(\mu\nu)}_{j'i'l',ijl}(\overline{a})
\end{aligned} \tag{17.8.7}$$

where $a = \boldsymbol{\theta} zh$ for all $h \in H$. Note that the set $\boldsymbol{M}^{\nu\mu} = \{M^{\nu\mu}_{ijl}\}$ closes under $h \in H$

whereas it is transformed to $M^{\mu\nu}$ under $\overline{a}$. Consequently, the combined row vector basis $[M^{\nu\mu}, M^{\mu\nu}]$ provides a representation $D(\overline{H}_{\overline{a}})$ defined by

$$D(\overline{h}) = \begin{bmatrix} P^{(\nu\mu)}(\overline{h}) & 0 \\ 0 & P^{(\mu\nu)}(\overline{h}) \end{bmatrix} \tag{17.8.8a}$$

$$D(\overline{a}) = \begin{bmatrix} 0 & P^{(\nu\mu)}(\overline{a}) \\ P^{(\mu\nu)}(\overline{a}) & 0 \end{bmatrix} \tag{17.8.8b}$$

where $P^{(\nu\mu)}$ and $P^{(\mu\nu)}$ are defined in (17.8.7).

To understand the algebraic properties of this representation $D(\overline{H}_{\overline{a}})$ introduced above, we may rewrite the set of matrices $P^{(\nu\mu)}(\overline{h})$ and $P^{(\nu\mu)}(\overline{a})$ defined in (17.8.7) formally as follows:

$$P^{(\nu\mu)}(\overline{h}) = S^{(\nu)}(h)^* \times S^{(\mu)}(h) \times D^{(\lambda)}(h) \tag{17.8.9a}$$

$$P^{(\mu\nu)}(\overline{a}) = \varepsilon_v[S^{(\mu)}(a)^* \mathbin{\tilde{\times}} S^{(\nu)}(a)] \times D^{(\lambda)}(zh) \tag{17.8.9b}$$

where $A \times B$ denotes an ordinary direct product whereas $A \mathbin{\tilde{\times}} B$ denotes a twisted direct product of matrices A and B defined by their matrix elements

$$A_{ab}B_{cd} = [A \times B]_{ac,bd} = [A \mathbin{\tilde{\times}} B]_{ac,db} \tag{17.8.9c}$$

The latter differs from the ordinary direct product in that the second set of subscripts is interchanged from that of the direct product. The dimensions of A and B are independent from each other for an ordinary direct product, but may be required to be equal for some properties of the twisted direct product; e.g. for the trace of $A \mathbin{\tilde{\times}} B$ (see below).

To facilitate the matrix manipulation of the representation $D(\overline{H}_{\overline{a}})$, we shall give here some of the algebraic properties of the twisted direct product defined by (17.8.9c) in comparison with those of the ordinary direct product:

(1) $\operatorname{tr} A \times B = \operatorname{tr} A \operatorname{tr} B$, but

$$\operatorname{tr}(A \mathbin{\tilde{\times}} B) = \operatorname{tr} AB \tag{17.8.10a}$$

(2) $(A \times B)(C \times D) = AC \times BD$, but

$$\begin{aligned} (A \times B)(C \mathbin{\tilde{\times}} D) &= (AC) \mathbin{\tilde{\times}} (BD) \\ (A \mathbin{\tilde{\times}} B)(C \times D) &= (AD) \mathbin{\tilde{\times}} (BC) \\ (A \mathbin{\tilde{\times}} B)(C \mathbin{\tilde{\times}} D) &= (AD) \times (BC) \end{aligned} \tag{17.8.10b}$$

(3) For the Hermitian adjoint, $(A \times B)^\dagger = B^\dagger \times A^\dagger$ and analogously

$$(A \mathbin{\tilde{\times}} B)^\dagger = (B^\dagger \mathbin{\tilde{\times}} A^\dagger) \tag{17.8.10c}$$

(4) If A and B are unitary, then both $A \times B$ and $A \mathbin{\tilde{\times}} B$ are unitary.

Here it has been assumed that the dimensions of matrices are the same as those involved in the ordinary matrix multiplication. By means of the algebraic properties of $A \mathbin{\tilde{\times}} B$ given above we can show that

$$\begin{aligned} P^{(\nu\mu)}(\overline{h})P^{(\nu\mu)}(\overline{a}) &= P^{(\nu\mu)}(\overline{h}\overline{a}) \\ P^{(\mu\nu)}(\overline{a})P^{(\nu\mu)}(\overline{h}) &= P^{(\mu\nu)}(\overline{a}\overline{h}) \end{aligned} \tag{17.8.11}$$

which will lead us to the proof that the representation of $\overline{H}_{\overline{a}}$ defined by (17.8.8) is indeed a unitary representation even though the matrix system $\boldsymbol{P}^{(\nu\mu)}$ is defined by the co-unirreps of H^z.

Now, according to selection rule 2 given in Section 6.10, the number n_1 of the linearly independent non-zero elements in the basis set $\{\boldsymbol{M}^{\nu\mu}, \boldsymbol{M}^{\mu\nu}\}$ for given ν and μ $(\neq \nu)$ equals the number of the identity representations contained in the representation $D(\overline{H}_{\overline{a}})$. This number n_1 is given by the sum of the characters of the representation over the group elements of $\overline{H}_{\overline{a}}$. Since $D(\overline{a})$ are off-diagonal from (17.8.8b), their traces are zero so that we have, from (17.8.8a) and (17.8.9a),

$$\begin{aligned} n_1^{(\nu\mu)} &= \frac{1}{2|H|}\sum_{h\in H} \operatorname{tr}[\boldsymbol{P}^{(\nu\mu)}(\overline{h}) + \boldsymbol{P}^{(\mu\nu)}(\overline{h})] \\ &= \frac{1}{|H|}\sum_{h\in H} R_e[\boldsymbol{\kappa}^{(\nu)}(h)^*\boldsymbol{\kappa}^{(\mu)}(h)]\chi^{(\lambda)}(h) \end{aligned} \qquad (17.8.12)$$

where R_e denotes the real part and $\boldsymbol{\kappa}^{(\nu)}(h) = \operatorname{tr} \boldsymbol{S}^{(\nu)}(h)$ and $\chi^{(\lambda)}(h) = \operatorname{tr} D^{(\lambda)}(h)$. Note here that the anti-unitary elements $a = a_0 h$ do not contribute to the above sum directly but do affect the result indirectly through the types of the co-unirreps $\boldsymbol{S}^{(\nu)}$ and $\boldsymbol{S}^{(\mu)}$, because, from (17.6.2),

$$\boldsymbol{\kappa}^{(\nu)}(h) = \begin{cases} \chi^{(\nu)}(h), & \text{for type a} \\ 2\chi^{(\nu)}(h), & \text{for type b} \\ \chi^{(\nu)}(h) + \chi^{(\overline{\nu})}(h), & \text{for type c} \end{cases}$$

where $\chi^{(\nu)}(h) = \operatorname{tr} \Delta^{(\nu)}(h)$ and $\chi^{(\overline{\nu})}(h) = \operatorname{tr} \Delta^{(\overline{\nu})}(h)$ for all $h \in H$ and analogous relations for $\boldsymbol{\kappa}^{(\mu)}$.

Identification of the non-zero elements requires the reduction of the representation $D(\overline{H}_{\overline{a}})$ defined by (17.8.8) by an appropriate similarity transformation as was discussed in Section 6.10. We shall not go into this reduction because it requires further explicit specification of the matrix system $\boldsymbol{M}^{\nu\mu}$ for a given specific problem.

17.8.3 Transitions between states belonging to the same co-unirrep

When $\nu = \mu$ in (17.8.1), the system of transition matrices becomes

$$M_{ijl}^{\nu\nu} = \langle \Phi_i^\nu, V_l \Phi_j^\nu \rangle \qquad (17.8.13)$$

The set closes under $\overline{H}_{\overline{a}}$, as can be seen from (17.8.7). For convenience, we may reproduce here the transformation properties of the basis Φ^ν under the three types of the co-unirreps $\boldsymbol{S}^{(\nu)}$ to the extent we need for the transformation of $\boldsymbol{M}^{\nu\nu}$. From (16.4.12)–(16.4.17), we have

$$\Phi^\nu = \begin{cases} [\psi^\nu, \phi^{\overline{\nu}}] \in S^{(\nu,\overline{\nu})} & (\nu \neq \overline{\nu}) \quad \text{for type c} \\ [\psi^\nu, \phi^\nu] \in S^{(\nu,\nu)} & \text{for type b} \\ \psi^\nu = \phi^\nu \in S^{(\nu+)} & \text{for type a} \end{cases} \qquad (17.8.14)$$

where ψ^ν and $\phi^{\overline{\nu}}$ are the bases of the unirreps $\Delta^{(\nu)}$ and $\Delta^{(\overline{\nu})}$ of the halving subgroup H, respectively. These are connected by an anti-unitary operator $a = a_0 h$ as follows. From (16.4.10b) and (16.4.11b),

$$a\psi^\nu = \phi^{\overline{\nu}} N(a), \qquad a\phi^{\overline{\nu}} = \psi^\nu \overline{N}(a) \qquad (17.8.15a)$$

where the transformation matrices $N(a)$ and $\overline{N}(a)$ are further related, from (16.4.11c), by

$$\overline{N}(a)N(a)^* = \Delta^{(\nu)}(a^2), \qquad N(a)\overline{N}(a)^* = \Delta^{(\overline{\nu})}(a^2) \tag{17.8.15b}$$

where $\overline{N}(a) = N(a)$ for type a and $\overline{N}(a) = -N(a)$ for type b from (16.4.14). First we consider the type c co-unirreps because the remaining cases can be treated as special cases of them.

17.8.3.1 Type c co-unirreps with $\nu \neq \overline{\nu}$

The matrix system (17.8.13) takes the form, using (17.8.14) for type c,

$$\begin{aligned} \boldsymbol{M}^{\nu\nu} &= \begin{bmatrix} \langle \psi^\nu, V\psi^\nu \rangle & \langle \psi^\nu, V\phi^{\overline{\nu}} \rangle \\ \langle \phi^{\overline{\nu}}, V\psi^\nu \rangle & \langle \phi^{\overline{\nu}}, V\phi^{\overline{\nu}} \rangle \end{bmatrix} \\ &\equiv \begin{bmatrix} V^{(11)} & V^{(12)} \\ V^{(21)} & V^{(22)} \end{bmatrix} \end{aligned} \tag{17.8.16}$$

In view of (17.8.15a) and (17.8.2b), we see that each set of the off-diagonal elements $V^{(12)} = \{V^{(12)}_{ijl}\}$ and $V^{(21)} = \{V^{(21)}_{ijl}\}$ spans an invariant subspace under $\overline{H}_{\overline{a}}$, while the sets of diagonal elements $V^{(11)} = \{V^{(11)}_{ijl}\}$ and $V^{(22)} = \{V^{(22)}_{ijl}\}$ are combined to form another invariant subspace of $\overline{H}_{\overline{a}}$. In the following we shall discuss the selection rules for each invariant subspace separately.

In the off-diagonal subspaces for type c, the invariant subspace

$$V^{(12)} = \langle \psi^\nu, V\phi^{\overline{\nu}} \rangle; \qquad \psi^\nu = \{\psi^\nu_i\}, \qquad \phi^{\overline{\nu}} = \{\phi^{\overline{\nu}}_i\} \tag{17.8.17a}$$

provides a unitary representation $D^{(12)}$ of $\overline{H}_{\overline{a}}$ defined, using (17.8.15a), by

$$\begin{aligned} D^{(12)}(\overline{h}) &= \Delta^{(\nu)}(h)^* \times \Delta^{(\overline{\nu})}(h) \times D^{(\lambda)}(h) \\ D^{(12)}(\overline{a}_0\overline{h}) &= \varepsilon_\upsilon(\overline{N}(a_0h)^* \mathbin{\tilde{\times}} N(a_0h)) \times D^{(\lambda)}(zh) \end{aligned} \tag{17.8.17b}$$

for all $h \in H$. It is a trivial matter to confirm that the matrix system $D^{(12)}$ defined by (17.8.17b) is indeed a matrix representation of $\overline{H}_{\overline{a}}$, if we use the algebraic properties (17.8.10b) of the twisted direct products and the transformation properties (17.8.15). Accordingly, the number of linearly independent matrix elements contained in the set $V^{(12)} = \{V^{(12)}_{ijl}\}$ is given by the character of $D^{(12)}(\overline{H}_{\overline{a}})$ as follows:

$$\begin{aligned} n_1^{(12)} &= \frac{1}{2|H|}\sum_{h\in H}\{\mathrm{tr}\, D^{(12)}(\overline{h}) + \mathrm{tr}\, D^{(12)}(\overline{a}_0\overline{h})\} \\ &= \frac{1}{2|H|}\sum_{h\in H}\{\chi^{(\nu)}(h)^*\chi^{(\overline{\nu})}(h)\chi^{(\lambda)}(h) + \varepsilon_\upsilon\chi^{(\nu)}((a_0h)^2)^*\chi^{(\lambda)}(zh)\} \end{aligned} \tag{17.8.17c}$$

where $\chi^{(\nu)} = \mathrm{tr}\,\Delta^{(\nu)}$, $\chi^{(\lambda)} = \mathrm{tr}\, D^{(\lambda)}$ and use of (17.8.10a) and (17.8.15b) has been made.

The second invariant off-diagonal space in (17.8.16) is spanned by the matrix system

$$V^{(21)} = \langle \phi^{\overline{\nu}}, V\psi^\nu \rangle \tag{17.8.18a}$$

It provides a representation $D^{(21)}(\overline{H}_{\overline{a}})$ that can be obtained from $D^{(12)}(\overline{H}_{\overline{a}})$ by the simultaneous interchanges $\nu \leftrightarrow \overline{\nu}$ and $N(a) \leftrightarrow \overline{N}(a)$. Accordingly, the number of linearly independent matrix elements contained in the set $\{V_{ijl}^{(21)}\}$ is given from $n_1^{(12)}$ of (17.8.17c), by

$$n_1^{(21)} = \frac{1}{2|H|}\sum_{h\in H}\{\chi^{(\overline{\nu})}(h)^*\chi^{(\nu)}(h)\chi^{(\lambda)}(h) + \varepsilon_v\chi^{(\nu)}((a_0h)^2)\chi^{(\lambda)}(zh)\} \tag{17.8.18b}$$

where[5] we have used $\chi^{(\overline{\nu})}(a^2)^* = \chi^{(\nu)}(a^2)$.

Since the numbers $n_1^{(12)}$ and $n_2^{(21)}$ are real, their sum can be expressed by $n_1^{(12)*} + n_2^{(21)}$ so that

$$n_1^{(12)} + n_1^{(21)} = \frac{1}{|H|}\sum_h\{\chi^{(\nu)}(h)\chi^{(\overline{\nu})}(h)^* R_e\chi^{(\lambda)}(h) + \varepsilon_v\chi^{(\nu)}((a_0h)^2)R_e\chi^{(\lambda)}(zh)\} \tag{17.8.20}$$

where R_e denotes the real part. In the special case in which the character $\chi^{(\lambda)}$ of $D^{(\lambda)}(H_z)$ is real, we have $n_1^{(12)} = n_1^{(21)}$.

The diagonal subspace for type c is spanned by the diagonal elements in (17.8.16):

$$V^{(11)} = \langle\psi^\nu, V\psi^\nu\rangle, \qquad V^{(22)} = \langle\phi^{\overline{\nu}}, V\phi^{\overline{\nu}}\rangle \tag{17.8.21}$$

The representation of $\overline{H}_{\overline{a}}$ based on $[V^{(11)}, V^{(22)}]$ is given, with use of (17.8.15a), by

$$D^{(11+22)}(\overline{h}) = \begin{bmatrix} D^{(\nu)}(\overline{h}) & 0 \\ 0 & D^{(\overline{\nu})}(\overline{h}) \end{bmatrix}$$

$$D^{(11+22)}(\overline{a}) = \begin{bmatrix} 0 & D^{(\overline{\nu})}(\overline{a}) \\ D^{(\nu)}(\overline{a}) & 0 \end{bmatrix} \tag{17.8.22}$$

where

$$D^{(\nu)}(\overline{h}) = \Delta^{(\nu)}(h)^* \times \Delta^{(\nu)}(h) \times D^{(\lambda)}(h)$$

$$D^{(\overline{\nu})}(\overline{h}) = \Delta^{(\overline{\nu})}(h)^* \times \Delta^{(\overline{\nu})}(h) \times D^{(\lambda)}(h)$$

$$D^{(\nu)}(\overline{a}) = \varepsilon_v[N(a)^* \tilde{\times} N(a)] \times D^{(\lambda)}(zh)$$

$$D^{(\overline{\nu})}(\overline{a}) = \varepsilon_v[\overline{N}(a)^* \tilde{\times} \overline{N}(a)] \times D^{(\lambda)}(zh) \tag{17.8.23a}$$

Here $a = a_0h = \boldsymbol{\theta}zh$.

The number of linearly independent matrix elements contained in the set $\{V^{(11)}, V^{(22)}\}$ is then given by

$$n_1^{(11+22)} = \frac{1}{2|H|}\sum_h\{|\chi^{(\nu)}(h)|^2 + |\chi^{(\overline{\nu})}(h)|^2\}\chi^{(\lambda)}(h) \tag{17.8.23b}$$

which is independent of the anti-unitary operator a, because $S^{(11+12)}(\overline{a})$ is off-diagonal. However, the existence of the $\chi^{(\overline{\nu})}(h)$ terms is a consequence of the anti-unitary operator $a \in H^z$.

[5] From (16.4.10a), $\Delta^{(\nu)}(a^{-1}ha)^* = N(a)^{-1}\Delta^{(\overline{\nu})}(h)N(a)$ and $(ah)^2 = h^{-1}(ha)^2h = a(ha)^2a^{-1}$ we obtain the following identities:

$$\chi^{(\nu)}(a^{-1}ha)^* = \chi^{(\overline{\nu})}(h), \qquad \chi^{(\overline{\nu})}(aha^{-1}) = \chi^{(\nu)}(h)^*$$

$$\chi^{(\nu)}((ah)^2) = \chi^{(\nu)}((ha)^2) = \chi^{(\overline{\nu})}((ah)^2)^* = \chi^{(\overline{\nu})}((ha)^2)^* \tag{17.8.19}$$

17.8.3.2 For a type b co-unirrep

This case will be treated by specializing the case of type c in Section 17.8.3.1 with the condition that $\nu = \overline{\nu}$ and $\overline{N}(a) = -N(a)$ for type b from (17.8.15b).

In the off-diagonal subspaces for type b, the invariant spaces are

$$V^{(12)} = \langle \psi^\nu, V\phi^\nu \rangle, \qquad V^{(21)} = \langle \phi^\nu, V\psi^\nu \rangle \tag{17.8.24a}$$

Note that $V^{(21)} = V^{(12)*}$. Thus, from (17.8.17b) with $\overline{N}(a) = -N(a)$, the matrix representations of $\overline{H}_{\overline{a}}$ for both cases are given by

$$\begin{aligned} D^{(12)}(\overline{h}) &= D^{(21)}(\overline{h})^* \\ &= \Delta^{(\nu)}(h)^* \times \Delta^{(\nu)}(h) \times D^{(\lambda)}(h) \qquad (= D^{(\nu)}(\overline{h})) \\ D^{(12)}(\overline{a}) &= D^{(21)}(\overline{a})^* \\ &= -\varepsilon_\upsilon[N(a)^* \tilde{\times} N(a)] \times D^{(\lambda)}(zh) \qquad (= -D^{(\nu)}(\overline{a})) \end{aligned} \tag{17.8.24b}$$

The numbers of linearly independent matrix elements for both cases of the off-diagonal space are given by

$$n_1^{(12)} = n_1^{(21)} = \frac{1}{2|H|}\sum_h \{|\chi^{(\nu)}(h)|^2\chi^{(\lambda)}(h) + \varepsilon_\upsilon\chi^{(\nu)}((a_0h)^2)\chi^{(\lambda)}(zh)\} \tag{17.8.24c}$$

where we have used (17.8.10a) and (17.8.15b).

In the diagonal subspace for type b, we have, from (17.8.23b) with $\nu = \overline{\nu}$,

$$n_1^{(11+22)} = \frac{1}{|H|}\sum_h |\chi^{(\nu)}(h)|^2\chi^{(\lambda)}(h) \tag{17.8.24d}$$

which is completely free from the anti-unitary operator $a \in H^z$.

17.8.3.3 For type a co-unirreps

The matrix system $V^{(\nu\nu)}$ is defined by

$$V^{(\nu\nu)} = \langle \psi^\nu, V\psi^\nu \rangle \tag{17.8.25a}$$

which may be regarded as a special case of $V^{(12)}$ with $\phi^{\overline{\nu}} = \psi^\nu$ in (17.8.17a). In this case, we have $\overline{N}(a) = N(a)$, so that the matrix representation of $\overline{H}_{\overline{a}}$ based on $V^{(\nu\nu)}$ is given, analogously to (17.8.23a), by

$$\begin{aligned} D^{(\nu)}(\overline{h}) &= \Delta^{(\nu)}(h)^* \times \Delta^{(\nu)}(h) \times D^{(\lambda)}(h) \\ D^{(\nu)}(\overline{a}) &= \varepsilon_\upsilon[N(a)^* \tilde{\times} N(a)] \times D^{(\lambda)}(zh) \end{aligned} \tag{17.8.25b}$$

The number of linearly independent matrix elements contained in $\{V_{ijl}^{\nu\nu}\}$ is given by

$$n_1^{(\nu\nu)} = \frac{1}{2|H|}\sum_h \{\chi^{(\nu)}(h)|^2\chi^{(\lambda)}(h) + \varepsilon_\upsilon\chi^{(\nu)}((a_0h)^2)\chi^{(\lambda)}(zh)\} \tag{17.8.25c}$$

17.8.4 Selection rules under a gray point group

The results obtained above for an anti-unitary group can be easily specialized to $\overline{H}_{\overline{\boldsymbol{\theta}}}$ associated with a gray group H^e. Since $\boldsymbol{\theta}$ commutes with every element $h \in H$, we have $\chi^{(\nu)}(h)^* = \chi^{(\overline{\nu})}(h)$, from (17.8.19). For transitions between states belonging to different co-unirreps, the general result for n_1 given by (17.8.12) holds without modification because it is independent of the anti-unitary operator a. Thus the transitions between the states belonging to the same co-unirreps will be discussed here.

17.8.4.1 For type c co-unirreps

From (17.8.17c), (17.8.18b) and (17.8.23b) we have

$$n_1^{(22)} = \frac{1}{2|H|}\sum_h \{[\chi^{(\nu)}(h)^*]^2 + \varepsilon_\upsilon \chi^{(\nu)}(e'h^2)^*\}\chi^{(\lambda)}(h)$$

$$n_1^{(21)} = \frac{1}{2|H|}\sum_h \{[\chi^{(\nu)}(h)]^2 + \varepsilon_\upsilon \chi^{(\nu)}(e'h^2)\}\chi^{(\lambda)}(h)$$

$$n_1^{(11+22)} = \frac{1}{|H|}\sum_h |\chi^{(\nu)}(h)|^2 \chi^{(\lambda)}(h) \tag{17.8.26a}$$

17.8.4.2 For type b co-unirreps

We have $\nu = \bar{\nu}$ so that $\chi^{(\nu)}(h) = \chi^{(\nu)}(h)^* = \text{real}$. The above results (17.7.26a) are simplified to

$$n_1^{(12)} = n_1^{(21)} = \frac{1}{2|H|}\sum_h \{[\chi^{(\nu)}(h)]^2 + \varepsilon_\upsilon \chi^{(\nu)}(e'h^2)\}\chi^{(\lambda)}(h)$$

$$n_1^{(11+22)} = \frac{1}{|H|}\sum_h |\chi^{(\nu)}(h)|^2 \chi^{(\lambda)}(h) \tag{17.8.26b}$$

17.8.4.3 For type a co-unirreps

We have from (17.8.25c)

$$n_1 = \frac{1}{2|H|}\sum_h \{[\chi^{(\nu)}(h)]^2 + \varepsilon_\upsilon \chi^{(\nu)}(e'h^2)\}\chi^{(\lambda)}(h) \tag{17.8.26c}$$

where $\chi^{(\nu)}(h)$ is again real as in the case of a type b unirrep. Note that n_1 equals $n_1^{(12)} = n_1^{(21)}$ for a type b co-unirrep.

The results of the present section can be applied to the selection rules for anti-unitary space groups.

Appendix
Character tables of the crystal point groups

1. Notation for point groups. A proper point group P of finite order is defined by two generators a and b and their product ab that satisfy a set of defining relations for the group P

$$a^n = b^m = (ab)^l = e \tag{A1}$$

where n, m and l are the orders of the respective elements. Let the symbol n stand for an n-fold rotation, then the set $\{a, b, ab\}$ characteristic to each P may be expressed by the set of their axis orders $\{n, m, l\}$ as follows:

$$C_n = \{n, 0, 0\}, \qquad D_n = \{n, 2, 2\}, \qquad T = \{2, 3, 3\}, \qquad O = \{4, 3, 2\}$$
$$Y = \{5, 3, 2\}$$

where the first generator a is referred to as the principal axis of rotation for each P. Let $\overline{1}$ be the inversion, then an improper rotation is expressed by $\overline{n} = \overline{1}n$, which is the n-fold rotation followed by the inversion. According to Section 5.5, an improper point group is isomorphic to a proper point group P or a rotation–inversion group $P_i = P \times C_i$, which is the direct product of a proper point group P and the group of inversion $C_i = \{1, \overline{1}\}$. Those improper point groups isomorphic to P are

$$C_{np} = \{\overline{2n}, 0, 0\}, \qquad C_{nv} = \{n, \overline{2}, \overline{2}\}, \qquad D_{np} = \{\overline{2n}, 2, \overline{2}\}$$
$$T_p = \{\overline{4}, 3, \overline{2}\}$$

See Table 5.7 for the Schönflies notation for these improper point groups. Since the inversion $\overline{1}$ commutes with any rotation, we have the isomorphisms ($\simeq$):

$$C_{np} \simeq C_{2n}, \qquad C_{nv} \simeq D_n, \qquad D_{np} \simeq D_{2n}, \qquad T_p \simeq 0$$

2. A unitary irreducible representation (unirrep) $D(P_i)$ of a rotation–inversion group $P_i = P \times C_i$ is given by the direct product unirrep $D(P_i) = D(P) \times D(C_i)$, where $D(P)$ and $D(C_i)$ are the unirreps of the groups P and C_i. Since there exist two unirreps for C_i classified by $D_g(\overline{1}) = 1$ and $D_u(\overline{1}) = -1$, each unirrep $D(P)$ gives rise to two unirreps $D_g(P_i)$ and $D_u(P_i)$ defined by

P_i	$p \in P$	$\overline{p} = \overline{1}p$	Bases
D_g	$D(p)$	$D(p)$	Even functions
D_u	$D(p)$	$-D(p)$	Odd functions

Thus the character table of each P_i is easily constructed if the character table of the corresponding P is given. Accordingly, we shall not give the character table of any P_i explicitly but simply write down the elementary bases of each unirrep of P_i.

3. Let ψ_1 and ψ_2 be bases of the unirreps D_1 and D_2 of a group G, respectively, then a basis of the direct product representation $D_1 \times D_2$ is given by the direct product bases $\psi_1 \times \psi_2$. If one of the unirreps D_1 and D_2 is one-dimensional, then the direct product $D_1 \times D_2$ is also a unirrep of G. In particular, let ψ_0 be a basis of the identity representation D_0, then the direct product basis $\psi_1 \times \psi_0$ provides an alternative basis for the unirrep $D_1 = D_1 \times D_0$. This method will be used extensively to correlate the elementary basis functions listed in the character table of each point group. To facilitate the construction of the set of symmetry-adapted linear combinations (SALCs) based on the correspondence theorem on basis functions belonging to a unirrep developed in Chapter 7, we have listed more than one elementary basis set for each unirrep in the character tables given below.
4. Let $\psi(x, y, z)$ be a basis of a unirrep $D(P)$ of a proper point group, then $\psi(\tilde{x}, \tilde{y}, \tilde{z})$ also belongs to $D(P)$: here $\tilde{x}$, $\tilde{y}$ and $\tilde{z}$ are the components of a vector product of two vectors $\boldsymbol{r} = (x, y, z)$ and $\boldsymbol{r}' = (x', y', z')$ given by

$$\boldsymbol{r} \times \boldsymbol{r}' = [\tilde{x}, \tilde{y}, \tilde{z}] = [yz' - zy', zx' - xz', xy' - yx']$$

which becomes the infinitesimal rotation if $\boldsymbol{r}' = \nabla = (\partial_x, \partial_y, \partial_z)$. An analogous statement holds, with some modification, for an improper point group isomorphic to a proper point group P. For example, see the character tables of the groups D_6, C_{6v} and D_{3p}.
5. The same set of notations is used for the unirreps of the groups which are mutually isomorphic. As a result, the present system of notation may have some minor differences from the notations of other authors; cf. Heine (1977) and Koster *et al.* (1963).
6. The symbols A and B are used for one-dimensional representations, E is used for a two-dimensional unirrep, and T for a three-dimensional one. More specifically, let $\chi(a)$ and $\chi(b)$ be the characters of the generators a and b of a point group defined by (A1). Then the symbol A stands for a unirrep with $\chi(a) = 1$ while B stands for one with $\chi(a) = -1$. Subscript 1 or 2 is attached to A and B for $\chi(b) = 1$ or $\chi(b) = -1$.
7. The vector unirreps of the point groups are constructed from the elementary spatial bases in Chapter 6. Let a double point group be defined by the defining relations

$$a^n = b^m = (ab)^l = e', \qquad e'^2 = e \tag{A2}$$

where e' is the 2π rotation. Then the integral and half-integral unirreps of the double point groups are constructed in Chapter 11 via the spinor bases: the former provide the single-valued (vector) unirreps of the point groups whereas the latter provide the double-valued unirreps of the point groups. These can be regarded as the projective unirreps belonging to the factor systems defined by the representation of 2π rotation: $D(e') = \pm 1$, where $+1$ is for the vector unirreps and -1 is for the double-valued unirreps. The characters of these two types of unirreps and their bases are tabulated for the crystal point groups. In each table, the two types of unirreps are separated by a broken line. Each class of a point group is represented by one typical element in the class, possibly a generator of the group, while the order of each class is denoted by the symbol # at the beginning of the table.

8. The unirreps of the uniaxial group $C_n = \{c_n^k;\ k = 0, 1, \ldots, n-1\}$ are given by (11.3.3b):

$$M_m(c_n^k) = \exp(-\mathrm{i}2\pi mk/n); \qquad m = 0, \pm\frac{1}{2}, \pm 1, \ldots, \pm\frac{n-1}{2}, \frac{n}{2} \tag{A3}$$

with the spinor bases

$$\phi(j, m) = \alpha^{j+m}\beta^{j-m}[(j+m)!(j-m)!]^{-1/2}; \qquad j > |m|$$

where $\alpha = \xi_1$ and $\beta = \xi_2$ in (10.3.1b). The product rule $M_{m1} \times M_{m2} = M_{m1+m2}$ helps one to find new bases via

$$\phi(j_1, m_1)\phi(j_1, m_2) \propto \phi(j_1 + j_2, m_1 + m_2) \tag{A4}$$

According to Theorem 10.4.1, when j is an integer, the spinor basis $\{\phi(l, m)\}$ can be replaced by a point basis $\{r^l Y_{l,m}(\vartheta, \varphi)\}$, where $\{Y_{l,m}\}$ are the spherical harmonics. Thus we have the correspondences under proper rotations

$$\begin{aligned} \phi(l, m) &\leftrightarrow (-1)^m P_l^m(z)(x + \mathrm{i}y)^m \\ \phi(l, -m) &\leftrightarrow P_l^m(z)(x - \mathrm{i}y)^m \end{aligned} \tag{A5}$$

where $P_l^m(z)$ is an even (odd) polynomial of z of the degree $l - |m|$. When j is a half integer $l + \frac{1}{2}$, the basis can be expressed linearly in the elementary spinor basis α or β as follows, from (A4),

$$\phi(l + \tfrac{1}{2}, m + \tfrac{1}{2}) \propto \phi(l, m)\alpha \qquad \text{or} \qquad \phi(l, m+1)\beta \tag{A6}$$

where $\phi(l, m)$ and $\phi(l, m+1)$ are expressed by the spatial variables. For a mixed basis like (A6), a rotation characterized by a given rotation vector $\boldsymbol{\theta}$ is defined by the direct product $S(\boldsymbol{\theta}) \times R(\boldsymbol{\theta})$ of the spinor transformation $S(\boldsymbol{\theta})$ and the spatial rotation $R(\boldsymbol{\theta})$. The correspondences (A5) fail under the inversion, because it acts only on the spatial variables.

9. The unirreps of the group D_n are described in Table 11.5, with the bases

$$\begin{aligned} \phi_+(j, m) &= 2^{-1/2}[\phi(j, m) + \phi(j, -m)] \\ \phi_-(j, m) &= -\mathrm{i}2^{-1/2}[\phi(j, m) - \phi(j, -m)] \end{aligned} \tag{A7}$$

When j is an integer l, we have, in view of (A5), the following correspondences:

$$\begin{aligned} \phi_+(l, m) &\leftrightarrow P_l^m(z)\,\mathrm{Re}\,(y - \mathrm{i}x)^m \\ \phi_-(l, m) &\leftrightarrow P_l^m(z)\,\mathrm{Im}\,(y - \mathrm{i}x)^m \end{aligned} \tag{A8}$$

where Re and Im are the real and imaginary parts, respectively.

Remark. The unirreps of D_n given by Table 11.5 have been determined assuming that $u_0 = 2_x$. In a case in which we take $u_0 = 2_y$ as for D_2, D_3 and D_6 in the character tables given below, the basis $\phi(j, m)$ in Table 11.5 should be replaced by $\mathrm{i}^{-m}\phi(j, m)$ to leave the unirreps unchanged. Accordingly, the correspondences (A8) should be modified by

$$\begin{aligned} \phi_+(l, m) &\leftrightarrow P_l^m(z)\,\mathrm{Re}\,(x + \mathrm{i}y)^m \\ \phi_-(l, m) &\leftrightarrow P_l^m(z)\,\mathrm{Im}\,(x + \mathrm{i}y)^m \end{aligned} \tag{A9}$$

Through the correspondences (A8) and (A9), all the point bases given in the character tables of D_n are obtained from the spinor bases given in Table 11.5. These are in accordance with Table 6.4 for the vector unirreps of D_n.

For convenience of tabulation, the half-integral unirrep E_m of D_n is defined by the spinor basis $[\alpha^{2m}, \beta^{2m}]$ for $u_0 = 2_x$ or $[\alpha^{2m}, \mathrm{i}^{2m}\beta^{2m}]$ for $u_0 = 2_y$: these are equivalent to the basis $[\phi_+(|m|, m), \phi_-(|m|, -m)]$ given in Table 11.5.

10. The character tables for the group T and O are from Tables 11.7 and 11.6, respectively.

The group C_i

#	1	1	
C_i	E	$\bar{1}$	Bases for $C_i = \{\bar{1}, 0, 0\}$
M_{0g}	1	1	$1, x^2, y^2, z^2, \tilde{x}, \tilde{y}, \tilde{z}$
M_{0u}	1	-1	x, y, z
$M_{\frac{1}{2}g}$	1	1	α, β
$M_{\frac{1}{2}u}$	1	-1	$z\alpha, z\beta$

1. $\tilde{x} = yz' - zy'$, $\tilde{y} = zx' - xz'$, $\tilde{z} = xy' - yx'$.
2. $\alpha \propto \phi(\frac{1}{2}, \frac{1}{2})$, $\beta \propto \phi(\frac{1}{2}, -\frac{1}{2})$.

The groups $C_2 \simeq C_m$

#	1	1		
C_2	E	2_z	Bases for $C_2 = \{2, 0, 0\}$	Bases for $C_m = \{\bar{2}, 0, 0\}$
C_m	E	$\bar{2}_z$		
M_0	1	1	$1, z, \tilde{z}, z^2$	$1, x, y, \tilde{z}, z^2$
$M_1 = M_1^*$	1	-1	$x, y, \tilde{x}, \tilde{y}$	$z, \tilde{x}, \tilde{y}$
$M_{\frac{1}{2}}$	1	$-\mathrm{i}$	$\alpha, M_0 \times M_{\frac{1}{2}}, M_1 \times M_{\frac{1}{2}}^*$	
$M_{\frac{1}{2}}^*$	1	i	$\beta, M_0 \times M_{\frac{1}{2}}^*, M_1 \times M_{\frac{1}{2}}$	

M^* is the complex conjugate representation of M.

The bases of the unirreps for the group $C_{2i} = C_2 \times C_i$

M_{0g}:	$1, \tilde{z}, z^2$	M_{0u}:	z
M_{1g}:	$\tilde{x}, \tilde{y}$	M_{1u}:	x, y
$M_{\frac{1}{2}g}$:	α	$M_{\frac{1}{2}u}$:	$z\alpha, x\beta, y\beta$
$M_{\frac{1}{2}g}^*$:	β	$M_{\frac{1}{2}u}^*$:	$z\beta, x\alpha, y\alpha$

The groups $D_2 \simeq C_{2v}$

#	1	1	1	1		
D_2	E	2_z	2_y	2_x	Bases for	Bases for
C_{2v}	E	2_z	$\overline{2}_y$	$\overline{2}_x$	$D_2 = \{2, 2, 2\}$	$C_{2v} = \{2, \overline{2}, \overline{2}\}$
$A,\ A_1$	1	1	1	1	$1, x^2, y^2, z^2, xyz$	$1, z, x^2, y^2, z^2$
$B_1,\ A_2$	1	1	-1	-1	$z, xy, \tilde{z}$	$xy, \tilde{z}, xyz$
$B_2,\ B_1$	1	-1	1	-1	$y, zx, \tilde{y}$	$zx, \tilde{y}, x$
$B_3,\ B_2$	1	-1	-1	1	$x, yz, \tilde{x}$	$yz, \tilde{x}, y$
$E_{\frac{1}{2}}$	2	0	0	0	$[\alpha, \mathrm{i}\beta]$	

1. Here A, B_1, B_2 and B_3 are used for D_2 because of the complete equivalence of the three two-fold axes. This is an exception to the general rule.
2. A_1, A_2, B_1 and B_2 are in accordance with the unirreps of D_n given in Tables 6.4 and 11.5 for $n = 2$ with $u_0 = 2_y$.

The bases of the unirreps for the group $D_{2i} = D_2 \times C_i$

A_g:	$1, x^2, y^2, z^2$	A_u:	$xyz, z\tilde{z}$
B_{1g}:	$xy, \tilde{z}$	B_{1u}:	z
B_{2g}:	$zx, \tilde{y}$	B_{2u}:	y
B_{3g}:	$yz, \tilde{x}$	B_{3u}:	x
$E_{\frac{1}{2}g}$:	$[\alpha, \mathrm{i}\beta]$	$E_{\frac{1}{2}u}$:	$z[\alpha, -\mathrm{i}\beta], x[\beta, \mathrm{i}\alpha], y[\beta, -\mathrm{i}\alpha]$

The groups $C_4 \simeq C_{2p}$

#	1	1	1	1		
C_4	E	4_z	2_z	4_z^3	Bases for	Bases for
C_{2p}	E	$\overline{4}_z$	2_z	$\overline{4}_z^3$	$C_4 = \{4, 0, 0\}$	$C_{2p} = \{\overline{4}, 0, 0\}$
M_0	1	1	1	1	$1, z, \tilde{z}, z^2, x^2 + y^2$	$1, \tilde{z}, z^2, x^2 + y^2, xyz$
$M_2 = M_2^*$	1	-1	1	-1	$x^2 - y^2, xy, xyz$	$z, x^2 - y^2, xy$
M_1	1	$-\mathrm{i}$	-1	i	$x + \mathrm{i}y, \tilde{x} + \mathrm{i}\tilde{x}$	$x - \mathrm{i}y, \tilde{x} + \mathrm{i}\tilde{y}$
M_1^*	1	i	-1	$-\mathrm{i}$	$x - \mathrm{i}y, \tilde{x} - \mathrm{i}\tilde{y}$	$x + \mathrm{i}y, \tilde{x} - \mathrm{i}\tilde{y}$
$M_{\frac{1}{2}}$	1	ω	$-\mathrm{i}$	ω^3	$\alpha, M_1 \times M_{\frac{1}{2}}^*$	
$M_{\frac{1}{2}}^*$	1	$-\omega^3$	i	$-\omega$	$\beta, M_1^* \times M_{\frac{1}{2}}$	
$M_{\frac{3}{2}}$	1	ω^3	i	ω	$\alpha^3, M_1 \times M_{\frac{1}{2}}, M_2 \times M_{\frac{1}{2}}^*$	
$M_{\frac{3}{2}}^*$	1	$-\omega$	$-\mathrm{i}$	$-\omega^3$	$\beta^3, M_1^* \times M_{\frac{1}{2}}^*, M_2 \times M_{\frac{1}{2}}$	

$\omega = \exp(-\mathrm{i}\pi/4)$.

The bases of the unirreps for the group $C_{4i} = C_4 \times C_i$

M_{0g}:	$1, \tilde{z}, z^2, x^2 + y^2$	M_{0u}:	z
M_{2g}:	$x^2 - y^2, xy$	M_{2u}:	xyz
M_{1g}:	$(x + iy)z, \tilde{x} + i\tilde{y}$	M_{1u}:	$x + iy$
M_{1g}^*:	$(x - iy)z, \tilde{x} - i\tilde{y}$	M_{1u}^*:	$x - iy$
$M_{\frac{1}{2}g}$:	α	$M_{\frac{1}{2}u}$:	$z\alpha, (x + iy)\beta$
$M_{\frac{1}{2}g}^*$:	β	$M_{\frac{1}{2}u}^*$:	$z\beta, (x - iy)\alpha$
$M_{\frac{3}{2}g}$:	$xy\beta$	$M_{\frac{3}{2}u}$:	$z\alpha^3, (x + iy)\alpha$
$M_{\frac{3}{2}g}^*$:	$xy\alpha$	$M_{\frac{3}{2}u}^*$:	$z\beta^3, (x - iy)\beta$

The groups $D_4 \simeq C_{4v} \simeq D_{2p}$

#	1	2	2	2	1			
D_4	E	4_z	2_x	2_{xy}	2_z			
C_{4v}	E	4_z	$\overline{2}_x$	$\overline{2}_{xy}$	2_z	Bases for	Bases for	Bases for
D_{2p}	E	$\overline{4}_z$	2_x	$\overline{2}_{xy}$	2_z	$D_4 = \{4, 2, 2\}$	$C_{4v} = \{4, \overline{2}, \overline{2}\}$	$D_{2p} = \{\overline{4}, 2, \overline{2}\}$
A_1	1	1	1	1	1	$1, z^2, x^2 + y^2$	$1, z, z^2, x^2 + y^2$	$1, z^2, x^2 + y^2, xyz$
A_2	1	1	−1	−1	1	$z, \tilde{z}, xy(x^2 - y^2)$	$\tilde{z}, xy(x^2 - y^2)$	$\tilde{z}, z(x^2 - y^2)$
B_1	1	−1	1	−1	1	$x^2 - y^2, xyz$	$x^2 - y^2$	$x^2 - y^2$
B_2	1	−1	−1	1	1	$xy, (x^2 - y^2)z$	xy, xyz	z, xy
E	2	0	0	0	−2	$[x, y], [\tilde{x}, \tilde{y}], [yz, -xz]$	$[y, -x], [\tilde{x}, \tilde{y}], [yz, -xz]$	$[x, -y], [\tilde{x}, \tilde{y}], [yz, -xz]$
$E_{\frac{1}{2}}$	2	$\sqrt{2}$	0	0	0	$[\alpha, \beta]$		
$E_{\frac{3}{2}}$	2	$-\sqrt{2}$	0	0	0	$[\alpha^3, \beta^3]$		

$[\tilde{x}, \tilde{y}](x^2 + y^2) \propto [yz, -xz]$; $\tilde{x} = y\partial_z - z\partial_y$, $\tilde{y} = z\partial_x - x\partial_z$.
$[\tilde{x}, \tilde{y}]z^2 \propto [yz, -xz]$.

Elementary bases of the unirreps for the group $D_{4i} = D_4 \times C_i$

A_{1g}:	$1, z^2, x^2 + y^2$	A_{1u}:	$\tilde{z}z, (x^2 - y^2)xyz$
A_{2g}:	$\tilde{z}$	A_{2u}:	z
B_{1g}:	$x^2 - y^2$	B_{1u}:	xyz
B_{2g}:	xy	B_{2u}:	$(x^2 - y^2)z$
E_g:	$[\tilde{x}, \tilde{y}], [yz, -xz]$	E_u:	$[x, y]$
$E_{\frac{1}{2}g}$:	$[\alpha, \beta]$	$E_{\frac{1}{2}u}$:	$z[\alpha, -\beta], [(x + iy)\beta, (x - iy)\alpha]$
$E_{\frac{3}{2}g}$:	$xy[\beta, \alpha], (x^2 - y^2)[\beta, -\alpha]$	$E_{\frac{3}{2}u}$:	$z[\alpha^3, -\beta^3], [(x + iy)\alpha, -(x - iy)\beta]$

The group C_3

#	1	1	1	
C_3	E	3_z	3_z^2	Bases for $C_3 = \{3, 0, 0\}$
M_0	1	1	1	$1, z, \tilde{z}, z^2, x^2+y^2, \cos(3\theta), \sin(3\theta)$
M_1	1	ω	$ω^2$	$x+iy, \tilde{x}+i\tilde{y}, (x-iy)^2$
M_1^*	1	$ω^2$	ω	$x-iy, \tilde{x}-i\tilde{y}, (x+iy)^2$
$M_{\frac{1}{2}}$	1	$-ω^2$	ω	$\alpha, M_1 \times M_{\frac{1}{2}}^*$
$M_{\frac{1}{2}}^*$	1	$-ω$	$ω^2$	$\beta, M_1^* \times M_{\frac{1}{2}}$
$M_{\frac{3}{2}} = M_{\frac{3}{2}}^*$	1	-1	1	$\alpha^3, \beta^3, M_1 \times M_{\frac{1}{2}}, M_1^* \times M_{\frac{1}{2}}^*$

$ω = \exp(-2\pi i/3)$; $\cos(3\theta) = x^3 - 3xy^2$, $\sin(3\theta) = y^3 - 3yx^2$.

The bases of the unirreps for the group $C_{3i} = C_3 \times C_i$

M_{0g}:	$1, z^2, \tilde{z}^2$	M_{0u}:	z
M_{1g}:	$(x-iy)^2, \tilde{x}+i\tilde{y}$	M_{1u}:	$x+iy$
M_{1g}^*:	$(x+iy)^2, \tilde{x}-i\tilde{y}$	M_{1u}^*:	$x-iy$
$M_{\frac{1}{2}g}$:	α	$M_{\frac{1}{2}u}$:	$z\alpha, (x+iy)\beta$
$M_{\frac{1}{2}g}^*$:	β	$M_{\frac{1}{2}u}^*$:	$z\beta, (x-iy)\alpha$
$M_{\frac{3}{2}g}$:	$\alpha^3, \beta^3, (\tilde{x}+i\tilde{y})\alpha$	$M_{\frac{3}{2}u}$:	$z\alpha^3, z\beta^3, (x+iy)\alpha, (x-iy)\beta$

The groups $D_3 \simeq C_{3v}$

#	1	2	3		
D_3	E	3_z	2_y	Bases for	Bases for
C_{3v}	E	3_z	$\overline{2}_x$	$D_3 = \{3, 2, 2\}$	$C_{3v} = \{3, \overline{2}, \overline{2}\}$
A_1	1	1	1	$1, z^2, x^2+y^2, y^3-3yx^2$	$1, z, z^2, x^2+y^2, y^3-3yx^2$
A_2	1	1	-1	$z, \tilde{z}, x^3-3xy^2$	$\tilde{z}, x^3-3xy^2$
E	2	-1	0	$[x, y], [\tilde{x}, \tilde{y}], [yz, -xz], [2xy, x^2-y^2]$	$[x, y], [\tilde{y}, -\tilde{x}], [xz, yz], [2xy, x^2-y^2]$
$E_{\frac{1}{2}}$	2	1	0	$[\alpha, i\beta]$	$[\alpha, \beta]$
B_1	1	-1	$-i$	$\alpha^3 + i\beta^3$	$\alpha^3 - \beta^3$
B_2	1	-1	i	$\alpha^3 - i\beta^3$	$\alpha^3 + \beta^3$

1. From Tables 6.4 and 11.5 with $u_0 = 2_y$ for D_2.
2. $[\partial_x, \partial_y](y^3 - 3yx^2) \propto [2xy, x^2-y^2]$.
3. $[\tilde{y}, -\tilde{x}]z \propto [x, y]$.

The bases of the unirreps for the group $D_{3i} = D_3 \times C_i$

A_{1g}:	$1, z^2, x^2+y^2$	A_{1u}:	$z\tilde{z}, y^3-3yx^2$
A_{2g}:	$\tilde{z}$	A_{2u}:	z, x^3-3xy^2
E_g:	$[\tilde{x}, \tilde{y}], [yz, -xz], [2xy, x^2-y^2]$	E_u:	$[x, y]$
$E_{\frac{1}{2}g}$:	$[\alpha, i\beta]$	$E_{\frac{1}{2}u}$:	$z[\alpha, -i\beta], [(x+iy)\beta, i(x-iy)\alpha]$
B_{1g}:	$z[(x+iy)\alpha+i(x-iy)\beta]$	B_{1u}:	$z(\alpha^3-i\beta^3), (x+iy)\alpha-i(x-iy)\beta$
B_{2g}:	$z[(x+iy)\alpha-i(x-iy)\beta]$	B_{2u}:	$z(\alpha^3+i\beta^3), (x+iy)\alpha+i(x-iy)\beta$

The groups $C_6 \simeq C_{3p}$

#	1	1	1	1	1	1		
C_6	E	6_z	3_z	2_z	3_z^2	6_z^5	Bases for	Bases for
C_{3p}	E	$\overline{6}_z$	3_z	$\overline{2}_z$	3_z^2	$\overline{6}_z^5$	$C_6 = \{6, 0, 0\}$	$C_{3p} = \{\overline{6}, 0, 0\}$
M_0	1	1	1	1	1	1	$1, z, z^2, \tilde{z}$	$1, z^2, \tilde{z}, \cos(3\theta), \sin(3\theta)$
$M_3 = M_3^*$	1	-1	1	-1	1	-1	$\cos(3\theta), \sin(3\theta)$	z
M_1	1	$-\omega^2$	ω	-1	ω^2	$-\omega$	$x+iy, (x+iy)z$	$(x+iy)z$
M_1^*	1	$-\omega$	ω^2	-1	ω	$-\omega^2$	$x-iy, (x-iy)z$	$(x-iy)z$
M_2	1	ω	ω^2	1	ω	ω^2	$(x+iy)^2$	$(x+iy)^2, x-iy$
M_2^*	1	ω^2	ω	1	ω^2	ω	$(x-iy)^2$	$(x-iy)^2, x+iy$
$M_{\frac{1}{2}}$	1	ρ	ρ^2	$-i$	ρ^4	ρ^5	$\alpha, M_1 \times M_{\frac{1}{2}}^*$	
$M_{\frac{1}{2}}^*$	1	$-\rho^5$	$-\rho^4$	i	$-\rho^2$	$-\rho$	$\beta, M_1^* \times M_{\frac{1}{2}}$	
$M_{\frac{3}{2}}$	1	$-i$	-1	i	1	$-i$	$\alpha^3, M_1 \times M_{\frac{1}{2}}$	
$M_{\frac{3}{2}}^*$	1	i	-1	$-i$	1	i	$\beta^3, M_1^* \times M_{\frac{1}{2}}^*$	
$M_{\frac{5}{2}}$	1	ρ^5	$-\rho^4$	$-i$	$-\rho^2$	ρ	$\alpha^5, M_2 \times M_{\frac{1}{2}}, M_3 \times M_{\frac{1}{2}}^*$	
$M_{\frac{5}{2}}^*$	1	$-\rho$	ρ^2	i	ρ^4	$-\rho^5$	$\beta^5, M_2^* \times M_{\frac{1}{2}}^*, M_3^* \times M_{\frac{1}{2}}$	

1. $\omega = \exp(-2\pi i/3)$; $\cos(3\theta) = x^3 - 3xy^2$, $\sin(3\theta) = y^3 - 3yx^2$.
2. $\rho = \exp(-\pi i/6)$; $\rho^4 = \omega$.

The bases of the unirreps for the group $C_{6i} = C_6 \times C_i$

M_{0g}:	$1, z^2, \tilde{z}$	M_{0u}:	z
M_{3g}:	$z(x^3-3xyz), z(y^3-3yz^2)$	M_{3u}:	x^3-3xy^2, y^3-3yx^2
M_{1g}:	$(x+iy)z$	M_{1u}:	$(x+iy)$
M_{1g}^*:	$(x-iy)z$	M_{1u}^*:	$(x-iy)$
M_{2g}:	$(x+iy)^2$	M_{2u}:	$(x+iy)^2z$
M_{2g}^*:	$(x-iy)^2$	M_{2u}^*:	$(x-iy)^2z$
$M_{\frac{1}{2}g}$:	α	$M_{\frac{1}{2}u}$:	$z\alpha, (x+iy)\beta$
$M_{\frac{1}{2}g}^*$:	β	$M_{\frac{1}{2}u}^*$:	$z\beta, (x-iy)\alpha$
$M_{\frac{3}{2}g}$:	$z(x+iy)\alpha$	$M_{\frac{3}{2}u}$:	$z\alpha^3, (x+iy)\alpha$
$M_{\frac{3}{2}g}^*$:	$z(x-iy)\beta$	$M_{\frac{3}{2}u}^*$:	$z\beta^3, (x-iy)\beta$
$M_{\frac{5}{2}g}$:	$(x+iy)^2\alpha$	$M_{\frac{5}{2}u}$:	$z\alpha^5, (x^3-3xy^2)\beta$
$M_{\frac{5}{2}g}^*$:	$(x-iy)^2\beta$	$M_{\frac{5}{2}u}^*$:	$z\beta^5, (x^3-3xy^2)\alpha$

The groups $D_6 \simeq C_{6v} \simeq D_{3p}$

#	1	2	3	3	1	2			
D_6	E	6_z	2_y	2_x	2_z	3_z			
C_{6v}	E	6_z	$\overline{2}_x$	$\overline{2}_y$	2_z	3_z	Bases for	Bases for	Bases for
D_{3p}	E	$\overline{6}_z$	2_y	$\overline{2}_x$	$\overline{2}_z$	3_z	$D_6 = \{6, 2, 2\}$	$C_{6v} = \{6, \overline{2}, \overline{2}\}$	$D_{3p} = [\overline{6}, 2, \overline{2}]$
A_1	1	1	1	1	1	1	$1, z^2, x^2+y^2$	$1, z, z^2, x^2+y^2$	$1, z^2, x^2+y^2, y^3-3yx^2$
A_2	1	1	-1	-1	1	1	$z, \tilde{z}$	$\tilde{z}$	$\tilde{z}, x^3-3xy^2$
B_1	1	-1	1	-1	-1	1	$y^3-3yx^2, (x^3-3xy^2)z$	y^3-3yx^2	$z\tilde{z}$
B_2	1	-1	-1	1	-1	1	$x^3-3xy^2, (y^3-3yx^2)z$	x^3-3xy^2	z
E_1	2	1	0	0	-2	-1	$[x, y], [\tilde{x}, \tilde{y}], [yz, -xz], [y\tilde{z}, -x\tilde{z}]$	$[x, y], [\tilde{y}, -\tilde{x}], [xz, yz], [y\tilde{z}, -x\tilde{z}]$	$[\tilde{x}, \tilde{y}], [yz, -xz]$
E_2	2	-1	0	0	2	-1	$[2xy, x^2-y^2]$	$[2xy, x^2-y^2]$	$[x, y], [2xy, x^2-y^2], [y\tilde{z}, -x\tilde{z}]$
$E_{\frac{1}{2}}$	2	$\sqrt{3}$	0	0	0	1	$[\alpha, \mathrm{i}\beta]$	$[\alpha, \beta]$	$[\alpha, \mathrm{i}\beta]$
$E_{\frac{3}{2}}$	2	0	0	0	0	-2	$[\alpha^3, -\mathrm{i}\beta^3]$	$[\alpha^3, \beta^3]$	$[\alpha^3, -\mathrm{i}\beta^3]$
$E_{\frac{5}{2}}$	2	$-\sqrt{3}$	0	0	0	1	$[\alpha^5, \mathrm{i}\beta^5]$	$[\alpha^5, \beta^5]$	$[\alpha^5, \mathrm{i}\beta^5]$

1. From Tables 6.4 and 11.5 with $u_0 = 2_y$ for D_6.
2. $[\partial_x, \partial_y](y^3 - 3yx^2) \propto [2xy, x^2 - y^2]$; $[\tilde{x}, \tilde{y}](x^2 + y^2) \propto [yz, -xz]$.
3. $[\partial'_x, \partial'_y]\tilde{z}\tilde{z} \propto [y\tilde{z}, -x\tilde{z}]$; $\tilde{z} = xy' - yx'$.

The bases of the unirreps for the group $D_{6i} = D_6 \times C_i$

A_{1g}:	$1, z^2$	A_{1u}:	$z\tilde{z}$
A_{2g}:	$\tilde{z}$	A_{2u}:	z
B_{1g}:	$(x^3 - 3xy^2)z$	B_{1u}:	$y^3 - 3yx^2$
B_{2g}:	$(y^3 - 3yx^2)z$	B_{2u}:	$x^3 - 3xy^2$
E_{1g}:	$[\tilde{x}, \tilde{y}], [yz, -xz]$	E_{1u}:	$[x, y]$
E_{2g}:	$[2xy, x^2 - y^2]$	E_{2u}:	$[2xy, x^2 - y^2]z$
$E_{\frac{1}{2}g}$:	$[\alpha, \mathrm{i}\beta]$	$E_{\frac{1}{2}u}$:	$z[\alpha, -\mathrm{i}\beta], [(x+\mathrm{i}y)\beta, \mathrm{i}(x-\mathrm{i}y)\alpha]$
$E_{\frac{3}{2}g}$:	$z[(x+\mathrm{i}y)\alpha, -\mathrm{i}(x-\mathrm{i}y)\beta]$	$E_{\frac{3}{2}u}$:	$z[\alpha^3, \mathrm{i}\beta^3], [(x+\mathrm{i}y)\alpha, \mathrm{i}(x-\mathrm{i}y)\beta]$
$E_{\frac{5}{2}g}$:	$[(x+\mathrm{i}y)^2\alpha, -\mathrm{i}(x-\mathrm{i}y)^2\beta]$	$E_{\frac{5}{2}u}$:	$z[\alpha^5, -\mathrm{i}\beta^5], (x^3 - 3xy^2)[\beta, \mathrm{i}\alpha], (y^3 - 3yx^2)[\beta, -\mathrm{i}\alpha]$

The group T

#	1	3	4	4	
T	E	2_x	3_{xyz}	3^2_{xyz}	Bases for $T = \{2, 3, 3\}$
A	1	1	1	1	$1, xyz, x^2+y^2+z^2$
A'	1	1	ω	ω^2	$u+\mathrm{i}v$
A''	1	1	ω^2	ω	$u-\mathrm{i}v$
T	3	-1	0	0	$[x, y, z], [yz, zx, xy], [\tilde{x}, \tilde{y}, \tilde{z}]$
$E_{\frac{1}{2}}$	2	0	1	-1	$[\alpha, \beta]$
$E'_{\frac{1}{2}}$	2	0	ω	$-\omega^2$	$A' \times E_{\frac{1}{2}}$
$E''_{\frac{1}{2}}$	2	0	ω^2	$-\omega$	$A'' \times E_{\frac{1}{2}}$

$\omega = \exp(-2\pi\mathrm{i}/3)$; $u = 2z^2 - x^2 - y^2$, $v = 3^{\frac{1}{2}}(x^2 - y^2)$.

The bases of the unirreps for the group $T_i = T \times C_i$

A_g:	$1, x^2+y^2+z^2$	A_u:	xyz
A'_g:	$u+\mathrm{i}v$	A'_u:	$(u+\mathrm{i}v)xyz$
A''_g:	$u-\mathrm{i}v$	A''_u:	$(u+\mathrm{i}v)xyz$
T_g:	$[yz, zx, xy], [\tilde{x}, \tilde{y}, \tilde{z}]$	T_u:	$[x, y, z]$
$E_{\frac{1}{2}g}$:	$[\alpha, \beta]$	$E_{\frac{1}{2}u}$:	$xyz[\alpha, \beta]$
$E'_{\frac{1}{2}g}$:	$(u+\mathrm{i}v)[\alpha, \beta]$	$E'_{\frac{1}{2}u}$:	$xyz(u+\mathrm{i}v)[\alpha, \beta]$
$E''_{\frac{1}{2}g}$:	$(u-\mathrm{i}v)[\alpha, \beta]$	$E''_{\frac{1}{2}u}$:	$xyz(u-\mathrm{i}v)[\alpha, \beta]$

The groups $O \simeq T_p$

#	1	6	8	6	3		
O	E	4_z	3_{xyz}	2_{xy}	2_z	Bases for	Bases for
$T_p\ (=T_d)$	E	$\overline{4}_z$	3_{xyz}	$\overline{2}_{xy}$	2_z	$O = \{4, 3, 2\}$	$T_p = \{\overline{4}, 3, \overline{2}\}$
A_1	1	1	1	1	1	$1, x^2+y^2+z^2$	$1, xyz$
A_2	1	-1	1	-1	1	$xyz, \tilde{x}\tilde{y}\tilde{z}$	$\tilde{x}\tilde{y}\tilde{z}$, $(y^2-z^2)(z^2-x^2)(x^2-y^2)$
E	2	0	-1	0	2	$[u, v]$	$[u, v]$
T_1	3	1	0	-1	-1	$[x, y, z]$, $[\tilde{x}, \tilde{y}, \tilde{z}]$	$[\tilde{x}, \tilde{y}, \tilde{z}]$, $[x(y^2-z^2), y(z^2-x^2), z(x^2-y^2)]$
$A_2 \times T_1 = T_2$	3	-1	0	1	-1	$[yz, zx, xy]$, $[\tilde{x}, \tilde{y}, \tilde{z}]xyz$	$[x, y, z], [yz, zx, xy]$
$E_{\frac{1}{2}}$	2	$\sqrt{2}$	1	0	0	$[\alpha, \beta]$	
$E'_{\frac{1}{2}}$	2	$-\sqrt{2}$	1	0	0	$A_2 \times E_{\frac{1}{2}}$	
Q	4	0	1	0	0	$E \times E_{\frac{1}{2}}$	

1. $u = 2z^2 - x^2 - y^2$, $v = 3^{1/2}(x^2 - y^2)$.
2. $[\tilde{x}, \tilde{y}, \tilde{z}]xyz = [x(y^2 - z^2), y(z^2 - x^2), z(x^2 - y^2)]$.
3. $[\partial_x, \partial_y, \partial_z]xyz = [yz, zx, xy]$.

The bases of the unirreps for the group $O_i = O \times C_i$

A_{1g}:	$1,\ x^2+y^2+z^2$	A_{1u}:	$xyz\tilde{x}\tilde{y}\tilde{z}$
A_{2g}:	$\tilde{x}\tilde{y}\tilde{z}(x^2-y^2)(y^2-z^2)(z^2-x^2)$	A_{2u}:	xyz
E_g:	$[u,\ v]$	E_u:	$[v,\ -u]xyz$
T_{1g}:	$[\tilde{x},\ \tilde{y},\ \tilde{z}]$	T_{1u}:	$[x,\ y,\ z]$
T_{2g}:	$[yz,\ zx,\ xy]$	T_{2u}:	$[x(y^2-z^2),\ y(z^2-x^2),\ z(x^2-y^2)]$
$E_{\frac{1}{2}g}$:	$[\alpha,\ \beta]$	$E_{\frac{1}{2}u}$:	$A_{1u} \times E_{\frac{1}{2}g}$
$E'_{\frac{1}{2}g}$:	$A_{2g} \times E_{\frac{1}{2}g}$	$E'_{\frac{1}{2}u}$:	$A_{2u} \times E_{\frac{1}{2}g}$
Q_g:	$E_g \times E_{\frac{1}{2}g}$	Q_u:	$E_u \times E_{\frac{1}{2}g}$

References

Chapter 1

Cartan, E. (1913), *Bull. Soc. Math. France* **41**, 53 and *The Theory of Spinors*, English translation, MIT Press, Cambridge, MA, 1966.

Kim, S. K. (1979a), A new method of matrix transformation I, matrix diagonalizations via involutional transformation, *J. Math. Phys.* **20**, 2153–8.

Kim, S. K. (1979b), A new method of matrix transformation II, general theory of matrix diagonalizations via reduced characteristic equations and its application to angular momentum coupling, *J. Math. Phys.* **20**, 2159–67.

Littlewood, D. E. (1950), *The Theory of Group Characters*, 2nd ed., Clarendon Press, Oxford.

Murnaghan, F. D. (1938), *The Theory of Group Representations*, Johns Hopkins Press, Baltimore.

Chapter 2

Brauer, R. and Weyl, H. (1935), Spinors in *n*-dimensions, *Am. J. Math.* **57**, 425–49.

Dirac, P. A. M. (1947), *The Principles of Quantum Mechanics*, 3rd ed., Clarendon Press, Oxford.

Kim, S. K. (1969), Involutional matrices based on the representation theory of *GL*(2), *J. Math. Phys.* **10**, 1225–34.

Kim, S. K. (1979a), *op. cit.*

Kim, S. K. (1979b), *op. cit.*

Kim, S. K. (1980a), The theory of spinors via involutions and its application to the representations of the Lorentz group, *J. Math. Phys.* **21**, 1299–313.

Kim, S. K. (1980b), Theory of involutional transformation applied to the Dirac theory of the electron 1, remarks on the Dirac plane wave, *J. Math. Phys.* **21**, 2282–5.

Kim, S. K. (1980c), Theory of involutional transformation applied to the Dirac theory of the electron II, remarks on the Dirac–Coulomb waves, *J. Math. Phys.* **21**, 2286–90.

Chapter 3

Altmann, S. L. (1977), *Induced Representations in Crystals and Molecules*, Academic Press, London.

Coxeter, H. S. M. and Moser, W. O. J. (1984), *Generators and Discrete Groups*, Springer-Verlag, New York.

Rotman, J. J. (1973), *The Theory of Groups, An Introduction*, Allyn and Bacon, Inc., Boston.

Chapter 4

Cartan, E. (1913), *op. cit.*

Littlewood, D. E. (1950), *op. cit.*

Kim, S. K. (1980a), *op. cit.*

Lomont, J. S. (1959), *Applications of Finite Groups*, Academic Press, New York.

Chapter 5

Coxeter, H. S. M. and Moser, W. O. J. (1984), *op. cit.*
Heine, V. (1977), *Group Theory in Quantum Mechanics*, Pergamon Press, Oxford.
Kim, S. K. (1983b), A unified theory of the point group III. Classification and basis functions of improper point groups, *J. Math. Phys.* **24**, 414–18.
Kroto, H. W., Heath, J. R., O'Brien, S. C., Carl, R. F. and Smalley, R. E. (1985), C_{60}: Buckminsterfullerene, *Nature*, **318**, 162–3.
Wyle, H. (1952), *Symmetry*, Princeton University Press, Princeton.
Zak, J., Casher, A., Glück, M. and Gur, Y. (1969), *The Irreducible Representations of Space Groups*, Benjamin, New York.

Chapter 6

Lyubarskii, G. Ya. (1960), *The Application of Group Theory in Physics*, Translated from the Russian, Pergamon Press, New York.
Wigner, E. P. (1962), *Group Theory and its Application to the Quantum Mechanics of Atomic Spectra*, 3rd Printing, Academic Press, New York.

Chapter 7

Cotton, F. A. (1990), *Chemical Applications of Group Theory*, 3rd ed., Wiley-Interscience, John Wiley and Sons, Inc., New York.
Herzberg, G. (1951), *Molecular Spectra and Molecular Structure, II. Infrared and Raman Spectra of Polyatomic Molecules*, D. Van Nostrand Co., Inc., New York.
Kim, S. K. (1981a), A new method of constructing symmetry-adapted linear combinations for a finite group I. SALC's of equivalent invariant functions, *Physica* A **105**, 577–92; II. Correspondence theorem and basis operators, *Physica* A **106**, 521–38.
Kim, S. K. (1981b), A new method of constructing the symmetry coordinates of molecular vibrations based on the correspondence theorem, *J. Math. Phys.* **22**, 2303–6.
Kim, S. K. (1986a), A new method of constructing the symmetry-adapted linear combinations based on the correspondence theorem and induced representations, *J. Math. Phys.* **27**, 365–9.
Wilson, E. B., Decius, J. C. and Cross, P. C. (1955), *Molecular Vibrations, The Theory of Infrared and Raman Vibrational Spectra*, McGraw-Hill Book Company, Inc., New York.

Chapter 8

Jansen, L. and Boon, M. (1967), *Theory of Finite Groups. Applications in Physics*, North-Holland, Amsterdam.
Kim, S. K. (1986a), *op. cit.*
Lomont, J. S. (1959), *op. cit.*
Miller, S. C. and Love, W. H. (1967), *Tables of Irreducible Representations of Space Groups and Corepresentations of Magnetic Space Groups*, Bruett Press, Boulder, CO.
Raghavacharyulu, I. V. V. (1961), Representations of Space Groups, *Can. J. Phys.* **39**, 830–40.

Chapter 9

Hamermesh, M. (1962), *Group Theory and its Application to Physical Problems*, Addison-Wesley, Reading, MA.
Wigner, E. P. (1962), *op. cit.*
Wybourne, B. G. (1974), *Classical Groups for Physicists*, John Wiley & Sons, New York.

Chapter 10

Biedenharn, L. C. and Van Dam, H. (Eds.) (1965), *Quantum Theory of Angular Momentum*, Academic Press, New York.
Condon, E. U. and Shortley, G. H. (1935), *The Theory of Atomic Spectra*, Cambridge University Press, Cambridge.
Dirac, P. A. M. (1974), *Spinors in Hilbert Space*, Plenum Press, New York.
Hamermesh, M. (1962), *op. cit.*
Kim, S. K. (1969), *op. cit.*
Kim, S. K. (1981c), On the projective representations of the point group, *Molec. Phys.* **43**, 1447–50.
Kim, S. K. (1983b), *op. cit.*
Van der Waerden, B. L. (1974), *Group Theory and Quantum Mechanics*, Springer-Verlag, New York.
Wigner, E. P. (1962), *op. cit.*

Chapter 11

Altmann, S. L. (1979), Double groups and projective representations, I. General theory, *Molec. Phys.* **38**, 489–511.
Brown, E. (1970), A simple alternative to double groups, *Am. J. Phys.* **38**, 704–15.
Kim, S. K. (1981b), *op. cit.*
Kim, S. K. (1981d), A unified theory on point groups and their general irreducible representations, *J. Math. Phys.* **22**, 2101–7.
Koster, G. F., Dimmock, J. O., Wheeler, R. G. and Statz, H. (1963), *Properties of the Thirty-two Point Groups*, MIT Press, Cambridge, MA.

Chapter 12

Bir, G. L. and Pikus, G. E. (1974), *Symmetry and Strain-induced Effects in Semiconductors*, John Wiley & Sons, New York.
Kim, S. K. (1983a), A unified theory of the point groups II. The general projective representations and their applications to space groups, *J. Math. Phys.* **24**, 411–13.
Kim, S. K. (1984c), A unified theory of point groups VI. The projective corepresentations of the magnetic point groups of infinite order, *J. Math. Phys.* **25**, 2125–7.
Lomont, J. S. (1959), *op. cit.*

Chapter 13

Kim, S. K. (1985), The plane rotations allowed for *d*-dimensional discrete lattices and their application to the Bravais lattices in three dimensions, *J. Math. Phys.* **26**, 2381–2.
Kim, S. K. (1986b), The 32 minimal general generator sets of 230 double space groups, *J. Math. Phys.* **27**, 1471–89.
Lyubarskii, G. Ya. (1960), *op. cit.*

Chapter 14

Bir, G. L. and Pikus, G. E. (1974), *op. cit.*
Bouchaert, L. P., Smoluchowski, R. and Wigner, E. (1936), Theory of Brillouin zones and symmetry properties of wave functions in crystals, *Phys. Rev.* **50**, 58–67.
Kim, S. K. (1983a), *op. cit.*
Miller, S. C. and Love, W. H. (1967), *op. cit.*
Zak, J., Casher, A., Glück, M. and Gur, Y. (1969), *op. cit.*

Chapter 15

Bir, G. L. and Pikus, G. E. (1974), *op. cit.*

Jones, H. (1975), *The Theory of Brillouin Zones and Electronic States in Crystals*, North-Holland, Amsterdam.

Koster, G. F., Dimmock, J. O., Wheeler, R. G. and Statz, H. (1963), *op. cit.*

Peierls, R. E. (1955), *Quantum Theory of Solids*, Clarendon Press, Oxford.

Chapter 16

Condon, E. U. and Shortley, G. H. (1935), *op. cit.*

Dimmock, J. O. (1963), Representation theory for nonunitary groups, *J. Math. Phys.* **4**, 1307–11.

Dimmock, J. O. and Wheeler, R. G. (1962), Symmetry properties of wave functions in magnetic crystals, *J. Phys. Chem. Solids*, **23**, 729.

Kim, S. K. (1983c), A unified theory of the point groups IV. The general corepresentations of the crystallographic and noncrystallographic Shubnikov point groups, *J. Math. Phys.* **24**, 419–23.

Kim, S. K. (1984b), General irreducibility condition for vector and projective corepresentations of antiunitary groups, *J. Math. Phys.* **25**, 197–9.

Kim, S. K. (1984c), *op. cit.*

Kubo, R. (1957), Statistical-mechanical theory of irreversible processes. 1. General theory and simple applications to magnetic and conduction problems, *J. Phys. Soc. Japan*, **12**, 570–86.

Onsager, L. (1931), Reciprocal relations in irreversible processes I. *Phys. Rev.* **37**, 405–26; II, **38**, 2265–79.

Shubnikov, A. V., Belov, N. V. *et al.* (1964), Translated from the Russian by Itzkoff, J. and Collob, J., Pergamon Press, Inc., New York.

Wigner, E. P. (1962), *op. cit.*

Chapter 17

Belov, N. V., Neronova, N. N. and Smirnova, T. S. (1955), *Trudy Inst. Kristallogr. Akad. Nauk SSR* **11**, 33; English translation in Shubnikov, A. V. and Belov, N. V., *et al.* (1964), *Colored Symmetry*, Pergamon Press, Inc., New York.

Bradley, C. J. and Cracknell, A. P. (1972), *The Mathematical Theory of Symmetry in Solids, Representation Theory of Point Groups*, Clarendon Press, Oxford.

Kim, S. K. (1983c), *op. cit.*

Kim, S. K. (1984a), A unified theory of the point groups V. The general projective corepresentations of the magnetic point groups and their applications to magnetic space groups, *J. Math. Phys.* **25**, 189–96.

Kim, S. K. (1984c), *op. cit.*

Kim, S. K. (1986c), The 38 assemblies of the general generator sets for 1421 magnetic double space groups, *J. Math. Phys.* **27**, 1484–9.

Miller, S. C. and Love, W. H. (1967), *op. cit.*

Opechowski, W. and Guccione, R. (1965), *Magnetic Symmetry in Magnetism*, edited by Rado, G. T. and Suhl, H., Academic Press, New York, Vol. 2A, pp. 105–65.

Appendix

Heine, V. (1977), *op. cit.*

Koster, G. F., Dimmock, J. O., Wheeler, R. G. and Statz, H. (1963), *op. cit.*

Index